AM-FM Broadcasting

Equipment, Operations, and Maintenance

AM-FM Broadcasting

Equipment, Operations, and Maintenance

by

Harold E. Ennes

HOWARD W. SAMS & CO., INC.
THE BOBBS-MERRILL CO., INC.
INDIANAPOLIS • KANSAS CITY • NEW YORK

FIRST EDITION

FIRST PRINTING—1974

International Standard Book Number: 0-672-21012-6
Library of Congress Catalog Card Number: 74-75274

Preface

Radio broadcasting is a highly specialized branch of electronics. Those who study for and take the examination for the FCC radiotelephone first-class operator license soon realize that their knowledge is incomplete when they assume the duties of actual broadcast operation. This text is intended for all who need a practical insight into the use of electronic circuitry as applied specifically to broadcasting.

The reader of this book must already have basic electronic training. However, advanced electronic training on a university level, while helpful, is not required.

The first ten chapters make up the engineering section of the text. This section is not intended as a design course for broadcast equipment; rather, it presents the engineering fundamentals needed by the chief engineer or maintenance technician, or trainee for these positions, of an a-m and/or fm broadcast station. Therefore, the design of networks for impedance-matching or bridging circuitry is included. A knowledge of mathematics at the high-school level is assumed, although a refresher course is included for reference.

The final four chapters are designed to answer the question, "I know the theory; now what do I do on the job?" The basic technical theory of radio broadcasting has remained essentially the same, but great advances have been made in the development of equipment and operations. These improvements have outmoded many techniques. Thus, the last chapters are devoted to a discussion of modern station operations.

In preparing the content, I have deliberately avoided tying down specific procedures to specific FCC rules. It is the duty of every engineer who is responsible for compliance with such rules (normally the chief engineer of the station) to subscribe to current FCC rules about broadcasting and to keep abreast of changes that occur. Therefore, whenever FCC rules are mentioned as examples, it must be understood that the presentation is for illustrative purposes only and is *not* necessarily the latest ruling of the Commission.

When maintenance and engineering personnel require a quick and easy reference, the extensive subject breakdown in the text and the complete index will prove helpful. Preventive maintenance and proof-of-performance testing are given full coverage.

An effort has been made to compile data which the broadcast engineer frequently needs but for which he must usually consult numerous sources. It is hoped that this volume will serve both as an informative text and as a useful reference source.

The author is indebted to the following for important contributions to this book: American Broadcasting Co.; CBS Laboratories; CCA Electronics Corp.; Electrodyne Co., Inc.; Gates Division Harris-Intertype Corp.; General Electric Co.; Marti Electronics, Inc.; Metron Instruments, Inc.; National Broadcasting Co.; RCA; Sansui Electric Co., Ltd.; Schafer Electronics Corp.; Sony Corp.; TelePro Industries, Inc.; and the 3M Co.

HAROLD E. ENNES

Contents

CHAPTER 4

CHAPTER 5

CHAPTER 6

CHAPTER 7

CHAPTER 8

CHAPTER 9

CHAPTER 10

CHAPTER 11

CHAPTER 12

CHAPTER 13

CHAPTER 14

APPENDIX A

APPENDIX B

CHAPTER 1

Orientation in AM-FM Broadcasting

This book is concerned with basic training in a-m and fm broadcast engineering, where the following definitions apply: The abbreviation a-m stands for "amplitude modulation," in which the transmitted rf carrier is fixed in frequency but is increased or decreased in amplitude (modulated) by the applied sound. The abbreviation fm stands for "frequency modulation," in which the rf carrier remains constant in amplitude, but the frequency is varied (modulated) by the applied sound. In this chapter we will first consider specifics of a-m transmission, then fm transmission. After the characteristics of each have been explored, we will examine studio and transmitter requirements common to both systems.

1-1. THE STANDARD A-M BROADCAST CHANNEL

The following three definitions describe the foundation for the standard a-m station assignments. *Standard broadcast station* means a broadcasting station licensed to operate on a channel in the frequency band of 535 to 1605 kilohertz (kHz) and to transmit radiotelephone emissions primarily intended to be received by the general public. *Standard broadcast band* means the band of frequencies extending from 535 to 1605 kHz. *Standard broadcast channel* means the band of frequencies occupied by the carrier and two sidebands (upper and lower) of a broadcast signal with the carrier frequency at the center. Channels are designated by their assigned carrier frequencies. There are 107 carrier frequencies assigned to standard broadcast stations, beginning at 540 kHz and continuing in successive steps of 10 kHz to 1600 kHz.

When a radio carrier is modulated in amplitude, the original carrier plus upper and lower sidebands exist. The extent of the sidebands depends on the highest frequency that modulates the carrier wave at the specific time. Thus, if the highest modulating frequency is 5000 Hz (5 kHz), the limits of the transmission bandwidth are the carrier frequency plus and minus

5 kHz. For example, a carrier of 540 kHz amplitude modulated by a 5-kHz tone would have an upper sideband of 545 kHz and a lower sideband of 535 kHz.

When a-m broadcasting was first established, 5 kHz was about the highest audio frequency that could be expected from either microphones or phono pickups. Telephone lines carrying program transmissions between studio and transmitter, and overland routes carrying network radio programs, were equalized to 5 kHz, which was entirely adequate to meet the needs of the times. Each a-m station was assigned a channel of 10 kHz (plus and minus 5 kHz), as described above.

For example, Fig. 1-1 shows a station with an assigned carrier frequency of 1000 kHz. By FCC definition, a station on 540 kHz has a lower sideband

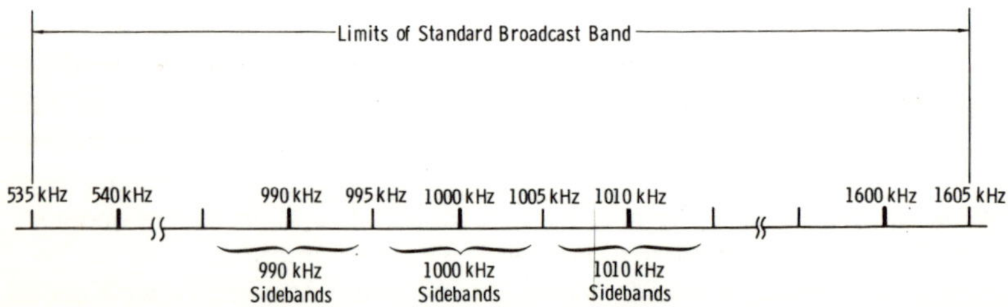

Fig. 1-1. Channels in standard broadcast band.

limit of 535 kHz, and a station on 1600 kHz has an upper sideband limit of 1605 kHz, thus fixing the assigned channel width at ±5 kHz. In Fig. 1-1, the 1000 kHz carrier, when modulated with a 5000 Hz tone, would have sidebands from 995 kHz to 1005 kHz, which would just meet the sideband limits from the adjacent (upper and lower) assigned carriers.

Modern technology allows the gamut of audible frequencies to be picked up, recorded, or reproduced. This covers a frequency range of 30 to 15,000 Hz. Practically all modern a-m broadcast transmitters are capable of an audio frequency response within 1 dB from 30 Hz to 12 kHz. This is entirely practicable in spite of the theoretical bandwidth limitation on a-m channels. In practice, certain minimum mileage separations must exist between stations on the same or adjacent channels. This is considered in greater detail in Chapter 9.

1-2. CLASSES OF STANDARD BROADCAST CHANNELS AND STATIONS

Table 1-1 presents a general outline of the assignment of stations to channels in the standard broadcast band in the United States. Caution: Always keep abreast of current FCC Rules and Regulations, which are subject to change. (NOTE: Some stations authorized under earlier allocation standards may not be operating in accord with Table 1-1.)

Clear Channels

A clear channel is one on which the dominant station or stations render service over wide areas. These stations are protected from objectionable interference within their primary service areas and over all or a substantial portion of their secondary service areas. Stations operating on these channels are classified as follows.

Class I Stations—A class I station is a dominant station operating on a clear channel and designed to render primary and secondary service over an extended area and at relatively long distances. Its primary service area is free from objectionable interference from other stations on the same and adjacent channels, and its secondary service area is free from interference except from stations on adjacent channels, and from stations on the same channel in accordance with the channel designation in current FCC rulings. The operating power must not be less than 10 nor more than 50 kW.

Class II Stations—A class II station is a secondary station that operates on a clear channel and is designed to render service over a primary service area, which is limited by and subject to such interference as may be received from class I stations. Whenever necessary, a class II station must use a directional antenna or other means to avoid interference with class I stations and with other class II stations.

Class II stations are divided into three groups as follows: A *class II-A* station is an unlimited-time class II station operating on one of the clear channels listed in Table 1-1 and assigned to a community within a state specified in current FCC rulings. A class II-A station must operate with power of not less than 10 kW at night nor more than 50 kW at any time. A *class II-B* station is an unlimited-time class II station other than those included in class II-A. A class II-B station must operate with power not less than 250 watts nor more than 50 kW. A *class II-D* station is a class II station operating daytime or limited time. A class II-D station must operate with power not less than 250 watts nor more than 50 kW.

Regional Channels

A regional channel is one on which several stations may operate with powers not in excess of 5 kW. The primary service area of a station operating on any such channel may be limited to a given field-intensity contour as a consequence of interference.

A *class III* station is a station that operates on a regional channel and is designed to render service primarily to a principal center of population and the surrounding rural area. Class III stations are subdivided into two classes. A *class III-A* station is a class III station that operates with power not less than 1 kW nor more than 5 kW and whose service area is subject to interference in accordance with current FCC rulings. A *class III-B* sta-

Table 1-1. Channels for A-M Stations

Freq (kHz)	Type of Channel*	Class of Station	Notes
540	C	II	Note 3
550	R	III	
560	R	III	
570	R	III	
580	R	III	
590	R	III	
600	R	III	
610	R	III	
620	R	III	
630	R	III	
640	C	I	Only One Class I Station
650	C	I	Only One Class I Station
660	C	I	Only One Class I Station
670	C	I & II	Only One Class I Station
680	C	I & II	
690	C	II	Note 1
700	C	I	Only One Class I Station
710	C	I & II	
720	C	I & II	Only One Class I Station
730	C	II	Note 2
740	C	II	Note 1
750	C	I	Only One Class I Station
760	C	I	Only One Class I Station
770	C	I	Only One Class I Station
780	C	I & II	Only One Class I Station
790	R	III	
800	C	II	Note 2
810	C	I & II	
820	C	I	Only One Class I Station
830	C	I	Only One Class I Station
840	C	I	Only One Class I Station
850	C	I & II	
860	C	II	Note 1
870	C	I	Only One Class I Station
880	C	I & II	Only One Class I Station
890	C	I & II	Only One Class I Station
900	C	II	Note 2
910	R	III	
920	R	III	
930	R	III	
940	C	I & II	
950	R	III	
960	R	III	
970	R	III	
980	R	III	
990	C	II	Note 1
1000	C	I & II	
1010	C	II	Note 1
1020	C	I & II	Only One Class I Station

Freq (kHz)	Type of Channel*	Class of Station	Notes
1030	C	I & II	Only One Class I Station
1040	C	I	Only One Class I Station
1050	C	II	Note 2
1060	C	I & II	
1070	C	I & II	
1080	C	I & II	
1090	C	I & II	
1100	C	I & II	Only One Class I Station
1110	C	I & II	
1120	C	I & II	Only One Class I Station
1130	C	I & II	
1140	C	I & II	
1150	R	III	
1160	C	I & II	Only One Class I Station
1170	C	I & II	
1180	C	I & II	Only One Class I Station
1190	C	I & II	
1200	C	I	Only One Class I Station
1210	C	I & II	Only One Class I Station
1220	C	II	Note 2
1230	L	IV	
1240	L	IV	
1250	R	III	
1260	R	III	
1270	R	III	
1280	R	III	
1290	R	III	
1300	R	III	
1310	R	III	
1320	R	III	
1330	R	III	
1340	L	IV	
1350	R	III	
1360	R	III	
1370	R	III	
1380	R	III	
1390	R	III	
1400	L	IV	
1410	R	III	
1420	R	III	
1430	R	III	
1440	R	III	
1450	L	IV	
1460	R	III	
1470	R	III	
1480	R	III	
1490	L	IV	
1500	C	I & II	
1510	C	I & II	
1520	C	I & II	
1530	C	I & II	

Table 1-1. Channels for A-M Stations—cont

Freq (kHz)	Type of Channel*	Class of Station	Notes
1540	C	I & II	
1550	C	I & II	
1560	C	I & II	
1570	C	II	Note 2
1580	C	II	Note 1
1590	R	III	
1600	R	III	

*C = Clear, R = Regional, L = Local

NOTES:

1. For Class I stations which will not deliver over 5 microvolts per meter ground wave or 25 microvolts per meter 10 percent time sky wave at any point on the Canadian border, and provided that such stations operating nighttime (i.e., during hours between sunset and sunrise at the location of the Class II station) are located not less than 650 miles from the nearest point on the Canadian border.
2. For Class II stations which operate daytime only, which will not deliver at any point on the Mexican border over 5 microvolts per meter ground wave, and which operate with no more than the following powers:
   (A) If not located within the areas specified below, 5 kilowatts.
   (B) If operating on any of the following frequencies within the following specified areas, no more than 1 kilowatt:
      (1) 800 kHz: less than 1319 kilometers (820 miles) from Ciudad Juarez, Chihuahua.
      (2) 1050 kHz: less than 998 kilometers (620 miles) from Monterrey, Nuevo Leon.
      (3) 1570 kHz: less than 998 kilometers (620 miles) from Ciudad Acuna, Coahuila.
3. For Class II stations which will not deliver a signal of more than 5 microvolts per meter ground wave or 25 microvolts per meter 10 percent sky wave at any point on the Canadian border, nor more than 10 microvolts per meter daytime or 50 microvolts per meter nighttime at any point on the Mexican border: Provided, that stations operating at night shall be located:
   (A) Within the continental United States including Alaska; and
   (B) Not less than 650 miles from the nearest point on the Canadian border; and
   (C) North of the parallel 35° N if west of the meridian 93° W, or north of the parallel 30° N if east of said meridian.

tion is a class III station that operates with power not less than 500 watts nor more than 1 kW nighttime and 5 kW daytime, and whose service area is subject to interference in accordance with current FCC rules.

Local Channels

A *local channel* is one on which several low-power stations operate. The primary service area of a station operating on any such channel may be limited to a given field-intensity contour as a consequence of interference. Such stations operate with power no greater than 250 watts nighttime and 1 kW daytime. The daytime power may be no greater than 250 watts if the station is located 100 kilometers (62 miles) or closer to the Mexican border, or in the area of the state of Florida south of 28° north latitude and between 80° and 82° west longitude.

A *class IV* station is a station operating on a local channel and designed to render service primarily to a city or town and the immediately surrounding suburban and rural areas. The power of a station of this class cannot be less than 250 watts, and it cannot be more than 250 watts nighttime or

1 kilowatt daytime. Its service area is subject to interference in accordance with current FCC rules. (Stations which were licensed to operate with 100 watts day or night prior to this ruling may continue to do so.)

Special Class II-A Station Rulings

Some of the FCC rulings applicable to class II-A stations are as follows:

No class II-A station shall be authorized unless at least 25 percent of its interference-free service area or at least 25 percent of the population residing therein receives no other interference-free nighttime primary service. Class II-A stations shall operate with a power of not less than 10 kilowatts nighttime.

The cochannel class I station shall be protected by the Class II-A station to its 0.1 millivolt per meter (mV/m) contour daytime and its 0.5 mV/m 50 percent sky-wave contour nighttime. All other stations of any class authorized on or before October 30, 1961 shall normally receive protection from objectionable interference from class II-A stations as provided in current FCC rules.

A class II-A station shall normally receive daytime protection to its 0.5 mV/m ground-wave contour and nighttime protection to the contour to which it is limited by the cochannel class-I station.

1-3. TECHNICAL DEFINITIONS FOR A-M BROADCASTING

The following terms, as used in this text, are defined as indicated.

Antenna current: The radio-frequency current in the antenna with no modulation.

Antenna power: The product of the square of the antenna current and the antenna resistance at the point where the current is measured.

Antenna resistance: The total resistance of the transmitting antenna system at the operating frequency and at the point at which the antenna current is measured.

Blanketing: That form of interference which is caused by the presence of a broadcast signal of one volt per meter (V/m) or greater intensity in the area adjacent to the antenna of the transmitting station. The 1 V/m contour is referred to as the *blanket contour,* and the area within this contour is referred to as the *blanket area.*

Combined audio harmonics: The arithmetical sum of the amplitudes of all the separate harmonic components. Root-sum-square harmonic readings may be accepted under conditions prescribed by the FCC.

Effective field: The root-mean-square (rms) value of the inverse distance fields at a distance of 1 mile from the antenna in all directions in the horizontal plane.

Grid modulation: Modulation produced by introduction of the modulating wave into any of the grid circuits of any tube in which the carrier-frequency wave is present.

High level modulation: Modulation produced in the plate circuit of the last radio stage of the system.

Last radio stage: The radio-frequency power-amplifier stage that supplies power to the antenna.

Low level modulation: Modulation produced in an earlier stage than the final.

Maximum percentage of modulation: The greatest percentage of modulation that may be obtained in a transmitter without producing in its output harmonics of the modulating frequency in excess of those permitted by the regulations.

Maximum rated carrier power: The maximum power at which the transmitter can be operated satisfactorily. It is determined by the design of the transmitter and the type and number of tubes used in the last radio stage.

Modulated stage: The radio-frequency stage to which the modulator is coupled and in which the continuous wave (carrier wave) is modulated in accordance with the system of modulation and the characteristics of the modulating wave.

Modulator stage: The last amplifier stage of the modulating wave which modulates a radio-frequency stage.

Operating power: The power that is actually supplied to the antenna.

Percentage modulation (amplitude): The ratio, expressed in percentage, of half the difference between the maximum and minimum amplitudes of the amplitude-modulated wave to the average amplitude.

Plate input power: The product of the direct plate voltage applied to the tubes in the last radio stage and the total direct current flowing to the plates of these tubes, measured without modulation.

Plate modulation: Modulation produced by introduction of the modulating wave into the plate circuit of any tube in which the carrier-frequency wave is present.

Service areas: The *primary service area* of a broadcast station is the area in which the ground wave is not subject to objectionable interference or objectionable fading. The *secondary service area* is the area served by the sky wave and not subject to objectionable interference. The signal is subject to intermittent variations in intensity. The *intermittent service area* is the area receiving service from the ground wave but beyond the primary service area and subject to some interference and fading.

1-4. FM BROADCAST CHANNELS

The fm broadcast band consists of that portion of the radio-frequency spectrum between 88 and 108 megahertz (MHz). It is divided into 100 channels of 200 kilohertz each. For convenience, the frequencies available for fm broadcasting (including those assigned to noncommercial educational broadcasting) are given numerical designations, which are shown in Table 1-2.

Table 1-2. FM Broadcast Channels

Frequency MHz	Channel No.	Frequency MHz	Channel No.	Frequency MHz	Channel No.
88.1	201	94.7	234	101.3	267
88.3	202	94.9	235	101.5	268
88.5	203	95.1	236	101.7*	269*
88.7	204	95.3*	237*	101.9	270
88.9	205	95.5	238	102.1	271
89.1	206	95.7	239	102.3*	272*
89.3	207	95.9*	240*	102.5	273
89.5	208	96.1	241	102.7	274
89.7	209	96.3	242	102.9	275
89.9	210	96.5	243	103.1*	276*
90.1	211	96.7*	244*	103.3	277
90.3	212	96.9	245	103.5	278
90.5	213	97.1	246	103.7	279
90.7	214	97.3	247	103.9*	280*
90.9	215	97.5	248	104.1	281
91.1	216	97.7*	249*	104.3	282
91.3	217	97.9	250	104.5	283
91.5	218	98.1	251	104.7	284
91.7	219	98.3*	252*	104.9*	285*
91.9	220	98.5	253	105.1	286
92.1*	221*	98.7	254	105.3	287
92.3	222	98.9	255	105.5*	288*
92.5	223	99.1	256	105.7	289
92.7*	224*	99.3*	257*	105.9	290
92.9	225	99.5	258	106.1	291
93.1	226	99.7	259	106.3*	292*
93.3	227	99.9	260	106.5	293
93.5*	228*	100.1*	261*	106.7	294
93.7	229	100.3	262	106.9	295
93.9	230	100.5	263	107.1*	296*
94.1	231	100.7	264	107.3	297
94.3*	232*	100.9*	265*	107.5	298
94.5	233	101.1	266	107.7	299
				107.9	300

*Class A channel.
NOTE: The frequency 108.0 MHz may be assigned to VOR test stations subject to the condition that interference is not caused to the reception of fm broadcasting stations, present or future.

The frequency band 88.1-91.9 MHz (channels 201-220) is for noncommercial educational fm broadcasting, except in Alaska, where the frequency band 88-100 MHz is allocated exclusively to government radio services and the nongovernment fixed service. However, the frequencies 100.1-107.9 MHz (channels 261 through 300, inclusive) are available for use by noncommercial educational stations in Alaska.

Except as provided in current and special FCC rulings, the frequencies shown by asterisks in Table 1-2 are designated as class A channels and are assigned for the use, in all zones, of class A stations only. A class A station is designed to render service to a relatively small community, city or town, and the surrounding rural area. (Most communities designated for fm broadcast service by the FCC were assigned class A channels.) A class A station may be authorized to operate with as much as 3 kW effective radiated power (erp) in the horizontal plane, with an antenna height above average terrain up to 300 feet. For improved reception and coverage, class A stations may also radiate a full 3 kW erp in the vertical plane.

Class B fm broadcast stations provide service to principal cities and larger communities. In general, these stations may radiate up to 50 kW erp in the horizontal plane (and also in the vertical plane), with antenna height up to 500 feet above average terrain.

The most powerful fm broadcast facility permitted by the FCC is the class C station. These stations are authorized to radiate up to 100 kW erp at an antenna height above average terrain up to 2000 feet. This applies to both the horizontal and vertical planes of polarization.

NOTE: For all classes, where the height of the antenna exceeds that specified, the maximum radiated power must be reduced. These conditions are explored more fully in Chapter 10.

The minimum required erp for a class A fm station is 100 watts. For a class B station, the minimum is 5 kW, and for a class C station it is 25 kW. No minimum antenna height above average terrain is specified.

1-5. TECHNICAL DEFINITIONS FOR FM BROADCASTING

As used in this text, the following terms related to fm broadcasting are defined as indicated.

Antenna height above average terrain: The average of the antenna heights above the terrain from 2 to 10 miles from the antenna for the eight directions spaced evenly for each 45° of azimuth, starting with true north. (In general, a different antenna height will be determined in each direction from the antenna.) The average of these various heights is considered the antenna height above the average terrain. In some cases, fewer than eight directions may be used. Where circular or elliptical polarization is employed, the antenna height above average terrain shall be based upon the height of the antenna which transmits the horizontal component.

Antenna power gain: The square of the ratio of the root-mean-square free-space field strength produced at 1 mile in the horizontal plane, in millivolts per meter for 1 kilowatt antenna input power, to 137.6 mV/m. This ratio should be expressed in decibels (dB). (If specified for a particular direction, antenna power gain is based on the field strength in that direction only.)

Center frequency: (1) The average frequency of the emitted wave when modulated by a sinusoidal signal. (2) The frequency of the emitted wave without modulation.

Cross talk: An undesired signal occurring in one channel and caused by an electrical signal in another channel.

Effective radiated power: The product of the antenna power (transmitter output power less transmission line loss) times (1) the antenna power gain or (2) the antenna field gain squared. Where circular or elliptical polarization is employed, the term effective radiated power is applied separately to the horizontal and vertical components of radiation. For allocation purposes, the authorized effective radiated power is the horizontally polarized component of radiation only.

Field strength: The electric field strength in the horizontal plane.

Fm broadcast band: The band of frequencies extending from 88 to 108 MHz, which includes those assigned to noncommercial educational broadcasting.

Fm broadcast channel: A band of frequencies 200 kHz wide and designated by its center frequency. Channels for fm broadcast stations begin at 88.1 MHz and continue in successive steps of 200 kHz to and including 107.9 MHz.

Fm broadcast station: A station employing frequency modulation in the fm broadcast band and licensed primarily for the transmission of radiotelephone emissions intended to be received by the general public.

Fm stereophonic broadcast: The transmission of a stereophonic program by a single fm broadcast station, utilizing the main channel and a stereophonic subchannel.

Free-space field strength: The field strength that would exist at a point in the absence of waves reflected from the earth or from other reflecting objects.

Frequency modulation: A system of modulation in which the instantaneous radio frequency varies in proportion to the instantaneous amplitude of the modulating signal (amplitude of modulating signal to be measured after pre-emphasis, if used) and the instantaneous radio frequency is independent of the frequency of the modulating signal.

Frequency swing: The instantaneous departure of the frequency of the emitted wave from the center frequency.

Left (or right) signal: The electrical output of a microphone or combination of microphones that are placed so as to convey the intensity, time, and location of sounds originating predominately to the listener's left (or right) of the center of the performing area.

Left (or right) stereophonic channel: The left (or right) signal as electrically reproduced in reception of fm stereophonic broadcasts.

Main channel: The band of frequencies from 50 to 15,000 Hz that frequency modulate the main carrier.

Multiplex transmission: The simultaneous transmission of two or more signals within a single channel. Multiplex transmission as applied to fm broadcast stations means the transmission of facsimile or other signals in addition to the regular broadcast signals.

Percentage modulation: The ratio of the actual frequency swing to the frequency swing defined as 100 percent modulation, expressed in percentage. For fm broadcast stations, a frequency swing of ±75 kHz is defined as 100 percent modulation.

Pilot subcarrier: A subcarrier serving as a control signal for use in the reception of fm stereophonic broadcasts. The pilot subcarrier has a frequency of 19 kHz.

Stereophonic separation: The ratio of the electrical signal caused in the right (or left) stereophonic channel to the electrical signal caused in the left (or right) stereophonic channel by the transmission of only a right (or left) signal.

Stereophonic subcarrier: A subcarrier having a frequency which is the second harmonic of the pilot subcarrier frequency and which is employed in fm stereophonic broadcasting.

Stereophonic subchannel: The band of frequencies extending from 23 to 53 kHz, and containing the stereophonic subcarrier and its associated sidebands.

1-6. GENERAL STUDIO REQUIREMENTS

Fig. 1-2 illustrates an example of studio and control-room layout that meets the requirements of the "medium-size" station, whether a-m, fm, or both. In this example, the transmitter is installed at a location remote from the studio.

Continuous operation is carried out in the main control room. The control console allows operation of all microphones, turntables, and cartridge and reel-to-reel tape machines, and it provides for switching of remote or network signals. It may handle operations of both the a-m and fm transmitters. Normally, all rack-mounted tape machines and possibly automation equipment may be controlled from the operating control console. If the transmitter(s) is remotely controlled, control and monitoring equipment is included, as shown at the right of the control console in the main control room.

The production control room normally handles all special recording facilities (cartridge and reel-to-reel tape machines), and it may also be used in hours of split am-fm programming. Many times, a complete "disc-jockey" show may be put together in this location, then played back at the regular program time from the main control room. Or delayed broadcasts from the studio, remote sources, or the network may be recorded here for future playback. A view of the ABC Radio Network production control operating center in Hollywood is shown in Fig. 1-3.

The record and tape library contains all the records (discs) and tapes used by the station. Turntables, tape equipment, and suitable monitoring facilities are installed in this room for the purpose of auditions and, in some cases, editing of tapes and programs.

1-7. GENERAL TRANSMITTER REQUIREMENTS

An example of typical transmitter equipment layout is shown in Fig. 1-4. When the transmitter is located remotely from the studio, the program is normally carried over telephone company (Telco) land lines, with a separate line to provide direct communication with the studio. In the case of an fm transmitter that is located atop a mountain, a radio-frequency

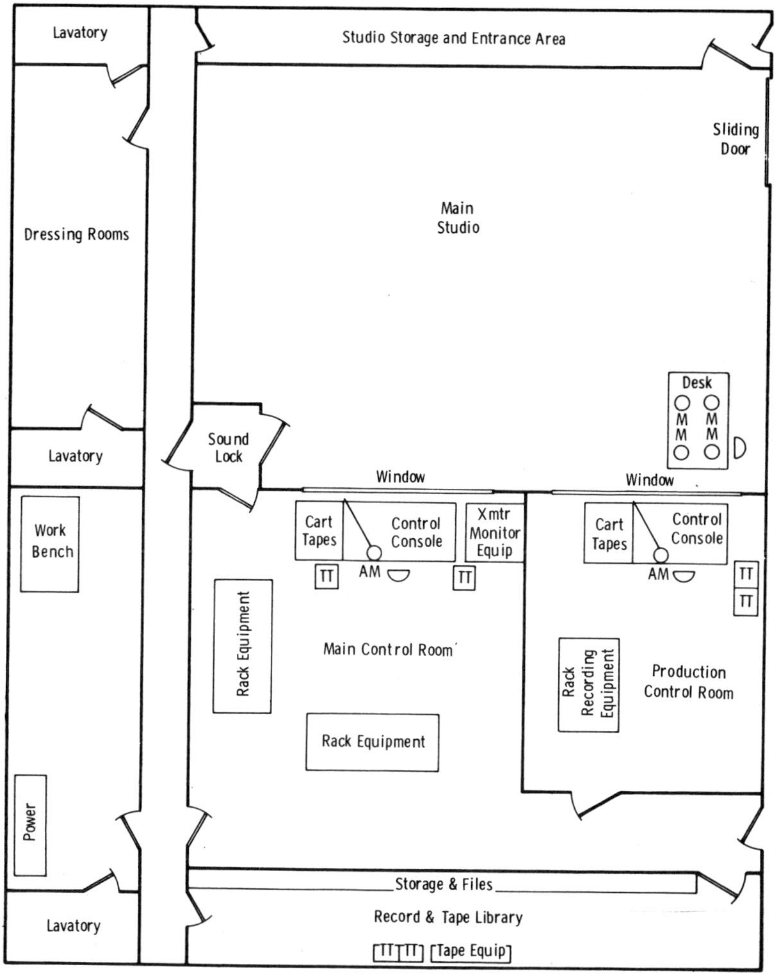

Legend: TT=Turntable
M=Microphone, desk or floor-stand
AM=Announce microphone

Fig. 1-2. Example of studio and control-room layout.

Courtesy American Broadcasting Co.

Fig. 1-3. ABC production control, Hollywood, California.

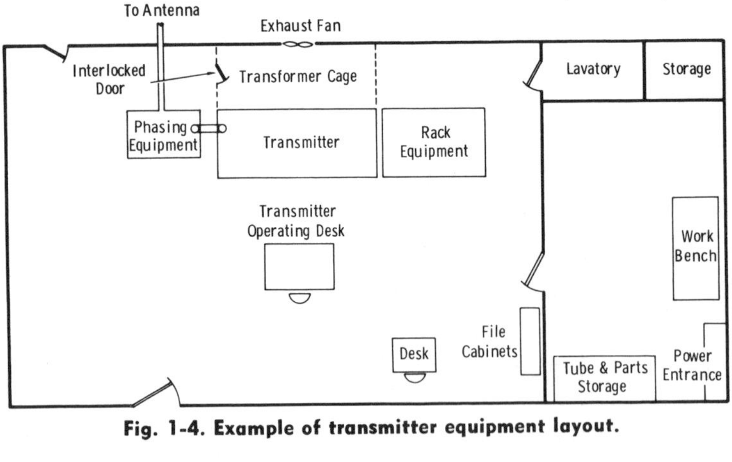

Fig. 1-4. Example of transmitter equipment layout.

studio-to-transmitter link (stl) is often employed. Program lines to a-m transmitters are usually equalized to 8 kHz or 10 kHz. Program lines to fm sites must be equalized to 15 kHz.

The rack equipment includes program amplifiers, power supplies, monitoring equipment, and, if authorized, remote-control equipment.

The transmitter control desk contains switches to control final-stage power adjustment and switches for "going directional" when required. It normally also houses metering equipment for antenna currents, remote indication of modulation percentage, and incoming audio level indicators. The phasing equipment provides the common-point feed to all towers; it is required only when a directional antenna system must be employed.

When a transmitter is operated by remote control, the following FCC rules must be satisfied:

(1) The equipment at the operating and transmitting positions shall be so installed and protected that it is not accessible to or capable of operation by persons other than those duly authorized by the licensee.

(2) The control circuits from the operating positions to the transmitter shall provide positive on and off control and shall be such that open circuits, short circuits, grounds, or other line faults will not actuate the transmitter, and any fault causing loss of such control will automatically place the transmitter in an inoperative position.

(3) A malfunction of any part of the remote-control equipment and associated line circuits resulting in improper control or inaccurate meter readings shall be cause for the immediate cessation of operation by remote control.

(4) Control and monitoring equipment shall be installed so as to allow the licensed operator at the remote-control point to perform all the functions in a manner required by the Commission's Rules.

EXERCISES

Q1-1. Define (A) amplitude modulation, (B) frequency modulation.

Q1-2. Give the frequencies covered by the standard broadcast band.

Q1-3. Give the lowest and highest assigned carrier frequencies in the standard broadcast band.

Q1-4. If an a-m transmitter on 890 kHz is modulated with a tone of 10 kHz, what is the (A) lower sideband and (B) upper sideband?

Q1-5. What are the minimum and maximum operating powers allowed by the FCC for (A) class I stations and (B) class IV stations?

Q1-6. Define the fm broadcast band in terms of frequency.

Q1-7. How wide is each fm channel?

Q1-8. When an fm station is authorized for a given maximum power, what else is specified?

Q1-9. What is a "pilot subcarrier" and when is it used?

Q1-10. Give the frequency band of a stereophonic subchannel.

CHAPTER **2**

Mathematics Review for Broadcast Engineering

A knowledge of fundamental mathematics is a necessity for a good understanding of broadcast engineering. A background equivalent to high school mathematics is assumed in this book. This chapter will serve as a review of fundamental relationships and as a reference for the remainder of the text. You can readily determine whether or not you need to study this chapter by taking the examination at the end of the chapter.

NOTE: Appendix A contains tables of common logarithms, natural trigonometric functions, and decibels.

2-1. NUMBERS

Real numbers when given in simple signless form such as 1, 2, 3, 30, etc., are automatically taken as *positive* values. But we know that it is also necessary to consider zero and *negative* numbers. See Fig. 2-1A. Zero is a reference or starting point. If we travel 2 miles east (to the right) from this starting point, then turn around and travel 4 miles west (to the left through the starting point), we can express the result as $+2-4=-2$ miles.

Be sure to visualize this relationship properly. We traveled in a positive direction for two miles, then turned around. If we had then traveled in a negative direction just two miles, we would have ended our trip at the starting or reference point (zero), although we had traveled a total of four miles. But we actually continued on through the starting point for a total of four miles in the negative direction, with the end result of arriving at the -2 mile point. This is a total excursion of six miles.

We can just as accurately plot the positive and negative values along the north-south (y) axis as along the east-west (x) axis, as shown in Fig. 2-1B.

(*A*) *X axis.*

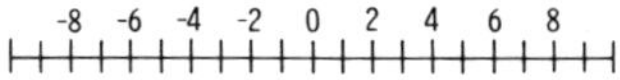

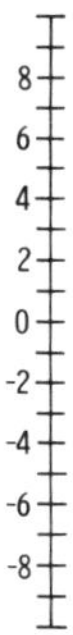

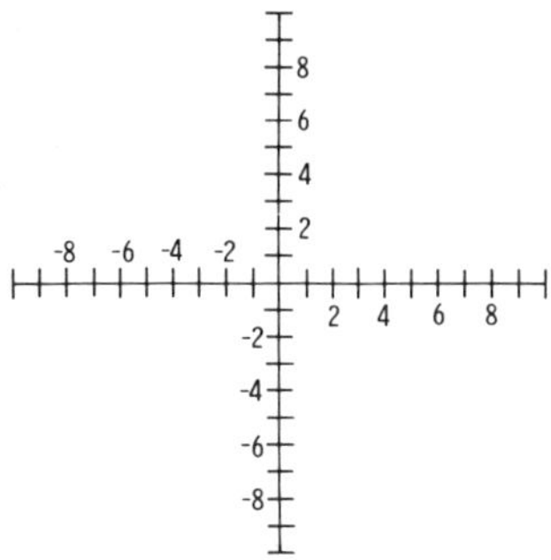

(*B*) *Y axis.* (*C*) *Combined axes.*

Fig. 2-1. Graphical representation of whole numbers.

Or we can combine Figs 2-1A and 2-1B to form the pattern of Fig. 2-1C. The directions of positive and negative values illustrated are those conventionally used.

The real number system consists of:

Zero (starting point)
Whole numbers (positive or negative)
Rational numbers (expressed as whole numbers or fractions)
Irrational numbers (which cannot be expressed as simple fractions)

An example of an irrational number is the square root of 2 ($\sqrt{2}$). This cannot be expressed as a simple fraction.

Whole numbers are also called "integral numbers." These are termed "even numbers" when exactly divisible by 2. When not exactly divisible by 2, they are called "odd numbers."

Adding and Subtracting Positive and Negative Numbers

Fig. 2-2 illustrates a basic electronic example of negative values. In Fig. 2-2A, a 3-volt and a 9-volt battery are connected in series aiding. A volt-

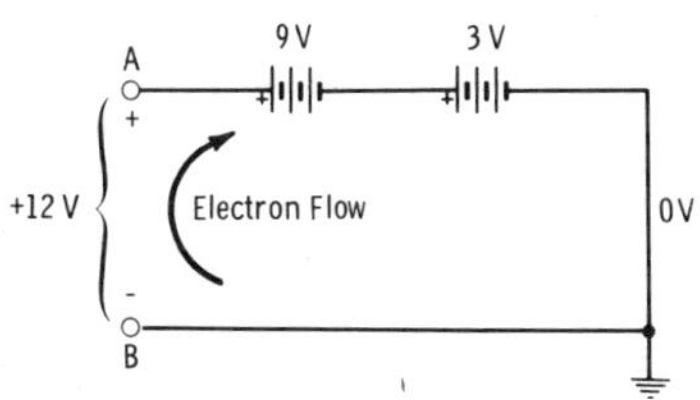

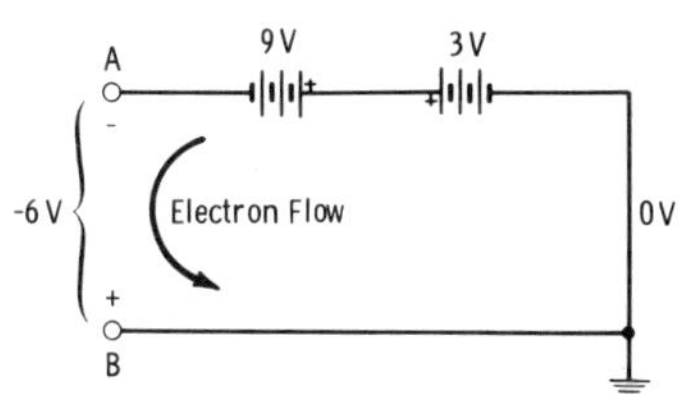

(*A*) *Batteries aiding.* (*B*) *Batteries opposing.*

Fig. 2-2. Basic example of negative values.

meter between points A and B shows point A to be 12 volts positive relative to reference (zero) point B $(9+3=12)$. It is just as accurate to say that point B is 12 volts negative relative to point A. In Fig. 2-1B, the 9-volt battery has been reversed. Now the voltmeter shows point A is 6 volts negative relative to reference point B $(-9+3=-6$ volts$)$. Or we can say that point B is 6 volts positive relative to point A. If we place a resistor between A and B in each circuit, the direction of electron flow between the terminals is in opposite directions in the two circuits.

Thus we have illustrated the basic rule of addition of positive and negative numbers. When we *add* two numbers of unlike sign, we simply *subtract* the smaller number from the larger number, and place the sign of the larger in front of the answer. When more than two numbers of unlike signs are to be added:

1. Add all positive numbers.
2. Add all negative numbers.
3. Subtract the smaller sum from the larger sum and affix the sign of the larger to the answer.

For example, add −2, 4, −8, 6, 9.
Solution:

$$\begin{array}{r} -2 \\ -8 \\ \\ \hline -10 \end{array} \qquad \begin{array}{r} 4 \\ 6 \\ 9 \\ \hline +19 \end{array}$$

so:

$$19+(-10)=+9$$

When adding only negative numbers, simply add all the numbers and affix the negative sign to the sum.

Now let us return momentarily to the travel example used in conjunction with Fig. 2-1A. In that example, we traveled a total of 6 miles and ended at the −2-mile point relative to the starting point. Now suppose we turn around again and travel 2 miles east. We arrive at our starting point (zero). What we have done essentially is to subtract −2 miles from our trip, which was exactly equivalent to *adding* a positive 2.

Thus, subtracting a negative number is equivalent to adding a positive number. The rule in subtraction is to change the sign of the *subtrahend* and add. For example, to subtract −10 from 6:

$$\begin{array}{rl} 6 & \text{Minuend} \\ -10 & \text{Subtrahend (Change sign to +.)} \\ \hline 16 & \text{Remainder} \end{array}$$

This would be expressed in equation form as:

$$6-(-10)=16$$

Fractions

Fig. 2-3 is a basic graphical representation of fractions. Proper fractions (values less than 1) are shown in Fig. 2-3A. Improper fractions (those totalling 1 or greater) are shown in Fig. 2-3B, which also indicates that improper fractions can be expressed as mixed numbers containing a whole number and a fraction.

The top number of a fraction is termed the *numerator*. This is an indicator of the number of fractional parts expressed. The bottom number of a fraction is termed the *denominator*. This is an indicator of the number of parts into which the whole unit has been divided.

Observe from Fig. 2-3A that smaller values are always to the left of 0 (the starting point) and larger values are always to the right of the starting point. The fraction $-1/4$ is larger than the fraction $-1/2$ because it lies to the right, as shown graphically. Conversely, the fraction $1/2$ is larger than the fraction $1/4$, again because it lies to the right on the graph.

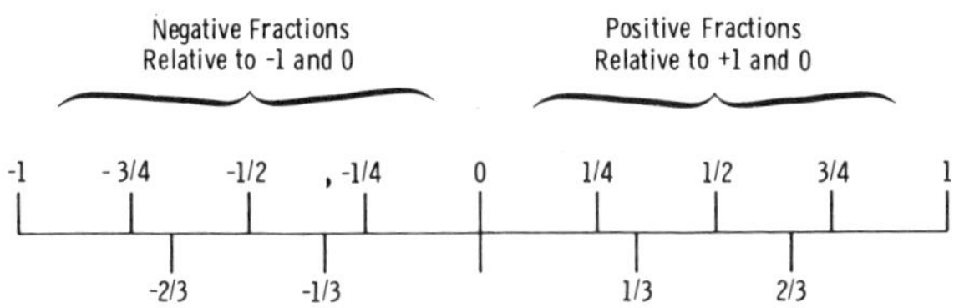

(*A*) *Proper fractions.*

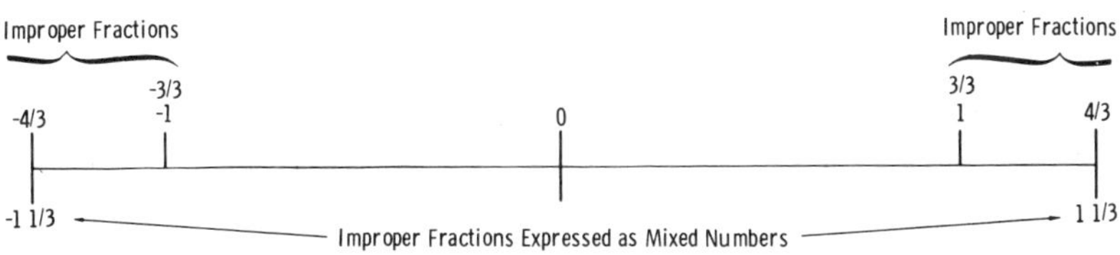

(*B*) *Improper fractions and mixed numbers.*

Fig. 2-3. Graphical representation of fractions.

Before we go further, note that the product (result of multiplication) can be written as (for example) $2 \times 6 = 12$ with the conventional multiplication sign. It can also be written $(2)(6) = 12$. We will use the latter form in this text.

The following basic rules for fractions apply:

1. The numerator and denominator can be *multiplied* or *divided* by the *same number* without changing the value of the fraction.

   For example, multiplying both the numerator and denominator of the fraction $2/3$ by 2 gives:

$$\frac{(2)(2)}{(3)(2)} = \frac{4}{6}$$

which is equivalent to ⅔.

Or, taking 4/6 and dividing both numerator and denominator by 2 gives:

$$\frac{4/2}{6/2} = \frac{2}{3}$$

2. To multiply a fraction by a given number, the numerator is multiplied by the number, or the denominator is divided by the number.

   For example,

$$\frac{4}{6}(2) = \frac{(4)(2)}{6} = \frac{8}{6} = 1\tfrac{1}{3}$$

or

$$\frac{4}{6/2} = \frac{4}{3} = 1\tfrac{1}{3}$$

3. To divide a fraction by a given number, the numerator is divided by the number, or the denominator is multiplied by the number.

   For example, 4/6 divided by 2:

$$\frac{4/2}{6} = \frac{2}{6} = \frac{1}{3}$$

or

$$\frac{4}{(6)(2)} = \frac{4}{12} = \frac{1}{3}$$

4. To add fractions:

   (A) Convert all the fractions to the lowest common denominator.
   (B) Add all the numerators.
   (C) Put that sum over the denominator.

   For example: ⅙ + ¼ + ½

To find the lowest common denominator (LCD), find the smallest number into which each denominator can be divided a whole number of times. In this example, 6 will go into 12 2 times, 4 will go into 12 3 times, and 2 will go into 12 6 times. Therefore 12 is the lowest common denominator.

Since 6 goes into 12 twice, then 1/6 is the same as 2/12. Since 4 goes into 12 3 times, 1/4 = 3/12. Since 2 goes into 12 6 times, 1/2 = 6/12. Then:

$$\frac{2}{12} + \frac{3}{12} + \frac{6}{12} = \frac{11}{12}$$

In the above example, the lowest common denominator was easy to find just by observation. Sometimes this is not the case; consider $1/5 + 1/4 + 1/2$.

To find this LCD, we use *factoring*. This simply means reducing to fundamental quantities. Since 5 and 1 are the only whole numbers that can be multiplied together to give 5, 5 is not factorable. Similarly, 2 is not factorable. But $4 = (2)(2)$. So we have:

$$5 = 5$$
$$4 = (2)(2)$$

The number 2 appears twice in 4, so $(2)(2)$ must be a part of the LCD, which must also include a 5. Thus:

$$\text{LCD} = (2)(2)(5) = 20$$

Then changing the fractions to the LCD:

$$1/5 = 4/20$$
$$1/4 = 5/20$$
$$1/2 = 10/20$$

Thus:

$$\frac{4}{20} + \frac{5}{20} + \frac{10}{20} = \frac{19}{20}$$

5. To convert a fraction to a decimal number, divide the denominator into the numerator. In the preceding example, the sum $4/20 + 5/20 + 10/20$ was found. These fractions can be converted to decimals as follows: $4/20 = 0.2$, $5/20 = 0.25$, and $10/20 = 0.5$. Then:

$$\begin{array}{l} 0.20 \\ 0.25 \\ \underline{0.50} \\ 0.95 \text{ total} \end{array}$$

This total stated in fractional form is ninety-five one-hundredths $(95/100)$. Dividing both numerator and denominator by 5 gives $19/20$, which was the answer obtained under rule 4 above.

6. To subtract one fraction from another fraction, (A) change both fractions to the same denominator as in rule 4 for addition, (B) subtract one numerator from the other numerator, and (C) place the remainder over the denominator. For example:

$$\frac{1}{4} - \frac{1}{5} = \frac{5}{20} - \frac{4}{20} = \frac{1}{20}$$

7. To multiply fractions, multiply the numerators and then multiply the denominators. For example, $(1/4)(1/5) = 1/20$. The product of

fractions always is less than either of the fractions. Take another example: $(3/5)(3/6) = 9/30$. Note that $9/30$ can be reduced to $3/10$ by dividing both numerator and denominator by 3.

8. To divide fractions, invert the divisor and then multiply. To keep your perspective on this rule, note that (for example) multiplying a number by $1/5$ is the same as dividing that number by 5. Thus $(3)(1/5) = 3/5$. Note that this is the same as dividing three by five $(3/5)$.

   For example, if $2/5$ is to be divided by $3/7$:

$$\frac{2}{5} \div \frac{3}{7} = \left(\frac{2}{5}\right)\left(\frac{7}{3}\right) = \frac{14}{15}$$

   Note that the divisor, $3/7$, was inverted $(7/3)$, and then the multiplication rule was used.

9. When a series of different operations must be performed, do them in the following sequence:

   1. Multiplications.
   2. Divisions.
   3. Additions.
   4. Subtractions.

   For example, solve

$$24 \div 6 - (4)(10) \div 5 + (20)(3) - 2$$

   Multiplications: $24 \div 6 - 40 \div 5 + 60 - 2$
   Divisions: $4 - 8 + 60 - 2$
   Additions: $64 - 8 - 2$
   Subtractions: $64 - 10 = 54$ (Answer)

2-2. PERCENTAGE

A percentage is a fractional value having a denominator of 100. For example, 1 percent is the same as $1/100$ or 0.01, 6 percent is the same as $6/100$ or 0.06, and 75 percent is the same as $75/100$ or 0.75.

If a power supply is said to be regulated within 1 percent of 280 volts, what is the allowable departure from 280 volts? Since 1 percent $= 0.01$, then $(280)(0.01) = 2.8$ volts (answer).

The regulation of a power supply is often computed in terms of the voltage difference between no-load and full-load conditions:

$$\%\ \text{Regulation} = \frac{E1 - E2}{E2}(100)$$

where,

E1 is the no-load voltage,
E2 is the full-load voltage.

For example, if E1 = 300 V and E2 = 290 V, then:

$$\% \text{ Regulation} = \frac{300 - 290}{290}(100)$$

$$= \frac{10}{290}(100)$$

$$= (0.0344)(100) = 3.44\%$$

2-3. EXPONENTS AND SQUARE ROOTS

The exponent of a number is a small figure placed at the upper right of the number. For example, 10^2 means that the number 10 is used as a factor two times, thus:

$$10^2 = (10)(10) = 100$$

Similarly,

$$10^3 = (10)(10)(10) = 1000$$

When the exponent is positive as in the above examples, the result is greater than unity, if the number being multiplied is greater than unity. The expression 10^2 is read "ten squared" or "ten raised to the second power"; 10^3 is read "ten raised to the third power"; etc.

When the exponent is negative, the reciprocal of the number (the number divided into 1) is raised to a positive power. Thus:

$$10^{-4} = \frac{1}{10^4} = \left(\frac{1}{10}\right)\left(\frac{1}{10}\right)\left(\frac{1}{10}\right)\left(\frac{1}{10}\right) = \frac{1}{10{,}000} = 0.0001$$

When the exponent is negative, the result is between zero and unity if the number to be raised to the power is greater than unity.

Unity (in terms of exponents) is any number raised to the zero power. Thus $10^0 = 1$, $2^0 = 1$, etc.

Addition of Exponents

Exponents are added in obtaining the product of like numbers raised to the same or different powers. Thus:

$$(2^2)(2^3) = 2^{2+3} = 2^5$$

or

$$(3^3)(3^6) = 3^{3+6} = 3^9$$

When unlike numbers are raised to powers and the product must be found, each number must be raised to the indicated power before multiplication. Thus the product of 2^2 and 3^4 is:

$$(2^2)(3^4) = (2)(2)(3)(3)(3)(3) = (4)(81) = 324$$

Subtraction of Exponents

Exponents are subtracted in the division of numbers raised to powers. Thus:

$$\frac{2^4}{2^2} = 2^{4-2} = 2^2 = 4$$

or

$$\frac{4^2}{4^3} = 4^{2-3} = 4^{-1} = \frac{1}{4}$$

or

$$\frac{10^6}{10^{-9}} = 10^{15}$$

But note that

$$\frac{2^3}{2^3} = 2^{3-3} = 2^0 = 1$$

(Any number raised to the zero power is equal to unity, or 1).

Multiplication of Exponents

Exponents are multiplied to find the power of a power. Thus $(3^3)^3 = 3^9$, $(3^3)^{-3} = 3^{-9}$, $(3^{-3})^2 = 3^{-6}$, $(4^{-3})^{-4} = 4^{12}$, etc.

Power of a Fraction

When a fraction is raised to a power, it can be written as the power of the numerator divided by the power of the denominator. For example:

$$\left(\frac{1}{2}\right)^2 = \frac{1^2}{2^2} = \frac{1}{4}$$

Roots

If the power of a number is a fraction, it can be written as the root of the number. Thus $16^{1/2} = \sqrt{16} = 4$. The symbol $\sqrt{\ }$ is always assumed to indicate the square root of the number, and the 2 is omitted. For any other root (such as the cube root, or $\sqrt[3]{\ }$) the order of the root (in this case 3) must be given. The symbol $\sqrt{\ }$ is called the *radical sign,* and the number under the radical is called the *radicand.*

Extracting Square Root

Extracting the square root of a number is relatively simple compared to extracting higher roots. These roots are obtained by using logarithms, as covered in Section 2-5.

How to extract the square root of a whole number is shown most easily by an example. Suppose it is desired to extract the square root of 48,400. Proceed as follows.

Step 1. Starting at the decimal point, divide the number into two-digit groups moving to the left: 4′84′00. Note that when the number of digits is an odd number, the leftmost digit is by itself.

Step 2. Find the largest number that when squared (multiplied by itself) will be less than or equal to the first group, in this case the single digit 4. Since 2 times 2 is 4, place the figure 2 immediately above the 4:

```
  2
√4′84′00
```

Step 3. The square of 2 is 4, so place this number under the first digit, and subtract:

```
  2
√4′84′00
  4
 ───
  0
```

Step 4. Bring the second group down with the remainder:

```
  2
√4′84′00
  4
 ────
   84
```

(Note that 4 − 4 is zero, so this is discarded, leaving 84).

Step 5. Double the number in the answer thus far and bring down the result as the first digit in a trial divisor for 84:

```
        2
      √4′84′00
        4
4     ) 84
```

Step 6. The second digit for this trial divisor must be such that if the total number making up the trial divisor is multiplied by the second digit, the product will be less than or equal to the remainder, in this case 84. We know that the trial divisor will become a number greater than 40 when a second digit is added. Thus 42 is the number, since 42 times 2 (the number added) is exactly equal to 84. So:

```
      2  2  0   (answer)
    √4′84′00
      4
42)  ─────
      84
      84
     ─────
      00′00
```

Thus in this example, the number added to 40 is 2, and 2 times 42 is 84. The number 2 is placed immediately above the ′84, and since the remainder of the total number is 00′00, a zero (0) is placed immediately above the ′00 under the radical sign.

In the above example, we extracted the square root of a perfect square. In the following example, the radicand is not a perfect square.

Problem: Extract the square root of 1410.

```
       3  7. 5
     √14′10.00
       9
  67)510
     469
 745)  4100
       3725
        375
```

The answer to the first decimal place is 37.5 with a remainder of 375. For greater accuracy (more decimal places), two more zeros would be brought down and the process continued.

The process is exactly the same in extracting the square root of a number less than unity. For example, find the square root of 0.00026. First, mark off the number in two-digit groups:

.00′02′60

Note that an extra cipher is added to the last digit to make a group of two.

```
       . 0  1  6  1
     √.00′02′60′00
            1
   26)     160
           156
  321)       400
             321
              79
```

The answer to four places is 0.0161 with a remainder of 79. To prove the answer, (0.0161)(0.0161) = 0.000259, or very nearly 0.00026.

2-4. SCIENTIFIC NOTATION

Scientific notation greatly simplifies computations that involve very large numbers. The process is simpler than would appear from a statement of the basic rule, which is: "Any number expressed in scientific notation is written as a decimal number between one and ten, multiplied by ten raised to the proper power."

Example 1: $5{,}000{,}000 = (5)(10^6)$
Example 2: $525 = (5.25)(10^2)$
Example 3: $0.000005 = (5)(10^{-6})$
Example 4: $0.00000024 = (2.4)(10^{-7})$

These examples illustrate that the decimal point is shifted as many places as necessary, and this result is multiplied by ten raised to a power equal to the number of places the decimal point was shifted. If the decimal point is moved to the left, the power of ten is positive. If the decimal point is moved to the right, the power of ten is negative.

Powers of ten from 10^6 to 10^{-6} and the conventional terminology are shown in Chart 2-1.Prefixes commonly used in electronics, along with their symbols and power-of-ten multipliers, are listed in Table 2-1.

Chart 2-1. Powers of Ten

10^6	10^5	10^4	10^3	10^2	10^1	10^0	.	10^{-1}	10^{-2}	10^{-3}	10^{-4}	10^{-5}	10^{-6}
						1	.						
					1	0	.	1					
				1	0	0	.	0	1				
			1	0	0	0	.	0	0	1			
		1	0	0	0	0	.	0	0	0	1		
	1	0	0	0	0	0	.	0	0	0	0	1	
1	0	0	0	0	0	0	.	0	0	0	0	0	1
One Million	One Hundred Thousand	Ten Thousand	One Thousand	One Hundred	Ten	One	Decimal Point	One Tenth	One Hundredth	One Thousandth	One Ten Thousandth	One Hundred Thousandth	One Millionth

In practice, the number need not be reduced to a quantity between one and ten if it is more convenient to do otherwise. In Example 1 above, the result could just as correctly be written as $(50)(10^5)$. Example 4 could just as accurately be written as $(24)(10^{-8})$, etc.

Adding and Subtracting Powers of Ten

To add or subtract numbers expressed in powers of ten, simply convert all numbers to the same power of 10, and keep this same power of 10 in the answer.

Table 2-1. Prefix Symbols

Prefix	Symbol	Multiplier	Notes
tera	T	10^{12}	one trillion
giga	G	10^{9}	one billion (1000 mega)
mega	M	10^{6}	one million
kilo	k	10^{3}	one thousand
hecto	H	10^{2}	one hundred
deka	D	10	ten
deci	d	10^{-1}	one tenth
centi	c	10^{-2}	one hundredth
milli	m	10^{-3}	one thousandth
micro	μ	10^{-6}	one millionth
nano	n	10^{-9}	millimicro (one thousandth of one millionth)
pico	p	10^{-12}	micromicro (one millionth of one millionth)
atto	a	10^{-18}	millionth of micromicro

Example 1: $3.28(10^3) + 1.14 + 12.5(10^2)$

$$\begin{aligned} 3.28(10^3) &= 3.28000(10^3) \\ 1.14 &= 0.00114(10^3) \\ 12.5(10^2) &= \underline{1.25000(10^3)} \\ \text{Total} &= 4.53114(10^3) \end{aligned}$$

Example 2: $3.28(10^3) - 1.14(10^2)$

$$\begin{aligned} 3.28(10^3) &= 32.80(10^2) \\ 1.14(10^2) &= \underline{-\ 1.14(10^2)} \\ \text{Total} &= 31.66(10^2) \end{aligned}$$

Multiplying Powers of Ten

To find the product of numbers when expressed as powers of 10, simply multiply the numbers and add the exponents of the tens. For example:

$$6.28(10^3) \times 3.14(10^2) = 19.7192(10^{3+2}) = 19.7192(10^5)$$

For negative exponents:

$$6.28(10^{-3}) \times 3.14(10^{-2}) = 19.7192(10^{-2-3}) = 19.7192(10^{-5})$$

For exponents with opposite signs:

$$6.28(10^3) \times 3.14(10^{-2}) = 19.7192(10^{3-2}) = 19.7192(10) = 197.192$$

Dividing Powers of Ten

To divide numbers expressed as powers of 10, simply divide the numbers and subtract the exponent of the denominator from the exponent of the numerator. For example:

$$\frac{6.28(10^3)}{3.14(10^2)} = 2(10^{3-2}) = 2(10) = 20$$

Sometimes it is more convenient to transfer the ten in the denominator to the numerator and change its sign:

$$\frac{6.28(10^3)(10^{-2})}{3.14} = 2(10^{3-2}) = 2(10) = 20$$

2-5. LOGARITHMS

Logarithms (often called "logs") enable us to simplify complex computations involving multiplication, division, powers, and roots. Logarithms reduce multiplication to a simple addition and division to a simple subtraction; raising to a power becomes a single multiplication, and extracting a root becomes a single division.

There are two generally used systems of logarithms, the *common* logarithm (most widely used and reviewed in this chapter) and the *natural* logarithm (not covered in this text). A natural logarithm is indicated by the notation $\log_\epsilon$, which designates a logarithmic base of 2.718. The common logarithm has a base of 10. The approximate relationship of the two systems is as follows:

$$\log_{10} \text{ (common log)} = 0.4343 \text{ times natural logarithm of number}$$
$$\log_\epsilon \text{ (natural log)} = 2.3026 \text{ times common logarithm of number}$$

The designation $\log_{10}$ is normally written simply as log, the base ten being assumed. Thus log 645 means the common logarithm of the number 645.

The common logarithm of any number is simply the power to which 10 must be raised to equal that number. For numbers that are integral powers of 10, the relationship is extremely simple, as evidenced by Table 2-2. When the number is not an integral power of 10, the log of that number consists of a whole number and a decimal fraction. The whole-number part of the logarithm is termed the *characteristic,* and the decimal-fraction part is termed the *mantissa.*

Table 2-2. Integral Powers of Ten

Number	Power of 10	Logarithm of Number (To the Base 10)
1000	10^3	3
100	10^2	2
10	10^1	1
1	10^0	0
0.1	10^{-1}	−1
0.01	10^{-2}	−2
0.001	10^{-3}	−3
0.0001	10^{-4}	−4

The Characteristic

Table 2-2 actually lists the characteristic of the number in the column headed "Logarithm of Number." This may be expanded as shown by Table 2-3. Note that the characteristic of a whole number has a positive value equal to one less than the number of digits before the decimal point, and the characteristic of a decimal fraction has a negative value equal to one more than the number of zeros after the decimal point.

Table 2-3. Characteristics of Numbers

Number	Characteristic
1 to 9	0
10 to 99	1
100 to 999	2
1000 to 9999	3
10,000 to 99,999	4
0.0001 to 0.0009	—4
0.001 to 0.009	—3
0.01 to 0.09	—2
0.1 to 0.9	—1

The Mantissa

The decimal-fraction part of a logarithm (the mantissa) is obtained from a table of common logarithms such as Table A-1 in Appendix A. To find the mantissa in the table, look under the left-hand column (N) and locate the first two digits of the number. Then in the column under the third digit of the number, the mantissa is given. For example, take the number 246. Locate 24 in column N. Then go across to the column numbered 6 and find the mantissa, 0.3909, which will be the decimal part of the logarithm of 246.

Suppose the number is a four-digit number such as 2468. You have already found the mantissa for 2460, which was 0.3909. To find the mantissa for 2468, use the *proportional parts* column to find the proportional part for 8. This is 14, so the mantissa is 0.3909 plus 0.0014, or 0.3923.

The Combined Characteristic and Mantissa (Logarithm of a Number)

The mantissa is always positive and determines the sequence of digits. It is the only value given by the log table. The characteristic may be either positive or negative; it determines the position of the decimal point.

You already know that the characteristic is one less than the number of digits to the left of the decimal point. Thus the log of 2468 = 3.3923. Now suppose you need to find the log of 0.02468. Again, look under the left-hand column (N) to locate the first two digits (24). Then in the column under the third digit of the number (6), read 3909. Then again using the proportional-parts column for 8, obtain the same mantissa as before, 0.3923. But now the characteristic is a negative 2. The negative characteristic is equal to the number of digits to the right of the decimal point to and including the first significant digit.

The preceding result could be written $0.3923-2$, but this form is impractical to use. Therefore, convert the characteristic to a positive number, for example: $8.3923-10$. This is to say that you add 10 (or a multiple thereof) to the negative characteristic, and compensate by indicating the subtraction of 10 from the entire logarithm.

Examples:

$\text{Log } 0.002468 = 7.3923 - 10$
$\text{Log } 0.2468 = 9.3923 - 10$
$\text{Log } 2.468 = 0.3923$
$\text{Log } 24.68 = 1.3923$

Antilogarithms

In practical problems it is necessary to find the antilogarithm (often called the "antilog"), which is the number corresponding to a given logarithm. Finding this number is the reverse of finding the logarithm. For example, since the log of 2468 is 3.3923, the antilog of 3.3923 is 2468.

To find the antilog of a number, locate the mantissa in the log table and write it down. Then place the decimal as indicated by the characteristic. For example, to find the antilog of 3.5453, first look up the mantissa (5453) to find the corresponding number, which is 351. A characteristic of 3 means that there will be 4 digits in the number, or 3510. So antilog $3.5453 = 3510$.

Examples:

$\text{Antilog } 2.5453 = 351$
$\text{Antilog } 1.5453 = 35.1$
$\text{Antilog } 0.5453 = 3.51$
$\text{Antilog } 7.5453 - 10 = 0.00351$

When you must find the antilogarithm of a logarithm the mantissa of which is not given exactly in the table, procede as follows:

Step 1. Compute the tabular difference between the next higher and next lower mantissas.

Step 2. Compute the difference between the given mantissa and the next lower mantissa, and divide this figure by the result of Step 1.

Step 3. Add the result of Step 2 to the significant figures corresponding to the next lower mantissa.

Step.4. Place the decimal as indicated by the given characteristic.

As an example, to find the antilog of 2.7376:

Step 1. Next higher mantissa 0.7380
Next lower mantissa 0.7372
Tabular difference 0.0008

Step 2. Given mantissa 0.7376
Next lower mantissa 0.7372
Tabular difference 0.0004

Then:

$$\frac{0.0004}{0.0008} = 0.5$$

Step 3. The significant figures corresponding to the next lower mantissa (0.7372) are 546. To this number, add the result of Step 2, giving 546.5.

Step 4. Since the given characteristic is 2, the antilog of 2.7376 is 546.5. If the antilog of 3.7376 is to be found, the result is 5465, etc.

To Multiply

To multiply numbers by the use of logarithms, add the logarithms and find the antilog of the sum.

Example 1: Multiply 246 by 788.

$$\begin{aligned} \text{Log } 246 &= 2.3909 \\ \text{Log } 788 &= \underline{2.8965} \\ \text{Sum } &= 5.2874 \end{aligned}$$

$$\text{Antilog } 5.2874 = 193{,}848$$

Example 2: Find (0.03) (0.05) (0.5).

$$\begin{aligned} \text{Log } 0.03 &= 8.4771 - 10 \\ \text{Log } 0.05 &= 8.6990 - 10 \\ \text{Log } 0.5 &= \underline{9.6990 - 10} \\ \text{Sum } &= 26.8751 - 30 \end{aligned}$$

$$\text{Antilog } (26.8751 - 30) = 0.00075$$

To Divide

To divide numbers by the use of logarithms, subtract the log of the divisor from the log of the dividend, and find the antilog of the difference.

Example: Divide 40,500 by 15.4.

$$\begin{aligned} \text{Log } 40{,}500 &= 4.6075 \\ -\text{Log } 15.4 &= \underline{-1.1875} \\ \text{Difference} &= 3.4200 \end{aligned}$$

$$\text{Antilog } 3.4200 = 2630$$

Powers

To raise a given number to any power, multiply the log of the given number by the power which the number is to be raised. Then find the antilog of the product.

Example: Find 26.4^3.

$$\begin{aligned} \text{Log } 26.4 &= 1.4216 \\ &\underline{\times 3} \\ \text{Product} &= 4.2648 \end{aligned}$$

$$\text{Antilog } 4.2648 = 18{,}400$$

NOTE: The answer 18,400 is as accurate as can be obtained with a four-place log table and is normally sufficiently accurate in practice. The actual value of 26.4^3 is 18,399.744. Six-place log tables are available for use when extreme accuracy is required.

Roots

To find a root of a number by the use of logarithms, divide the log of the given number by the index of the root, and find the antilog of the quotient.

Example: Find $\sqrt[3]{2.09}$.

$$\begin{aligned} \text{Log } 2.09 &= 0.3201 \\ 0.3201 \div 3 &= 0.1067 \end{aligned}$$

$$\text{Antilog } 0.1067 = 1.279$$

2-6. ALGEBRA

In ordinary arithmetic, numbers are given in equations. In algebra, letters are used basically to express general conditions. Let us express certain of the preceding logarithmic relationships in algebraic form as an example:

$$\text{Log } (a)(b) = \log a + \log b$$

$$\text{Log} \frac{a}{b} = \log a - \log b$$

A basic tool in algebra is the ability to change an expression from one form to another more suitable for the problem at hand. The basic rules are: if $A = B/C$, then $B = AC$, and $C = B/A$. Thus when we consider the basic rule of Ohm's law that states that voltage is equal to the product of amperes and ohms ($E = IR$), we know that $I = E/R$ and that $R = E/I$.

Another type of expression is the ratio or proportion:

$$\frac{A}{B} = \frac{C}{D}$$

which is read "A is to B as C is to D." Cross multiplying gives AD = BC, and from this:

$$A=\frac{BC}{D};\qquad B=\frac{AD}{C};\qquad C=\frac{AD}{B};\qquad D=\frac{BC}{A}$$

A basic formula for tuned circuits is:

$$f=\frac{1}{2\pi\sqrt{LC}}$$

This formula (relationship) is true for resonance no matter what numerical values are involved. Thus if you know the two values (inductance L and capacitance C), you can compute the frequency of circuit resonance.

But suppose you have a given capacitance and must compute the value of inductance required to resonate with that capacitance at a given frequency. In this case, the two known values are the frequency and capacitance, and the inductance must be found.

First remove the radical by squaring both sides of the equation:

$$f^2=\frac{1}{4\pi^2LC}$$

Then solve for L:

$$L=\frac{1}{4\pi^2f^2C}$$

If the inductance is known and the capacitance is unknown, the formula becomes:

$$C=\frac{1}{4\pi^2f^2L}$$

The basic relationship for ac impedance is:

$$Z=\sqrt{R^2+X^2}$$

where the reactance (X) is actually $X_L - X_C$. Then

$$Z^2=R^2+X^2$$

and

$$R=\sqrt{Z^2-X^2}$$

and

$$X=\sqrt{Z^2-R^2}$$

2-7. GEOMETRY AND TRIGONOMETRY

We will make here a basic distinction between geometry and trigonometry that is most important to the electronic engineer. The most valuable use of geometry for our purpose in this text is to find the hypotenuse of a right triangle when the dimensions of the two legs are known. As used in this text, trigonometry is a tool for determining the angles of a right triangle and their relationship to the sides of the triangle.

In Fig. 2-4A, a force (A) of a 3 units acts at right angles to a force (B) of 4 units. Since there is a difference in direction between the two forces, we cannot add them either arithmetically or algebraically. The solution (the magnitude of the resultant force) must be obtained by geometric addition.

Fig. 2-4B is the graphical representation of completing the rectangle so that the geometric sum of A and B can be represented. Ovbiously, this sum (C) is simply the hypotenuse of a right triangle. Fig. 2-4C illustrates the conventional representation.

RULE: The hypotenuse of a right triangle is equal to the square root of the sum of the squares of the other two sides.

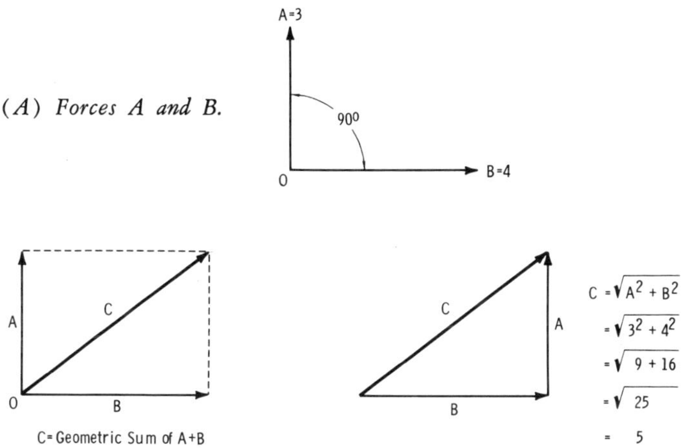

(*A*) *Forces A and B.*

(*B*) *Completing rectangle.*

(*C*) *Geometric addition.*

Fig. 2-4. Resultant of two forces at right angles.

Thus the resultant of forces A and B in Fig. 2-4 is 5, as indicated in Fig. 2-4C. Note that this solution gives the answer in magnitude only; the angle of the resultant force is determined by trigonometric methods. If the two sides are equal, the hypotenuse forms an angle of 45° with either leg (halfway between the two legs). Since force B in Fig. 2-4 is greater than force A, the resultant lies closer to the larger force, B. Hypotenuse C is a *vector.* Vector analysis (Section 2-8) uses both geometry and trigonometry.

Fig. 2-5 illustrates additional basic geometric relationships useful to the electronic engineer.

Fig. 2-6 illustrates some of the terminology used with trigonometric

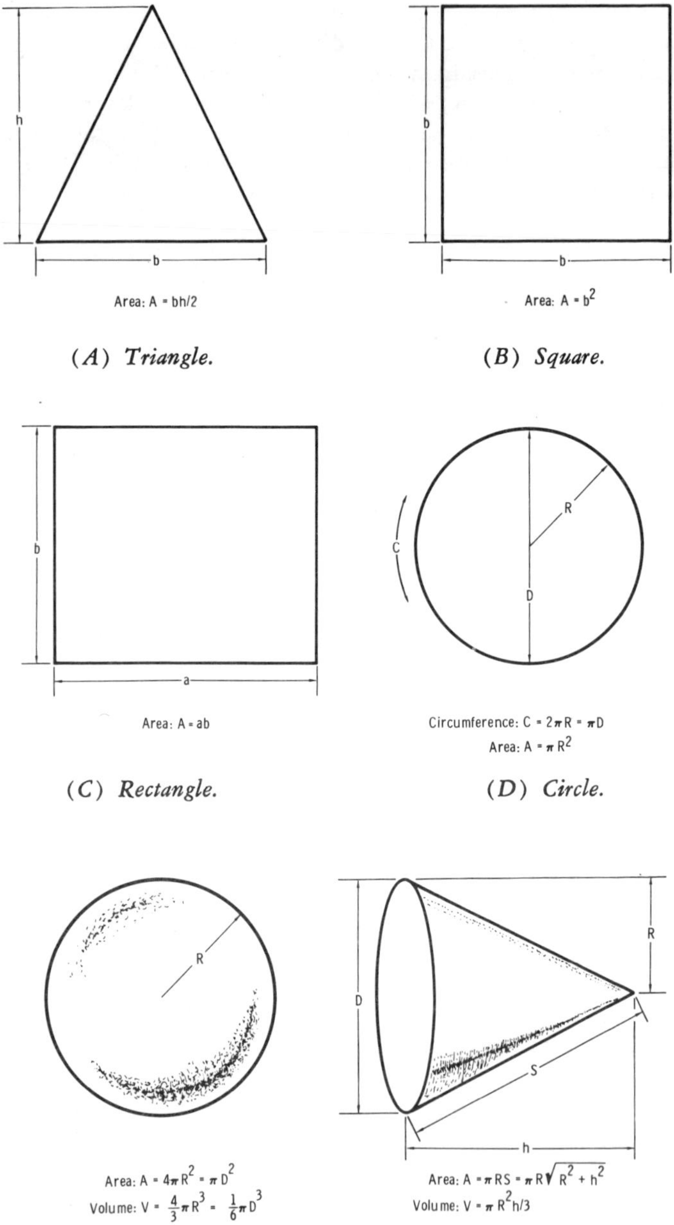

(*A*) *Triangle.* (*B*) *Square.*

(*C*) *Rectangle.* (*D*) *Circle.*

(*E*) *Sphere.* (*F*) *Cone.*

Fig. 2-5. Basic geometric relationships.

C = Hypotenuse
B = Side Opposite Angle b and Side Adjacent to Angle a
A = Side Adjacent to Angle b and Side Opposite Angle a
a = Acute Angle Formed by the Hypotenuse and the Altitude Leg (B)
b = Acute Angle Formed by the Hypotenuse and the Base Leg (A)

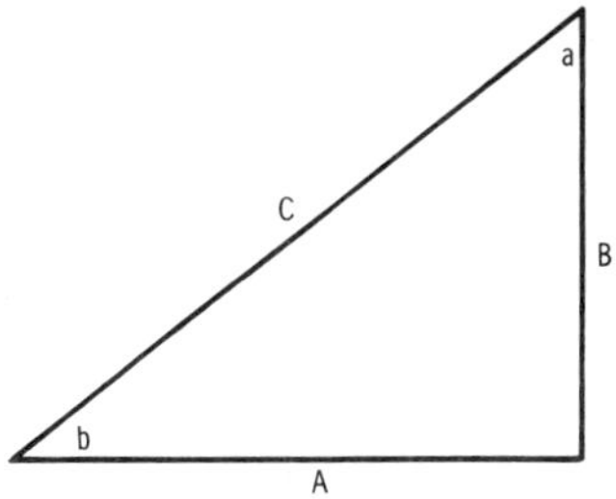

Fig. 2-6. Terminology applicable to trigonometric functions.

Table 2-4. Trigonometric Formulas

Known Values	Formulas for Unknown Values of				
	A	**B**	**C**	**∠b**	**∠a**
A & B	——	——	$\sqrt{A^2 + B^2}$	arc tan $\frac{B}{A}$	arc tan $\frac{A}{B}$
A & C	——	$\sqrt{C^2 - A^2}$	——	arc cos $\frac{A}{C}$	arc sin $\frac{A}{C}$
A & ∠b	——	A tan ∠b	$\frac{A}{\cos \angle b}$	——	90° — ∠b
A & ∠a	——	$\frac{A}{\tan \angle a}$	$\frac{A}{\sin \angle a}$	90° — ∠a	——
B & C	$\sqrt{C^2 - B^2}$	——	——	arc sin $\frac{B}{C}$	arc cos $\frac{B}{C}$
B & ∠b	$\frac{B}{\tan \angle b}$	——	$\frac{B}{\sin \angle b}$	——	90° — ∠b
B & ∠a	B tan ∠a	——	$\frac{B}{\cos \angle a}$	90° — ∠a	——
C & ∠b	C cos ∠b	C sin ∠b	——	——	90° — ∠b
C & ∠a	C sin ∠a	C cos ∠a	——	90° — ∠a	——

The expression "arc sin" or "$\sin^{-1}$" indicates "the angle whose sine is. . . ." Similarly, "arc tan" or "$\tan^{-1}$" indicates "the angle whose tangent is . . ." , etc.

functions. Table 2-4 applies to this figure. Table A-2 in Appendix A gives the natural sines, cosines, tangents, and cotangents of angles.

> NOTE ON USE OF TABLE OF TRIGONOMETRIC FUNCTIONS: To find values for angles from 0° to 45°, use the headings at the top of the table and the degree listings in the left-hand column. For angles from 45° to 90°, use the headings at the bottom of the table and the degree listings in the right-hand column. Read the degree listings in the right-hand column from bottom to top; thus the 10′ listing directly above 89° signifies 89° 10′.

The trigonometric functions of an angle are normally abbreviated as follows:

Sine = sin

Cosine = cos

Tangent = tan

Cotangent = cot

The basic relationships are:

$$\sin = \frac{\text{opposite side}}{\text{hypotenuse}}$$

$$\cos = \frac{\text{adjacent side}}{\text{hypotenuse}}$$

$$\tan = \frac{\text{opposite side}}{\text{adjacent side}}$$

$$\cot = \frac{\text{adjacent side}}{\text{opposite side}}$$

Angles can be formed in any of the four quadrants that make up a complete circle, or 360°. In trigonometric tables, only angles up to 90° are given. Thus the circle is divided into four 90° quadrants as shown by Fig. 2-7. For an angle between 90° and 180°, the trigonometric function is equal in magnitude to the same function of an angle equal to 180° minus the angle in question. This angle will then be between 0 and 90° (that is, the angle is measured from the x-axis). For an angle in the third quadrant subtract 180°, and in the fourth quadrant subtract the angle from 360°. In any case, the result is an angle between 0 and 90°.

Assume in Fig. 2-6 that A = 4 and B = 3. What is the magnitude of C, and what angle does it make with the x-axis? From geometry, we know that

$$C = \sqrt{4^2 + 3^2} = \sqrt{25} = 5$$

This result gives the magnitude of C but does not specify angle b.

From trigonometry, we know that we can find angle b by any or all of the relationships listed above (sine, cosine, tangent, or cotangent).

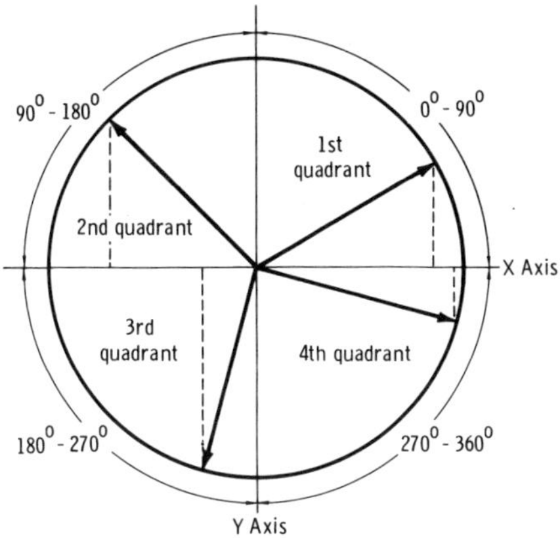

Fig. 2-7. Quadrants of a circle.

$$\sin b = \frac{3}{5} = 0.60$$

Then b = 36° 50′ (from trig table by reading to the nearest sine value listed, which is 0.5995).

$$\cos b = \frac{4}{5} = 0.80$$

Then b = 36° 50′ (from trig table to nearest listed cosine value, or 0.8004)

$$\tan b = \frac{3}{4} = 0.75$$

Then b = 36° 50′ (from trig table to nearest listed tangent value, or 0.7490). (The 36° 50′ is read "36 degrees, 50 minutes.")

The sum of the interior angles of a triangle is 180°. Therefore the sum of angles a and b in Fig. 2-6 is 90°. When angle b is known (36° 50′), angle a is found by subtracting angle b from 90°. In this example, 90° − (36° 50′) = 53° 10′.

Note that if any one angle and one side are known, the other side (or hypotenuse) can be found by the use of algebra on the angle relationships given above. Thus, for example, since

$$\sin = \frac{\text{opposite side}}{\text{hypotenuse}}$$

then

$$\text{hypotenuse} = \frac{\text{opposite side}}{\sin}$$

or

$$\text{opposite side} = (\sin)(\text{hypotenuse})$$

2-8. VECTORS

We have been using vectors in geometry and trigonometry. It is now possible to review vectors in greater detail. Fig. 2-8A shows vectors (A and B) which are not at right angles. These vectors may be resolved into their components along the x and y axes as illustrated in Fig. 2-8B. Then:

$$\text{Sum of x components} = R_x = A_x + B_x$$
$$\text{Sum of y components} = R_y = A_y + B_y$$

The geometric sum of the resultant x and y components (R_x and R_y) yields the resultant vector R at an angle r with the x axis. This is shown in Fig. 2-8C, which illustrates the graphical solution to this problem. Line AR is drawn parallel to vector B, and line BR is drawn parallel to vector A. The intersection determines the magnitude of resultant vector R.

To solve on paper without drawing a graph, the magnitude of the resultant is:

$$R = \sqrt{R_x^2 + R_y^2}$$

or actually

$$R = \sqrt{(A_x + B_x)^2 + (A_y + B_y)^2}$$

As an example, assume:

$$A = 8 \text{ and angle a} = 20° \ (\text{written } 8 \angle 20°)$$
$$B = 6 \text{ and angle b} = 40° \ (\text{written } 6 \angle 40°)$$

From the fundamental trigonometric relationships:

$$\cos a = \frac{A_x}{A}$$
$$A_x = A \cos a$$

Similarly,

$$\sin a = \frac{A_y}{A}$$
$$A_y = A \sin a$$

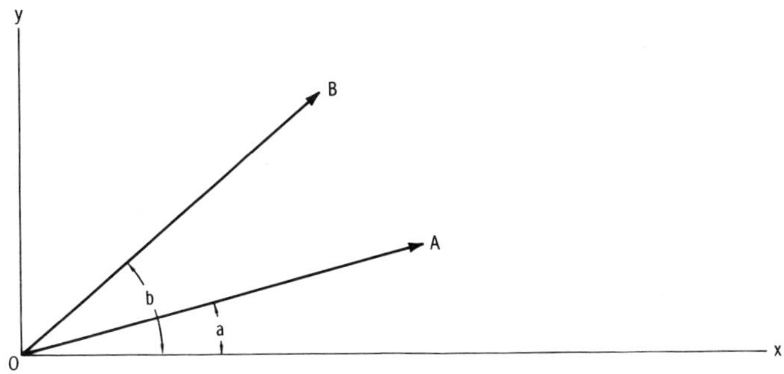

(A) Vectors not at right angle.

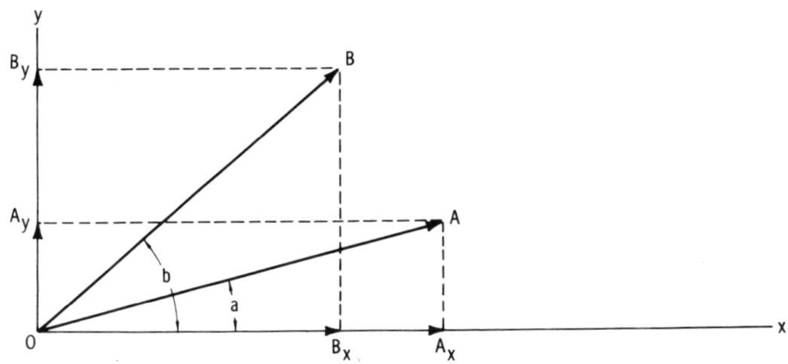

(B) Components along x and y axes.

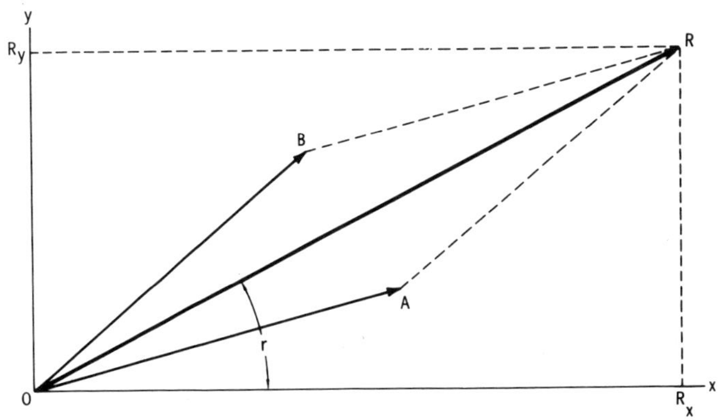

(C) Graphical solution of problem.

Fig. 2-8. Basic vector addition.

Therefore

$$A_x = A \cos 20° = 8(0.9397) = 7.5$$
$$A_y = A \sin 20° = 8(0.3420) = 2.7$$

Then finding the x and y components of vector B:

$$B_x = B \cos b = 6 \cos 40° = 6(0.7660) = 4.6$$
$$B_y = B \sin b = 6 \sin 40° = 6(0.6428) = 3.9$$

Collecting the x and y components:

$$R = \sqrt{(7.5 + 4.6)^2 + (2.7 + 3.9)^2} = \sqrt{190} = 13.8$$

Note from the above that $R_x = 7.5 + 4.6 = 12.1$ and that $R_y = 2.7 + 3.9 = 6.6$. Therefore

$$\tan r = \frac{R_y}{R_x} = \frac{6.6}{12.1} = 0.545$$

then

$$r = 28° \, 40' \text{ (approx)}$$

Thus the complete solution is written:

$$R = 13.8 \, \angle 28° \, 40'$$

You can find the magnitude of vector R directly in terms of the magnitudes of vectors A and B and trigonometric functions of angles a and b by the following relationship:

$$R = \sqrt{(A \cos a + B \cos b)^2 + (A \sin a + B \sin b)^2}$$

Then value of tan r is

$$\tan r = \frac{A \sin a + B \sin b}{A \cos a + B \cos b}$$

Now see Fig. 2-9. Given the magnitudes and angles of vectors A and B, find R and angle r.

$$R = \sqrt{(8 \cos 20° + 20 \cos 120°)^2 + (8 \sin 20° + 20 \sin 120°)^2}$$

Since vector B is in the second quadrant, find the angle it makes with the negative x axis and attach the correct sign:

$$\cos 120° = -\cos (180 - 120) = -\cos 60°$$

then:

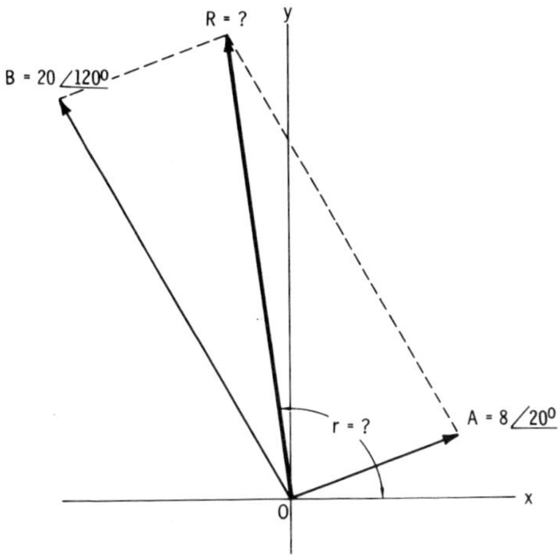

Fig. 2-9. Example of vector addition with vectors in first and second quadrants.

$$\begin{aligned} R &= \sqrt{[8(0.9397) + 20(-0.5)]^2 + [8(0.3420) + 20(0.866)]^2} \\ &= \sqrt{[7.52 + (-10)]^2 + [2.74 + 17.32]^2} \\ &= \sqrt{(-2.48)^2 + (20)^2} \\ &= \sqrt{6.15 + 400} = \sqrt{406} = 20.1 \end{aligned}$$

The magnitude of R is thus 20.1. The tangent of angle r is:

$$\tan r = \frac{20}{-2.48} = -8.06$$

and the corresponding angle from the table is approximately 82° 55′. But angle r is in the second quadrant, and the angle found in the table is the angle between R and the negative x axis. Angle r is then 180° − (82° 55′), or 97° 05′. The solution to the problem is therefore $R = 20.1 \angle 97° 05'$.

2-9. COMPLEX NOTATION (OPERATOR j)

In the study and measurement of propagation factors and broadcast antennas, it is necessary to understand the combined effect of magnitude and phase. This is most conveniently visualized in vector form. The length of a vector represents the magnitude, and the direction (with respect to a reference axis) represents the phase. The combined result may then be represented by one resultant line.

For example, a circuit employs a resistance of 3 ohms and a coil with inductive reactance (at the frequency specified) of 4 ohms. From fundamental theory, the impedance (Z) is the square root of the sum of the resistance squared and the reactance squared:

$$Z = \sqrt{R^2 + X^2} = \sqrt{9 + 16} = \sqrt{25} = 5 \text{ ohms}$$

While this gives the total impedance, it does not express the resultant phase angle of the total impedance relative to a purely resistive circuit. It is this phase angle that must be considered in the proper phasing and tuning of antenna systems.

There are two basic notations for expressing a vector quantity. These are the polar form (just described in Section 2-8) and the rectangular form (to be explored in this section). The rectangular form involving the j operator is a simple expression for impedance that shows the resistive and reactive components directly. In the example of 3 ohms resistance and 4 ohms inductive reactance, the rectangular form is simply $3 + j4$.

Inductance is considered a positive reactance; thus it is represented as 4 units *up* along the y axis (Fig. 2-10). If the reactance is capacitive (negative reactance), the four units are taken *down* along the y axis, and the rectangular form is $3 - j4$. The resistance of 3 ohms is represented as 3 units along the x axis.

The $+j$ operator produces 90° counterclockwise rotation of any vector to which it is applied, without changing the magnitude of that vector. The $-j$ operator produces 90° clockwise rotation of any vector to which it is applied, without changing the magnitude of that vector. Successive applications of the $+j$ operator to a vector produce 90° steps in sequence in the counterclockwise direction. Similarly, successive applications of the $-j$ operator to a vector move that vector 90° clockwise for each application. This is tabulated in Table 2-5.

The addition of quantities in rectangular form is illustrated by the following examples:

Example 1: Add $3 + j4$ to $3 - j4$

$$\begin{array}{r} 3 - j4 \\ 3 + j4 \\ \hline 6 - \ 0 \end{array}$$

Example 2: Add $3 + j4$ to $3 + j4$

$$\begin{array}{r} 3 + j4 \\ 3 + j4 \\ \hline 6 + j8 \end{array}$$

One quantity may also be subtracted from another; for example, $3 + j2$ may be subtracted from $3 + j4$:

$$\begin{array}{r} 3 + j4 \\ -3 - j2 \\ \hline 0 + j2 \end{array}$$

Or, $3 - j2$ may be subtracted from $3 + j4$:

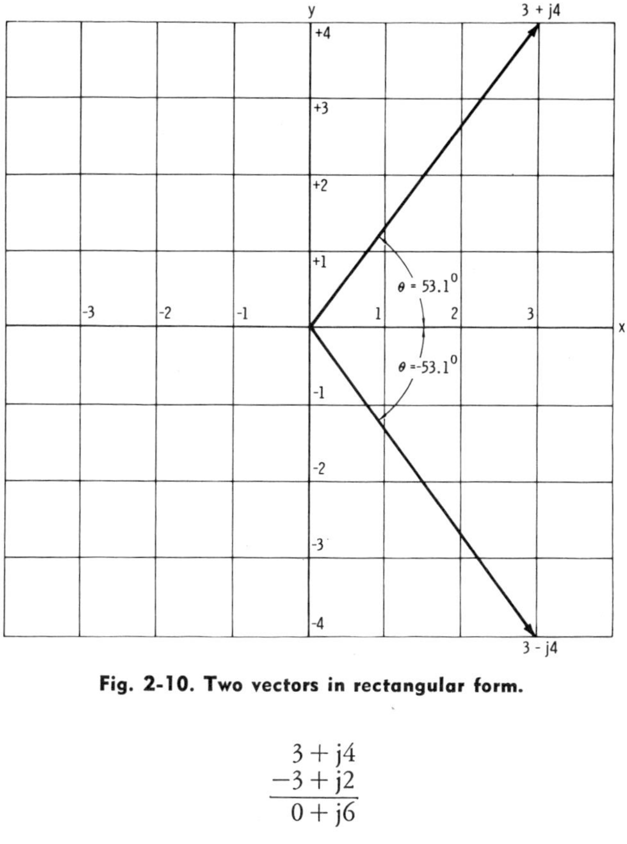

Fig. 2-10. Two vectors in rectangular form.

$$\begin{array}{r} 3 + j4 \\ -3 + j2 \\ \hline 0 + j6 \end{array}$$

The quantities may also be multiplied or divided, but it is more convenient to change from rectangular to polar form first (Fig. 2-10). The polar form of $3 + j4$ is $5 \angle 53°$. The magnitude (5) is simply the vector addition of $3 + j4$, or the hypotenuse of the right triangle whose other two

Table 2-5. Vector Rotation Due to j Operator

Operator	Direction of Rotation	Amount of Rotation
j	Counterclockwise	90°
j^2	"	180°
j^3	"	270°
j^4	"	360°
$-j$	Clockwise	−90°
$(-j)^2$	"	−180°
$(-j)^3$	"	−270°
$(-j)^4$	"	−360°

sides are 3 and 4. The hypotenuse is equal to the square root of the sum of the squares of the other two sides.

The rectangular form can be converted to the polar form by use of the following trigonometric relationships:

$$\sin\theta = \frac{\text{opposite side}}{\text{hypotenuse}}$$

$$\cos\theta = \frac{\text{adjacent side}}{\text{hypotenuse}}$$

$$\tan\theta = \frac{\text{opposite side}}{\text{adjacent side}}$$

The angle of the polar form may be determined first, even before the magnitude. Note that (in this example) the ratio of the opposite and adjacent sides of the right triangle containing the angle θ is already known to be 4/3, or 1.33. Then the angle whose tangent is 1.33 (from trigonometric tables) is approximately 53°.

If the rectangular form is $3 - j4$, the polar form is $5 \angle -53^\circ$.

Numbers in polar form may be multiplied conveniently. The numbers representing the magnitudes of the vectors in polar form are multiplied, and their corresponding angles are added algebraically. For example:

$$(5 \angle 53^\circ)(5 \angle -53^\circ) = 25 \angle 0^\circ$$

Vectors in polar form may also be divided conveniently. To do this, divide the vector magnitude in the numerator by the vector magnitude in the denominator; then algebraically subtract the angle in the denominator from the angle in the numerator. For example:

$$\frac{10 \angle +20^\circ}{5 \angle -10^\circ} = 2 \angle +30^\circ$$

Polar form can be converted to rectangular form with the following relationship:

$$A \angle \theta = A\cos\theta + jA\sin\theta$$

For example, convert $5 \angle +53^\circ$ to rectangular form. From trigonometric tables, cos 53° is 0.6, and sin 53° is 0.8. Then:

$$\begin{aligned} 5 \angle +53^\circ &= 5(0.6) + j5(0.8) \\ &= 3 + j4 \end{aligned}$$

If angle θ is negative, the relationship becomes:

$$A \angle -\theta = A\cos\theta - jA\sin\theta$$

The resultant rectangular form of $5 \angle -53^\circ$ is therefore $3 - j4$.

2-10. THE DECIBEL AND VOLUME UNIT

Understanding the terminology and correct usage of signal generators, meters, and measuring devices used in broadcasting is vitally important in testing and maintenance procedures. Therefore, this entire section is devoted to the units of measurement and their application.

When sound is increased in magnitude, the *loudness* is said to increase; the impression to the brain is roughly proportional to the logarithm of the ratio of the acoustical powers of the two sound levels. Loudness is a complex function dependent on many variables and is covered more fully in Chapter 11.

For example, suppose that a speaker driven with 1 watt has its driving power increased to 2 watts. It is meaningless to say that the power was increased by 1 watt unless it is also stated that the original power was 1 watt. What is important is that the power was doubled. The ear interprets this as a certain change in loudness; but the same degree of change is perceived with an increase of only 1/2 watt if the original power was 1/2 watt, or with an increase of 2 watts if the original power was 2 watts.

The common logarithm of the ratio of two powers is an expression of their relationship in *bels:*

$$\text{bels} = \log_{10}\left(\frac{P_2}{P_1}\right)$$

where,

P_1 is the reference power,

P_2 is the power being compared to P_1.

The bel is too large a unit for practical use in radio work, so a unit equal to one-tenth of a bel, the *decibel* (dB), is commonly used. Therefore the difference in level between P_1 watts and P_2 watts is given by:

$$\text{dB} = 10\log_{10}\left(\frac{P_2}{P_1}\right)$$

To avoid cumbersome computations, tables and graphs are normally used (see Table A-3 in Appendix A for the conversion of ratios to decibels). Note that a power ratio of 2/1 is 3.01 dB; this is normally stated as 3 dB.

Zero dB (0 dB) may designate any convenient reference level. Although it is normally based on the ratio between two powers, the decibel can also be used to indicate absolute power, provided the reference level (*zero level*) is specified. In the past, so-called standard reference levels have been variously specified as 1, 6, 10, 12.5, and 50 milliwatts (mW). The 1 mW reference level is most widely used today, but the practicing engineer will occasionally find 6 and 12.5 mW referred to as 0 dB. The term *dBm* is used to indicate that zero level is 1 mW. Note that power levels expressed

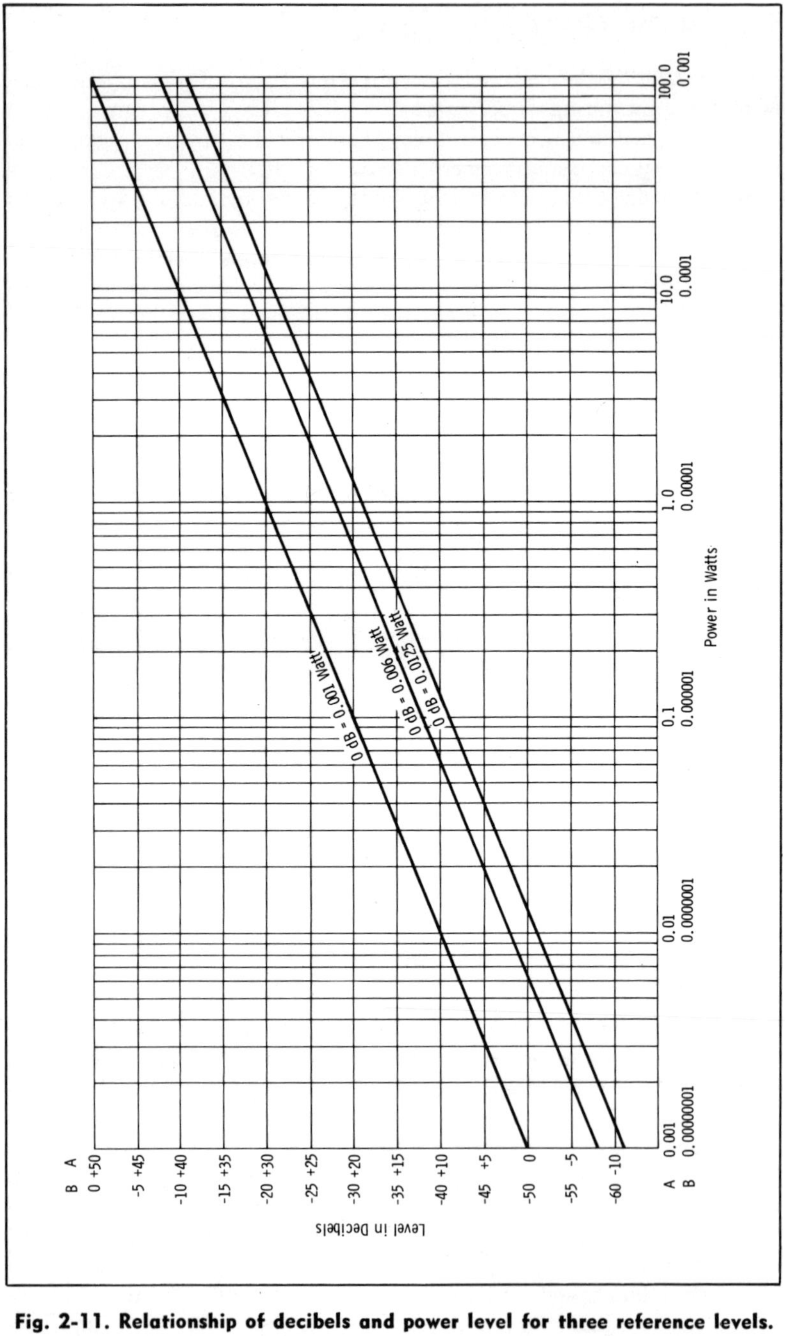

Fig. 2-11. Relationship of decibels and power level for three reference levels.

in decibels are independent of impedance values. Fig. 2-11 shows a graph of decibels versus power for three reference levels.

Note from Fig. 2-11 that:

1. To convert from 1 mW reference to 6 mW, add −7.78 dB.
2. To convert from 1 mW reference to 12.5 mW add −10.97 dB.
3. To convert from 6 mW reference to 1 mW add +7.78 dB.
4. To convert from 12.5 mW reference to 1 mW add +10.97 dB.

With any reference level, a plus sign indicates so many "decibels up"; a minus sign indicates so many "decibels down." The statement of a *power ratio* in decibels is independent of reference level or impedance. The statement of *absolute power* in decibels is meaningless unless the reference level is stated.

When decibels are related to voltage or current, the value of impedance must be taken into account, since the voltage across or the current through an impedance depends on the impedance as well as the power level:

$$E = \sqrt{WR}$$

where,

E is the voltage across the impedance,
W is the power in watts,
R is the impedance (purely resistive) in ohms.

Power is proportional to the square of the voltage or the current. When a number is squared, the logarithm of that number is doubled; therefore, when decibels are used to express voltage ratios, the following relation applies:

$$dB = 20 \log_{10} \left(\frac{E_2}{E_1}\right)$$

For current ratios, the relation is:

$$dB = 20 \log_{10} \left(\frac{I_2}{I_1}\right)$$

Table A-3 (Appendix A) lists decibels relative to voltage or current ratios as well as power ratios. Note that for a given ratio of power, the number of decibels is one-half the value for the same ratio of voltage or current.

The vtvm is a convenient tool that is often used in gain or loss measurements. Fig. 2-12 shows the dBm-to-volts relationship for the impedances given.

Volume Units and VU Meters

The VU meter is a standardized instrument intended for the monitoring of radio-program content. Since the power in program signals is constantly

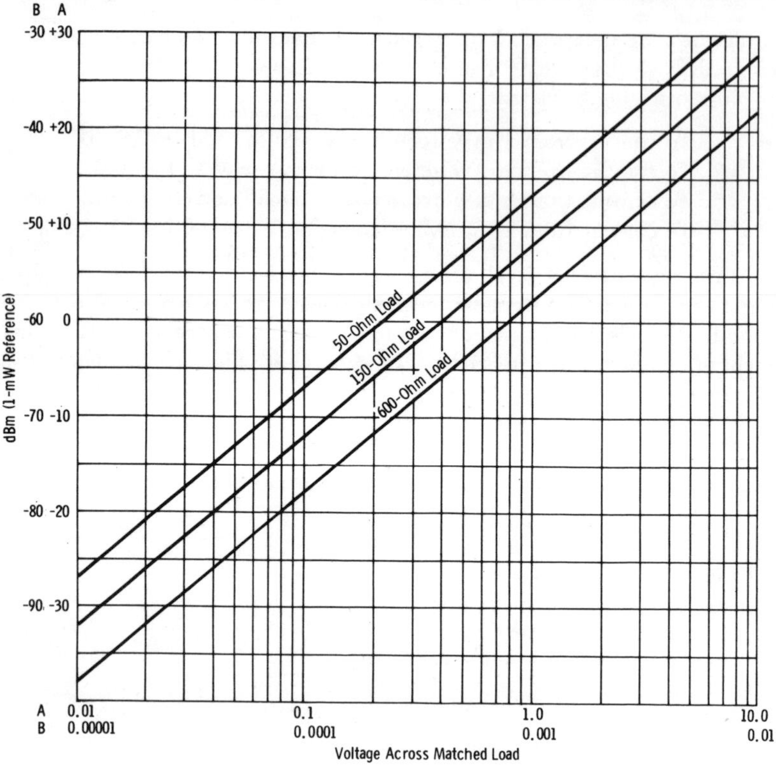

Fig. 2-12. Relationship of dBm and load voltage.

fluctuating, the meter reading must be standardized as to whether it is peak, rms, or average, and the meter movement must have specified ballistic characteristics, such as speed of response and damping.

The standardized VU meter employs a full-wave dry-disc rectifier. Its dynamic characteristics are such that if a sinusoidal voltage in the frequency range concerned and of such amplitude as to give reference deflection (under steady-state conditions) is suddenly applied, the pointer will reach 99 percent of reference in 0.3 second (within 10 percent). The pointer will then overshoot reference deflection by a minimum of 1 percent and a maximum of 1.5 percent.

Fig. 2-13 illustrates the conventional external circuitry involved with the use of the standard VU meter for monitoring transmission levels. In practice, the signal level is such that the range of the meter must be extended. The impedance of the meter itself varies with the voltage across the meter terminals and, therefore, must be isolated by a resistance network. The ballistic characteristics described are dependent on the meter load impedance, which must be 3900 ohms. The decibel reading of the calibrated, variable meter multiplier (C) plus the scale reading of the

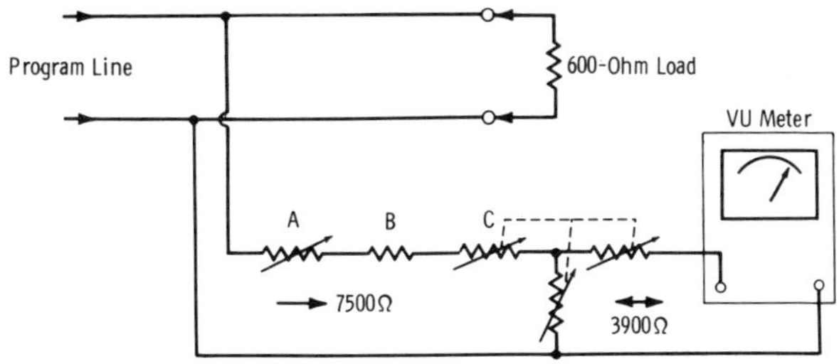

Fig. 2-13. External circuit of standard VU meter.

meter yield a measurement of the transmitted level. The complete network contains the following components:

A. Zero adjuster, approximately 800 to 1000 ohms.
B. Fixed resistor, approximately 3200 ohms, selected so that with A at midposition, $A + B = 3600$ ohms
C. Meter multiplier, "T" attenuator, 3900 ohms input and output.

The meter input impedance, as seen by the program line, is 7500 ohms, except when a 1-mW position is provided. This is for a test position only and is not used in program monitoring.

The volume unit implies a complex wave, a waveform with a higher peak-to-rms ratio than a sine wave. When the VU meter is used to measure steady, single-frequency sine-wave signals, the reading should be referred to as so many dBm. Volume units should never be used to indicate the level of a sine-wave signal.

The volume unit has as its reference a steady-state condition, however. This reference is 1 mW of sine-wave power at 1 kHz through a resistance of 600 ohms. When the instrument (having an internal impedance of approximately 3600 ohms) and an external calibration potentiometer are connected across such a circuit, the indication should be 0 VU. However, when the meter is bridged across a 600-ohm program circuit as shown in Fig. 2-13 (total bridging impedance of 7500 ohms), the maximum sensitivity of the meter is +4 dBm for 0 VU. Most program lines are fed with a +8 dBm level. In this case, the external multiplier is set to +8, and the meter indicates 0 VU at +8 dBm.

The VU meter is, by definition, properly calibrated only when connected across 600 ohms. When it is connected across any impedance other than 600 ohms, the reading must be corrected by adding $10 \log_{10} (600/Z)$, where Z is the actual impedance in ohms. The EIA and FCC specifications for noise and distortion measurement require that a meter with standard VU characteristics be used. In this case, the meter on the measuring equipment reads an actual 1 mW of sine-wave power in 600 ohms for zero reference.

Always remember the basic differences in terminology between program and test signals, which may be summarized as follows: *Reference volume* is that strength of program signals that causes 0 VU (or 100 percent) deflection under the conditions described. This definition is arbitrary because the complex nature of the waveform makes a definition in fundamental terms impossible. *Reference level* is that steady-state condition in which there is 1 mW of a 1000-Hz sine wave in a 600-ohm impedance across which the meter indicates 0 reference level. This should be termed 0 dBm. However, when the meter is connected per standard practice according to Fig. 2-13, the maximum sensitivity is +4 dBm for 0 VU deflection. The actual level is the setting of the multiplier in decibels plus the meter reading. Some multiplier arrangements allow a 1-mW position for test purposes only; this position taps down on the attenuator for a lower multiplier resistance.

Decibels In Practice

The first consideration is feeding the output of the sine-wave signal generator to the input of the system or device being tested. (Actual techniques are covered in Chapters 13 and 14. Only the proper reading of decibels is of concern here.) Fig. 2-14 shows a typical arrangement. The generator dBm meter is always loaded by the constant impedance of the output transformer. The actual output is the reading of the meter minus the setting of the calibrated attenuator. For example, to feed a microphone preamplifier, the generator gain might be adjusted to give a meter reading of +15 dBm and the attenuator set to 65 dB. The actual output is then $15 - 65 = -50$ dBm, a typical value for the input circuit of a microphone preamplifier.

The generator output-transformer secondary is then adjusted (usually by means of a switch connected to transformer taps) to match the load, which is the input of the preamplifier. This is normally 600, 150, or 50 ohms. The actual dBm input to the device is independent of the value of the load. Zero dBm sets the reference level at 1 mW in any load.

The voltage developed across the load and the current through the load are dependent on the value of the load in ohms. For example, 0 dBm (1 mW) results in 0.774 volt across 600 ohms, 0.387 volt across 150 ohms, and 0.224 volt across 50 ohms (Fig. 2-12).

Provided the generator output is matched to the load impedance, no conversion in meter reading (in dBm) is necessary when feeding the device or system, regardless of the input impedance. This is simply a practical application of Ohm's law, but it has resulted in some confusion in practice. The explanation of why the power remains the same regardless of impedance should be reviewed, as follows.

When the impedance is reduced from 600 to 150 ohms, the ratio is 4/1. Since the turns ratio of the output transformer is the square root of the impedance ratio:

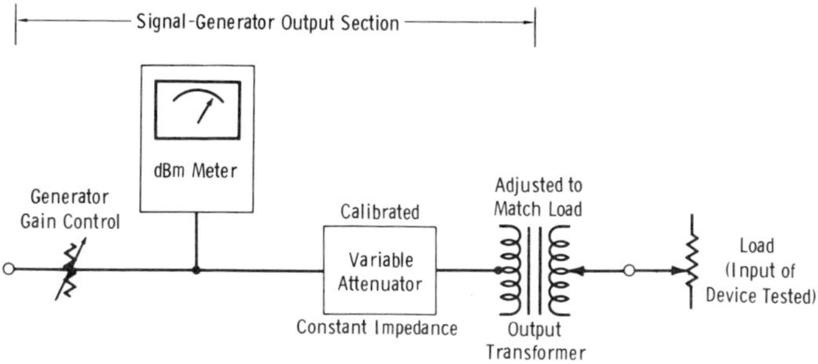

Fig. 2-14. Connection of signal generator to system being tested.

$$\text{Turns ratio} = \sqrt{600/150} = \sqrt{4} = 2 \text{ to } 1$$

Since the voltage developed is in direct proportion to the turns ratio, just one-half the voltage is developed across 150 ohms as for 600 ohms at any reference level. However, calculation shows that the same power exists in the load for either case. Therefore the level indicated on the generator dBm meter (minus the attenuator setting) is the actual level at the system input, provided the impedances are matched.

When the load does not match the output impedance of the generator, it is convenient to measure the voltage gain and then convert to decibels. The generator is adjusted to the closest match available, and the following calculation performed:

$$E_L = E_{OC}\frac{R_L}{R_L + R_O}$$

where,

E_L is the voltage across the external load,
E_{OC} is the open-circuit voltage (twice the voltage that would appear across a load equal to R_O),
R_L is the external resistance load,
R_O is the output impedance of the generator.

For example, assume that the external load is 250 ohms, the generator output impedance is set at 150 ohms, and the generator output is −10 dBm. From Fig. 2-12 the voltage across a 150-ohm load would be 0.125 volt, so E_{OC} in this example is 0.25. Substituting in the formula gives:

$$E_L = 0.25\frac{250}{250 + 150}$$

$$= 0.25\frac{250}{400} = 0.156 \text{ volt}$$

Note from Fig. 2-11 that −10 dBm represents a power of 0.1 mW. The power (E^2/R) in the 250 ohm load is:

$$\frac{0.156^2}{250} = \frac{0.0243}{250} = 0.097 \text{ mW}$$

This is less than 0.5 dBm from the -10 dBm reference of 0.1 mW, an error that normally may be disregarded. Unless the mismatch is 2/1 or greater, any correction factor usually can be ignored except for the most precise measurements.

As mentioned previously, noise and distortion measurements are made with a meter that has standard VU characteristics to meet EIA and FCC specifications. This VU meter is properly calibrated only when reading across 600 ohms. For example, assume it is necessary to measure the gain of an amplifier with 150 ohms input and output. Further assume that the gain of the amplifier is 40 dB. If the input is to be -20 dBm, the output should be $+20$ dBm. As described, the generator should feed -20 dBm into the amplifier, and no conversion factor is involved. However, the output must be measured with a meter calibrated in dBm for 600 ohms. It has been noted that the impedance ratio between 600 ohms and 150 ohms is 4/1, which, for a given power, corresponds to a voltage ratio of 2/1. A 2/1 voltage ratio is equal to 6 dB (Fig. 2-12 or Table A-3). Therefore, the meter reading is 6 dB low across 150 ohms, and the correction factor is $+6$ dB. Then the meter at the output of the amplifier should read $20-6$, or $+14$ dBm. The actual output is then $14+6=20$ dBm. Fig. 2-15 gives the correction factor for impedances from 10 ohms to 10,000 ohms, for the reference 0 dBm $=$ 1 mW in 600 ohms.

Note that with the usual broadcast-type signal generator and measuring equipment no correction factor (in dBm) is involved at the system input. But if the measurement is made across other than 600 ohms at the output, the proper correction factor must be applied to the reading.

There are many applications in practice in which the standard VU meter is used across impedances other than 600 ohms. It is important to under-

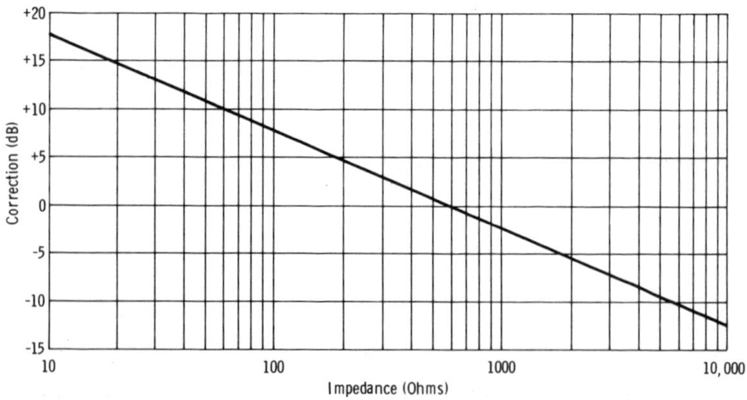

Fig. 2-15. Correction factor for VU meter across various impedances.

stand the proper interpretation for maintenance and level setting. For example, it may be desirable to monitor all inputs to a crossbar switcher where all the inputs are 150 ohms. Assume the proper level at this point is +10 dBm. If the standard minimum insertion loss multiplier is used, the level across 600 ohms would be +4 dBm for 0 VU deflection. However, the correction factor for 150 ohms is +6 dB. Therefore, 0 VU deflection now indicates 0 dBm +4 dB +6 dB, or +10 dBm in 150 ohms.

Now further assume that an external test decibel meter which does not employ the program-line bridging network is used, and the reference level of 1 mW in 600 ohms is stated on the scale of the meter. In this case, the 150-ohm crossbar input level should indicate +4 dBm (0 dBm with the external multiplier set on +4). Then with the +6-dB correction factor, the actual level is +10 dBm.

The preceding information is important to the installation engineer and to the maintenance department. Once the VU meters are installed in a system, the operating engineer is not interested in absolute levels; he only needs to see that the program level is maintained at the zero reference level on peaks.

Always bear in mind that when a VU meter peaks at 0 VU or 100 percent on program material, actual instantaneous peaks will occur at well over 1 mW in 600 ohms. This peak factor of the average program wave is generally taken to be 10 dB over the peak-to-rms value of a sine wave. For this reason, a unit or system is sometimes tested and measured with a sine-wave power of 10 dB above the program operating reference. The Bell Telephone test board commonly feeds tones at 10 dB over the program operating level when measurements for distortion and cross talk are being made. This is important to the operator who may be monitoring the incoming network line for purposes of setting level on network circuits. When there is any doubt, the local test board concerned should be contacted to ascertain the level being transmitted.

Microphone Output in Decibels

A microphone output may be expressed in terms of either voltage or power. Since the output is obviously dependent on the magnitude of excitation, the reference is normally made to either 1 or 10 dynes per square centimeter. (A level of 0.0002 dyne/cm^2 is considered to be the threshold of audibility.) All microphone output ratings are expressed at a single stated frequency.

When the output rating is given in terms of voltage, the reference is 1 volt (open circuit), usually at 1 dyne per square centimeter. The expression is abbreviated dBV, which indicates decibels with a reference of 1 volt as 0 dB. Thus if a microphone specification sheet states the output as −60 dBV, this indicates that an open-circuit voltage of 60 dB below 1 volt is generated with a sound pressure of 1 dyne/cm^2. When the microphone is connected to a matched load, the voltage is decreased to one-half the open circuit voltage, or 6 dB less. The effective output is then −66 dBV. Note in Table A-3 that −60 dB indicates 0.001 of the reference voltage, or 1 millivolt (open circuit). An additional 6 dB cuts this value in half, giving an effective 0.5-mV signal.

Most broadcast-type microphones are rated in terms of power output (dBm) at a stated sound pressure. Typical ratings are from −50 to −65 dBm at a sound pressure of 10 dynes/cm^2. Note that 10 dynes/cm^2 is 50,000 times the pressure at the threshold of hearing, 0.0002 dynes/cm^2. A voltage ratio of 10,000/1 is 80 dB. The ratio of 50,000 to 10,00 is 5, which gives an additional 14 dB and a final level of 94 dB above threshold level. This is in the upper region of the average program sound pressure encountered in practice. (See Fig. 2-16, which is a graph of sound level ranges in dynes/cm^2 relative to decibels, where 0.0002 dyne/cm^2 is 0 dB at 1000 Hz.)

The EIA (formerly RETMA) microphone system rating is also a power rating, but it is a ratio in decibels, relative to 1 mW and 0.0002 dyne/cm^2, of the power available from the microphone to the square of the undisturbed sound field pressure in a plane progressive wave at the microphone position. (This simply specifies the axis of the microphone relative to the sound front.)

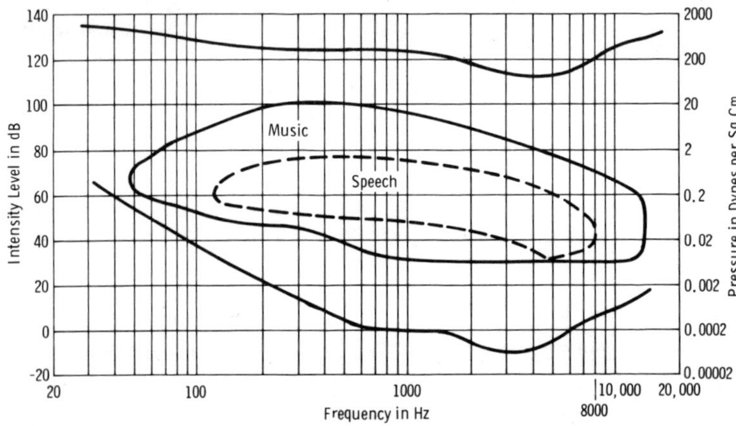

Fig. 2-16. Sound-level ranges.

The EIA system rating is given by:

$$G_M = \left(20 \log_{10} \frac{E}{P} - 10 \log_{10} R_{MR}\right) - 50 \text{ dB}$$

where,

G_M is the microphone system rating (sensitivity) in dBm,
E is the open-circuit voltage generated by the microphone,
P is the sound pressure in dynes per square cm,
R_{MR} is the microphone rating impedance.

Table 2-6. Microphone Impedance Ratings

Nominal Mic Z	Rating Impedance (R_{MR})
19-75 Ohms	38 Ohms
75-300 Ohms	150 Ohms
300-1200 Ohms	600 Ohms

Microphone rating impedances (R_{MR}) for broadcast-type units are given in Table 2-6.

Note that you can convert readily from the EIA microphone system rating to the effective output rating in dBm. It is only necessary to consider the difference in acoustical pressure between 0.0002 dyne/cm^2 and 1 or 10 dynes/cm^2. The ratio of 1 to 0.0002 is 5000/1, or 74 dB. The ratio of 10 to 0.0002 is 50,000/1, or 94 dB, as discussed previously. Thus to convert from a G_M rating to effective output, for 1 dyne/cm^2 add 74 dB and for 10 dynes/cm^2 add 94 dB. For example, if a certain microphone has a G_M rating of -150 dBm, the effective output level at 1 dyne/cm^2 is $-150+74=-76$ dBm. At 10 dynes/cm^2, the effective output level is $-150+94=-56$ dBm.

A natural question that may occur at this point is why the voltage ratio is used in expressing pressure ratios in terms of decibels and then also is used in adding to a power level in dBm. Actually, when a microphone is connected to an unloaded input transformer, its output cannot be expressed in power since no appreciable power is delivered. The effective output level of the microphone is simply that level which, when added to the amplifier power gain in dB, gives the correct output level from the amplifier in dBm.

The effective output level of a microphone connected to a matching impedance is given by:

$$P_O = 1000 \frac{{E_G}^2}{4\,R_M}$$

where,

P_O is the output level in milliwatts,
E_G is the open-circuit output in volts,
R_M is the nominal microphone impedance.

The power in milliwatts can then be converted into dBm.

Radio-Frequency Applications of Decibels

The field strength of a radio-frequency wave is conveniently stated in terms of dBu, where 0 dBu = 1 microvolt per meter (μV/m). Note that this simply involves a voltage ratio. Power is conveniently stated in terms of dBk, where 0 dBk = 1 kW. This simply involves a power ratio.

There are three primary advantages of using decibels to express field strength (volts) and power (watts):

1. Antenna power gain in dB may be added directly to transmitter power level in dBk.
2. Antenna field gain in dB may be added to received field strength in dBu. That is, an increase of 1 dB at the transmitter antenna results in an increase of 1 dB in received field strength.
3. Transmission-line losses in dB may be subtracted directly from transmitter power output in dB above or below 1 kW.

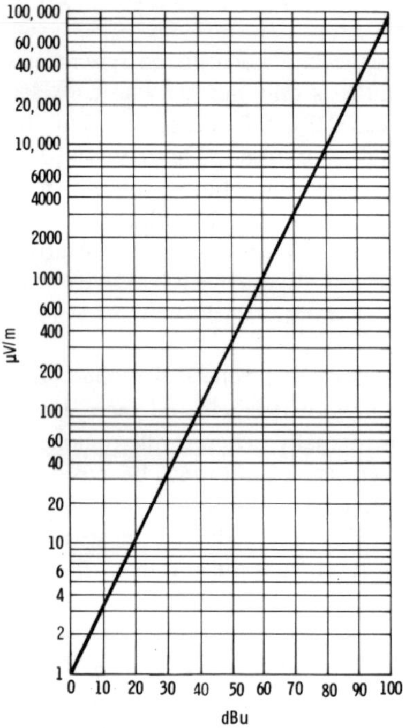

Fig. 2-17. Graph relating dBu to microvolts per meter.

For convenience, Fig. 2-17 is a graph relating dBu to actual microvolts per meter. Fig. 2-18 is a graph relating dBk to actual watts of power. Fig. 2-19 relates percent efficiency to decibels of attenuation and is particularly useful for handling transmission-line characteristics in the calculation of effective radiated power.

To apply decibels to some familiar figures in broadcasting, consider the 3.16 mV/m required to be delivered by an fm station over the principal city to be served. This can be expressed as 70 dBu. The outer limit of service for an fm transmitter is 50 μV/m, or 34 dBu. Note from the graph in Fig. 2-18 that since 0 dBk = 1 kW (1000 watts), 0.1 kW (100 watts) = −10 dBk and 10 kW (10,000 watts) = +10 dBk.

Effective radiated power (erp) is a measure of the actual signal strength radiated by the station. It depends on the transmitter power output, trans-

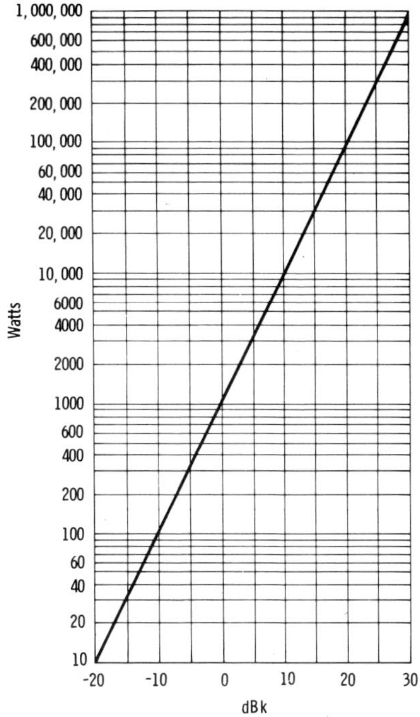

Fig. 2-18. Graph relating dBk to power in watts.

mission-line attenuation, and antenna gain. For example, assume that the transmitter power output is 10 kW, the transmission-line efficiency (for type, frequency, and length of line required) is 79 percent, and the antenna power gain is 3.2. Note that if the conventional computation of erp is made, it is necessary to take 79 percent of 10,000 and then multiply by 3.2 to obtain the erp (25,280 watts). However, the computation can be carried out as follows:

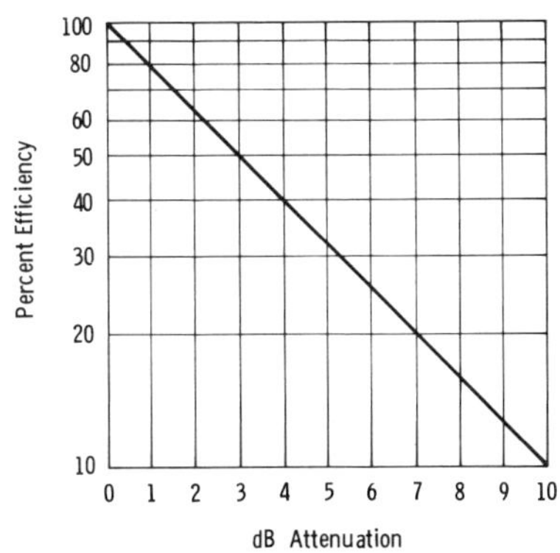

Fig. 2-19. Graph relating percent efficiency and dB attenuation.

10 kW = +10 dBk
79 percent efficiency = 1 dB attenuation (Fig. 2-19)
Power gain of 3.2 = 5 dB (Table A-3)

Then:

$$+10 - 1 + 5 = +14 \text{ dBk}$$

Note the simple addition and subtraction for the erp in terms of dBk. This is obviously still more simplified when the transmission-line attenuation and the antenna gain are specified in decibels, as some manufacturers list them.

2-11. THE CURVE AND THE GRAPH

Graphs (such as those of Figs. 2-17, 2-18, and 2-19) are widely used in electronics for a pictorial presentation of algebraic and/or geometric functions. On a graph, adjacent points may be joined by straight lines, or a sufficient number of points may be plotted to permit drawing a smooth curve. A common example of the latter is a graph of the amplitude-versus-frequency response of an amplifier or complete system.

Graph Paper

Graph paper is available in a wide variety of forms, including rectangular-coordinate, logarithmic, polar-coordinate, and many special-purpose types. Fig. 2-20 is a basic example of the rectangular coordinate system. Four points are plotted in Fig. 2-20B. For point A, x = 6 and y = −4; for B, x = 3 and y = 3; for C, x = −5 and y = 6; for D, x = −7 and y = −8. Conversely, if point A is known, the x and y components of A may be found by extending lines from A to the corresponding points on the x and y axes.

When a plot must be made covering a wide range of values, either log-log scales or semilog scales are employed. "Log-log" means that both the x and y axes are marked with logarithmic scales. On semilog paper, one scale is logarithmic and the other is linear. For example, the usual scale used for plotting amplitude-versus-frequency response is logarithmic along the x axis and linear along the y axis. (For an example, refer to Fig. 5-5 in Chapter 5.)

An example of polar-coordinate paper is illustrated in Fig. 2-21. Note that the scale is marked in degrees both clockwise and counterclockwise. For directional-antenna patterns, the reference line is north (0°), and the angles are measured clockwise.

Smith Charts

Interpretation of some measurements made on transmission lines and waveguides may be made on a Smith chart. This form of transmission-line

Quadrant	X	Y
First	+	+
Second	-	+
Third	-	-
Fourth	+	-

(*A*) *Signs of coordinates.*

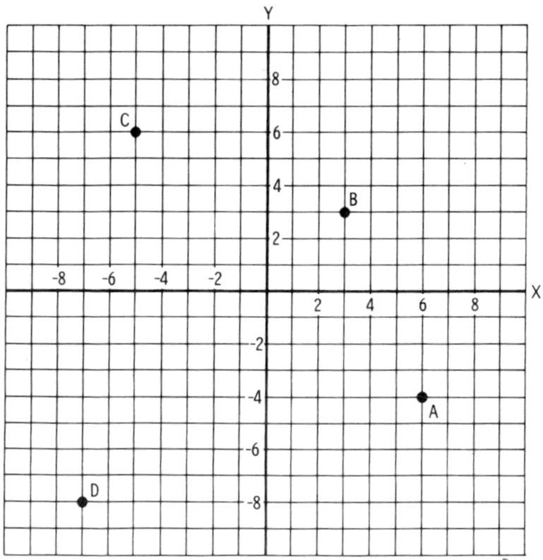

(*B*) *Plotting of points.*

Fig. 2-20. Rectangular coordinates.

chart is a polar plot, consisting of a system of impedance coordinates, superimposed upon which is another system of lines representing loci of constant standing-wave ratio and constant distance along the line. The chart may be used for the interpretation of the measured values of vswr and the location of the voltage minima in terms of equivalent input impedance (or admittance) circuits, or in terms of reflection coefficients. It is also useful for determining the effect of a discontinuity or a change in characteristic impedance, and for solving impedance-matching problems.

Point X at the center of the chart (Fig. 2-22) is the origin for a polar plot of the reflection coefficient. The angle is indicated on the circular scale around the rim, and the magnitude is indicated by the radial distance measured outward from the center on a scale graduated linearly from 0 to 1. Circular arcs, such as A, extending from point Y to the outer rim are the loci of all reflection coefficients that correspond to normalized impedances having equal reactive parts; for arc A, the impedances have a reactance of 1.0. Circles such as B represent reflection coefficients corresponding to normalized impedances having equal resistive parts; for circle B this resistance is 0.8.

In traversing a standing-wave circle (S) with a standing-wave ratio of 2.0, the resistance axis is crossed at two points, one giving a very high resistance and the other a very low resistance. These correspond to the voltage maxima and minima, respectively, observed on a standing-wave detector. On the basis of this, the Smith chart can be used in connection with the standing-wave detector to determine unknown load impedances. For ex-

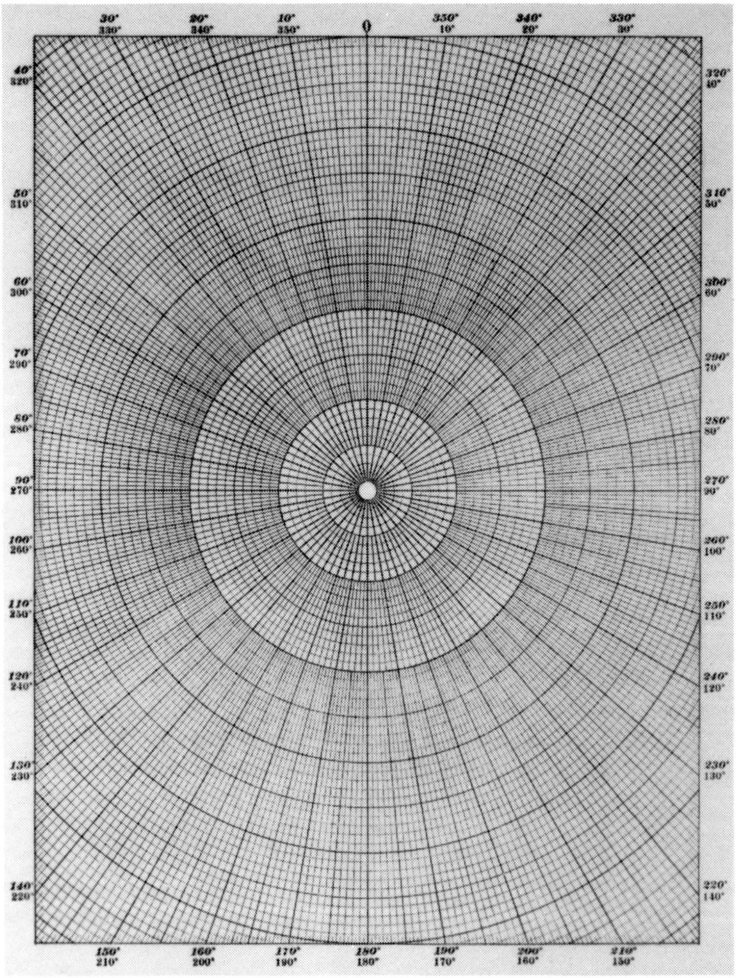

Fig. 2-21. Example of polar-coordinate paper.

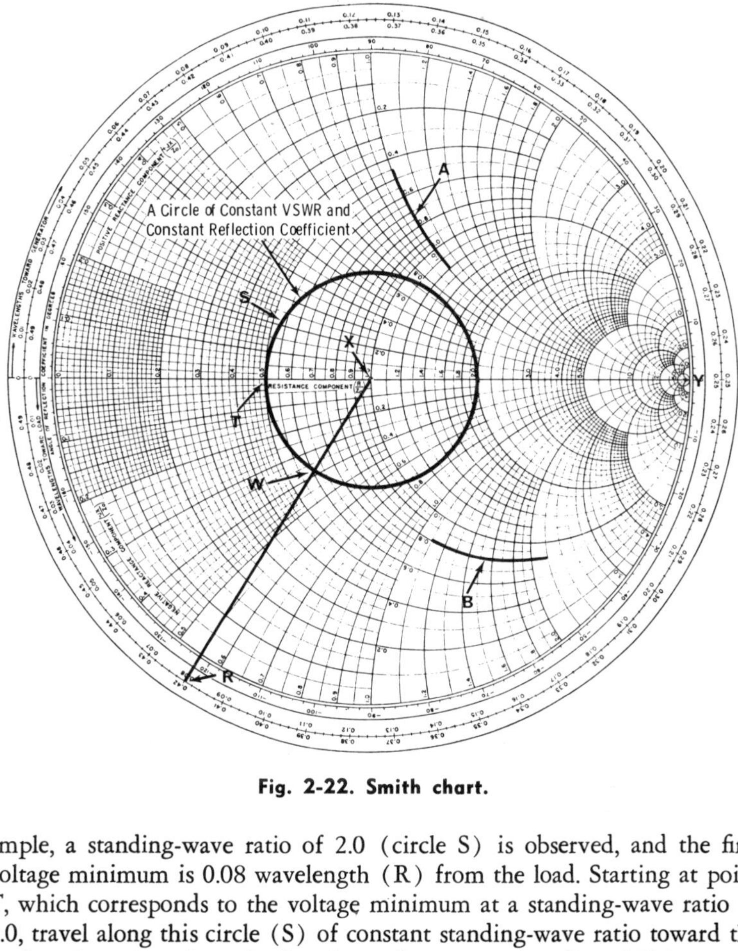

Fig. 2-22. Smith chart.

ample, a standing-wave ratio of 2.0 (circle S) is observed, and the first voltage minimum is 0.08 wavelength (R) from the load. Starting at point T, which corresponds to the voltage minimum at a standing-wave ratio of 2.0, travel along this circle (S) of constant standing-wave ratio toward the load to the point where the line drawn between points R and X intersects circle S. At this point of intersection (W) read the coordinates of the load impedance, which in this example are 0.62 − j0.38. Multiplying these numbers by the characteristic impedance of the standing-wave detector gives the actual impedance of the terminating load.

EXERCISES

Q2-1. Add: 4, 7, −10, 8, −6, 9

Q2-2. Add: 21, 5, −30, 2, −16

Q2-3. Subtract −16 from 5.

Q2-4. Multiply 4/5 by 3.

Q2-5. Divide 9/32 by 3.

Q2-6. Solve: $1/7 + 1/3 + 1/2$

Q2-7. Solve: $1/4 - 1/6$

Q2-8. $(2/3)(1/5) = ?$

Q2-9. $(3/4)(6/10) = ?$

Q2-10. Divide 2/3 by 3/5.

Q2-11. $\frac{30}{4} - \frac{(5)(3)}{3} + (45)(4) - 5 = ?$

Q2-12. The no-load voltage output of a power supply is 48 volts, and the full-load voltage output is 47.5 volts. What is the percentage regulation?

Q2-13. $10^0 = ?$
$4^0 = ?$
$2^0 = ?$

Q2-14. $(5^{-4})(5^3)(5^2)(5^{-9}) = ?$

Q2-15. (A) $(4^5)^3 = ?$
(B) $(4^{-5})^{-3} = ?$
(C) $(4^{-5})^3 = ?$
(D) $(4^2)^{-6} = ?$

Q2-16. $(3/4)^2 = ?$

Q2-17. $\sqrt{69{,}573} = ?$

Q2-18. $\frac{4(10)^3(5)(10^{-12})(2)(10^6)}{5(10^7)(4)(10^{16})} = ?$

Q2-19. (A) Log 1000 = ?
(B) Log 100 = ?
(C) Log 10 = ?
(D) Log 1 = ?
(E) Log 0.0001 = ?

Q2-20. $\text{Log}_{10}\ 54.65 = ?$

Q2-21. Multiply 1.24 by 246 using logarithms.

Q2-22. Divide 961 by 224 using logarithms.

Q2-23. $638^5 = ?$ Use logarithms.

Q2-24. If $X = AB$
$B = ?$
$A = ?$

Q2-25. In Fig. 2-8, assume $A = 10 \angle 30^\circ$ and $B = 5 \angle 60^\circ$. What are the magnitude (R) and angle (r) of the resultant vector?

Q2-26. Convert $20 \angle 30^\circ$ to rectangular form.

Q2-27. Convert 30 + j50 to polar form.

Q2-28. One dBm = ?

Q2-29. If the input to a transmission line is 150 watts and you measure 100 watts at the output end of the line, what is the loss in decibels?

Q2-30. What is the power gain of an amplifier rated at 40 dB gain?

CHAPTER **3**

Semiconductors and Logic for Broadcast Engineers

Discrete transistor circuitry and integrated circuits (ICs) are commonplace in modern broadcast installations. This is true not only for audio amplifiers but for control circuitry such as switch gear, transmitter remote control, and automation facilities as well. Low-power rf stages in transmitters are now almost completely solid state.

> NOTE: The reader should have basic solid-state training at least equivalent to the author's *Workshop in Solid State*.[1]

The fundamentals of semiconductor circuitry most important to the am-fm broadcast technician will be reviewed in this chapter. As developed in the above-mentioned reference, we will refrain from the equivalent-circuit approach, which involves *h* parameters and complex mathematical analysis, and emphasize a much simpler circuit analysis with practical applications. We will then study the principles of logic circuitry most often found in various applications in am-fm broadcast installations.

3-1. BASIC PARAMETERS OF SOLID-STATE AMPLIFIERS

There are certain practical parameters that are common to all solid-state amplifiers. If the maintenance technician is sufficiently familiar with these characteristics, he knows almost exactly what to expect in waveform and voltage analysis, regardless of what make of amplifier or control circuit he is troubleshooting.

Table 3-1 lists the symbols used in this study, and Chart 3-1 gives the basic parameters common to all circuit configurations—common emitter,

---

[1]Harold E. Ennes, *Workshop in Solid State* (Indianapolis: Howard W. Sams & Co., Inc., 1970).

Table 3-1. Symbols Used in This Book

α	Alpha. Common-base short-circuit current gain.
A_i	Current gain.
A_p	Power gain.
A_v	Voltage gain.
β	Beta. Common-emitter short-circuit current gain.
f_{hfb}	Alpha cutoff frequency (common base).
f_{hfe}	Beta cutoff frequency (common emitter).
f_{max}	Maximum frequency of oscillation.
C_{ce}	Collector-emitter capacitance.
C_{eb}	Collector-base capacitance.
C_{in}	Input capacitance.
Ge	Germanium.
g_m	Transconductance.
h_{FE}	dc short-circuit current gain.
h_{fe}	ac (signal) short-circuit current gain.
I_B	dc base current.
I_b	ac (signal) base current.
I_C	dc collector current.
I_c	ac (signal) collector current.
I_{CO}	Collector leakage current (cutoff current).
I_E	dc emitter current.
I_e	ac (signal) emitter current.
r_e	Small-signal emitter resistance.
R_E	Emitter resistor.
R_{EB}	Emitter-base junction resistance (assume 4 ohms average).
R_f	Feedback resistance.
R_G	Generator resistance.
R_{in}	Input resistance.
R_L	Load resistance.
R_S	Source resistance.
r_{tr}	Transresistance.
Si	Silicon.
V_B	Base voltage (dc).
V_{BB}	Base supply voltage.
V_{BE}	Base-to-emitter voltage (dc).
V_C	Collector voltage (dc).
V_{CC}	Collector supply voltage.
V_{CE}	Collector-to-emitter voltage (dc).
V_E	Emitter voltage (dc).
V_{EE}	Emitter supply voltage.
Z_g	Generator impedance.
Z_{in}	Input impedance.
Z_o	Output impedance.
Z_s	Source impedance.

Chart 3-1. Basic Parameters Common to All Configurations

Base Current	$I_B = I_E - I_C = \dfrac{I_C}{h_{FE}}$ or $\dfrac{I_E}{h_{FE}} - I_{CO}$
Collector Current	$I_C = I_E - I_B = \alpha I_E = h_{FE} I_B$
Collector Power	$P_C = V_{CE} I_C$
Emitter Current	$I_E = I_B + I_C$ (Total Current)
Small-Signal Emitter Resistance	$r_e = \dfrac{26}{I_E}$ where, I_E is emitter current in mA.
Transresistance	$r_{tr} = r_e + R_{EB} + R_E$ where, r_e is the small-signal emitter resistance in ohms, R_{EB} is assumed to be 4 ohms, and R_E is the unbypassed external emitter resistance.
Transconductance	$g_m = \dfrac{1}{r_{tr}} = \dfrac{I_E}{26}$ where, I_E is emitter current in mA.
Bandwidth	$f_{hfe} = \dfrac{f_{hfb}}{h_{fe}}$, or $f_{hfb} = h_{fe} f_{hfe}$ where, f_{hfe} is the beta cutoff frequency (3-dB point), f_{hfb} is the alpha cutoff frequency (3-dB point), h_{fe} is $\beta = \dfrac{\alpha}{1 - \alpha}$
Upper Frequency Limit	$f_u = \dfrac{g_m}{6.28C_t}$ where, f_u is upper frequency limit (unity gain) in MHz, g_m is transconductance in micromhos, and C_t is total capacitance in pF.
Input Capacitance	$C_{in} = \dfrac{g_m}{6.28 f_{hfb}}$

Chart 3-2. Parameters of Common-Emitter Circuit

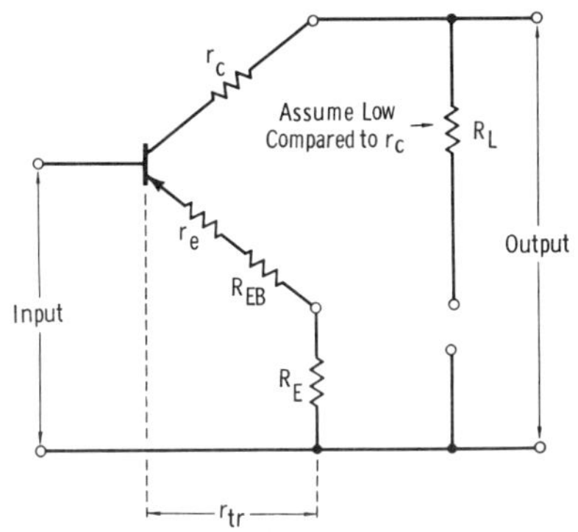

Input Impedance	$Z_{in} = h_{fe} r_{tr}$
Load Impedance	$Z_L = R_L$ in parallel with input impedance of next stage.
Current Gain	$A_I = \frac{\Delta I_C}{\Delta I_B} = h_{fe}$ where, $h_{fe} = \beta = \frac{I_c}{I_b} = \frac{\alpha}{1 - \alpha}$
Voltage Gain	$A_v = \frac{\Delta V_C}{\Delta V_B} = \frac{Z_L}{r_{tr}} = g_m Z_L$
Power Gain	$A_p = \frac{V_{out} I_{out}}{V_{in} I_{in}} = \beta \frac{Z_L}{r_{tr}}$
Typical Values (Single Stage)	Z_{in} 500-1500 ohms Z_L 1k-50k A_v 100-1000 A_i 25-100 A_p 25-70 dB

common base, and common collector (emitter follower). Charts 3-2, 3-3, and 3-4 give the specific parameters for the common-emitter, common-base, and common-collector configurations, respectively.

The Common-Emitter Amplifier

The *common-emitter* circuit (Fig. 3-1) is by far the most widely used in amplifier design. The *common-base* circuit is most often used to match a low impedance to a high impedance, and the *common-collector* (emitter-follower) circuit is most often used to match a high impedance to a low impedance.

The design parameters for Fig. 3-1 apply to a stage intended for linear

Chart 3-3. Parameters of Common-Base Circuit

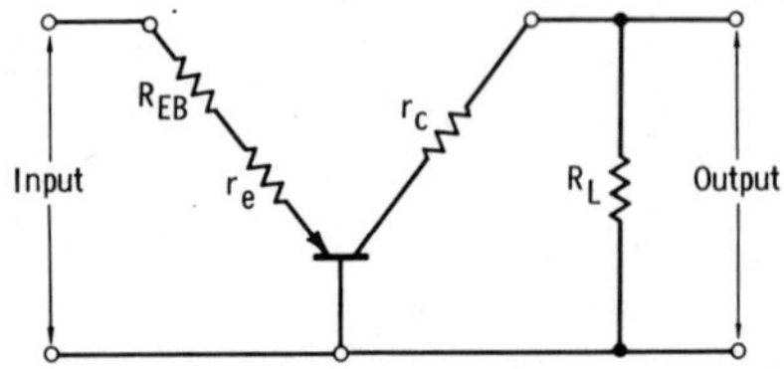

Input Impedance	$Z_{in} = r_{tr}$
Load Impedance	$Z_L = R_L$ in parallel with input impedance of following stage.
Current Gain	$A_i = \alpha = \dfrac{\beta}{1+\beta}$ (In practice, α is 0.95 to 0.995, or approximately 1.)
Voltage Gain	$A_v \approx \dfrac{Z_L}{r_{tr}} = g_m Z_L$
Typical Values (Single Stage)	Z_{in} 5-150 ohms Z_L 100k-500k A_v 100-1500 A_i Less than 1 (slightly) A_p 20-30 dB

(The common-base circuit is used primarily to match a low impedance to a high impedance.)

Chart 3-4. Parameters of Common-Collector Circuit

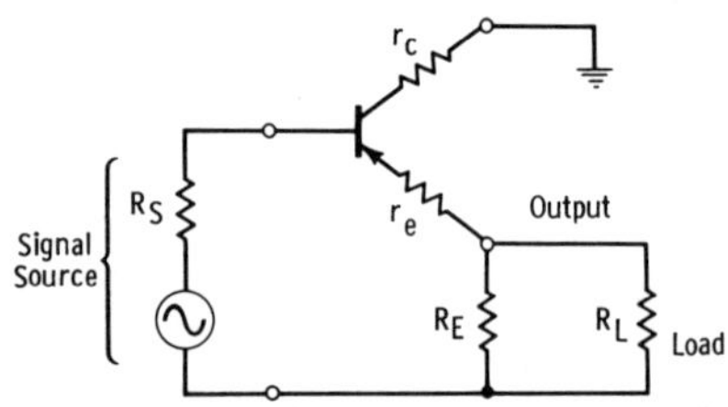

Input Impedance	$Z_{in} = (\beta + 1)Z_L$ where, Z_L is R_L in parallel with R_E.
Output Impedance	$Z_{out} = \dfrac{R_s}{\beta + 1}$ where, R_s is the output impedance of the signal source.
Current Amplification	$A_i \approx \beta$
Voltage Amplification	A_v = Less than unity
Typical Values (Single Stage)	Z_{in} 1k-500k Z_{out} 2-1000 ohms A_v Less than 1 A_i 25-100 A_p 10-35 dB

(The common-collector circuit is used primarily to match a high impedance to a low impedance.)

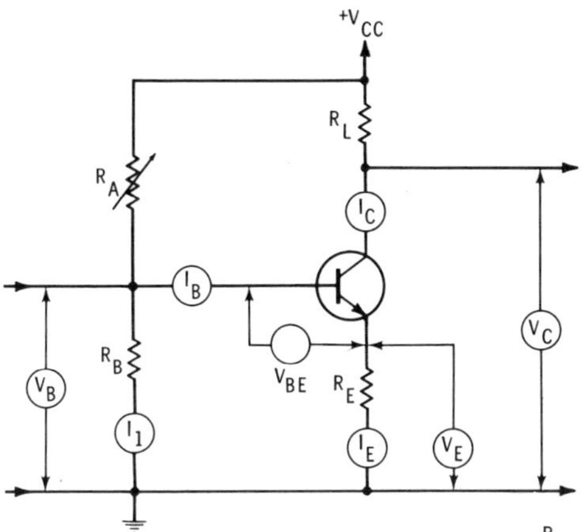

(1) $R_L > 5R_E$

(2) $I_1 > 5I_B$

(3) R_A Adjusted for $V_C \cong 1/2\, V_{CC}$

(4) I_B Dependent Upon (3)

(5) $R_B = V_B/I_1 \cong$ 5 to 10 Times R_E

(6) $R_E = \dfrac{V_E \text{ Desired } (>5V_{BE})}{I_E \text{ Desired (From Signal Mode)}}$

(7) $V_B = V_E + V_{BE}$, or $\dfrac{R_B}{R_A + R_B} V_{CC}$

(8) $V_E > 5V_{BE}$

(9) $V_C \cong 1/2\, V_{CC}$

(10) $A_V \cong R_L / R_E$ (With R_E Unbypassed and $I_E > 1$ mA)

Fig. 3-1. Design parameters for common-emitter amplifier.

amplification, as revealed by the fact that the quiescent collector voltage (V_C) is fixed to the approximate midpoint between ground and full collector supply voltage (V_{CC}). Some form of stabilizing feedback is normally provided, and we will examine this basic characteristic now.

Emitter Feedback

Emitter feedback is, very simply, the result of the unbypassed emitter resistor, R_E. Since the signal induced in useful collector load R_L must pass through R_E, the signal is degenerated. This type of feedback is a form known as *series feedback,* and it increases the input impedance. Before we go further, it is important to examine in more detail this most common form of feedback stabilization.

First, let us put down applicable relationships for the circuit of Fig. 3-1 for future reference:

$$V_{BE} = 0.2 \text{ to } 0.3 \text{ V for germanium transistors} \qquad \text{(Eq. 3-1)}$$

$$V_{BE} = 0.6 \text{ to } 0.7 \text{ V for silicon transistors} \qquad \text{(Eq. 3-2)}$$

$$I_B = \frac{I_E}{h_{FE}} - I_{CO} \qquad \text{(Eq. 3-3)}$$

$$I_C = \alpha I_E \quad \text{(Eq. 3-4)}$$

$$I_C = I_E - I_B \quad \text{(Eq. 3-5)}$$

$$I_C = h_{FE} I_B \quad \text{(Eq. 3-6)}$$

$$Z_{in} = h_{FE} r_{tr} \quad \text{(Eq. 3-7)}$$

$$Z_{load} = R_L \text{ in parallel with following-stage input } Z \quad \text{(Eq. 3-8)}$$

$$A_i = h_{FE} \quad \text{(Eq. 3-9)}$$

$$A_v = \frac{R_L}{r_{tr}} \quad \text{(Eq. 3-10)}$$

$$A_v = g_m R_L \quad \text{(Eq. 3-11)}$$

$$A_p = (A_i)^2 \frac{R_L}{\beta r_{tr}} = A_i A_v = \beta \frac{R_L}{r_{tr}} \quad \text{(Eq. 3-12)}$$

Now study the ten points listed on Fig. 3-1. They show how the design engineer arrives at stable operating parameters for a linear (class-A) amplifier.

In point 6, it is stated that the emitter current desired is derived from the intended operating mode. For example, a low-level stage (such as an audio preamplifier) must be operated at very low emitter (hence base and collector) currents in the interest of low noise level. As the signal level is brought higher and higher in amplitude, more and more dc current must be employed to accommodate the higher peak-to-peak signal swings.

The main problem in stabilization is to control the current gain (β, or h_{FE}) of the common-emitter circuit. The actual value of β differs with different transistors of the same type, and it also varies with temperature changes. Consequently, practical circuit design requires a *controlled beta* (controlled current gain) so that this parameter will remain fixed under varying operating temperatures and with necessary transistor replacements. Thus, if a certain transistor type has a "minimum h_{FE}" of 20, the circuit is designed to limit beta to no more than 20 by means of fixed resistance ratios.

We can now examine a practical single-stage common-emitter linear amplifier and see how such a design is made. Let us analyze the circuitry in Fig. 3-2 for practice in determining what we should expect in dc voltage measurements, and to see how it meets the design requirements of Fig. 3-1.

Step 1 is to determine the base voltage, V_B. We can do this by figuring the current in R_B, or we can use the second relationship of item 7 in Fig. 3-1. In either case, $V_B = +5$ volts.

Then, since the transistor is a silicon (Si) type:

$$V_E = +5 - 0.6 = +4.4 \text{ volts}$$

(Note that this value is greater than 5 times V_{BE}, as called for in Fig. 3-1.)

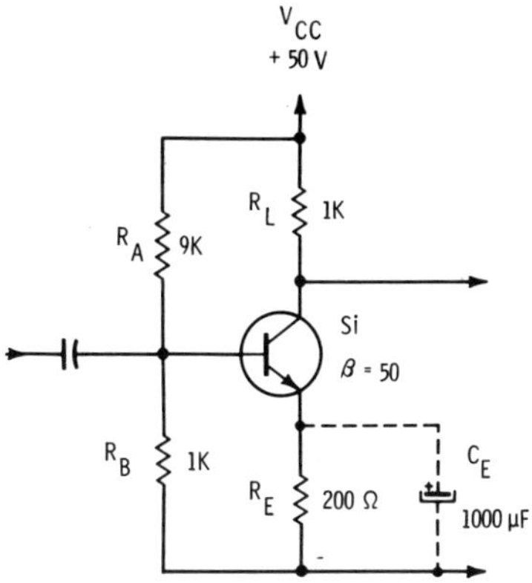

Fig. 3-2. Class-A common-emitter amplifier circuit.

Since the emitter voltage is +4.4 volts, the emitter current (from Ohm's law) must be:

$$I_E = \frac{4.4}{200} = 22 \text{ mA}$$

For rapid analysis, assume the same current in the collector as in the emitter:

$$I_C = 22 \text{ mA}$$

Then the dc voltage drop across R_L is:

$$V_{RL} = (0.022)(1000) = 22 \text{ volts}$$

and:

$$V_C = 50 - 22 = +28 \text{ volts}$$

(Note that this value is close to $1/2\ V_{CC}$.)

For the given transistor, $h_{FE} = \beta = 50$. Then, from equation 3-3 (ignoring I_{CO}):

$$I_B = \frac{0.022}{50} = 0.44 \text{ mA}$$

Note that current I_1 is 5 mA and therefore meets the requirement of being more than 5 times the base current (I_B).

With R_E bypassed (C_E connected):

$$A_i = h_{FE} = 50 \text{ (from specification sheets)}$$

$$r_{tr} = \frac{26}{22} + 4 = 5 \text{ (approx)}$$

$$A_v = \frac{R_L}{r_{tr}} = \frac{1000}{5} = 200$$

$$Z_{in} = (50)(5) = 250 \text{ ohms (from equation 3-7)}$$

Note that since R_E is bypassed for the signal, the signal is not degenerated. Therefore, the full value of beta is in effect, with the exception of the slight shunting action of bias resistors R_A and R_B.

Now consider the case in which R_E is not bypassed by C_E. The first change to note is the effect on input impedance Z_{in} of the transistor alone. When R_E was bypassed, Z_{in} equalled 250 ohms. But when R_E is not bypassed, r_{tr} becomes 205 ohms, or approximately 200 ohms. Then:

$$Z_{in} = (50)(200) = 10{,}000 \text{ ohms}$$

Now it is evident that the value of R_B essentially fixes the value of the new input impedance. Since R_B is only one tenth of the transistor input impedance, the current gain will be reduced drastically.

For practical circuit analysis, the following relationship of currents and resistances is sufficiently accurate:

$$\frac{I_e}{I_{in}} = \frac{R_B}{R_E}$$

This equation says that the emitter current is to the input current as base resistor R_B is to emitter resistor R_E. From this basic relationship, we can derive a useful equation for quickly evaluating current gain in a degenerative amplifier. Since collector current (for rapid estimates) can be considered to be the same as emitter current, then substituting I_c (collector current) for I_e (emitter current) in the above basic equation:

$$\frac{I_c}{I_{in}} = \frac{R_B}{R_E}$$

Note that the ratio of I_c (collector current) to I_{in} (input current) is an expression for current gain (A_i). So:

$$A_i = \frac{R_B}{R_E} \text{(for an unbypassed } R_E\text{)} \qquad \text{(Eq. 3-13)}$$

Note also that when an unbypassed R_E is much greater than $r_e + R_{EB}$, voltage amplification (A_v) can be expressed:

$$A_v = \frac{R_L}{R_E} \text{ (for an unbypassed } R_E\text{)} \qquad \text{(Eq. 3-14)}$$

This equation is valid so long as $r_e + R_{EB}$ is very small in value compared to R_E and can be ignored.

Based on the above equations, the new current gain with R_E unbypassed is:

$$A_i = \frac{R_B}{R_E} = \frac{1000}{200} = 5$$

and the new voltage gain is:

$$A_v = \frac{R_L}{R_E} = \frac{1000}{200} = 5$$

Other Basic Negative-Feedback Circuits

Although emitter degeneration (series feedback) is almost universally used, other forms of degenerative feedback, which may or may not be used along with emitter degeneration, are often encountered. We will explore briefly the fundamentals of such circuitry.

The basic arrangement for collector-base feedback (termed *shunt feedback* is shown in Fig. 3-3. For most practical circuits, the resistance of R_f is small enough that we can use these approximate gain formulas for quick-analysis purposes:

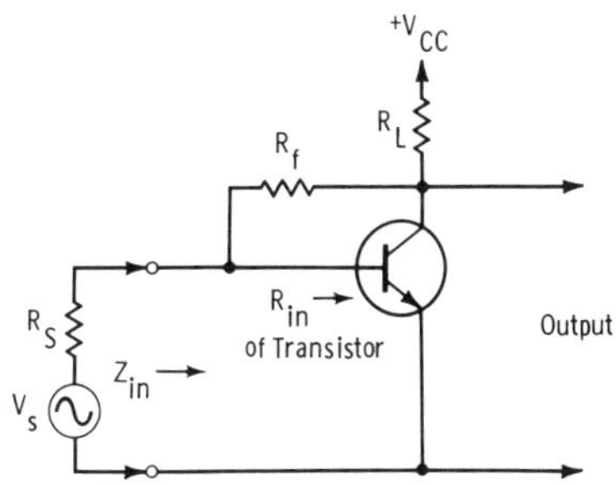

Fig. 3-3. Circuit with shunt feedback.

$$A_i = \frac{R_f}{R_L} \qquad \text{(Eq. 3-15)}$$

$$A_v = \frac{R_f}{Z_{in}} \qquad \text{(Eq. 3-16)}$$

The quantity Z_{in}, the effective input impedance of the stage, is approximately:

$$Z_{in} = \frac{R_L + R_f}{\beta R_L + R_f} R_{in} \qquad \text{(Eq. 3-17)}$$

where R_{in} is the input impedance without feedback.

Note that shunt feedback *decreases* the effective Z_{in}, whereas series feedback (such as that provided by an unbypassed emitter resistor) *increases* the effective Z_{in}.

NOTE: The gain formulas given above are adequate for quick analysis of most circuits because R_f is usually relatively low in value to meet the requirement of "adequate feedback." Somewhat more nearly exact relationships are as follows:

$$A_i = \frac{R_f}{R_f + \beta R_L} \beta$$

$$A_v = \frac{R_f}{R_L + R_f} \beta \frac{R_L}{R_{in}}$$

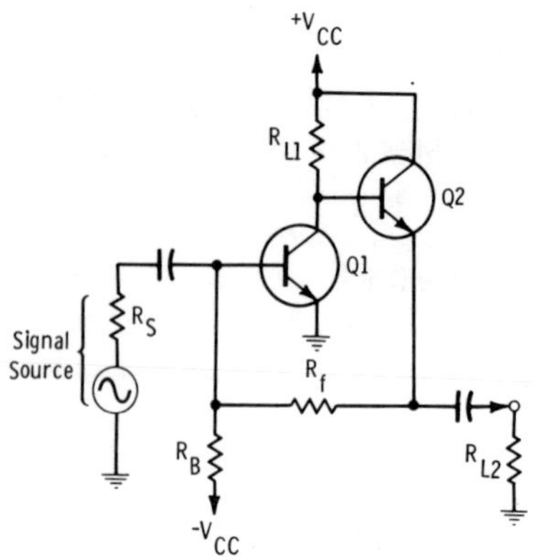

Fig. 3-4. Diagram of a feedback pair.

Fig. 3-4 shows a "feedback pair." This circuit is gaining universal acceptance as a stable, wideband amplifier with inherent ac and dc stability. This type of circuit has low input and output impedance. For quick analysis, its voltage gain can be calculated from the expression:

$$A_v = \frac{R_f}{R_s} \qquad \text{(Eq. 3-18)}$$

where R_s is the output resistance of the signal source. This formula is sufficiently accurate for use in normal applications of the circuit with low input and output impedances. The voltage gain is almost universally held to a value less than 10.

Feedback Gain-Impedance Relationships

We should be certain of a "sharp focus" on impedance-to-gain relationships. For example, series feedback affects voltage gain much more than current gain, and shunt feedback affects current gain much more than voltage gain. These facts always bring up the question, "If you cut current gain, how can voltage gain remain the same?" Or, "If current gain is not affected, why does voltage gain decrease?" This discussion should answer these questions.

See Fig. 3-5A. This circuit is without negative feedback. Assume that:

$A_i = h_{fe} = 50$
Input signal current $= 10\ \mu A$
Effective $Z_{in} =$ 1000 ohms (resistive)
$R_L =$ 2000 ohms

Then:

Signal current in $R_L = (50)\ (10\ \mu A) = 500\ \mu A.$
Signal voltage in $Z_{in} = IR = (10\ \mu A)\ (1000) = 0.01$ volt
Signal voltage in $R_L = (500\ \mu A)\ (2000) = 1$ volt

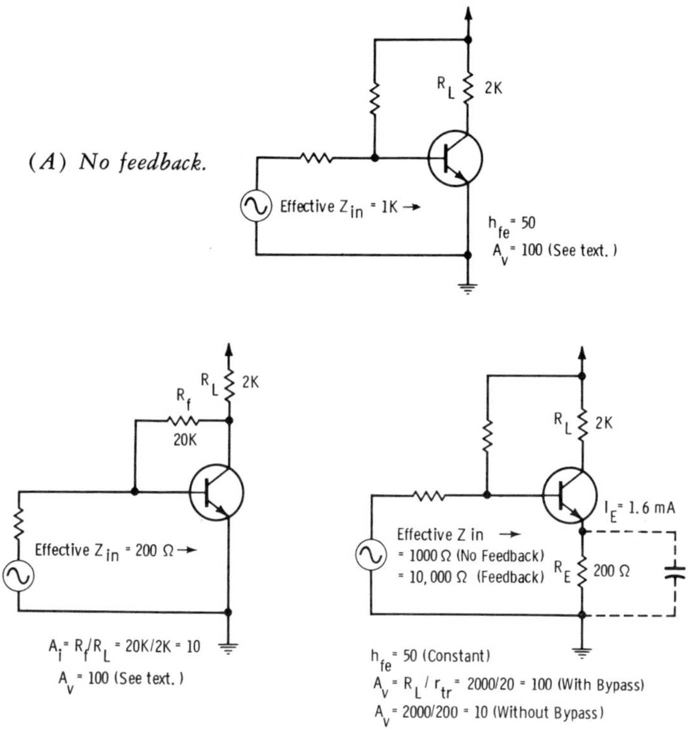

(A) No feedback.

(B) Shunt feedback. *(C) Series feedback.*

Fig. 3-5. Comparison of shunt and series feedback.

$$A_v = \frac{V_{out}}{V_{in}} = \frac{1}{0.01} = 100$$

Now see Fig. 3-5B. Shunt feedback is used, and:

A_i is cut to 10 (from 50).
Signal input current is maintained at 10 μA (for reference).
R_{in} is reduced (by shunt feedback) from 1000 ohms to about 200 ohms (from equation 3-17).

Now:

Signal current in $R_L = (10)(10\ \mu A) = 100\ \mu A = 0.1$ mA.
Signal voltage in $R_L = (0.1\ mA)(2000\ \Omega) = 0.2\ V = 200$ mV.
Signal voltage in $R_{in} = (10\ \mu A)(200\ \Omega) = 0.002\ V = 2$ mV.

$$A_V = \frac{200\ mV}{2\ mV} = 100 \text{ (the same as for the circuit of Fig. 3-5A).}$$

Note that although the output signal voltage was reduced in amplitude for the same signal input current, the voltage *amplification* remained the same for the circuit of Fig. 3-5B as for that of Fig. 3-5A. The current amplification is limited to 10 (which will be below the minimum beta of the transistor used) so that voltage output is stabilized for changes in h_{fe}.

As a matter of interest, the voltage gain calculated from the more complete equation given in the note above is:

$$A_V = \frac{20K}{2K + 20K}(50)\frac{2K}{1K} = 90$$

The "quick-analysis" value is within about 10 percent of this result.

Fig. 3-5C shows the series-feedback arrangement of the common-emitter circuit. Whether bypassed or not, R_E has no effect on the current-gain parameter of the transistor. But we already know that it drastically affects the voltage gain of the circuit.

Assume that R_E is bypassed and that the signal input is 10 microamperes. Then:

$$\text{Signal current in } R_L = (50)\ (10\ \mu A) = 500\ \mu A$$

$$\text{Signal voltage in } R_L = (500\ \mu A)\ (2000\ \Omega) = 1 \text{ volt}$$

$$\text{Signal voltage in } R_{in} = 10\ \mu A\ (h_{fe}r_{tr}) = 10\ \mu A\ (50)\left(\frac{26}{1.6} + 4\right)$$

$$= 10\ \mu A\ (1000\ \Omega) = 0.01 \text{ volt}$$

$$A_v = \frac{1}{0.01} = 100$$

Now assume that R_L is not bypassed (series feedback to signal):

$$R_{in} = 50\ (20 + 200) = 10{,}000 \text{ ohms (approx)}$$

$$V_{in} = (10\ \mu A)\ (10{,}000\ \Omega) = 0.1 \text{ volt}$$

$$\text{Signal current in } R_L = 50\ (10\ \mu A) = 500\ \mu A\ (h_{fe} \text{ not changed})$$

$$\text{Signal current in } R_L = (500\ \mu A)\ (2000\ \Omega) = 1 \text{ volt}$$

$$A_V = \frac{1}{0.1} = 10 \quad \left(\text{Same as } \frac{R_L}{R_E} = \frac{2000}{200} = 10\right)$$

It has been pointed out that h_{FE} is the dc beta (common emitter) and that h_{fe} is the ac, or signal, beta. This difference can be emphasized by studying Fig. 3-6.

The circuit of Fig. 3-6A is equivalent to a single stage that has a 5K collector load and is feeding a following circuit with a 1.5K input impedance. At low frequencies, the load resistance is effectively 5000 ohms. Assume $h_{FE} = h_{fe} = 50$. The output voltage swing for a base-current swing of (for example) 3 microamperes is:

$$V = (3\ \mu A)\ (50)\ (5000\ \Omega) = 750 \text{ mV}$$

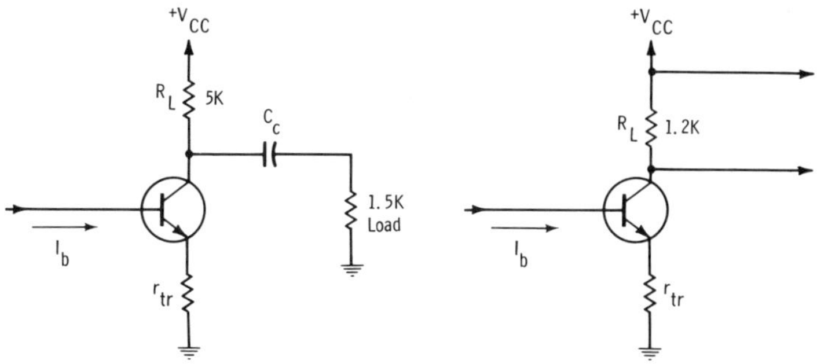

(*A*) *Dc load.* (*B*) *Ac load.*

Fig. 3-6. Circuit equivalents for ac and dc gain.

At high frequencies (for which the reactance of C_C can be neglected), the effective load becomes 5K in parallel with 1.5 K, or about 1.2K (Fig. 3-6B). Now the output-voltage swing for a 3-microampere base-current swing is:

$$V = (3\ \mu A)\ (50)\ (1200\ \Omega) = 180\ mV$$

This example simply emphasizes that with h_{fe} equal to h_{FE}, the actual voltage swing is much less for h_{fe} due to the difference in the load the transistor sees.

The Common-Collector (Emitter-Follower) Circuit

The emitter follower is used extensively for applications in which a low output impedance is desired. Fig. 3-7 shows an output stage driving a 75-ohm coaxial line, as used in certain forms of control circuitry. The 8 volts on the base is developed across the collector load of Q1. The emitter measures 8.5 volts, so the voltage drop across R_L is $20 - 8.5 = 11.5$ volts. Therefore:

$$I_E = \frac{11.5}{470} = 24\ mA$$

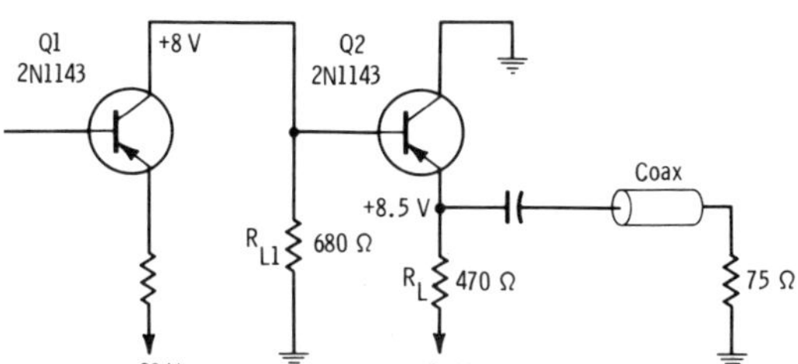

Fig. 3-7. Typical emitter follower.

The transistor has a typical β of 50; therefore:

$$I_B \doteq \frac{I_E}{h_{FE}} = \frac{0.024}{50} = 480\ \mu A$$

It can be seen that a current gain (480-μA base current to nearly 24-mA collector current) occurs in the emitter follower just as in the common-emitter amplifier (Fig. 3-8A). But the *voltage* gain is approximately unity. The output signal is of the same amplitude as the input signal, less a very small signal-voltage drop across the base-emitter junction. The voltage gain of this circuit can be explained as follows:

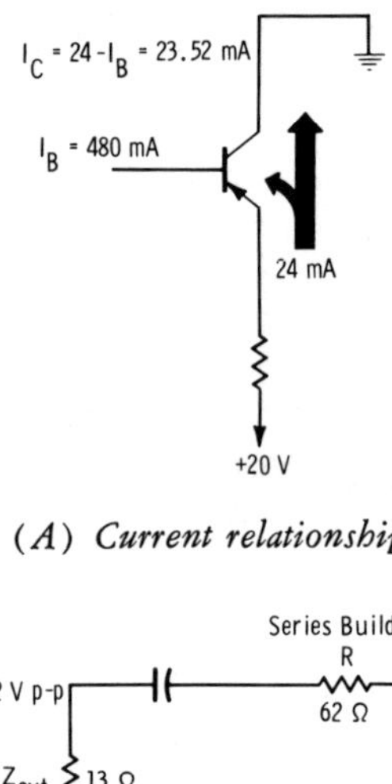

(*A*) *Current relationship.*

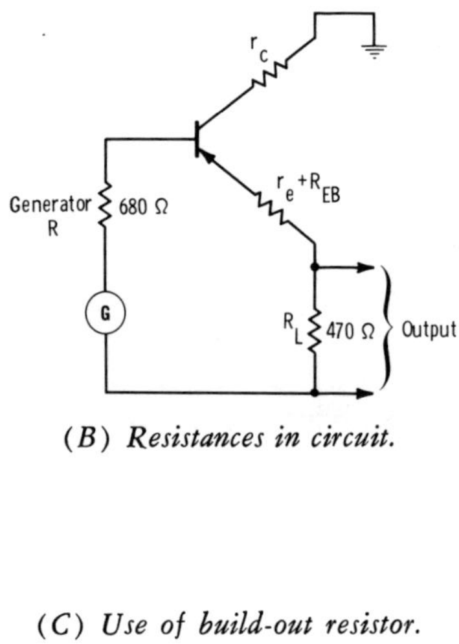

(*B*) *Resistances in circuit.*

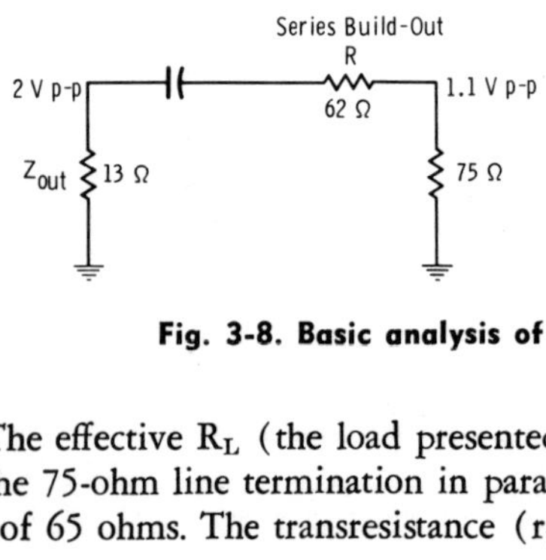

(*C*) *Use of build-out resistor.*

Fig. 3-8. Basic analysis of emitter-follower stage.

The effective R_L (the load presented to the signal voltage) in this case is the 75-ohm line termination in parallel with 470 ohms, for an effective R_L of 65 ohms. The transresistance (r_{tr}) is:

$$r_{tr} = \frac{26}{I_E} + R_{EB} + R_L$$

$$= 1\ (\text{approx}) + 4 + 65 = 70\ \text{ohms}$$

Then, since g_m is $\frac{1}{r_{tr}}$ 1/70, or 0.014:

$$A_v = g_m R_L = (0.014)(70) =) 0.98$$

Or, we can say that, since R_L is much greater than $r_e + R_{EB}$:

$$A_v = \frac{R_L}{R_E} \text{ (approx)}$$

In this case, $R_E = R_L$, so:

$$A_v = \frac{R_L}{R_L} = 1 \text{ (approx)}$$

In practice there are many cases in which it is not possible to measure any amplitude difference between the signals at the base and emitter of a common-collector stage. For this analysis, we can delete R_{EB} from the g_m computation. The gain may normally be assumed to be unity for quick circuit analysis. Note also that the dc parameters were used in the above analysis, so we are considering only low frequencies (and dc) at the present.

Study Fig. 3-8B, and observe that r_c appears in parallel with the sum of r_e, R_{EB}, and R_L. Since r_c can be assumed to be 1.5 megohms or more, it need not be considered for quick circuit analysis. What is important is to note that the actual input impedance and the output impedance are somewhat interdependent. This is true because the input load is part of the output, and the output load is part of the input. The high intrinsic value of r_c is not serving as isolation between input and output as it does for the other configurations; it now forms a parallel resistance only.

Since the emitter follower is used primarily as an impedance transformer, the input and output impedances are most important characteristics to know. While conventional treatments result in highly complex formulas, we can use the following rules of thumb that are fairly accurate in practice:

$$Z_{in} = (\beta + 1) R_L \qquad \text{(Eq. 3-19)}$$

$$Z_{out} = \frac{R_S}{\beta + 1} \qquad \text{(Eq. 3-20)}$$

The input resistance of the stage in Fig. 3-8 is:

$$Z_{in} = (50 + 1)\, 65 = 3300 \text{ ohms (approx)}$$

This resistance has a reasonably small loading effect on the 680-ohm collector load of Q1.

$$Z_{out} = \frac{680}{50 + 1} = 13 \text{ ohms (approx)}$$

The internal output impedance (sending-end impedance) of an emitter follower can be made quite low, sufficiently low to drive 2-ohm switching buses directly. This low impedance also is the reason why a series "build-out" resistor is usually used (Fig. 3-8C). For the receiving end (75 ohms) to "see" a 75-ohm impedance, the 13-ohm output would require a series resistance of 62 ohms. Note also that this arrangement forms a voltage divider, and a 2-volt signal across R_L is dropped to 1.1 volts across the 75 ohms.

In practice, as the effective load impedance increases, the input impedance increases. As the generator impedance increases, the output impedance increases. Providing the generator impedance (previous collector R_L) is not extremely low (so long as it is greater than 10 times R_L, as is usually the case), the formulas given above will prove sufficiently accurate for practical purposes.

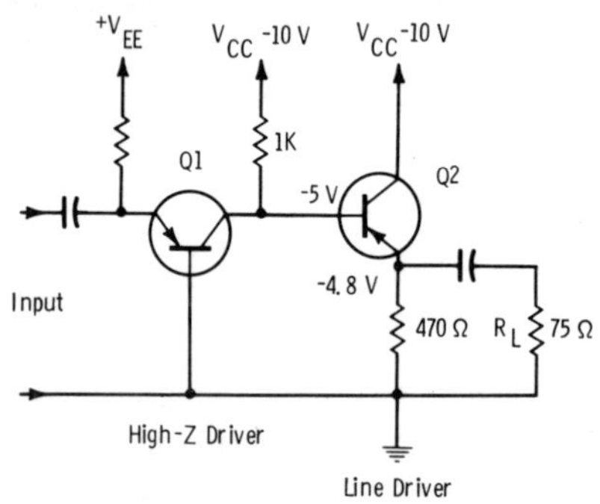

Fig. 3-9. Emitter-follower line driver.

Figs. 3-9 through 3-12 illustrate the basic applications of the common-collector configuration. In Fig. 3-9, the output impedance of Q1 is essentially the collector load of 1000 ohms. For class-A operation, we expect a quiescent collector voltage of $1/2\ V_{CC} = -5$ volts. Assume germanium transistors with a beta of 50. Then the common-collector line driver, Q2, has:

$$R_L = \frac{470\ (75)}{470 + 75} = 65 \text{ ohms (effective load)}$$

$$Z_{out} = \frac{1000}{50 + 1} = 20 \text{ ohms (approx)}$$

$$Z_{in} = (50 + 1)\ 65 = 3300 \text{ ohms (approx)}$$

Now study Fig. 3-10A. In this case, the emitter-follower line driver is *driven* by an emitter follower. The 1K load is now in the emitter circuit of Q1 rather than the collector circuit. Let us see what happens. (Assume Q1 sees a source resistance of 1000 ohms, and β is 50 for both transistors.)

The output impedance of Q1 is:

$$Z_{out} = \frac{1000}{50 + 1} = 20 \text{ ohms (approx)}$$

Then the output impedance of Q2 is:

$$Z_{out} = \frac{20}{50 + 1} = 0.4 \text{ ohm (approx)}$$

The equivalent circuit of the output stage is shown in Fig. 3-10B. Note that to feed a 75-ohm load, a build-out resistance of 75 ohms is needed so that the effective internal output impedance becomes 75 ohms. The build-out and load combine to form a 2:1 voltage divider. We must therefore expect to find the signal voltage at the emitter of Q2 to be twice as great as the signal voltage across the 75-ohm load. This condition is normal. For a 1-volt peak-to-peak signal to appear across the 75-ohm load, a 2-volt peak-to-peak signal must be available at the Q2 emitter.

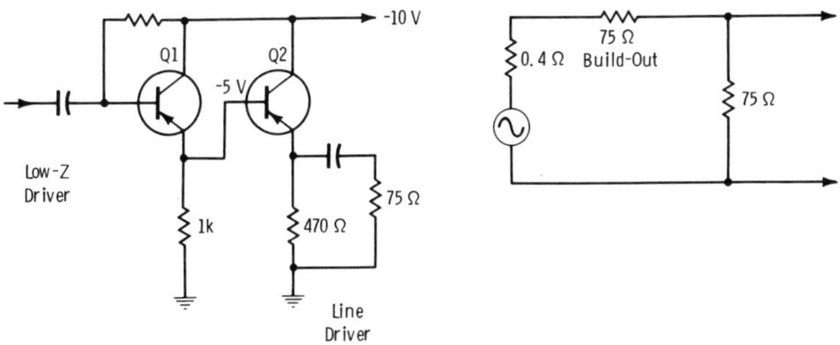

(*A*) *Schematic diagram.* (*B*) *Equivalent of output stage.*

Fig. 3-10. Cascaded emitter followers.

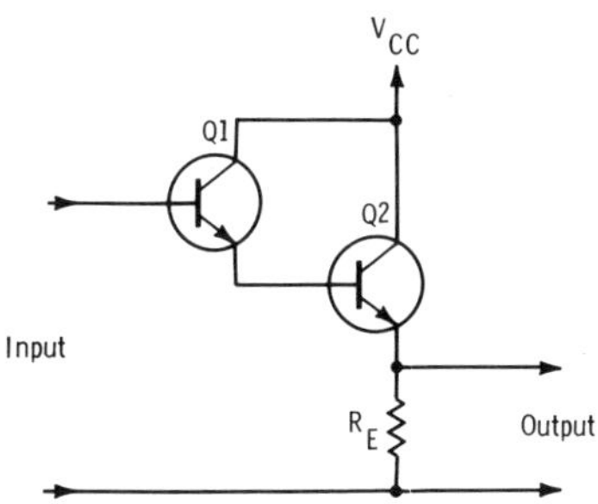

Fig. 3-11. Darlington circuit.

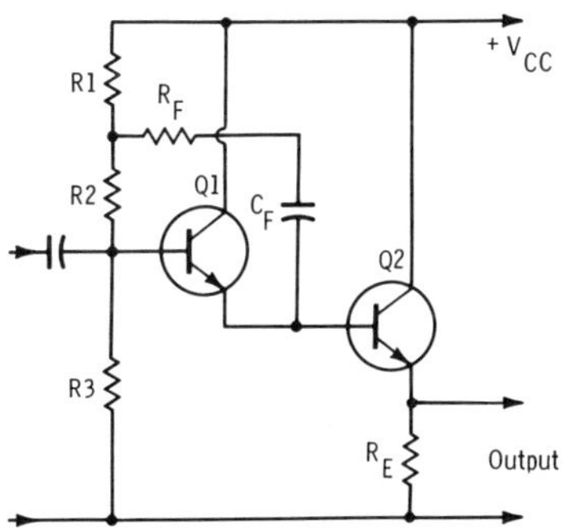

Fig. 3-12. Circuit with bootstrapping.

In solid-state circuit applications requiring a very high input impedance, the emitter follower is quite naturally chosen. The Darlington emitter follower shown in Fig. 3-11 consists of two current amplifiers in cascade to provide large current gain and very high input impedance. The input impedance is raised due to the effective product of the betas (h_{fe}) of both transistors. That is, since the base currents have been reduced by the product of the current gains, the input impedance is raised accordingly.

The input impedance of the Darlington current multiplier is limited by the shunting effect of the Q1 collector resistance and capacitance. A technique for reducing this shunting effect is to employ positive shunt feedback, termed *bootstrapping.* In Fig. 3-12, the value of R_F is such as to produce at the junction of R1 and R2 a voltage which is in phase with the input voltage. The signal-voltage drop across the bias resistors is thereby reduced, which is equivalent to raising the resistance of the input network. Bootstrapping is used in other than Darlington circuits for the same reason —to reduce the effective loading of the input biasing networks.

The Field-Effect Transistor

The field-effect transistor (FET) is a relatively new device, but it has gained substantially in practical applications. As with any device, it has advantages and disadvantages when compared to the conventional transistor.

For the conventional transistor, there are two major differences in operation compared to the vacuum tube; the input circuit must be forward biased (the tube is actually reverse biased), and the output circuit must be reverse biased. So essentially we have a current-operated device (transistor) compared to a voltage-operated device (tube). This is simply the most convenient way to contrast the two devices.

Now see Fig. 3-13A. In this *junction* type of FET, a pn junction is employed for the *gate* (control) electrode. The sole function of this gate is

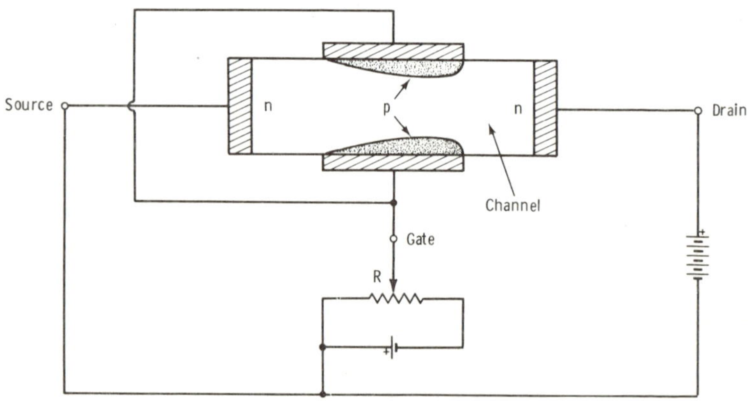

(A) Simplified diagram.

(B) Schematic symbols.

Fig. 3-13. Fundamentals of junction FET.

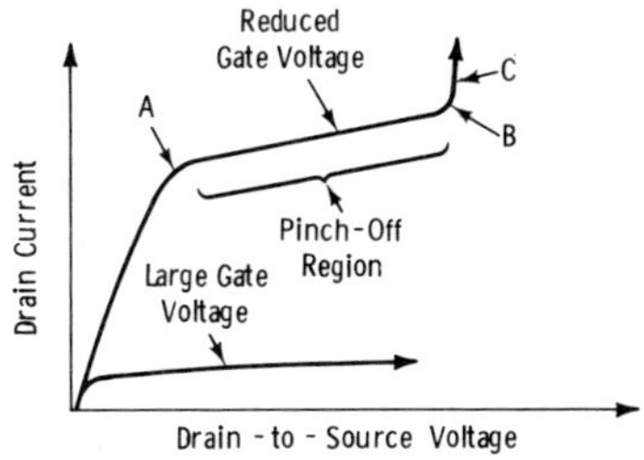

Fig. 3-14. Typical FET drain-current characteristic.

to provide a control element analogous to the grid of a vacuum tube. The FET also has a *source* (analogous to the cathode) and a *drain* (analogous to the plate). Note that the gate receives a voltage which is the reverse bias neccessary to control majority-carrier current in the *channel.*

When the gate voltage is increased, the fields set up by the junction barriers cause a reduction in the number of majority carriers flowing through the channel from source to drain (Fig. 3-14). As the gate voltage is reduced, drain current increases. The area in which the drain voltage has a relatively small effect on drain current (between points A and B in Fig. 3-14) is termed the *pinch-off region* of operation. As the drain-to-source voltage is increased, or the gate voltage goes to zero bias or a slight forward bias, the drain current can increase excessively, and the FET can be damaged by the resultant heating of the junction (part C of the curve in Fig. 3-14).

Just as with the vacuum tube, the input impedance of the FET in the common-source configuration is very high due to the reverse bias on the gate. The FET is advantageous to use where very high input impedances are required in a single stage of amplification. The device is extremely delicate in handling and servicing, but this disadvantage is rapidly being overcome by "protected gate" design.

Schematic symbols for the FET are shown in Fig. 3-13B. Figs 3-15A and 3-15B present a circuit drawn for an n-channel and a p-channel FET. The gate input impedance is normally around 10 megohms, and a fixed resistor (R_G) is used so that the input signal voltage (v_i) is developed across a known load for all frequencies concerned.

Typically, the signal at the source (v_s) will be between 0.6 and 0.75 of v_i when R_S is not bypassed. Resistor R_S supplies the proper bias to the gate and (for class-A operation) fixes the drain voltage at about 1/2 of V_{DD}. The drain signal output (v_d) is inverted.

Just as $\beta = 50$ is a good "average" for the conventional transistor, so the transconductance (g_m) for the proper operating point of the average FET can be assumed to be $g_m = 500$ micromhos, and $A_v = g_m R_L$ (approx) without feedback.

NOTE: In an FET, g_m varies with drain current just as the beta of a conventional transistor varies with collector current. Therefore, it is subject

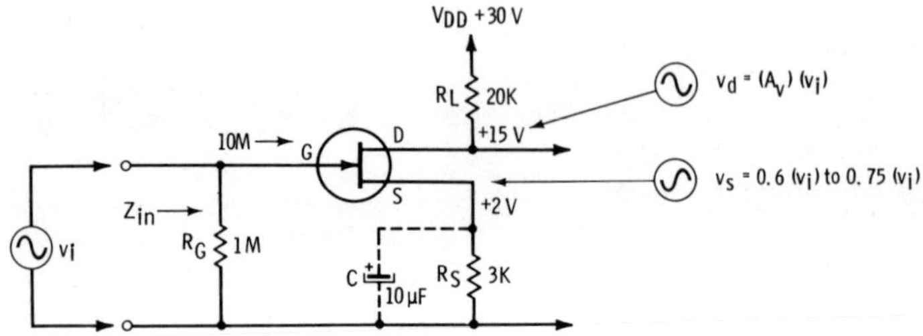

(*A*) *N-channel.*

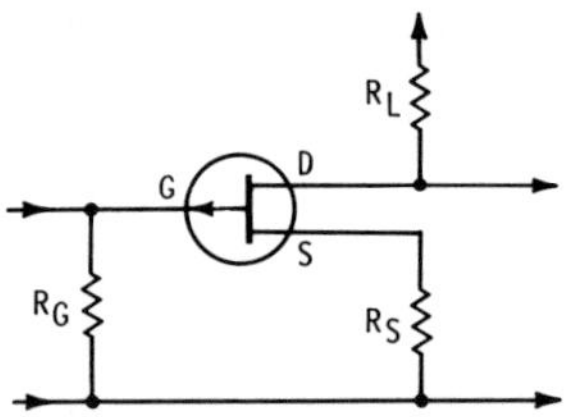

(*B*) *P-channel.*

Fig. 3-15. Amplifier circuits using FETs.

to wide variation, and data sheets must be consulted to determine g_m at a specified drain current.

The voltage amplification with source feedback (C not connected) is:

$$A_v = \frac{g_m R_L}{1 + g_m R_s} \text{ (approx)}$$

For the circuit of Fig. 3-15A, assume that $g_m = 500$ micromhos. Then the gain without feedback is:

$$A_v = (0.0005)(20{,}000) = 10 \text{ (approx)}$$

With C not used (source feedback):

$$A_v = \frac{0.0005\ (20{,}000)}{1 + 0.0005\ (3000)} = 4 \text{ (approx)}$$

Fig. 3-16. shows symbols for three types of metal-oxide-semiconductor FET (MOSFET). These transistors are treated further in applicable portions of following chapters.

The characteristics of FETs may be summarized as follows:

1. High impedances permit use of vaccum-tube biasing techniques.
2. Very good thermal stability.
3. Extremely low feedback capacitance.
4. Low noise figure (nf = 3.5 dB, typical). A noise figure of 5 dB is considered very good for conventional transistors.

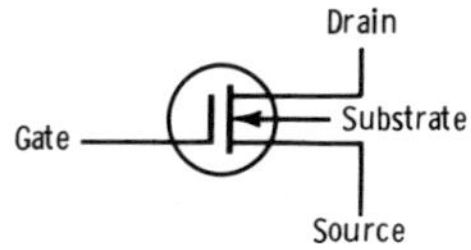

(*A*) *N-channel MOSFET symbol.*

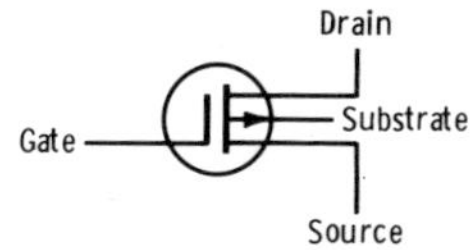

(*B*) *P-channel MOSFET symbol.*

(*C*) *Dual-insulated-gate MOSFET.*

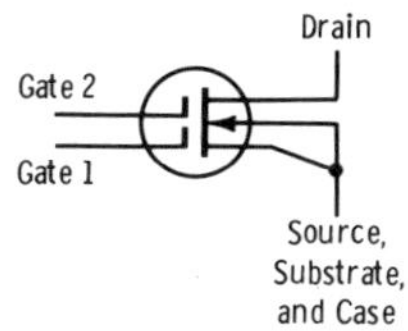

Fig. 3-16. Schematic symbols for MOSFETs.

Class-B Audio Operation of Transistors

Class-B push-pull power amplifiers are used mainly in equipment that must provide high power output and high power efficiency. Fig. 3-17A shows a simplified circuit of a class-B amplifier. The emitter-base junctions are zero biased. In this circuit, each transistor conducts on alternate half cycles of the input signal. The output-signal half cycles are combined in the secondary of the output transformer. Maximum efficiency is obtained even during idling periods (no input signal), because neither transistor conducts during these periods.

An indication of the output current waveform for a given signal current input can be obtained by considering the dynamic transfer characteristic for the amplifier. It is assumed that the two transistors have identical dynamic transfer characteristics. The characteristic for one of the transistors is shown in Fig. 3-17B. The variation in output (collector) current is plotted against input (base) current under load conditions. Since two transistors are used, the overall dynamic transfer characteristic for the push-pull amplifier is obtained by placing two of the curves back-to-back. The two curves are shown back-to-back *and combined* in Fig. 3-17C. Note that the zero lines of the two curves are lined up vertically to reflect the zero bias current. Note that severe distortion occurs at the crossover point (the point where the signal passes through zero). This crossover distortion becomes more severe with low signal input currents. Crossover distortion can be eliminated by using a small forward bias on both transistors of the push-pull amplifier.

A class-B push-pull amplifier with a small forward bias applied to the base-emitter junctions is shown in Fig. 3-18A. A voltage divider is formed by resistors R2 and R1; the voltage developed across R1 supplies the base-emitter bias for both transistors.

A study of the dynamic transfer characteristic curve of the amplifier demonstrates the elimination of crossover distortion. In Fig. 3-18B, the dynamic transfer characteristic curves of the two transistors are placed back-to-back (but not combined) for zero-base-current bias conditions. The dash lines indicate the zero-signal point for each curve when forward bias is applied to the transistors. To obtain the overall characteristic (Fig. 3-18C), the two graphs are shifted until the dash lines meet; then the individual curves (dotted lines) are combined to form the solid curve.

Resistance-capacitance coupling to the input of a class-B push-pull amplifier cannot be used satisfactorily without special considerations. In the class-B push-pull amplifier, one transistor conducts during one half cycle

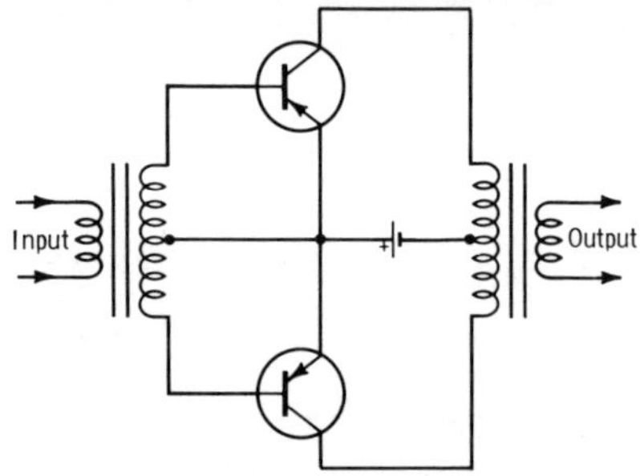

(*A*) *Schematic diagram.*

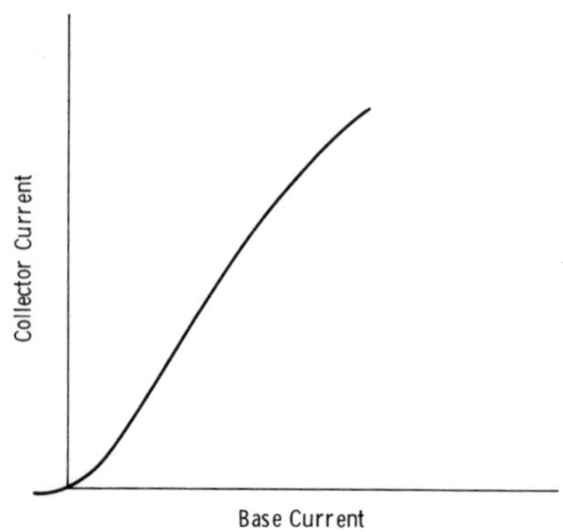

(*B*) *Single transfer characteristic.*

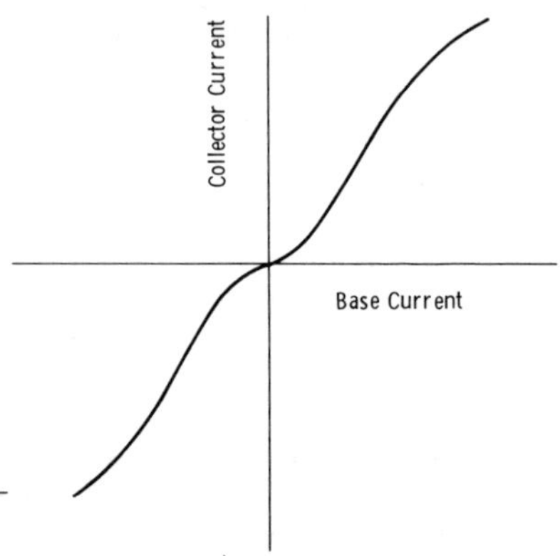

(*C*) *Combined transfer characteristic.*

Fig. 3-17. Basic class-B amplifier.

of the input signal; the other transistor remains nonconducting during this time, except for the small forward bias applied to minimize crossover distortion. Refer to Fig. 3-19A. Assume that the input signal causes transistor Q1 to conduct. Electrons (solid-line arrows) leave the right-hand plate of coupling capacitor C1, enter the base-emitter junction of transistor Q1, flow to the emitter ground connection, go through ground to the junction of resistors R1 and R2, and pass through resistor R1 to the left-hand plate of C1. (Resistor R1 represents the output resistance of the previous

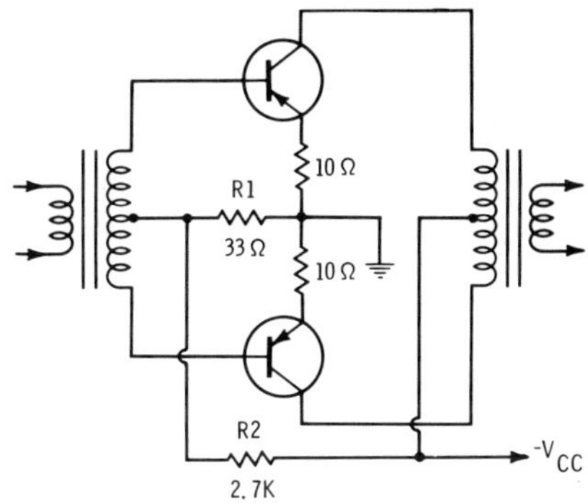

(*A*) *Schematic diagram.*

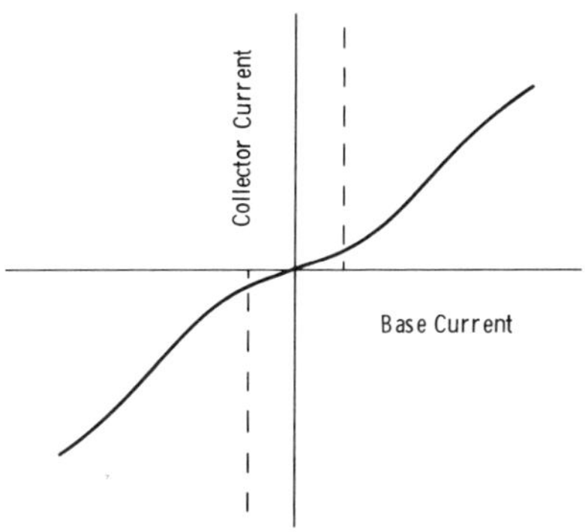

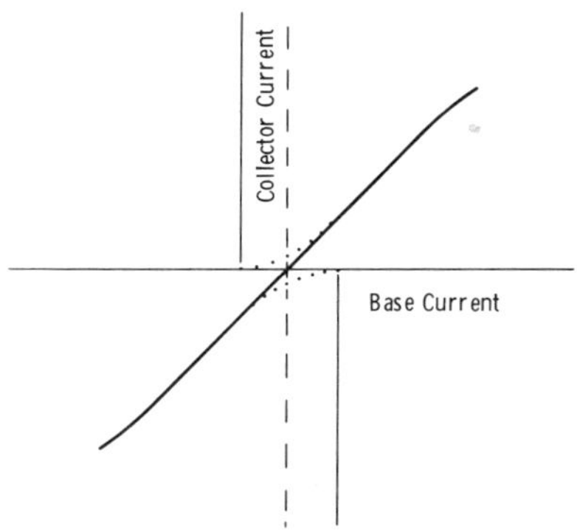

(*B*) *Uncombined transfer curves.* (*C*) *Combined transfer curve.*

Fig. 3-18. Use of forward bias in class-B amplifier.

amplifier.) Capacitor C1 charges rapidly through this low-resistance path. However, C1 cannot *discharge* through the emitter-base junction of Q1; for practical purposes, the junction represents an open circuit to electron flow *from emitter to base.* Electrons on the left-hand plate of capacitor C1 must flow through resistor R1, through ground to the emitter connection of Q1, and through resistors R6 and R3 to the right-hand plate of C1. (The discharge path is shown by the dash-line arrows.) Capacitor C1 discharges slowly because it must discharge through R3; normally the resistance of R3 is made large to avoid shunting of signal current around the Q1 base-emitter junction. The discharge current through R3 develops a reverse-bias voltage with the polarity indicated. The reverse bias can cause *class-C* operation with resultant severe distortion of the signal. (The discharge current through resistor R6 does not result in a reverse bias because the battery current in the opposite direction through this resistor maintains a forward bias.)

The advantages of RC coupling (economy and better frequency response) can be retained by replacing resistor R3 with a diode, as shown in

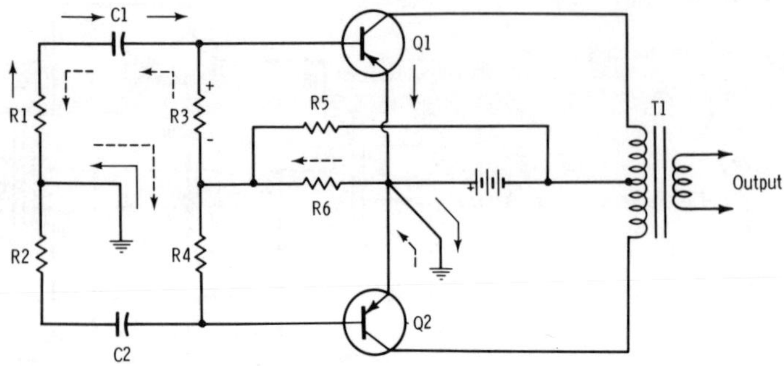

(*A*) *Input resistors.*

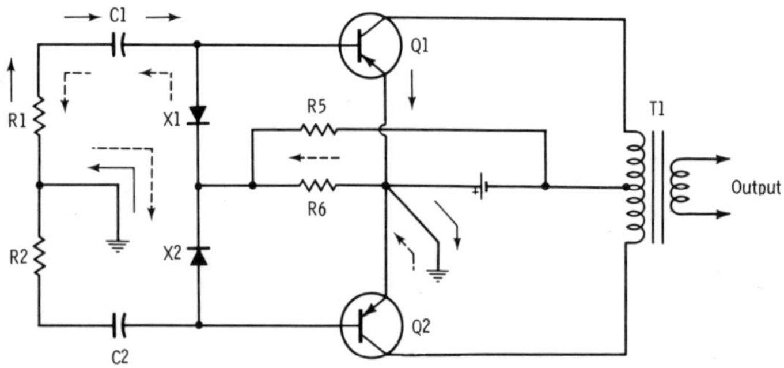

(*B*) *Input diodes.*

Fig. 3-19. Class-B amplifier with RC input coupling.

Fig. 3-19B. Diode X1 is reverse biased and acts as a high-value resistance when the input signal has the polarity that causes the emitter-base junction of Q1 to conduct. This high resistance prevents the shunting of signal current around the emitter-base junction. The charge path (solid-line arrows) for capacitor C1 is the same as that for the circuit shown in Fig. 3-19A. In the discharge path (dash arrows), resistor R3 is now replaced by diode X1, which is forward biased and has negligible resistance during the discharge period. Capacitor C1 discharges rapidly, therefore, and reverse bias of the emitter-base junction is avoided.

Note: The secondary windings of any class-B driver transformer (input to class-B stage) should be bifilar wound to obtain tight coupling and thereby minimize leakage inductance. Otherwise, "ringing" may occur in the crossover region as a result of the "kickback" energy stored in leakage inductance.

Now let us consider some power computations. By definition:

$$A_p = \frac{P_{out}}{P_{in}} = \frac{I_o^2 R_L}{I_{in}^2 R_{in}} \qquad \text{(Eq. 3-21)}$$

Since, for a small load resistance, the ratio I_o/I_{in} is the same as beta (common-emitter circuit), the power gain (which can be expressed in terms of the square of the current gain times the ratio of output to input resistance) can be expressed:

$$A_p = \beta^2 \frac{R_{c\text{-}c}}{R_{b\text{-}b}} \qquad \text{(Eq. 3-22)}$$

where,

$R_{c\text{-}c}$ is the collector-to-collector load resistance,
$R_{b\text{-}b}$ is the base-to-base input resistance.

With this information, observe again Fig. 3-18A. Assume that the power-supply voltage is −12 volts, the impedance of the output-transformer primary is 300 ohms, center tapped, and the impedance of the input-transformer secondary is 2000 ohms, center tapped. Assume a nominal beta of 50.

Then from Equation 3-22:

$$A_p = (50)^2 \frac{300}{2000} = 2500(0.15) = 375$$

What is the actual power output? The rule of thumb is:

$$P_o = \frac{2\,V_{CE}^2}{R_{c\text{-}c}} \qquad \text{(Eq. 3-23)}$$

In this equation, V_{CE} is the collector-to-emitter voltage at *no signal.* Remember that in a class-B stage, this voltage is very close to the power-supply potential (close to current cutoff); it may be taken as about 85 percent of V_{CC}. Thus, in this example, assume $V_{CE} = -10$ volts. Then:

$$P_o = \frac{2(-10)^2}{300} = \frac{200}{300} = 666 \text{ mW}$$

But remember that the power in the secondary depends on transformer efficiency, which normally may be assumed to be about 75 percent. So the actual power delivered by the secondary is (0.75) (666 mW) = 500 mW (approx).

Since the power gain is 375 and the power output (collector-to-collector) is 666 mW, then the required input power is:

$$P_{in} = \frac{P_{out}}{A_p} = \frac{666 \text{ mW}}{375} = 1.8 \text{ mW (approx) base-to-base}$$

An efficiency of 75 percent must also be assumed for the driver transformer, so the driver power must be:

$$P_{driver} = \frac{\text{P into input}}{0.75} = \frac{1.8 \text{ mW}}{0.75} = 2.4 \text{ mW}$$

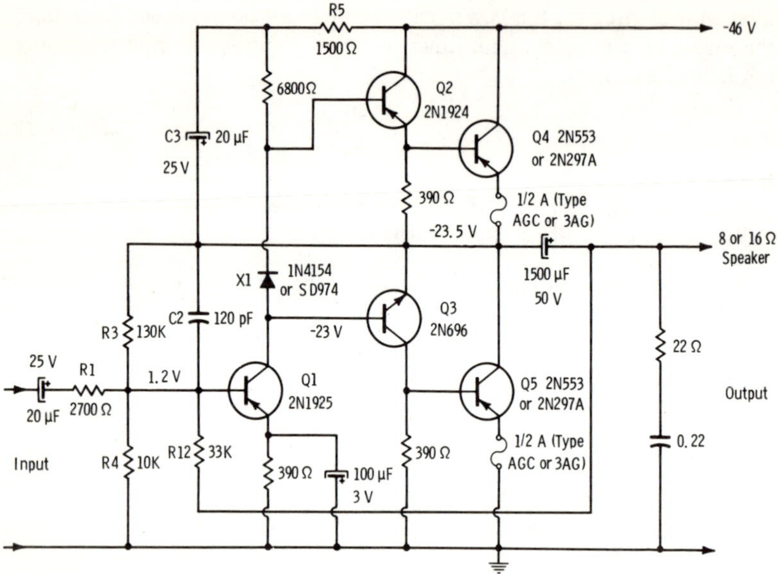

(*A*) *Amplifier schematic*

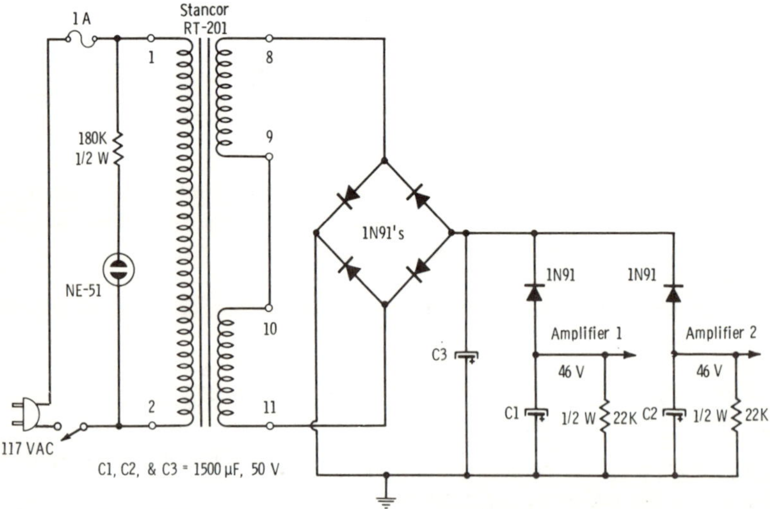

(*B*) *Power-supply schematic.*

Fig. 3-20. "Single-ended push-pull" amplifier.

A good performance check of any amplifier is to determine how faithfully it reproduces a square wave. When a transformer is used, it must be physically large for good low-frequency response, and it must be of high-quality design and construction throughout. Therefore, a great deal of effort has been expended in developing "transformerless" push-pull circuitry. In tube circuits, a number of tubes must be used in parallel to obtain the necessary current to develop power in low-impedance devices such as the voice coils of loudspeakers. Power transistors, however, lend themselves ideally to this application since they are inherently low-voltage, high-current devices. Fig. 3-20A shows an amplifier of the "single-ended push-pull" output class. (The information relative to Fig. 3-20 was furnished by the General Electric Company, Semiconductor Products Department, and is used with permission.)

Note that the bases of Q2 and Q3 are driven in phase. Transistor Q2 is used in an emitter-follower stage (no phase inversion), whereas Q3 is in a common-emitter stage (phase inversion). When a negative signal is applied to the base of Q4, this transistor draws current, which must pass through the speaker because the simultaneous positive signal at the base of Q5 holds Q5 near cutoff. When the signal polarity reverses, Q4 is cut off and Q5 conducts. Thus we have a "single-ended push-pull" output stage. As the diagram shows, its input is strictly direct-coupled, resulting in excellent low-frequency response.

The circuit also has the advantage of dc feedback for temperature stabilization of all stages. This feedback system stabilizes the voltage division across power-output transistors Q4 and Q5. Transistors Q2 and Q3 also operate class-B in the Darlington connection to increase the current gain. Using an npn transistor for Q3 gives the required phase inversion for driving Q5 and also makes possible the advantage of push-pull emitter-follower operation from the output of Q1 to the load. Emitter-follower operation makes possible lower inherent distortion and low output impedance.

Transistors Q4 and Q5 have a small forward bias of 10 to 20 mA to minimize crossover distortion and to operate the output transistors in a more favorable beta range. This bias is set by the voltage drop across the 390-ohm resistors that shunt the inputs to Q4 and Q5. Transistors Q2 and Q3 are biased at about 1 mA (to minimize crossover distortion) with the voltage drop across the silicon diode (X1). Junction diodes have a temperature characteristic similar to that of the emitter-base junction of a transistor. Therefore, this diode also gives compensation for temperature variation of the emitter-base resistance of Q2, Q3, and Q4. These resistances decrease with increasing temperature; thus the decrease in forward voltage drop of approximately 2 millivolts/degree centigrade of the diode provides some temperature compensation.

Transistor Q1 is a class-A driver with an emitter current of about 3 mA. Negative feedback to the base of Q1 lowers the input impedance of this stage. A source impedance higher than the input impedance is needed so that the feedback current will pass into the amplifier rather than into the source; resistor R1 limits the minimum value of source impedance. The value of R3 permits about one-half the supply voltage to appear across transistor Q5.

About 11 dB of positive feedback is applied across R5 by way of C3. This bootstrapping action helps to compensate for the unsymmetrical output circuit and permits the positive peak signal swing to approach the amplitude of the negative peak. This positive feedback is offset by about the same magnitude of negative feedback through R3 to the base of Q1. The net amount of negative feedback is approximately 14 dB resulting from the connection of R12 from the output to the input. In addition, there is the local feedback inherent in the emitter-follower stages. The value for feedback capacitor C2 was chosen for optimum square-wave response. (The value was selected to provide short rise time and minimum overshoot.)

A 1/2-ampere fuse is used in the emitter circuit of each output transistor for protection and to provide local feedback. (The type of fuse used has a dc resistance of about 1 ohm.) This local feedback increases the bias stability of the circuit and also improves the declining frequency response of Q4 and Q5 at the upper end of the audio spectrum. Because of reduced transistor efficiency above 10 kHz, care should be used when checking the amplifier for maximum continuous sine-wave output at these frequencies. If continuous power is applied for more than a short time, the resultant heating may raise the transistor current enough to blow the 1/2-ampere fuses. Actual performance of the amplifier does not suffer, since the power level in music and speech declines as the frequency increases beyond about 1 to 2 kHz.

The speaker system is shunted by a 22-ohm resistor in series with a 0.22-μF capacitor. This network is used to compensate for the continued rise of the amplifier load impedance and its accompanying phase shift beyond the audio spectrum.

The overall result of using direct coupling and ample degeneration is an amplifier with low distortion, good bandwidth, and output impedance of about 1 ohm for good speaker damping. The performance specifications for the amplifier of Fig. 3-20A include: power response at 1 watt, flat from 30 Hz to 15 kHz and down 3 dB at 50 kHz; total harmonic and IM distortion at 1 watt both less than 1 percent. At 7 watts the IM distortion is less than 2 1/2 percent, and the total harmonic distortion is less than 1 percent measured at 50 Hz, 1 kHz, and 10 kHz. Performance is about the same for both 8- and 16-ohm loads.

This amplifier is capable of about 8 watts of continuous output power with 1-volt-rms input, or 10 watts of music power into 8 or 16 ohms when used with the power supply of Fig. 3-20B. This power supply has diode decoupling, which provides excellent separation (80 dB) between the two stereo amplifier channels.

Power transistors Q4 and Q5 should each be mounted on an adequate heat radiator, such as is used for the output in an automobile radio, or on a 3″ × 3″ × 3/32″ aluminum plate that is insulated from the chassis.

> NOTE: Further practical applications of solid-state devices in microphone preamplifiers, pickup preamplifiers, light-controlled attenuators, attenuator coupling, and audio agc amplifiers are covered in Chapter 6 of this text. Radio-frequency applications are covered in Chapters 8, 9, and 10.

3-2. THE ANALOG-DIGITAL RELATIONSHIP

Anyone now involved in broadcast engineering must immediately become familiar with analog and digital circuitry. The reader totally unfamiliar with the fundamental "language" of modern systems may be pleasantly surprised at how easy the subject matter can become.

You may not be familiar with an *analog computer,* but you are already familiar with *analog display* devices. See Fig. 3-21. Current through the meter coil deflects the needle across a continuous scale to a point which is proportional to the voltage being measured by the voltmeter. When you look at the meter scale, your brain converts the reading to a series of digits as closely as you can interpret the pointer position, in this example +5.2 volts. The accuracy of this "digital readout" is influenced by the basic meter accuracy, and your interpretation is influenced by parallax and mental factors.

One method of obtaining a direct digital readout (digital voltmeter) is illustrated in Fig. 3-22. At the start of the measurement cycle, a ramp-voltage generator is activated. The ramp voltage and the voltage being measured are applied to an input comparator; at the instant these voltages become equal, a coincidence circuit generates a start pulse which opens a gate. This allows clock pulses to enter a counter. A second comparator circuit senses when the ramp voltage has reached zero. The output of this zero (or ground) comparator generates a stop pulse which closes the gate. The counter stops, and the resultant count of the clock pulses is displayed on in-line indicating tubes. The accuracy of such time-interval counters can be made extremely good compared to the process of Fig. 3-21. The

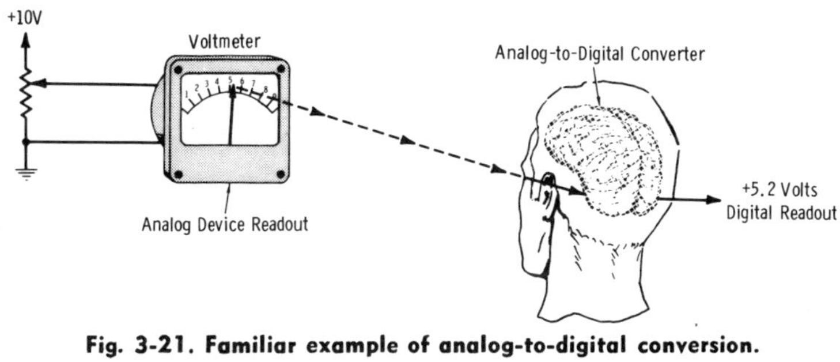

Fig. 3-21. Familiar example of analog-to-digital conversion.

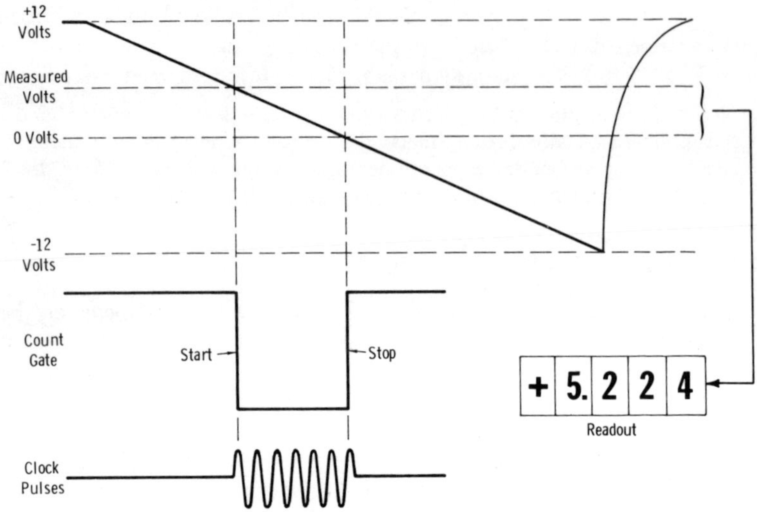

Fig. 3-22. Time-to-voltage conversion for digital voltmeter readout.

+5.224-volt indication of the digital voltmeter would be only a guess in the process of Fig. 3-21.

The applications of digital circuitry in broadcasting cover much more than increased accuracy of voltage readings, which is not all that important. Test generators, audio and rf switching systems, frequency and modulation monitors, and audio tape systems have rapidly become "digitalized." But to keep the subject in focus, we must retain the analog-to-digital and the digital-to-analog relationship, which will be stressed in this chapter.

The analog computer is so termed because it can draw an analogy to things in nature. The simple circuit of the voltmeter in Fig. 3-21 can say that the voltage is the force on a body, the resistance is the value of acceleration, and the current is the mass of the body. Since voltage is the product of current and resistance (Ohm's Law), and force is the product of acceleration and mass (Newton's law), the two problems are analogous and can be solved by the same circuit.

The analog computer is a continuous-scale device, able to draw a smooth curve. The digital computer can solve the same problems, but in discrete steps of integers or fractions. The analog computer works with continuous values, whereas the digital computer works with discrete, individual digits. In fact, the digital device recognizes only two digits, zero and one. From this simple off-on relationship, any numerical value can be "read out."

3-3. THE CONVENTIONAL DECIMAL NUMBER

The conventional decimal number system is represented by the diagram in Fig. 3-23. The *radix* is 10. This simply means that digits to the left of the decimal point are integers in powers of ten, and those to the right of the decimal point are fractions in powers of ten.

For example, 3486.412 is read as three thousand four hundred eighty six and four hundred twelve thousandths. It is also 3000+400+80+6+0.4+ 0.01+0.002. Or it can be $3(10^3)+4(10^2)+8(10^1)+6(10^0)+4(10^{-1})+1(10^{-2})+2(10^{-3})$. In working with the conventional decimal numbering system, we do not normally convert the numbers in this

	Thousands	Hundreds	Tens	Ones (Units)	Point	Tenths	Hundredths	Thousandths
	+	+	+	+	·	+	+	+
Power of 10	10^3	10^2	10	10^0		10^{-1}	10^{-2}	10^{-3}
Decimal Equivalent	1000	100	10	1		0.1	0.01	0.001
	Integral Part					Fractional Part		

Fig. 3-23. Decimal number system.

fashion since it merely complicates an otherwise fast reading. But now we are going to use a numbering system which has only two digits, 0 and 1, which can be used in combinations to represent any number. This is termed the *binary* system.

3-4. THE BINARY SYSTEM

In the binary system, the radix is 2. All numbers are constructed on the basis of powers of 2. For example, 1001 is $1(2^3)+0(2^2)+0(2^1)+1(2^0)$. (NOTE: $2^0=1$.) The first term $[1(2^3)]$ is simply the 1 of the first digit of the number (1001) times 2 raised to the power which represents the number of digits remaining in the original number (3). To find the decimal equivalent:

$$1(2^3)=8$$
$$0(2^2)=0$$
$$0(2^1)=0$$
$$1(2^0)=1$$

Adding: $8+0+0+1=9$, the decimal equivalent of binary 1001.

Binary Addition

From the above, construct the binary table as follows. Start by writing down four zeros:

0000 (Make this equal to zero.)
0001 (Add 1.)

0001 (This equals 1 in the decimal system.)

Now add another 1 to get a decimal value of 2.

0001
0001

When 1 is directly under 1 in the binary system, the sum is zero, and 1 is carried to the next column to the left. If there is already a 1 in that column, the sum is again zero and the 1 carries to the left again, etc. So:

0001
0001

0010 (This equals 2 in the decimal system.)

Now add another 1:

0010
0001

0011 (This is 3 in the decimal system.)

Add another 1:

0011
0001

0100 (This is 4 in the decimal system.)

Table 3-2. Binary-Decimal Equivalents (Whole Numbers)

Binary	Decimal
0000	0
0001	1
0010	2
0011	3
0100	4
0101	5
0110	6
0111	7
1000	8
1001	9
1010	10

The binary equivalents of the decimal numbers 0 to 10 are given in Table 3-2.

The radix in the binary system is 2 for both whole numbers and fractions. For example, binary 0.010 is $0(2^{-1}) + 1(2^{-2}) + 0(2^{-3}) = 0 + (1/2)^2 + 0 = 0 + 1/4 + 0 = 0.250$. Some binary fractions and their decimal equivalents are given in Table 3-3.

Table 3-3. Binary-Decimal Equivalents (Fractions)

Binary	Decimal
0.000	0
0.001	0.125
0.010	0.250
0.011	0.375
0.100	0.500
0.101	0.625
0.110	0.750
0.111	0.875

It has already been noted that the digits on the left of the binary point are coefficients of increasing *positive* powers of 2, with 2^0 adjacent to the binary point. The digits on the right of the point are coefficients of increasing *negative* powers of 2, with 2^{-1} adjacent to the point.

For another example, the decimal equivalent of binary 100101.01 is found as follows:

$$
\begin{array}{rcr}
\textit{Binary} & & \textit{Decimal} \\
1(2^5) & = & 32 \\
+0(2^4) & = & 0 \\
+0(2^3) & = & 0 \\
+1(2^2) & = & 4 \\
+0(2^1) & = & 0 \\
+1(2^0) & = & 1 \\
+0(2^{-1}) & = & 0 \\
+1(2^{-2}) & = & 0.25 \\
\hline
100101.01 & = & 37.25
\end{array}
$$

The binary method requires only two digits, 0 and 1. All numbers can be expressed by combinations of these digits. Since it is necessary to work with "powers of two," Table 3-4 is very convenient.

With a number system employing only 0 and 1, an open circuit can signify 0, and a closed circuit can signify 1. So an open switch or relay, a blocked tube or transistor, an extinguished lamp, etc., may represent 0. If the operation is reversed, the device says 1.

In Table 3-2, binary numbers are indicated to four places. Let us see how far we can go with four-place binary numbers in the decimal system. Starting with 1010 (binary 10), successively add binary 1 (0001) and obtain 1011 (or 11), 1100 (12), 1101 (13), 1110 (14) and 1111 (15). Adding 0001 again gives 10000 (binary 16), which has five places. This tells us that four "gates" (on-off devices) will handle up to the equivalent of 15 in the decimal system.

It is quite easy to convert any binary number to its decimal equivalent if Fig. 3-24 is mentally visualized. Take as an example binary 101100.101, and correlate Fig. 3-25 with Fig. 3-24. In Fig. 3-25A, the positional

Table 3-4. Positive Powers of Two

Power of 2	Decimal Equivalent
2^0	1
2^1	2
2^2	4
2^3	8
2^4	16
2^5	32
2^6	64
2^7	128
2^8	256
2^9	512
2^{10}	1024
2^{11}	2048
2^{12}	4096

Weight of Each Digit	Thirty-Twos	Sixteens	Eights	Fours	Twos	Ones (Units)	Point	Halves	Quarters	Eighths	Sixteenths	Thirty-Seconds	Sixty-Fourths
	+	+	+	+	+	+	.	+	+	+	+	+	+
Power of 2	2^5	2^4	2^3	2^2	2^1	2^0		2^{-1}	2^{-2}	2^{-3}	2^{-4}	2^{-5}	2^{-6}
Decimal Equivalent	32	16	8	4	2	1		1/2 0.5	1/4 0.25	1/8 0.125	1/16 0.0625	1/32 0.03125	1/64 0.015625

Integral Part — Fractional Part

Fig. 3-24. Binary numbers in powers of 2 and decimal equivalents.

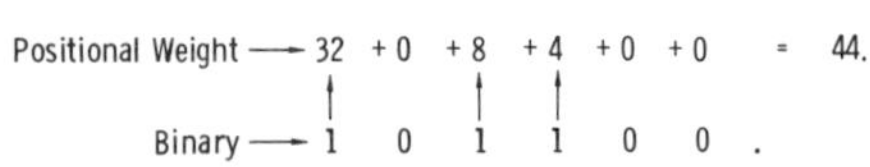

(A) Integral part.

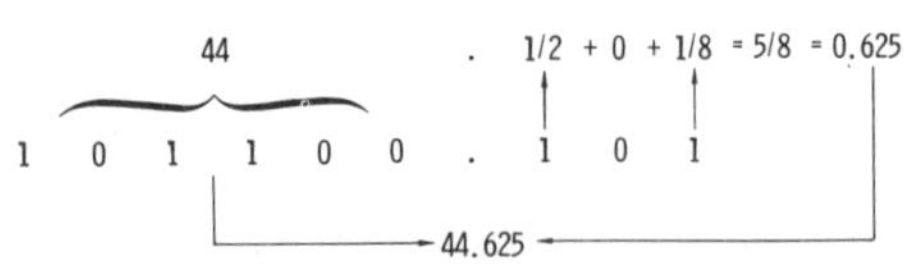

(B) Fractional part.

Fig. 3-25. Conversion of binary number to decimal form.

weights of the digits (top of Fig. 3-24) are added to obtain 44 for the integral part of the decimal number. In Fig. 3-25B, the fractional part is converted to obtain the total decimal number 44.625 for the 101100.101 binary number.

Complements in Binaries

We are now in a position to study the addition of positive and negative numbers in binary form. For example, consider the addition of 0101 and −0010. Note that in decimal form this is 5 − 2. The binary technique is: change the sign of the negative number, *complement* this number, and add the result to the positive number.

To complement a binary number, simply change all zeros to ones, change all ones to zeros, and *then add one*. Thus, to complement 0010:

$$\begin{array}{r} 1101 \\ +1 \\ \hline 1110 \end{array}$$

Then add the 1110 to 0101:

$$\begin{array}{r} 0101 \\ +1110 \\ \hline 10011 \end{array}$$

Now *discard* the overflow digit in the answer and *replace it with either a plus or minus sign.* (NOTE: The overflow digit is the digit in the answer one place to the left of the leftmost digit in the two numbers added. In this example, the leftmost 1 in 10011 is the overflow digit.) *If the digit is a 1,* the sign is plus. If the overflow digit is a *zero,* the sign is *minus.* Thus the answer in the example is +0011. This is the binary notation for decimal 3, which satisfies the condition 5 − 2 = 3.

When a negative number is added to a smaller positive number, the overflow digit is always zero, and the answer is always negative. When the result is negative, the number must be *recomplemented* to obtain the correct solution. For example: Add −1000 (decimal −8) to 0011 (decimal 3). First find the complement of 1000:

$$\begin{array}{r} 0111 \\ +1 \\ \hline 1000 \end{array}$$

Now add this to 0011:

$$\begin{array}{r} 1000 \\ +0011 \\ \hline 01011 \end{array}$$

Since the overflow digit is 0, discard it and substitute a negative sign, giving −1011. Now recomplement:

$$\begin{array}{r} 0100 \\ +1 \\ \hline 0101 \end{array}$$

So the answer is -0101. This is decimal number -5, which results from adding -8 to $+3$ in the decimal system.

Binary Subtraction

Binary subtraction is exactly the same as addition of positive and negative binary numbers. For example, subtract 0100 (decimal 4) from 1010 (decimal 10). First find the complement of 0100:

$$\begin{array}{r} 1011 \\ +1 \\ \hline 1100 \end{array}$$

Then add:

$$\begin{array}{r} 1010 \\ +1100 \\ \hline 10110 \end{array} = +0110$$

Note that 0110 is decimal 6, which is the result of subtracting decimal 4 from decimal 10.

Binary Multiplication, Digital Methods

Digital computers multiply by over-and-over addition and divide by over-and-over subtraction. A 5-MHz flip-flop can perform five million such operations per second.

Over-and-over addition involves the use of two registers, the *accumulator* and the *multiplier*. The accumulator starts with all zeros, and the multiplier keeps account of the number of additions performed. To illustrate, let us start with conventional numbers and multiply 462×232:

	Accumulator	Multiplier
1.	000000	232
2.	000462	
3.	000462	231
4.	000462	
5.	000924	230
6.	004620	
7.	005544	220
8.	004620	
9.	010164	210
10.	004620	
11.	014784	200
12.	046200	
13.	060984	100
14.	046200	
15.	107184 (Answer)	000

The above table shows the results of the following operations: In Step 1, zeros are recorded in the accumulator column, and the multiplier (232) is recorded in the right column. In Step 2, the multiplicand is recorded. Step 3 is the first addition, $000 + 462 = 462$. Since one addition has now been made, the multiplier is reduced by one, to give 231. In Step 4, the multiplicand is recorded again. In Step 5, $462 + 462 = 924$. The multiplier register is again reduced by one to give 230. The first-order digit of the multiplier is now zero, so in the next step, move the multiplicand one place to the left. (Now the multiplicand will be added *ten* times in each step.) In Step 7, add Steps 5 and 6 to get 5544, and reduce the multiplier by ten, to give 220. In Step 8, the multiplicand of Step 6 is repeated. In Step 9, the sum of 5544 and 4620 is 10164. Put it down and again reduce the multiplier by ten to get 210. In Step 10, repeat the multiplicand as in Step 8. In Step 11, the total of Steps 9 and 10 is 14784. The multiplier is now 200. Now the second-order digit is zero, so in the next step again move the multiplicand one place to the left. (Now the multiplicand will be added 100 times in each step.) In Step 13, record the new sum in the accumulator column, and reduce the multiplier by 100 to get 100. In Step 14, repeat the same multiplicand. In Step 15, the total is 107184; this is the answer, since the multiplier has been reduced to zero.

Now multiply binary 1011 times binary 11010:

	Accumulator	Multiplier	Operations
1.	000000000	11010	Move multiplicand one place left.
2.	000010110		Add.
3.	000010110	11000	Move multiplicand two places left.
4.	001011000		Add.
5.	001101110	10000	Move multiplicand one place left.
6.	010110000		Add.
7.	100011110	00000	This is the answer.

Converting Decimal Numbers to Binary Numbers

Figs. 3-24 and 3-25 show how easy it is to convert binary numbers to decimal numbers. It is also necessary to be able to convert decimal numbers to binary numbers. In Fig. 3-26A the table of Fig. 3-24 has been extended to a higher decimal value (power of 2). In the example of Fig. 3-26, decimal 745.21 is converted to its binary equivalent out to four places to the right of the binary point.

Step 1 (Fig. 3-26B) is to write down the decimal number. Step 2 is to write down the largest whole power of 2 (in decimal value) that can be subtracted from the number. This is 512, so a 1 is placed under the decimal value of 512 in the bottom row of Fig. 5-26A. Step 3 is the remainder, and Step 4 is to find the largest whole power of 2 that can be subtracted from 233.21. This is 128, so a 1 is placed under 128. The process is repeated

← Extend as Far as Needed →   Point

Powers of 2	2^{10}	2^9	2^8	2^7	2^6	2^5	2^4	2^3	2^2	2^1	2^0	•	2^{-1}	2^{-2}	2^{-3}	2^{-4}
Decimal Value	1024	512	256	128	64	32	16	8	4	2	1	•	0.5	0.25	0.125	0.0625
Binary for 745.21	0	1	0	1	1	1	0	1	0	0	1	•	0	0	1	1

(*A*) *Tabulation of conversion of 745.21 to binary form.*

Step		Notes	Operation in Table Above
1.	745.21	Number To Be Converted	
2.	—512.00	Largest Whole Power of 2 That Can Be Subtracted	Place a 1 under 512.
3.	233.21	Remainder	
4.	—128.00	Largest Whole Power of 2	Place a 1 under 128.
5.	105.21	Remainder	
6.	—64.00	Largest Whole Power of 2	Place a 1 under 64.
7.	41.21	Remainder	
8.	—32.00	Largest Whole Power of 2	Place a 1 under 32.
9.	9.21	Remainder	
10.	—8.00	Largest Whole Power of 2	Place a 1 under 8.
11.	1.21	Remainder	
12.	—1.00	Largest Whole Power of 2	Place a 1 under 1.
13.	0.210	Remainder	
14.	—0.125	Largest Whole Power of 2	Place a 1 under 0.125.
15.	0.0850	Remainder	
16.	—0.0625	Largest Whole Power of 2	Place a 1 under 0.0625.
17.	Stop for this example, and place a 0 under all vacant numbers in bottom row above.		

(*B*) *Steps in converting decimal 745.21 to binary equivalent.*

Fig. 3-26. Conversion of decimal number to binary form.

through all remaining steps. Thus decimal 745.21 = binary 1011101001.-0011.

3-5. BINARY CODES

Binary numbers can become very long in representing decimal numbers. In practice, not only numbers but also letters and various symbols are represented. For convenience, certain terms are used to identify parts of these binary representations. The term *bit* means a binary digit (*BI*nary digi*T*). The term *character* means a group of bits. The term *word* refers to the total number of bits required for a particular system. A word may be defined either by the total number of bits or the total number of characters. For example, a certain system may use a 192-bit word. A character may be divided into (for example) groups of 4 bits. Thus this system may also be defined as using a 48-character word.

Digital systems primarily use instructions in binary form. However, humans use decimal numbers and alphabetic letters. Various codes have been devised to facilitate communication between human operators and digital units.

Pure Binary and BCD Codes

Observe the table of Fig. 3-26A. This shows the pure binary code, which uses the exact positional weight (defined in Fig. 3-24) of each binary digit as the weight value.

See Fig. 3-27. The binary coded decimal (bcd) code employs four binary *bits* per *character* and the weight scheme of 8, 4, 2, 1. Each character then has the decimal value that the four bits represent. Note from Fig. 3-27 that the decimal equivalent is simply the binary-number sum expressed in decimal form. Thus 1001 = 9, which is 8 + 1.

A 4-bit number can have values from zero to fifteen (0000 to 1111). Normally, however, only sufficient combinations are used in the bcd code to express all 10 decimal symbols (0 to 9). To express decimal numbers greater than 9, a separate 4-bit group is used for each digit. For example:

92 = 1001 0010
920 = 1001 0010 0000
591 = 0101 1001 0001

NOTE: A digital system addressed with the 4-bit number using values from 0 to 15 is termed pure binary. When the values used are 0 to 9, it is termed bcd.

The USASCII Code

The digital system must be able to work with alphabetic characters as well as numeric characters; this is termed an *alphanumeric* code. Attempts have been made at standardizing on a universal code. The USASCII (USA

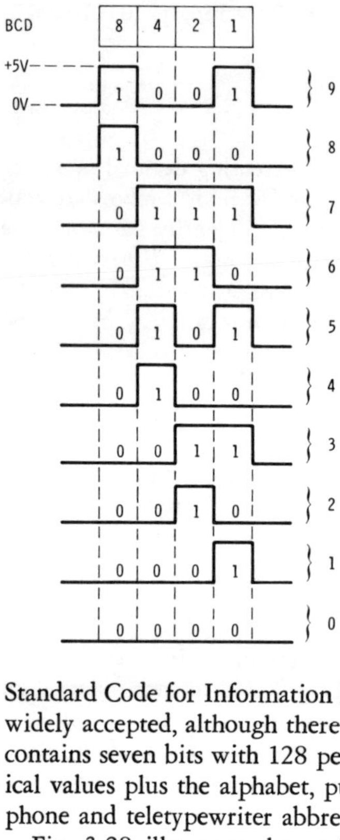

Fig. 3-27. Binary-coded decimal characters.

Standard Code for Information Interchange) has been found to be the most widely accepted, although there are many others in everyday use. This code contains seven bits with 128 permutations. The seven bits allow all numerical values plus the alphabet, punctuation marks, symbols and special telephone and teletypewriter abbreviations to be expressed.

Fig. 3-28 illustrates the entire code. Observe the column headed 100 (column 4). Capital letters of the alphabet from A to O are listed in order in this column. This information is decoded by using the four digits shown on the left and adding the three digits at the head of the column. For capital A this is:

$$A = 100\ 0001$$

Similarly, for the small letter "a" (column 6):

$$a = 110\ 0001$$

The last four binary digits express the decimal number in 8, 4, 2, 1 bcd form. Therefore, the USACII code is compatible with systems designed to use the 8,4,2,1 bcd code.

Octal Coded Binary

The octal system is a common numbering system within the digital area; using it makes it possible to avoid long strings of zeros and ones. The positional weighting is based on powers of eight. The octal system converts readily to binary form and vice versa because the radix of the octal system (8) is an integral power of 2 ($8 = 2^3$).

Take the binary word 101001.01, which contains 8 digits. Since the octal system has 2^3 as the primary base. the word can be arranged in groups of three bits each; this converts to the equivalent octal number. See Fig. 3-29.

BITS				b_7 b_6 b_5 →	0 0 0	0 0 1	0 1 0	0 1 1	1 0 0	1 0 1	1 1 0	1 1 1
b_4 ↓	b_3 ↓	b_2 ↓	b_1 ↓	ROW \ COL	0	1	2	3	4	5	6	7
0	0	0	0	0	NUL	DLE	SP	0	@	P	\	p
0	0	0	1	1	SOH	DC1	!	1	A	Q	a	q
0	0	1	0	2	STX	DC2	"	2	B	R	b	r
0	0	1	1	3	ETX	DC3	#	3	C	S	c	s
0	1	0	0	4	EOT	DC4	$	4	D	T	d	t
0	1	0	1	5	ENQ	NAK	%	5	E	U	e	u
0	1	1	0	6	ACK	SYN	&	6	F	V	f	v
0	1	1	1	7	BEL	ETB	'	7	G	W	g	w
1	0	0	0	8	BS	CAN	(	8	H	X	h	x
1	0	0	1	9	HT	EM	)	9	I	Y	i	y
1	0	1	0	10	LF	SUB	*	:	J	Z	j	z
1	0	1	1	11	VT	ESC	+	;	K	[	k	{
1	1	0	0	12	FF	FS	,	<	L	\	l	¦
1	1	0	1	13	CR	GS	-	=	M	]	m	}
1	1	1	0	14	SO	RS	.	>	N	^	n	~
1	1	1	1	15	SI	US	/	?	O	_	o	DEL

Fig. 3-28. USASCII encoding chart.

The word 101001.01 has been arranged in groups of three bits each. Note that when this is done, the corresponding octal digit for each group is determined by the 4, 2, 1 weight factor. Thus the octal number has only 3 digits compared to 8 or 9 binary digits. This is shown by rows A and B of Fig. 3-29.

When converting from octal to decimal form, it is most convenient to use the octal-to-binary conversion as an intermediate step. This is shown by rows B and C of Fig. 3-29. Then the binary number is converted to decimal form in the normal manner as shown by rows C and D.

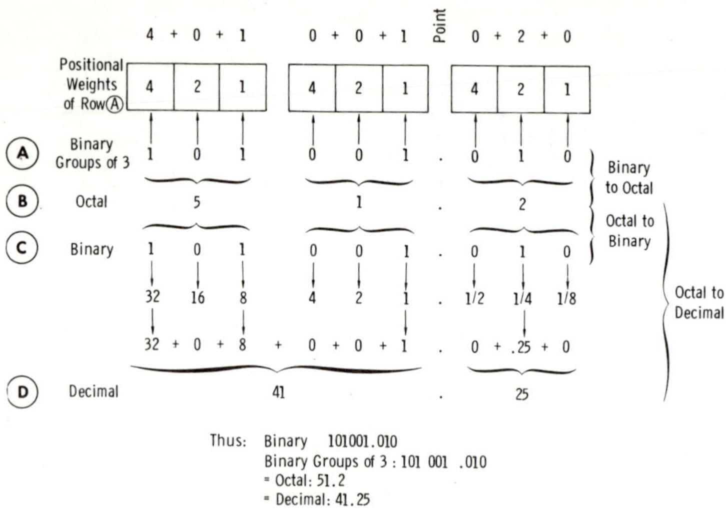

Fig. 3-29. Binary-to-octal-to-decimal conversion.

3-6. BASIC BOOLEAN ALGEBRA

Boolean algebra has become an important link in understanding electronic computer and logic systems. This algebra is concerned with elements that have only two possible stable states and no unstable states. It is useful in representing switching circuits, because a switch can be in only one of two possible stable states, "closed" or "open" (Fig. 3-30). The symbol 0 (zero) can represent an open switch, and 1 (one) can represent a closed switch. From the outset remember this: The 0 and 1 *do not represent numbers;* they are shorthand for representing the presence or absence of a conducting path.

0 = Open circuit, or nonconducting path

1 = Closed circuit, or conducting path

Fig. 3-31A shows a simple circuit with two switches in series. Obviously, both switches must be closed to complete the circuit. If either switch A or B is open, the state is 0. If both are closed, the state is 1. A *truth table* for this circuit as shown in Fig. 3-31B; this is a list of all the possible states of the switches (columns A and B) and the corresponding states of the

Fig. 3-30. Basic logic states.

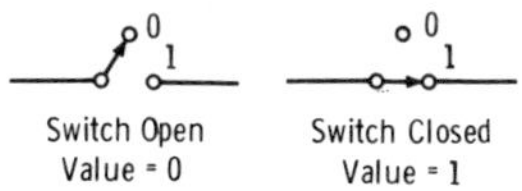

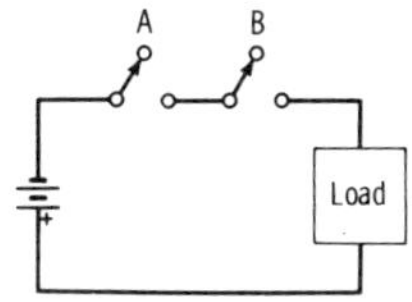

(*A*) *Basic function.*

Truth Table (AND)

Condition	A	B	f (A, B) = AB
1	0	0	0
2	1	0	0
3	0	1	0
4	1	1	1

(*B*) *Truth table.*

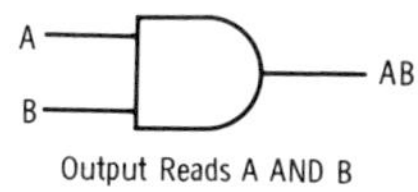

(*C*) *Diagram symbol.*

Fig. 3-31. Fundamentals of an AND circuit.

output function of A and B (in the last column). Condition 1 is with A and B open (0 state), so function f(A,B) is also 0. Conditions 2 and 3 have only one switch closed (state 1) at a time, so the function of A and B is still 0. Condition 4 is with both A and B in the closed (1) state, and therefore f(A,B) = 1. This is an AND (coincidence) circuit. Both A and B must be in the 1 state for a current to exist in the load. The symbol AB is read "A *and* B." It *does not* mean the product A × B as in conventional algebra.

Shown in Fig. 3-31C is the diagram symbol for the AND circuit. When this symbol is used without further identification, it should be recognized as an AND circuit. When a regular block is used, it should be labelled AND.

The significance of Fig. 3-32 should now be apparent, except for one thing. The expression A + B is read "A OR B," not "A plus B." This is an OR circuit. The truth table of Fig. 3-32B can be seen to be true if you think of the logic involved. The expression A + B means A OR B, which is function f(A,B). So if A or B takes the value of 1 (Switched closed), then f(A,B) must equal 1. Compare this to the logic of the AND circuit (Fig. 3-31).

If 0 is an open circuit and 1 is a closed circuit, what happens when a signal resulting from a given state goes through a phase reversal? See Fig. 3-33A, which shows an AND circuit feeding the base of a transistor. The AND-circuit output is A AND B, so if both inputs are 1's, the base is in a

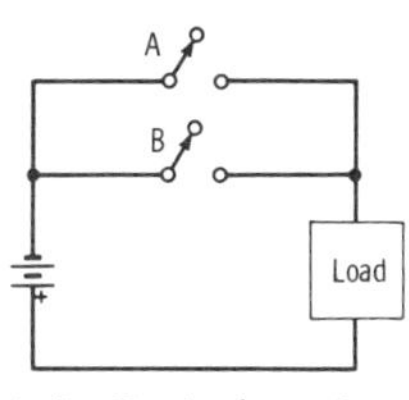

(*A*) *Basic function.*

Truth Table (OR)

Condition	A	B	f (A, B) = A+B
1	0	0	0
2	1	0	1
3	0	1	1
4	1	1	1

(*B*) *Truth table.*

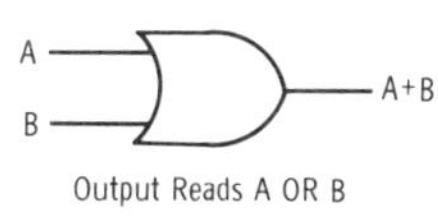

(*C*) *Diagram symbol.*

Fig. 3-32. Fundamentals of an OR circuit.

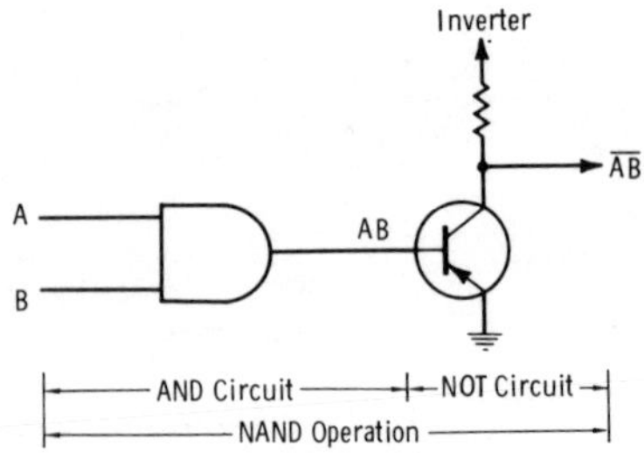

(*A*) *Diagram.*

Truth Table (NAND)

A	B	AB	f (A, B) = $\overline{AB}$
0	0	0	1
1	0	0	1
0	1	0	1
1	1	1	0

(*B*) *Truth table.*

Fig. 3-33. Fundamentals of a NAND circuit.

1 state (condition 4 of the truth table, Fig. 3-31B). This signal is inverted at the collector, so the output is now a 0 state, or NOT AB, (signified by a line over AB, or $\overline{AB}$). The entire circuit is now termed a NAND circuit. This simply means the circuit provides an inverted AND, or NOT AND, operation, which is to say an AND circuit with phase reversal.

Note that the first three columns of the NAND truth table (Fig. 3-33B) are a copy of the AND truth table in Fig. 3-31B. But the inverted output changes all 0's to 1's, and the 1 to a 0. Note that the whole quantity AB, not the separate variables of individual switch states, is complemented (reversed).

Fig. 3-34 shows the same kind of logic applied to an OR circuit. When the phase is reversed, the state becomes NOT A OR B (designated $\overline{A+B}$). Compare the resulting truth table (Fig. 3-34B) with the one in Fig. 3-33B.

For Figs. 3-35 through 3-40, a positive signal is defined as 1. Here is the way to look at these circuits:

Fig. 3-35 shows a *diode-logic* (DL) circuit. If a positive pulse is applied to either input A or B, A OR B will appear at the output.

Fig. 3-36 is also a diode logic circuit, but in this case the diodes are forward biased through their respective resistors. If a positive pulse is applied to only one of the inputs, the other diode is still conducting, providing a low-impedance bypass to any signal change. So the voltage at the output remains low. If positive pulses are applied to *both* inputs simultaneously, the diodes become open switches, and the output rises to the R_L supply voltage. This is an AND (coincidence) circuit.

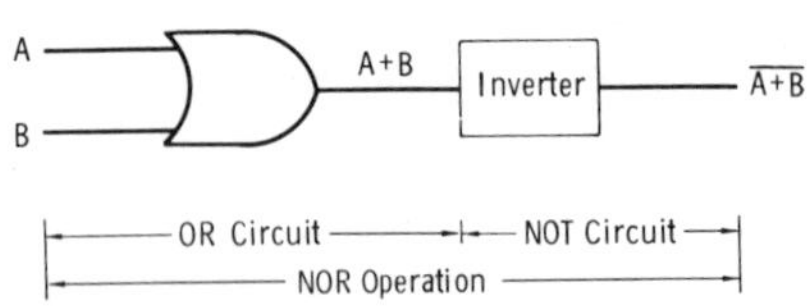

(*A*) *Diagram.*

Truth Table (NOR)

A	B	A+B	f(A, B) = $\overline{A+B}$
0	0	0	1
1	0	1	0
0	1	1	0
1	1	1	0

(*B*) *Truth table.*

Fig. 3-34. Fundamentals of a NOR circuit.

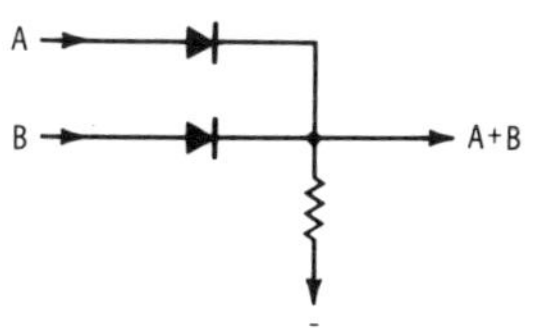

Fig. 3-35. Diode-logic OR circuit.

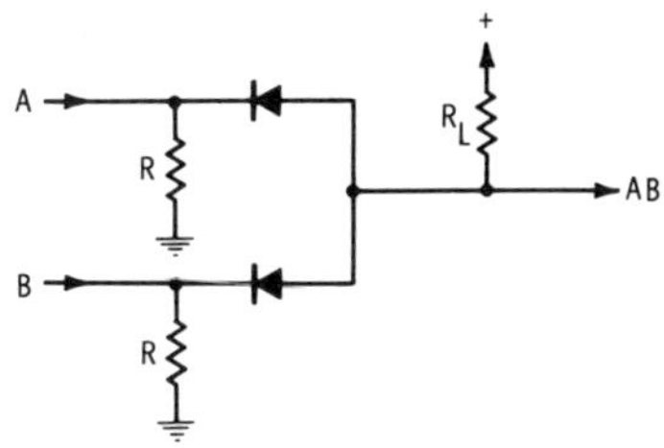

Fig. 3-36. Diode-logic AND circuit.

A *direct-coupled-transistor-logic* (DCTL) circuit is shown in Fig. 3-37. A positive pulse applied to only one of the inputs still leaves the other transistor without forward bias. Therefore, the switch is still open, and no change occurs at the output. A positive pulse must be applied to *both* inputs simultaneously to close the switch. Since phase inversion occurs, the circuit is a NOT AND, or NAND, circuit.

Another DCTL circuit is shown in Fig. 3-38, but in this case a positive pulse appearing at either input will appear inverted at the output. So this is a NOT OR (NOR) circuit. Circuits using only transistors are also termed TTL for *transistor-transistor logic.*

The diagram in Fig. 3-39 represents a *diode-transistor-logic* (DTL) circuit. In this case, an OR circuit is followed by a NOT circuit (NOR operation).

The *resistor-transistor-logic* (RTL) circuit in Fig. 3-40 is the same as the circuit of Fig. 3-39 except that resistors are used instead of diodes. Sometimes capacitors are used around the resistors (to enhance switching speed). When this is done, the circuit is termed *resistor-capacitor-transistor logic* (RCTL).

Instruction manuals for modern solid-state equipment are filled with the above terminology. Regardless of how complex the circuits may look at first glance, they work on these principles.

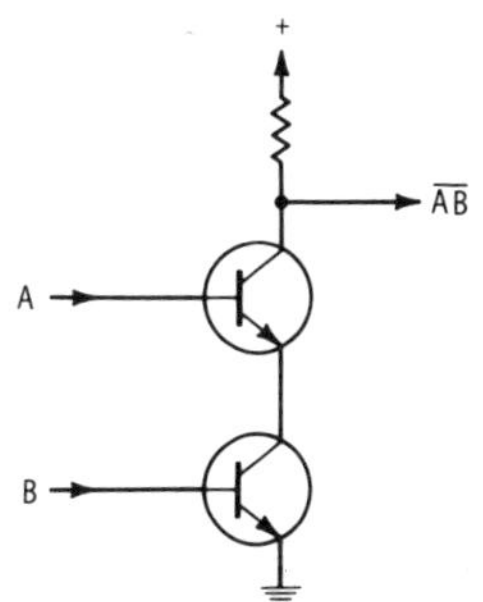

Fig. 3-37. Direct-coupled-transistor-logic NAND circuit.

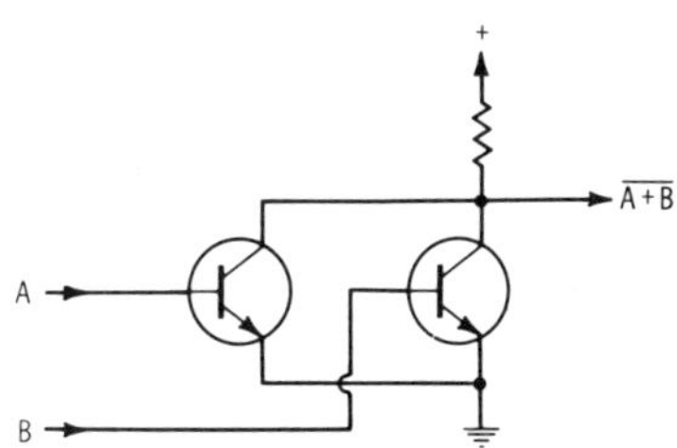

Fig. 3-38. Direct-coupled-transistor-logic NOR circuit.

3-7. LOGIC-SCHEMATIC NOTATION

You will need to become well acquainted with the term "logic," and know how to read "schematics" of logic systems. It makes no difference in what field you are engaged, electronic equipment is absorbing an ever-increasing amount of "logic."

With such gear, servicing as we have known it will not exist. Some present applications (and a great number of upcoming broadcast and commercial applications) use *chips* or *integrated circuits,* (a single chip can replace a fantastic number of transistors) on printed-circuit or etched-wiring boards. Does this mean transistor studies are all in vain, except as a transition from vacuum tubes through transistors to chips? Certainly not. Study and experience with transistors will make the new field of microminiaturization a "natural." And please rule out any thought that new sys-

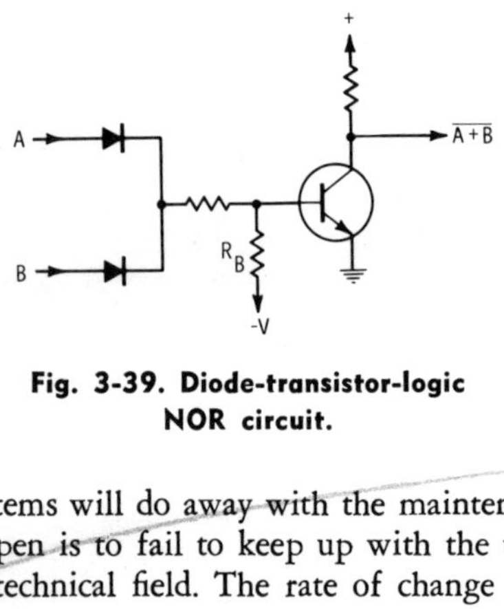

Fig. 3-39. Diode-transistor-logic NOR circuit.

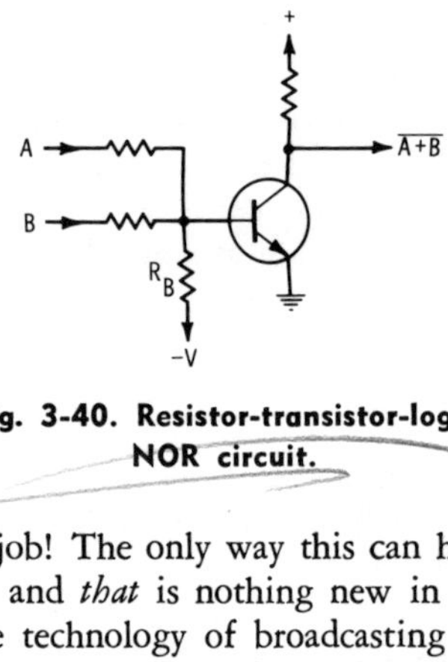

Fig. 3-40. Resistor-transistor-logic NOR circuit.

tems will do away with the maintenance job! The only way this can happen is to fail to keep up with the times; and *that* is nothing new in the technical field. The rate of change in the technology of broadcasting requires technical personnel to spend a greater proportion of time in acquiring new knowledge to solve problems. Continuing education, therefore, is no longer incidental to the job, but an essential part of it.

The digital concept recognizes only two numbers, which are "conditions": 1 and 0. *Positive logic* means that a 1 (high level) represents the true or more positive level and a 0 (low level) represents the false or less positive level. This is the logic most often used. *Negative logic* (sometimes used) means that the voltage level assigned to logical 1 is *negative* with respect to the voltage level assigned to logical 0. Unless stated otherwise, we will be concerned with positive logic in this text.

"Schematics" of logic systems are quite simply block or flow diagrams with certain symbology that tells what is expected of the circuit. It is not possible to service a single "stage" of a microminiature block; you must know how to check for the trouble and then replace a single plug-in "block" on a board, or an entire card of blocks in a multiple-card module.

Fig. 3-41A shows the familiar series of pulses making up a square wave. Now simply add an L (for low, or negative-going, pulses) or an H (for high, or positive-going, pulses) to each portion of the waveform. Fig. 3-41B lists the nomenclature applied in logic systems. Column 1 indicates a "high-level" or a "low-level" point. Column 2 repeats this in numerical form. Column 3 says the "high" level can be a plus voltage and the "low" level can be ground, or zero volts. Column 4 shows that the high level can be zero volts (common or ground) and the low level can be a minus voltage. In either case, the high level is positive relative to the low level. Column 5 says that "1" can be "true" and "0" can be "false." This can be reversed, depending upon application. Further columns would be limited only by the imagination; they could be "John" or "Mary," "go" or "no-go," or some other combination.

Fig. 3-42A shows the symbol for an AND (coincidence) gate. (Remember that any number of inputs could exist, not just two.) The input waveforms and corresponding output waveform are shown in Fig. 3-42B. Note

(*A*) *Pulses.*

Column	1	2	3	4	5	
	Hi	1	+	0	True	etc.
	Lo	0	0	-	False	

(*B*) *Nomenclature.*

Fig. 3-41. Pulse nomenclature for logic systems.

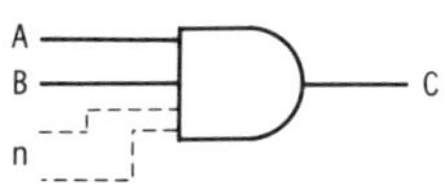

(*A*) *Symbol.*

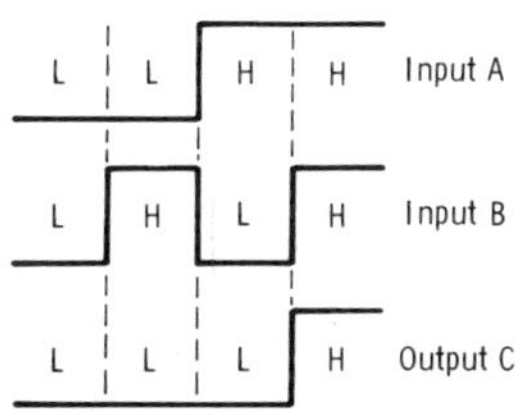

(*B*) *Waveforms.*

Input		Output
A	B	C
L	L	L
L	H	L
H	L	L
H	H	H

(*C*) *Truth table.*

Fig. 3-42. Coincidence (AND) gate.

a significant fact here: When both the A and B inputs are low, the output at C is also low, or zero. But the *same* output condition exists when inputs A and B have pulses of *opposite polarity* applied. The significance of this fact is that the coincidence of two *positive* pulses at the inputs is the "reference" operation; it is the *only* operation which results in a change of output waveform.

Fig. 3-42C is the corresponding truth table. This table is much more convenient than drawing out waveforms for A, B, and C. Again, it shows that the reference operation (the only condition causing a change in output level) is for two coincident positive pulses (or high, or 1, etc.) to be present at the inputs.

Fig. 3-43A shows the symbol for a NAND gate (NOT AND, or AND gate with inversion). Note the small circle at the output end of the symbol; this circle indicates signal inversion. From the waveforms of Fig. 3-43B, observe that coincident high-level pulses must occur at the input to change the output to the low level. Note that column C of the truth table of Fig. 3-43C is just opposite to that of Fig. 3-42C.

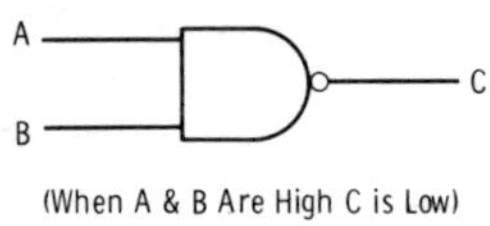

(*A*) *Symbol.*

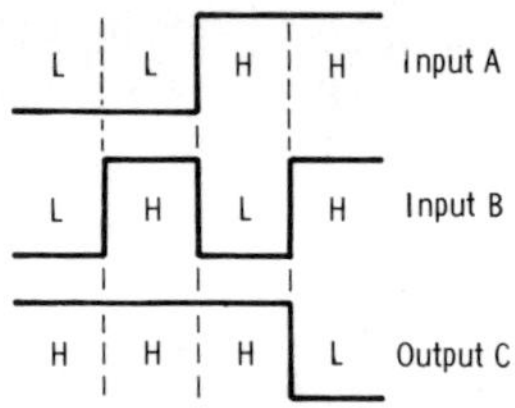

(*B*) *Waveforms.*

Input		Output
A	B	C
L	L	H
L	H	H
H	L	H
H	H	L

(*C*) *Truth table.*

Fig. 3-43. NAND gate.

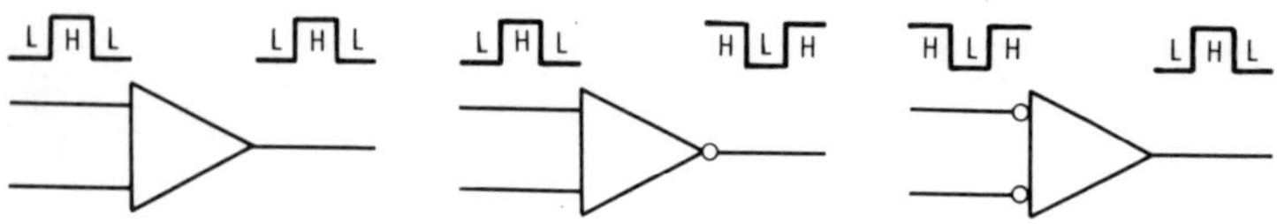

(*A*) *Noninverting stage.* (*B*) *High-level activated.* (*C*) *Low-level activated.*

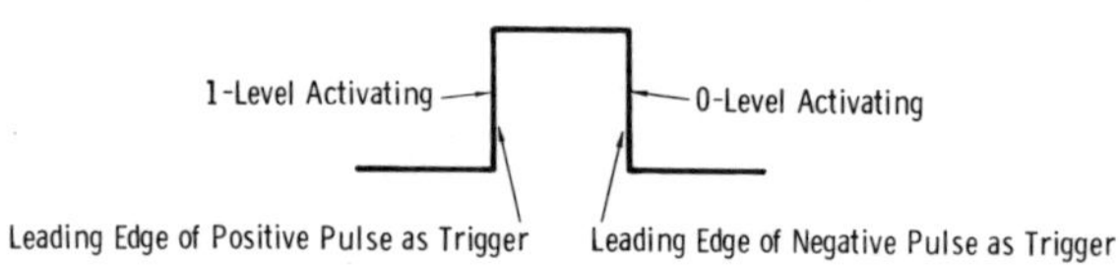

(*D*) *Triggering waveform.*

Fig. 3-44. Inversion in logic stages.

Fig. 3-44A shows the "triangle" representation of any stage that is noninverting. (NOTE: In logic circuitry, this symbol is normally replaced by the block symbol of an AND or OR circuit.) In Fig. 3-44B, the circle at the output means that the low-level output is the reference. This symbol represents a high-level-activated device, because the high level is required to change the output to a low level. In Fig. 3-44C, the circles are at the inputs. This is a low-level-activated device; low-level inputs are required for a high-level output.

Fig. 3-44D illustrates that a 1 (high-level) activation is the leading edge of a positive pulse. This is the same as the trailing edge of a negative pulse. Also, a zero (low-level) activation is the leading edge of a negative pulse.

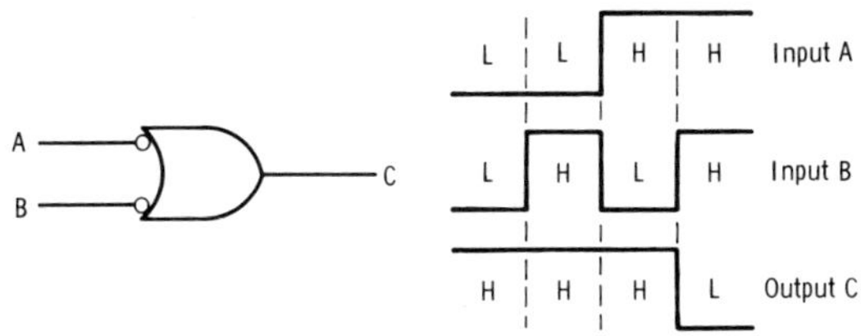

(*A*) *Symbol.* (*B*) *Waveforms.*

Input		Output
A	B	C
L	L	H
L	H	H
H	L	H
H	H	L

(*C*) *Truth table.*

Fig. 3-45. OR gate with phase inversion.

This is the same as the trailing edge of a positive pulse. These relationships should help to keep the comprehension of "levels" in proper perspective.

Fig. 3-45A shows the most common symbol for the OR gate with phase inversion added; therefore a NOR gate is indicated. The circles at the inputs designate a low-level-activated device; if either input A or B is *low,* the output will be *high.* Fig. 3-45B shows the waveforms, and Fig. 3-45C is the more convenient truth table for this circuit function.

Fig. 3-46A shows a block with the labels "SS," "30 μs," "0," and "1." The SS means single-shot (monostable) multivibrator, the 30 μs indicates the unstable (timing) period, and the two outputs (one from each side of the multivibrator) are designated as the 0 output and the 1 output. Some schematics say "OS" (for one-shot), which is the same as "SS" (single-shot). In addition, note that the input line is terminated in a circle. This circle indicates that the input is at low level for the reference polarity of output signals.

Now study the waveforms of Fig. 3-46B. Prior to time t_1, the input is high, the 0 output is high, and the 1 output is low. At time t_1, the input swings low. This action triggers the single-shot multivibrator into its unstable state (which in this example has a duration of 30 μs); thus the 0 side transfers from high to low, and the 1 side transfers from low to high. This condition prevails for the built-in timing period of 30 μs, and then at t_2 the outputs reverse polarity again. Now remember that the single-shot

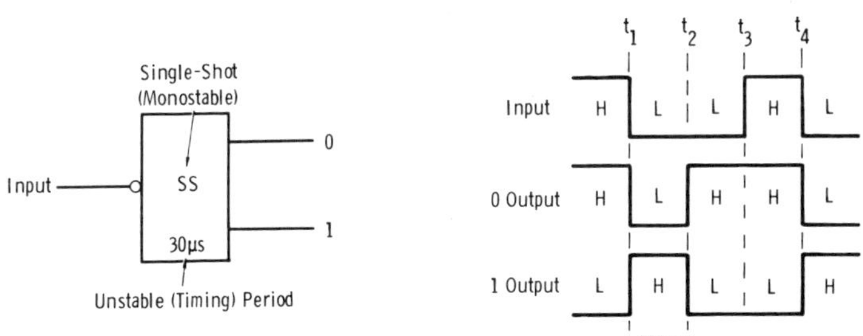

(*A*) *Symbol.* (*B*) *Waveforms.*

Fig. 3-46. Monostable multivibrator.

must wait for another "low" (negative-going) trigger before it can operate again. Thus from t_2 to t_4 the outputs remain unchanged. At t_4, the timing period starts again (negative, or low, input pulse).

Note carefully that only during the timing period, which was initiated by the negative-going transistion, is the 1 output really 1 and the 0 output really zero (or low). This is the reference pulse mentioned before.

Another basic block of digital circuitry is the *flip-flop.* This is a bistable multivibrator which remains in its last state until an input causes it to change states. Usually, the input trigger pulse is differentiated and then applied to a diode to polarize the pulse so that only the positive-going or negative-going edge causes the bistable circuit to respond. Sometimes a dc shift rather than a pulse is used.

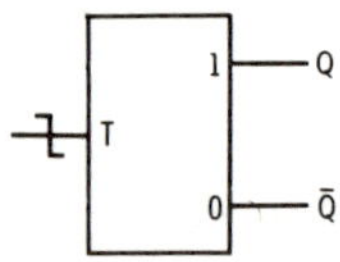

⎿ = Input responds to negative-going signal only.

(*A*) *Symbol.*

Given Present State		State After Trigger Pulse	
Q	Q̄	Q	Q̄
0	1	1	0
1	0	0	1

(*B*) *Truth table.*

Fig. 3-47. Fundamentals of toggle flip-flop.

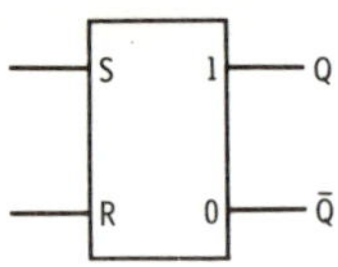

Note: S = Set Pulse
R = Reset (or Clear) Pulse

(*A*) *Symbol.*

Input		Output	
R	S	Q	Q̄
0	0	No Change	
0	1	1	0
1	0	0	1
1	1	Not Allowed	

(*B*) *Truth table.*

Fig. 3-48. Fundamentals of set-reset flip-flop.

For example, note the symbology for the *toggle* flip-flop of Fig. 3-47A. Every time a negative-going edge of a pulse at input T occurs, the bistable changes state. Recall that a multivibrator can have two outputs 180° apart from opposite sides. Thus, if a negative-going trigger arrives when Q=0 and $\bar{Q}$=1, then Q flips to 1 and $\bar{Q}$ to 0.

With the toggle flip-flop, there is no predetermined state for the two outputs when the circuit is first turned on. Thus the state of the outputs following application of a trigger pulse cannot be predicted unless the present state is known. The circuit simply changes states each time a negative-going transition is applied.

The *set-reset* bistable (Fig. 3-48) overcomes this problem. This circuit (also termed RS flip-flop) has two inputs and the usual two complementary outputs. As indicated by the truth table (Fig. 3-48B), a 1 input to the set terminal makes the 0 output ($\bar{Q}$) a 0 and the 1 output (Q) a 1. A 1 input

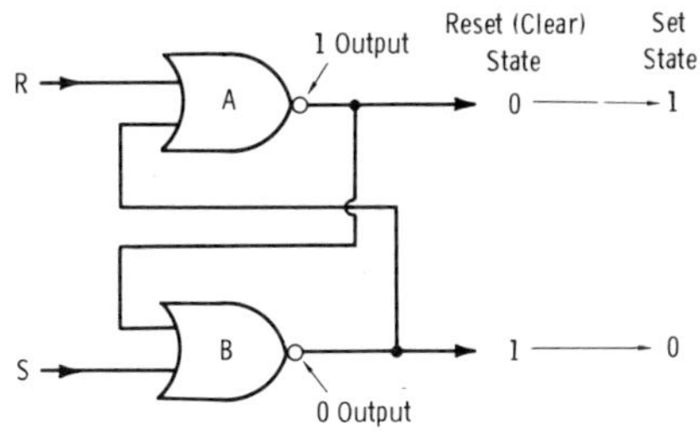

Fig. 3-49. NOR gates cross-connected to form set-reset flip-flop.

at the reset (clear) terminal reverses the state: the Q output becomes 0 and the $\bar{Q}$ output becomes 1. Zero signals on both inputs do not change the state. If both inputs should receive simultaneous signals (1's), the next state cannot be predicted and is ambiguous. Thus simultaneous inputs are commonly termed "not allowed" or "forbidden" combinations. This simply says that the device cannot be in both states simultaneously. The RS flip-flop cannot be used in logic situations which include the possibility of simutaneous set and reset inputs.

This is best understood by going momentarily to Fig. 3-49. Two NOR gates cross-connected as shown form a flip-flop. When power is applied, opposite states will exist; we will arbitrarily say these are a 1 output for B and a 0 output for A. The 0 output of A at the input of gate B becomes a 1 at the output, and the 1 output of gate B at the input of gate A becomes a 0 (phase inversion of NOR gate).

Now assume a 1 appears at the S (set) terminal. This 1 becomes a 0 at the output of gate B and drives the output of gate A to 1. Thus a set input has set the significant output to the normal 1 and the previous 1 output to 0. A 1 input to R will now reset the device to the previous state. Follow this action again from the truth table for Fig. 3-48B. Note that the circuit is predictable for three of four input conditions.

Flip-flops may be *clocked* or *unclocked.* In the unclocked flip-flop just discussed, the outputs respond to the inputs as the inputs change. In the clocked flip-flop, a clock input must exist at the time the inputs change for the outputs to respond.

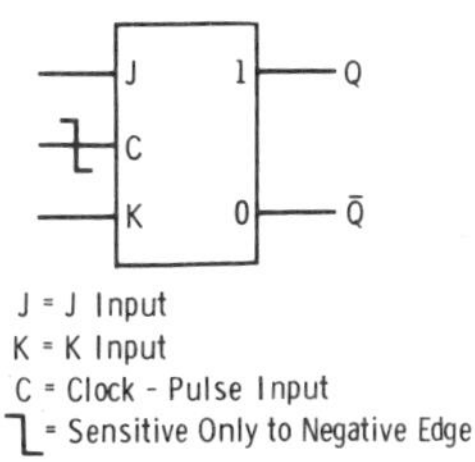

(*A*) *Symbol.*

Input		Output	
J	K	Q	$\bar{Q}$
0	0	No Change	
0	1	0	1
1	0	1	0
1	1	Complement	

(*B*) *Truth table.*

Fig. 3-50. Fundamentals of clocked JK flip-flop.

One of the most popular logic elements is the JK flip-flop (Fig. 3-50). There are no ambiguous states. When a 1 is applied to the J input, the 1 output (Q) is 1 and the 0 output ($\overline{Q}$) is 0. (When the flip-flop is clocked, the clock must be present.) When a 1 is applied to K, the 1 output flips to 0 and the 0 output flips to 1. When a 1 is applied to both the J and K inputs, the flip-flop switches to its complement state. Sometimes two or more J and K inputs exist. One J and one K input may be tied together for use as a clock input.

Special forms of gates are used in logic functions, and these gates have special symbols. Fig. 3-51A indicates two NAND gates with outputs paralleled. The symbol indicating that this circuit actually performs as an OR circuit is shown at the junction of outputs f1 and f2. This is termed a *wired* OR or sometimes a *phantom* OR. The truth statement for this circuit is: If f1 is true OR f2 is true, the output is true.

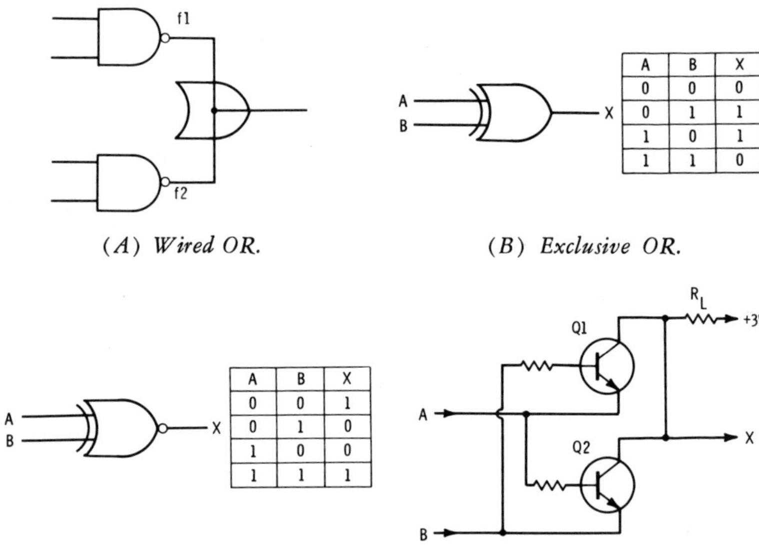

A	B	X
0	0	0
0	1	1
1	0	1
1	1	0

A	B	X
0	0	1
0	1	0
1	0	0
1	1	1

(*A*) *Wired OR.* (*B*) *Exclusive OR.*

(*C*) *Exclusive OR with phase inversion.* (*D*) *Circuit diagram of exclusive OR gate.*

Fig. 3-51. Special forms of gates.

A circuit which produces a true output only when the input states are *not* identical is termed an *exclusive* OR gate (Fig. 3-51B). Note from the truth table that a 1 (high) output is produced if only one input is 0 (low). If both inputs are of like polarity (either 0's or 1's), the output is 0, or low.

The complement of this function is shown in Fig. 3-51C. A 0 output is obtained only if the inputs are of opposite polarity. To make this clear,

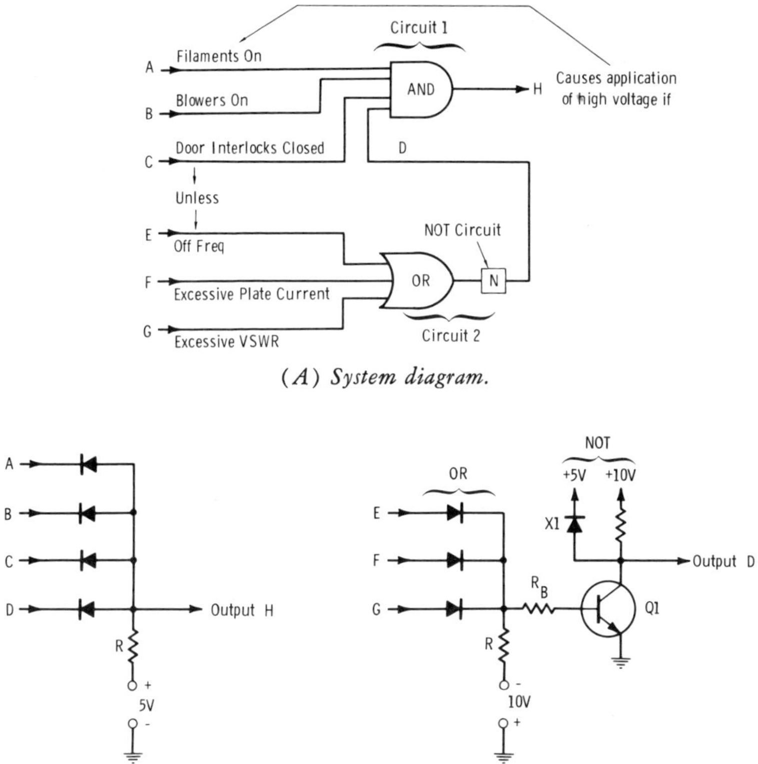

(*A*) *System diagram.*

(*B*) *Circuit 1.* (*C*) *Circuit 2.*

Fig. 3-52. Logic circuits for transmitter high voltage.

study the discrete resistor-transistor logic (RTL) circuit illustrated in Fig. 3-51D. If both inputs are 0's or 1's, both transistors have zero-biased base-emitter junctions, and neither can conduct. Thus output X sees the full value of supply voltage (high level) since there is no current in R_L.

Since the A input is tied to the emitter of Q1 and the base of Q2, and the B input is tied to the emitter of Q2 and the base of Q1, *unlike* polarity (opposite logic levels) will turn on one of the transistors. The resultant voltage drop across R_L will send output X to a low level. Note again that this is the complementary function of the circuit in Fig. 3-51B, as indicated by the small circle at the output of the logic symbol in Fig. 3-51C.

A simple example of how logic schematics are drawn is shown in Fig. 3-52. Output H will apply high voltage to the final transmitter stage if:

Filaments are on
Blowers are on
Door interlocks are closed

but not if:

Exciter is out of frequency limits
Final-stage plate currents are excessive when power is applied
VSWR is excessive when power is applied

These conditions mean that output H will be logic 1 when all the inputs (A, B, C, and D) to the AND circuit are logic 1. Inputs A, B, and C are logic 1 if the corresponding conditions are satisfied. The NOT output (which provides input D) will be 1 if inputs E, F, and G are all logic 0. If any one of these inputs becomes logic 1, the OR output is logic 1. Since this output is complemented (inverted) by the NOT circuit, output D becomes 0, and high voltage is not applied because output H of the AND gate is logic 0.

Here is a brief review of the action of circuit 1 and circuit 2 in Fig. 3-52: In circuit 1 (AND gate, Fig. 3-52B), if all inputs are at logic 1, the resulting positive potential at the diode cathodes holds the diodes at cutoff, and there is no current through resistor R. Therefore the potential at H is at plus 5 volts, or logic 1. If any one of the inputs goes to logic 0, the associated diode cathode is essentially returned to ground, and the diode conducts. Therefore the voltage at H drops to ground potential (minus the small forward voltage drop across the diode), and the output voltage becomes essentially 0 (logic 0). Thus, *all* the inputs must be at logic 1 to have logic 1 at the output.

In circuit 2 (Fig. 3-52C), consider the OR gate and NOT gate separately. If *any* of the inputs to the OR gate is at logic 1, the associated diode will conduct, and the voltage at the junction of R and R_B will be essentially the positive voltage applied to the diode. This back-biases the other diodes, and the OR-gate output goes to logic 1. If all inputs are at logic 0 (ground), the voltage at the R-R_B junction goes to zero minus the small forward voltage drop across the diodes.

Now consider the NOT circuit in Fig. 3-52C. When the base of Q1 goes positive (indicating a logic 1 from the OR circuit, which in turn indicates a faulty-condition input), the transistor is saturated and output D goes to essentially ground potential. So the AND circuit will not apply high voltage to the transmitter, since input D is at logic 0. When the Q1 base is returned to ground (all inputs to the OR circuit at logic 0), the transistor is cut off, and output D goes to plus 5 volts, or logic 1. Then if inputs A, B, and C are logic 1, output H causes application of the high voltage.

There is just one thing that might not be apparent immediately; what is the purpose of X1 in the collector circuit of Q1? This diode simply prevents the logic-1 output from exceeding 5 volts. When Q1 is not conducting, X1 becomes forward biased and clamps the logic-1 output to plus 5 volts. When Q1 is saturated, X1 is reverse biased and effectively out of the circuit. This technique is used when it is desirable to match all input levels to the AND gate.

3-8. THE OPERATIONAL AMPLIFIER

Before taking up the subject of analog-to-digital and digital-to-analog conversions, it is necessary to be familiar with a basic block used in such conversions, the operational amplifier. First study the negative-feedback amplifier of Fig. 3-53. The base-emitter current varies only slightly with relatively large input signal voltages due to the fact that R_{in} in series with the base is very high relative to the forward base-emitter resistance. Thus the input source as seen from the transistor base is essentially a constant-current source. The input impedance is essentially the value of R_{in}, and the transistor base is very near ground potential (about −0.6 volt for a silicon pnp transistor).

Fig. 3-53. Pnp negative-feedback amplifier.

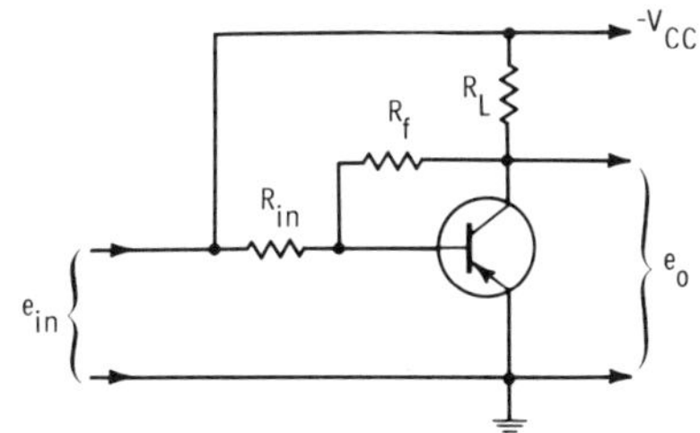

The ratio of R_f (feedback resistance) to R_{in} is greater than unity, and for practical purposes the gain (A) of the stage may be considered fixed by this ratio, independent of transistor characteristics. The feedback voltage through R_f is, of course, 180° out of phase with the input voltage and is degenerative. Thus $A = R_f/R_{in}$, which simply fixes the ratio of negative-feedback voltage to input voltage. This type of amplifier is sometimes used in computers to multiply the dc input by a constant, which is the fixed stage gain.

An operational amplifier is a direct-coupled (dc) feedback amplifier that has very high gain, high input impedance, and low output impedance. It normally takes the configuration of Fig. 3-54. It is usually arranged so that the input signal is referenced to ground (Fig. 3-54A) and the output signal is referenced to zero volts dc with respect to ground. The zero dc level at the output is established by operating the amplifier between two power sources of opposite polarity to ground (Fig. 3-54B), with the output terminal connected to the point which is at zero dc voltage with respect to ground.

As indicated by Fig. 3-54A, the operational amplifier is represented on schematic diagrams by a triangular symbol with terminals for external connections shown, but no internal details included. The output terminal has a plus sign (when indicated at all), and one input terminal has a minus sign and the other input terminal a plus sign. The minus input terminal is

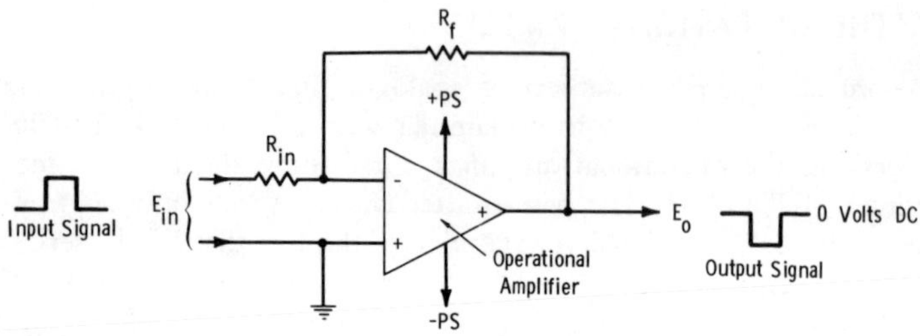

(*A*) *Inverting input referenced to ground.*

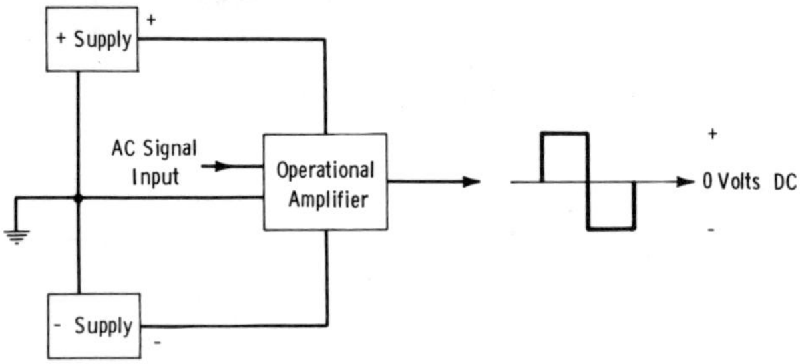

(*B*) *Method of obtaining dc ground at output.*

Fig. 3-54. Fundamentals of operational amplifier.

termed the *inverting* input (output signal inverted from input polarity), and the plus input terminal is termed the *noninverting* input (output signal same polarity as input). With the noninverting input terminal at ground, the no-signal input voltage at the inverting input of the amplifier is at ground dc potential also.

The two inputs of an operational amplifier comprise a *differential input*. One input is inverted and the other not inverted. If the *same signal* is applied to both inputs, the output is $E1 - E2 = 0$. If different signals are applied, the output amplitude is the *difference* between the input amplitudes; the output depends on actual amplitude difference, width difference, or phase difference. The use of the differential amplifier is covered in applicable portions of following chapters.

At present, we are concerned with the application typified by Fig. 3-54A. Actual voltage amplification is the ratio of R_f to R_{in}. Thus:

$$E_o = -E_{in}\frac{R_f}{R_{in}}$$

NOTE: The negative sign simply indicates that the output is of opposite polarity from the input.

As previously stated, the gain of an operational amplifier is extremely high. When operated as an open loop (no inverse feedback), it can become merely a switch with (typically) an input signal of only 1 to 5 millivolts. See Fig. 3-55A. The saturation limits at the ends of the linear transfer curve are termed the *common-mode input-voltage limits.* (*"Common mode"* means one input is at ground or other reference potential, as distinguished from differential input, described earlier.)

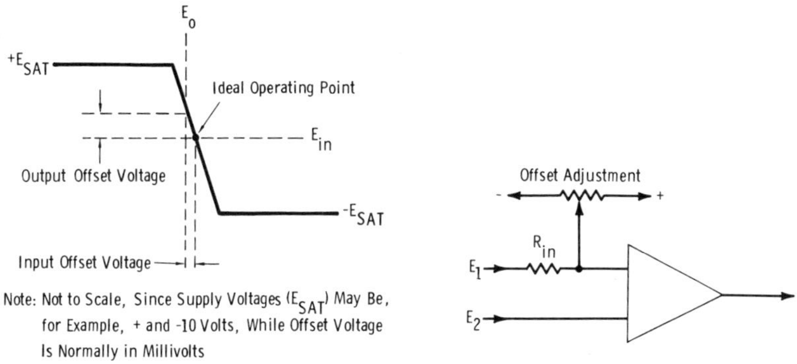

(*A*) *Transfer characteristic.* (*B*) *Offset compensation.*

Fig. 3-55. Open-loop operational-amplifier operation.

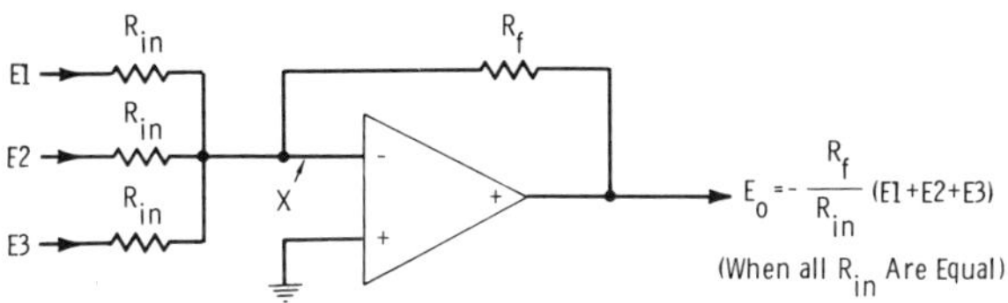

Fig. 3-56. Voltage summer.

Also indicated in Fig. 3-55A is the "ideal operating point." In the ideal amplifier, the output voltage is zero when the input voltages are zero. In practice, the characteristic is shifted, or *offset* slightly when the input voltage is zero. Therefore, the technician will normally find an offset-voltage adjustment, as illustrated in Fig. 3-55B. This control is adjusted to balance out the offset.

If a number of voltage sources, E1, E2, E3, etc., are connected through individual resistors of the same value to the inverting input (Fig. 3-56), the circuit acts as a *voltage summer.* As stated previously, feedback brings the voltage at the inverting input of the amplifier (point X) to the same level as the noninverting terminal, which in this circuit is ground. Each input signal voltage produces an output the same as if it were the only input present. For example, for E1:

$$E_o = -E_1 \frac{R_f}{R_{in}}$$

The total voltage output therefore is the sum:

$$E_o = -\frac{R_f}{R_{in}}(E1 + E2 + E3 + \ldots E_n)$$

R_{in} need not be the same for all inputs. In that case, the gain for each input voltage is given by the ratio of R_f to the corresponding input resistor.

When the feedback element is changed from a resistance to a capacitance, an integrator or ramp generator is formed. The circuit of Fig. 3-57A is an inverting amplifier with a feedback capacitor C_f, and a switch across the capacitor. In practice, the switch is normally a transistor electronic switching device clocked at a given rate.

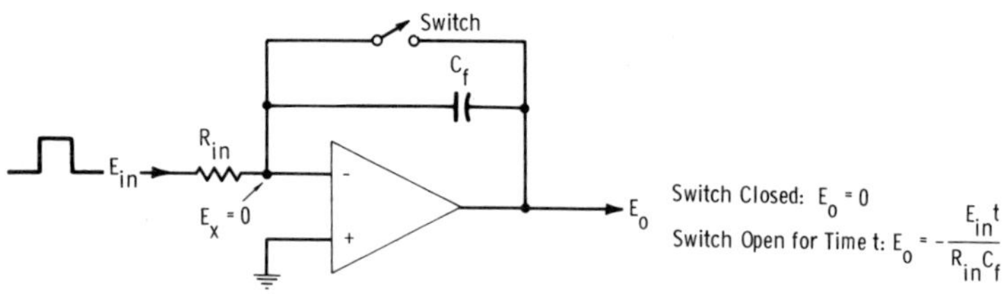

(*A*) *Diagram.*

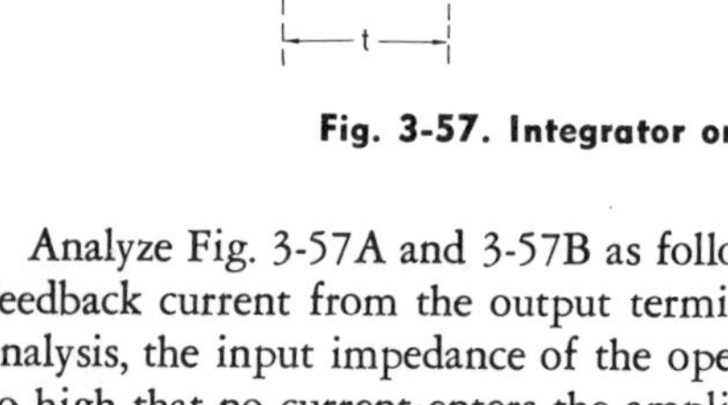

(*B*) *Output waveform.*

Fig. 3-57. Integrator or ramp generator.

Analyze Fig. 3-57A and 3-57B as follows: When the switch is closed, all feedback current from the output terminal passes into R_{in}. (For practical analysis, the input impedance of the operational amplifier may be assumed so high that no current enters the amplifier.) The voltage at E_x is zero, or the same as the noninverting input, which is at ground. Therefore the output voltage (E_o) is zero because the switch is a direct short between input and output.

When the switch is opened, feedback current from the output terminal must enter the capacitor, and the current through C_f must enter R_{in}. This degenerative current prevents the amplifier input voltage from changing instantaneously with E_{in}. The input current and feedback current combine to maintain a constant charging current through C_f, producing a linear voltage change across the capacitor. This is to say that the charging current is constant and is equal to E_{in}/R_{in}.

The slope of the output waveform is dependent on time constant $R_{in} C_f$. Since the input side of the capacitor (E_x) remains at ground potential, the entire voltage change must appear at the side of the capacitor connected to the output terminal. Since E_{in} is applied to the inverting input, E_o must have opposite polarity. Consequently the *rate of change* (slope) of the output voltage ramp equals $-E_{in}/R_{in}C_f$.

The voltage across the capacitor is:

$$E_c = \frac{It}{C_f}$$

where,

E_c is the capacitor voltage,
t is the time (seconds) the switch is open,
C_f is the feedback capacitance in farads,
I is the charging current.

It was stated previously that:

$$I = \frac{E_{in}}{R_{in}}$$

where,

R_{in} is the input resistance in ohms.

Substituting for I in the equation for E_c:

$$E_c = \frac{E_{in}t}{R_{in}C_f}$$

But:

$$E_c = -E_o$$

Therefore:

$$E_o = -\frac{E_{in}t}{R_{in}C_f}$$

If the input signal (E_{in}) is a steady dc voltage while the switch is open, E_o changes at a constant rate of $-E_{in}/R_{in}C_f$ volts per second. The output waveform for a positive input (Fig. 3-57B) therefore is a ramp having a slope equal to this rate. The ramp starts at zero volts and peaks at a value of $-E_{in}t/R_{in}C_f$ volts, where t is the length of time the switch is open. The output voltage, of course, levels off if this time is sufficiently long that the amplifier reaches saturation.

It can be seen that the effect of the amplifier is to transform E_{in} into an output voltage having a rate of change (slope) proportional to E_{in} with a constant of proportionality of $-1/RC$. Since this electrical operation is analogous to the mathematical operation called integration, the circuit is sometimes termed an *integrator*.

3-9. DIGITAL-TO-ANALOG CONVERSION

A common method of converting digital information to analog information involves the use of an operational amplifier with a feedback loop consisting of series-arranged resistors across which are relay contacts or electronic switches. The switches across the individual resistors can be either open or closed, depending on the 0's and 1's of the digital data, changing the effective feedback resistance and resulting in an analog output at the amplifier output terminal.

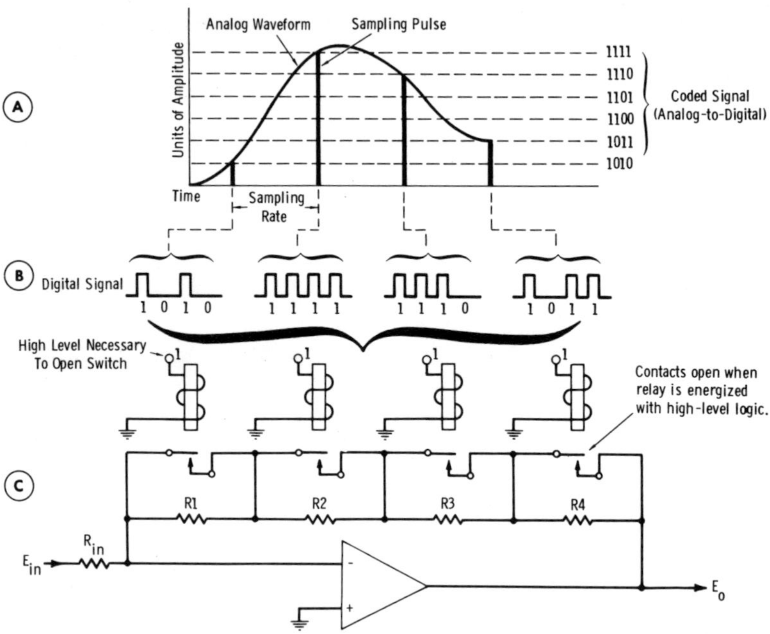

Fig. 3-58. Digital-to-analog conversion using operational amplifier with variable feedback controlled by relays.

Fig. 3-58A represents an analog waveform which is sampled to result in pulses corresponding to the instantaneous voltage amplitude at the time of sampling. This is termed pulse-code modulation (pcm). A coding device divides the sampled signals into incremental amplitude levels (right-hand side of Fig. 3-58A) and develops corresponding digital pulses (Fig. 3-58B).

The operational amplifier of Fig. 3-58C contains a feedback loop with a series array of selected resistors, R1 through R4. Each resistor is bypassed with a switch which is open or closed. Although relays are shown for clarity, a transistor switch is often used. If all contacts are closed, the feedback path has zero resistance (a direct short between output and input), and no voltage gain is produced ($A = R_f/R_{in}$). When a 1 binary input is applied, the switch opens, and resistance is inserted in the feedback path, producing voltage gain. Therefore, E_o represents the analog value of a digital function.

The four bits of each sampling time are fed individually to each of the switches. Take the first sample, 1010. The first 1 (left-most digit) is fed to the first switch; the 0 (second bit) is fed to the switch for R2; the third bit (1) is fed to the switch for R3; and the next 0 is fed to the switch for R4. The process is repeated for the next word. The first bit (left-most digit) is termed the *most significant digit,* since it has the greatest weighting value (2^3, or 8), whereas the last digit is termed the *least significant digit,* since $2^0 = 1$. The resistance values of R1, R2, R3, and R4 are weighted accordingly to give the correct amount of gain. Note that E_{in} can be a reference dc voltage which "traces a curve" at the output in accordance with the initial analog function.

In the example of Fig. 3-58, a four-digit word is used, and four switches are required. A switch is needed for each binary input digit, so a ten-digit word would require ten switches and associated resistors. The greater the number of switches, the smaller the possible incremental difference in output voltage levels can be, and the higher the *resolution* can be.

Another method of digital-to-analog conversion is a series of transistor switches termed a *digilog* which feeds a resistance network termed a *ladder-adder.* The switches simply divide ground between plus and minus reference voltages in the resistor network so that the applied voltages are varied in analog proportion to the digital signals which cause the switching.

3-10. ANALOG-TO-DIGITAL CONVERSION

One type of analog-to-digital conversion was shown in Fig. 3-22, and another type in A and B of Fig. 3-58. In either case, analog inputs are converted, or *encoded,* into digital outputs. The *comparison method* is the most common type of analog-to-digital conversion.

The comparison method of converting analog information to digital information is illustrated basically in Fig. 3-59. The familiar sine wave is shown to represent an analog voltage, which may be seen to be analogous to rotation of a potentiometer between plus and minus values through a zero reference voltage. (The input source could just as easily be the secondary of a transformer with the primary connected to a microphone, phono pickup, synchro system, etc.) The potentiometer provides one input to a voltage-comparator circuit. Each position of the potentiometer at any instant is termed the *analog address.*

The other input to the voltage comparator is a linear ramp voltage from the function generator. The ramp is initiated from a start signal which also enables a gate circuit to allow continuous-running clock pulses to pass to the digital counter (flip-flops). As long as there is a difference in magnitude between the analog and function-generator inputs to the comparator, clock pulses at a constant repetition rate are permitted to reach the counter. When the linearly rising ramp voltage reaches equality with the analog address, a stop signal is generated by the comparator; this signal disables the gate and blocks the clock pulses from the counter. The number of pulses accumulated in the counter during the comparison time interval is proportional to the amplitude of the analog input voltage at the time of address, and is the desired digital information.

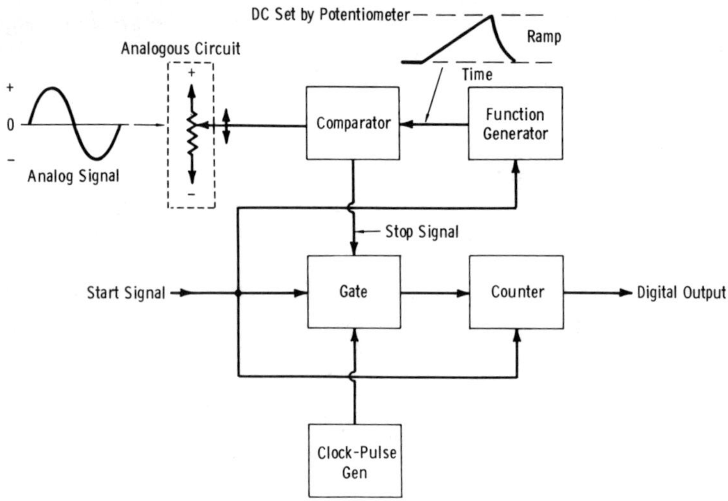

Fig. 3-59. Comparison method of analog-to-digital conversion.

3-11. COUNTERS

A basic block in digital systems is the binary counter, which is simply a series of flip-flops. Although most flip-flops are in the form of ICs, discrete transistors are sometimes used for this circuitry. In any event, it is desirable that the reader have a good comprehension of the action of the binary counter, and this is most readily gained by study of discrete circuitry.

Figs. 3-60A and 3-60B show two identical circuits, not interconnected. From the base voltage-divider arrangement, it is obvious that each circuit would be conducting. (The base voltage is a little less than 1/2 of the −25-volt supply voltage, in a forward-bias polarity.) Analyze the two extremes of operation as follows.

Saturation: Since the emitter resistance is 1/10 of the collector resistance, the voltage at the emitter would be about −2.5 volts. Since there is negligible voltage drop across the transistor in saturation, the voltage at the collector is also approximately −2.5 volts.

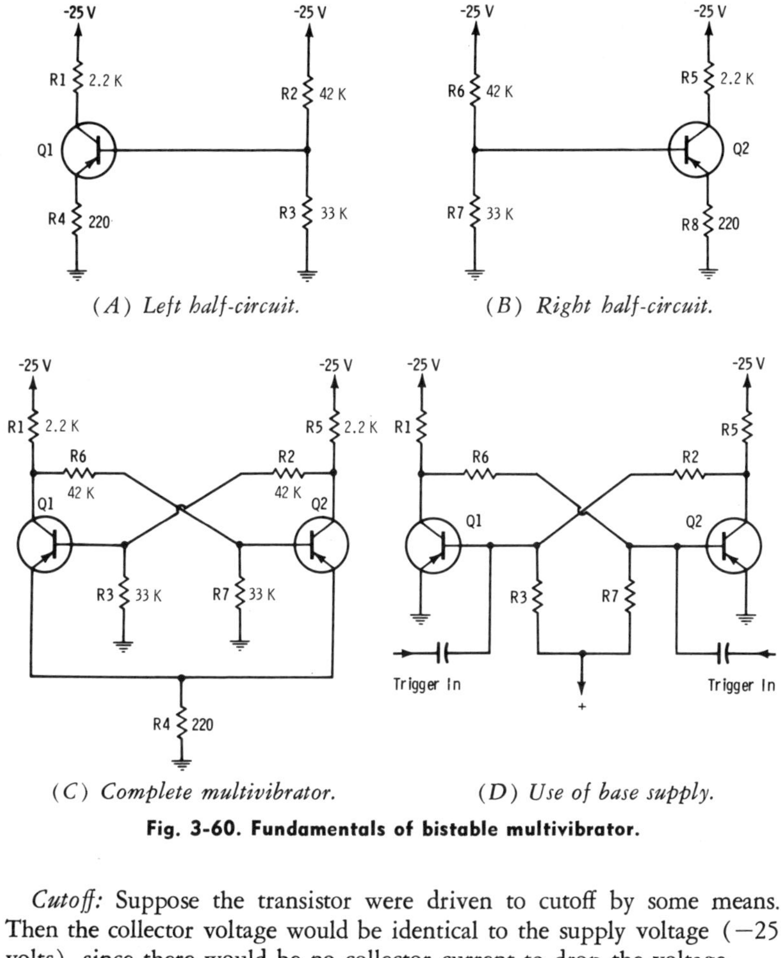

(*A*) *Left half-circuit.* (*B*) *Right half-circuit.*

(*C*) *Complete multivibrator.* (*D*) *Use of base supply.*

Fig. 3-60. Fundamentals of bistable multivibrator.

Cutoff: Suppose the transistor were driven to cutoff by some means. Then the collector voltage would be identical to the supply voltage (−25 volts), since there would be no collector current to drop the voltage.

Now suppose it is possible to switch the top end of R2 between the −25 volts and −2.5 volts. Further suppose that the emitter voltage of Q1 can be made a constant −2.5 volts. With this condition, since the base of Q1 will be something under 1/2 of −2.5 volts (actually about −1 volt), the base is positive relative to the emitter, and the transistor cannot conduct.

Fig. 3-60C shows the individual circuits of Figs. 3-60A and 3-60B cross-coupled. The emitters of Q1 and Q2 are now returned to ground through a common resistor, R4. The "top" ends of R2 and R6 are each connected to the collector of the opposite transistor.

For analysis, since both circuits are identical, we must arbitrarily assume one transistor or the other in conduction. Assume Q1 is in saturation. Then its emitter and collector are at −2.5 volts, and the top of R6 is also at −2.5

volts. Note that since the emitters are now common, both emitters will be at −2.5 volts; consequently, Q2 is at cutoff, from the previous analysis. Since Q2 is cut off, its collector is at −25 volts, the top of R2 is also at −25 volts, and therefore Q1 is truly in saturation. The circuit is "flipped"; in the absence of any change, one transistor is saturated and the other is cut off.

Suppose now that base triggering is used. A trigger applied to either base will start a chain reaction; since pnp transistors are being used, the trigger must be positive to the base of the on transistor, or negative to the base of the off transistor. Assume, for example, that a positive-going trigger is applied to the on transistor, Q1. Transistor Q1 will be driven toward cutoff, its collector will swing negative (less voltage drop across R1), and the base of Q2 will become more negative because of the cross coupling through R6. So Q2 starts to conduct; the drop through R5 swings the Q1 base positive (toward ground from −25 volts), further cutting off Q1. The regenerative action continues (even though the trigger pulse is very short in duration) until Q1 is completely cut off and Q2 is saturated. The circuit is "flopped"; the reverse conditions from the previous mode now exist. In the absence of any change, the circuit will remain in this mode.

When the bases are returned to ground, a common emitter resistor must be used to make the emitter of the off transistor slightly negative. Thus the less negative base can keep the transistor cut off. The emitter resistor can be deleted when the bases are returned to a slight reverse bias, as shown in Fig. 3-60D. This circuit works the same as that of Fig. 3-60C, except that the off transistor is held off by the external reverse bias.

In the common-emitter circuit of Fig. 3-60C, emitter triggering may be used. When emitter triggering is not used, emitter resistor R4 usually is bypassed with a capacitor to avoid degeneration of the signal voltage.

Further note that in the foregoing example a positive trigger pulse to the base of Q1 was used to turn the transistor off. To turn it on again, a pulse of the opposite polarity would need to be applied to the Q1 base, or a pulse of the same polarity would need to be applied to the Q2 base. But suppose an emitter trigger is used. Assume Q1 is saturated (Q2 off) and the trigger is negative-going. Since the base of Q2 is already positive relative to the emitter, a negative pulse at the emitter will only momentarily increase the reverse bias, and hence will not affect the operation of Q2. But Q1 is conducting because its base is negative relative to its emitter. When the negative trigger pulse arrives at the common emitter, Q1 is driven toward cutoff by the reduced forward bias, and regeneration takes care of the rest of the action. Transistor Q1 becomes cut off, and Q2 becomes saturated; the multivibrator has changed state.

Now another negative pulse arrives. Since Q1 is already cut off, it is not affected. But Q2 is driven toward cutoff, and again regeneration causes the circuit to flop back to the original state. So a chain of negative pulses (or, as should be evident, a chain of positive pulses) is all that is required. No "steering" of pulses with regard to polarity is required in emitter triggering. The negative emitter pulse has no affect on the off transistor but affects the on transistor. If the pulses are positive, they will not affect the transistor which is already on, but will affect the off transistor. Remember also that this entire description concerned pnp transistors. For npn transistors, all polarities are reversed. Be sure to visualize the action for npn bistable multivibrators.

Observe again Fig. 3-60D. A trigger at the base of Q1 changes the state of the multivibrator. To change the state again, a trigger of the opposite polarity at the Q1 base would be required, or a trigger of the same polarity at the Q2 base. But we normally will be concerned with a given polarity of trigger pulses for a bistable multivibrator, so let us see how this is done.

See Fig. 3-61A. This bistable multivibrator is identical to the circuit of Fig. 3-60D except for the method of triggering. Negative pulses from one input source are applied to the bases of both transistors. Since this is base triggering for pnp transistors, a negative pulse will drive the off tran-

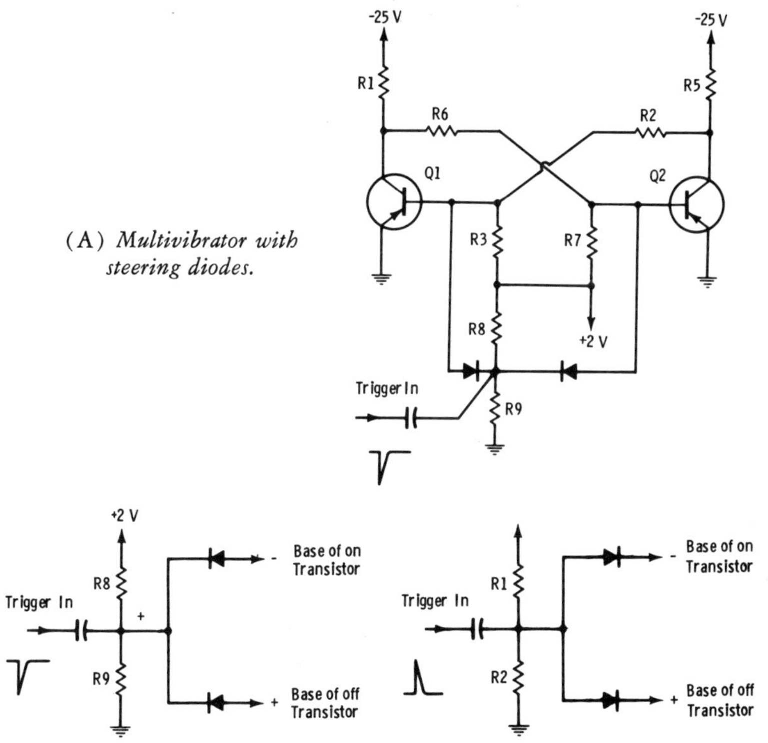

(A) *Multivibrator with steering diodes.*

(B) *Steering diodes for negative pulses.*

(C) *Steering diodes for positive pulses.*

Fig. 3-61. Fundamentals of pulse steering.

sistor on. See Fig. 3-61B for analysis. Resistors R8 and R9 form a voltage divider from the +2-volt reverse-bias supply so that the cathodes of the diodes are at a slight positive (reverse) potential. The base of whichever transistor is on will be negative, so the corresponding diode is reverse biased, and this circuit remains open (no effect.) The base of the off transistor is held positive by the reverse-bias supply, so its diode is turned "on" by the negative pulse. Hence, the off transistor receives the negative pulse, which turns it on. Regeneration completes the action; the circuit flips and reverses the polarity of the output voltage. The next negative pulse flops the circuit back to the original state. The diodes are termed *steering diodes* for obvious reasons. Fig. 3-61C shows the diode arrangement for positive pulse steering.

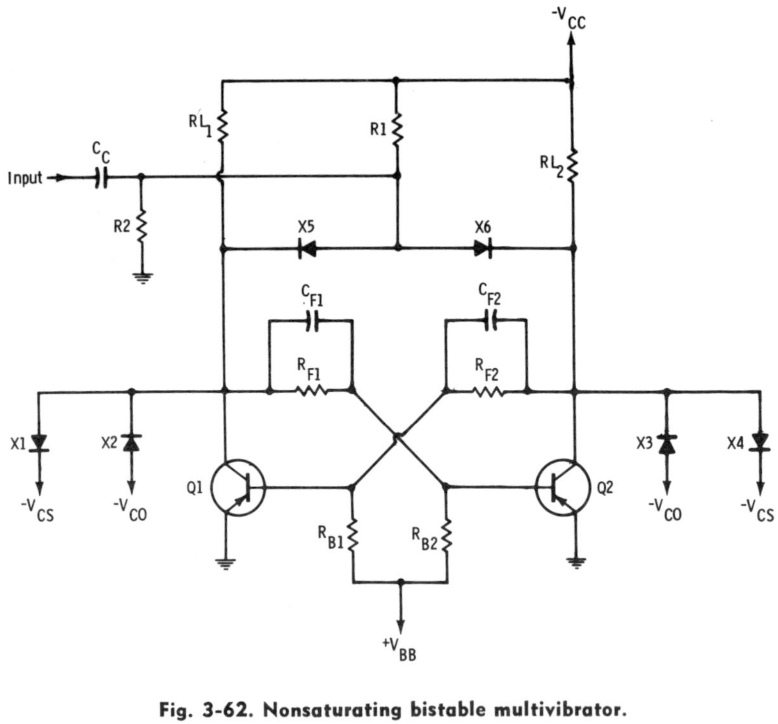

Fig. 3-62. Nonsaturating bistable multivibrator.

For analysis of positive-pulse steering, see Fig. 3-61C, and use it in conjunction with Fig. 3-62. At the moment, we are confining the analysis of Fig. 3-62 to triggering only (diodes X5 and X6). This arrangement is always termed *collector triggering,* but remember that the collector pulse is actually cross-coupled to the opposite transistor base. This cross coupling is what changes the state of the multivibrator.

Since the trigger pulse is positive and the transistors are pnp, the trigger will drive the on transistor off. The steering is evident from Fig. 3-61C when you bear in mind that the base of the on transistor is tied to the collector of the off transistor (Fig. 3-62).

The circuit of Fig. 3-62 is a nonsaturating bistable multivibrator. The multivibrator is held out of both saturation and cutoff by means of biased diodes. This decreases the storage time and increases the pulse-repetition-rate capability.

Diodes X1 and X4 with bias source V_{CS} are used for saturation clamping. Diodes X2 and X3 with bias source V_{CO} are used for cutoff clamping. The circuit is so designed that without clamping it would both saturate and cut off. Clamping the upper and lower limits of the output permits more reliable substitution of transistors when necessary. The analysis of the circuit is as follows:

Saturation clamping: Collector diodes X1 and X4 have their cathodes returned to a slight negative voltage, for example −2 volts. When the transistor associated with one of the diodes is not in saturation, the anode of the diode is considerably more negative than the cathode, and the diode is effectively an open circuit. When the transistor is driven toward saturation, the voltage at the collector attempts to reach ground potential (more positive than the base) and forward-bias the base-collector junction (the required condition for saturation). But as soon as the collector reaches a voltage slightly less negative than the diode cathode potential, the diode becomes forward biased, and the collector is held at the V_{CS} supply voltage. Thus the transistor cannot saturate.

Cutoff clamping: Diodes X2 and X3 have their anodes returned to a reverse bias which is somewhat less than the collector supply voltage. When the transistor nears cutoff, the diode is forward biased, and the collector is held at $-V_{CO}$. This action prevents the transistor from being cut off.

> NOTE: Clamping diodes hold the operation in the more linear portion of the output characteristic curve. Thus the average power dissipation is increased, and to avoid transistor damage the designer must choose a load line such that operation of the circuit falls within the maximum permissible power dissipation.

Fig. 3-63 illustrates what a binary circuit does. Note that binary 1 has one output pulse for every two input pulses. Then binary 2 again has one output pulse for every two pulses from binary 1. So the output of binary 2 is at one-fourth of the original frequency. Stated another way:

$$\text{Total division} = 2^n$$

where,

n is the number of binary scalers (number of bistable multivibrators or flip-flops).

A series of binaries can be made to count an odd number, such as the 525 count required in sync generators, by means of reset pulses corresponding to an odd count.

We can now examine in greater detail the readout of a binary counter. Fig. 3-64 illustrates four flip-flops in cascade, forming a $2^4 = 16$ scaler. The capacity of a counter to count to a given numerical value is termed the *modulus* of the counter.

The series of input pulses (waveform 1 in Fig. 3-64B) is applied to the first flip-flop (FF1), which is assigned the *least significant* binary number. The output of FF4 has the *most significant* (highest order of 2, or 2^3) binary number. The reader must bear in mind that when we write a binary number, the *least significant* number is on the right, the reverse of the readout on the block diagram of Fig. 3-64A. In reading the 8-4-2-1 binary output, recall that $2^0 = 1$, $2^1 = 2$, $2^2 = 4$, and $2^3 = 8$; hence, the binary weight of FF1 is 2^0. (Review Fig. 3-24.) The resulting binary count is shown at the bottom of Fig. 3-64 with the digits arranged from the bottom (waveform 5) to the top (waveform 2) so that the most significant number (waveform 5) is in the proper position.

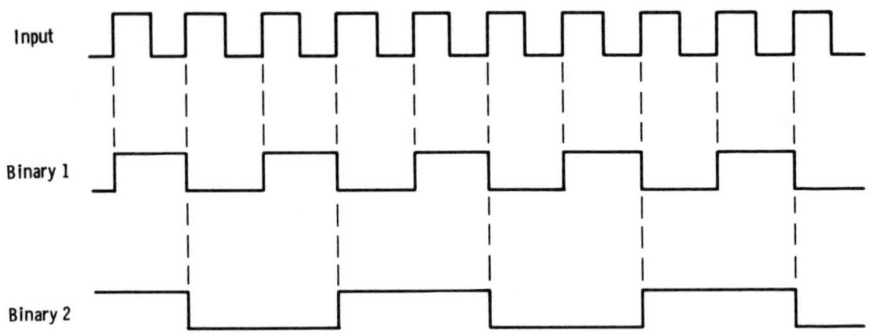

Fig. 3-63. Waveforms of binary circuits.

Before analysis of Fig. 3-64, bear in mind the basic facts about a reset-set (RS) flip-flop. When the flip-flop is the set state (for positive logic), the 1 side of the circuit reads 1 (high level) and the 0 side reads 0 (low level). If the circuit is in the reset (clear) position, the 1 side reads 0 and the 0 side reads 1. When the clear (reset) side changes from the 0 state to the 1 state, (1 to 0 on the set side), a pulse is sent to the input of the flip-flop of next higher order, as shown by the block diagram of Fig. 3-64A. Analyze Fig. 3-64 as follows:

Prior to pulse 1: All four flip-flops are in the clear position; thus each output reads 0, and the total counter output is 0000.

Pulse 1: Flip-flop 1 changes to its set position, producing at 1 at its set output. However, neither FF2 nor any of the following stages receives a signal because FF1 produces a pulse at its R (clear) output only when changing from set to reset (clear). Note that the counter output is now 0001.

Pulse 2: The second pulse changes FF1 from set to reset, producing a 0 at the set output and a 1 at the clear output. The R output pulse changes FF2 from the clear to the set position (the S output of FF2 now reads 1). The

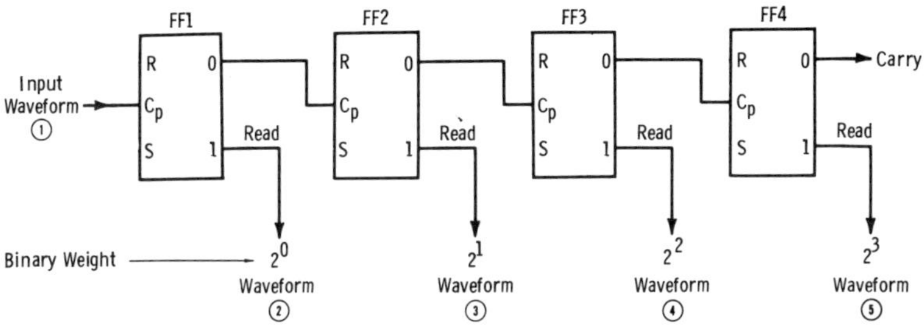

(A) Block diagram.

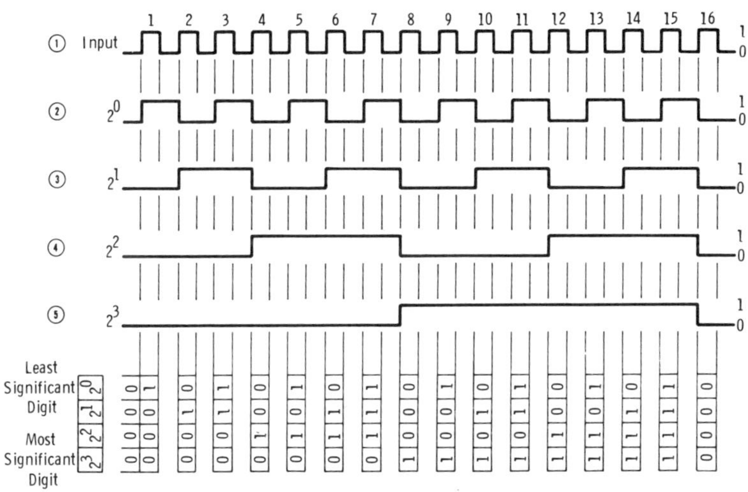

(B) Waveforms.

Fig. 3-64. Divide-by-sixteen circuit.

clear output of FF2 is 0, and the remaining flip-flops are not affected. Thus the total counter output is now 0010, or binary 2.

Pulse 3: The third pulse changes FF1 to the set state again, but the following flip-flops remain in the previous state. Thus the output is now 0011.

Pulse 4: The fourth pulse changes FF1 and FF2 to the clear position and FF3 to the set position. The output is now 0100.

By continuing this process, the circuit will count up to 15 pulses and then return to clear (0000) on the sixteenth pulse. Table 3-5 shows the resulting truth table for the counter. Note that this analysis has developed a binary-to-decimal conversion table for numbers through the maximum decimal number (15) possible with a four-digit binary.

Thus the highest possible binary number in the output of this counter is 1111, or binary 15. The modulus of any counter is the fixed capacity of the counter plus one. The modulus of this counter then is $15 + 1 = 16$. The addition of another flip-flop would make the modulus 32.

A practical parallel to the above situation can be drawn by considering that most automobiles have odometers with a decimal readout capacity of 99,999.9 miles. Since 99,999.9 is the maximum readout, recycling of the indicator to 00000.0 represents 100,000.0. Thus it is apparent that the modulus of any counter is the fixed capacity of the counter plus one.

There are many varieties of counter, but the technique described above is basic to all types.

3-12. MEMORY, OR STORAGE

Some instruction books may refer to a physically separate magnetic tape unit as a *storage* device and a magnetic-core or drum unit in the same sys-

Table 3-5. Truth Table for Divide-by-Sixteen Counter

Input Pulse	2^3	2^2	2^1	2^0	Decimal Equivalent
0	0	0	0	0	0
1	0	0	0	1	1
2	0	0	1	0	2
3	0	0	1	1	3
4	0	1	0	0	4
5	0	1	0	1	5
6	0	1	1	0	6
7	0	1	1	1	7
8	1	0	0	0	8
9	1	0	0	1	9
10	1	0	1	0	10
11	1	0	1	1	11
12	1	1	0	0	12
13	1	1	0	1	13
14	1	1	1	0	14
15	1	1	1	1	15
16	0	0	0	0	0

tem as a *memory* device. Basically, however, the terms memory and storage can be used interchangeably.

A memory unit may consist merely of a series of capacitors that store a charge for a required period of time to be used for readout. Or it may consist of an *accumulator* which stores data in the form of static dc voltages at the outputs of flip-flop circuits. A memory could store information on a length of magnetic tape or on the magnetic surface of a drum or disc. Magnetic core matrices are often employed as memories. More recently, integrated-circuit memory devices have become quite common.

A *magnetic core* is a tiny ring of ferromagnetic material, typically 0.050 inch in outside diameter, 0.030 inch in inside diameter, and 0.015 inch thick. It is capable of retaining saturation in either of two states, depending on the direction of the applied current, which are arbitrarily assigned the values 0 and 1.

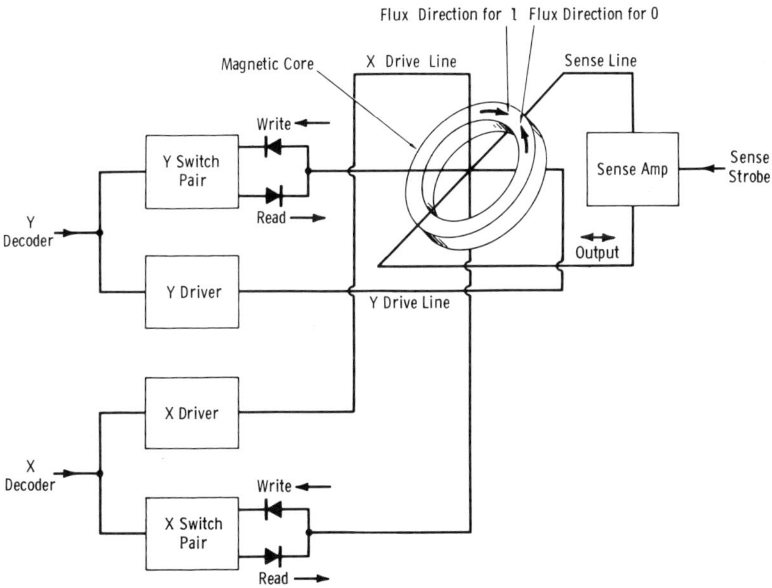

Fig. 3-65. Basic magnetic-core schematic.

See Fig. 3-65. The X and Y drive lines (explained later) supply the current for core magnetization. A current in one direction establishes a magnetization in the core in a given direction. Reversing the current direction reverses the direction of the flux field and the core magnetization. The write-in and read-out functions determine the current direction as shown in Fig. 3-65.

The X and Y drive lines constitute two separate windings on the core. To sense the state of a core, a third winding termed the *sense* winding is used. The two wires (X and Y) are used for write selection, and by reversing the direction of the current, the same two wires are used for read. The write current places the core in a 1 state. The read current shifts the state of the core, and the sense wire picks up the signal voltage generated by the flux change of the core in shifting from 1 to 0. Since the core state stored before sensing is destroyed during readout, this type of memory is referred to as a *destructive-readout* memory. With this type of memory, if the stored data is to be used again, the readout pulse is immediately followed by a restore (or rewrite) pulse to return the core to the original state.

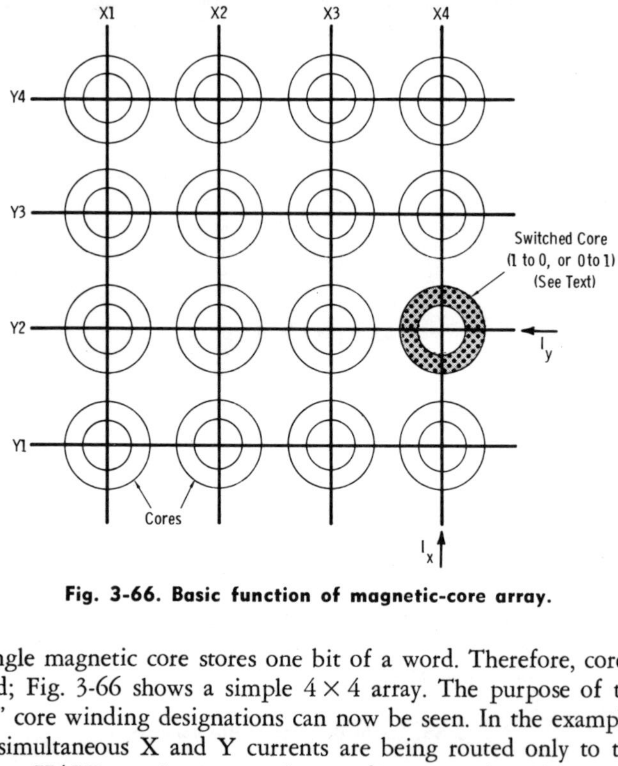

Fig. 3-66. Basic function of magnetic-core array.

A single magnetic core stores one bit of a word. Therefore, core *arrays* are used; Fig. 3-66 shows a simple 4 × 4 array. The purpose of the "X" and "Y" core winding designations can now be seen. In the example illustrated, simultaneous X and Y currents are being routed only to the core related to X4,Y2 on the rectangular-coordinate matrix. Although the X current is being routed to the other three cores in the X4 line, and the Y current also feeds the other cores on the Y2 line, a single current alone will not change the state of the core. If the X and Y (simultaneous) current applied is a write current and core X4,Y2 is in a 0 state, it is shifted to a 1 state. If the core is already in a 1 state, no change takes place. Then for a read current, the core changes state (1 to 0), and the resultant voltage is picked up by the sense winding. Any core in the matrix is read in a similar manner by application of a read pulse to the associated X and Y winding of the core. If the core has not been switched to a 1 by a previous write current, it contains a 0. Application of a read pulse to the X and Y windings of that core does not switch the direction of magnetic intensity, and no pulse is induced in the sense winding. The read current can only switch a 1 to a 0, not vice versa. Similarly, as previously stated, a write current can only switch a 0 to a 1. If a 1 already exists, no change occurs.

NOTE: In some complex system arrays, a large number of "stacked" arrays are necessary. The X and Y driver windings of each array are in series or parallel with every corresponding X and Y winding on other array assemblies. Some 1's and some 0's are normally written in one vertically stacked column of cores. When the X and Y wires are energized to write a 1, the intended cores in the stack must be selected. This is done by running an *inhibit* wire in parallel with either the X or Y line, adding a fourth winding to the core. A current is made to pass through the inhibit wire in the opposite direction to that in the adjacent (X or Y) drive line. This cancels the magnetic field to prevent writing a 1 into any core that should be a 0.

As previously mentioned, integrated-circuit memories are rapidly appearing in new digital systems. One example is the RCA CD4005 or CD4005D (identical except for physical construction) memory IC. These are 16-word, 1-bit-deep, nondestructive-readout (ndro) memories consisting of 16 flip-flop storage cells arranged in an X-Y matrix.

Each storage cell contains a flip-flop which consists of two cross-coupled complementary-symmetry MOS (metal oxide semiconductor) inverter stages and two bidirectional gates. Each gate consists of two n-channel MOS transistors which are connected to perform the gating to and from each cell. The functional diagram is shown in Fig. 3-67A, and Fig. 3-67B illustrates the cell-arrangement diagram.

Addressing of a particular cell is accomplished by the simultaneous application of high voltage levels on the X-Y coordinates of that cell, just as for the magnetic-core system. A 1 is written into an addressed cell by the application of a low voltage level on the D1 line and a high voltage level on the D0 line. Conversely, a 0 is written into the addressed cell by forcing the D0 line low and the D1 line high. Nondestructive readout, or sensing, at the Q-sense return line of an addressed cell is accomplished by providing sense current by means of an externally forced high-level voltage on the D1 line. The D0 line is also maintained at a high voltage level during readout.

Sensing can also be accomplished by external logic-level monitoring of the D1 and D0 lines. In this mode of sensing, logic circuitry that inhibits the writing operation during readout is provided, and the sense line must be connected to V_{SS} (the source voltage supply).

NOTE: For positive logic, the low level (0) is the negative source voltage, and the high level (1) is the positive drain voltage. Similarly, for voltage

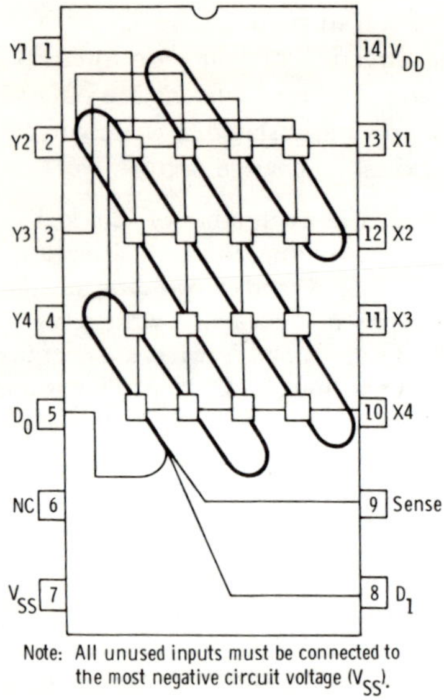

(*A*) *Functional diagram.*

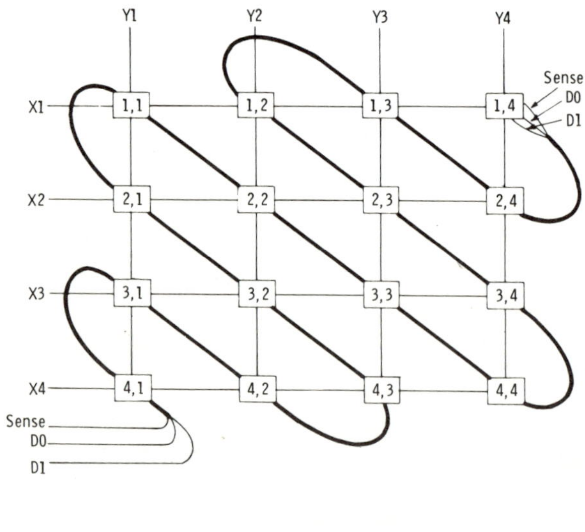

(*B*) *Cell arrangement.*

Courtesy RCA

Fig. 3-67. Integrated-circuit memory.

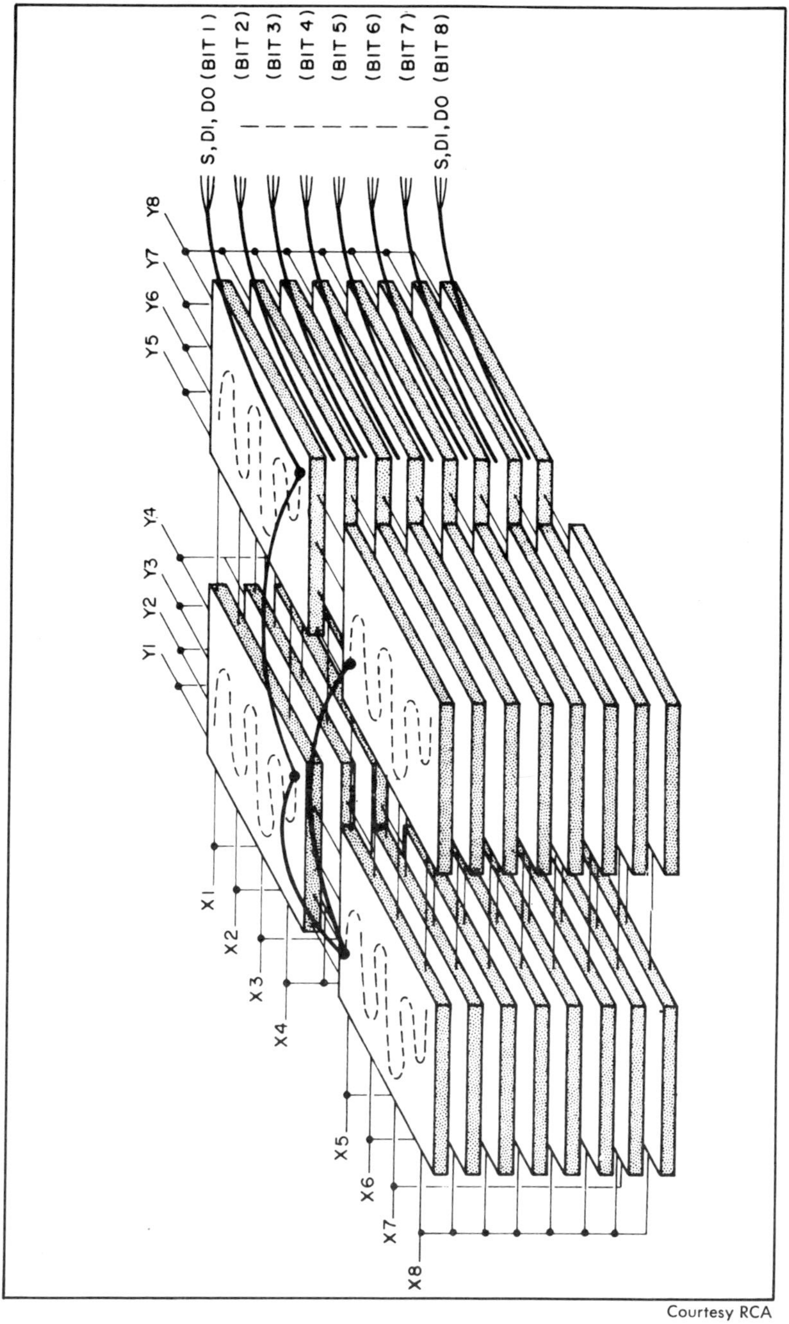

Courtesy RCA

Fig. 3-68. 64-word × 8-bit memory using 32 packages.

sensing, the low level is −1.5 volts, and the high level is −0.9 volt. For current sensing, the low level is essentially zero current, and the high level is 500 μA.

Expansion of this memory beyond the 16-word, one-bit capacity is accomplished by the utilization of any number of packages. Fig. 3-68 shows how 32 packages can be interconnected to produce a 64-word × 8-bit memory. Word capacity is increased by horizontal expansion in the X and Y directions (increasing the number of cell locations). The sense lines in each plane (each plane represents one bit) are made common, either by direct connection or through gates. Likewise, the D0 lines are made common and the D1 lines are made common in the same manner. Bit expansion is accomplished by an increase in the number of horizontal planes. The corresponding X and Y terminals in each plane are connected together (directly or through gating).

As an example of operation, consider writing the 8-bit binary word 10111011 into word location X3,Y4. The least significant bit is bit 1, on the far right (adjacent to the binary point).

1. The X3 and Y4 inputs are set to a high level.
2. The D0 and D1 inputs are set to high (H) and low (L) levels as follows:

Bit Number	*D0*	*D1*
1	H	L
2	H	L
3	L	H
4	H	L
5	H	L
6	H	L
7	L	H
8	H	L

To read this same word out of word location X3,Y4:

1. X3 and Y4 are again set to a high level
2. Current from an external source is supplied on the D1 line for each bit (D1 is set to a high level)
3. All D0 lines are maintained at a high voltage level

All sense leads associated with cells in which a 1 was stored will carry approximately 500 microamperes; those cells containing a 0 will pass virtually zero current through their respective D1 leads.

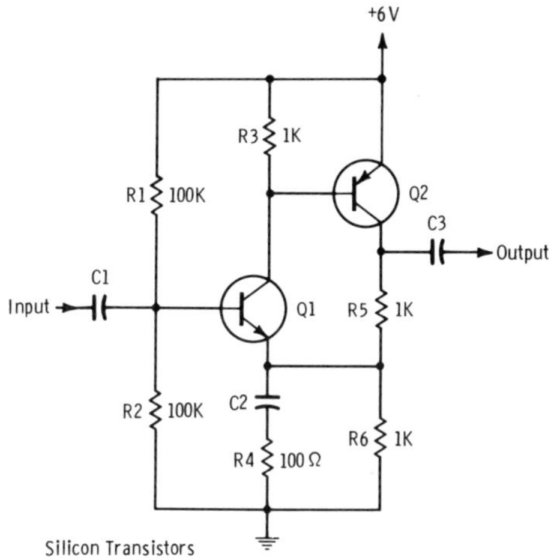

Fig. 3-69. Circuit for question Q3-1.

EXERCISES

Q3-1. Make a dc analysis of the circuit in Fig. 3-69. What voltages would you expect to read at the bases of Q1 and Q2, the collectors of Q1 and Q2, and the emitter of Q1? Assume silicon transistors.

Q3-2. Convert decimal 16 to: (A) pure binary, (B) bcd, (C) octal.

Q3-3. Convert decimal 11 to (A) binary and (B) octal.

Q3-4. Convert decimal 568 to (A) binary and (B) octal.

Q3-5. Convert binary 101001.01 to decimal.

Q3-6. Read aloud the expressions: (A) AB, (B) A + B.

Q3-7. Define the terms: (A) DTL, (B) RTL, (C) RCTL, (D) TTL.

Q3-8. Prepare the truth table for circuit 1 (AND) of Fig. 3-52A.

Q3-9. Prepare the truth table for the OR portion of circuit 2, Fig. 3-52A.

Q3-10. Prepare the truth table for the NOT portion of Fig. 3-52A.

Q3-11. Does a series of cascaded binaries always count to 2^n, where n is the number of binary circuits?

CHAPTER 4

The Fundamentals of Transducers

A transducer converts one type of power to another. A microphone converts alternate condensations and rarefactions of air (sound waves) into corresponding electrical variations. A phono pickup converts mechanical motion from recorded tracks on a record to corresponding electrical variations. A magnetic pickup head converts alternating magnetic fields on the tape to corresponding electrical variations. A loudspeaker converts electrical variations to a mechanical movement, causing corresponding air vibrations that can be detected by the ear.

4-1. THE MICROPHONE

In dealing with any mechanical or electrical component, there is always an "ideal" which the designer must use as a criterion of performance. An ideal microphone for most purposes should have an output voltage independent of frequency, and a uniform response for any one direction of a sound wave. In this ideal case, the instrument would deliver to the audio amplifier power whose frequency, phase, and amplitude components would be exactly the same as the original acoustical energy.

There are two basic types of microphones: the pressure type, and the pressure-gradient type. Pressure types, in turn, are quite varied, including carbon, condenser, crystal, moving-coil (dynamic), etc. There is only one pressure-gradient instrument, the *velocity* type, incorporating a metallic ribbon suspended in a magnetic field.

Each different type of microphone exhibits a different response pattern (Fig. 4-1). The pattern simply represents the amplitude response for varying positions of the sound sources about the microphone. A *nondirectional* pattern is illustrated in Fig. 4-1A. If a constant-intensity sound source is moved around the microphone at a constant distance, the amplitude of the electrical impulses from the microphone will also remain constant. In Fig.

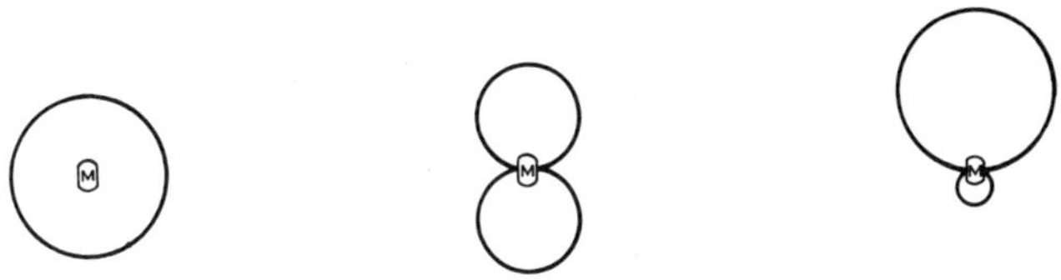

(*A*) *Nondirectional.* (*B*) *Bidirectional* (*C*) *Unidirectional.*

Fig. 4-1. Three fundamental microphone patterns.

4-1B, the response of the microphone decreases as the angle of the sound source approaches 90° from either face of the instrument. This microphone is sensitive from front and rear while theoretically "dead" at both sides, resulting in a *bidirectional* pattern. Still another fundamental type is illustrated in Fig. 4-1C; it is sensitive only to sounds originating in front of the microphone and is said to have a *unidirectional* response. Briefly, it may be stated that pressure microphones are nondirectional, velocity microphones are bidirectional, and a combination of the two is unidirectional.

These response patterns are measured in an acoustically "dead" room. The condition is equivalent to an open space where there is absolutely no reflection of sound waves. The effect of room acoustics on the action of the microphone will be discussed in Chapter 11. It will also be shown that response patterns vary somewhat for different frequencies.

Impedance Values and Cable Lengths

Just as there is a need for a wide variety of response patterns, there is also a need for a difference in the designs for permanent or portable use, and a number of impedance values for use in the complex fields of radio and sound.

The output of a microphone will either be high impedance working directly into the grid circuit of the input amplifier stage, or low impedance using a 30-, 150-, or 250-ohm transformer. This low impedance works into the input transformer of an amplifier. Some microphones built for various applications have a tapped output transformer with screwdriver-adjustable switch to select either high- or low-impedance output.

In general, high-impedance microphones are used for general communications, amateur radio, some pa installations, and some home recorders. They are especially applicable if the length of the microphone cable is kept under 25 feet.

The effect of the length of a microphone cable is directly related to the impedance and the application of the sound system. The metallic shielding in a cable forms a capacitance with the inner conducting wires, and this is added to the capacitance between the wires themselves. The effect is equivalent to placing a capacitor across the line; the higher the impedance used the greater will be the shunting effect across the line with consequent loss of high-frequency response. Cable length, therefore, definitely must be considered when high-impedance microphones are used. The effect of any length of cable is negligible for low-impedance systems.

Fig. 4-2 shows the signal loss caused by extra cable length over 20 feet. This is based on a good microphone cable with a low capacitance of 0.0007 μF per 20-foot length. For a cable length of 200 feet, the frequency response at 5000 Hz would be 15 dB lower than for a length of 20 feet. Thus, broadcasting stations, recording studios, and special pa installations

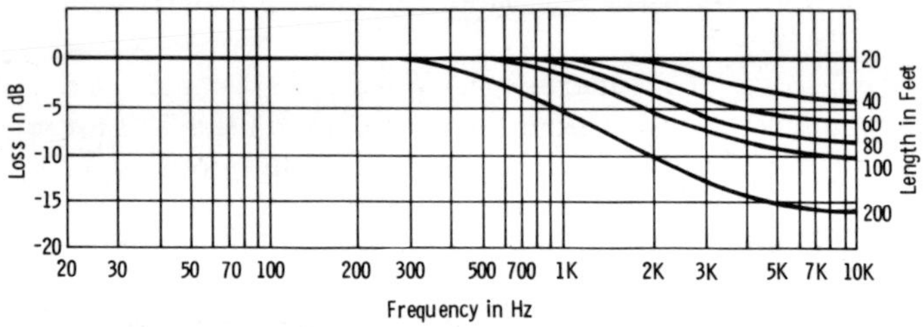

Fig. 4-2. Signal loss versus cable length (high-impedance input).

in which long cable lengths are a necessity always use low-impedance microphones. Low impedance must also be used in any form of communications where the operator works at a point remote from the microphone amplifier, since medium and high frequencies are important for crispness and intelligibility of speech.

Output and Noise Levels

Every electrical amplifier has a noise level that depends largely on the thermal noise generated in the input stage of the amplifier and which becomes larger as the bandwidth is increased. Thermal noise is generated in such microphones as the carbon and some condenser varieties where a voltage is applied to the microphone elements, but there is also noise in all other types due to certain fluctuations in the air pressure from thermal velocities of air molecules. Just as the noise level of a particular amplifier limits the lowest signal volume to be amplified, the thermal noise in the air places a lower limit on the atmosphere when used as a sound-energy medium. In fact, the noise component of air acting on the diaphragm of an extremely sensitive microphone is about equal to the thermal noise of a high-quality amplifier. This noise, however, is just about equal to the threshold of hearing of the most sensitive ears, between 1000 and 5000 Hz. Should any microphone (other than a carbon type) noticeably raise the noise level at the output of a good amplifier, a fault within the microphone, cable, or input circuit is indicated (unless a stray ac field is being picked up).

High-impedance microphones and carbon microphones intended for general communications use have a somewhat higher output level than low-impedance units. The output levels of low-impedance microphones are always given as open-circuit or unloaded transformer input ratings. It is quite important in most low-impedance installations that the microphone be worked into an unloaded circuit. The response of ribbon and combination-type microphones may be seriously affected by a reflected resistance load on the mechanical constants of the moving elements. Fig. 4-3 shows typical microphone input circuits. The input is said to be "unloaded" since, although the transformer primary matches the microphone impedance at a given frequency (usually 400 Hz), no loading resistance is used.

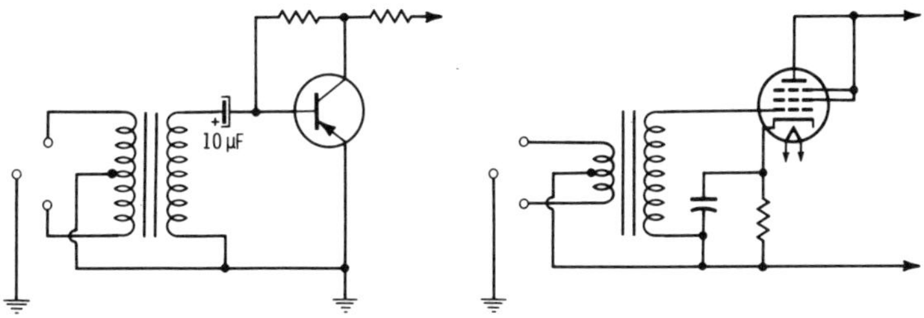

(*A*) *Transistor amplifier.* (*B*) *Vacuum-tube circuit.*

Fig. 4-3. Microphone input circuits.

Volume-Handling Capability

Another limiting factor of a microphone is the maximum sound pressure that can be handled without distortion. This is particularly true of pressure-type units. It is the function of the diaphragm, which is connected to the moving element inside, to move physically in accordance with the applied sound pressure, thus providing a voltage image of the sound. There is, however, an elastic limit to this movement. Suppose that the diaphragm is connected to a moving coil as in the inductor (or dynamic) microphone. Fig. 4-4A shows what happens when an extremely loud sound pressure, such as shouting, occurs close to the unit. At point A, the sound is still increasing, but the elastic limit of the diaphragm has been reached, and it has come to rest. This means that the coil has also come to rest, and the voltage

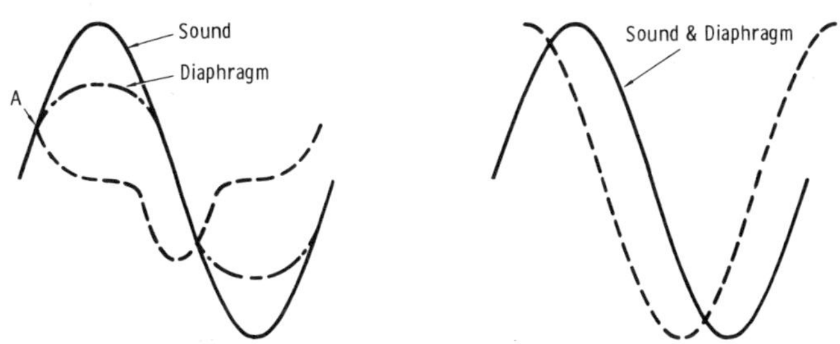

(*A*) *Action under excessive pressure.* (*B*) *Action with normal loading.*

Fig. 4-4. Movement of diaphragm in dynamic microphone.

output has dropped to zero, even though the air pressure is still increasing on the diaphragm. The diaphragm then returns at high velocity as the pressure subsides, and a surge in voltage is created; the distortion is shown by the dashed voltage line. This kind of waveshape results in *blasting* and contains a number of distortion components and false harmonics. Fig. 4-4B shows how the voltage is generated in a normal microphone that is not overloaded; the voltage follows the shape of the sound pressure exactly.

The Pressure Microphone

The pressure microphone, known variously as the moving-conductor, inductor, and dynamic microphone, is one of the most popular types in all fields of radio and sound. The electrical output results from the motion of a conductor (usually a small coil) in a magnetic field. The magnets are of the permanent type and require no external voltage.

There is actually a slight technical difference between a moving-coil microphone and an inductor type, although the operating principles are identical. The moving-coil type, as its name implies, utilizes a small coil in the magnetic field. In the inductor type, the diaphragm actuates a straight conductor that is suspended between the magnets.

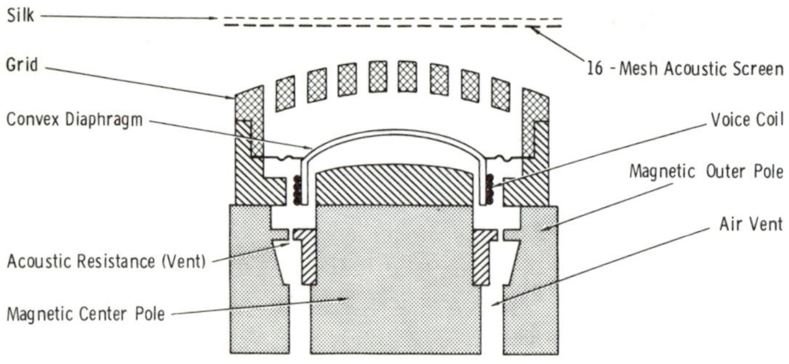

Fig. 4-5. Basic diagram of moving-coil microphone.

Fig. 4-5 shows the basic construction of a moving-coil unit. Motion of the diaphragm causes the small coil to move back and forth in the magnetic field in accordance with the actuating sound pressures. In practice, the small voltage thus generated is applied across a transformer, the primary of which matches the impedance of the coil. A varying audio voltage appears across the secondary of the transformer; this secondary has an impedance of 50, 150, or 250 ohms or high impedance depending on the application desired. The mass resistance on the opposite end from the coil is used to balance the movement of the diaphragm.

Without a sort of resistance control or damping, however, a diaphragm and coil assembly of this kind would have a frequency response somewhat similar to that of curve A in Fig. 4-6. Curve B is the idealized response

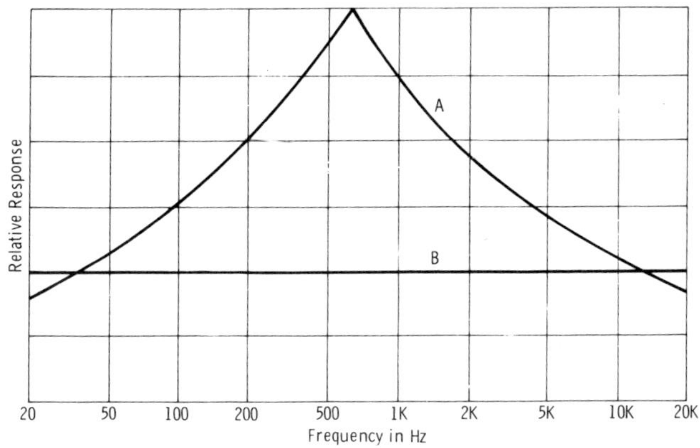

Fig. 4-6. Dynamic-microphone response curves.

curve of a microphone and indicates the necessity for some kind of resistance control.

Damping is accomplished in several ways. Fig. 4-7 shows an assembly with a tube (A) and an air slit (B). Air slits overcome the mass reactance of the diaphragm at high frequencies (which reduces high-frequency response) resulting in an increase of over an octave at the high end of the scale. An increase in low-frequency response is obtained by the action of the tube and its associated air cavity. The addition of these features is designed to maintain uniform diaphragm velocities at all frequencies.

Most coils used in dynamic microphones are wound with aluminum ribbon or wire; aluminum provides a maximum ratio of conductivity to mass. When ribbon is used, it is wound edgewise on the form. Diaphragms of the dynamic and inductor types are made of either aluminum alloys, styrol, Bakelite, or paper.

The Pressure-Gradient Microphone

The pressure-gradient, velocity, or (more commonly) ribbon microphone employs the moving-conductor principle in which a ribbon is suspended

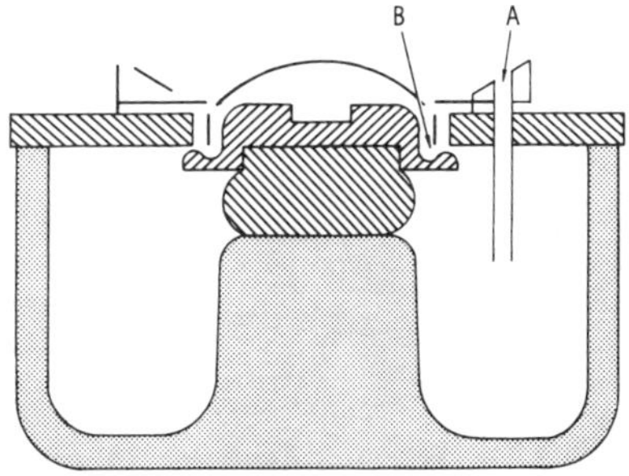

Fig. 4-7. Means of damping (A) and compensation for mass reactance (B) at high frequencies.

so that it vibrates freely in a magnetic field. This ribbon constitutes its own diaphragm; since it is not housed in any closed type of case and is exposed to the air on its two active sides, there are no adverse effects of cavity resonance, diaphragm resonance, or pressure doubling.

Fig. 4-8 illustrates the basic construction of a ribbon microphone. The ribbon is made of a thin corrugated aluminum strip (aluminum alloy to minimize mass) and is suspended between the poles of a permanent magnet as shown. Movement of the ribbon caused by passing sound waves causes the magnetic lines of force to be cut transversely, inducing a corresponding voltage between the two ends of the ribbon. In order that the ribbon constitute a mass reactance over the usable range, its resonant frequency is made lower than the lowest frequency to be reproduced. This eliminates peaks.

The ribbon moves through the magnetic field in the direction of diminishing pressure as a function of the difference in sound pressure between the front and back of the ribbon. Thus, under certain conditions, it is said to correspond in motion to the particle velocity in a sound wave, hence the term velocity microphone. This rate of pressure change with distance is called *pressure gradient.*

The ribbon can respond only along the axis perpendicular to its surface and is "live" to sound waves only from either face; it is "dead" from the side (90° from the zero axis). Hence, this microphone is, by the nature of its design and construction, a bidirectional instrument.

It should be understood that although the pressure of a sound wave in air may be independent of frequency, the pressure gradient, or rate of pressure change with distance, is not. Therefore, the mechanical impedance of a ribbon is *proportional to the frequency* of the actuating sound waves. In practice, this is only important where the spherical character of low-fre-

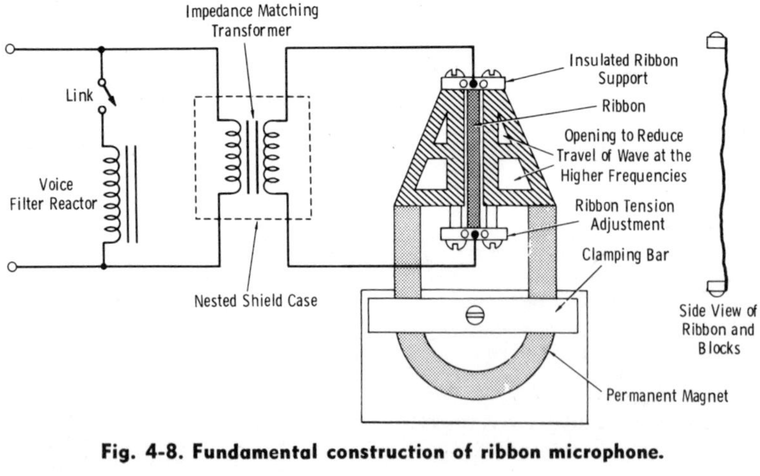

Fig. 4-8. Fundamental construction of ribbon microphone.

quency speech waves very close to the mouth will cause a greater pressure on the ribbon than the higher-frequency components of speech. Thus, when a person speaks into a ribbon microphone at close range, there is a definite "boominess" resulting from an accentuation of the lower frequencies. This is overcome by use of a *speech trap* when the microphone is to be used for speech only. It consists of an adjustable tap on the output transformer which raises the impedance of the ribbon from its extremely low value to 50, 150, or 250 ohms, and lowers the open-circuit reactance of the transformer such that low-frequency response is decreased. The tap is adjustable by means of a slotted switch on the bottom of the microphone case and is usually marked "music" and "voice."

Fig. 4-9. RCA SK-46 velocity microphone.

Courtesy RCA

A typical ribbon microphone, the RCA Model SK-46, is illustrated in Fig. 4-9. It is a later version and is less than half the size of the earlier RCA Model 44-BX. Actually, there are two models of this later version, the SK-35 and the SK-46. They are similar except for the close-talking feature of the SK-35. This unit is particularly suited for remote broadcasts in factories, busy offices, or other noisy locations. The lips of the speaker should be less than one inch from the front screen of the SK-35 in this application. The SK-46 is also used in studio applications, where it is especially suitable

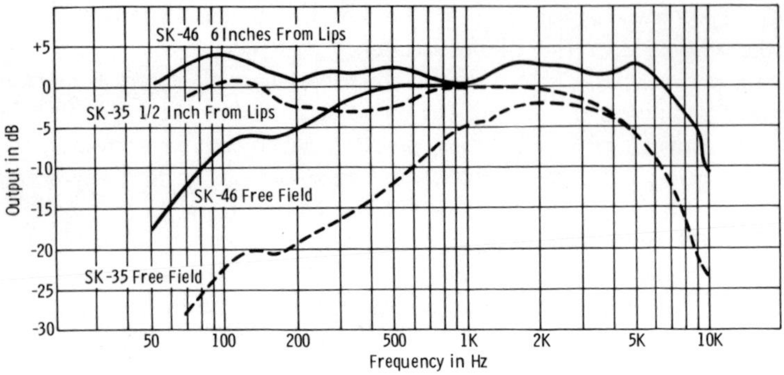

Fig. 4-10. Frequency response of SK-35 and SK-46 microphones.

for across-the-table discussions. The frequency-response curves for the ribbon microphones discussed are shown in Fig. 4-10.

Patterns of Response

Fig. 4-11 illustrates the diaphragm of a pressure-type microphone being excited by sound waves. In Fig. 4-11A, the sound originates in front of the instrument; as the wavefront passes, the diaphragm is brought from rest to the position shown by the dashed line. This results from the pressure of the condensed air and is the principle of a pressure-type microphone. In Fig. 4-11B, the sound is coming from the back of the microphone, but the pressure of the wavefront causes the diaphragm to move inward just as before. In Fig. 4-11C, the sound originates at the side, and the point of sound pressure causes the diaphragm to move inward once again. Hence, it may be seen that the diaphragm always moves in the same direction regardless of the initial direction of the traveling wavefront, resulting in a nondirectional response pattern. This type of microphone can be made semidirectional by orientation of the case and by the use of baffles. However, it is not a truly directional microphone as are other types.

Fig. 4-12 illustrates the action in a velocity microphone when the ribbon is actuated by sound waves. Waves on either face of the instrument (0° or 180°) cause the ribbon to move, although in opposite directions, for a given wavefront. Sound coming from the side (90° to axis of face) cause equal and opposing pressures on the sides of the ribbon, resulting in a theoretically zero movement of the ribbon. Hence, the microphone has essentially a "figure eight," or bidirectional, response.

The unidirectional microphone is just what its name implies: an instrument that by design is live on only one face, and dead toward the sides and rear. It is useful in broadcasting and recording studios. The unidirectional pattern may be obtained in a number of ways. The first method used was the combination pressure and pressure-gradient principle that is still widely employed. Primarily, it consists of a pressure element, usually a moving

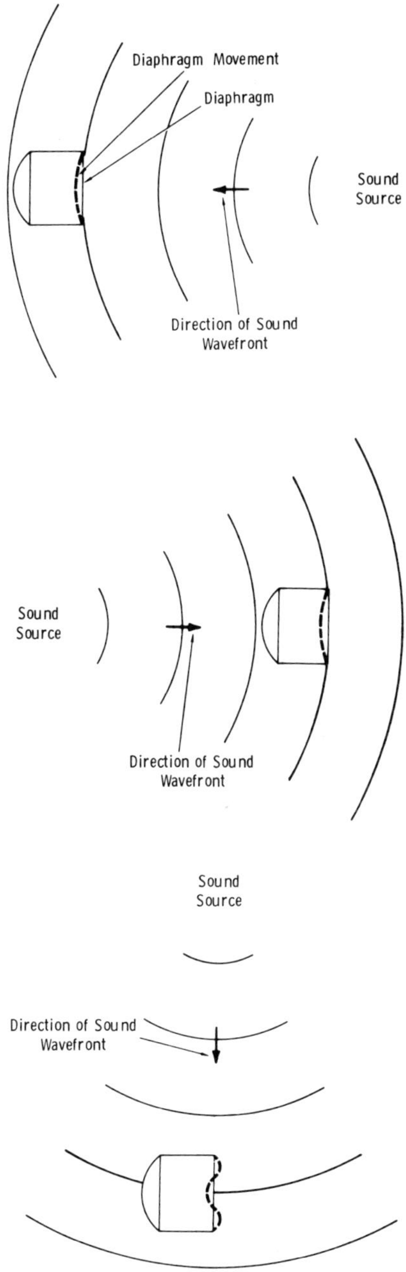

(*A*) *Sound from front.*

(*B*) *Sound from rear.*

(*C*) *Sound from side.*

Fig. 4-11. Action of pressure microphone.

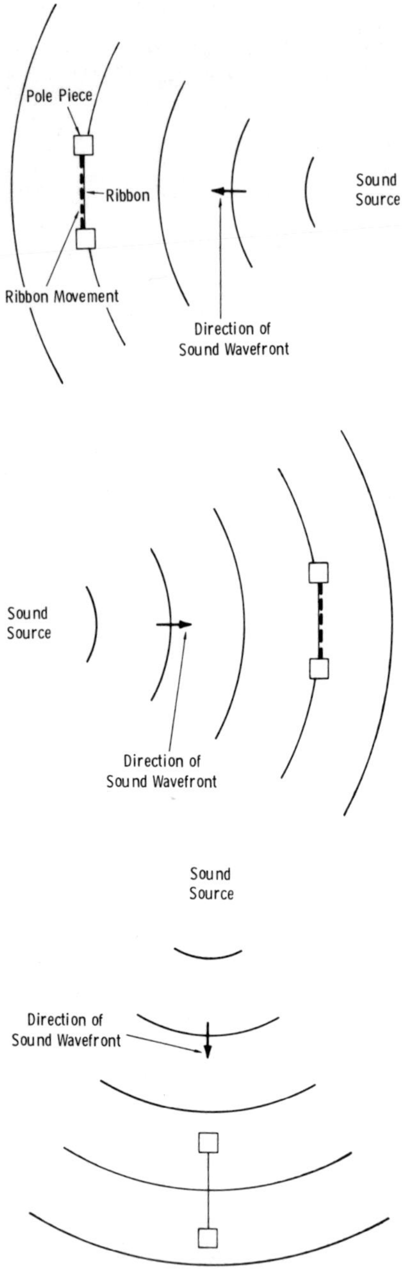

(*A*) *Sound from front.*

(*B*) *Wavefront from rear.*

(*C*) *Pressure from side.*

Fig. 4-12. Action of ribbon microphone.

coil, and a ribbon element connected in series. They are phased so that the outputs of the separate elements reinforce each other for sound coming from a given direction relative to the face of the instrument, and they cancel each other for sounds coming from the sides and rear. This will be clarified by the following brief review of pressure and pressure-gradient element action.

Look again at Fig. 4-11A and compare it with Fig. 4-12A, noting that the diaphragm and its associated coil assembly and the ribbon in the velocity microphone respond in the same direction to the traveling wavefront. Therefore, if these two elements were connected in series and properly phased, the output of the microphone would be large in amplitude with sound from this direction. Now compare Fig. 4-11B and 4-12B. The pressure element moves in the same direction as before, whereas the ribbon element moves in the opposite direction. With the same wire connections as before, the voltage generated by the ribbon will oppose that of the moving coil. It is obvious, then, that sound emanating from this direction will be effectively canceled out at the microphone transformer, resulting in zero response. Sound from the side will result in zero response from the ribbon element, but will still actuate (to a modified extent) the pressure element. This is why the typical unidirectional microphone has a much wider angle of pickup to the front compared with the velocity (bidirectional) microphone.

Effect of Angle on Frequency Response

In the hypothetical instance of the "ideal" microphone case mentioned earlier, the instrument should be uniformly directional at all frequencies; however, such is not true in practice. High frequencies travel in a relative beam-like configuration compared to the low frequencies, and a number of design and construction compromises must be made. The use of a nondirectional microphone, for example, oriented so as to become semidirectional is possible only for the higher frequencies which are sufficiently deflected by the dimensions of the case. At lower frequencies, approximately 1000 Hz or less, the microphone is nondirectional in any position. Such an orientation is useful in reducing high-frequency feedback in public-address installations associated with broadcasts which require a nondirectional microphone.

It is well known that high-frequency response falls off with increasing angles from the zero axis, or from a line drawn through the exact center of the diaphragm. Whenever the wavelength of a sound wave is short compared to the width of the diaphragm, several points of unequal pressure exist which cause an irregular movement. Thus, it becomes obvious that for best high-frequency pickup at increasing angles to the face, the diaphragm must be smaller than the wavelength of the highest frequency considered. In practice, this is impractical due to the extremely low output that would be obtained from such an instrument, and a compromise must be made. It will be noted, therefore, that all microphones are directional at high frequencies and have wider angles of equal response areas at the middle and lower frequencies. A typical frequency-directivity response pattern is shown in Fig. 4-13.

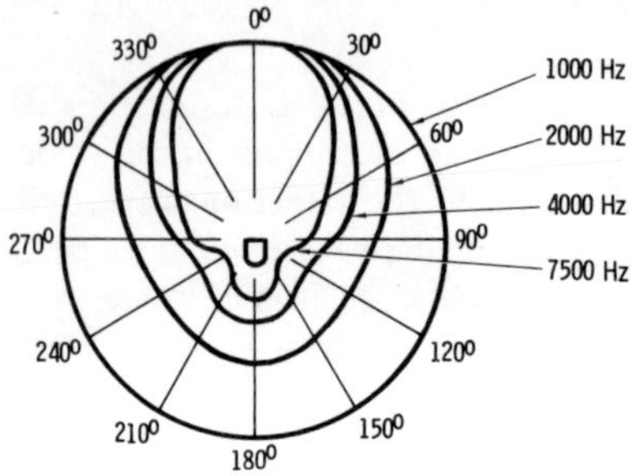

Fig. 4-13. Frequency dependence of directivity.

The Unidirectional Microphone

The feature of directivity is a useful characteristic for discriminating between wanted and unwanted sound. The applications are endless. Public-address engineers must guard against feedback. The communications engineer is interested, generally, only in a voice immediately at one side of the instrument; the same is true for the amateur operator. Recording and broadcasting technicians are concerned with the same general principles of wanted and unwanted sounds, and all shades in between for proper balance of various sources.

The first approach to achieving a truly directional instrument was the combination pressure and pressure-gradient principle. This effectively nulls one loop of a bidirectional pattern, making the microphone sensitive only to sounds from one side. Since the ribbon element is the most practical for a pressure-gradient microphone, and the moving-coil or dynamic principle has proven highly effective (due to light weight and freedom from temperature effects) for the pressure microphone, these elements were used in the unidirectional instrument.

A simple connection of the two elements in series is only the beginning, however, of obtaining a practical unidirectional response. This is because the magnitude of response for a given sound pressure and the phase of the output voltage differ over the usable frequency range, particularly at low and high frequencies. Thus, certain deviations in design and construction from the conventional ribbon and dynamic units are apparent in the microphone.

First, the diaphragm of the dynamic unit must be as close as possible to the ribbon in order to minimize phase difference. At the same time, there must be no disturbance of normal operation due to the proximity of the units. Thus, the housing of the dynamic unit must be streamlined to have minimum effect on the close ribbon element; and the ribbon assembly must

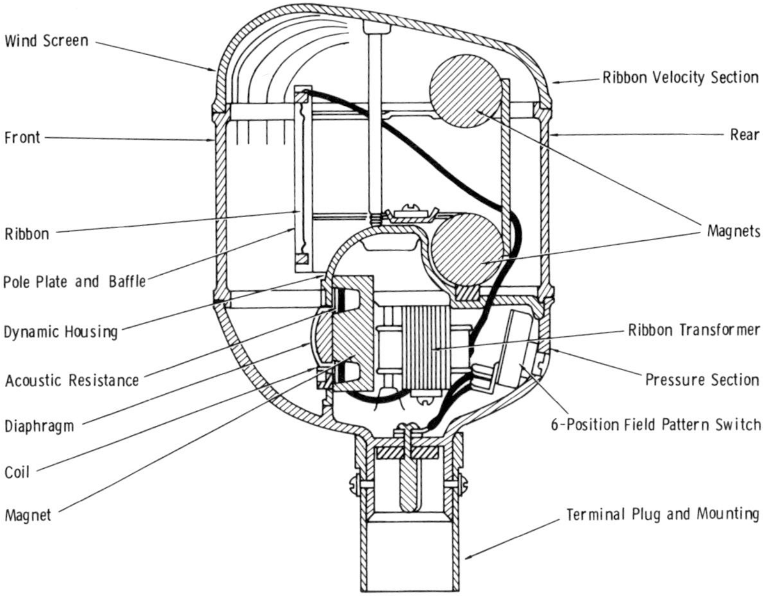

Fig. 4-14. Construction of cardioid microphone.

be designed with a wider-than-normal air gap requiring heavier permanent magnets.

Second, since the two elements have essentially the same magnitude and phase response at middle frequencies but may deviate considerably at low and high frequencies, an electrical equalization or corrective network must be used in the outputs of the two elements before their combination at the output transformer.

The Western Electric 639B cardioid microphone (Fig. 4-14) was a typical combination ribbon and dynamic instrument. The ribbon element in this microphone was redesigned to meet the above requirements and also to reduce flutter caused by wind when used outdoors. This ribbon is about half the length of a normal ribbon, somewhat narrower, but greater in thickness. This thickness results in a stiffness about ten times that normally employed in a velocity microphone with greater stability and consequent reduction of noise caused by wind.

The lower part of the microphone case contains the dynamic unit and housing, in which is located the ribbon transformer, electrical equalizer, and variable pattern switch. This switch permits use of the microphone either as a dynamic unit alone, ribbon unit alone, or both in series to achieve the unidirectional response pattern. With the switch in the D (dynamic) position, the response is nondirectional. In the R (ribbon) position, the ribbon is used alone, and the usual bidirectional pattern is obtained. To obtain the cardioid or unidirectional response, the switch is turned to the C position, which utilizes both elements through the electrical equalization circuit.

There are several more modern means of constructing a microphone with unidirectional response other than the combination ribbon and moving-coil system just described. Their basic action, however, is based on the same general principles.

Fig. 4-15 illustrates a method in which a ribbon element, such as commonly employed in velocity microphones, acts as a pressure-operated device. It is suspended in the usual fashion in the magnetic field of a permanent magnet, but differs from the velocity type in that the face is exposed to the atmosphere whereas the rear side is terminated in an acoustic resistance. Thus the back of the element is enclosed so that it presents an infinite impedance to sound, resulting in a microphone that is pressure-operated with nondirectional characteristics, except at high frequencies.

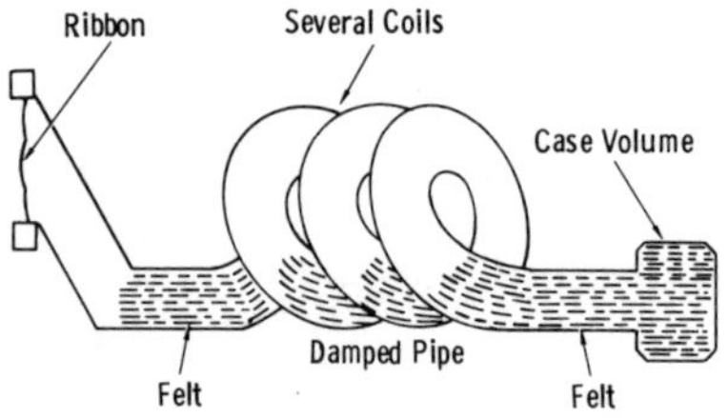

Fig. 4-15. Infinite impedance for producing unidirectional characteristics.

By combining this principle with that of the pressure-gradient ribbon action, a unidirectional response pattern is obtained. Fig. 4-16 illustrates this method, where one continuous ribbon is used, with the upper half acting as a velocity (pressure-gradient) microphone and the lower half acting as a ribbon pressure device terminated in the rear with a folded pipe (the acoustic resistance). The pipe is usually damped with tufts of felt.

Due to the length of pipe behind the pressure section, the velocity of the pressure ribbon leads the pressure in the sound waves at the low audio frequencies. To compensate for this characteristic, a cloth screen is placed in front of the velocity section to introduce a corresponding phase shift in the velocity ribbon. Phase shift at high frequencies is minimized by using the same ribbon element for both units and suitable geometrical configurations of the field structure.

The design of the RCA Type 77-D microphone is somewhat similar to the principle described with an added means of varying the acoustic impedance presented to the ribbon. A switch varies the area of an adjustable opening in the labyrinth connector. When the opening is so large that the back of the ribbon is entirely exposed to the air (as in the ordinary velocity microphone), the acoustic impedance is zero, and the response pattern is bidirectional. When this aperture is completely closed, the acoustic im-

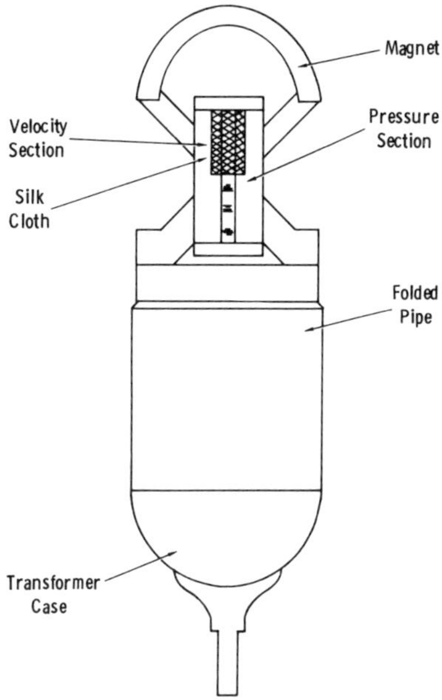

Fig. 4-16. Unidirectional microphone.

pedance is infinite, and the microphone becomes nondirectional in the manner of the usual pressure-operated device. By varying the size of the opening, a great variety of response patterns can be obtained. Fig. 4-17 shows typical patterns ranging from bidirectional to nondirectional.

A more recent RCA development is the BK-5A Uniaxial microphone, illustrated in Fig. 4-18. The term Uniaxial is derived from the fact that the

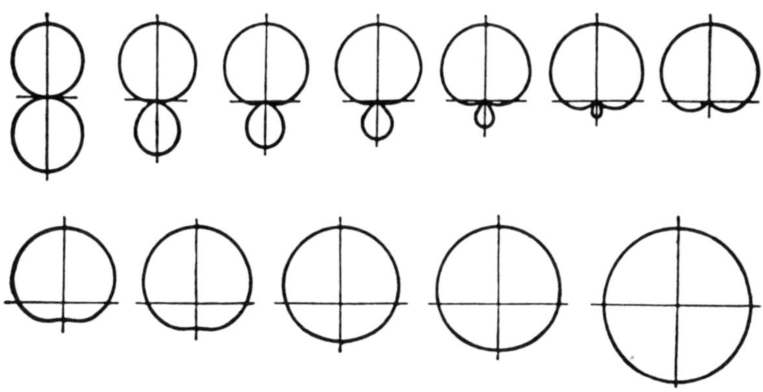

Fig. 4-17. Directional patterns produced by altering area of damping-tube opening.

direction of maximum sensitivity coincides with the major axis of the unit. The directional properties in the midfrequency region are essentially cardioid with an 18-dB front-to-back ratio (Fig. 4-19). The high-frequency directional properties have been improved over the Type 77-D, both in pickup angle and in front-to-back ratio. The pattern of the BK-5A is not adjustable.

As described, the Type 77-D employs one (adjustable) port at the back of the ribbon, which in combination with the acoustic labyrinth controls the directional properties. In the Type BK-5A there are two ports, one placed at each end of the ribbon. The ports are covered with an acoustically controlled cloth to form the proper opening impedance. The movable element is a ribbon with a horn and screen assembly in front and the acoustic connector and labyrinth behind. The port impedance and the acoustic resistance of the labyrinth along with the physical separation of the front and rear pickup points give this unit its unidirectional characteristic. The acoustic labyrinth is a pipe about 31 inches long folded into the form of a cylinder. It is damped along its length to eliminate resonances and acts as a pure acoustic resistance over a substantial part of the frequency range.

The electrical circuit consists of the corrugated aluminum-foil ribbon, a line-matching transformer (brought to a terminal board for adjustment to 30-, 150-, or 250-ohm circuits), and a response-compensation reactor. A switch incorporated with this reactor allows selection of three response characteristics, music (M), voice 1 (V1), and voice 2 (V2) (Fig. 4-20).

Courtesy RCA

Fig. 4-18. RCA BK-5A microphone.

4-2. THE TURNTABLE AND PICKUP

The broadcast turntable (Fig. 4-21) is designed to handle all commercial disc recordings. A preamplifier for the pickup cartridge is usually mounted within the turntable housing. Although designs vary, the following list outlines a typical turntable facility:

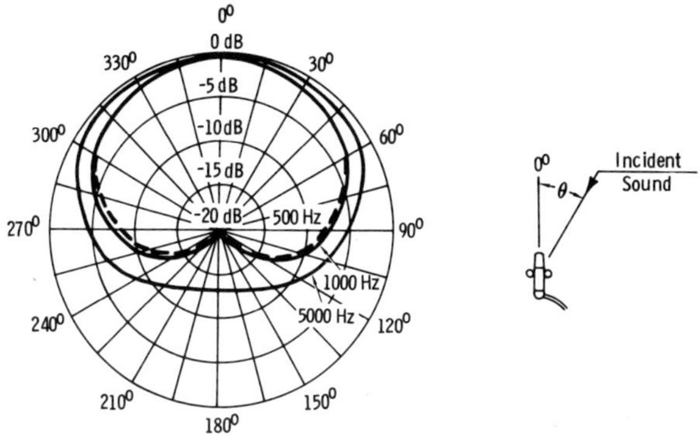

(*A*) *About vertical axis.*

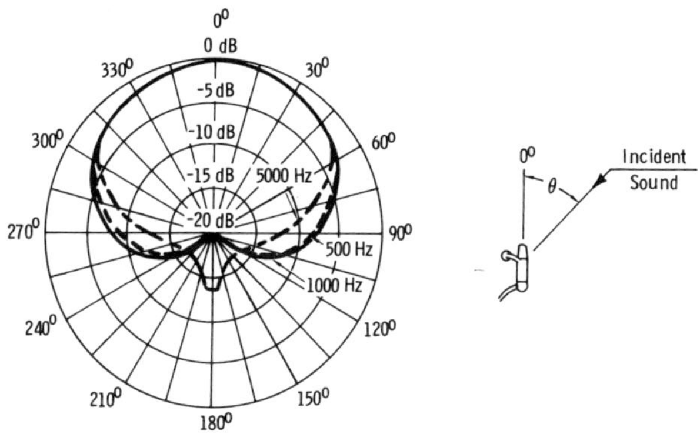

(*B*) *About horizontal axis.*

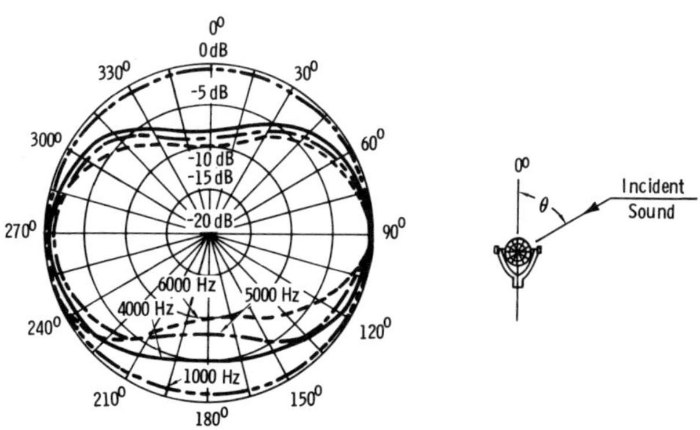

(*C*) *About longitudinal axis.*

Courtesy RCA

Fig. 4-19. Directional characteristics of BK-5A microphone.

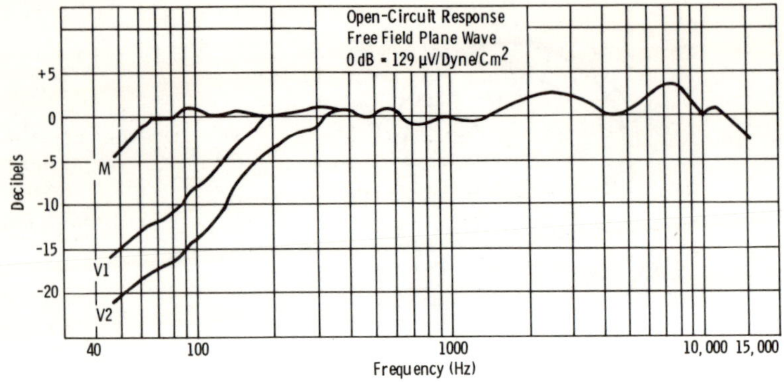

Courtesy RCA

Fig. 4-20. Effect on frequency response of BK-5A music/voice switch.

1. Drive Mechanism: Usually a capstan-to-idler rim drive or a belt drive, powered by a 1/100-horsepower 1800-rpm synchronous motor (or a 4-pole motor) with a three-step pulley (for three speeds) coupled to the motor shaft.
2. Controls
   A. Speed Selector Switch: A three-position control linked to a cam which allows the three rubber idlers to engage, one at a time, between the motor pulley and the turntable rim. The speed of the turntable is determined by the ratio of diameters between the motor pulley and the turntable rim.

Courtesy Broadcast Electronics, Inc.

Fig. 4-21. Broadcast turntable.

B. Off-On Switch: Usually operates a mercury switch (for quiet operation) which energizes the motor and simultaneously engages or disengages the rubber idlers.

C. Equalizer: A circuit which is usually incorporated in the preamplifier with a switch shaft extending through the turntable mounting board for convenience. Terminology varies considerably, but the following results are achieved:

(1) Normal: Provides standard NAB/RIAA equalization. (Recordings are pre-emphasized with complementary de-emphasis on playback to achieve a higher over-all signal/noise ratio.)

(2) Plus Highs: Results in greater high-frequency response (less de-emphasis) for use where some of the high-frequency response has been lost due to wear, but signal/noise ratio is still good.

(3) Minus Highs: Less high-frequency response for elimination of higher-frequency noise where required.

The turntable preamplifier and equalizer are more fully described in Chapter 6.

The most important characteristics of the turntable and pickup along with the associated preamplifier are wow, flutter, hum, rumble, and frequency response. The use of test records and general maintenance procedures are included in Chapter 13.

How the Pickup Cartridge Works

One of the most popular types of pickups used in broadcasting today is the *variable reluctance* pickup. When the signal recorded in the lateral variations of the record grooves vibrates the stylus from side to side, currents are set up in the magnetic poles. This generates an electrical signal which corresponds to the signal which originally caused the lateral groove variations of the recording. The stylus is mounted so that it is insensitive to vertical variations, isolated from external mechanical vibration, and subject to a minimum of surface noise.

Either a 2.5- or 3-mil stylus (1 mil = 1/1000 inch) is used for 78-rpm discs; a 1-mil stylus is required for microgroove (33 1/3 rpm) recordings. Some units employ plug-in cartridges that can be easily removed and replaced.

The Stereo Disc and Pickup

Stereo is covered later in this book. Suffice it to say here that two separate signals and amplifying channels are required, one each for the left and right listening areas.

The modern stereo system, including stereo disc recording, is compatible. This means that the stereo recording can be played back in a conventional monaural manner as well as providing two channels for stereo systems. The two channels are provided in a single groove on the stereophonic disc recording. The basic concept of stereo disc recording is illustrated in highly simplified form in Fig. 4-22.

The recording process is represented in Fig. 4-22A. The left and right pickup area signals are fed to their respective cutting-stylus drive coils. These coils are arranged to provide modulation axes in quadrature to each other and at 45° angles relative to the record surface. Fig. 4-22C shows the groove resulting from a signal applied only to the right (1) or left (2) coil. Note in each case that a vertical component results which identifies the difference between channels. If identical signals are applied to both channels in phase (3), a lateral groove with no difference component results. If identical signals are applied to both channels with a 180° phase difference, only vertical (difference) modulation results (4 in Fig. 4-22C).

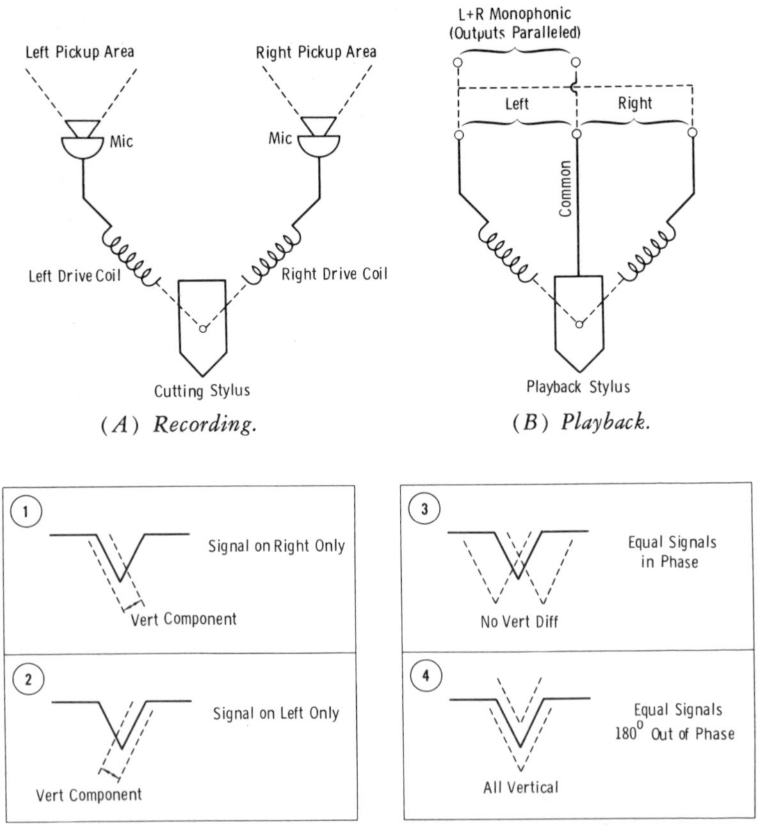

(*A*) *Recording.* (*B*) *Playback.*

(*C*) *Grooves.*

Fig. 4-22. Principles of stereo disc recording.

In essence, this is a lateral recording of the sum of the left and right signals, and a vertical recording of the difference between the channels. The stereo playback cartridge (Fig. 4-23A) has a single stylus and three terminals, left, right, and common. For a stereo system, the left and common terminals feed the equalizer for the left channel, and the right and common terminals feed the equalizer for the right channel. The left and right terminals can be paralleled to feed a single monophonic channel. Also, the lateral grooves make the disc compatible for a conventional monophonic pickup cartridge, since this represents the sum (L + R) signal, which is essentially what any monophonic system reproduces. *A word of*

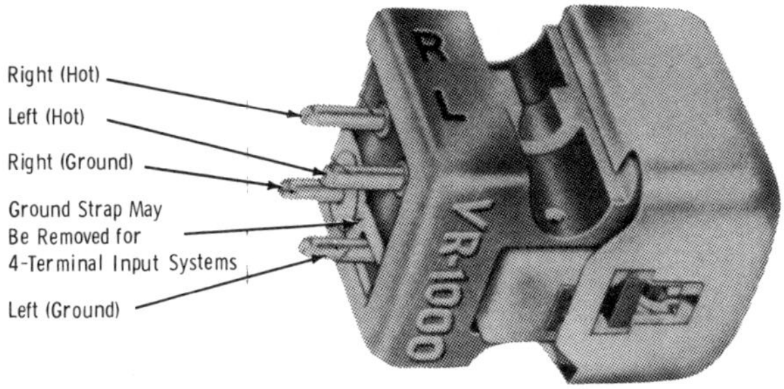

(A) GE VR-1000 cartridge.

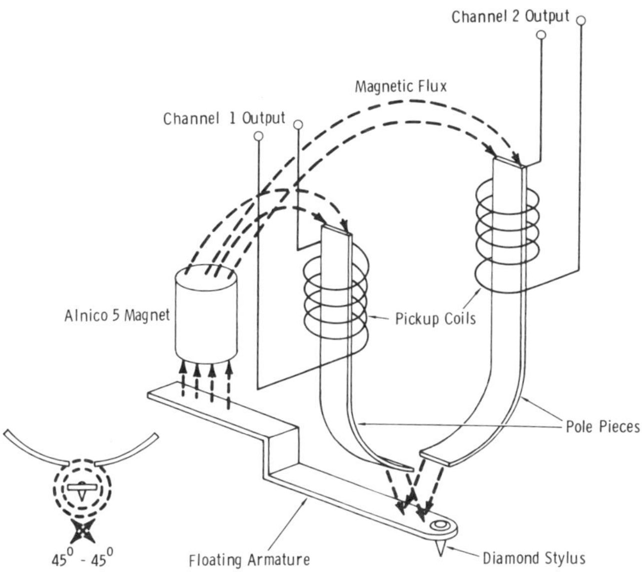

(B) Operating principles.

Courtesy General Electric Co.

Fig. 4-23. Stereo pickup cartridge.

caution: Just any monophonic pickup is not suitable for playing a stereo recording on a monophonic system. The cartridge must have good vertical compliance to avoid damage to the disc and to track the vertical modulation without causing distortion in the signal.

Fig. 4-23A illustrates the GE stereo variable-reluctance cartridge, type VR-1000. The principle of operation is shown in Fig. 4-23B. A linear relationship between stylus motion and output voltage is achieved by placing the magnet over the end of the floating armature, thereby magnetically saturating the special iron. The armature design causes a circular magnetic field to envelop the stylus gem. These circular lines of force surrounding the stylus move with the stylus as it traces the recorded groove. A lateral motion of the stylus varies the *proportion* of flux between the two pole pieces. A vertical motion of the stylus varies the *total* flux in the magnetic circuit. Thus the motion of the stylus by the recorded grooves is translated into left-channel and right-channel outputs in accordance to the original recording.

The stereo pickup stylus has a diameter of 0.7 or 0.5 mil, contrasted to the 1-mil stylus used for monophonic recordings. Some systems employ readily replaceable plug-in cartridges for stereo, microgroove, and 78-rpm discs.

The Quadraphonic (4-Channel) Disc

Quadraphonic (4-channel) systems are fully discussed in Chapter 7. At this time, we are concerned only with the nature of the disc and the stylus action involved.

At the time of this writing, the entire status of quad sound is in a state of flux. There are two basic systems involved, the matrixing technique and the discrete system. The matrixing system has various names: the SQ system (CBS), the QS system (Sansui), and the Stereo-4 system (Electro-Voice).

The matrixing system essentially mixes four separate channels into a composite that is recorded on the disc. It is encoded into a stereo fm transmitter. A decoder at the receiver separates the four channels to be reproduced by four speakers. The same program can be reproduced on standard mono or stereo receivers, just as color television signals can be received in black and white on monochrome receivers.

In the quad technique, the four reproducers are basically located left-front and right-front, left-rear and right-rear. Briefly, the matrixed quad disc has left-front and right-front channels recorded in the two side walls of the grooves along the 45° lines as in normal stereo records. In addition, a *helical* groove cut is imposed by circular motion of the cutting stylus, clockwise for the left-rear modulation and counterclockwise for the right-rear modulation. In playback, the stylus does not actually rotate clockwise or counterclockwise; the single stylus motion is a complex function that reproduces the four individual signals that can be decoded and separated into the four separate speaker feeds. (Encoding and decoding are covered in Chapter 7.)

The discrete disc employs a 30- to 40-kHz subcarrier that tells the playback cartridge which speaker the sound should go to at any instant. Again, the two groove walls at 45° angles to the vertical (90° to each other) carry the conventional stereo signal information. In addition, a *difference* signal is added to the *sum* signal in each groove wall. These difference signals are frequency and phase modulated on the subcarrier. The demodulator at the receiver sorts the sum and difference signals into the proper four amplifier channels.

4-3. THE MAGNETIC TAPE HEAD

The complete magnetic tape system is covered in Chapter 5. We are concerned at this time with the head and its associated components.

A magnetic field is created when an electric current passes through a conductor. When a magnetizable substance is brought within this field, the small atomic regions or *domains* of the material become oriented so that their axes are aligned with the field of force about the conductor, and the material exhibits a magnetic field of its own. When the current is stopped or the magnetized substance is withdrawn from the field, the material retains a magnetic force to an extent depending on the *remanence* of the material.

In magnetic tape recording, actual steel is no longer used as the magnetic substance. Instead, a magnetic coating is placed on a plastic base that is one-quarter inch wide (for most audio applications) and about two thousandths of an inch thick. The actual coating is about half a thousandth of an inch deep.

Fig. 4-24A illustrates the basic action of a magnetic tape system. As the signal is fed into the recorder head, the tape passes at a constant speed across the very small gap of the recorder head core. This gap is usually 0.001 to 0.002 inch, but may be as small as 0.00005 inch (0.05 mil). Thus at any given instant, a tiny magnet of this dimension is formed, and the recorded signal consists of a string of these minute longitudinal magnets whose flux and polarity correspond to the signal current in the head at the instant each magnet was formed.

The magnetic properties of a signal recorded on magnetic tape may be illustrated as shown in Fig. 4-24B. The sound track here has been made visible by means of a method that is similar to the mapping of fields about a bar magnet by using iron filings. In this instance, the recorded tape was passed through a suspension of carbonyl iron in heptane. Carbonyl iron consists of particles much smaller than ordinary iron filings. It may be observed by this illustration that the tape actually consists of a string of magnetic fields which correspond to the amplitude variations of the original signal that was recorded.

Basic Principles

Fig. 4-25 shows the usual mechanical arrangement of a tape recorder. This is known as the *capstan drive system* and is used in all tape recorders where reliable and flutter-free operation is mandatory. The capstan is driven by the drive motor and is weighted with a flywheel to assure speed regulation. The tape is held firmly against the capstan by the pressure roller and tension. The supply reel containing the tape is at the left; it is actuated by the rewind motor to rewind the tape after recording. The tape is threaded across the erase, record, and playback heads, around the capstan, and onto the take-up reel. This reel is driven by the forward, or take-up, motor. The speed of rotation may be adjusted either by variable speed motors or by varying the size of the capstan. The tape transport is more fully described in Chapter 5.

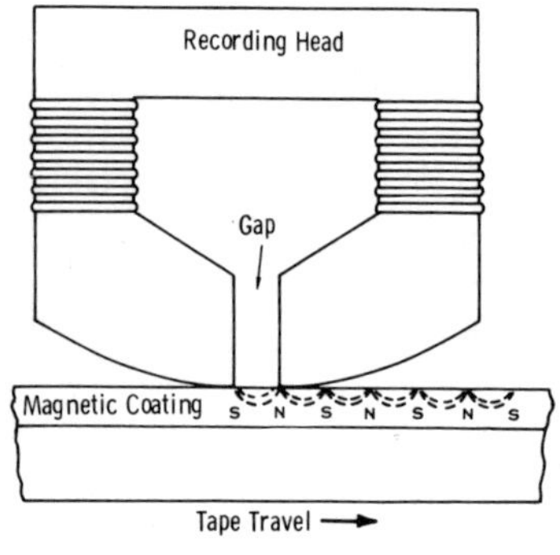

(*A*) *Magnetizing action of recording head.*

Courtesy Ampex Corp.

(*B*) *Word "tape" recorded at 30 in/s, full track.*

Fig. 4-24. Magnetic recording principles.

The wavelength of the recorded signal for a given frequency depends on tape speed. For a 15,000-Hz tone on a tape moving at 15 inches per second, one full wavelength is 0.001 inch. On playback, if the head gap is equal to the recorded wavelength, there is complete cancellation of the magnetic field. The gap should be no more than one-half the wavelength of the highest frequency to be recorded, or in this case 0.0005 inch. For a gap of 0.001 inch, it would be necessary to run the tape at 30 inches per second to cover a frequency range up to 15,000 Hz (so that the recorded wavelength at this frequency would be 0.002 inch).

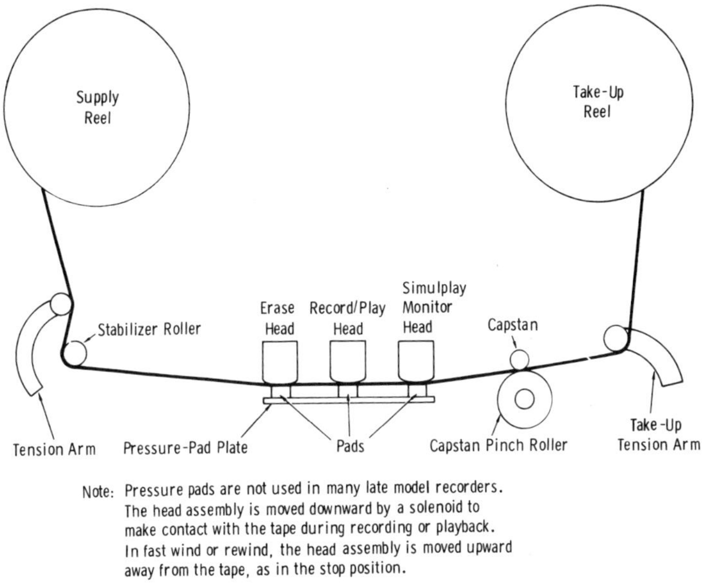

Fig. 4-25. Basic parts of reel-to-reel magnetic tape recorder.

The Necessity for Bias

Remember from magnetic theory that the extent of magnetization depends on the strength of the magnetic field. Thus, the amount of magnetism retained depends on the magnitude of the original magnetic field for a substance of given remanence (ability to retain magnetic force).

Fig. 4-26 shows a graph of the amount of remanence versus the strength of the signal field. This curve may be compared in effect to the plate current of a vacuum tube as plotted against grid voltage. It is obvious that the straight portions of the curve which would result in an undistorted signal output are very small. Fig. 4-27 shows the type of magnetic characteristic curve necessary for undistorted output.

Since there is no practical straight portion of a magnetization curve such as the one in Fig. 4-26, dc bias is out of the question when using a tape of zero magnetization as the operating axis. To overcome this difficulty, a high-frequency bias is used. Since this bias current is far above the highest audible frequency to be used, it is called *supersonic bias.* This voltage is *not* modulated by the signal current, but rather they are mixed together to form the pattern shown in the illustration.

The audio component on the plus side of the zero magnetization axis results in the distorted output waveform shown in the upper portion of Fig. 4-26, where the a loop is larger than the b loop. The audio component on the minus side results in the lower distorted waveform, where the c loop

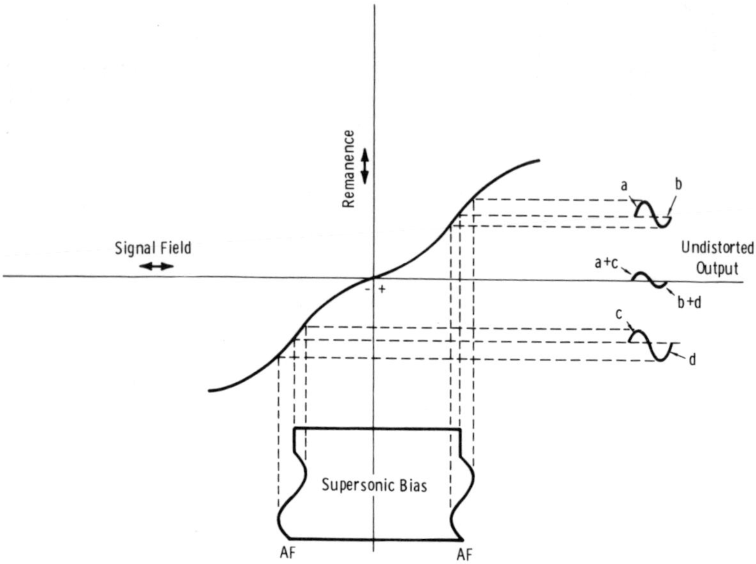

Fig. 4-26. Remanence versus signal strength for magnetic tape coatings.

is smaller than the d loop. Since these complete loops are combined on the axis of operation, the resulting output signal is undistorted as shown. That is, the algebraic sum of a plus c equals the algebraic sum of b plus d when the bias is of the proper magnitude for a particular recording head and magnetic tape.

The exact frequency of the bias does not seem to be critical, as long as it is high enough not to beat with the harmonics of the audio signal or subcarriers (such as fm stereo). In practice, it is customary to make the

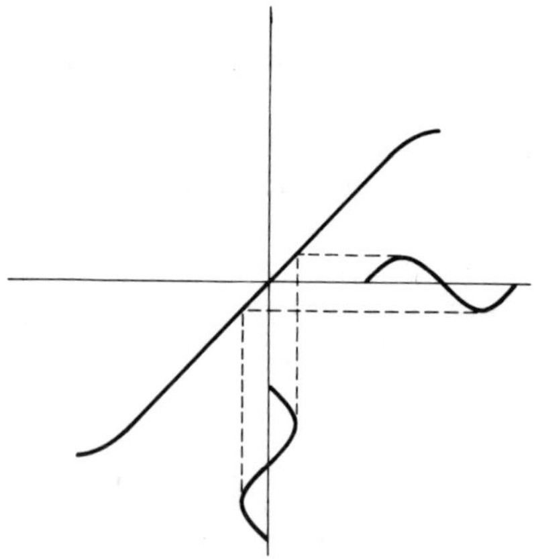

Fig. 4-27. Linear (ideal) remanence curve.

bias frequency five times the highest audio frequency to be reproduced plus 5 kHz. Thus, a tape recorder with a frequency range up to 5000 Hz would use a bias frequency of five times 5, or 25 kHz, plus 5, equaling 30 kHz. A recorder system utilizing the full range up to 15,000 Hz would, therefore, require a bias frequency of at least 80 kHz. Commercial tape recorders utilize bias frequencies ranging from 30 to 100 kHz. Higher bias frequencies seem to give better signal-to-noise ratio on the average than lower bias frequencies, and most manufacturers are now using the high frequencies even for home-type recorders.

Effect of Bias Magnitude

Bias magnitude, or amount of bias used (measured by bias-current readings or ampere-turns of bias), is an important characteristic of magnetic recording. Too little bias results in high noise and distortion levels. Too much bias results in loss of high frequencies. In practice, however, bias is adjusted for minimum noise and distortion. Any resultant loss of highs is compensated for by increasing the tape speed or by equalization procedures.

The 3M Company, one of the leading researchers in magnetic-tape characteristics, has conducted exhaustive tests on the effects of bias with its particular sound tapes. As a basis from which to start, tests were run first on the effect of bias current on the output that could be obtained with no more than 1 percent third-harmonic distortion. Measurement of the third-harmonic distortion was chosen in preference to total distortion, since the total measurement would be affected by noise level. Measuring the third-harmonic distortion theoretically relates the measurement only to the magnetic properties of the medium. Even-harmonic distortion has been found to be negligible when using magnetically neutral tape (ac neutralized tape instead of dc-biased or saturated tape) with a supersonic bias of good waveform.

The curves of Fig. 4-28 show the results of the manufacturer's tests on the effect of bias. The actual number of bias ampere-turns is dependent on the particular recording head used, but the effects of changes of bias are substantially the same with any head. (Ampere-turns are the product of the current through the coil in amperes times the number of turns.) The optimum bias in this test for No. 111 sound tape results from 2.4 ampere turns.

The effects of bias on frequency response are shown in Fig. 4-29. It is seen that raising the bias results in severe loss of high-frequency response (probably due to a partial erasing action of the stray bias field). The recording, of course, is unequalized.

It may also be observed from the above studies that using the optimum value of bias to obtain maximum output at no more than 1 percent third-harmonic distortion results also in the best high-frequency response relative to low frequencies. It is also known, however, that the noise level may

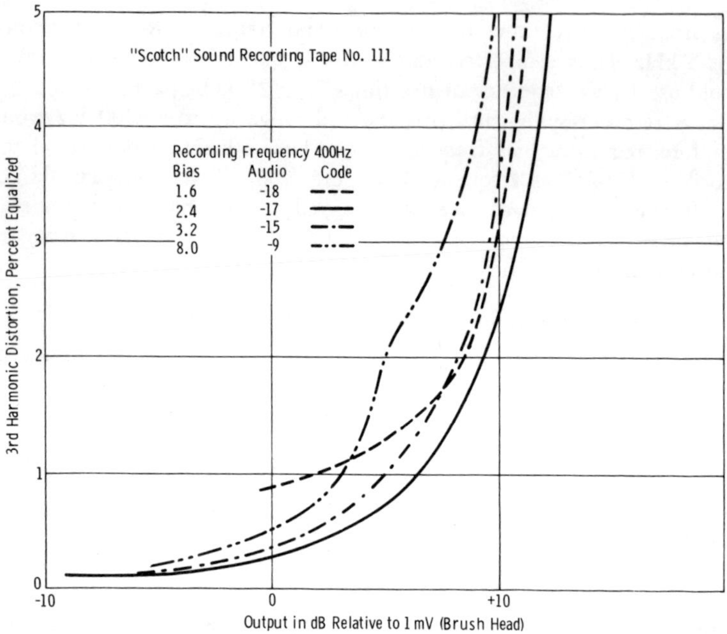

Courtesy 3M Co.

Fig. 4-28. Effect of bias on distortion.

be decreased by using higher values of bias, resulting in a somewhat better signal-to-noise ratio. The final design must be a compromise for the particular purpose of the recorder, and, as mentioned before, high frequencies may be brought up if necessary by using appropriate equalization, by using very small air gaps in the recording head, or by increasing the speed of the tape movement.

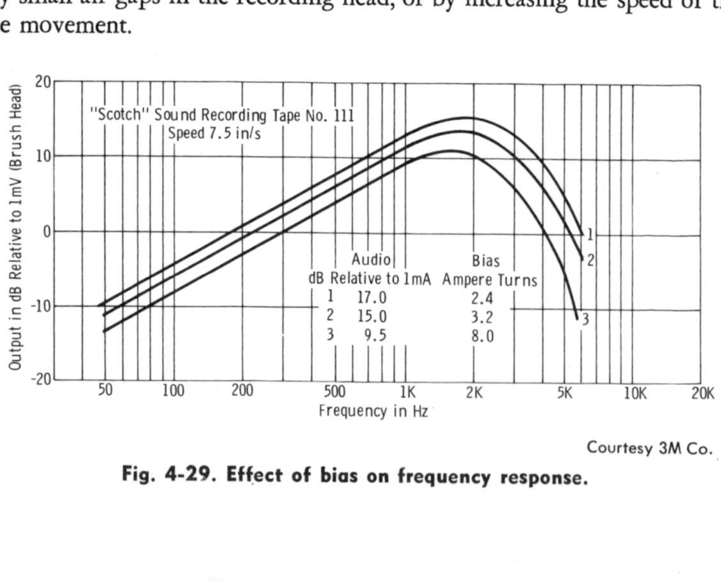

Courtesy 3M Co.

Fig. 4-29. Effect of bias on frequency response.

Effect of Tape Speed

The effect of speed on the recorded wavelength and consequent high-frequency response has been mentioned. The influence of recording-head gap size was also shown, and these factors should be kept clearly in mind for a comprehensive picture of magnetic recording.

Fig. 4-30 shows the effect of speed when the optimum value of bias is used. It may be observed that the highest speed of 18 inches per second results in best high-frequency response. The "high speed" has now been standardized at 15 inches per second.

Frequency response is dependent on several factors other than those mentioned. These may be enumerated briefly as follows:

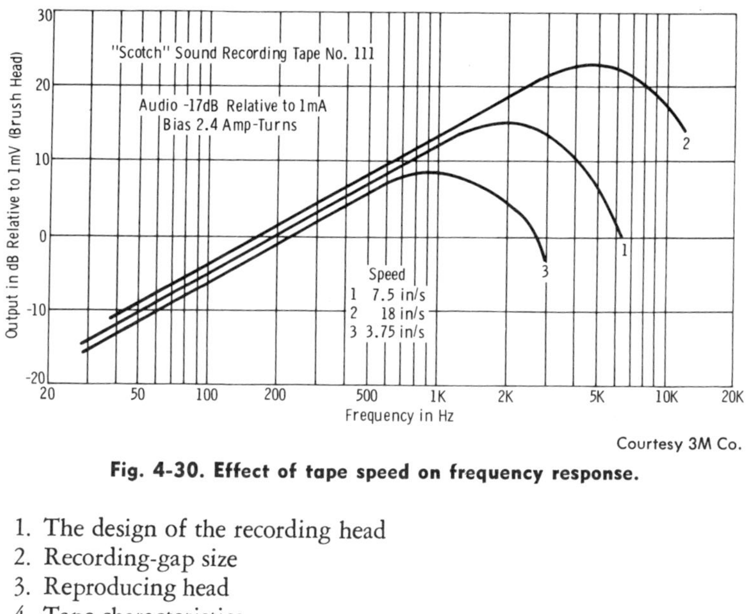

Courtesy 3M Co.

Fig. 4-30. Effect of tape speed on frequency response.

1. The design of the recording head
2. Recording-gap size
3. Reproducing head
4. Tape characteristics
5. Azimuthal orientation of heads
6. Characteristics of associated amplifiers
7. Tape contact against pole pieces

Importance of Tape Contact

In magnetic tape recorders, it is very important that the tape be held firmly against the smooth surface of the pole pieces and across the gap. The effect of poor contact between the tape and the playback pole pieces becomes increasingly noticeable at higher frequencies. Fig. 4-31A shows the attenuation caused by various separations of tape and playback head. In these tests, a tape was recorded with various pure tones and then reproduced on a good system with proper contact. Then this tape was reproduced several times with paper shims of different thicknesses used to separate the tape and pole pieces by known amounts. The level was then recorded for each frequency and separation and compared with the original levels. The separation values are given in mils. Data on the frequency scale are plotted for a tape speed of 7.5 in/s, but it will be remembered from the previous discussion that the wavelength effect is the important characteristic. In other words, for a given separation, the same attenuation will result at 5000 Hz (7.5 in/s), at 10,000 Hz (15 in/s), at 2500 Hz (3.75 in/s), etc. For this reason, the wavelength scale is shown at the top of the graph and is more significant, being true for any tape speed.

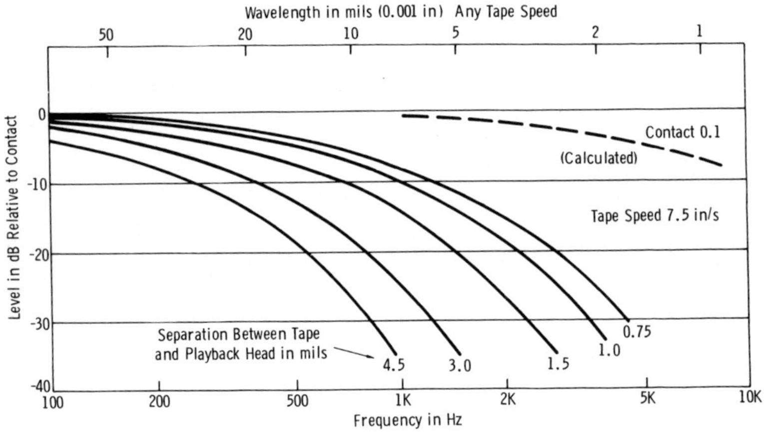

(A) Function of frequency for various separations.

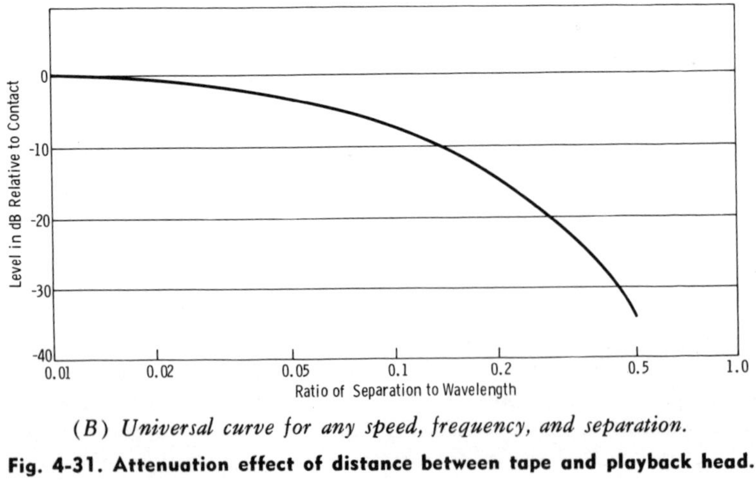

(B) Universal curve for any speed, frequency, and separation.

Fig. 4-31. Attenuation effect of distance between tape and playback head.

Fundamentally, attenuation depends on the ratio of the separation to the wavelength. The same attenuation results from a 10-mil wavelength and 1-mil separation as for a 45-mil wavelength and 4.5-mil separation. Thus, Fig. 4-31B shows the amount of attenuation against the ratio of separation to wavelength.

It is evident from these results that very small separations can cause serious trouble in attaining comparatively good high-frequency response. The pole pieces must be clean, smooth, and well machined. The tape must be smooth and flexible, with adequate tension or pressure. Plastic-base tape is superior to a similar paper-base tape; in addition, the noise level is somewhat lower. Later tapes use a polyester or acetate backing.

Table 4-1 lists typical magnetic tape-head specifications. Note from the top two listings (full-track standard record/play heads) that a broad variation in efficiency can exist between different types or makes of heads.

Table 4-1. Typical Magnetic Tape-Head Specifications

Head Description and Function	Gap Spacer (mils)	Record Current (rms μA)	Output (rms mV)
Full Track Track Width 0.240 in Record/Play	0.2	650	0.5
Full Track Track Width 0.240 in Record/Play	0.2	150	2.4
2-Track Mono Track Width 0.08 in Track 0.04 in Off Center Record/Play	0.1	80	1.2
2-Track Stereo Track Width 0.08 in Center-Center Track Spacing 0.16 in Record/Play	0.1	60	1.8
2-Track Stereo Track Width 0.08 in Center-Center Track Spacing 0.16 in Play Only	0.05		2.2
Quadrasonic Sound 4-Track, 4-Channel Track Width 0.037 in Center-Center Track Spacing 0.071 in Record/Play	0.1	34	1.4

NOTE: Above characteristics are an average employing 3M tape and 1-kHz tone recorded 12 dB below tape saturation.

EXERCISES

Q4-1. Can a nondirectional microphone be made directional by "aiming" it toward the desired source?

Q4-2. What is the "normal" response (pattern) of an unmodified pressure-gradient microphone?

Q4-3. How many terminals are there on the normal stereo phono pickup, and how are they used?

Q4-4. Can a stereo phono pickup be used in regular mono broadcasting?

Q4-5. Does a phono pickup for quadrasonic sound employ a single or multiple stylus?

Q4-6. How wide is the normal audio tape?

Q4-7. What is the normal track width recorded on a magnetic tape employing full-track recording?

Q4-8. What is the normal track width recorded on each track of a 2-track stereo tape, and what is the center-to-center track spacing?

Q4-9. What is the track width on a 4-track, 4-channel quadrasonic tape head, and what is the center-to-center track spacing?

Q4-10. Why is ultrasonic bias used in a magnetic tape head?

CHAPTER 5

The Magnetic Tape System

Tape cartridges and reel-to-reel magnetic tape contribute a considerable share of modern broadcast schedules. The nature of the medium lends itself readily to automation in varying degrees. Fig. 5-1 illustrates program automation facilities including four reel-to-reel tape machines. Broadcast automation systems are covered in Chapter 7. We will consider only the individual reel-to-reel and cartridge tape systems in this chapter.

5-1. REEL-TO-REEL MAGNETIC TAPE SYSTEMS

The magnetic tape head was basically covered in the preceding chapter. The basic parts of the transport system were shown in Fig. 4-25 in that chapter.

The Tape Transport

Referring again to Fig. 4-25, note that the tape leaves the supply reel to pass over a tension arm to wrap around a stabilizing roller. This roller sometimes is attached to a flywheel to eliminate any tendency to flutter. The tape then passes in contact with the erase, record/play, and simulplay (monitor) heads to the capstan and capstan pinch roller and then across the take-up tension arm to the take-up reel. If tape breakage should occur, either tension arm thus released stops the transport.

Some machines use individual and separate record and reproduce heads, with the reproduce head serving for monitoring during recording. Other (2-channel) equipment employs four heads: two-track erase, two-track record, quarter-track reproduce, and two-track reproduce. The two-track or quarter-track reproduce operation in this case is selected by positioning a switch on the head assembly. Some special-purpose reel-to-reel recorders such as those used for logging at a speed of 1⅞ inches/second do not employ an erase head. The tape must be erased with a bulk eraser external to the machine.

Three motors are normally employed in professional machines, two torque motors for the supply and take-up reels, and a multispeed hysteresis-synchronous motor for driving the capstan. In the record or reproduce mode, the pinch roller moves to engage the tape tightly against the rotat-

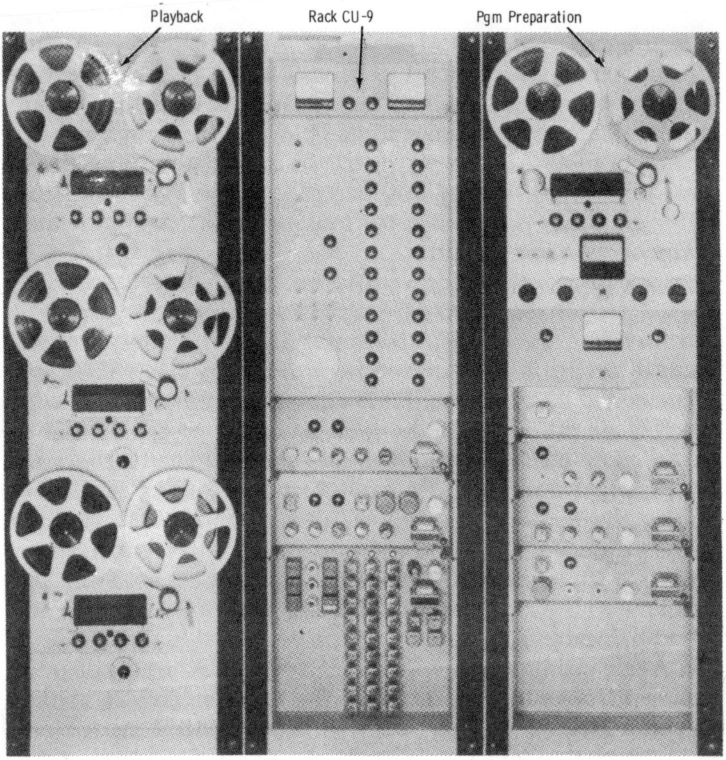

Courtesy Schafer Electronics Corp.

Fig. 5-1. Program automation equipment.

ing capstan shaft. This capstan and pinch roller combination exerts full control over the tape speed. The supply and take-up reel motors act only to maintain proper tape tension. The supply-reel motor actually applies a slight backward pull on the tape as it unwinds. The take-up reel motor applies sufficient forward torque to maintain correct tape tension on that side of the capstan assembly. Electromechanical brakes (usually mounted on the torque motors) stop the supply and take-up reels evenly to prevent "spilling" the tape.

Overall Electronics

Fig. 5-2 is a block diagram of a typical 2-speed magnetic tape system. The most common speeds for normal use are 3¾ and 7½ inches per second. The 7½ and 15 in/s equipment normally is used only where a large amount of production requiring copies (dubs) is scheduled.

Record Path—The program input is fed through a record level control to an amplifier coupled to an emitter follower. The emitter follower is employed wherever high input and low output impedances are desirable. Pre-equalization then occurs in series through selectable equalizers depending on the speed selected. Low input and output impedances are normally used for equalization circuitry, so another emitter-follower stage follows the equalizer. A constant-current amplifier (very high internal resistance) is used to drive the recording head through a bias filter. This filter prevents the bias current from interacting with the constant-current amplifier and vice versa. The bias oscillator also supplies the erase head for recording, and the bias magnitude to the record head is varied by means of an adjustable trimmer capacitor.

Reproduce Path—The reproduce head feeds an amplifier and an emitter follower. Post equalization normally provides parallel circuitry to the reproduce channel. The VU meter can monitor either the reproduce level or the recording level at the point shown by means of the input-reproduce switch. The record calibrate control is normally a factory adjustment which can be checked periodically by the maintenance technician. This control is adjusted so that with normal sine-wave input level, the VU meter on the machine reads 0 VU when a vtvm measurement at the output terminals reads +4 dB. (Review Section 2-10, Chapter 2).

The complementary-symmetry emitter follower used to drive the line output provides a low-impedance, single-ended push-pull output signal current with a high degree of linearity.

NOTE: Fig. 5-2 illustrates a system in which separate record and reproduce heads are employed. On machines where the same head is used for playback and for recording (record/play head), the head is relay-switched between the record output and play (reproduce) input depending on the mode of operation selected.

Equalizers

The need for equalization for a magnetic tape medium should be apparent after a review of Figs. 4-29 and 4-30 (Chapter 4). Note that, for example, if the output level at 200 Hz is 0 dB, the output at 400 Hz will be about +6 dB. This is a rise of 6 dB per octave (each time the frequency is doubled, a change of one octave has occurred). At low frequencies, there is a small *rate of change* across the gap; hence there is very low induced magnetic field energy. At higher frequencies, a greater rate of change occurs across a given gap size; hence there is more induced energy. At still

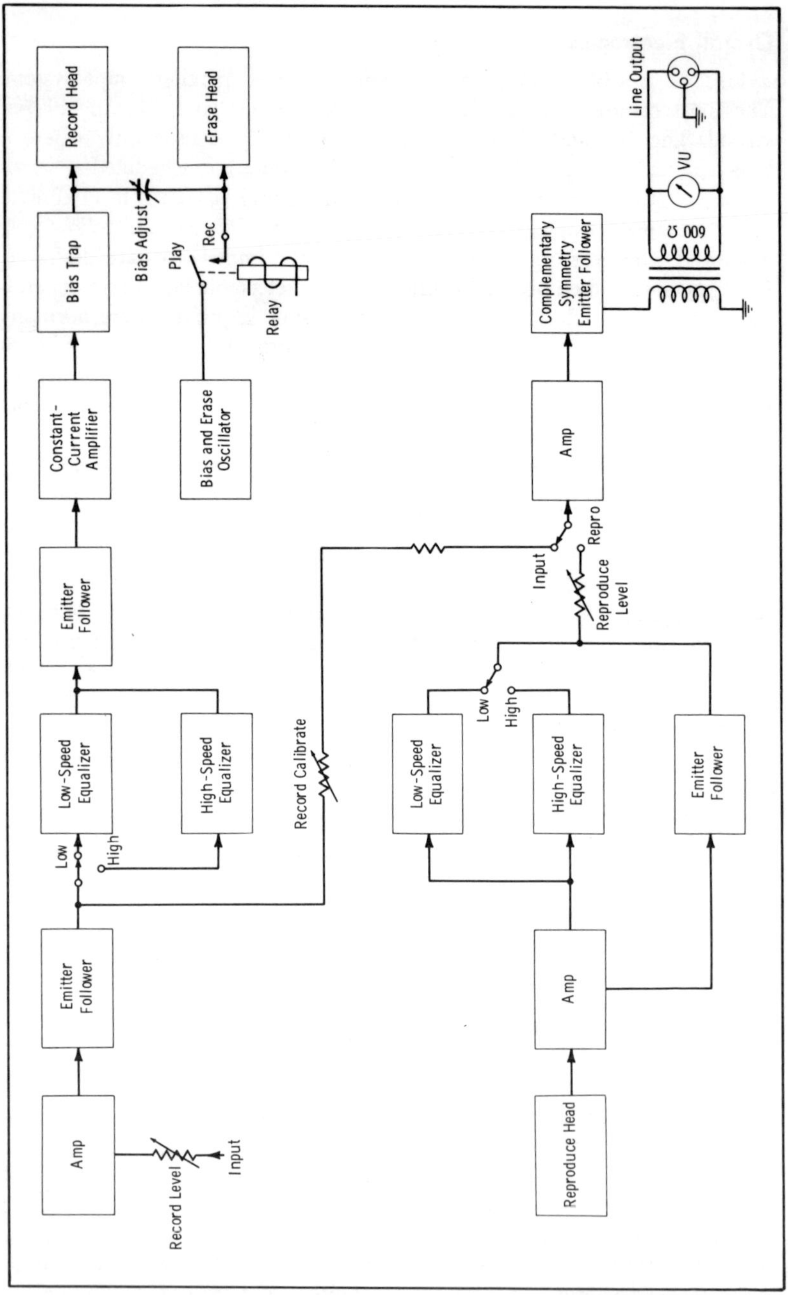

Fig. 5-2. Typical two-speed magnetic tape system.

higher frequencies, several additional factors appear. There is a loss in the tape from tape-coating thickness factor. Also, as the wavelength approaches the gap size, cancellation of energy begins to occur. At the high-frequency limitation, the response falls quite rapidly.

The recorded wavelength is:

$$\text{Recorded wavelength} = \frac{\text{Velocity}}{\text{Frequency}}$$

The recorded wavelength for a 15-kHz tone at a tape speed of 7.5 in/s is:

$$\text{Wavelength} = \frac{7.5}{15{,}000} = 0.0005 \text{ inch} = 0.5 \text{ mil}$$

If the magnetic tape head has a gap size of 0.2 mil, the 0.5 mil recorded wavelength is a little more than twice as long as the gap width. The ratio of wavelength to gap size must be at least two to one as a practical limit so that the output can be adequately pre-emphasized for good reproduction and signal-to-noise ratio. Losses in the magnetic core structure cause the signal to begin to decrease before this point.

To examine the low-frequency problem, look at the response curves of Figs. 4-29 and 4-30 in the reverse direction. If the response at 1000 Hz is +10 dB, the response one octave lower (500 Hz) will be +4 dB. Another octave lower (250 Hz), the response is −2 dB. Still another octave lower (125 Hz), the response is −8 dB. The limit of how far down the response can go and still be usable is set by the residual noise level encountered in the tape, head, and machine electronics of any particular system as a whole.

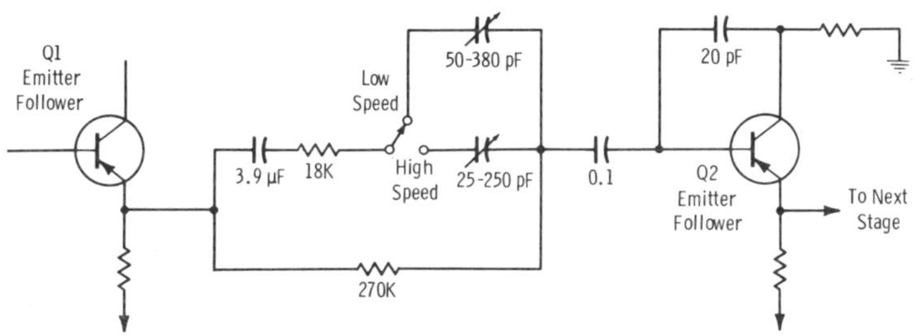

Fig. 5-3. Typical pre-equalization in two-speed tape system.

It is obvious that equalizers must be used to bring both the low- and high-frequency regions of the audio range (30-15,000 Hz) into line with the middle frequencies. In practice, this involves both pre-equalization (recording mode) and post-equalization (reproduce mode), as shown in Fig. 5-2. In addition, for multispeed transports, separate equalizers must be used for each speed employed.

Fig. 5-3 illustrates a typical pre-equalization network for a two-speed magnetic tape system running at either 3¾ or 7½ in/s. The additional fixed equalization provided by the small capacitance between collector and base of Q2 is usually found in most recorders.

Fig. 5-4 shows typical circuitry for post-equalization. Note that the circuitry is identical for both speeds, but that the arrangement permits proper adjustment of equalization for either speed without the need of such critical adjustments each time the speed is changed. Resistors R3 and R4 provide high-frequency equalization, and R1 and R2 provide low-frequency equalization. This is a shunt type of circuit between the emitter of Q3 and the emitter of Q1 (low-impedance sources).

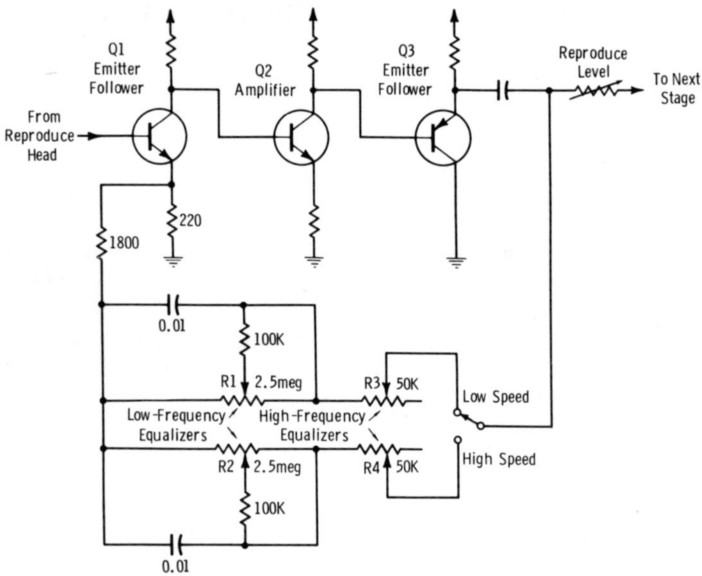

Fig. 5-4. Simplified diagram of typical post-equalization circuitry.

Fig. 5-5 presents the NAB (National Association of Broadcasters) standard pre- and post-equalization curves used by the broadcast industry. Note that pre-equalization provides boost at both the low and high ends of the frequency range and that post-equalization provides a complementary curve. The overall result is a practically flat frequency-response curve with maximum signal-to-noise ratio.

Record-Head Drivers

Fig. 5-6 illustrates a typical constant-current record-head amplifier of the type used in professional recorders. Transistor Q4 acts as a bootstrap on Q3. Over the audio range, the collector of Q3 works into an impedance value which is sufficiently high compared to the record head to provide constant-current characteristics.

The audio signal is fed to the record head through a bias trap. At the head, it is mixed with the bias frequency (normally 60 to 100 kHz). The bias trap is simply a parallel tuned circuit which is resonant at the bias frequency. The values shown in Fig. 5-6 are for a bias frequency of 100 kHz.

The Erase and Bias Oscillator

A typical erase and bias oscillator is shown in Fig. 5-7. It is a conventional flip-flop working into a tuned circuit. Capacitor C1 across adjustable transformer T1 sets the frequency of oscillation. The activating dc voltage is applied only when the system is placed in the record mode. The bias symmetry control adjusts the current division between Q1 and Q2 so that a symmetrical waveform can be obtained.

The Erase and Bias Output Network

See Fig. 5-8. The network providing the feed to the erase and record heads must be well shielded, as are the heads themselves. Capacitor C1 allows adjustment of the erase current to the erase head, which may require anywhere from about 125 to 325 milliamperes, depending on the construction. Similarly, R1 provides adjustment of the magnitude of the supersonic bias current to the record head. Construction of these heads varies widely, and the amount of bias current required is more accurately given in ampere-turns. The bias current varies in practice between different types of heads and may cover a range of 4 to 70 milliamperes.

NOTE: A special form of magnetic tape recording for adding reverberation (electronically) to live studio productions is described in Chapter 11.

5-2. THE SELF-CONTAINED TAPE CARTRIDGE SYSTEM

The tape cartridge is preloaded with magnetic tape for immediate recording or playback. The compact cartridge slips into position in seconds, and no threading is involved. A typical cartridge employing a continuous loop of lubricated tape drawn from the center is shown in Fig. 5-9. As the tape is used, it is wound on the outer part of the loop. This loop principle does not require take-up or rewind motors; only the capstan motor and pinch roller are required for drive.

Two general types of loops are used in the tape cartridge. These are the *continuous* loop in which the beginning and end are spliced together in a straightforward manner, and the *mobius* loop, in which a half-twist is imparted to the beginning of the tape before it is spliced to the end. The mobius loop permits dual half-track recording to double the recording time.

The tape made for continuous loops is lubricated on the backing side only. A mobius loop requires double-coated tape with lubrication on both sides and a Mylar base in the middle.

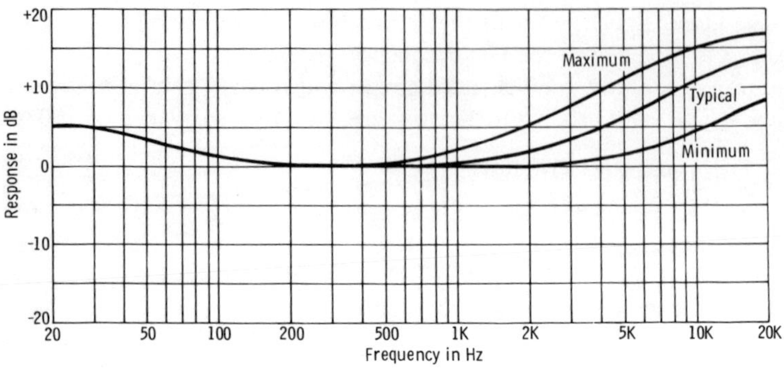

(A) 3¾ in/s record.

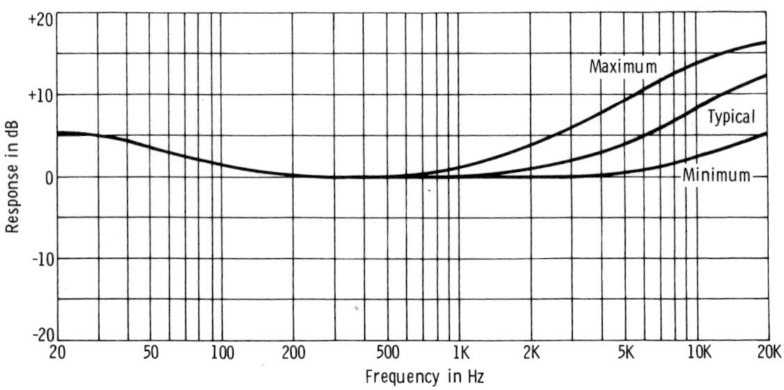

(B) 7½ in/s record.

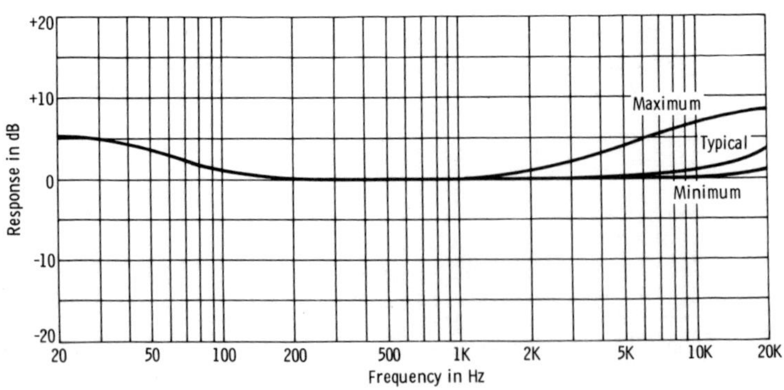

(C) 15 in/s record.

Fig. 5-5. Standard pre-equalization

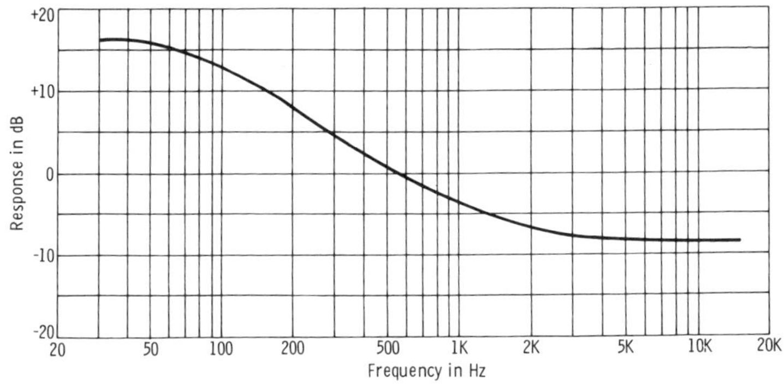

(*D*) *3¾ in/s reproduce.*

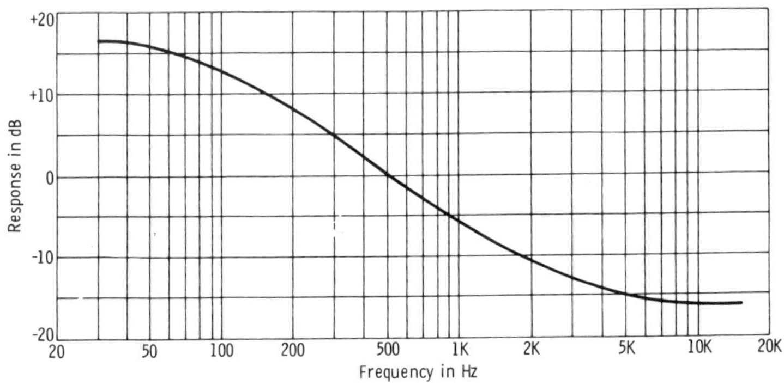

(*E*) *7½ in/s reproduce.*

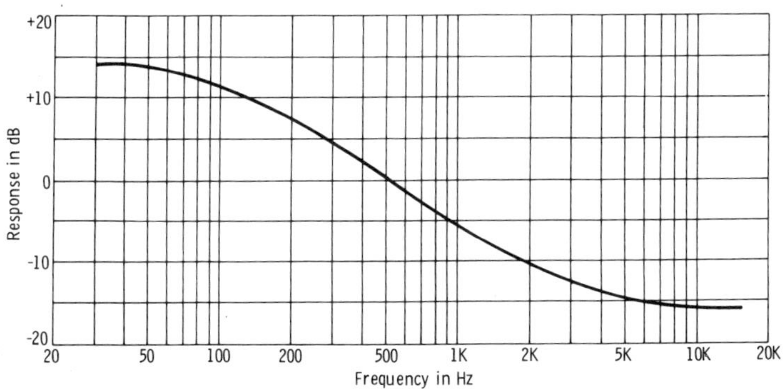

(*F*) *15 in/s reproduce.*

and post-equalization curves.

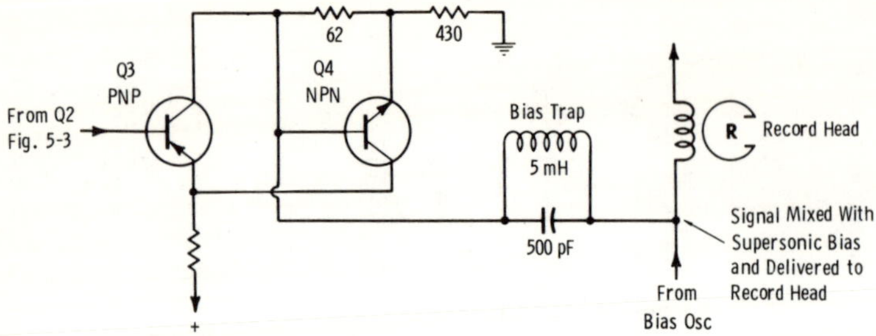

Fig. 5-6. Constant-current record-head amplifier.

Erase heads are not normally used on the cartridge tape deck. The tape must be erased on an external bulk eraser.

When the cartridge is placed on the tape deck, a lever is moved to pivot the pressure (pinch) roller from below the deck position to hold the tape securely against the capstan. Pressure pads within the cartridge hold the tape flat and firmly against the record/play and cue heads.

Tape Cartridge Terminology

Following is a tabulation of accepted terminology as applied to tape cartridge equipment:

Cue track: The lower track in a two-track cartridge monophonic system, and the lowest track in a three-track cartridge stereophonic system.

Program Track: The upper track in a two-track cartridge monophonic system, and the upper track (left channel) and middle track (right channel) in a three-track cartridge stereophonic system.

Head A: The magnetic (reproducing) head nearest the capstan.

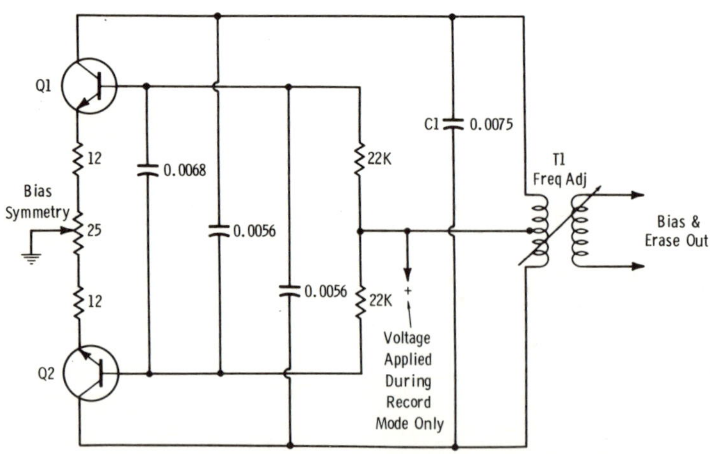

Fig. 5-7. Typical erase and bias oscillator.

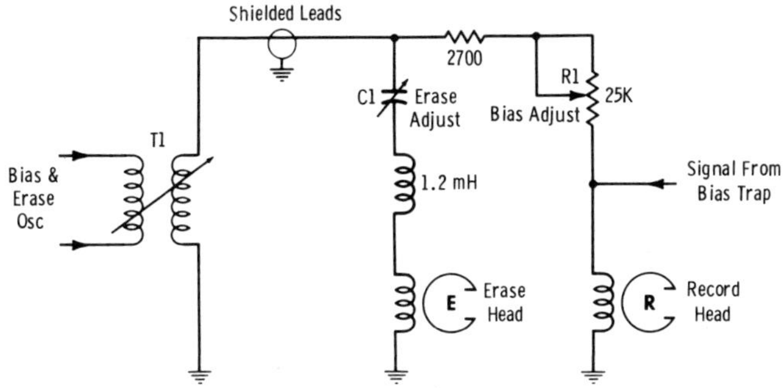

Fig. 5-8. Feeds to erase head and record head.

Head B: The magnetic (recording) head adjacent to head A.

Primary Cue Tone: A 1000-Hz tone recorded on the cue track and used for placing the tape in the cued position. The stop tone.

Secondary Cue Tone: A 150-Hz tone recorded on the cue track and used as the "end of message" signal. This tone is generally used for starting other equipment in an automation system.

Tertiary Cue Tone: An 8000-Hz tone recorded on the cue track and used as desired. This tone has been known in the past as an auxiliary or trip cue tone.

Cartridge Size A: The size A cartridge was previously known as Series 300, or the smallest size.

Cartridge Size B: The size B cartridge was previously known as Series 600.

Cartridge Size C: The size C cartridge was previously known as Series 1200.

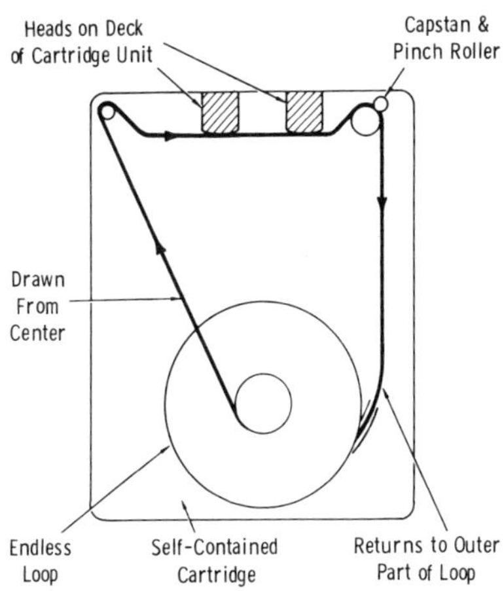

Fig. 5-9. Path of tape in endless-loop cartridge.

The tape speed is 7½ inches per second. Spot announcements or program material can be as short as one second or as long as 31 minutes. When a recording is started, the 1000-Hz tone is automatically placed on the cue track. In many operations, the spot or program material simply plays through the capacity of the particular cartridge and is automatically stopped by the 1000-Hz tone and is then recued for another play. Thus the 1000-Hz tone is the *primary cue tone* which serves as the cued position, or stop position. If more than one spot or production is to be placed on one cartridge, the stop button is depressed after the first segment, and the additional recording is added so that the primary cue tone will again serve as the cued or stop position. Upon completion of the final segment, the tape is allowed to continue until stopped by the first primary cue tone.

Note from the list of terminology above that two additional tones are available. Sometimes it is desirable to insert secondary and tertiary cue tones after the message(s) has been completely recorded. This permits monitoring and accurate placement of the tones, especially where they are employed to start other equipment in an automation system.

Some tape cartridge equipment is arranged so that the record unit and playback unit comprise separate systems. In other equipment, both systems

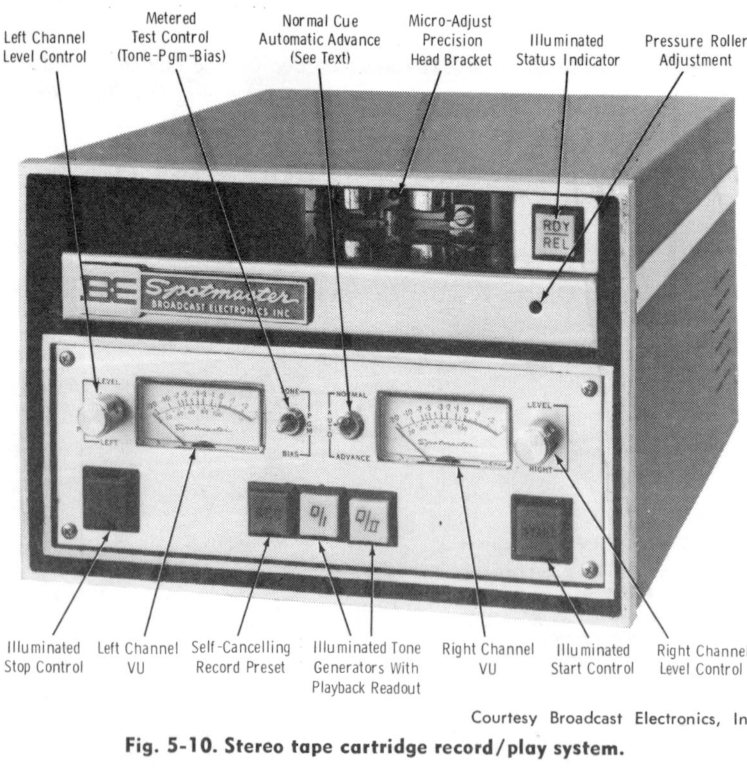

Courtesy Broadcast Electronics, Inc.

Fig. 5-10. Stereo tape cartridge record/play system.

are housed in the same cover. Fig. 5-10 illustrates a stereo record/play unit with separate level controls for left and right channels. Note the cue switch (three position) marked Normal-Auto-Advance. The normal position allows normal cueing as previously described. The automatic cue provides automatic end of message and rapid advance to the next segment on the tape. The advance position provides manual fast forward with automatic stop on the 1000-Hz primary tone.

Basic Electronics of Cartridge Tape Systems

Fig. 5-11 shows a simplified block diagram of the basic recording circuitry. The program recording circuitry is similar to that of conventional

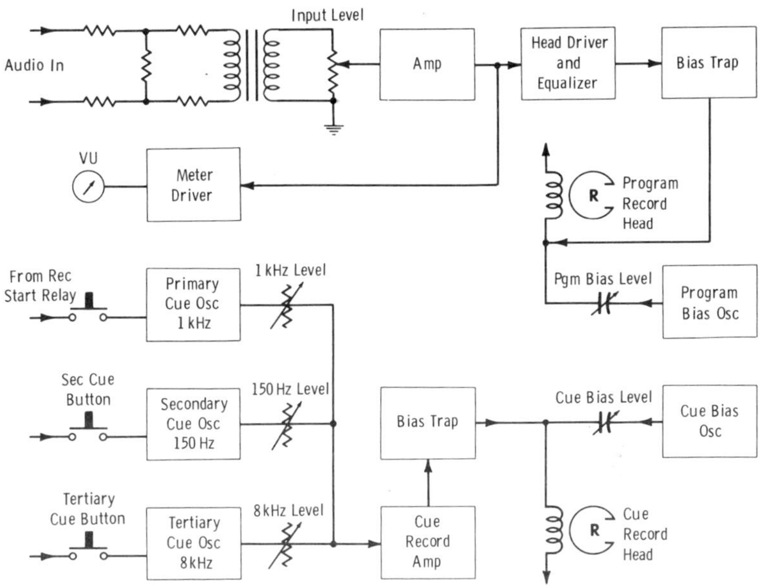

Fig. 5-11. Typical recording electronics of tape cartridge system.

reel-to-reel magnetic recording, described previously. The basic additions are those associated with the cue-track recording head. When recording is started, the primary cue oscillator is automatically started by a momentary contact associated with the record relay. The secondary and tertiary cue oscillators (optional equipment) are activated by relays associated with their respective momentary-contact push buttons. The cue tones are mixed through individual level controls and fed (through a bias trap) to the cue record head, which also receives the cue bias frequency.

Fig. 5-12 is a block diagram of typical playback electronics for tape cartridge systems. The small signal output of the playback head is amplified and fed through a playback level control to the line output driver. The pickup from the cue playback head is amplified through a broadband amplifier and fed through a cue sensitivity control (level control) to individual (tuned) amplifiers, each of which activates a relay. The primary cue tone trips the machine to stop. The secondary and tertiary cue tones activate relays the normally open and/or normally closed contacts of which appear on a terminal board to be carried to any control point desired.

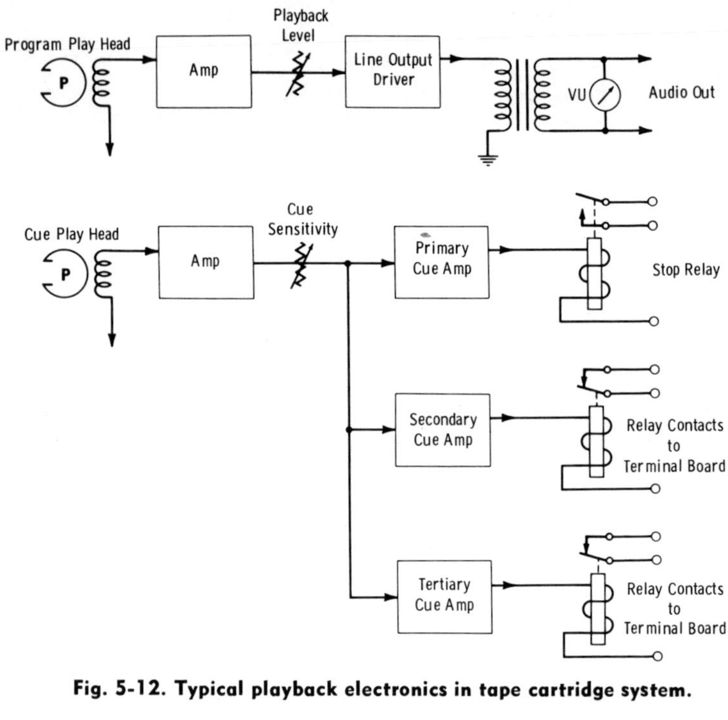

Fig. 5-12. Typical playback electronics in tape cartridge system.

The Playback Cue Amplifier

The cue-tone oscillators in the recording unit are conventional LC stabilized oscillators working from a well regulated dc supply. The playback cue-tone amplifiers (Fig. 5-13) are sensitive only to their individual tone frequencies. In the quiescent state, Q1, since its emitter resistance (R1) is greater than its collector load resistance (R2), is in a low-gain operation. Also, Q2, since it has no forward bias in the quiescent state, is cut off. The combination of L1 and C1 forms a series-resonant circuit at the individual tone frequency. Thus, on the arrival of a tone that is at the frequency of resonance of the series-tuned emitter return, the emitter circuit of Q1 becomes a very low impedance to ground and allows the transistor stage to become a very-high-gain amplifier. The resulting large (negative) excursions coupled to the base of Q2 saturate that transistor so that the relay

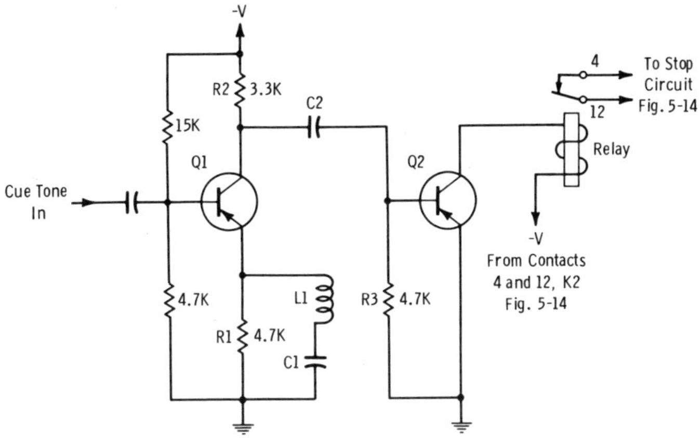

Typical Values

Freq	L1	C1	C2
1 kHz	500 mH	0.047 µF	0.22 µF
150 Hz	500 mH	2.0 µF	5.0 µF
8 kHz	5 mH	0.068 µF	0.1 µF

Fig. 5-13. Typical cue amplifier.

circuit is completed to ground and is thus activated, opening the normally closed contacts. After the short tone duration, Q2 again cuts off. The time constant of C2 and R3 is made long compared to one cycle of the tone frequency so that Q2 does not operate as a rapid switch during the individual cycles of the tone.

Typical Control Circuitry

Fig. 5-14 will enable the reader to become familiar with the important basic control functions of a typical tape cartridge system. In this example, the record and play units are separated.

The playback-unit control functions will be examined first. When the main power on-off switch is closed, indicator lamp I1 (usually a neon lamp) indicates that primary power has been applied and the system is ready for use. When a tape cartridge has been placed in the holder properly, the stop lamp is illuminated through closed contacts 2 and 10 of relay K1. Stop switch S2 is a normally closed switch and is momentarily opened by pushing the switch. This switch is in series with contacts 4 and 12 of K3. Note also that when the cartridge is in place to operate switch S3, power is applied to the capstan motor. Thus the motor is running at constant speed before the capstan solenoid engages the pinch roller and tape.

When start switch S1 is depressed, dc is applied to the start lamp and relay K1. This relay is latched on by contacts 9 and 5 (now closed) on K1.

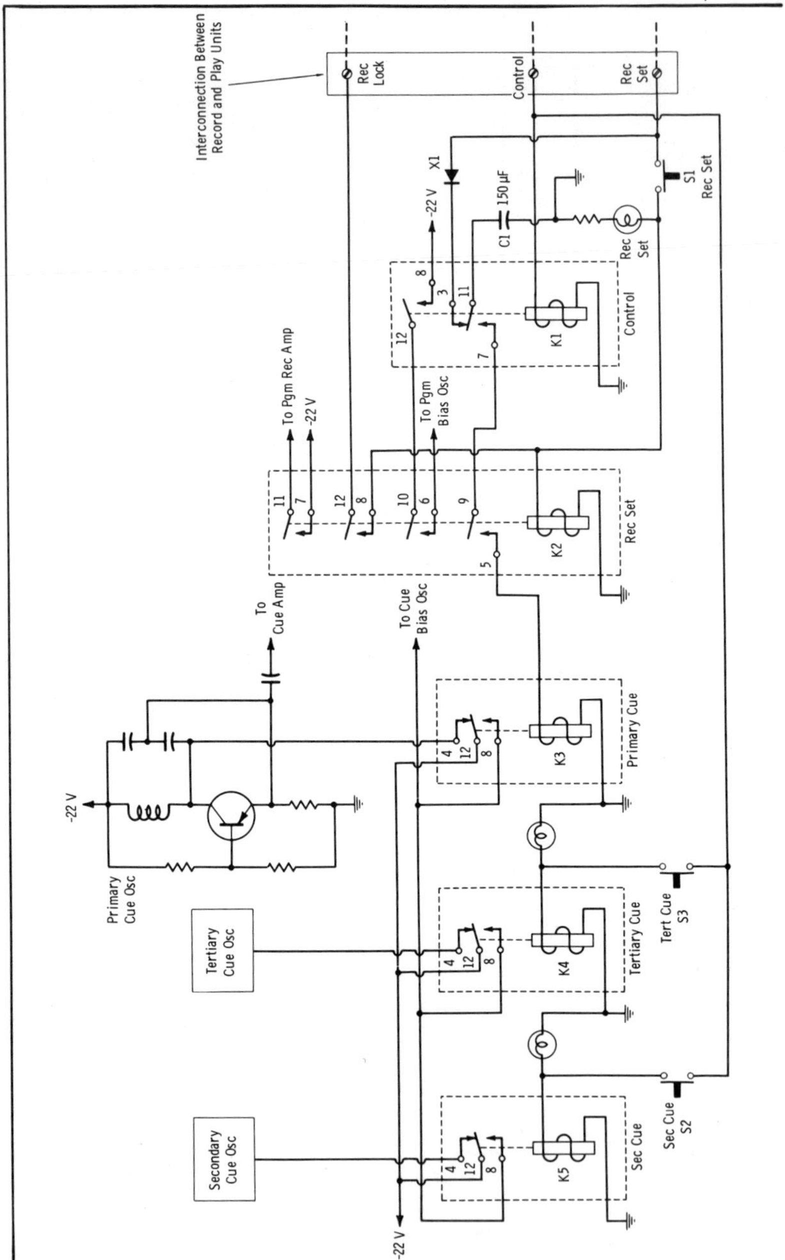

(*A*) *Record section.*

Fig. 5-14. Typical

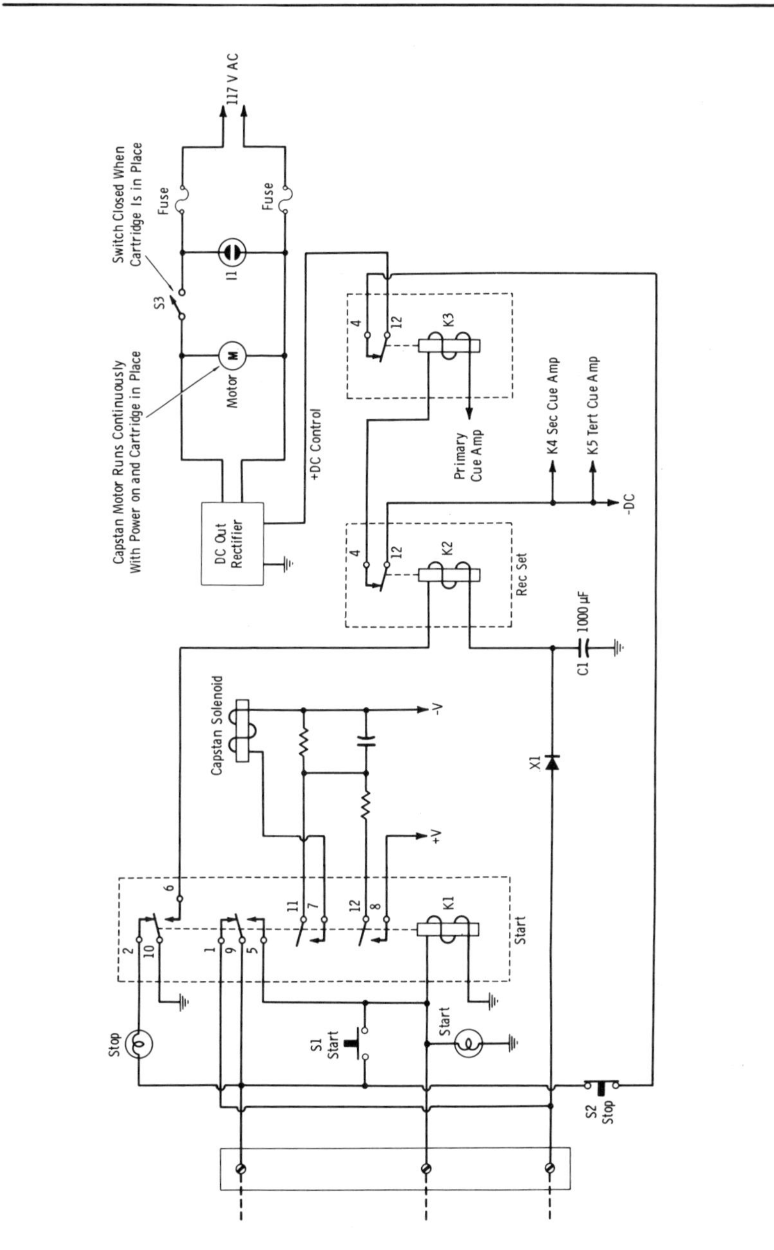

(*B*) *Playback section.*

control circuitry.

Contact pairs 8 and 12 and 7 and 11 energize the capstan solenoid, engaging the pinch roller with the tape to move it instantly at the proper speed. The grounded swinger arm (10) is moved to contact 6, which causes the stop indicator to go off and completes the return for record set relay K2. Therefore C1, which has charged through X1, is now able to discharge through K2 to operate that relay momentarily and open contacts 4 and 12 to prevent operation of K3.

Relay K3 is in the collector circuit of the primary cue amplifier (Fig. 5-13). Until this relay is operated by the primary cue tone (stop cue), contacts 4 and 12 of K3 are closed in series with the closed contacts of the stop switch, S2.

When the 1-kHz cue tone occurs, relay K3 is energized as described previously, opening contacts 4 and 12 of K3 and removing power from K1, thus stopping the transport. All contacts return to their previous state, and the stop lamp is again illuminated. Note that the action is the same as if the manual stop switch (S2) had been depressed.

We can now procede to the record unit. When record set switch S1 is depressed, power is applied to the record set lamp and relay K2. This relay is latched on by contacts 8 and 12 of K2. Contacts 7 and 11 close to apply dc to the program record amplifier. Contacts 6 and 10 close, but the circuit is not completed until control relay K1 is energized. Similarly, contacts 5 and 9 close but without circuit completion for the same reason.

Note that in the present mode, capacitor C1 in the record unit is charged to the applied voltage through diode X1 and normally closed contacts 3 and 11 of control relay K1.

When the start button on the playback unit is depressed, the tape moves as described before, and control relay K1 in the record unit is energized by contacts 9 and 5 of K1 in the play unit. Capacitor C1 in the record unit discharges through contacts 7 and 11 of K1, contacts 9 and 5 of K2, and primary cue relay K3. This momentarily closes contacts 8 and 12 of K3 to apply dc to the cue bias oscillator and removes the primary cue oscillator inhibit voltage (described below) by opening contacts 4 and 12.

All of the cue-tone oscillators work on the same principle as the automatic primary cue-tone oscillator shown in Fig. 5-14. When contacts 4 and 12 on K3 are closed, the collector of the oscillator is essentially returned to ground through the low internal impedance of the dc supply. When K3 operates, this ground is removed, and the cue tone is passed to the cue recording head simultaneously with the cue bias frequency. As soon as C1 is discharged, relay K3 again opens and the tone is removed. The secondary and tertiary cue oscillators (when used) are operated by momentary push buttons S2 and S3; this action also energizes the cue bias oscillator.

The unit will remain in the record mode until automatically stopped by the primary cue tone as described before for normal playback.

NOTE: Special types of tape recorders (such as those used for artificial reverberation) as well as operating techniques for all tape systems are covered in Chapter 11. Maintenance procedures are outlined in Chapter 13.

EXERCISES

Q5-1. Do all tape recording systems employ an erase head?

Q5-2. On a reel-to-reel tape recorder, does the torque of the take-up reel motor have any control over the tape speed?

Q5-3. What governs the exact tape speed?

Q5-4. What are the two most common tape speeds for reel-to-reel tape systems employed in am-fm broadcasting?

Q5-5: Why must low-frequency equalization as well as high-frequency equalization be employed in a tape recorder?

Q5-6. What is the standard tape speed for the tape cartridge system?

Q5-7. In a tape cartridge system, give the standard frequency for:
(A) The primary cue tone.
(B) The secondary cue tone.
(C) The tertiary cue tone.

Q5-8. Where is the cue track on a cartridge tape?

Q5-9. Give the arrangement of program tracks on a stereo cartridge tape.

Q5-10. In a tape cartridge system, why is the capstan motor forced to run before the actual start of tape movement?

CHAPTER 6

The Monaural Studio and Control Room

In this chapter, we will study the studio and control facilities for the single channel (monaural) method of broadcasting as practiced by all a-m stations and by monaural fm stations. The term "single channel" should not be misinterpreted. This simply means that all microphones and/or tape or disc channels may be mixed as desired, but there is no separation of sound information as to direction of the source.

Transmitter remote control and studio automation facilities are covered in Chapter 7, although such facilities are just as commonly used for monaural transmission as for stereo or quadraphonic transmission.

6-1. STUDIOS (PRODUCTION)

It must be emphasized here that the majority of broadcast stations today have no concern with production-type studio acoustics, being primarily concerned with disc reproduction and the various tape-system signal sources. However, some stations are heavily scheduled in producing programs involving musical pickups, dramatic skits for special events or commercial spots, local panel discussions, etc. These are most often recorded on tape for later use. To provide a useful text, we will cover production facilities and techniques as well as the more common practice of prerecorded sound sources.

The endeavor to realize high-fidelity transmission of broadcast programs is not new; it has been the goal of at least some engineers since the earliest days of broadcasting. The realization of overall high-fidelity service, however, includes the receiving set in the home, and it has not been until recently that the "average" set in the medium-price market was worthy of the extraordinary efforts of some broadcasters to render high-fidelity service. Conversely, it is apparent that with a good receiver, noticeable differences in fidelity characteristics of different stations within the range of the receiving position now may be noticed by the critical listener.

If the present state of development in broadcast amplifier equipment is taken as the sole criterion, then high-fidelity transmission is truly here. Frequency response is within 2 dB of the 1000-Hz reference from 30 to 15,000 Hz, and is limited only by wire-line connecting links in a-m installations or not at all in fm installations. The noise level at the antenna of the transmitter is at least 60 dB below 100-percent modulation, and the dynamic-range capability is at least 40 dB for a-m and 70 dB for fm. Unfortunately, however, the actual existence of high fidelity depends on many factors other than the af and rf amplifiers associated with the installation. These amplifiers, according to the ideas of some, form the heart of the transmission system insofar as high fidelity is concerned. Actually, they are merely a link in the chain of necessary functions of broadcasting a program and are no more important to fidelity than the other parts of the overall system, as Fig. 6-1 demonstrates.

In order to focus attention on the possible weak links, by eliminating the amplifiers, there remain: program and talent; production technicians responsible for pickup technique; the studio; program producers and announcers; microphones; mixing and switching circuits; control-room, master-control, and transmitter operators; wire lines; feeder systems and matching units; antennas; and the limitations set by channel bandwidths subject to government regulations. This presents quite a formidable list, and each item is recognizably inferior in performance to the modern amplifier associated with the broadcast installation. To those familiar with broadcasting, however, it may be shown that the weakest links and those which cause most concern at the present time are the studio itself, operating personnel, wire lines, and bandwidth limitations.

The limitations set by wire-line transmission are not serious if considered in relation to the allowable 10-kHz channel of the standard broadcast

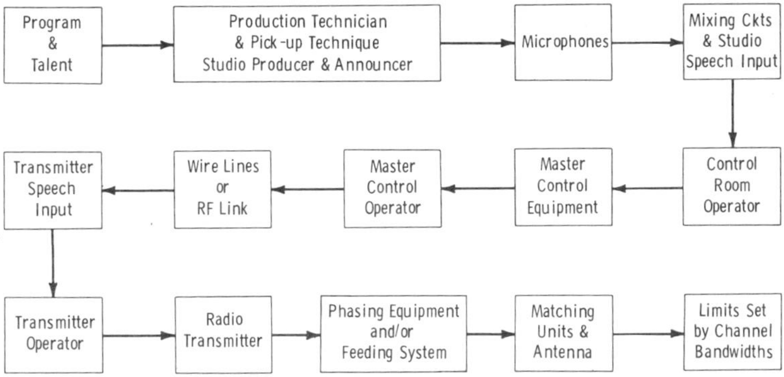

Fig. 6-1. Links in the chain involved in putting a program on the air.

installation. Formerly, all lines were equalized to 5000 Hz, which is, theoretically, the highest frequency of any effective strength tolerable if adjacent-channel interference is to be prevented. On the other hand, insofar as the relatively small primary coverage area is concerned, the frequency range of modern a-m transmitters (12,000 Hz), when utilized, allows a marked improvement, with class-B and -C service areas suffering from increased cross talk and interference. Although this situation is a deplorable one, it requires little discussion, in that the problem is primarily one under the control of the FCC.

Thus, there remain two factors to be considered, studio design and operating personnel. It is obvious that the broadcaster could possess high-fidelity equipment from microphone to antenna and still not provide high-fidelity service. In the final analysis, the outcome of any program for a given equipment installation depends entirely on the ability of the technical staff responsible for the operating technique. Realizable dynamic range, for instance, which is a highly important factor in high-fidelity transmission, is rarely utilized by station operators. It should be stated, however, that this is not entirely the fault of operators, but is due rather to a combination of factors including an incomplete correlation between the philosophy of dynamic range and compression amplifiers, inadequate visual monitoring indicators for wide dynamic range, and a confusion of ideas existent among pesonnel as to the amount of dynamic range tolerable in the home receiver for various types of program content. With fm transmission, this problem is even more important. Technical operations are fully explored in Chapter 11 of this text.

In general, the broadcast studio must meet the following requirements:

1. Freedom from noise, internal or external
2. Freedom from echoes
3. Diffusion of sound, providing a uniform distribution of sound energy throughout the microphone pickup area
4. Freedom from resonance effects
5. Reverberation reduction such that excessive overlapping of successive sound energy of speech articulation or music does not occur
6. Sufficient reverberation such that emphasis of speech and musical overtones is provided to establish a pleasing effect as judged by the listener

In the earliest days of broadcasting, the foremost problems encountered were quite naturally noise and echoes, since studios were simply rectangular rooms with conventional windows and ordinary walls. The first design steps were taken to treat the walls acoustically to prevent echoes and "flutter," and to cover the windows with the same acoustical material. This sufficed for a certain era in broadcasting, provided the operator with control over echoes, and practically isolated the microphone from factory whistles, fire sirens, etc. At that time, this type of studio was entirely adequate to satisfy the fidelity requirements of the program transmission possible with the associated equipment; indeed the electronic amplification of broadcast programs was so much better than the acoustic phonograph that the general public thought of the radio as a realization of true high-fidelity reproduction.

With the advent of the dynamic speaker, microphone improvements, and higher-power and wider-band amplifiers, the scope of fidelity possibilities began to broaden considerably. Signal-to-noise ratio was improved, and higher volumes could be handled in the receiver without distortion, resulting in a greater dynamic-range capability, but at the same time adding to the burden of studio design, since extraneous noises picked up at the studio were now more noticeable in the home receiver. This fact led to the "floating-studio" type of construction.

The following period saw many phenomenal improvements in broadcast equipment in general, such as 100-percent modulation of the transmitter with greatly reduced distortion, improvement in syllabic transmission characteristics, reduction of spurious frequencies and ripple level, greatly reduced noise levels in switching and mixing circuits, and nonmicrophonic tubes. Yet, strangely enough, studio design remained nearly stagnant over a period of six or seven years, except in isolated cases.

The broadcast engineer found himself faced with many apparent difficulties in rectangular studios. The big factor in a room with parallel walls is the excessive acoustical treatment necessary to overcome the effect of echoes, as mentioned previously. This has resulted, in the past, in extreme high-frequency attenuation and a lack of liveness such that the brilliance of musical programs was completely lacking. The loudness intensity for a given reading on the volume indicator is very low for a studio of this type in comparison with that obtained from a modern studio.

This effect obviously leads into complex operational difficulties, requiring a lower peaking of voice in relation to music to obtain a comparable loudness intensity in the receiver. Furthermore, in this type of studio a number of microphones must be used for a group of performers, since, if a single microphone is employed, a lack of reinforcement of harmonics and overtones of the instruments results in a thin sound lacking in body.

Another difficulty resulting from parallel-wall construction is that the angle of incidence of the wavefronts remains the same no matter how many reflections occur. Due to the acoustical treatment, this reflection (to any great extent) occurs only at the lower frequencies, and the nodes have marked regions of coincident reinforcement, resulting in resonance effects at the lower frequencies; thus conditions that would result in diffuse sound distribution are reduced. It becomes obvious that items 3, 4, and 6, given earlier in the requirements for good acoustics, are lacking in studios of rectangular design. In addition, the high-frequency response so necessary to brilliancy is reduced, effective dynamic range is inadequate, and operational difficulties are numerous. Thus, it is apparent that the studio becomes the weakest link in the high-fidelity chain in the great majority of broadcast installations today. Exceptions, of course, are the main network studios and a few independent stations more production conscious than the main body of independent broadcasters. It is certainly obvious that the large-scale expansion of fm service makes necessary a revolutionary education in studio requirements for the independent station operator, when concerned with local production.

From the foregoing discussion, the difficulties to be overcome may be listed as follows:

1. Lack of sound diffusion
2. Resonance at low frequencies
3. Insufficient reverberation for music
4. High-frequency absorption
5. Critical and multiple microphone placement
6. Operational complexities

The size and dimensions of the studio constitute a certain problem in studio design since an optimum volume per musician in the studio exists. Reduced to practice, however, this problem becomes one of simply proportioning the studio for a maximum number of musicians expected. This is possible because no difficulty exists in obtaining a good pickup of a small group in a studio designed for a great number of musicians; conversely, because a small room cannot conveniently be "aurally" enlarged, a large band in a small studio presents a difficult problem. Portable hard flats are often used in large studios to enclose a small group of musicians, thus providing the optimum dimensions required for good pickup of a given number of performers.

High-frequency absorption, particularly at frequencies over 5000 Hz, is relatively great. The absorption of sound by air at these frequencies is actually greater than the surface absorption of the studio, even under normal conditions of temperature and relative humidity. It is not possible to construct a studio having a reverberation time of over 1.2 seconds at 10,000 Hz even with theoretical zero absorptivity of the acoustical treatment. By distributing the reflector surfaces in proximity to the musical instruments, a maximum of diffused, polyphased high-frequency sound will exist at the microphone without being attenuated injuriously by space behind the instruments. A minimum number of microphones for adequate pickup is necessary under these conditions.

Fig. 6-2 illustrates one method of breaking up parallel surfaces in the pickup area. Both walls and ceilings are interspersed with rounded surfaces to enhance musical sounds without parallel resonance effects. Fig. 6-3 shows Johns-Manville Transite acoustical panels as used in the general-purpose type of studio.

Fig. 6-4 shows the sound-absorbing characteristics of three materials developed in the acoustical research department of Johns-Manville. By

Courtesy National Broadcasting Co.

Fig. 6-2. NBC studio A.

proportioning the amount or adjusting the orientation of these three materials in a studio, the time-frequency curve can be adjusted to any desired contour. This type of studio has the advantage of unlimited pickup area,

Courtesy Johns-Manville

Fig. 6-3. Acoustical panels used in a studio.

but it has the disadvantage of being affected by the size of the studio audience. A great difference in reverberation time exists between a vacant studio and one occupied by a large group of people.

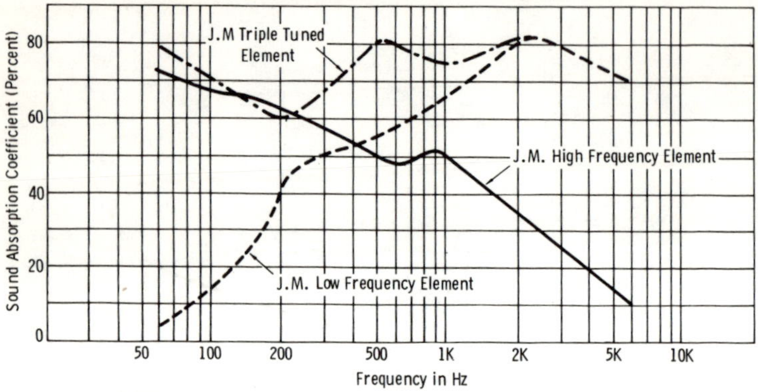

Fig. 6-4. Sound absorbing characteristics of three acoustical materials.

6-2. MATCHING AND BRIDGING PADS AND TECHNIQUES

Due to the complex nature of control and signal routing in typical broadcast systems, fixed attenuation pads are widely used. These devices should be understood before the reader undertakes the study of the main operations center.

The configurations of the two most common types of fixed pads employed in broadcast circuits are shown in Table 6-1. These pads provide constant-impedance loss circuitry generally specified in decibel values of attenuation. The T pad is used where one side of the circuit is grounded or provides a common return. The H pad is used where balanced-to-ground circuitry is involved (the most usual condition).

When the input and output impedances are equal, the series resistors are equal in value. When the input and output impedances differ, the series resistors are of different values on the input and output sides so that the impedance of the pad matches the impedance into which each side works. This arrangement is termed a *taper pad.* All pads must be constructed of noninductive elements so that no frequency discrimination occurs.

Table 6-1 lists the nearest EIA values of resistors to be used when the input and output impedances are equal to 600 ohms. Note the multiplying factors to be used when the equal impedances are other than 600 ohms. Observe that the values of the series resistors in the H pad are essentially one-half the values of those in the T pad, since the two sides of the circuit must total the desired amount; R2 is the same in either case.

Due to shunt capacitance effects across large resistance values, it is impractical to obtain attenuation of more than 40 dB in one pad. When greater attenuation is desirable, pads are connected in tandem.

Table 6-1. Design Data for Fixed Pads

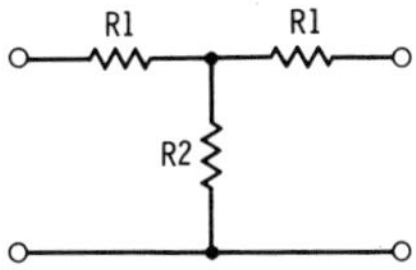

T Pad

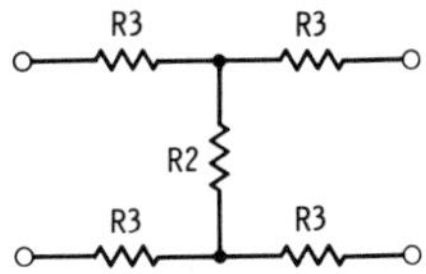

H Pad

For case where $Z_{in} = Z_{out} = 600$ ohms. For other than 600 ohms (but equal impedances), multiply all resistance values by factor $Z_x/600$ (0.40 for 250 ohms, 0.25 for 150 ohms, 0.083 for 50 ohms.

EIA Resistor Values Nearest to Exact Values

Loss (dB)	R1	R2	R3
0.5	18	10K	8.2
1.0	36	5.1K	18
2	68	2.7K	36
3	100	1.8K	51
4	130	1.2K	68
5	160	1K	82
6	200	820	100
7	220	680	110
8	270	560	130
9	300	470	150
10	300	430	160
11	330	360	160
12	360	330	180
13	390	270	200
14	390	240	200
15	430	220	200
16	430	200	220
17	470	180	220
18	470	150	240
19	470	130	240
20	510	120	240
22	510	100	270
24	510	75	270
26	560	62	270
28	560	47	270
30	560	39	270
32	560	30	300
34	560	24	300
36	560	18	300
38	560	15	300
40	560	12	300

Taper Pads

It frequently is necessary to use fixed pads which match unequal impedances. One common example is the isolation pad used to feed the transmitter line (Fig. 6-5). Program lines for fm, which are used for wide-band, low-noise service, are often designed to be terminated at both ends in 150 ohms rather than the more common 600-ohm impedance. The pad is used to provide a constant load at all frequencies to the line amplifier, and the repeat coil minimizes induction pickup along the line. The repeat coil is normally operated as a 1:1 transformer, although taps are available for other arrangements. A Faraday screen is used between windings to provide an electrostatic shield which prevents capacitive coupling.

Fig. 6-6A shows the notation used with T pads designed for unequal input and output impedances. Fig. 6-6B illustrates the H-pad notation. Note that, as before, the series resistance values of the H pad are one-half those of the T pad for a given decibel loss. This notation is different from that used with equal-impedance pads because the series resistors must have different values in the input and output arms due to the requirement of matching unequal impedances.

Step 1: Determine the minimum loss by means of the graph in Fig. 6-7. This minimum loss is related to the impedance ratio. For example, the impedance ratio of 600/150 is 4/1. The pad must be designed to the nearest decibel loss of an even number (multiple of 2) above this minimum loss value. In the case of a 4/1 ratio, the graph designates approximately 12 dB as the minimum loss. In practice, the design should be for at least a 14-dB loss.

Step 2: To calculate the resistor values, two K factors are required; these may be obtained from Table 6-2. Calculate R1, R2, and R3 as follows:

$$R1 = \frac{(Z1 + Z2)K_1 + (Z1 - Z2)}{2}$$

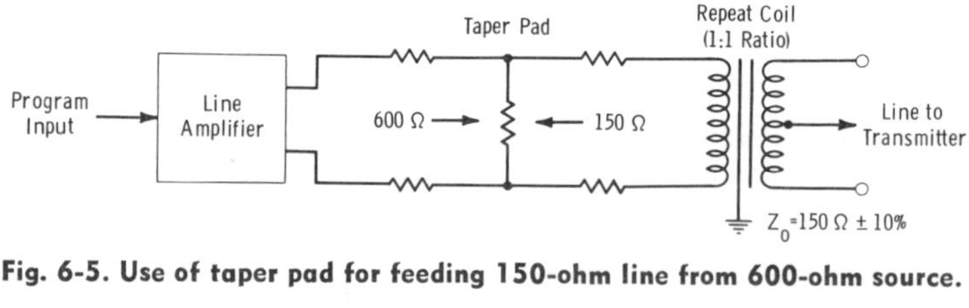

Fig. 6-5. Use of taper pad for feeding 150-ohm line from 600-ohm source.

(A) Notation for T pad. *(B) Notation for H pad.*

Fig. 6-6. Notation for pads used to match unequal impedances.

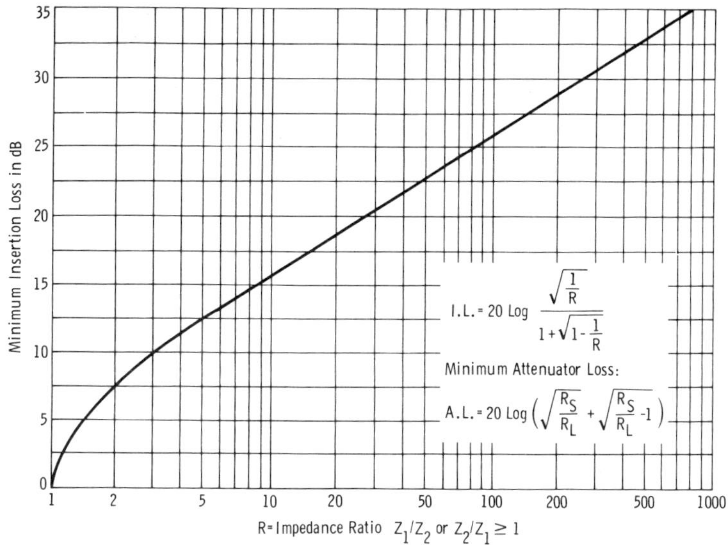

Fig. 6-7. Minimum-loss chart.

NOTE: For H pad, divide the result again by 2, per Fig. 6-6B.

$$R2 = \frac{(Z1 + Z2)K_1 - (Z1 - Z2)}{2}$$

NOTE: For H pad, divide the result again by 2, per Fig. 6-6B.

$$R3 = \frac{Z1 + Z2}{2K_2}$$

Table 6-2. Voltage Ratio Values for a Given Loss in Decibels

dB	K_1	K_2	dB	K_1	K_2	dB	K_1	K_2	dB	K_1	K_2
1	0.057	0.115	14	0.667	2.404	27	0.914	11.188	40	0.980	50.237
2	0.114	0.232	15	0.697	2.720	28	0.923	12.484	41	0.982	56.079
3	0.171	0.352	16	0.726	3.075	29	0.931	14.091	42	0.984	63.230
4	0.226	0.447	17	0.752	3.468	30	0.938	15.734	43	0.985	70.583
5	0.280	0.609	18	0.766	3.907	31	0.945	17.744	44	0.987	78.792
6	0.331	0.747	19	0.798	4.398	32	0.950	19.810	45	0.988	88.836
7	0.382	0.897	20	0.818	4.952	33	0.956	22.339	46	0.990	100.165
8	0.430	1.055	21	0.835	5.555	34	0.960	24.939	47	0.991	111.813
9	0.476	1.233	22	0.852	6.262	35	0.965	27.121	48	0.992	126.070
10	0.519	1.422	23	0.867	7.013	36	0.968	31.393	49	0.993	140.729
11	0.560	1.634	24	0.880	7.868	37	0.972	35.397	50	0.994	158.672
12	0.598	1.863	25	0.893	8.870	38	0.975	39.515			
13	0.634	2.122	26	0.904	9.977	39	0.978	44.555			

Take a specific example of a 600/150-ohm pad which should provide a 15-dB loss. From Table 6-2, K_1 is 0.697 and K_2 is 2.72. Then:

$$R1 = \frac{(600 + 150)(0.697) + (600 - 150)}{2} = \frac{973}{2} = 486 \text{ ohms}$$

$$R2 = \frac{(600 + 150)(0.697) - (600 - 150)}{2} = \frac{72}{2} = 36 \text{ ohms}$$

$$R3 = \frac{600 + 150}{2(2.72)} = \frac{750}{5.44} = 135 \text{ ohms}$$

Thus for a T pad (to the nearest EIA values):

$$R1 = 470 \text{ ohms}$$
$$R2 = 36 \text{ ohms}$$
$$R3 = 130 \text{ ohms}$$

For an H pad (to the nearest EIA values):

$$R1 = 240 \text{ ohms}$$
$$R2 = 18 \text{ ohms}$$
$$R3 = 130 \text{ ohms}$$

NOTE: The R numbers above refer to the configurations of Fig. 6-6.

Bridging Pads

A bridging pad provides a high impedance to the bus to be bridged so that the characteristic impedance is not disturbed, and a matching impedance is seen by the amplifier or unit used. A bridging pad has already been seen in relation to the standard VU meter (Fig. 2-13).

In practice, a bridging impedance must be at least 10 times the impedance of the circuit to be bridged. This is so that there is no practical attenuation on the bridged circuit itself due to loading. Thus, for a 600-ohm bus, a bridging circuit should present an impedance of at least 6000 ohms to the bus. Consideration must be given to the total number of units or amplifiers likely to be bridged across the circuit, and the bridging impedance of each unit is therefore made as high as possible considering the gain of the amplifier. The gain must be sufficient to make up the bridging loss.

Some amplifiers designed specifically for isolation purposes provide the needed high-impedance input for bridging applications. However, when a standard amplifier input (such as 150 or 600 ohms) must be used, a bridging pad must be inserted between the bus and the amplifier (Fig. 6-8A). The shunt resistor (R2) is made equal to the input impedance of the amplifier.

Although the bridging arrangement shown in Fig. 6-8A may be used for either a balanced or unbalanced amplifier, the use of a bridging transformer

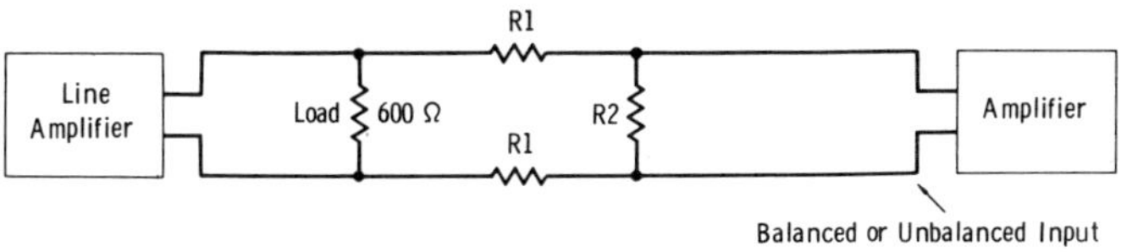

(*A*) *Bridging pad.*

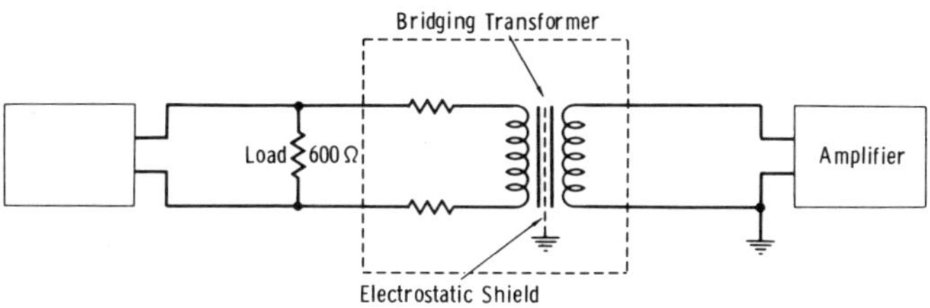

(*B*) *Isolation transformer.*

Fig. 6-8. Bridging arrangements.

with an electrostatic shield is preferred for bridging to an unbalanced circuit (Fig. 6-8B). The loss introduced by such a bridging coil is typically 20 dB.

Table 6-3 shows the two basic types of bridging circuits (unbalanced and balanced), with the values of bridging resistance used for the listed

Table 6-3. Value of R1 for Bridging 600-Ohm Circuit

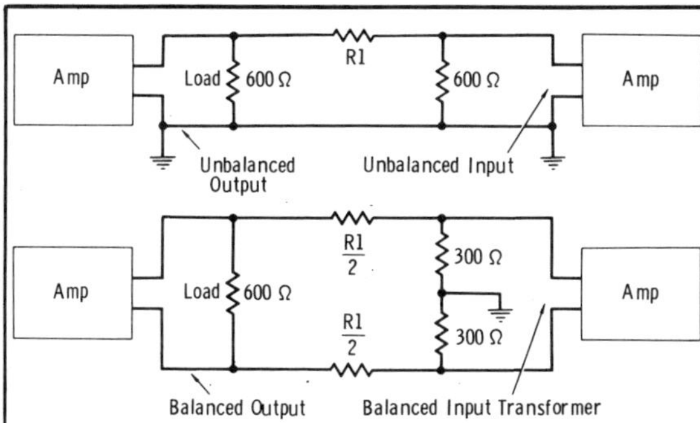

Loss dB	R1 in Ohms
40.1	30,000
38.5	25,000
36.8	20,000
34.2	15,000
30.7	10,000
28.3	7500
24.9	5000
20.8	3000

(CAUTION: The given dB loss is accurate only when bridging a 600-ohm circuit.)

attenuations when the circuit to be bridged is 600 ohms. Bridging resistances down to 3000 ohms have been included. Recall from a previous statement that the minimum bridging resistance should be at least 10 times the impedance of the circuit to be bridged; for 600 ohms, this would be 6000 ohms. The actual loss in the circuit to be bridged may be computed as follows:

$$\text{Loss in dB} = 20 \log \frac{2B_r + R}{2B_r}$$

where,

B_r is the bridging input of the pad,
R is the impedance of circuit bridged.

Thus for the 3000-ohm bridge mentioned:

$$\begin{aligned} dB &= 20 \log \frac{6000 + 600}{6000} \\ &= 20 \log 1.1 \\ &= 0.82 \text{ dB, or almost a 1-dB loss in the 600-ohm circuit.} \end{aligned}$$

Ideally, the loss in the circuit to be bridged should be under 2 percent. The problem encountered is not so much the 0.82-dB loss in the bridged circuit, but rather the poor isolation between the 600-ohm circuit and the bridged amplifier. Any form of noise pickup or oscillation that can occur in the input stage of an amplifier would "cross-talk" onto the 600-ohm signal. Also, no more bridging on this line could be accomplished without lowering the total bridging resistance to an excessively low value.

Combining Pads

It is frequently necessary to employ branching or combining networks so that more than one feed may be obtained from a single source. For example, a program line might be branched to a monitor line, a test panel, and the program line to the transmitter. Basic arrangements employed for three branches from a single source are illustrated in Fig. 6-9. Build-out resistors (R_B) are used to provide isolation between the branch feeds and the source. The value of the build-out resistors is calculated as follows:

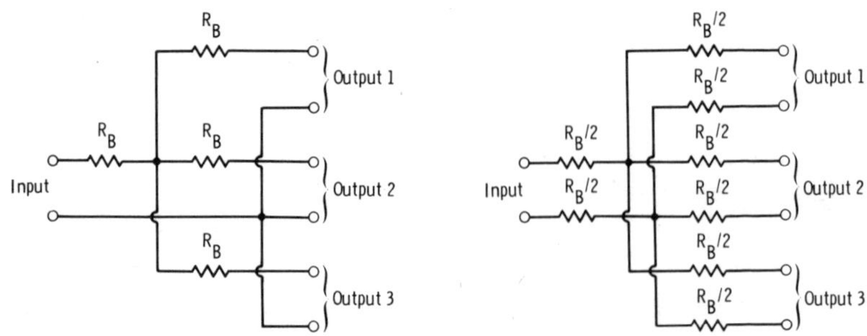

(*A*) *Unbalanced.* (*B*) *Balanced.*

Fig. 6-9. Combining networks.

$$R_B = \frac{N-1}{N+1} Z$$

where,

R_B is the build-out resistor,
N is the number of branch output circuits,
Z is the circuit impedance.

NOTE: Combining pads are used only between sources of equal impedances.

For a 600-ohm line branching to three circuits of 600 ohms:

$$R_B = \frac{3-1}{3+1} 600$$
$$= 0.5\ (600)$$
$$= 300 \text{ ohms}$$

The loss through any two branches is 20 log $(N-1)$, where N is the total number of circuits. This total includes the source circuit. Thus for the circuits shown in Fig. 6-9:

$$\text{dB loss} = 20 \log (4-1)$$
$$= 20 \log 3$$
$$= (20)(0.4771)$$
$$= 9.54 \text{ dB}$$

The loss is the same for either balanced or unbalanced branching pads. Note, however, that the value of R_B is halved for the balanced circuit.

For convenience, Table 6-4 lists the losses for branching pads with one input and up to 10 outputs, when working between impedance values of 600 ohms.

Table 6-4. Loss (dB) for Single-Input, Multiple-Output Pads (Impedance 600/600 Ohms)

Number of Outputs	Loss (dB)
2	6.0
3	9.5
4	12.0
5	14.0
6	15.6
8	18.1
10	20.0

6-3. ATTENUATORS (FADERS OR MIXERS)

Variable attenuators serve as *faders* or *mixers* in the audio system. They are generally mounted on the operating console. Like the fixed pads described above, the mixer provides a constant input and output impedance, regardless of the amount of attenuation introduced into the path. Two basic physical constructions are used, the rotary type and the vertically-operated type.

The constant input and output impedance of the variable attenuator is achieved by simultaneously increasing the series resistance and decreasing the shunt resistance as the control is turned toward maximum attenuation (and vice-versa for rotation toward minimum attenuation). This is true for all but the simple ladder-type attenuator, which is described first below. The ladder type does not provide a complete match at extremes of rotation, and this is its main disadvantage, along with a minimum insertion loss of 6 dB.

The Unbalanced Ladder

Due to its mechanical simplicity which requires only one row of contacts (Fig. 6-10), the unbalanced ladder attenuator is quite commonly used as a sound mixer. A ladder network is simply a number of cascaded pi resistor sections combined to supply the required terminal impedances and attenuation.

When the ladder-type fader is employed, even though the arm is rotated fully clockwise (minimum attenuation) to the contact at R2, the insertion loss is 6 dB. This minimum insertion loss results from the parallel resistors and the resistor (R4) in the slider-arm circuit. The minimum loss of 6 dB holds true only when the device is operating between like impedances. For example, if the input impedance is 600 ohms and the output impedance is 150 ohms, the resulting impedance ratio of 4 to 1 introduces an additional attenuation of approximately 11 dB (Fig. 6-7).

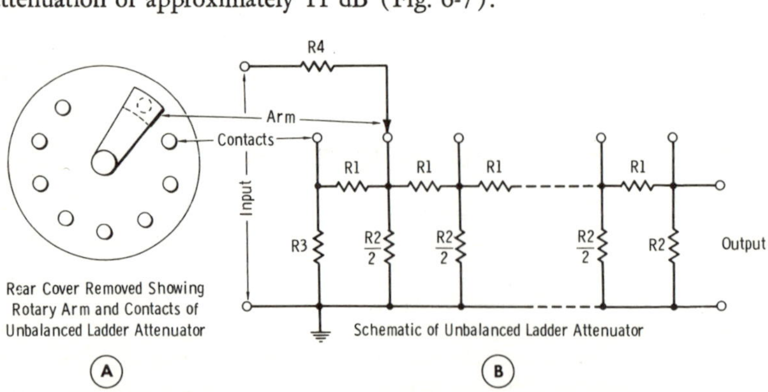

Fig. 6-10. Principles of unbalanced ladder attenuator.

Fig. 6-11 illustrates an unbalanced ladder attenuator with a built-in cueing circuit added. As shown, in the extreme attenuation position (maximum counterclockwise) a connection is made to feed the incoming signal to a terminal-board lug for an external cue amplifier. Such attenuators are often provided for turntable and tape inputs to allow proper cue-up of the source, or to fade in the signal at a given cue point without the addition of more switches.

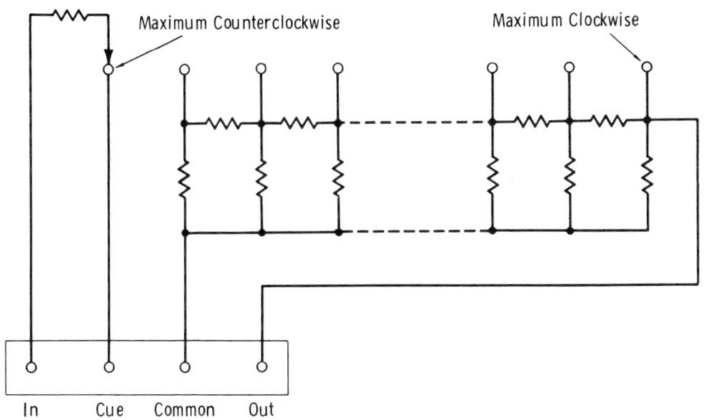

Fig. 6-11. Unbalanced ladder attenuator with cueing position.

The T Attenuator

The *bridged T* attenuator (Fig. 6-12A) is another unbalanced type of fader popularly used in sound mixers. It requires just two rows of contacts as contrasted to the three rows necessary for the *straight T* of Fig. 6-12B. Note that in either case, when the control is turned to the maximum clockwise position (arms on last contact to right) the input is connected directly to the output (with maximum shunt resistance); hence, zero insertion loss can be obtained when the device is operated with a 1:1 impedance ratio. When the straight T attenuator is used, the arm consists of three separate leaves, each of which makes contact with its individual row.

Balanced Attenuators

Fig. 6-13 presents schematics of the three basic types of balanced attenuators. The balanced ladder (Fig. 6-13A) is simply two unbalanced ladder networks inserted in the two sides of the line and coupled together on a common shaft. Like the unbalanced ladder, this attenuator has a minimum insertion loss of 6 dB.

The *bridged H* attenuator (Fig. 6-13B) is simply two bridged T networks mounted on a common shaft. Each section requires two rows of contacts; hence two units are mounted in tandem and operated by the common shaft. The front and rear units are securely fastened together within a totally enclosed dust cover. By depressing release springs located on the

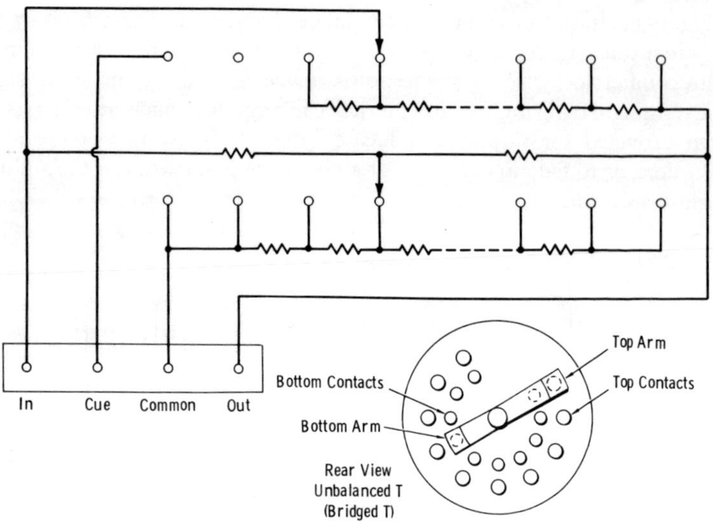

(*A*) *Bridged T.*

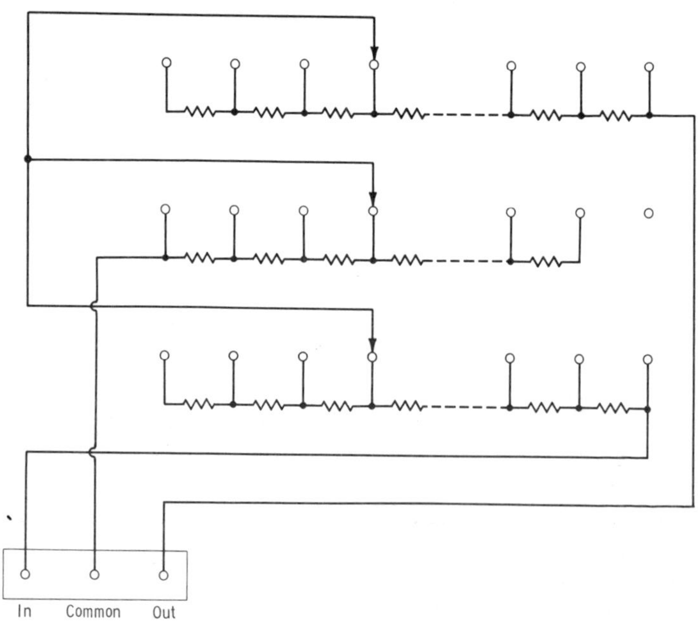

(*B*) *Straight T.*

Fig. 6-12. Types of T attenuators.

sides, the rear section can be removed to expose the contacts of both sections for cleaning.

The *balanced H* attenuator (Fig. 6-13C) consists of two straight T units mounted in tandem. Three variable resistors are used to form the T network in each section, and the sections are connected to form a balanced H. Each section requires three rows of contacts.

The bridged H and balanced H attenuators provide zero insertion loss when operated with a 1:1 impedance ratio, just as is true for the unbalanced T attenuators.

Low-Level Mixer Circuits

Low-level mixing is often employed in the simpler control consoles (six channels or fewer) such as those sometimes used in remote locations or as subassemblies in larger installations. For six-channel mixers, the T-type attenuator is normally used to avoid the extra 6-dB minimum insertion loss of the simpler ladder type. Four-channel mixers often employ the ladder type for channel mixers and a T type for the master level control (when used).

Low-level mixing simply means the input signals pass through the mixing network prior to amplification. Fig. 6-14 shows the basic principles of low-level mixing circuits. This example is an unbalanced parallel type of mixer with a 1:1 impedance ratio. Most larger control consoles employ high-level mixing, which means an individual attenuator is used between amplifiers for each channel.

As given in Fig. 6-14, the minimum loss for this type of mixing circuit is 20 log N, where N is the number of channels. If the ladder-type mixer is used, the insertion loss of 6 dB must be added to this computation. The data of Fig. 6-14 are for a 1:1 impedance ratio such as 600:600, 150:150, etc. The value of the build-out resistor (R) is dependent on both the source impedance (which in this case is also the output impedance) and the number of channels employed. The table of values of R in Fig. 6-14 is a "quick computation" list for reference.

Variable attenuators are purchased from a manufacturer, but the user must know how to combine these networks when building his own gear, such as portable sound mixers or an auxiliary mixing panel. The data of Fig. 6-14 (for 1:1 impedance ratio) and Fig. 6-15 (for taper attenuators between unlike impedances) are required for this purpose.

The diagram in Table 6-5 gives the basic design parameters for allowable mixer loss to a given amplifier or preamplifier. For convenience in application, Table 6-5 lists the actual mixer loss for the number of channels indicated in the left column.

An impedance-matching transformer sometimes is used in the position of the master gain control of Figs. 6-14 and 6-15 to feed a following amplifier. The master gain control then follows this amplifier to feed the program line amplifier.

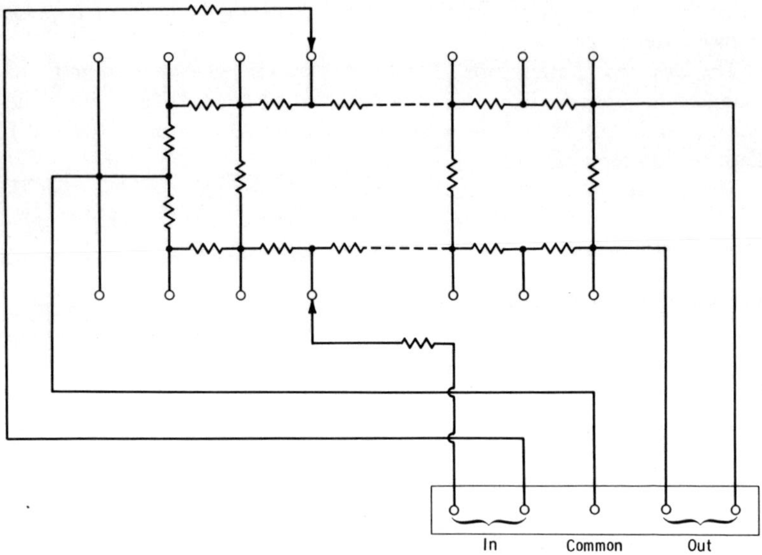

(*A*) *Balanced ladder.*

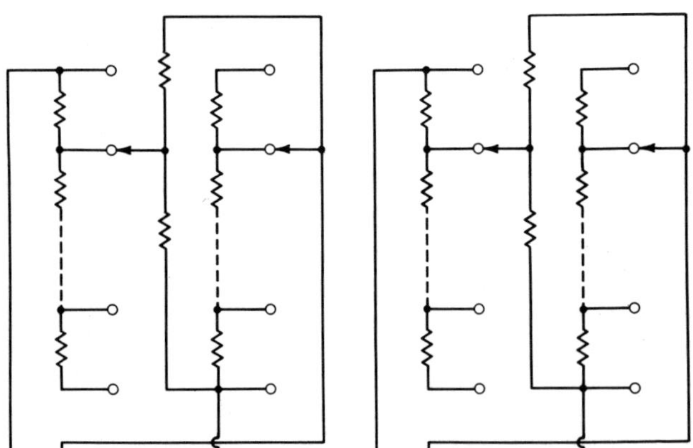

(*B*) *Bridged H.*

Fig. 6-13.

A side view of a typical vertical-type attenuator is shown in the drawing of Fig. 6-16. Sometimes the in-out and common terminals are brought to a receptacle to allow plug-in of units.

6-4. EQUALIZERS

Equalization as applied to tape recording and playback has already been discussed. Based on the same principle are equalizer networks for long lines (studio to transmitter, network distribution, remote installations, etc.)

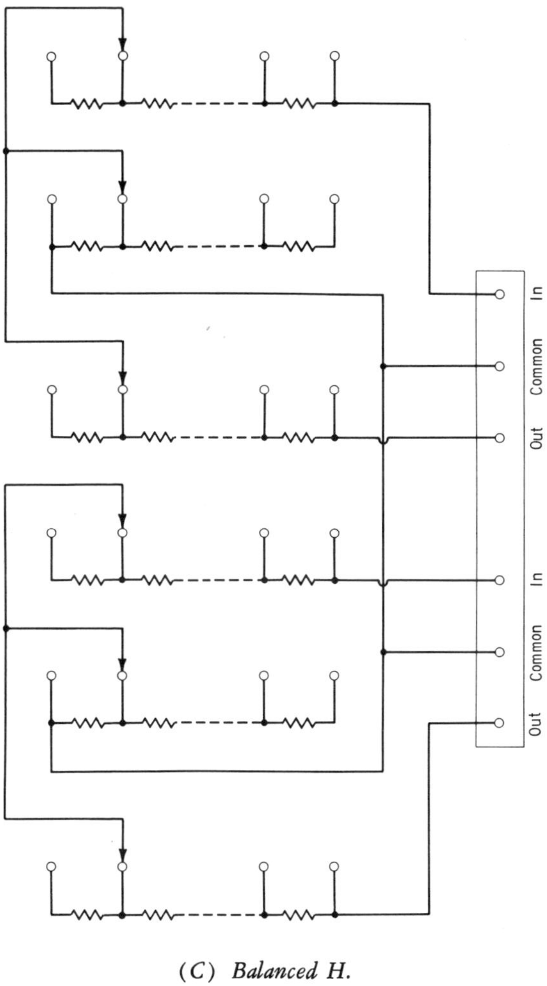

(*C*) *Balanced H.*

Balanced attenuators.

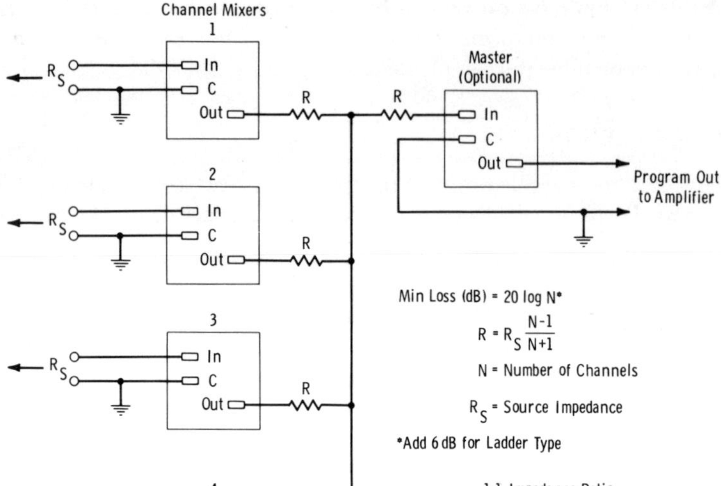

1:1 Impedance Ratio

No. of Channels	2	3	4	5	6
Value of R	$\frac{R_S}{3}$	$\frac{R_S}{2}$	$\frac{3R_S}{5}$	$\frac{2R_S}{3}$	$\frac{5R_S}{7}$

Fig. 6-14. Low-level mixing circuit, 1:1 impedance ratio.

Table 6-5. Mixer Loss

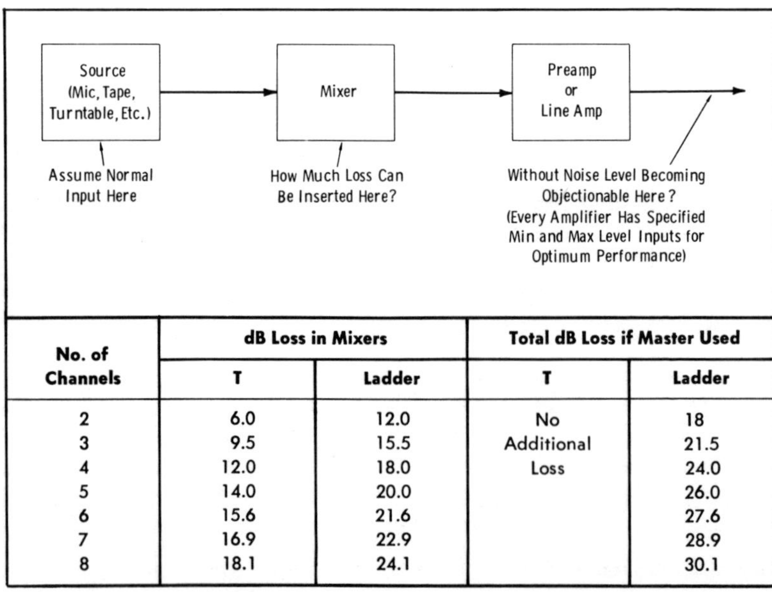

No. of Channels	dB Loss in Mixers		Total dB Loss if Master Used	
	T	Ladder	T	Ladder
2	6.0	12.0	No Additional Loss	18
3	9.5	15.5		21.5
4	12.0	18.0		24.0
5	14.0	20.0		26.0
6	15.6	21.6		27.6
7	16.9	22.9		28.9
8	18.1	24.1		30.1

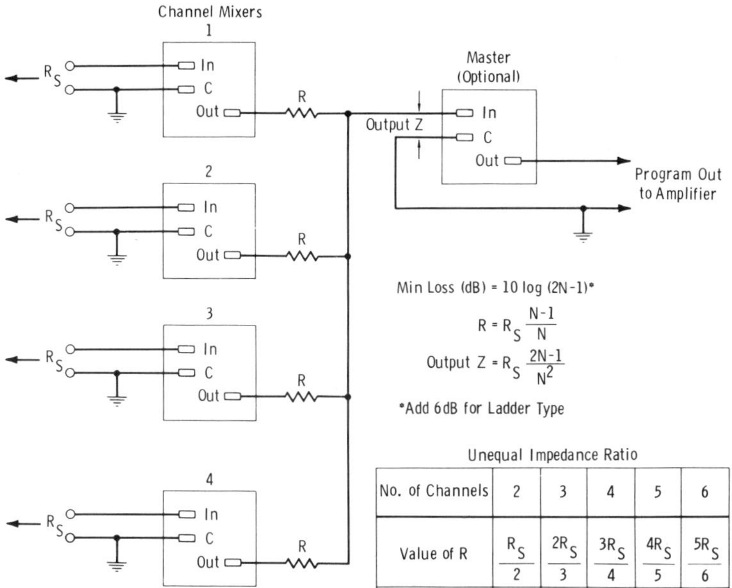

Min Loss (dB) = 10 log (2N-1)*

$$R = R_S \frac{N-1}{N}$$

$$\text{Output } Z = R_S \frac{2N-1}{N^2}$$

*Add 6dB for Ladder Type

Unequal Impedance Ratio

No. of Channels	2	3	4	5	6
Value of R	$\frac{R_S}{2}$	$\frac{2R_S}{3}$	$\frac{3R_S}{4}$	$\frac{4R_S}{5}$	$\frac{5R_S}{6}$
Output Z (To Master or Repeat Coil)	$\frac{3R_S}{4}$	$\frac{5R_S}{9}$	$\frac{7R_S}{16}$	$\frac{9R_S}{25}$	$\frac{11R_S}{36}$

Fig. 6-15. Low-level mixing circuit, unequal impedances.

and special equalizers used in production-type equipment for special effects, frequency emphasis or de-emphasis for echo use, etc.

Line Equalizers

Telephone lines used for the distribution of broadcast signals have a gradual rolloff off high-frequency response due to series inductance and shunt capacitance. Equalizers are installed by the telephone company at the receiving end of the line or cable, such as at the transmitter for studio-to-

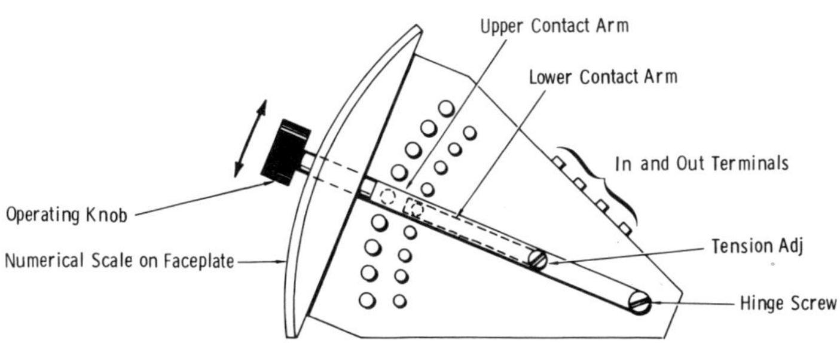

Fig. 6-16. Side view of typical vertical attenuator.

transmitter paths, or incoming network or remote lines at the studio. It is also desirable to have an adjustable line equalizer at the studio for use on remote pickups for which the lowest-priced line service may be requested for a "one-time" pickup (which means the line is not equalized by the telephone company).

Although most commercial line equalizers contain multiple sections for critical equalization, Fig. 6-17 illustrates the principle of operation. The LC network is normally made to resonate at a frequency slightly above the highest frequency of concern, such as 8 kHz for an a-m line or 15 kHz for an fm line. In noncritical remote pickups where only voice signals are involved, this frequency can be made about 5 kHz. The value of adjustable resistance R depends on the length of the line. The longer the line, the more equalization is required; thus a longer line needs less resistance than a shorter line. This is to emphasize that the variable resistance determines the magnitude of equalization; if R is made a short circuit, the maximum equalization is in effect. As the frequency is increased, the impedance of the shunt circuit increases; hence the response on the line is increased.

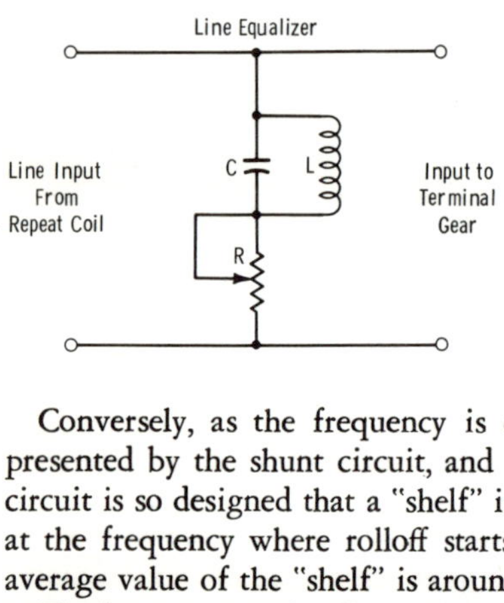

Fig. 6-17. Simplified representation of line equalizer.

Conversely, as the frequency is decreased, less effective impedance is presented by the shunt circuit, and lower frequencies are attenuated. The circuit is so designed that a "shelf" is provided such that equalization starts at the frequency where rolloff starts on the particular line involved. The average value of the "shelf" is around 1000 Hz.

The best way to visualize a line equalizer is to consider it as an attenuator at lower frequencies in the passband, with characteristics such that the response across the entire passband is flat. This explains the inevitable insertion loss of the equalizer, which then obviously depends on length of line, wire size, and frequency band to be equalized. The average value of this insertion loss in practice lies between 30 and 40 dB. A booster or line amplifier with a resistive input terminating network to load the equalizer is required to compensate for this loss.

Production Equalizers

Professional recording, production-type broadcasting, and sound-reinforcement systems incorporate many types of special equalizers. Fig. 6-18

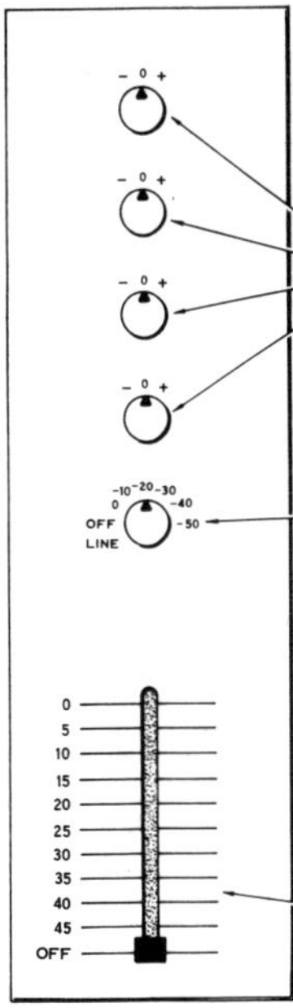

Fig. 6-18. Typical control module for individual channel.

illustrates a typical control-panel module for an individual channel in such a system. Quite often these modules are of the plug-in variety for maximum versatility of application in custom consoles. These may be used with or without echo equipment, which has wide application in special productions (Chapter 11).

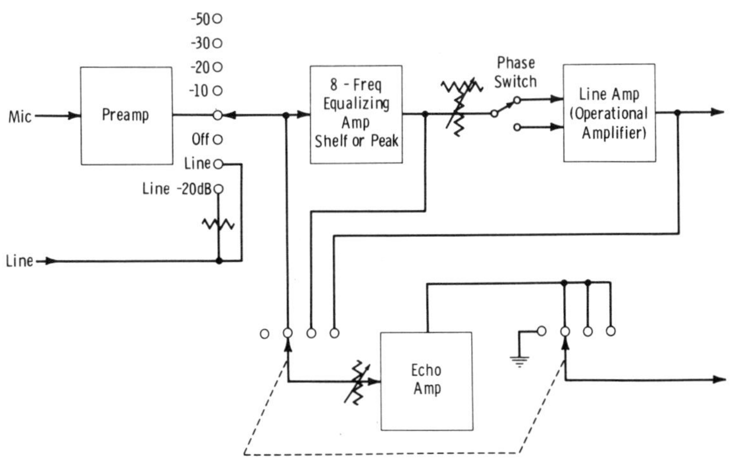

Fig. 6-19. Line-level or microphone input, 8-frequency equalization in Electrodyne system.

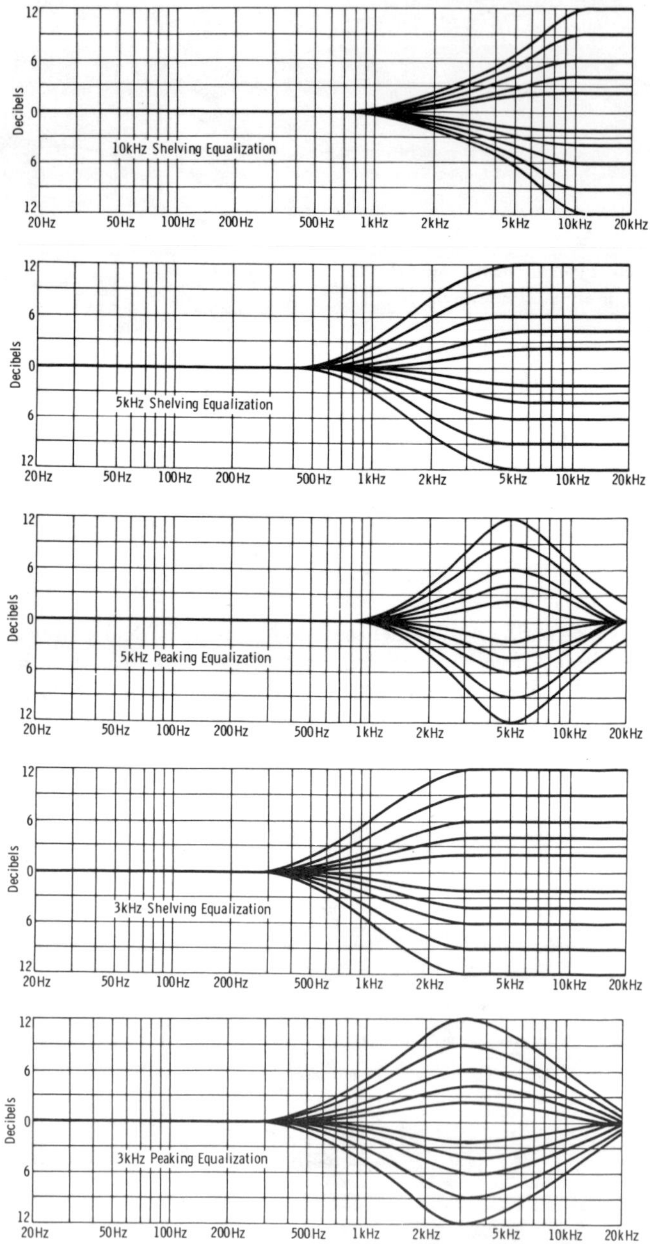

Fig. 6-20. Shelving and peaking

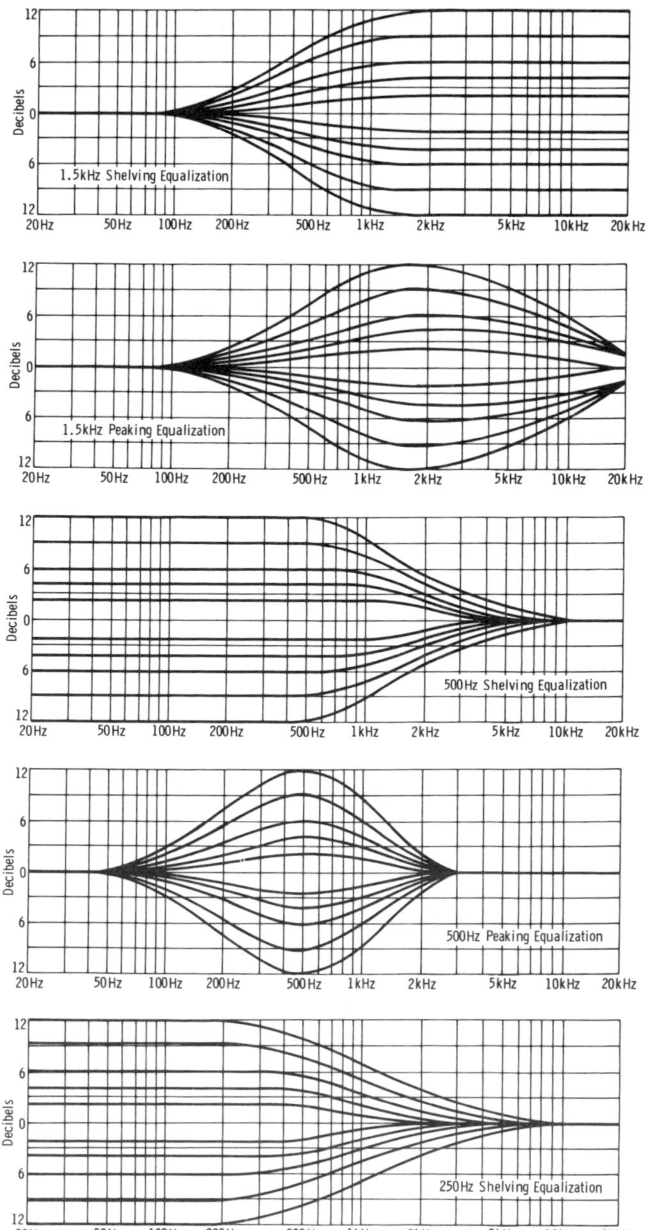

curves of Electrodyne system (continued on next page).

A block diagram of the Electrodyne system, which employs control modules of this type, is shown in Fig. 6-19. The Electrodyne Model 711L combines a low-noise microphone preamplifier; low- and high-frequency equalizers; and echo, cue, and program amplifiers into a single plug-in module. Front-panel controls allow 12-dB boost and attenuation and reciprocal equalization curves for four low frequencies and four high frequencies. Three of the high-frequency curves and two of the low-frequency curves can be selected as shelving or peaking curves (Fig. 6-20). The low-frequency equalization points are selectable at 40, 100, 250, or 500 Hz. High-frequency equalization points are selectable at 1.5, 3, 5, or 10 kHz. The 1.5-kHz position is included to permit control over the critical range required for dialogue and vocal enhancement. The input selector lever (a dual concentric switch) allows selections between microphone and line. There are two positions for the line, one with a 20-dB pad to compensate for high-level input signals. A microphone-preamplifier gain knob is provided, allowing up to 50 dB of gain in the microphone position. Other positions provide 10, 20, 30, and 50 dB of attenuation.

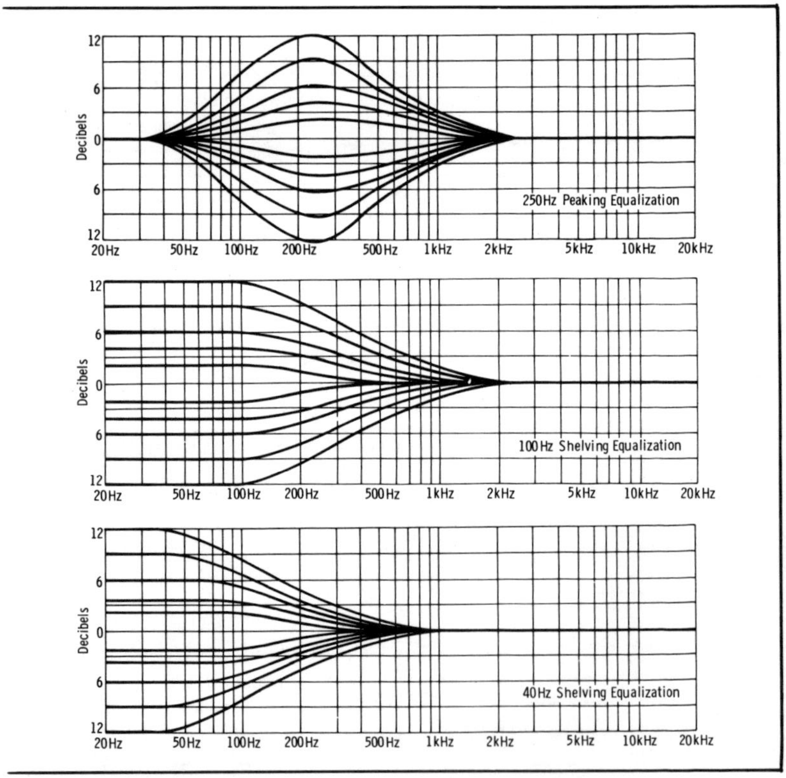

Fig. 6-20. Shelving and peaking curves of Electrodyne system—cont.

The −50 dB position is desirable for handling the high signal levels of modern condenser microphones. This position allows signal levels as high as +18 dBm. A phase-reversal push-button switch is provided to give 180° phase shift to the incoming signal by switching between the inverting and noninverting inputs of an operational amplifier. One echo send pot and selector switch gives the operator a choice of echo send from ahead of the attenuator, ahead of the attenuator but after equalization, or after the attenuator and equalization.

NOTE: The design of equalization networks in general involves advanced study beyond the scope of this text. Equalizers are normally purchased from the manufacturer after applications and specifications are formed by the user. However, for those readers interested in all types of equalizer design, the following text is recommended: Howard M. Tremaine: *Audio Cyclopedia* (Indianapolis: Howard W. Sams & Co., Inc., 1969).

6-5. THE CONTROL CONSOLE

The operations console may be quite complex in terms of the number of circuits and control functions. However, modern centers are designed and installed to achieve an easily operated setup that allows as nearly foolproof switching as possible along with flexibility of functions. Briefly, the general requirements are as follows:

1. Amplifiers are needed for stepping up the minute electric energy produced by turntable pickups, tape heads, and microphones; these are termed preamplifiers. High-level amplifiers are then required to make up losses in such control circuits as the variable attenuators. Isolation amplifiers are required to feed such points as monitoring speakers so that no interaction occurs on program and rehearsal or recording lines.
2. Switching and mixing arrangements on the control console allow selection of the proper program source and blending of individual inputs for desired program "balance."
3. Facilities are provided for auditioning or rehearsing a program to follow, or for recording on tape for use at a later time. These facilities must not "cross-talk" on the regular program line.
4. Inputs and outputs of amplifiers may be normaled through jack panels to allow rapid rerouting of the signal in case of trouble in any one amplifier or channel, or for flexibility in feeding signals to any desired point.
5. Incoming and outgoing lines may be normaled through jack panels to permit receiving or transmitting the signal in any desired path.

Fig. 6-21 illustrates one type of modern commercially built console. This completely transistorized center employs 24 illuminated touch-control keys which allow a total of 45 inputs, replacing the more conventional switches and knobs spread across the control board. Eight large control knobs (faders or attenuators) are provided, one for each of eight mixing channels. A fully interlocked cue-intercom system is also incorporated.

Some control consoles employ self-contained amplifiers and power supplies. Others provide only the mixing controls, switches, push buttons and VU meters, with all amplifiers and power supplies interconnected from rack-mounted units. In all cases, reel-to-reel and cartridge tape equipment as well as the turntables are external units from the control console.

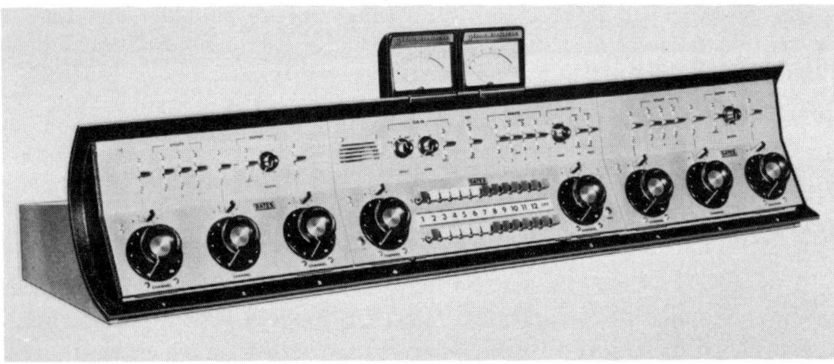

Courtesy Gates Division Harris-Intertype Corp.

Fig. 6-21. Dual-channel console employing transistor circuitry.

Network centers and larger independent stations sometimes use a facility known as a producer's console in conjunction with the individual studio control board. This panel contains a talk-back (intercom), production timing facilities, and appropriate signaling lights.

In large installations, a master control room may be used as the central switching facility. Each individual studio control is linked to the master control room by audio cables, relay control cables, tally-light (signaling) cables, and intercom lines. The audio cable connects the studio output terminals to master-switcher relay contacts; the relay-control cable connects push-button switches and supervisory lamps in the individual studio switcher (through turnkey switches in the master) to the appropriate dc control circuits. Any studio console can be fed to a given number of outgoing channels—the turnkey switches permit the master-control operator to assign control of any channel to any studio console. Push-button switches and lamps in the master switcher are connected to the control circuits of all relays, allowing switching to be performed either by the master operator or studio operator.

Fig. 6-22 is a block diagram of the CCA Executive 8-fader, dual-function control console for a-m or fm use. The 8-fader arrangement provides for up to 19 inputs. Nine inputs are for microphones, two for turntables, five for high-level lines switchable to two faders, and three for tape lines. Any single channel can be switched to the audition or program channel. Output lines are: program (1), muted speaker (4), intercom (5), headphones (2), low-level production channel (1), and high-level (+18 dBm) production channel (1).

Three-position selector switches for the microphone and tape channels are mounted immediately above their corresponding attenuators. The channel 6 position is normally used for the control-room microphone. The turntable, tape, and remote-line faders contain a cue provision as described earlier in this chapter.

The two remote (or network) faders have five high-level inputs available which can be switched to either of the two channels. It is impossible for the same input to be connected to both faders simultaneously; thus "segueing" (fading out one source while fading in the next source) between the two channels can be accomplished. When a remote-line input selector switch is in the center position, the program is fed back to that remote channel for cue or public-address amplifier feed at the remote pickup point.

The console monitor amplifier has facilities to select the program, audition (production), or external lines as a driving source. It also has facilities to drive the normal program or production lines. Thus it can serve as an emergency program line amplifier or a production line amplifier and simultaneously drive four speakers. The VU meter can be switched to the program line, the production line, or an external source.

Speaker Muting Circuitry

Whenever a microphone is turned on, the monitoring speaker in the same studio must be cut off simultaneously to avoid acoustical feedback. This is accomplished by a control circuit which is tied through a set of contacts on the microphone switch and which operates a muting relay for the studio speaker. The speaker is thereby disconnected, and a resistor of equivalent impedance is placed across the monitor bus to provide the same termination. If this is not done, other speakers connected to the same monitor bus would exhibit varying sound levels as speakers were turned off and on.

While some speaker muting arrangements employ only relay contacts actuated from the microphone key, others, such as the one in the CCA Executive described above, incorporate transistor switches for the relays. Fig. 6-23 illustrates the basic action of a two-studio system for purposes of discussion.

When the line-voltage ac is applied to the console, the +24-volt power supply is on, and all transistors (Q1 and Q2 in Fig. 6-23) are fully conducting due to the base current through the 22K resistors. Therefore, the associated relays are energized, the monitor speakers are on, and the on-air lights are off. In Fig. 6-23, it is assumed that a microphone is turned on in Studio A. When a microphone key is closed, the base of the corresponding transistor is shorted to ground, turning that transistor off and de-energizing the associated relay. This turns the on-air light on and mutes the monitor speaker by opening the high side of the circuit and substituting the back load resistor (R1 in this example) for the former speaker load resistance. Thus the signal level on the monitor bus is maintained the same in other studios whether another studio monitor speaker is on or off.

With the transistor arrangement, there is virtually no "click" in the system as muting is accomplished, since there is a total of only about 1 mA that turns the relays on and off. Diodes across the relays are used to damp the system against counter emf's from the relay coils.

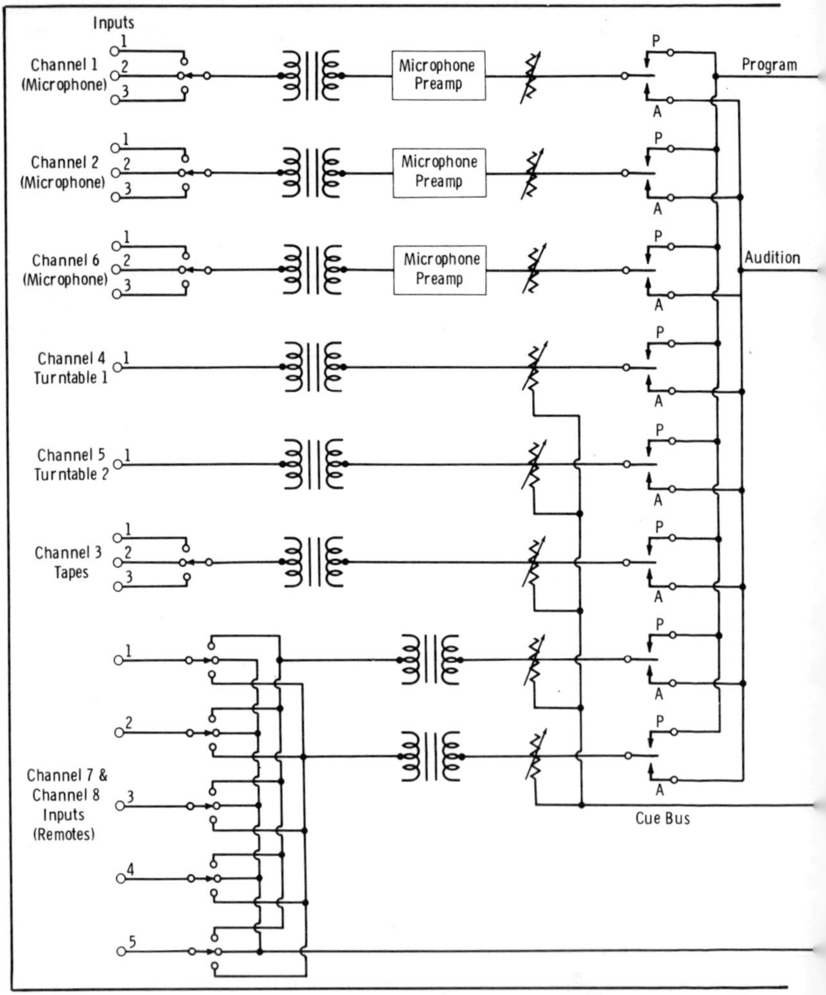

Fig. 6-22. Simplified diagram of CAA Executive

Talk-Back System

The CCA console previously described is designed to provide front-panel talk-back to auxiliary channels 9 and 10. From Fig. 6-24, it can be seen that there is a transistor which is normally off. Turning the talk-back switch to studio A, B, C, or remote turns the transistor on by applying forward bias to its base. This operates the talk-back relay. The talk-back relay reverses the cue bus so that the front-panel speaker serves as a microphone. The output of this speaker goes through the talk-back switch, then the talk-back relay, then the cue amplifier, back again through the talk-back

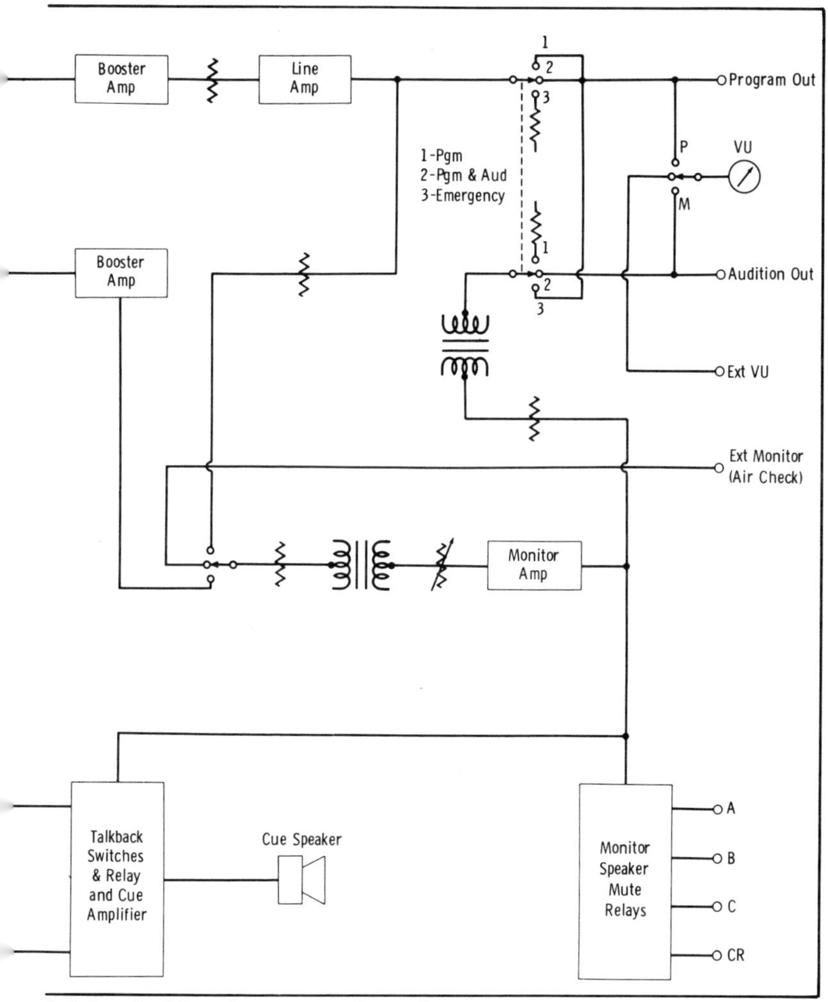

8-fader, dual-function, monaural console.

relay, then to the talk-back keys front-panel switch, and then to the appropriate studio speaker selected by the talk-back key. It should be noted that the monitor amplifier normally feeds the program channel to the auxiliary lines when the switches are in the center position.

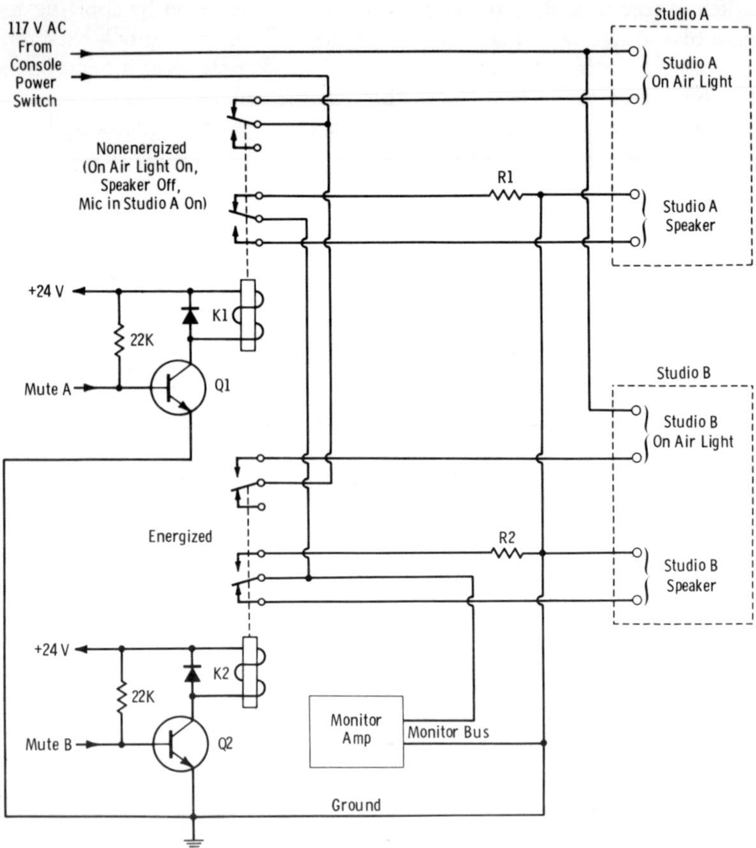

Fig. 6-23. Speaker muting and on-air light arrangement.

Electronic Attenuators and Switches

It now remains to become familiar with the latest type of electronic console gain controls; when these controls are used, audio itself is not brought into the control console except for VU-meter or monitoring purposes. In Fig. 6-25A, the light-controlled attenuator and light source are in one module which is placed in the audio circuitry. The control potentiometer is in the audio control console, which may be at some distance from the audio racks. Note in this case that the slider arm rotates between plus and

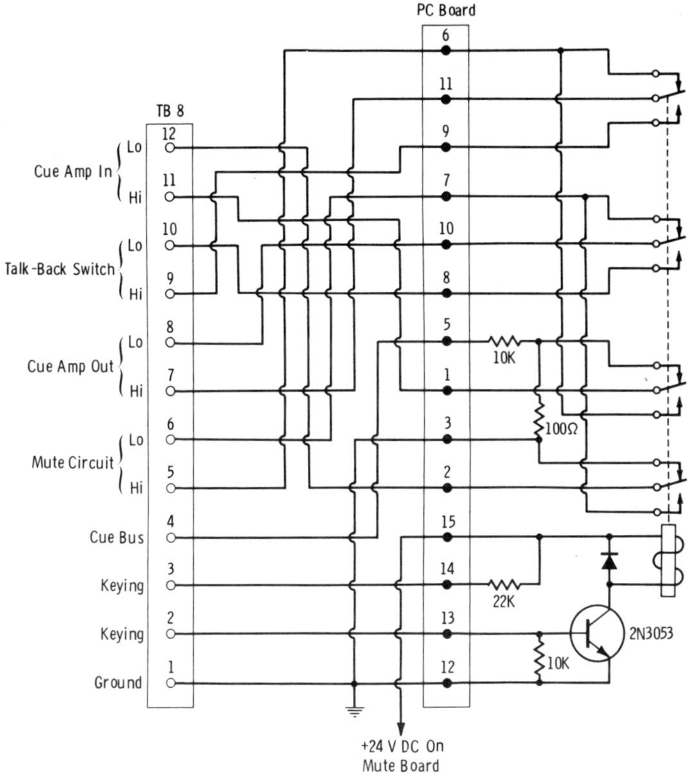

Fig. 6-24. Talk-back board for CCA consoles.

minus voltages so that the constant impedance of the attenuator is maintained just as in mechanical faders.

Fig. 6-25B illustrates the basic principle of these controls. The cells are normally of the cadmium-sulfide type, and the intensity of the light or lights (located adjacent to the cell) changes the cell resistance, thereby affecting the gain of the audio path in which the cell is placed. This provides a noise-free control over a wide dynamic range without transients or contact chatter. The resistance *decreases* as the lamp intensity *increases.* A current-limiting resistor may be found in series with the lamp, or a transistor constant-current source may be used. The control current is then linear with voltage over a stated range. Fig. 6-25C shows typical terminal arrangements.

The cell resistance at maximum light intensity may vary between about 50 and 150 ohms. The dark resistance may be as high as 100 megohms or more depending on type. A reaction time of 5 to 20 milliseconds is typical.

A defective unit is indicated by an open circuit between the control-circuit (lamp) terminals. Normal resistance at the lamp terminals as

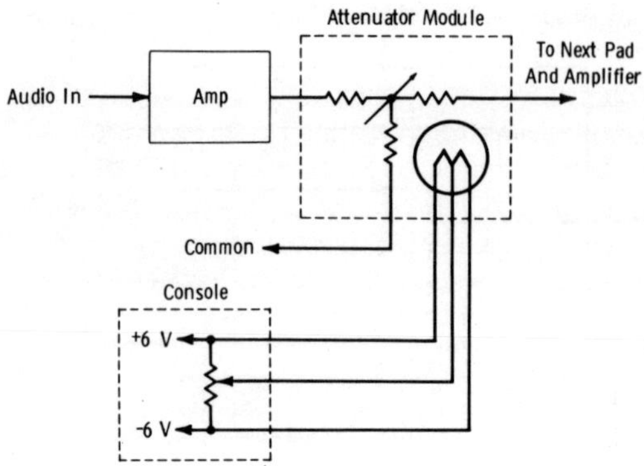

(A) Typical application.

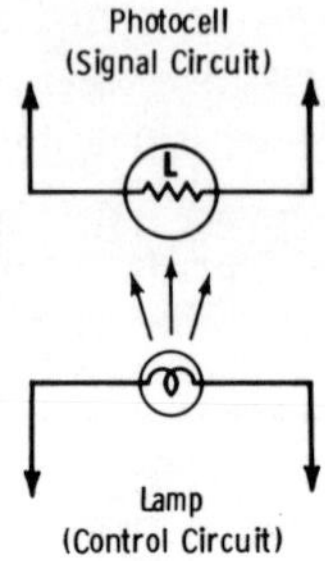

(B) Principle of attenuator.

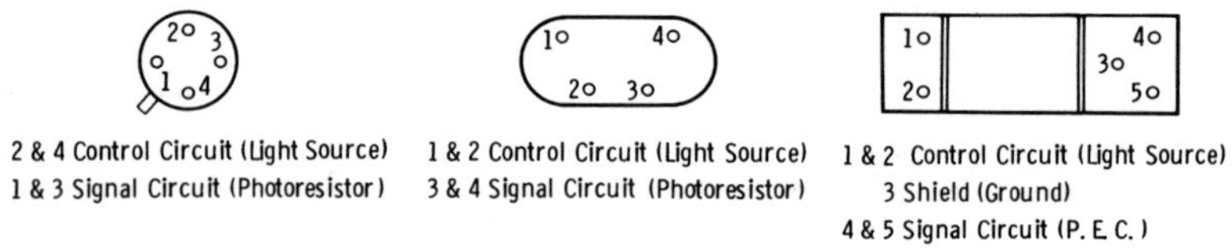

(C) Terminal arrangements.

Fig. 6-25. Light-controlled attenuators.

(A) FET as voltage-variable attenuator.

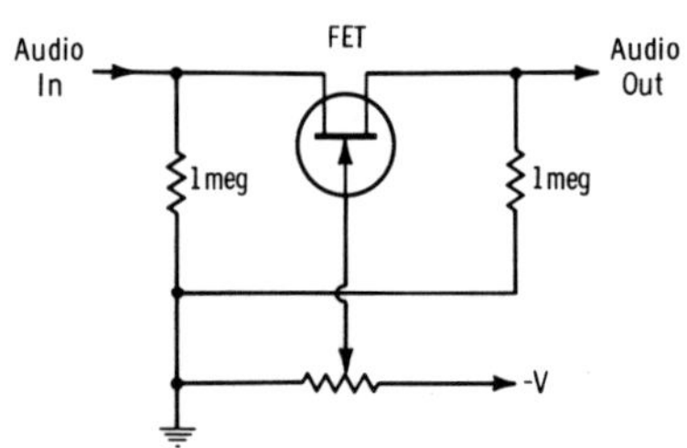

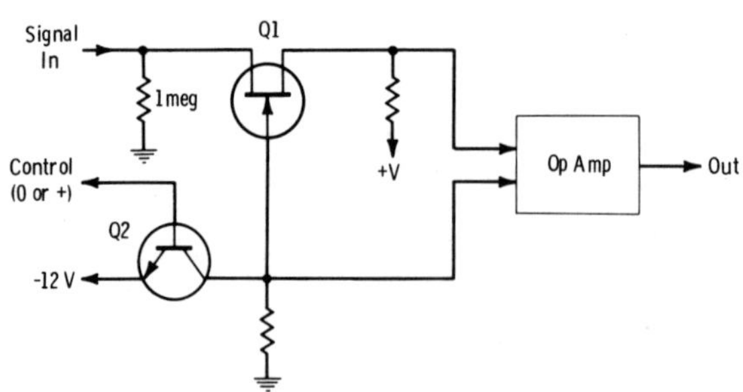

(B) Typical FET switch circuit.

Fig. 6-26. Attenuators using FETs.

checked with an ohmmeter is around 100 to 200 ohms, depending on the voltage rating of the lamp.

Fig. 6-26A illustrates another type of electronic attenuator, in which an FET is employed as a voltage-variable resistance. In this application, the drain-source voltage is biased below pinch-off. The series resistance is about 10^{10} ohms when the gate is at maximum negative voltage, decreasing linearly to about 240 ohms with the gate at ground potential.

Fig. 6-26B shows a typical FET circuit used for audio switching in place of relay or key contacts. When the control voltage on the base of Q2 is zero, or low, Q2 is off and forms an open switch. Transistor Q1 is turned fully on, presenting a series impedance of about 240 ohms. When the control voltage on the Q2 base goes toward 1 volt (high), Q2 saturates, forming a closed switch to apply −12 volts to the gate of Q1, turning the FET

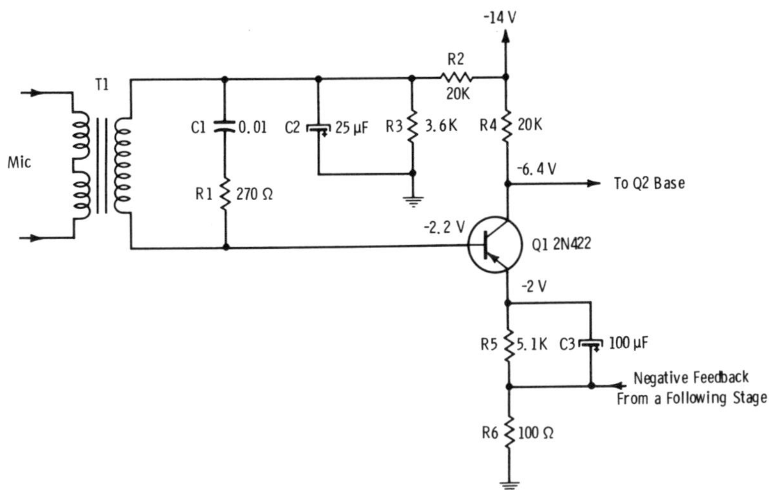

(A) Input stage.

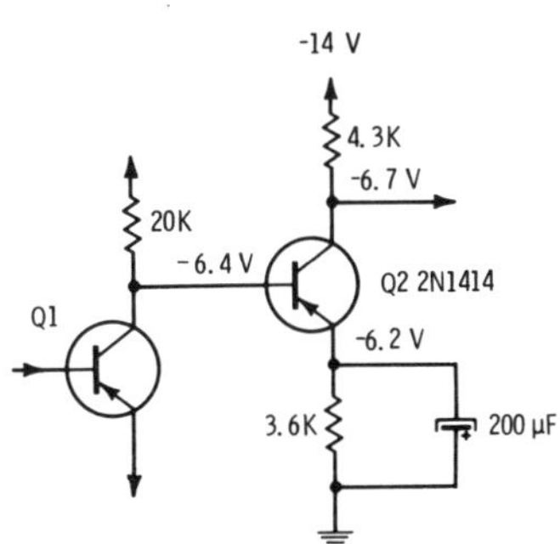

(B) Second stage.

Fig. 6-27. Microphone preamplifier circuits.

off. In this condition, the series resistance presented to the signal is about 10^{10} ohms. Switches using FETs lend themselves readily to the logic portions of automation systems.

6-6. MICROPHONE PREAMPLIFIERS

A common microphone input circuit for broadcast application is shown in Fig. 6-27A. Broadcast microphones are always of low impedance, so the first requirement (low source resistance) for a low noise factor is met. Second, the microphone normally has fairly long cables, so a transformer input is used to minimize hum and other extraneous pickup.

The primary of T1 is balanced to eliminate hum and noise pickup. The secondary is operated unloaded, which is conventional for low-impedance microphones. Note that the base of Q1 is series-fed from the secondary of T1; this is the type of feed often used in solid-state broadcast applications. This technique minimizes any tendency toward load change with signal variation and results in maximum input gain (maximum transducer gain) from T1. The network R1-C1 across T1 is used to roll off frequencies above the audio passband; therefore, it is sometimes stated that the network provides high-frequency stabilization of the circuit.

Since V_{CE} and the emitter current must be kept very low for a low noise factor, high values of R_L (R4) and R_E (R5-R6) are invariably found in the first stage of a preamplifier.

Now we will go through the dc analysis (operating point) for this stage. Voltage divider R2-R3 provides about −2.2 volts to the base of Q1. Since the 2N422 is a germanium transistor, the emitter voltage will be about −2 volts. So the emitter current is 2/5.2K = 0.38 mA (approx).

Assuming the collector current is the same as the emitter current, the collector voltage is $-[14-(0.38)(20)] = -6.4$ volts, so $V_{CE} = -6.4 - (-2) = -4.4$ volts.

If you ignore negative feedback and the input impedance of the following stage, what voltage gain would you expect?

$$r_{tr} = \frac{26}{0.38} + 100 = 168$$

and

$$A_v = \frac{R_L}{r_{tr}} = \frac{20000}{168} = 120 \text{ (approx)}$$

Fig. 6-27B shows a typical second stage into which the circuit of Fig. 6-27A might be coupled. Since the emitter current of this second stage is 6.2/3.6K = 1.7 mA:

$$r_{tr} = \frac{26}{1.7} + 4 = 19 \text{ (approx)}$$

and

$$Z_{in} = (h_{fe})(r_{tr}) = (40)(19) = 750 \text{ ohms (approx)}$$

Now the effective R_L of the first stage is 20K in parallel with 750 ohms, or about 720 ohms. Therefore the actual A_V of the first stage (still ignoring any negative feedback) is:

$$A_V = \frac{R_L}{r_{tr}} = \frac{720}{168} = 4.3$$

In practice, negative feedback (which is usually applied by an arrangement similar to that in Fig. 6-27A) can bring the voltage gain down to unity or less (depending on the type of feedback).

This analysis has been made to emphasize a point: The first stage of a preamplifier is usually operating at a very low gain, sometimes an actual voltage loss, in the interest of maximum fidelity and low inherent noise.

A slightly different approach is taken in the RCA BA-72A preamplifier. As shown by the simplified schematic of Fig. 6-28, this preamplifier consists of an unloaded input transformer (T101), a two-stage negative-feedback amplifier (Q1 and Q2), a two-stage output amplifier (Q3 and Q4), and an output transformer (T102). The Q1 collector load (R5) is 10K, the Q2 collector load (R13) is 2K, and the bypassed emitter resistor of Q2 (R14) is 200 ohms.

The high side of the input-transformer secondary is connected directly to the base of transistor Q1. This transistor operates in a common-emitter circuit with its collector connected to the base of Q2 through capacitor C6. Transistor Q2 also operates in a common-emitter circuit. The emitter resistor of Q2 is completely bypassed by capacitor C8, making the input impedance of Q2 very low. The low input impedance causes a larger signal current in the base of Q2, which results in a higher signal voltage developed across R13, the collector load resistor.

Negative feedback current passes from the collector of Q2 to the emitter circuit of Q1 through C9 and R9. Capacitor C5 bridges R9 to improve the signal response at high frequencies. The feedback is used mainly for stabilizing the gain and reducing distortion. As shipped from the factory, the 46-dB terminals are not strapped, and the gain of the preamplifier is 40 dB. If 46-dB gain is desired, a strap is provided which shunts R4 across R7, thereby reducing the inverse feedback voltage developed in the emitter circuit of Q1.

The voltage developed across R13 is direct-coupled to the base of Q3, which operates as an emitter follower. The emitter of Q3 is coupled by C12 to the primary winding of the output transformer, T102. The voltage developed across R15 is applied through C10 to the base of Q4. The collector output of Q4 is direct-coupled to the emitter of Q3, where it then passes through C12 to the output transformer. In the Q3-Q4 circuit, a current increase in Q3 causes a decrease in Q4. In effect, Q4 is a variable

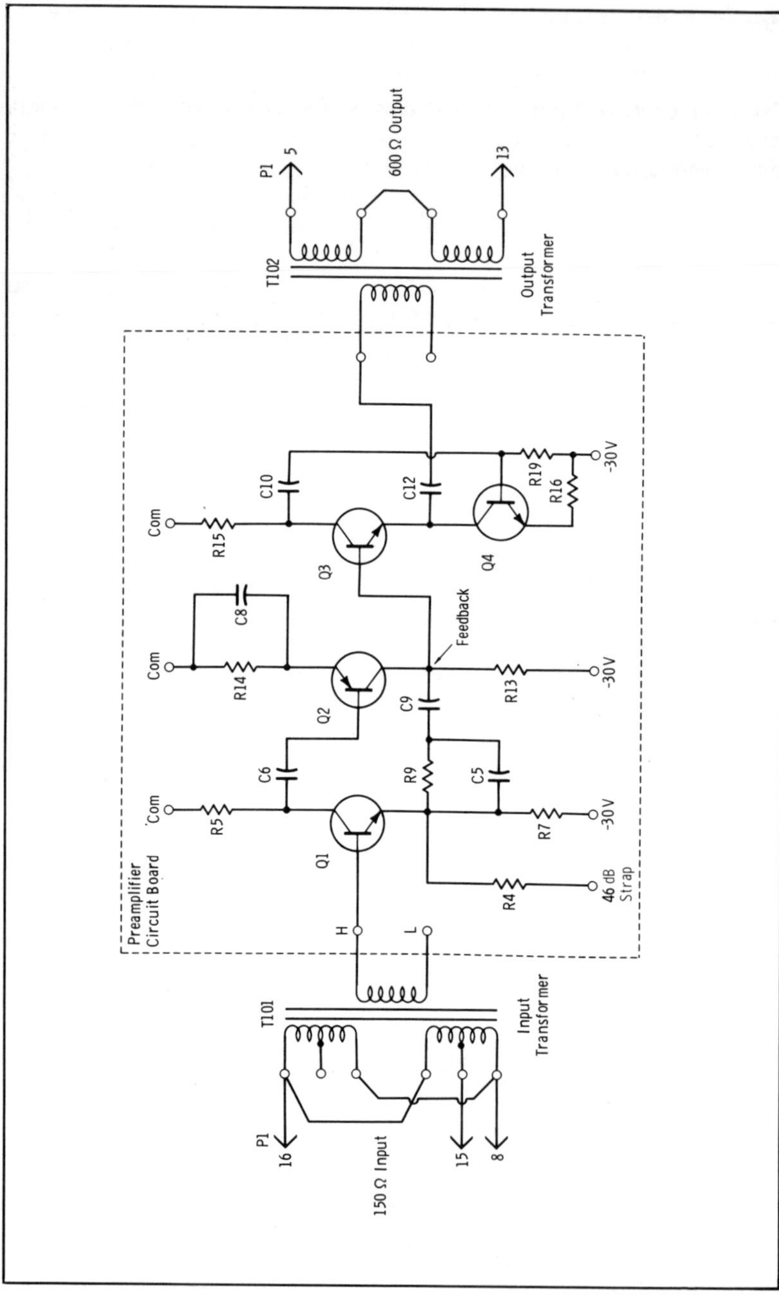

Fig. 6-28. Simplified schematic diagram of RCA BA-72A preamplifier.

emitter load for emitter follower Q3, which results in increased current in output transformer T102.

Note that in the description of the negative feedback circuit above, it was stated that C5 (620 pF) across R9 (6190 ohms) improves the high-frequency response. Capacitor C9 (15 μF) has no frequency discrimination. Since this is *negative* feedback, as the frequency is increased with decreased reactance of C5, more inverse feedback is provided, *decreasing* the high-frequency gain while reducing distortion. The highest frequency of concern in the audio passband is 20 kHz. But a rapid roll-off above this frequency must not occur since high harmonic distortion results in distortion of the fundamental frequencies. The reactance of C5 at the second harmonic of 20 kHz (40 kHz) is about 6200 ohms, or approximately equal to the resistive value of R9 (6190 ohms). In this manner, excellent frequency response at very low harmonic distortion is obtained. For the same reason (obtaining high-frequency stabilization), a 430-pF capacitor is shunted across the input transformer secondary (not shown on simplified schematic). Without such stabilization, supersonic oscillations can occur, and this would result in high total harmonic distortion.

6-7. AMPLIFIER INPUT LOADING

The input transformer of a professional audio amplifier generally provides for a source impedance of either 150 or 600 ohms. It may be adjusted for either impedance by transformer taps. In actuality, for an unloaded input transformer, the input impedance is higher than the source impedance for all frequencies from 20 to 20,000 Hz. An amplifier can never be

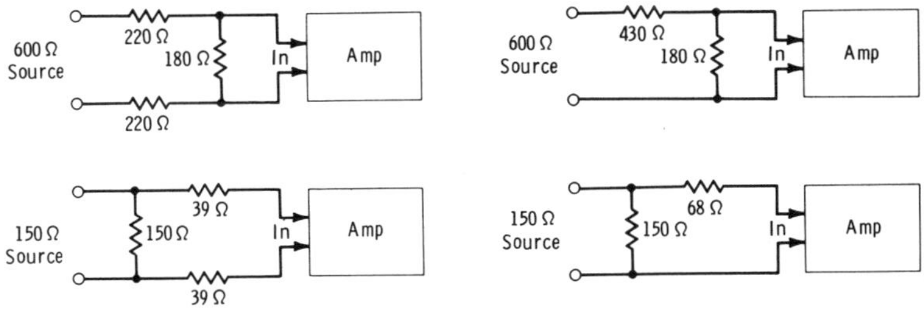

(*A*) *Balanced input.* (*B*) *Unbalanced input.*

Fig. 6-29. Matching pads for amplifier driven by a preceding amplifier.

fed from the output transformer of a previous amplifier (such as using the amplifier as a booster) without resistive loading of the input to provide a constant impedance across the passband. The unloaded input transformer is used only when driven by a microphone or low-impedance source such as a phono pickup.

Minimum-loss (6 dB) matching pads for this purpose are shown in Fig. 6-29. Additional pads may be required to reduce the level at the amplifier input terminals to its specified maximum value.

6-8. TURNTABLE PREAMPLIFIERS

In practice, it is impossible to consider the stylus, pickup head, and preamplifier separately since each depends on the other. Even recording characteristics themselves must be included in the overall discussion. There are two basic methods of recording, the constant-velocity and the constant-amplitude methods.

Constant-Velocity Recording

Constant velocity refers to the maximum transverse velocity of the stylus tip at the mean axis. Since this is held constant as the frequency changes, the peak amplitude is inversely proportional to the frequency (Fig. 6-30A). The maximum slope of the curve is the same for all frequencies.

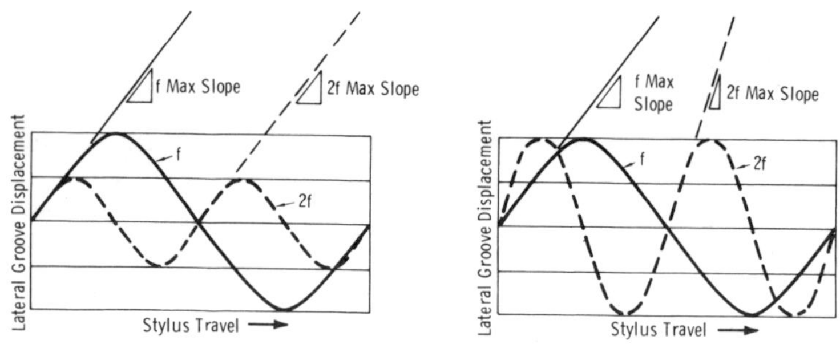

(*A*) *Constant-velocity recording.* (*B*) *Constant-amplitude recording.*

Fig. 6-30. Two basic methods of recording.

This method of recording is not suitable over a wide frequency range. For example, over a range of 8 octaves (each octave doubles the frequency) the ratio of maximum to minimum amplitudes is 256 to 1. This results in an impractical variation in peak amplitudes of the recorded grooves.

Constant-Amplitude Recording

The peak amplitude is held constant (for constant power output) as the frequency changes; therefore, the maximum slope is proportional to the frequency (Fig. 6-30B). This method is satisfactory for low frequencies but is not suitable for large amplitudes at the highest frequencies. This is due to excessive transverse velocity of the needle tip, which produces distortion in both recording and reproduction.

Combination Recording

Constant-amplitude recording is optimum for low frequencies; constant-velocity recording is quite satisfactory over a limited range including the medium and high frequencies. Therefore, practical recording systems employ an approximation to constant-amplitude recording at low frequencies and an approximation to constant-velocity recording for the medium- and high-frequency range. The crossover from constant amplitude to constant velocity normally occurs near 500 Hz. Any difference in the crossover frequency requires a different playback equalization for a flat frequency response.

Since the maximum slope of the displacement curve in the constant-amplitude mode is proportional to the frequency (Fig. 6-30B), an increased voltage is obtained with an increase in frequency. The voltage increases 6 dB per octave. The "knee" of the crossover region is rounded off; that is, there is no sharp demarcation between the constant-amplitude and constant-velocity characteristics. The recording then enters the constant-velocity region and high-frequency preemphasis.

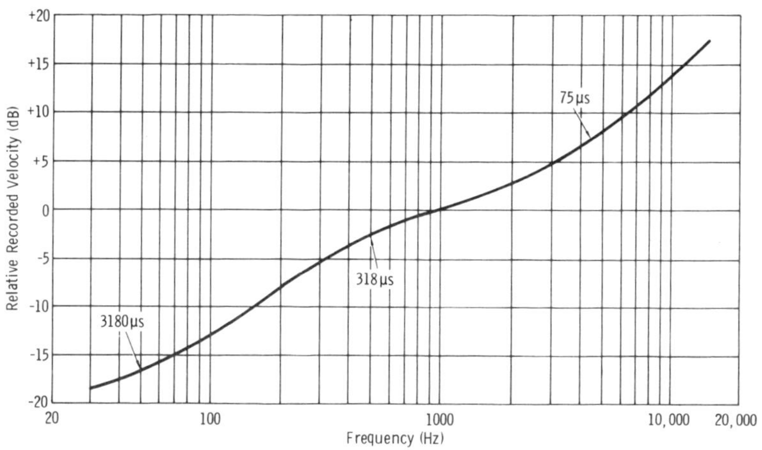

Fig. 6-31. EIA standard lateral disc-recording characteristic.

In practice, both the low- and high-frequency regions undergo certain equalization procedures which modify the recording characteristics just described. For standardization, the slope of the curve in decibels per octave is normally stated as a time constant. For example, a simple RC network comprising a tone control is standardized by the RC time constant. Fig. 6-31 illustrates the EIA standard lateral disc-recording characteristic. Table 6-6 tabulates the recording and complementary playback curves. This characteristic has been adopted by NAB and RIAA.

Table 6-6. Lateral Disc Characteristics (EIA, NAB, and RIAA)

Frequency (Hz)	Recording Characteristic	Reproducing Characteristic
15,000	+17.17	−17.17
14,000	+16.64	−16.64
13,000	+15.95	−15.95
12,000	+15.28	−15.28
11,000	+14.55	−14.55
10,000	+13.75	−13.75
9000	+12.88	−12.88
8000	+11.91	−11.91
7000	+10.85	−10.85
6000	+ 9.62	− 9.62
5000	+ 8.23	− 8.23
4000	+ 6.64	− 6.64
3000	+ 4.76	− 4.76
2000	+ 2.61	− 2.61
1000	0	0
700	− 1.23	+ 1.23
400	− 3.81	+ 3.81
300	− 5.53	+ 5.53
200	− 8.22	+ 8.22
100	−13.11	+13.11
70	−15.31	+15.31
50	−16.96	+16.96
30	−18.61	+18.61

The standard recording characteristic is specified as the algebraic sum of the ordinates (expressed in decibels) of three individual curves that conform to the admittances of the following three networks:

1. A parallel LR network having a time constant of 3180 μs.
2. A series RC network having a time constant of 318 μs.
3. A parallel RC network having a time constant of 75 μs.

Note from the curve of Fig. 6-31 that these three time constants specify the equalization of the frequencies up to the crossover frequency of 500 Hz, the rolloff of the crossover point itself, and the high-frequency pre-emphasis.

The overall combination of the stylus, pickup head, arm, and preamplifier equalizer must provide the proper complementary reproduction characteristic given in Table 6-6. Then the overall response will be "flat."

Circuit Analysis of Typical Phonograph Preamplifier

Fig. 6-32 shows a typical preamplifier for a magnetic phonograph pickup or tape head. This circuit was chosen purposely to explain how one would go about analysis of the dc operating point. Note that the base of Q1 is biased from a current that is directly proportional to the emitter current

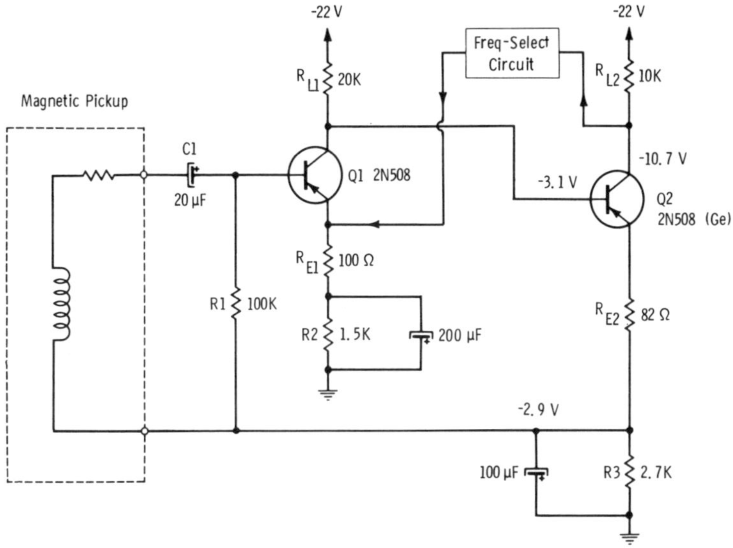

Fig. 6-32. Preamplifier for pickup.

of Q2. This connection stabilizes voltage and current bias points for changes in operating temperature and transistor h_{FE}.

This circuit would be difficult to analyze for dc voltage operating point except for the basic rule of thumb that class-A amplifiers have a collector voltage somewhere near one-half the collector supply voltage. (This rule applies to the second and following stages of a preamplifier, *never* to the first stage.) If we start with the assumption of −11 volts at the Q2 collector, we can see how close we would come to the actual voltage readings shown, as follows:

Since this is a class-A amplifier, assume the Q2 collector voltage is one-half the collector supply voltage, or −11 volts. Then for Q2:

$$I_C = \frac{11\text{ V}}{10\text{K}} = 1.1\text{ mA}$$

$$I_E = 1.1\text{ mA (approx)}$$

$$V_E = (-1.1\text{ mA})\ (2700\text{ ohms}) = -3.0\text{ V}$$

$$V_B = -3.0\text{ V} + (-0.2\text{ V})$$

$$= -3.2\text{ V (2N508 is germanium)}$$

Now, for Q1:

$$V_C = -3.2\text{ V}$$

$$\text{Drop across } R_{L1} = -22\text{ V} - (-3.2\text{ V}) = 18.8\text{ V}$$

$$I_C = 18.8/20\text{K} = 0.94\text{ mA}$$

$$I_E = 0.94 \text{ mA (approx)}$$
$$V_E = (-0.94 \text{ mA})(1600 \text{ ohms}) = -1.5 \text{ V}$$
$$V_B = -1.5 + (-0.2) = -1.7 \text{ V}$$

Note also the similarity of this circuit to those of Fig. 6-27A and Fig. 6-27B combined. The second stage has a relatively low input impedance.

Response-curve equalization is sometimes located in the negative-feedback circuit, as in Fig. 6-32. Fig. 6-33 shows a typical feedback network for RIAA/NAB playback characteristics. Note that the capacitors across the resistors provide more feedback at high frequencies than at lower frequencies. Since this is a degenerative circuit, high-frequency response is reduced to provide the proper playback response. This method eliminates the need to load a magnetic cartridge with the proper resistance for high-frequency compensation. Just as in the case of a microphone, a magnetic phonograph pickup is best operated into a relatively high-impedance circuit with proper equalization following the input.

The output stage of a broadcast preamplifier normally feeds a 150-ohm or 600-ohm fader. Fig. 6-34A shows a typical emitter-follower output circuit. To review the analysis of this stage, assume a 600-ohm fader, a 720-ohm source resistance, and a β of 50. Then:

$$Z_{out} = \frac{720}{50+1} = 14 \text{ ohms (approx)}$$

and:

$$R2 = 600 - 14 = 586 \text{ ohms (nearest EIA value} = 560 \text{ ohms)}$$

Fig. 6-34B shows the typical output stage when a transformer is used. The negative feedback provided by R_f reduces the output impedance as well as the input impedance. The output-transformer primary is always loaded with a fixed resistor to insure a fixed load on the collector circuit.

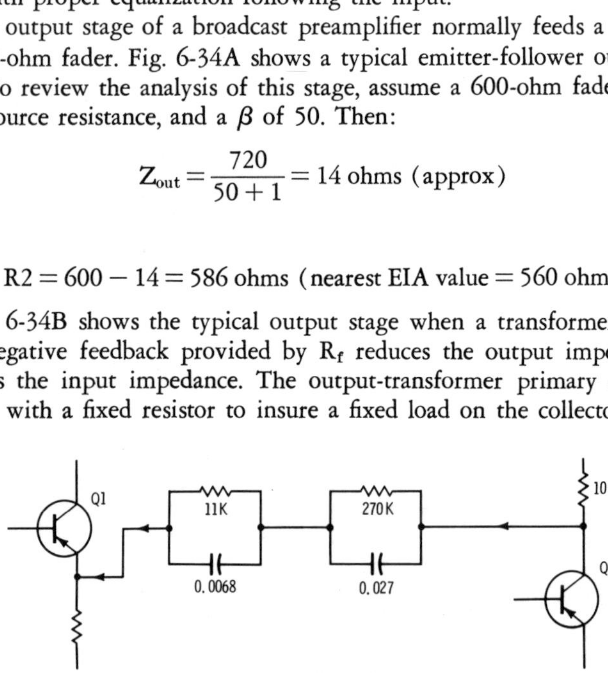

Fig. 6-33. RIAA/NAB feedback network.

6-9. AUDIO SYSTEM TECHNOLOGY

This section will describe the overall audio system, including program (line) and monitor amplifiers and the fundamentals of automatic audio

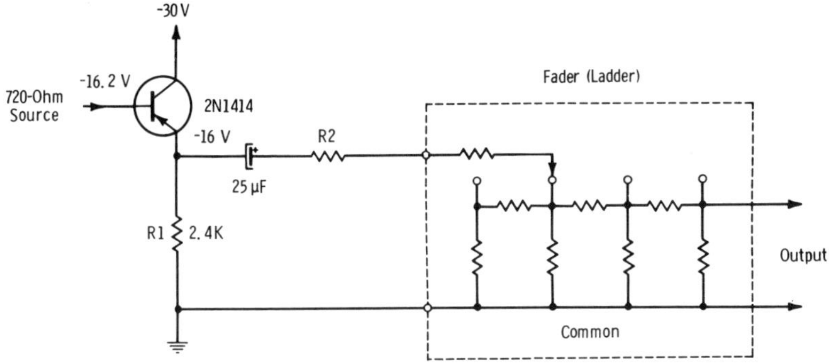

(A) Emitter follower.

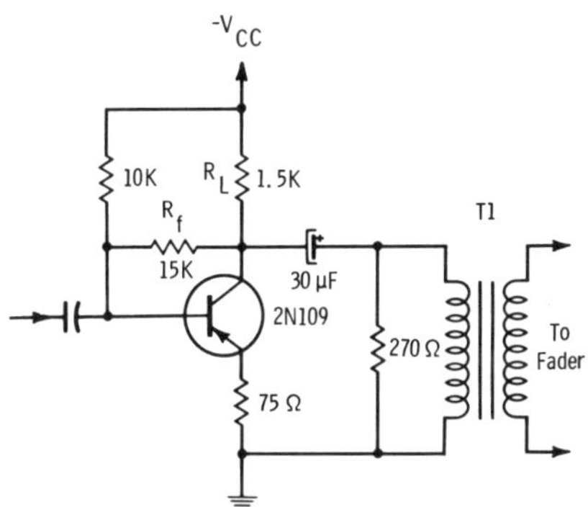

(B) Use of transformer.

Fig. 6-34. Attenuator coupling.

signal processing systems. The simple installation of Fig. 6-35 will serve for basic discussion. Circuit impedances of 150 ohms are typical up to the line feed of 600 ohms, but all the impedances could be 600 ohms. Conversely, many lines now have 150-ohm impedances, and all circuits in the studio might be 150 ohms.

A typical level at the microphone input is −50 dBm. The gain of the preamplifier is 40 dB, so the output into the mixing circuitry is −10 dBm. Typical operating mixer loss is 18 dB, giving −28 dBm at the mixer output. Since the maximum input level to most line amplifiers is around −32 to −35 dBm, a 10-dB pad loads the mixer output to provide a line-amplifier input level of −38 dBm. In this example, the line amplifier is adjusted to provide +18 dBm into the master-gain attenuator. A typical loss setting here is 10 dB, providing +8 dBm to the line-isolation pad. The VU-meter multiplier is set so that a level of +8 VU at this place in the system gives 0 deflection on the meter. (At this point, it should be emphasized that the reader must have a good working grasp of VU, dB, and dBm relationships. Review Section 2-10, Chapter 2).

Typical Program (Line) Amplifier

A simplified schematic of the RCA BA-43 program amplifier is shown in Fig. 6-36. This unit consists of an input transformer and a two-stage feedback preamplifier coupled through an emitter follower to a four-stage output amplifier.

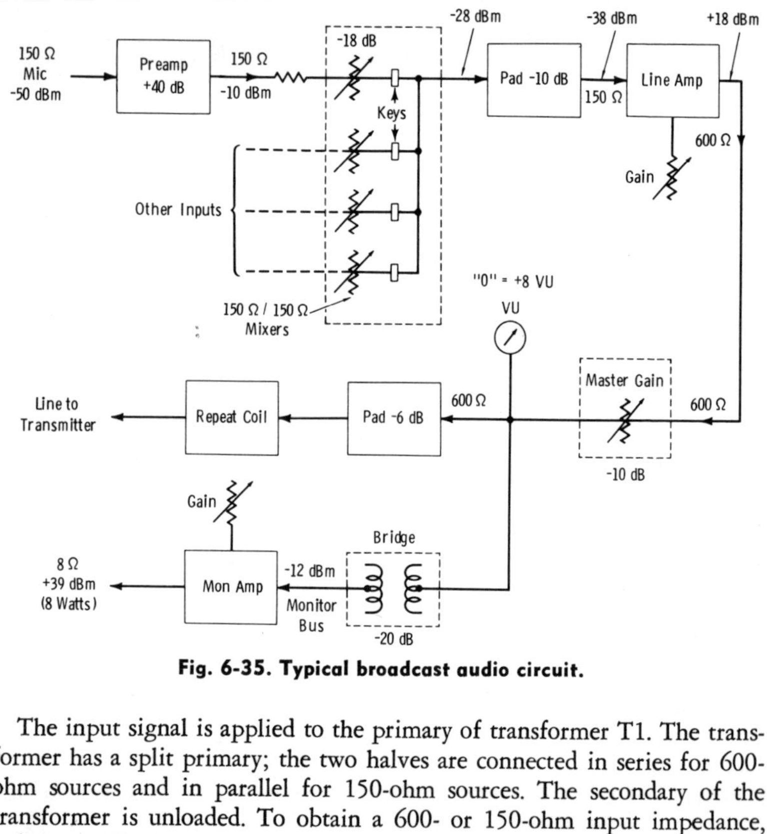

Fig. 6-35. Typical broadcast audio circuit.

The input signal is applied to the primary of transformer T1. The transformer has a split primary; the two halves are connected in series for 600-ohm sources and in parallel for 150-ohm sources. The secondary of the transformer is unloaded. To obtain a 600- or 150-ohm input impedance, a 600- or 150-ohm resistor must be connected across the primary. The signal from the secondary of T1 is applied through capacitor C2 to the base of Q1. The amplified signal from the collector of Q1 is applied to the base of Q2 and is again amplified. The signal from the collector of Q2 is applied through resistor R9 to the base of emitter follower Q3. Negative feedback current from Q3 to Q1 passes through resistor R7. The output of the preamplifier passes through C5 to gain control R8 and then to terminal 9 of connector P1.

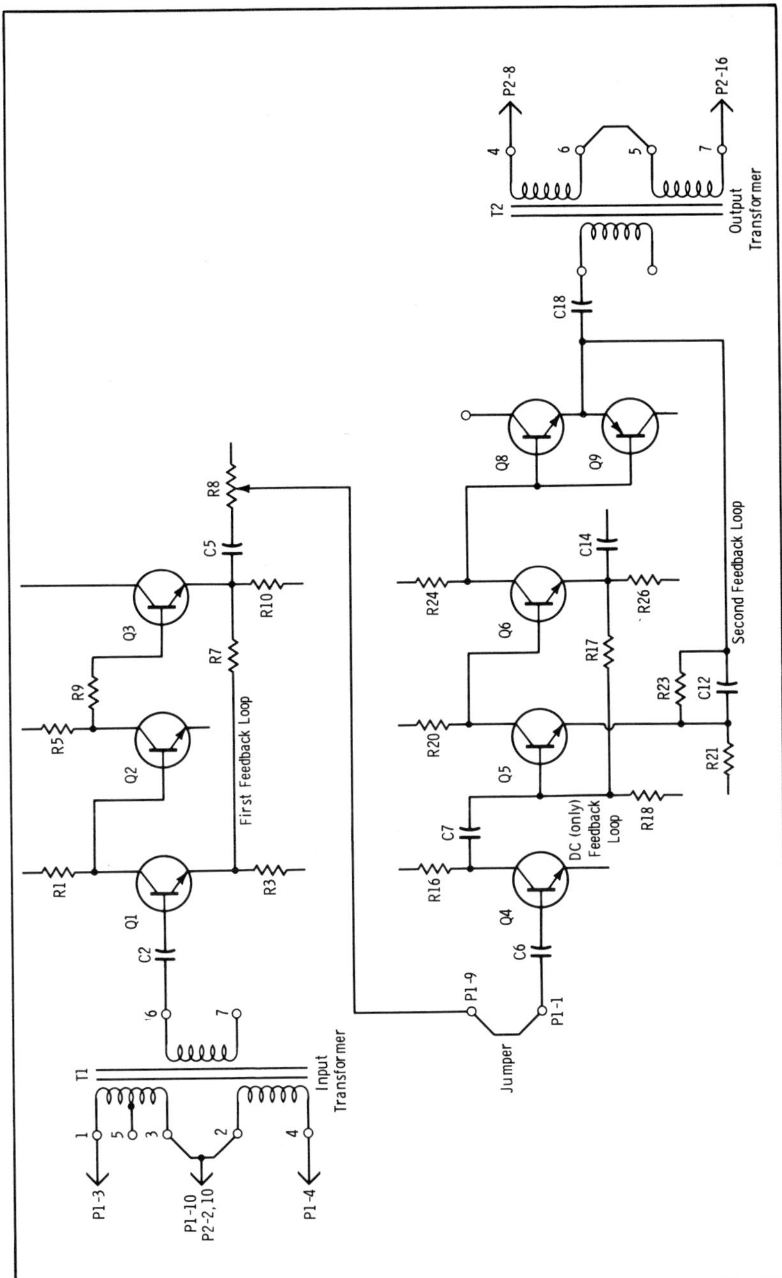

Fig. 6-36. Simplified schematic of RCA BA-43 program amplifier.

The operation of the output-amplifier portion is as follows: The input signal goes from terminal 1 of connector P1 to the base of Q4, is amplified, and is coupled to the base of Q5 through capacitor C7. The signal is amplified by Q5 and Q6, and the output from the collector of Q6 is applied to the bases of Q8 and Q9 which operate as a complementary-symmetry output stage. The output from the emitters of Q8 and Q9 is applied through C18 to output transformer T2. A dc (only) feedback loop exists through R17 between the Q6 emitter and the Q5 base. A second feedback loop in the output amplifier is formed by R23 and C12 between the output and the Q5 emitter. The result is a frequency response which is better than ±1 dB from 20 to 20,000 Hz at the maximum output level of +30 dBm. At this output level, the output noise level is less than −43 dBm, giving a signal-to-noise ratio of 73 dB, at a total harmonic distortion of 0.5 percent maximum.

Automatic Level-Control Systems

The trend in modern broadcast techniques is toward the use of automatic level-control equipment. Manual gain riding, a burdensome task at best, has the disadvantage of inhibiting the station from operating at its highest potential power. Every system is limited to a level which results in 100-percent modulation of the transmitter. Since maximum use must be made of the authorized power, and since highest signal-to-noise ratio is obtained for high degrees of modulation, it is desirable to maintain the *average* level of modulation as high as practicable. In addition, complete automation requires the use of automatic level riding to maintain high modulation levels without exceeding FCC modulation limits.

For automatic systems, some form of *agc* (automatic gain control) amplifier is normally employed at the studio. This amplifier has the ability to hold a relatively constant average output level over a wide range of input levels; this ability far exceeds that which can be obtained with manual gain riding. However, since this gain-control function must be performed in an inaudible manner, it is designed to be relatively slow acting when compared to the capability of electronic circuitry to act in fractions of a microsecond. These relatively slow attack and release times are not sufficient to prevent fast-acting signal peaks from overmodulating the transmitter. Therefore a *limiter* amplifier to control the faster-acting signal peaks is normally used at the transmitter.

Fig. 6-37A shows a block diagram of the fundamental audio amplitude limiter. Fig. 6-37B shows a block diagram of the fundamental agc system. The differences between these circuits will be discussed in the following paragraphs.

Limiter (or compressor) amplifiers normally employ only the closed-loop arrangement of Fig. 6-37A. This loop is degenerative, acting to reduce the gain as the input signal rises above a given level. For instance, an input-signal increase of 10 dB might result in an increase at the output of

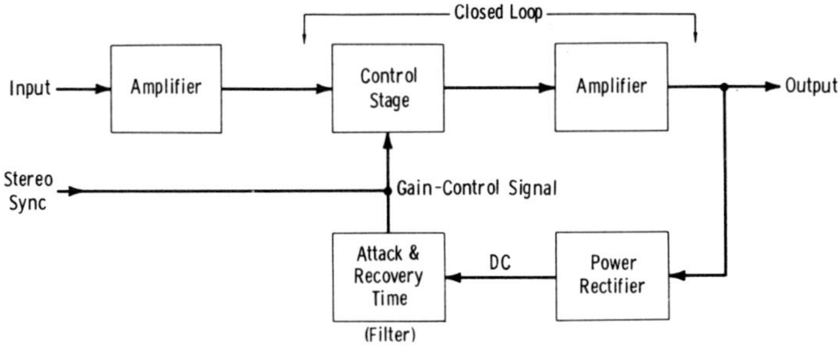

(*A*) *Limiter amplifier.*

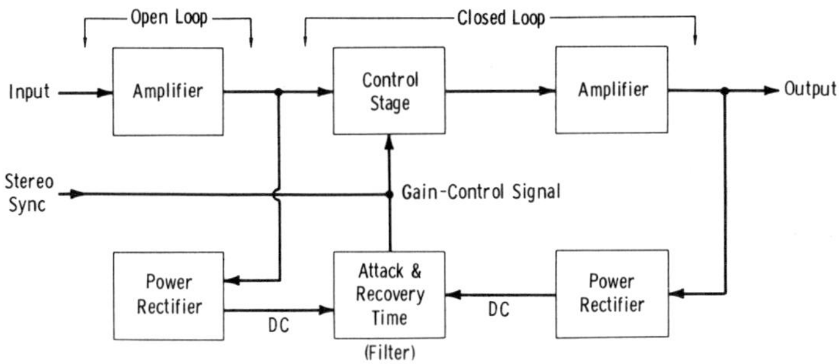

(*B*) *Agc amplifier.*

Fig. 6-37. Audio level-control devices.

only 1 dB. The gain reduction (in this example 9 dB) is the amount the signal has been limited or compressed.

In addition to gain reduction on high signals, the agc amplifier (Fig. 6-37B) allows a gain expansion over a limited range of lower-level signals. This combination of expansion and limiting, using both the open-loop and closed-loop control, maintains high uniform average levels without overmodulation and (ideally) without raising the noise level during dead sound intervals.

When agc is used in stereo applications, an additional factor is involved. It is important to maintain the original amplitude difference between channels so that maximum separation effect is obtained. Using a separate agc amplifier in each channel results in equal left and right volume, and some of the stereo positioning is lost. This problem is solved by tying the agc amplifiers together so that both have the same dc gain-control signal. This maintains relative gains, and the amplitude difference is preserved.

The most common types of gain-controlling devices are those that have variable amplification, those that exhibit a variable impedance, or those that exhibit both methods combined. For example, amplification can be made quite dependent on base bias; this bias can be supplied by the dc gain-control signal.

The power rectifier in the blocks of Fig. 6-37 is typified by the circuit of Fig. 6-38. It is a filtered dc amplifier driven from the signal-rectifier diode. In some cases, only the base-emitter junction is used as the rectifier at the expense of the power gain.

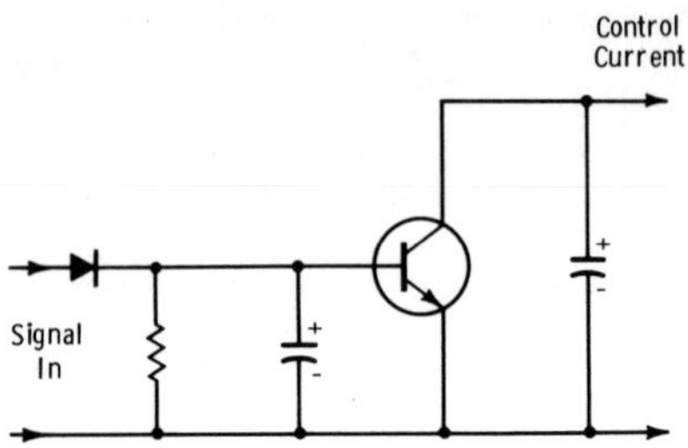

Fig. 6-38. Power rectifier circuit diagram.

Fig. 6-39A shows an example of amplification dependent on base bias. The transistor, biased in the active region, has its base voltage modified by the agc voltage from the power rectifier. Note from Fig. 6-39B that the *slope* of the curve (ratio of change in collector current to change in base-emitter voltage) is affected. As the signal increases, a larger negative agc voltage is applied to drive the transistor toward cutoff. The maximum gain reduction available in this type of circuit is about 15 dB if serious nonlinear distortion is to be avoided.

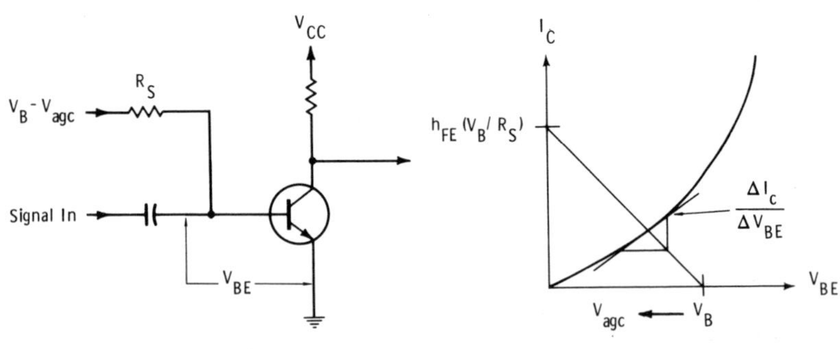

(*A*) *Circuit diagram.* (*B*) *Variation of gain.*

Fig. 6-39. Base-bias control of gain.

In Fig. 6-40 is an example of variable impedance in parallel with the signal path. When the current through a diode increases, the diode resistance decreases, and vice versa. The input current divides between the amplifier input and the diode resistance, R_D. When the signal is increased, the agc voltage increases and R_D decreases, lowering the overall stage gain. With this type of circuit, the signal source must have a high resistance (R_S), and the amplifier input impedance must be high also.

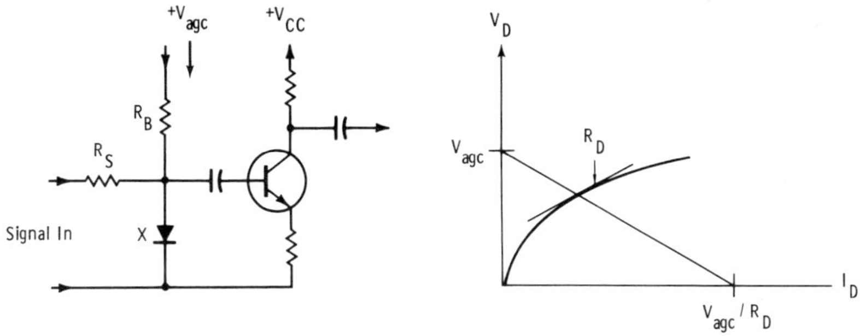

(*A*) *Circuit diagram.* (*B*) *Diode resistance.*

Fig. 6-40. Variable-impedance gain control.

Fig. 6-41 shows a two-terminal shunt arrangement. If a large agc voltage is developed (large signal), Q2 will have maximum base current and will saturate. Capacitor C2 is then returned to ground and provides a short circuit to the signal. As less agc is developed, Q2 assumes a higher shunt impedance (because its base current is decreased), and the gain is increased. Most of the agc action occurs in the region near Q2 saturation, where the slope of the output characteristic is changing most rapidly.

The RCA BA-45 AGC Unit

The RCA BA-45 agc unit contains expansion and compression circuits located on separate etched wiring boards. The BA-45 is designed for use with the BA-43 program amplifier and provides either expansion or compression, or both, by means of independent switches mounted behind the front panel. With both expansion and compression disabled, the BA-43 program amplifier will perform as a normal program amplifier with approximately 10 dB less gain than with a BA-43 operating alone. Refer to Figs. 6-42 and 6-43 for the following description.

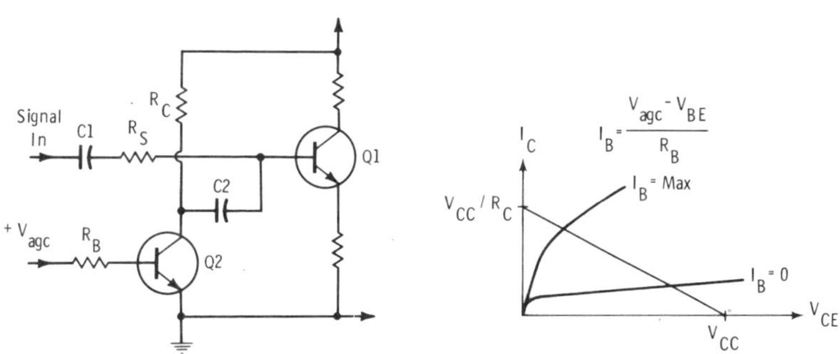

(*A*) *Circuit diagram.* (*B*) *Transistor operation.*

Fig. 6-41. Variable impedance with transistor.

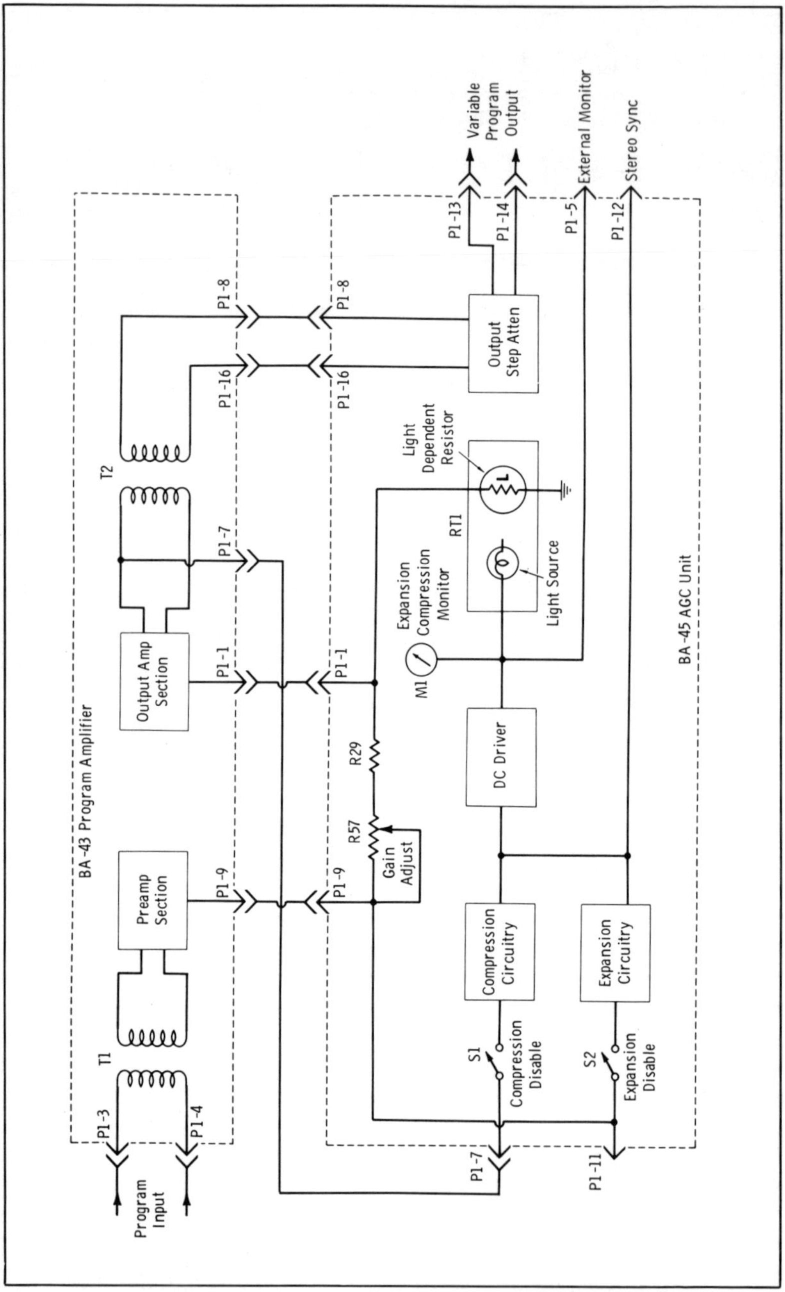

Fig. 6-42. Simplified interconnection diagram of BA-43 and BA-45 units.

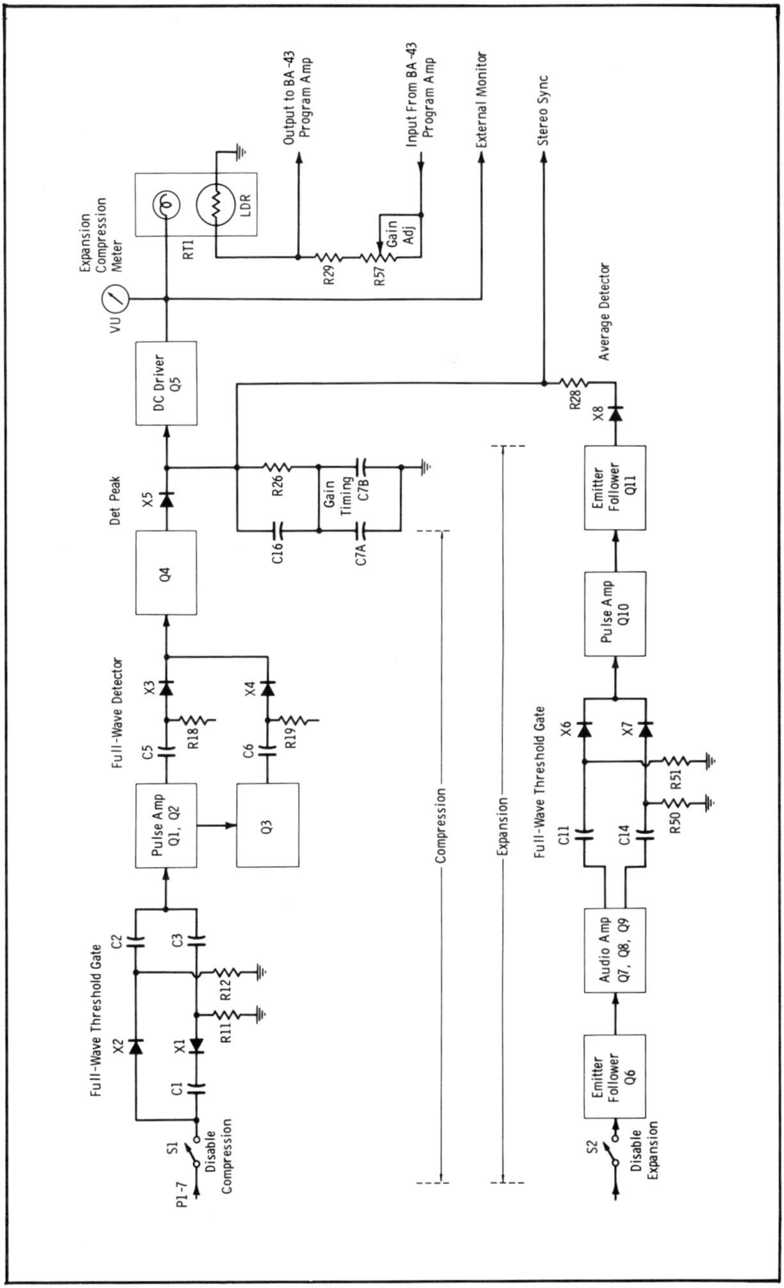

Fig. 6-43. Simplified block diagram of RCA BA-45 audio agc unit.

The expansion circuitry consists of an emitter follower, audio amplifiers, a full-wave threshold gate, a pulse amplifier, an emitter follower, and an average detector. The audio signal to the expansion circuitry is applied from the preamplifier section of the BA-43 program amplifier through terminals 9 and 11 of connector P1 and disable/expansion switch S2 to the base of emitter follower Q6. The audio signal from the emitter of Q6 is applied to the base of Q7 and is further amplified by audio amplifiers Q7, Q8, and Q9. The outputs from the emitter and collector of Q9 are applied to a full-wave threshold gate consisting of X6, X7, and associated circuitry. This gate sets the threshold level for expansion. The resulting pulses from the full-wave threshold gate are applied to the base of pulse amplifier Q10 and then to the base of emitter follower Q11. The signal from the emitter of Q11 is applied to the average detector, X8 and R28. This average detector provides a relatively slow attack and release time of

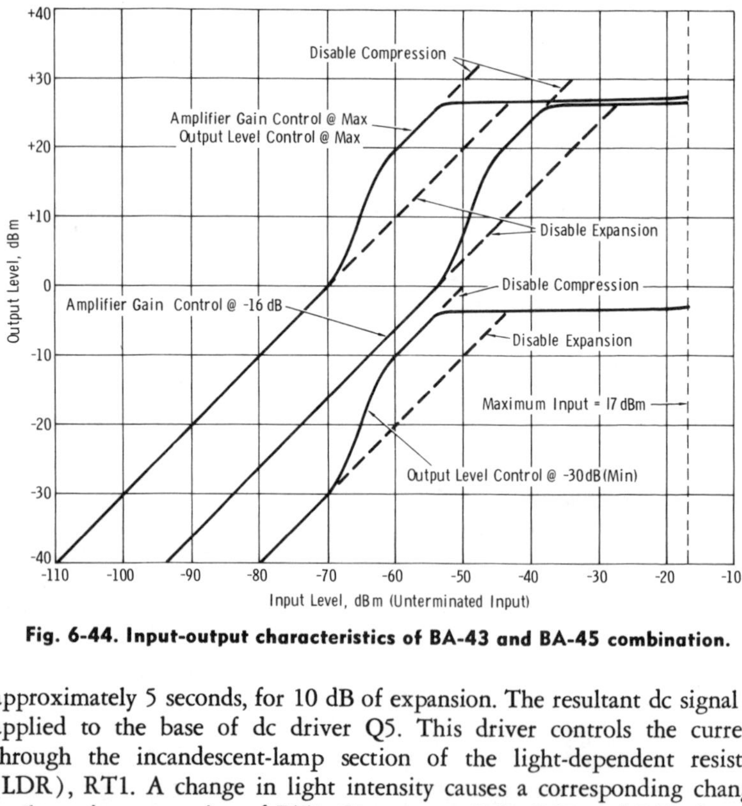

Fig. 6-44. Input-output characteristics of BA-43 and BA-45 combination.

approximately 5 seconds, for 10 dB of expansion. The resultant dc signal is applied to the base of dc driver Q5. This driver controls the current through the incandescent-lamp section of the light-dependent resistor (LDR), RT1. A change in light intensity causes a corresponding change in the resistance section of RT1. Components R29, R57, and RT1 form a voltage divider which varies the audio signal level to the output-amplifier section of the BA-43 program amplifier. See Figs. 6-44 and 6-45.

The compression circuitry consists of a full-wave threshold gate, pulse amplifiers and a phase inverter, a full-wave detector, an emitter follower, and a peak detector. The compression circuitry shares the dc driver and the LDR with the expansion circuitry.

The audio input to the compression circuitry is applied from the output-amplifier section of the BA-43 through terminal 7 of connector P1 and disable/compression switch S1 to the full-wave gate which consists of X1, X2, and associated circuitry. This gate sets the threshold level for compression. The resulting pulses from the full-wave threshold gate are applied to

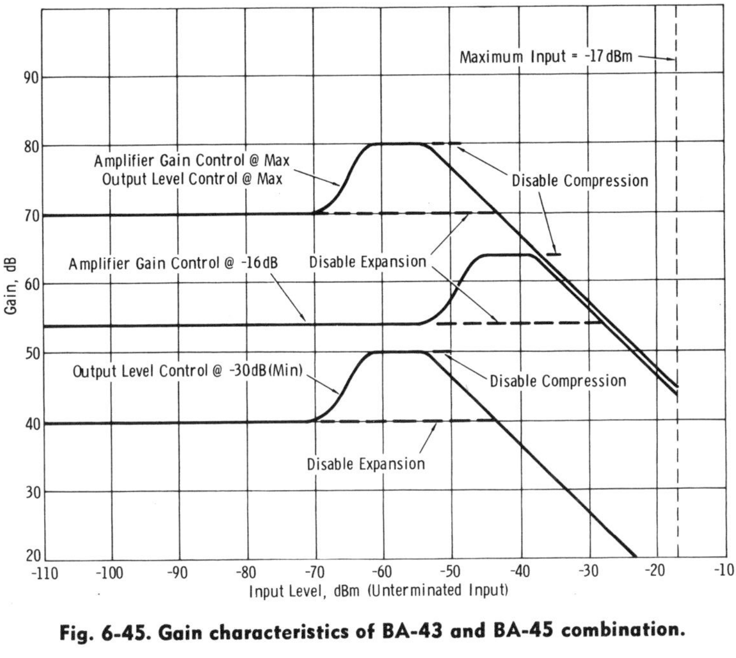

Fig. 6-45. Gain characteristics of BA-43 and BA-45 combination.

pulse amplifiers Q1 and Q2. One output from the collector of Q2 is applied through C5 to the full-wave detector which consists of X3, X4, and associated circuitry. Another output from the collector of Q2 is inverted by Q3 and then applied through C6 to the full-wave detector. The resulting pulses are applied to the base of emitter follower Q4. The output from the Q4 emitter is applied to the peak detector which consists primarily of X5, C7A, and C7B. Components X5, C7A, C7B, and Q4 make up a fast charging circuit which provides the compression circuitry with a rapid attack time (15 milliseconds for 10 dB of compression). When the audio signal drops below the compression threshold, X5 becomes an open circuit, and C7 then discharges through a higher impedance at the slower rate of 3 seconds for 10 dB of compression. The dc signal from the peak detector is applied to the base of emitter follower Q5. Transistor Q5 operates in the manner described above.

Monitor-Speaker Distribution

It is sometimes necessary to employ several speakers in a studio or other location, particularly where the area is large and the acoustics are influenced by the number of persons present. Fig. 6-46 shows a typical monitor output

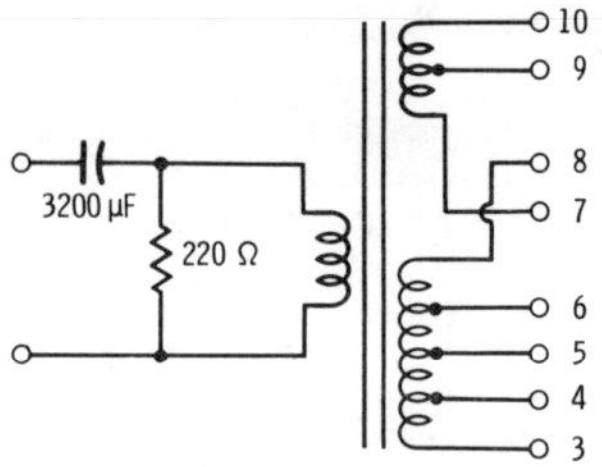

Impedance (Ohms)	Transformer Connections	Notes
4	Connect to 3-4	
8	Connect to 3-5	
16	Connect to 3-6	
150	Connect to 3-10	Jumper 3-7, 8-10
600	Connect to 3-10	Jumper 7-8
70 - Volt Line	Connect to 3-10	Jumper 7-8

Fig. 6-46. Typical monitor-amplifier output transformer.

transformer with connections provided to meet a variety of applications. Fig. 6-47 reviews the basic principles of matching multiple speakers to the monitor-line feed. Fig. 6-47A illustrates how two 16-ohm speakers may be connected in parallel to the 8-ohm monitor output terminals. Each speaker receives half the total power on the line. Fig. 6-47B shows two 8-ohm speakers connected in series to the 16-ohm monitor output terminals. Again, each speaker receives half the total line power. Any combination may be used: two 4-ohm speakers could be placed in series to feed from the 8-ohm line; two 8-ohm speakers could be paralleled to feed from the 4-ohm line, etc.

Due to the relatively large amount of current required for a given power, low-impedance monitor lines (such as 4 to 16 ohms) are limited to a short

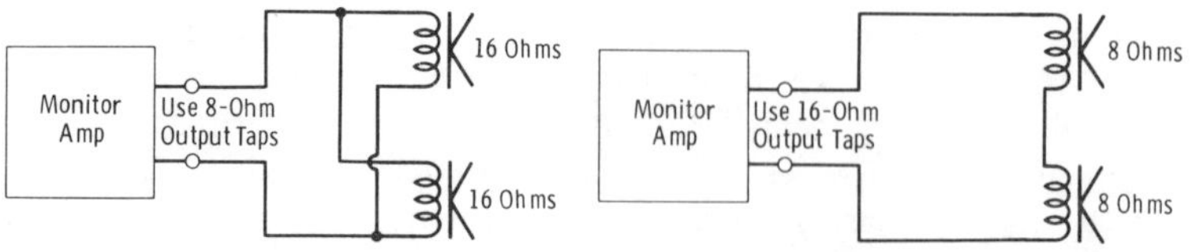

(*A*) *Parallel feed.* (*B*) *Series feed.*

Fig. 6-47. Matched-impedance feed of multiple speakers.

length of run for a practical wire size. When monitor lines must be long, lines of 150- or 600-ohm impedance are generally used. Table 6-7 gives the pertinent values of wire size and maximum feed length for the various impedances.

Table 6-7. Maximum Length in Feet for Monitor Lines

Wire Size	4 Ohms	8 Ohms	16 Ohms	150 Ohms	600 Ohms
14	125	250	450	1000	5000
16	60	150	300	775	3000
18	50	100	200	425	2000
20	25	50	100	275	1500

For very long monitor runs such as might be encountered at field events and associated pa coverage, a *constant-voltage* line, generally of the 70-volt level, is usually used, as shown in Fig. 6-48. With this technique, various speakers may be switched on or off without substituting back-load resistors and with little effect on the volume from other speakers. Speakers are connected across the distribution line through transformers that reduce the audio line voltage to the desired speaker level.

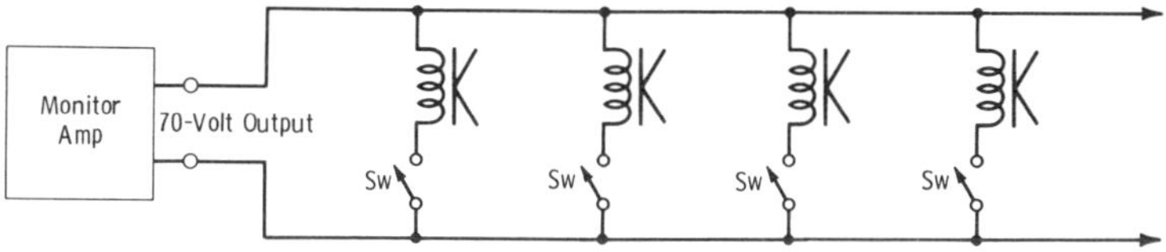

Fig. 6-48. 70-volt speaker distribution line.

In actuality, the audio line voltage is strictly dependent on the output regulation characteristics of the monitor amplifier driving the line. Amplifiers which provide for a 70-volt line feed should have output regulation such that the audio voltage has a maximum variation of 1.5 dB from no load to full load. The full-load requirement must not exceed the amplifier capability.

Reviewing Fig. 6-46, note that the 600-ohm line feed and the 70-volt line feed are identical. This really means that the transformer illustrated is probably associated with a 10-watt monitor amplifier. Thus with 8 watts feeding 600 ohms, the audio voltage is about 70 volts. For 50 watts of audio power on the line, a 100-ohm output would be used. In practice, monitor amplifiers designed for 70-volt line feeds have the 70-volt tap designated on the output transformer so that no computation is necessary regardless of the power rating of the amplifier. In practice, the actual audio line voltage varies with the signal. The 70-volt designation simply means the output voltage for maximum power.

In multiple speaker installations, it is usually desirable to be able to control the volume of one or more of the individual speakers. Fig. 6-49A shows the normal method used for a 70-volt line. When the monitor audio is distributed at voice-coil impedances, the T pad (Fig. 6-49B) or L pad (Fig. 6-49C) may be used. The L pad presents a constant impedance in one direction only (the side toward the monitor line).

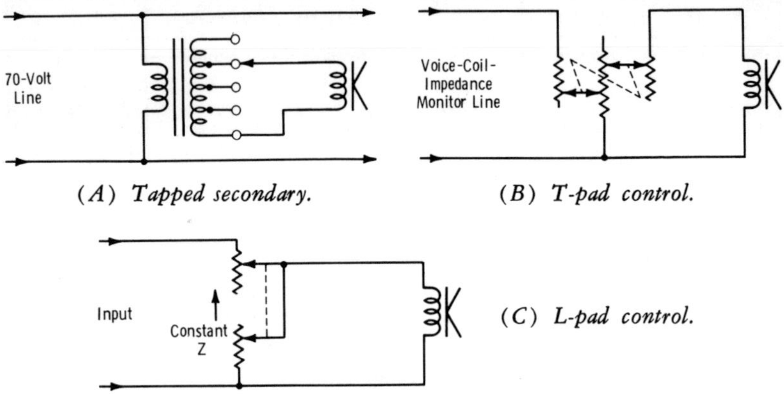

Fig. 6-49. Control of volume of individual speakers.

Proper Grounding Techniques

Hum, noise, and cross talk in any given installation are highly dependent on the technique of grounding. Every rack of equipment should have the grounds heavily strapped to a common point on that rack, and all rack grounds should be connected with good copper strapping to a common system ground point.

Fig. 6-50 gives the basic principle of microphone wiring. The cable from the studio wall outlet to the microphone should be a shielded twisted pair with flexible rubber covering for ease of handling and minimum shock noise if the cable needs any movement during use. The shield wire must not contact ground at any point in the system until it encounters the first preamplifier in the control room.

In extremely large installations, two grounding systems are employed: an equipment chassis ground, and a separate signal ground. In this case, the microphone-circuit ground is carried to the separate signal ground.

A cable with three wires plus a shield is often used for microphone circuits. In this event, the shield and third wire are tied together to form a more reliable ground connection over long periods of possible rough handling in service.

For jack fields, the sleeve or frame should preferably be mounted on a copper strap carried across the width of the panel, to provide convenient connection to the system or signal ground.

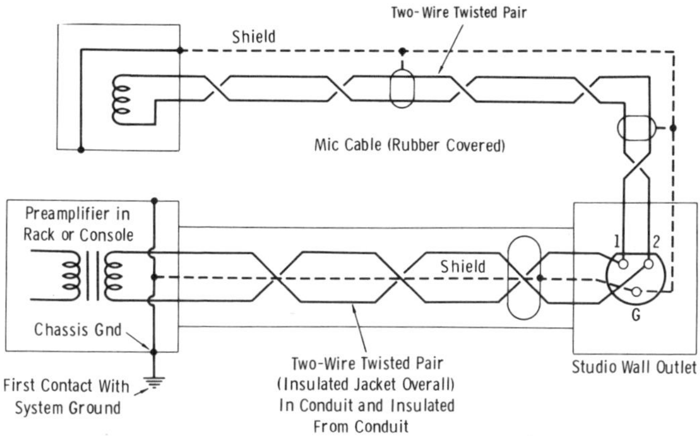

Fig. 6-50. Proper grounding of a microphone circuit.

In system or rack wiring, careful attention should be given the routing of cableforms carrying widely different signal levels. Three level specifications should be used, as follows:

1. Low level: Below −10 dBm, such as microphone circuitry
2. Intermediate level: −10 dBm to +20 dBm, such as rack amplifiers, boosters, and special effects (echo, reverberation, etc.)
3. High level: above +20 dBm, such as monitor lines, where +40 dBm is common

Cableforms of these three levels should be separated from each other by a distance of at least 2 inches.

All balanced mixing circuits employ center taps on the potentiometers, which may or may not need grounding. A test for best signal-to-noise (and signal-to-hum) ratio will determine whether this is necessary.

EXERCISES

Q6-1. Assume a balanced 600-ohm line carries a signal level of +10 dBm. This must feed an amplifier of 150-ohm balanced input impedance. The specified maximum input level for this amplifier is −22 dBm. Design the proper pad for this purpose.

Q6-2. What is the minimum insertion loss of a mixer potentiometer?

Q6-3. What is the minimum insertion loss of a ladder attenuator?

Q6-4. What is the minimum insertion loss of a:
(A) T attenuator.
(B) H attenuator.

Q6-5. A preamplifier with a fixed gain of 40 dB is rated as having a noise level of −80 dBm referred to the output. What is the minimum input level required to maintain a signal-to-noise (s/n) ratio of 60 dB?

Q6-6. If the preamplifier of Q6-5 is rated at a maximum output level of +18 dBm, what is (A) the maximum input level (B) the s/n ratio at maximum input level?

Q6-7. It is necessary to bridge a 600-ohm line carrying a 0-dBm level. An amplifier available for the bridging circuit has a balanced 600-ohm input and must have an input level of −25 dBm to obtain the required output. How could this bridging be done with resistors?

Q6-8. Assume you must build a four-microphone low-level T mixer to include a master gain control. The microphones are 150 ohms. What is:

(A) The value of R (see Fig. 6-14).

(B) The input level to the amplifier following the master gain control if the microphones give a −60-dBm output?

Q6-9. What s/n ratio would you expect from the amplifier of Q6-5 in the circuit of Q6-8?

Q6-10. Assume four 4-ohm speakers are connected in series across an 8-watt, 16-ohm monitor line. How many watts does each speaker receive?

CHAPTER **7**

The Stereo Control Room, Quadraphonic Systems, and Automation

A new era in broadcasting was ushered in with the FCC approval of fm stereophonic broadcasting in April, 1961. This system provides high-fidelity stereophonic sound while preserving compatible monophonic reception for existing single-channel receivers. Provision is also made to preserve SCA (storecasting) service simultaneously with stereo programming.

Stereocasting provides a fair approximation of listening in person. In nature, the spacing between our ears enables us to localize directions from which sound emanates. Part of this effect is the result of a few milliseconds delay of a sound wavefront from ear to ear. Higher frequencies have wavelengths near or less than the ear-to-ear spacing, resulting in phase differences. The combination provides depth and is missing in monophonic transmission, in which all microphones are mixed into a single channel.

The stereo system provides this depth by means of separate left and right channels that are reproduced by corresponding left and right speakers. The "center" is provided by effect only, as when a single sound source is of equal intensity and in phase on both channels. This is like monophonic reproduction by two or more speakers.

The number one requirement is *compatibility*. This means that:

1. A standard receiver can reproduce a monophonic output from a stereocast, without loss of original program information.
2. A stereo receiver can reproduce a monophonic program without loss of original program information.

The two separate audio channels are encoded into a composite signal at the fm transmitter, as fully explained in Chapter 10. In this chapter, only the studio terminal of the complete stereo system will be discussed, along with existing and proposed quadraphonic systems.

The most obvious difference is that two separate channels, left and right, are involved for a single stereo program. For live pickup, one or more microphones are employed for the left channel and one or more for the right channel. Reel-to-reel or cartridge tapes for stereo employ two-track tape and two heads. Stereo disc recordings are reproduced by means of the conventional stereo pickup head feeding a stereo preamplifier which has left and right-channel outputs.

7-1. THE STEREO CONSOLE

Fig. 7-1 illustrates the CCA Ultimate stereo 10-channel control console. Fig. 7-2 is a highly simplified block diagram of the console. The designation by symbols of the buses near the center of the diagram is as follows:

PL: Program left
PR: Program right
AL: Audition left
AR: Audition right
C: Center (Feeds combined left and right channels)

Input circuits include six microphones switchable to two faders, two turntables, ten high-level lines switchable to four faders, one external monitor input, three remote lines, and three cartridge lines. A description of each of the channels follows.

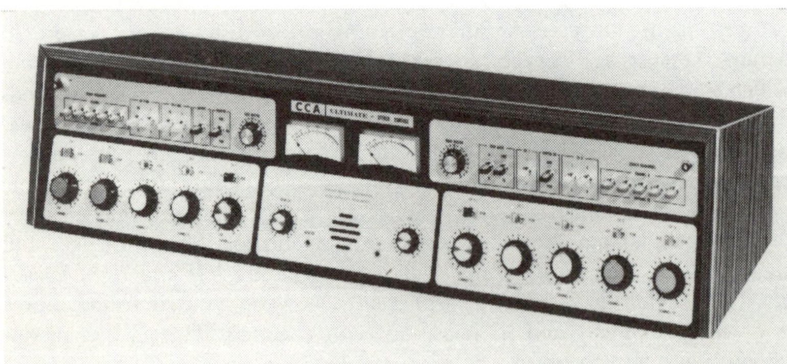

Courtesy CCA Electronics Corp.

Fig. 7-1. CCA Ultimate stereo control console.

Channels 1 & 2 (Stereo Auxiliaries)

Channels 1 and 2 are high-level stereo channels that have five high-level stereo inputs available which can be switched to either of the two stereo channels. It is impossible for the same input to be connected to both faders simultaneously, and thus segueing between the two channels with the five high-level inputs can be accomplished. Each stereo fader is of Altec manufacture with removable covers for cleaning; each contains a cue position. The output of each channel is switched to either the stereo audition or program lines.

Channels 3 & 4 (Microphone Channels)

Channels 3 and 4 each have available three inputs which in turn are used to drive monophonic preamplifiers that can be used to feed either the left or right amplifier or both in parallel. This selection can be achieved by means of a front-panel switch.

These inputs are nominally 200 ohms and are designed to accept input levels of −50 dBm. The three inputs are fed to the input selector switch, which in turn selects an appropriate input to drive the preamplifier. The output of the preamplifier is controlled by a fader which is so identified on the front panel of the console. The output of this fader, by the appropriate switch position identified as center-channel switching, can be directed to either the left, right, or center channel. From the output of this channel-selection switch, the signal is then fed to the audition-program switch. This switch is located just above the individual channel gain control.

Channels 5 & 6

Channels 5 and 6 are high-level stereo channels recommended to be used as turntable inputs for both the conventional left and right turntables. In order to achieve the optimum in signal-to-noise ratio, it is suggested that equalized preamplifiers for the turntables be installed in the turntable housings.

Channel 7

Channel 7 is a high-level stereo channel that has three switchable inputs and is suggested as a channel for stereo cartridges. It contains a cue position, and its output may be switched to either the audition line or the program line.

Channel 8

Channel 8 is available to be used as a remote high-level monophonic channel. It contains facilities for selecting one of three inputs and feeding the output of the fader to either the left, right, or center channel. This channel also contains facilities for talk-back.

Channel 8 in the console has essentially the same characteristics as channels 3 and 4 except for the fact that it does not have a preamplifier. It accepts a high-level input and feeds a transformer. This transformer isolates the source from the console. The output of the transformer is fed to a front-panel fader, then to the left-center-right switch, and then finally to the audition-program switch.

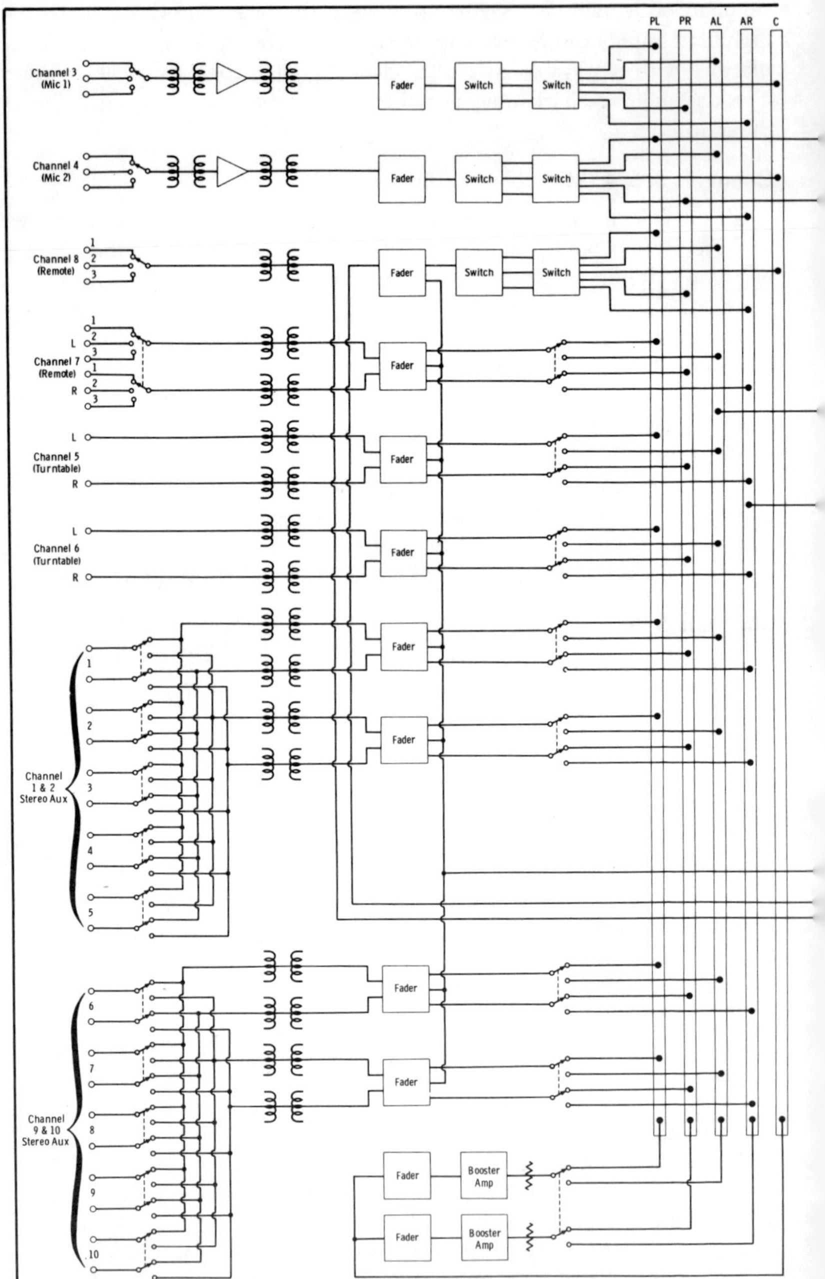

Fig. 7-2. Simplified block diagram,

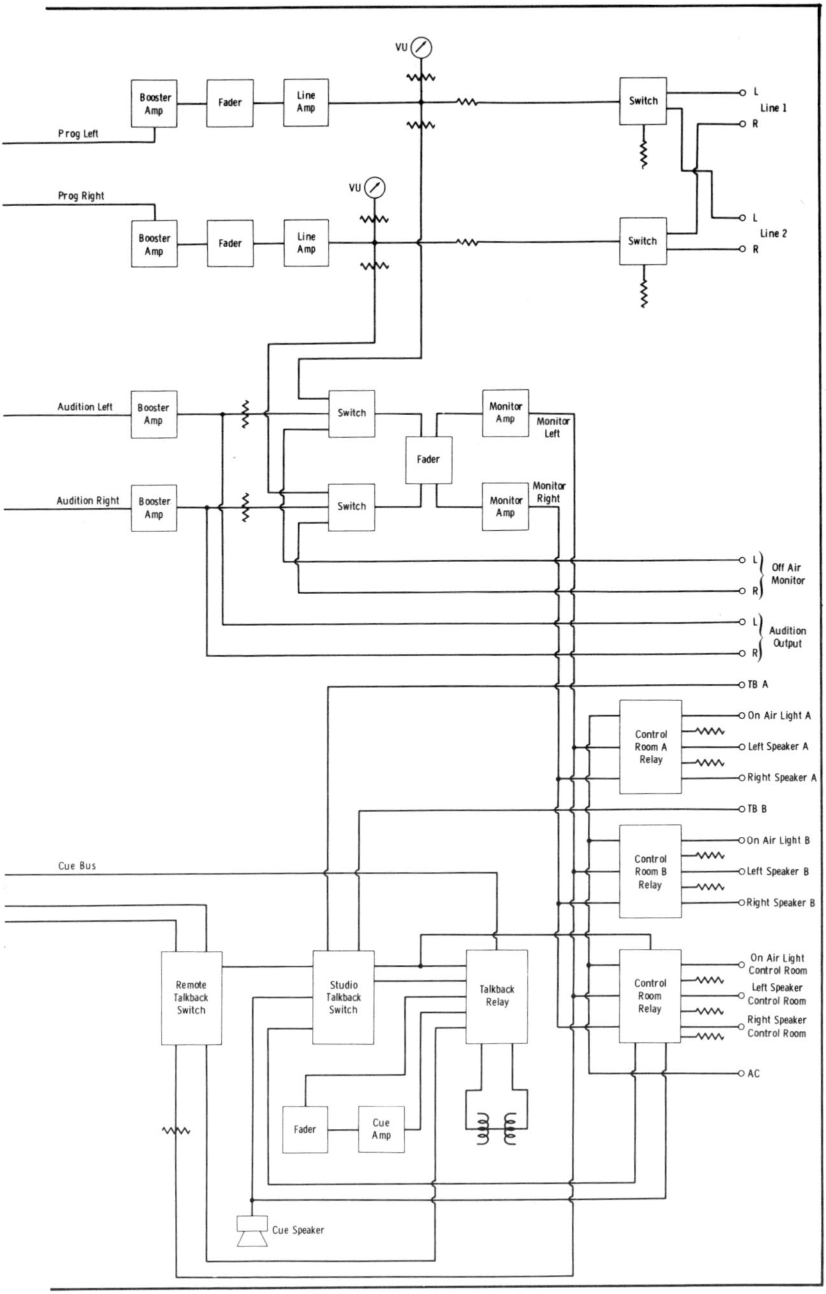

CCA Ultimate stereo control console.

Channels 9 & 10

Channels 9 and 10 are similar in construction and philosophy to channels 1 and 2. They are high-level stereo channels with facilities to select any one of five high-level stereo inputs for either of the two channels. It is impossible for both channels to be fed with the same input simultaneously. This philosophy provides a foolproof method of switching between two channels without causing an attendant drop in the audio level of the program.

Isolated High-Level Inputs

Transformers are used in every high-level input. This prevents problems associated with ground loops.

Program Channel

Assuming that the audition-program channel switch is in the program position, the output of the channel is connected to the program line. In the case of the monophonic inputs, when the switch is in the center-channel position, both the left and right program lines are fed in parallel. In the left position, only the left channel is fed, and in the right position only the right channel is fed. In the stereo mode, both the left and right booster amplifiers are fed.

The outputs of the booster amplifiers are then used to feed the front-panel master faders, both left and right, which in turn feed the line amplifiers. The outputs of the line amplifiers are available to feed either line 1 or line 2; in the center position, the console is self-terminated. Thus it is possible to disconnect the console from its terminating lines and still evaluate its performance by having it self-terminated.

Audition Channel

With the audition-program switch in the audition position, the outputs of the individual channels are fed to the audition booster amplifiers. The outputs of these booster amplifiers are used to feed an external set of terminals or, through the appropriate monitor selector switch, the monitor amplifiers. The audition channel does not have any individual master controls. The output level is controlled by the individual dual channel masters. With normal settings of the individual channel controls, −10 dBm will appear at the audition output terminals.

Cue Channel

Other than channels 3 and 4 (microphones), every fader in the console (8 faders) contains a switch position that connects the input signal to the cue bus. The cue bus normally feeds the cue amplifier through the contacts of the talk-back relays. Thus, the cue bus cannot inadvertently feed a speaker when the microphone is live.

The cue speaker is driven from the cue amplifier through the talk-back relay contacts and the control-room mute relay contacts. Under the condition of talk-back, the cue-amplifier input and output circuits are reversed, and it is used to amplify the output from the front-panel speaker which, in the talk-back position, serves as an audio source rather than a termination. The cue amplifier has capabilities of achieving 8 watts output, and its module is identical to that of the monitor amplifier.

A front-panel jack is used for monitoring with earphones. The cue amplifier is related to the left channel only.

Stereo Monitoring

The monitoring system has facilities for monitoring, on a stereo basis, the main channel, the audition channel, and external off-air programs. By the use of a front-panel selector switch, the monitor amplifier can be fed from the program line, the audition channel, or an external audio input. It contains a stereo fader for both the left and right channels, and its output is connected to three sets of speaker systems through the speaker-mute relay panel. It also has facilities for driving the front-panel monitor phone. It is designed basically to feed low-impedance devices; thus the monitor phone should have an impedance of approximately 8 ohms, and the speaker impedance should be approximately 24 ohms (three speakers in parallel across the 8-ohm monitor output: one each in studio A, studio B, and the control room). It is capable of producing 8 watts output to both the left and right channels under normal drive conditions.

Auditioning Without Monitor Amplifier

The console contains booster amplifiers in its audition lines. Thus, it is possible to audition a number of program sources and obtain an output level great enough that the monitor amplifier is not required. This addition of booster amplifiers in the audition channel adds considerable versatility.

Center Channel Control and Switching

The output of the center channel can be switched to either the audition or program channel, and the levels available to both the left and right channels are controllable. Controllable center channels are available not only from the low-level microphone channels but also from a high-level channel.

Channels 3, 4, and 8 contain facilities for feeding the outputs of the preamplifiers and the independent high-level inputs to either the left, right, or center channel, as the result of the appropriate selection of the individual channel selector switch above these channels. When the switch is placed in the center-channel position, the output of the switch is connected to two individual potentiometers that are mounted on the tray behind the front panel. These controls are preadjusted to achieve equal output level on both lines. They drive individual Altec 1578A booster amplifiers. The outputs of these booster amplifiers are used to feed an audition-program switch for the center channel. By turning the switch to the audition position, both the left and right audition channels are driven; when it is placed in the program position, the left and right line amplifiers are driven.

Switchable Output Lines

The console has front-panel facilities to switch the output to a second pair of stereo lines. This can serve as an emergency output in the event that the terminal equipment driven by the normal output becomes defective and the alternative output could be used to drive the proper termination for the console.

Microphone Preamplifier

The console contains two microphone preamplifiers. From Fig. 7-3, it can be seen that the preamplifier consists of two individual modules. They are an rf filter and an Opamp Lab 350P preamplifier. Both of these modules are located on the shelf behind the control panel.

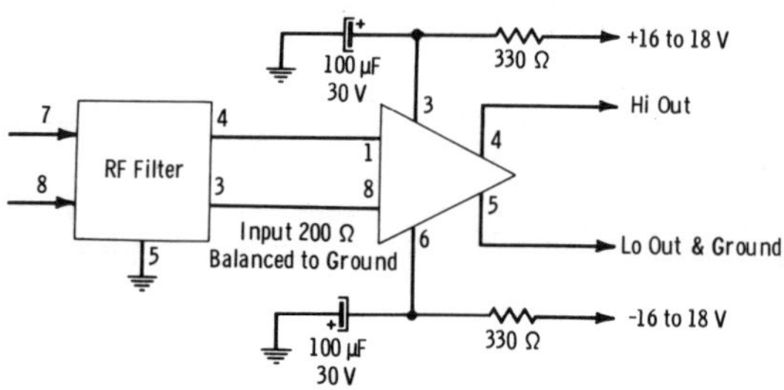

Fig. 7-3. Preamplifier for microphones.

The purpose of the rf filter is to minimize rf overloading of the preamplifier. This feature is particularly important when the installation is close to the transmitter, where a large rf field may exist.

The Opamp Lab 350P operational amplifier has a fixed gain of 50 dB. The input impedance is fixed at about 200 ohms and is electrically balanced to ground. The amplifier is capable of a +12-dBm output before clipping occurs, which means the input signal may be as large as −38 dBm.

Equivalent input noise is less than −120 dBm in the band of 30 Hz to 15 kHz. Being an operational amplifier with a bipolar power supply, the amplifier is insensitive to power-supply noise.

Program Line Amplifier

The line amplifier (Fig. 7-4) is built around an Opamp 425, which has an open-loop gain of 20,000. The UTC A11 and Altec 15356 transformers provide impedance matching and a balanced line in and out. The gain of the unit may be adjusted by changing the 15K feedback resistor. Since it is a bipolar operational amplifier, the unit is insensitive to power-supply

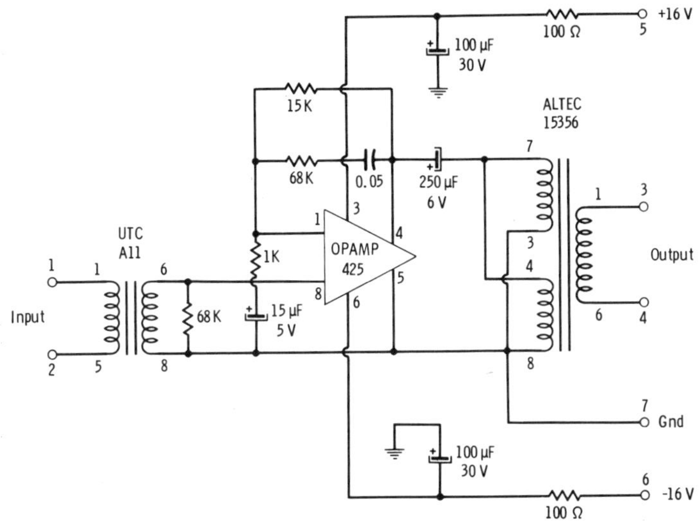

Fig. 7-4. Circuit of line amplifier.

noise. With an input level of -30 dBm, the amplifier can achieve a distortion-free output of $+14$ dBm.

Both a left and a right line amplifier are used. These units are of a plug-in variety and are located on the rear base of the cabinet.

7-2. STEREO MICROPHONE INPUTS

Fig. 7-5 illustrates the microphone input circuitry of the stereo console described above. Note the arrangement of the microphone switch which selects the microphone from studio A, control room (CR), or studio B. The switches are shown in the A position. Note also that a set of contacts on each switch runs to the program-audition switch so that this switch must be moved from the center position to select either program or audition before the speaker-muting and on-air light circuitry is activated.

For stereo broadcasting, it is extremely important that the left and right program sources be in phase with each other for correct aural perspective. When inputs are connected to a stereo console, color coding of the microphone pins and microphone cables must be observed so that they are connected to corresponding terminals on the terminal board of the console.

All microphone manufacturers give the polarity of their particular models for correct phasing. For example, Fig. 7-6A shows this information for the E-V Model 666 microphone. Positive pressure on the generating element results in positive voltage on the red cable lead (pin 1) relative to the black cable lead (pin 2). For the Shure Model SM 58 (Fig. 7-6B), positive pressure produces positive voltage at pin 2 (blue) with respect to pin 3 (red).

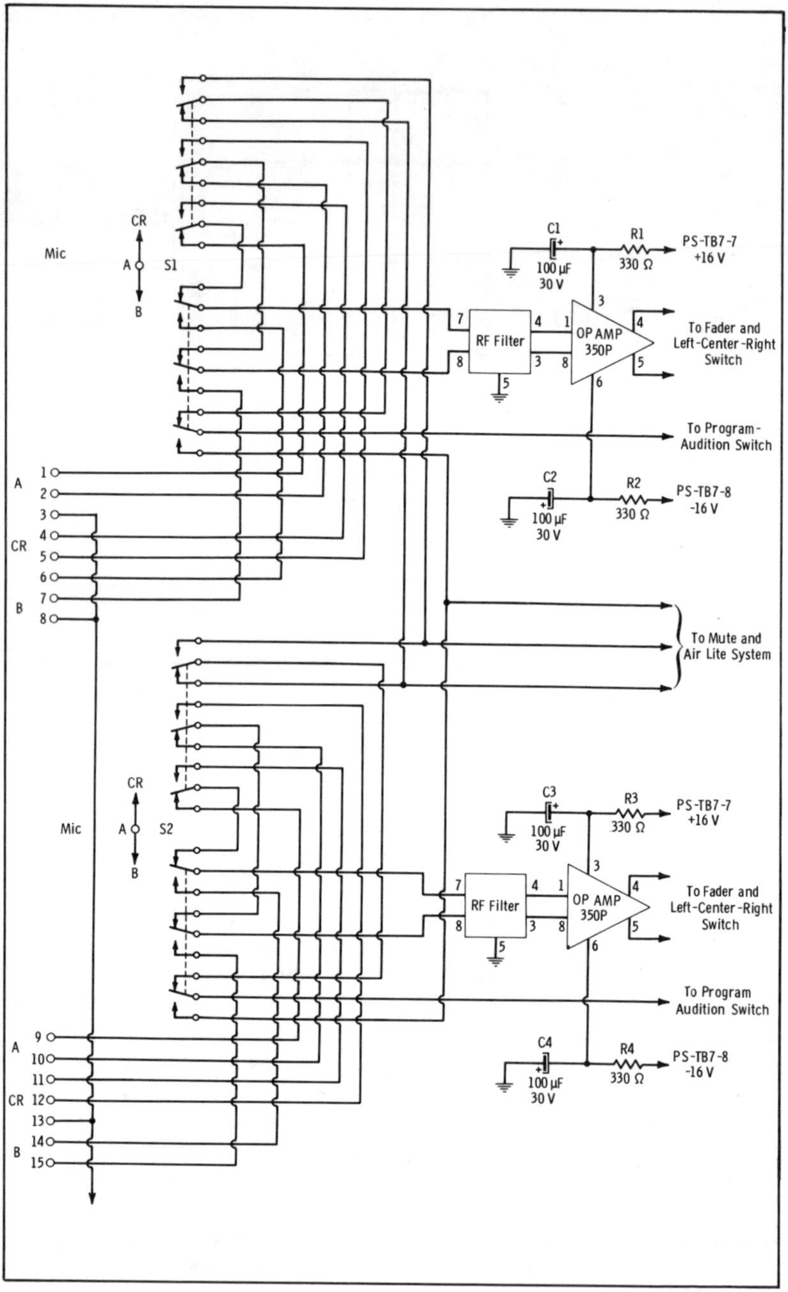

Fig. 7-5. Microphone input circuitry.

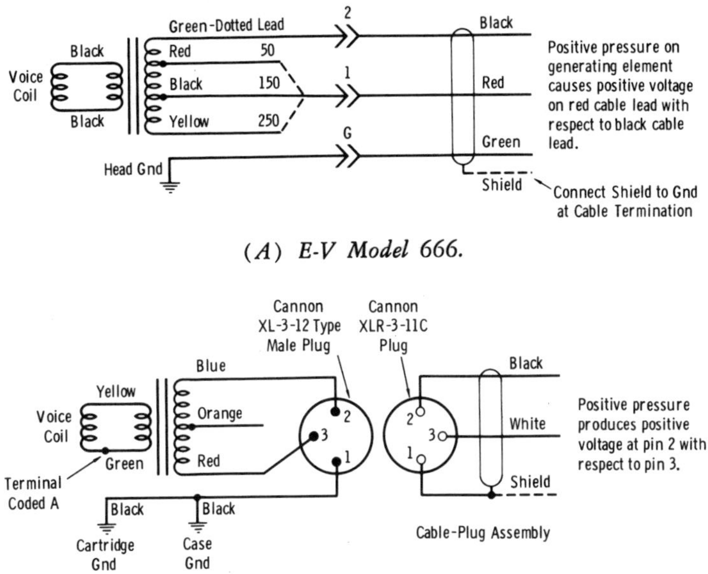

Fig. 7-6. Examples of microphone wiring.

The above example illustrates the amount of thought that the stereo technician must give to the installation and usage of microphones. Review Fig. 6-50 of Chapter 6. The studio wall microphone outlets are wired in a consistent manner into the rack or console preamplifiers. In Fig. 6-50, pins 1 and 2 are the signal leads, and pin 3 serves as the ground carry-through. This means that if you employ different makes and models of microphones, you must adapt the microphone cable for proper phasing of all microphones.

The color coding of shielded two-wire cable is black-white-shield or black-red-shield. The color coding of shielded three-wire cable is usually black-red-green-shield. Let us now review how to handle different types of microphones for the installation of Fig. 6-50, with either two-wire or three-wire shielded microphone cable (plug end for wall outlet).

For the EV Model 666 (Fig. 7-6A) and shielded two-wire cable with black and white wires, connect black to pin 1, white to pin 2, and shield to pin 3. This carries out the usual power-line color coding in which black is always "hot" with respect to white. For two-wire shielded cable with red and black wires, connect red to pin 1, black to pin 2, and shield to pin 3. For three-wire shielded cable with black, red, and green wires, connect red to pin 1, black to pin 2, green and shield to pin 3.

For the Shure Model SM 58 (Fig. 7-6B) and shielded two-wire cable with black and white wires, connect black to pin 1 at the wall-outlet plug end of the cable and to pin 2 at the microphone end of the cable. Connect white to pin 2 at the wall-outlet plug end and to pin 3 at the microphone end of the cable. Connect the shield to pin 3 at the wall-outlet plug end and to pin 1 at the microphone end of the cable. For two-wire shielded cable with black and red wires, connect red to pin 1 at the wall-outlet end of the cable and to pin 2 at the microphone end of the cable. Connect black to pin 2 at the wall-output plug end of the cable and to pin 3 at the microphone end of the cable. Connect the shield to pin 3 at the wall-output plug end and to pin 1 at the microphone end of the cable.

Note that when different-type microphones may employ the same type (usually XL) of connectors for both ends of a cable, the cables *must be identified* to maintain proper phasing. The same attention must be given to microphone extension cables.

> NOTE: In many instances, the microphone head receptacle (when used) may be rewired at the microphone end for the system configuration. In other instances, all microphone cables are wired the same, and short identifiable adapter cords are inserted for a particular microphone. Microphones with heads that have a cord running from internal wiring can, of course, have the cord-plug wiring modified to suit any system configuration.

Upon initial installation of a control console, the wiring from the studio microphone outlet receptacles into the preamplifiers must be made consistent. For example, in Fig. 7-5, if the studio A microphone is wired so that the positive terminal is connected to terminal 1, the positive terminal of all other outlets must go to terminal 9 of the second switch, etc.

7-3. WHY QUADRAPHONIC?

We have just two ears, and the question inevitably arises in the newcomer's mind as to the necessity for four-channel sound reproduction. In monophonic broadcasting, even though two microphones might be used spaced apart the same distance as the average set of human ears, there is no information in the signal as to directional properties of the two microphones. Regardless of the number of microphones employed in the pickup, the information is the same in all reproducers regardless of the number of speakers used. Certainly you can get "surround sound" in monophonic reproduction by using four speakers—left front, right front, left rear, and right rear. But there still is no directional information in the single-channel sound from the four speakers.

In stereo broadcasting, two channels designated "left" and "right" are employed. Two separate bits of information are available to the stereo receiver so that these two channels are reproduced through two sets of speakers, imparting directional sound perspective to the produced signal.

When you attend a live sound performance, it is true that your two ears do pick up certain magnitudes of direct sound from left and right. But almost 75 percent of the sound you actually hear is reflected from the side and rear walls, the ceiling, and even objects and people. Again, even though a number of microphones are split into two groups for "left" and "right" signal information, and even though these microphones are arranged to pick up reflected (ambient) sound, only two bits of information are present for the stereo receiver to interpret. The direct and ambient sound are channeled into just two groups (left and right), and there is no separate information to distinguish direct from ambient sound.

The problem (and even the solution) is very much like that of color television. Even though a monochrome camera looks at the colors in the original scene, there is only brightness information in the signal, which carries no color information for use by a color receiver. The color information must be separately encoded (onto a subcarrier) and interleaved into the tv-channel bandwidth employed for regular monochrome broadcasting, so that the color receiver can use this information to reproduce the original scene, while not interfering with reproduction of a color signal by a monochrome receiver in black and white.

Thus in quadraphonic broadcasting we are concerned with four separate channels of information, left and right front and left and right rear. All of this extra information must be contained within the regular channel bandwidth of the station. It must be compatible in that regular monophonic or stereophonic receivers can reproduce the total signal without loss of any important information.

At the time of this writing, there are two basic systems of quad sound, the matrix system and the discrete system. The basic action of these two systems will be compared in the next section.

7-4. BASIC COMPARISON OF MATRIX AND DISCRETE QUAD SYSTEMS

Simply stated, the *discrete* system is the transmission of four separate, isolated sound channels to four speakers. The *matrix* system takes the four separate channels and reduces them to two by an encoding process at the studio end. At the receiver end, a decoder is employed to convert the two channels back into four for application to four separate speakers.

The matrix system actually requires no encoding equipment at the studio end if previously matrixed discs or tapes are used. (Four-channel tapes require an encoder.) The additional matrix (encoder) is required only if the station produces its own live quad sound, quad discs, or tapes. Otherwise the matrixed signal from the quad source is simply handled in the conventional two stereo channels. However, the quad decoder is required for monitoring and, of course, at the receiver end. The only additional cost, therefore, to the broadcaster is that involved in obtaining the quad discs themselves and the decoder required for proper monitoring of the quad signal, plus the additional speakers.

The matrix system does not require FCC approval since there is no change in standards involved (no modifications of transmitters or bandwidths are needed). The discrete system requires some form of carrier multiplexing and, at the time of this writing, has not been authorized by the FCC except on an experimental basis. The Dorren Quadraplex (discrete quad) has been demonstrated experimentally by KIOI-FM (San Francisco) using a subcarrier at 76 kHz. Current FCC rules require that the last 25 kHz of the 100 kHz alloted to each station be free of signals to serve as a guard band. The rules further state that the radiated signal must be down at least 25 dB between 120 and 240 kHz from the assigned carrier. Measurements at KIOI-FM have shown their experimental signal down 42 dB at 100 kHz even when using the 76-kHz subcarrier, and the FCC has been petitioned for approval of this discrete quad system.

Discrete quad can be termed a 4-4-4 system; four individual channels are separately conveyed to a recorder and/or transmission system and reproduced on four correlated speakers. Matrixed quad can be termed a 4-2-4 system; four original sound channels are reduced to two channels to a recorder and/or transmission system (in this case a conventional stereo transmitter) and then decoded into four separate speakers.

There is another system in common use "hidden" in the stereo signal; this system does not involve the transmission end at all. It can be termed a 2-2-4 system which involves only a decoder at the stereo receiver end for adding two additional speakers to the existing stereo pair.

It was stated previously that in stereo broadcasting microphones may be placed in positions that receive not only direct sounds from left and right front, but also reflected sounds from the walls and ceiling. The 2-2-4 decoder system uses this principle in a unique way to achieve additional sound perspective, admittedly of a pseudo nature, where realism as such need not necessarily be required or desirable.

The decoder employs the principle that front signals may be essentially in phase, left and right signals independently phased in left and right components, and rear signals of random phase specifying reflected sound components. Such decoders are usually designed so that they are not limited to the 2-2-4 application, but are used with the matrixed quad broadcasting signal when available from a station within range of the receiver (4-2-4 application).

7-5. THE MATRIX SYSTEM

A *matrix* is simply a cross-connected voltage divider, involving phase splitters to obtain 90° and 180° phase relationships where necessary in the voltage division. It must be understood at the outset that the matrixing for quad sound is done at the studio to reduce four discrete channels to two channels for transmission. Normal stereo matrixing is then carried out at the fm stereo transmitter as covered in Chapter 10.

At least four matrixing techniques have been developed, but most of them are sufficiently compatible that a given decoder can be used for the four separate speaker feeds. There are certain basic requirements that must be satisfied that are quite adequately listed by P. Scheiber[1] as follows:

1. Basic four-channel performance:
   A. The ability to record sounds occurring at any point in 360°, and to reproduce each sound from the correct location in playback.
   B. Nondegradation of signal quality, including noise, frequency distortion, and nonlinear distortion as consistent with highest standards in the state of the art.
2. Compatibility:
   A. Four-channel compatibility: Nonobsolescence of playback equipment, using standard components and construction wherever possible.
   B. Stereo compatibility: The ability to reproduce the four-channel program on all standard two-channel stereo equipment, with all sounds in the four-channel program heard in their proper left-right positions.
   C. Monaural compatibility: Monaural playback possible on all standard equipment, without losing or altering the relative level of any sound in the four-channel program.
3. Economy:
   A. Adaptability to standard practices for software manufacture.
   B. Full playing time within a given format, as compared with the equivalent stereo recording.
   C. Usable with all major recording media and, preferably, broadcast.

A rather confusing situation exists in that two basically dissimilar matrix systems employ the letters "SQ" and "QS" as their system designations. We will describe the CBS-Sony SQ system, then the Sansui QS system, below. (Sansui has mentioned that they might change their designation to "SS.")

The SQ Matrix System

The basic SQ matrix system employed by Columbia Records for cutting quad discs is shown in Fig. 7-7. Fig. 7-7A represents the encoding matrix which supplies the required phase shift within two degrees over the audio band of 20 to 20,000 Hz. The $-j$ and $+j$ terms in the matrix equations signify how the signal components to the two rear channels are shifted in phase by opposite angles of 90° to result in a 180° phase relationship. Fig. 7-7B shows the basic decoder matrix, which is complementary to the encoder matrix of Fig. 7-7A.

---

[1]P. Scheiber, "Four Channels and Compatibility," *Journal of the Audio Engineering Society,* April 1971.

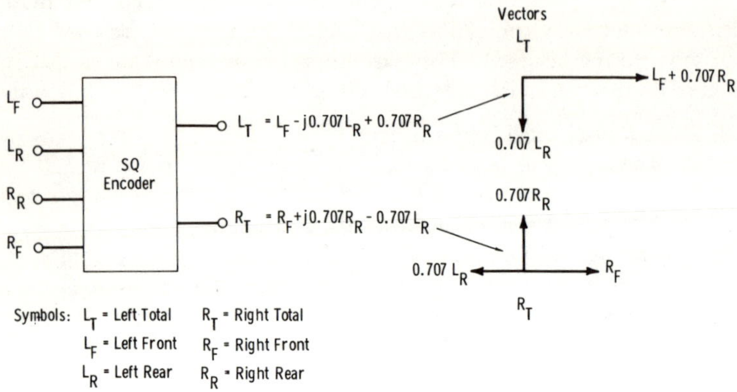

(*A*) *Encoder.*

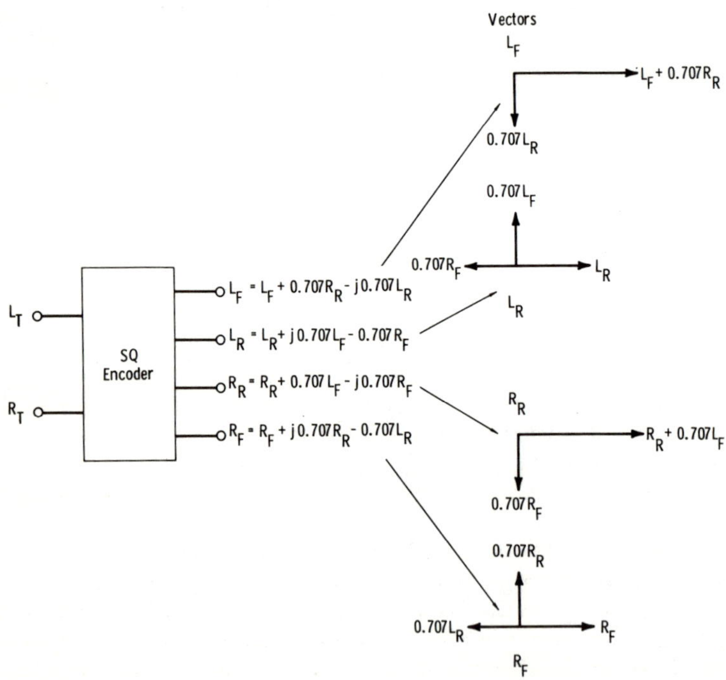

(*B*) *Decoder.*

Fig. 7-7. The basic SQ matrix system.

The separation between left and right front speakers and left and right rear speakers is approximately as good as ordinary stereo disc separation. However, it may be observed from the matrix equations and the vector diagrams of Fig. 7-7 that some crossover of information occurs between the front and rear channels. With the conventional (basic) decoder matrix of Fig. 7-7B, the front-back separation is only about 3 dB. A special matrix *blend decoder* has been developed which trades off some of the separation in the left-right channels (both front and rear) to gain better front-back separation. Columbia Records says that many variations of cross-coupling ratios are practical using the same type of passive circuitry. Their most efficient blend decoder (in terms of front-back channel separation) at the time of this writing has the following specified characteristics:

Left-front–right-front channel separation: 14 dB
Left-front–right-rear channel separation: 8 dB
Center-front–center rear separation: 7 dB

Either the basic matrix or the blend matrix decoder operates with passive circuitry composed of fixed phase-shift and combining networks. In contrast, the SQ *Logic Decoder* is an active device employed in conjunction with a basic matrix circuit. Input logic circuitry monitors the incoming signals before they reach the matrix network but does not otherwise affect the decoding matrix. This matrix feeds the four individual channel amplifiers which, in the Logic system, are voltage-controlled variable-gain devices. Analysis of the various amplitude and phase relationships of the input signals by the logic network results in a derivation of certain command functions such that appropriate individual control voltages are fed to each of the variable-gain amplifiers. The automatic gain variation in each amplifier changes dynamically as a function of the disc modulation. The manufacturers claim this technique makes it possible to isolate a signal in any channel with an arbitrary or selectable degree of cross talk to the other channels.

A factor in the SQ system which critics of the entire matrixed technique consider a drawback is the lack of a "rear center" channel, as illustrated in Fig. 7-8. This shows the vector sum for a sound source in the center of the rear area that would produce equal signal amplitudes in LR and RR when LF and RF theoretically receive no signal. In this case, the output of the encoder is essentially zero. Obviously, this cancellation depends on many variables such as frequency, phase differences, and precision of levels reaching the two rear-channel pickups. Critics of the SQ system, including some matrix manufacturers, claim that this equal-energy distribution often occurs in mix-down of multiple-channel tape recordings into two or four tracks, and that lack of directionality and loss of sound localization results. It is known that CBS instructs the recording engineers to avoid placing soloists or any important bit of information in the rear center of the pickup area.

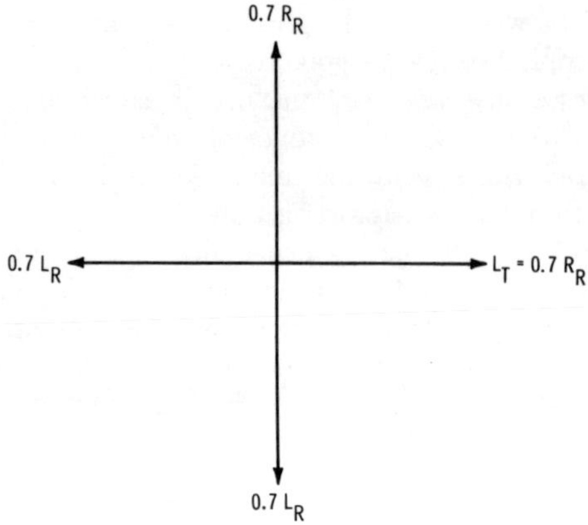

Fig. 7-8. Lack of "rear center" channel.

Another interesting situation exists when all channels receive identical signals. The resultant encoder output is depicted in Fig. 7-9, which shows that most of the sound energy would be concentrated in the left front channel.

RIAJ Standards on Matrix System Disc Recording

Before proceding with a discussion of the QS system, it will be helpful to present extracts from "Standard of the Engineering Committee, RIAJ (Record Industry Association of Japan) on Regular Matrix System Disc Recording." These extracts follow.

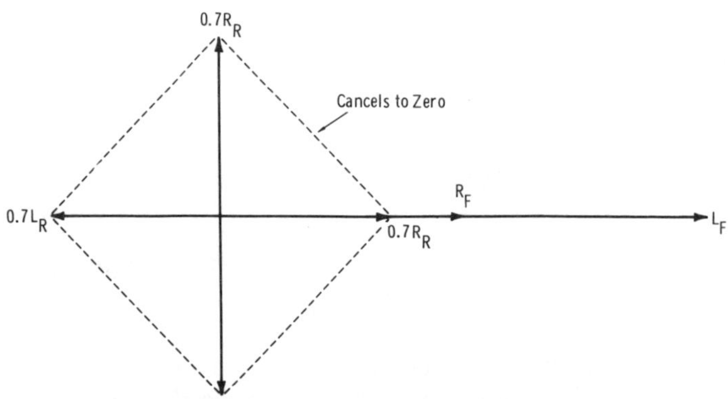

Fig. 7-9. Encoder output when $L_F = R_F = L_R = R_R$.

1. SCOPE OF APPLICABILITY

This standard shall apply to commercially marketed regular matrix system disc recordings. JIS regulations set forth under S. 8502 (Disc Recording) shall apply to all aspects of such recordings not covered by this standard.

2. RECORDING SYSTEM

The sound groove of the regular matrix system disc recording shall be modulated by two signals, left and right, in two directions at 90° to each other and at 45° to the record surface. These two signals shall be converted from multiple original signals in accordance with the regulations given under subsection 2.1. The left signal shall be recorded in the wall of the groove which is closer to the center axis of the record, and the right signal in the opposite wall.

If the two signals are in phase with each other and identical in quantity, they shall be recorded in such a manner that they can be reproduced by the movement of a reproducing stylus tip in directions parallel to the record surface and lateral to the sound groove.

2.1. Conversion of Signals

The two signals that modulate the sound groove shall consist of one left signal and one right signal converted from multiple original signals. The conversion of original signals into these two signals shall basically be achieved in the manner described below.

2.1.1. Front and Back Signals

A signal originated at the front center shall be converted into a left signal and a right signal which are mutually in phase and of identical quantity. A signal originated at the back center shall be converted into a left signal and a right signal which are out of phase with each other by 180° but of identical quantity.

2.1.2. Left and Right Signals

A signal originated on the left-hand (right-hand) side of the front and back centers shall be converted so that the left (right) signal is of greater quantity than the right (left) signal.

2.1.3. Center Signal

A signal originated at the center of the original sound field shall be converted so that the left and right signals are of identical quantity but so that the former has a phase lead of 90° relative to the latter.

2.2. Relationship of Direction of Sound Groove Modulation to Sound Source Direction

The relationship of the direction of the modulation of the sound groove to the direction of the corresponding sound source in the original sound field shall, in principle, be such that the angular direction of the former is half the angular direction of the latter (Fig. 7-10).

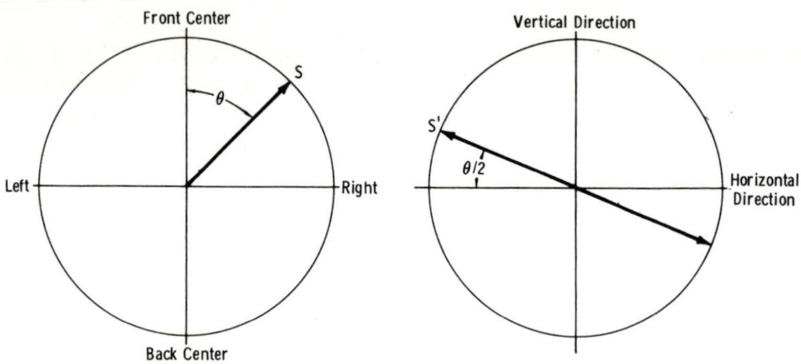

(*A*) *Direction of sound source* (*S*).

(*B*) *Direction of sound groove modulation* (*S'*).

Fig. 7-10. Directions of sound source and sound groove modulation.

ELABORATION

FOREWORD

The Engineering Committee of the Record Industry Association of Japan has compared and examined the various matrix system disc recordings being marketed by various manufacturers to date. Results of such studies have ascertained that all of them, with the exception of the SQ matrix system, are based fundamentally on one and the same system, that they are encoded similarly, and that they possess satisfactory compatability with one another. Hence the same committee hereby standardizes them as "regular matrix system disc recordings."

NOTE: At the time of this writing, Electro-Voice is providing a modification to its own decoder for use with the CBS-Sony SQ matrix system.

1. SCOPE AND APPLICABILITY

This standard governs only those aspects which are peculiar to the regular matrix system disc recording. All other aspects, such as its physical dimensions and quality, shall be regulated by JIS. S. 8502 (Disc Recording).

The regular matrix system disc recording which this standard regulates encompasses all matrix system disc recordings that are cut by converting the information of sound source directions into linear modulations of a spiral sound groove.

2. RECORDING SYSTEM

So as to ensure compatibility with two-channel stereo playback, this standard is formulated in compliance with the stereophonic recording system stipulated under JIS. S. 8502.

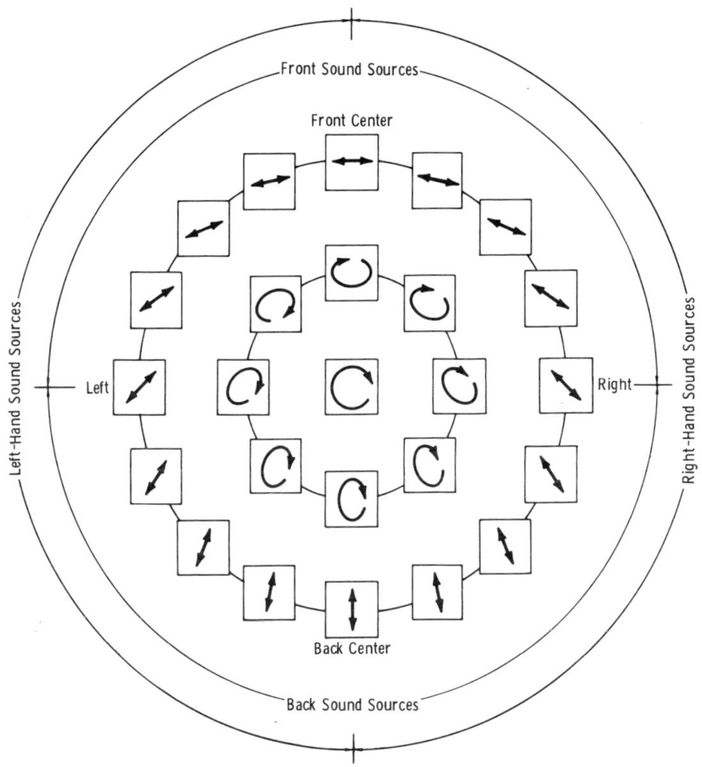

Note: As the sound source moves closer to the center of the sound field, the locus approaches an ellipse. When it is at the dead center, the locus assumes a perfect circle.

Fig. 7-11. Direction of sound groove modulation.

Thus the regular matrix system disc recording manufactured to this standard, when and if reproduced by regular two-channel stereo playback equipment, does not impair the relative sound image and sound volume balance between the left and right channels.

3. RELATIONSHIP OF DIRECTION OF SOUND GROOVE MODULATION TO SOUND SOURCE DIRECTION

The relationship of the direction of a sound source in the original sound field to the direction of the modulation of the sound groove on the regular matrix system disc recording is set forth in Fig. 7-11.

The term "direction of a sound source in the original sound field" is used to describe the direction of a sound source intended at the time of recording, whereas the term "direction of the modulation of the sound groove" is used to describe the locus of the vibration of a cutting stylus tip.

To reproduce the regular matrix system disc recording in more than two channels, it is thus possible to place three or more loudspeakers freely, depending upon the matrixing parameter of the decoder used (including a speaker matrix type).

Basic Principles of QS Encoding-Decoding System

The ability to encode the original 360° without any loss and mislocalization of sound sources was achieved by the adaptation of ±90° phase shifters and by setting disc-cutting vector angles (θ) between four channels at 22.5° in an ordinary disc groove. Fig. 7-12 presents a basic block diagram of the Sansui QS system.

Unlike conventional four-channel matrixing circuitry, this encoder phase-shifts the left and right rear channels ±90° instead of using an ordinary 180° phase inverting method to achieve a reverse phase relationship between these channels. This puts the four encoded channels in an ideal phase relationship, as shown in Fig. 7-13A. Signals are not cancelled in the encoder, and any information from any direction in the original sound field can be encoded.

The above-mentioned phase-shift technique allows a unique action at the decoder. The rear channels are phase shifted in a manner opposite to that in the encoder; namely, the left rear is shifted by −90° and right rear by +90°. Thus the reverse-phase relationship between the rear channels is reconverted to an in-phase one, as depicted by Fig. 7-13B.

The function can be explained in terms of vector angles in the disc groove. The encoder outputs will be:

$$L = (L_F + jL_R)\cos\theta + (R_F + jR_R)\sin\theta$$
$$R = (R_F - jR_R)\cos\theta + (L_F - jL_R)\sin\theta$$

The above equations show that there is no loss of information in the encoding process by adapting j-phase (±90°) shifters in the manner used here.

Now assume a signal of amplitude 1 enters all channels simultaneously in this system:

$$L = (\cos\theta + \sin\theta) + j(\cos\theta + \sin\theta) = 1.30 + j1.30$$
$$R = (\cos\theta + \sin\theta) - j(\cos\theta + \sin\theta) = 1.30 - j1.30$$

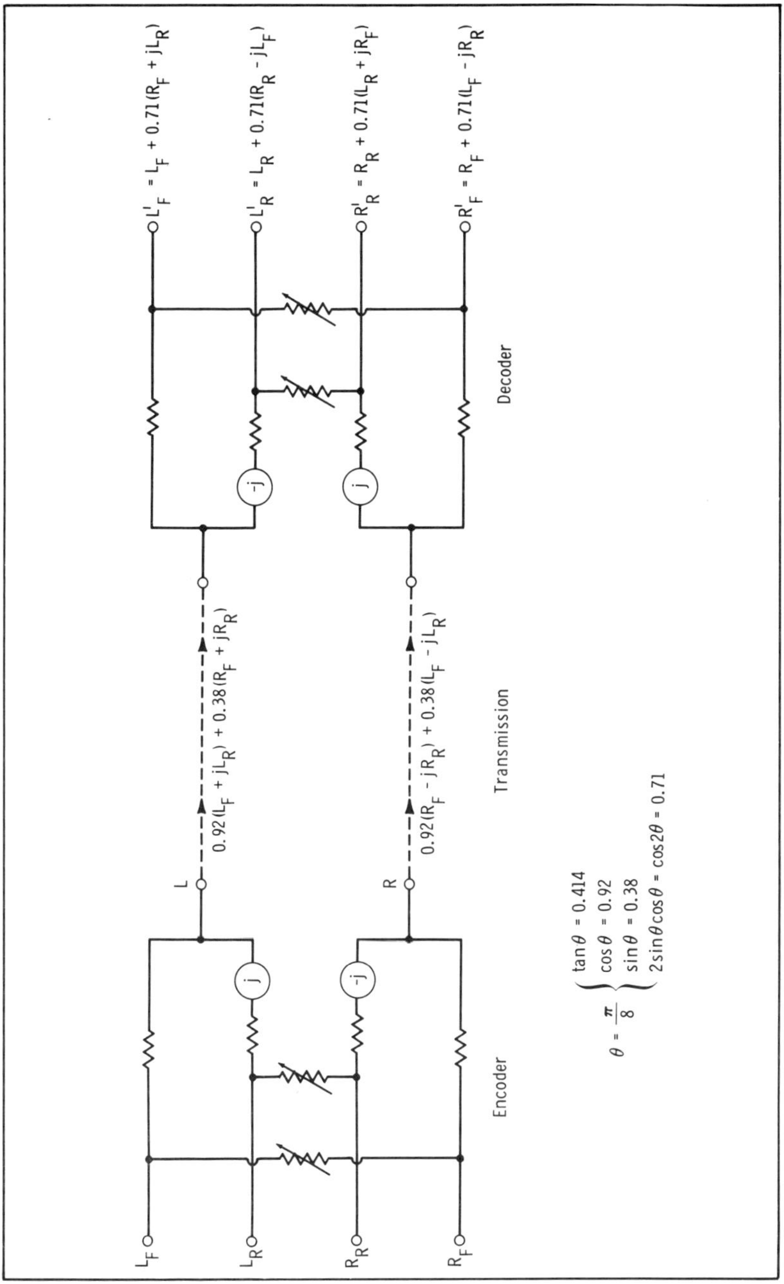

Fig. 7-12. Block diagram of QS coding system.

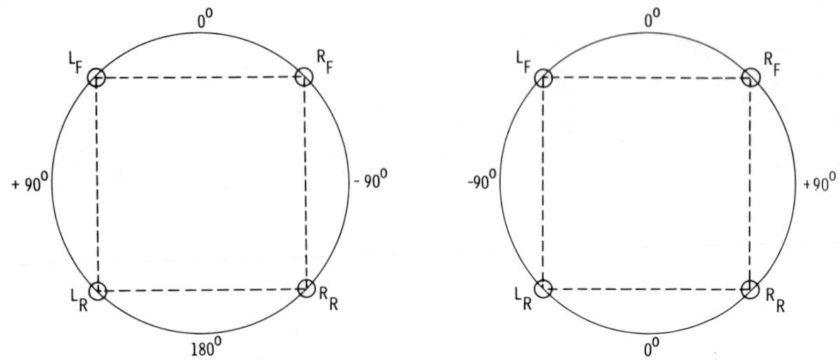

(*A*) *Encoder output.* (*B*) *Decoder output.*

Fig. 7-13. Phase relationship between channels.

Thus the encoded output signals of L_T and R_T are identical in level but have a phase difference of 90°. After decoding, identical levels exist in all channels, with the two rear channels phase shifted by j.

Next assume a signal of amplitude 1 enters L_R and R_R, with zero signal in L_F and R_F:

$$L = j(\cos\theta + \sin\theta) = j1.30$$
$$R = -j(\cos\theta + \sin\theta) = -j1.30$$

Thus the encoded outputs in this instance are signals 180° out of phase. When reconverted in the decoder, they are equal in-phase signals, localizing the sound at rear center.

Blending Coefficient for the Encoder-Decoder

In determining the value of the interchannel blending coefficient (for a matrix encoder), careful consideration must be given to the types of program sources available. Even conventional two-channel stereo sources are available in variety, including:

1. Those utilizing only two (left and right) monophonic channels independently (no intermediate phantoms).
2. Those with sound images localized at three monophonic channels, i.e., left, center, and right.
3. Multiple-track sources with sound images localized at multiple points on a line between left and right channels.
4. Sound-field recordings without any distinctive sound images, such as those of a church organ.

Four-channel program sources can also be classified in a similar manner, and this fact must be carefully weighed in determining the blending coefficient. The blending quantity among four channels of information in the encoding process must in no way restrict the flexibility.

In a four-channel stereo system, the information contained in each channel must be treated equally. This can be accomplished only when the vector angles among the four channels are identical, i.e., when they are all $\pi/8$ $(2\theta=\pi/4)$. In this condition:

A. The cross talks among adjacent channels are equally 3 dB. Thus, the four channels are reproduced uniformly to obtain a square sound field (Fig. 7-14).

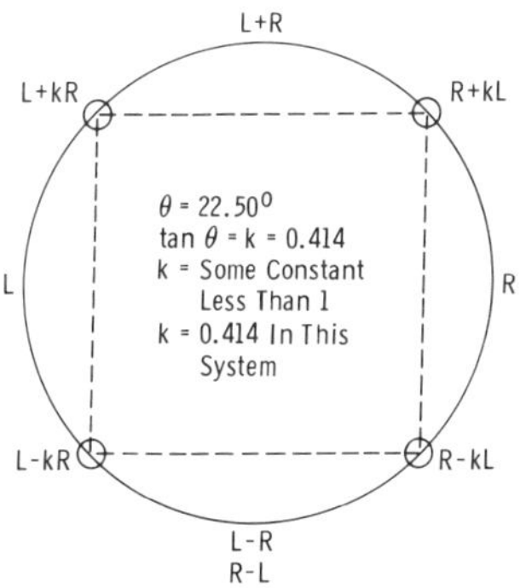

Fig. 7-14. Ideal four-channel reproduction of sound field.

B. Equal volume balance is attained among the four channels, so that distinct sound images can be positioned in any direction inside the square sound field (Fig. 7-15).
C. The encoder and decoder can be allowed to blend an identical quantity of information into adjacent channels.
D. Programs encoded by an encoder in which $\theta=22.5°$ can be decoded by a decoder with a different vector angular value without losing much of their four-channel effect.
E. Conversely, a decoder in which $\theta=22.5°$ is able to reproduce programs encoded by an encoder with a different vector angular value without losing much of their four-channel effect.

To clarify the points made above, Figs. 7-15, 7-16A, and 7-16B show the sound pressure response in different channels to visualize how the sound images are formulated in reproduction. Fig. 7-15 illustrates eight visual patterns of the phase, cross-talk, and directional relationships among the channels. It is noted that the QS decoder allows distinct psychoacoustic images of direction to be formed in the same directions as those in the original sound field.

Fig. 7-16A shows the sound pressure response of a sound source located in the left front channel and indicates the reproduced sound image is shaped as a symmetrical pattern accurately directed from the crossing of polar coordinates toward the point where the sound source is located. Fig. 7-16B shows the same situation with respect to a sound source located at

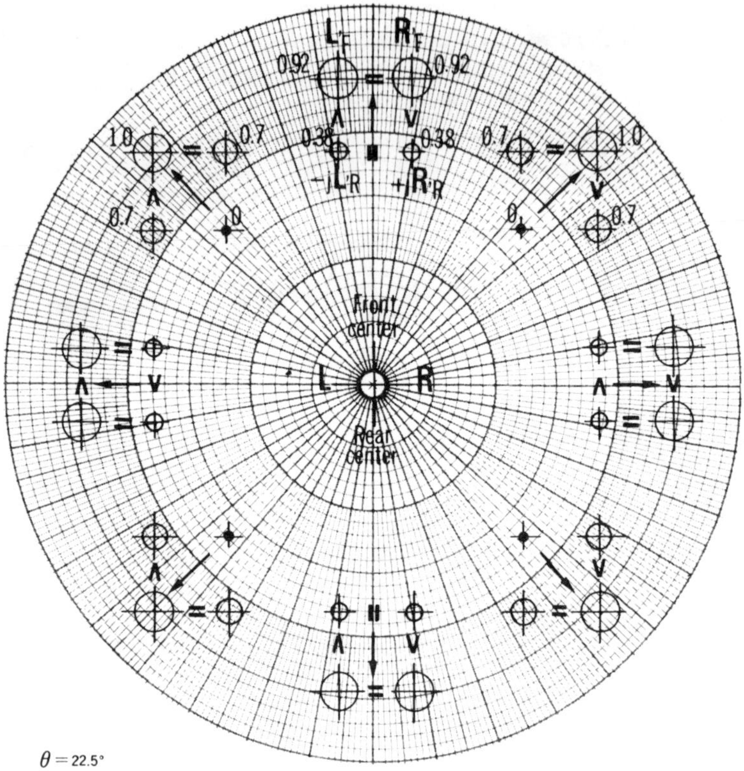

$\theta = 22.5°$
= ; In phase
|| ; Out of phase
ΛV; "j" phase
← ; Direction of Sound Source

Courtesy Sansui Electric Co., Ltd.

Fig. 7-15. Sound pressure response patterns and phase relationships among channels in reproduced sound field.

the front center. While all four speaker systems emit sound in this case, the cross talk to the opposite center direction is −7.7 dB, enabling a distinctively well balanced psychoacoustic field to be created among the four channels with the accurate directionality toward the front center.

Let us now consider what would happen if vector angle θ were not 22.5°. First, if it were smaller than 22.5° (Fig. 7-17A), the separation between the left front and rear and between the right front and rear would be impaired. This would result in a vertically short oblong sound field, making it impossible to obtain equal separations among the four channels.

If the angle were larger than 22.5° (Fig. 7-17B), the separation between the left and right front and between the left and right rear would deteriorate, impairing the volume balance among the channels.

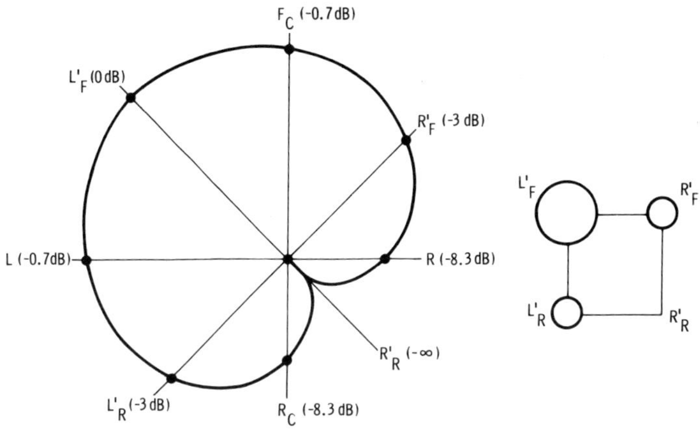

(*A*) *Left front input.*

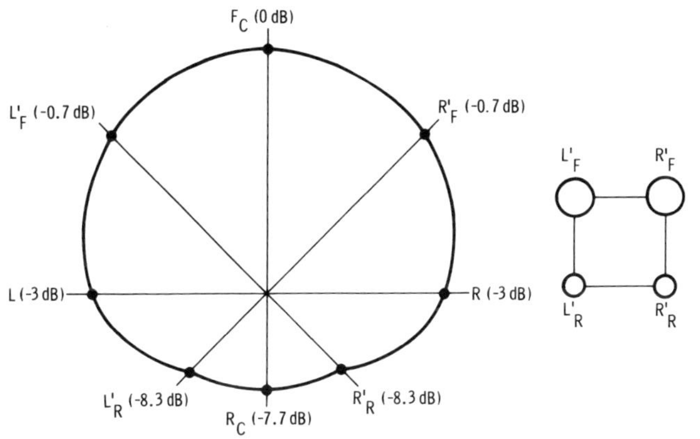

(*B*) *Front center input.*

Fig. 7-16. Sound pressure response patterns.

Thus it can be seen that the value of vector angle θ has a serious effect on the directional characteristics of the reproduced sound images and the shape of the reproduced sound field. This is relatively inconsequential in a 2-2-4 conversion, but if the vector angle were different in the encoder and decoder, it is obvious that the original sound field pattern would not be faithfully reproduced.

Phase Modulation

In a live (original) sound field, the indirect sound components converge upon the human ear from innumerable directions and with infinitely varying phase differences. Such a sound field, therefore, cannot be reproduced

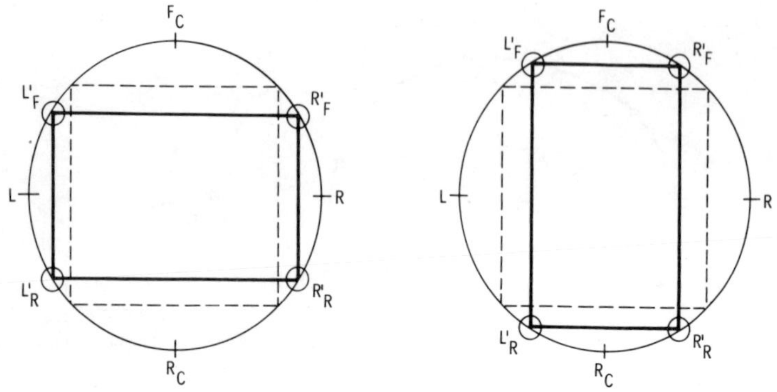

(*A*) *Angle less than 22.5°.* (*B*) *Angle greater than 22.5°.*

Fig. 7-17. Reproduced sound field simulation when angle θ is not 22.5°.

faithfully simply by adding one or two more speaker systems, although this will make possible a considerable improvement in the tonal quality over conventional two-channel stereo. Furthermore, the sound sources (direct sounds) themselves keep changing their positions and strengths every moment, so that the relative relationship between the direct and indirect sounds becomes infinitely complex. To reproduce the original sound field faithfully, it is therefore necessary to reproduce the direct and indirect sound components with continuously varying phases.

The Sansui QS 4-channel synthesizer decoder adopts a phase modulator circuit (Fig. 7-18) for this purpose. This circuit "phase modulates" the

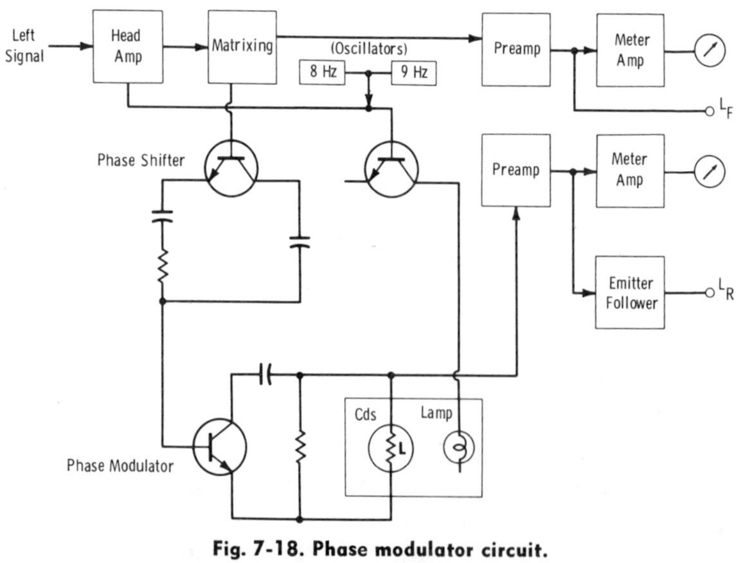

Fig. 7-18. Phase modulator circuit.

rear-channel signals, which then meet with the nonmodulated direct sound components from the front speaker systems in the air to enhance the presence of the reproduced sound. Phase modulation in this case produces infinitesimal time and frequency differences between the indirect sound components and the nonmodulated direct sound components. Not only is the presence of the reproduced sound greatly improved by this process, but the dynamic range of "pulsive" notes is substantially expanded. In designing the system, Sansui engineers repeated actual listening tests for a long period of time to determine the optimum frequency range and depth of phase modulation.

While various circuit designs are possible for such a phase modulator circuit, the Sansui QS 4-channel synthesizer decoder adopts a phase-shift frequency modulation system, whereby the resistance component is altered in circuits similar to the phase-shifter circuit. Modulating signals are a mixture of the middle- and high-frequency components of the audio input signal and the signals from an oscillator constructed of two generators, permitting the left and right channels to be modulated by different signals. If a sine wave is applied (for the purpose of testing, etc.), the resultant phase modulation would acquire a periodic nature. But in the case of an ordinary music source, the modulating signals are random signals for an optimum phase-modulation effect.

Panning Technique

The blending coefficient can be varied electrically by the use of *panpots* on the mixing console, as illustrated by Fig. 7-19. The two pairs of encoder inputs (L1 and R1, and L2 and R2) are each independently manipulated.

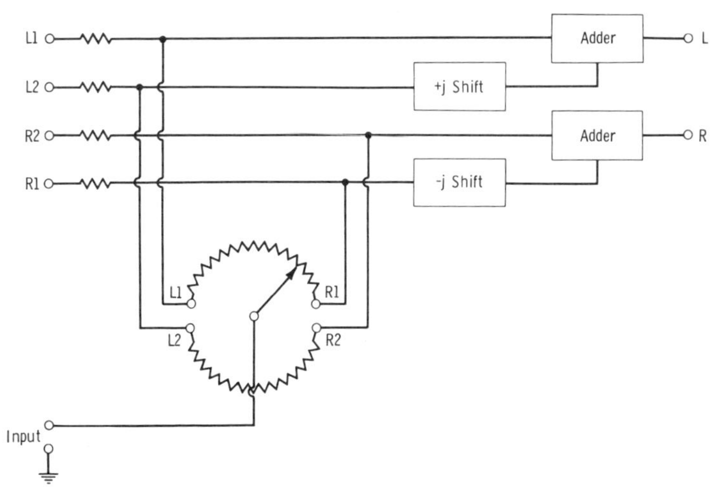

Fig. 7-19. Panning circuit.

By this technique it is possible to shift the focus in any direction at the recording end; speaker systems can be positioned anywhere on the playback side. Thus this coding system permits the localization of sound images at any point in a full 360° both in recording and playback, not only in a four-channel setup but in any other multichannel setup.

7-6. THE DISCRETE FOUR-CHANNEL SYSTEM

The discrete system, as implied by the term, involves the transmission of information in four separate channels throughout the system. Proponents of this method take exception to the "psychoacoustic" matrix system, claiming that true quad sound cannot be a compromise from four distinct channels representing left and right front and left and right rear. The discrete system therefore requires some form of carrier multiplexing of a stereo fm transmitter to carry the two additional channels of information.

We will first study the discrete system in its initial closed-circuit form, starting with the quad disc and ending with the four separate speaker systems. We will then describe briefly the currently proposed techniques of broadcasting discrete quad sound through a modified fm stereo transmitter.

The RCA-Panasonic-JVC discrete system, currently termed the CD-4 system, will be described, since this is the most popular at the time of book preparation. The term CD-4 is derived from *C*ompatible *D*iscrete *4*-Channel, and this designation may or may not be retained by RCA for future production models.

Signal relationships of the four channels are as follows:

Channel 1: Left front
Channel 2: Left rear
Channel 3: Right front
Channel 4: Right rear

Matrixing is employed in the discrete system, but the function is different than that of the matrix system previously described. The recording matrix circuit accepts the four discrete signal sources and produces four discrete signal outputs as follows:

Channel 1 plus Channel 2. $(L_F + L_R)$ Channel 3 plus Channel 4. $(R_F + R_R)$	SUM CHANNELS
Channel 1 minus Channel 2. $(L_F - L_R)$ Channel 3 minus Channel 4. $(R_F - R_R)$	DIFFERENCE CHANNELS

Very briefly, the subtracted channels frequency and phase modulate a 30-kHz carrier which is then mixed with the sum channels. This mixed signal is fed to a recording amplifier which drives a specially designed cutting head for the quad disc. The quad disc actually contains the modulated carrier and its sidebands with a frequency response up to 45 kHz.

The CD-4 Recording System

Fig. 7-20A shows a basic block diagram of the method of recording a CD-4 disc. The sum of channel 1 plus channel 2 ($L_F + L_R$) is recorded in the left wall of the groove (Fig. 7-20B), and the sum of channel 3 and channel 4 ($R_F + R_R$) is recorded in the right wall. The difference signals frequency and phase modulate a 30-kHz carrier which is also recorded on the disc, in the manner illustrated by Fig. 7-20B.

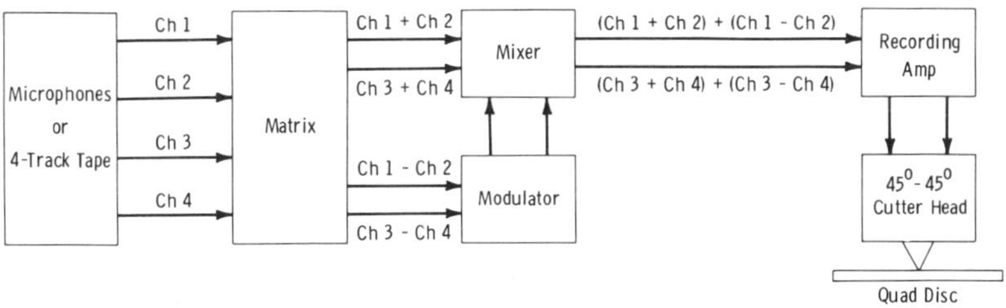

(A) Block diagram.

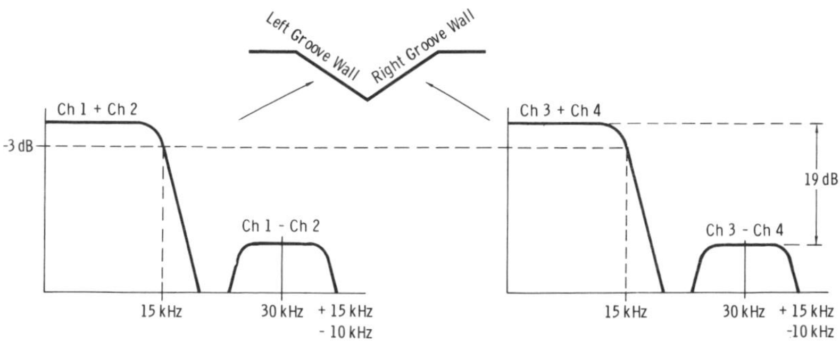

(B) Groove modulation.

Fig. 7-20. CD-4 disc recording system.

The CD-4 Reproducing System

Fig. 7-21A is a basic block diagram of the reproducing system. The pickup cartridge is a special high-compliance type with a stylus pressure of about 1.5 grams and a frequency response of 20 Hz to 45 kHz. Fig. 7-21B shows details of the demodulator block in Fig. 7-21A. The front-to-rear separation for this system is said to be about 25 dB, or very nearly equal to conventional stereo left-right separation.

When the CD-4 disc is played on a conventional stereo system, the stereo pickup is not sensitive to the 30-kHz carrier, and the sum signals are reproduced in the two channels as for normal stereo recordings. When the

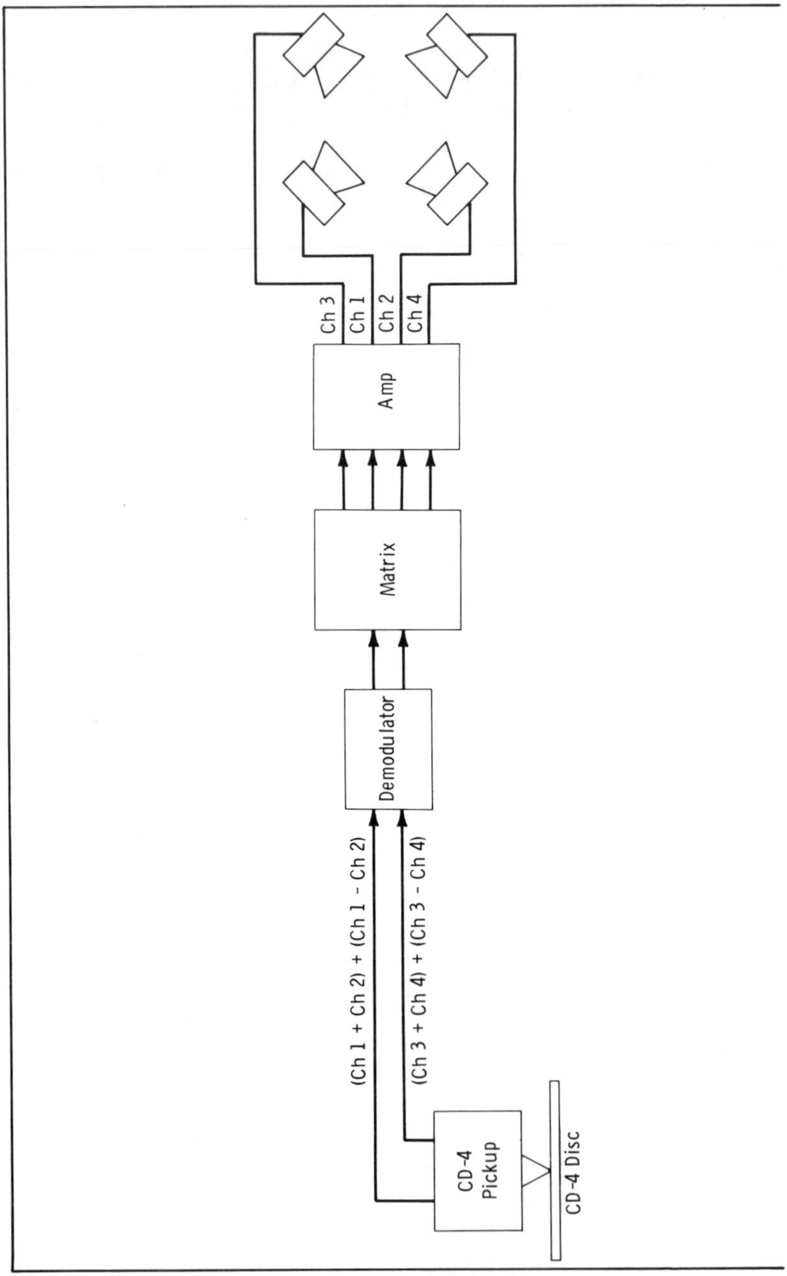

(A) Basic block diagram.

Fig. 7-21. CD-4

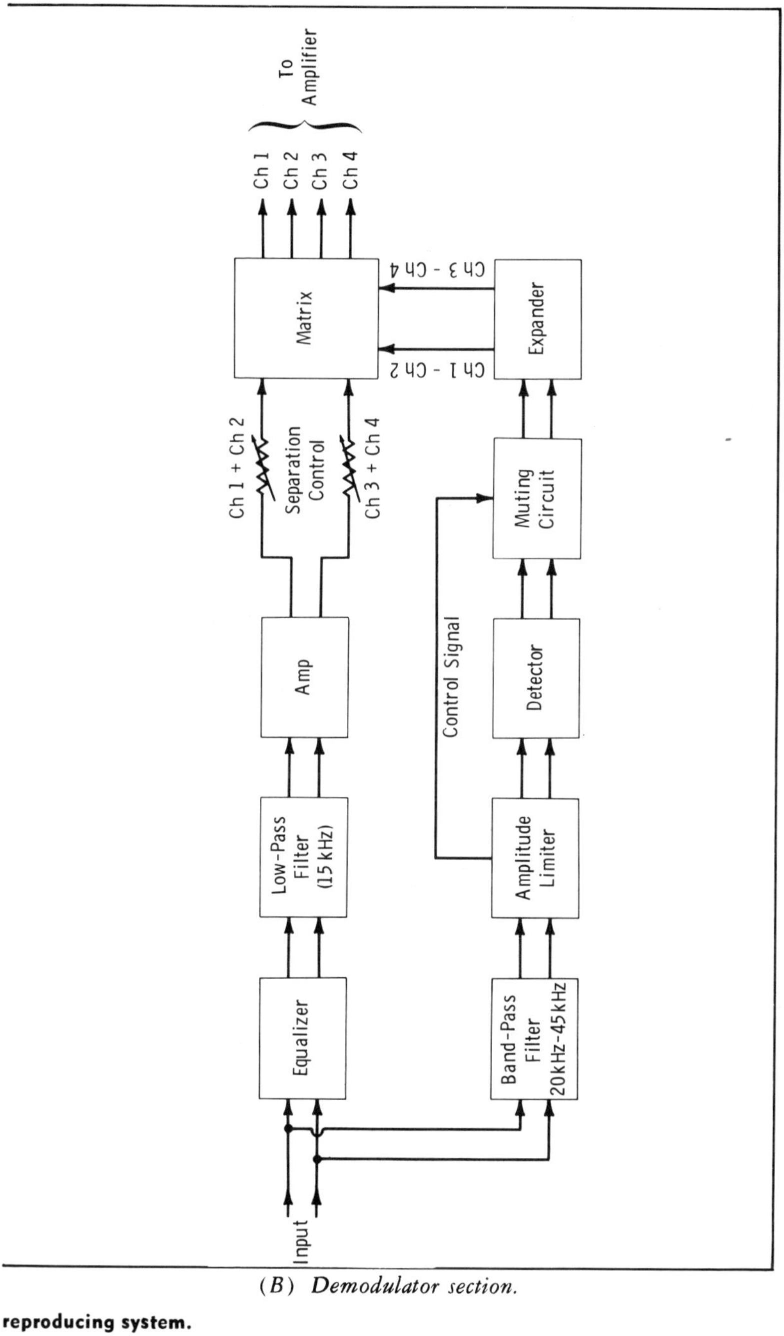

(*B*) *Demodulator section.*

reproducing system.

quad disc is played monophonically, the sum of all four signals is reproduced. For either stereo or monophonic reproduction, the demodulator is switched out of the circuit.

Getting Discrete Quad on the Air

As mentioned previously, the broadcasting of discrete four-channel sound sources involves some form of carrier multiplexing technique which requires special FCC approval. At the time of this writing, only experimental broadcasting has been authorized, with no approval of any specific discrete system.

Fig. 7-22 shows the multiplexed carrier frequencies of three proposed discrete quad systems. McMartin's TDM (time-division-multiplex) system places the subcarrier at the presently assigned SCA frequency of 67 kHz. The space is time-shared by sampling at the rate of 19 kHz locked to the stereo pilot frequency. The total bandwidth of 16 kHz provides 8 kHz for the left-rear and right-rear channels. Amplitude modulation is used in this time-shared system.

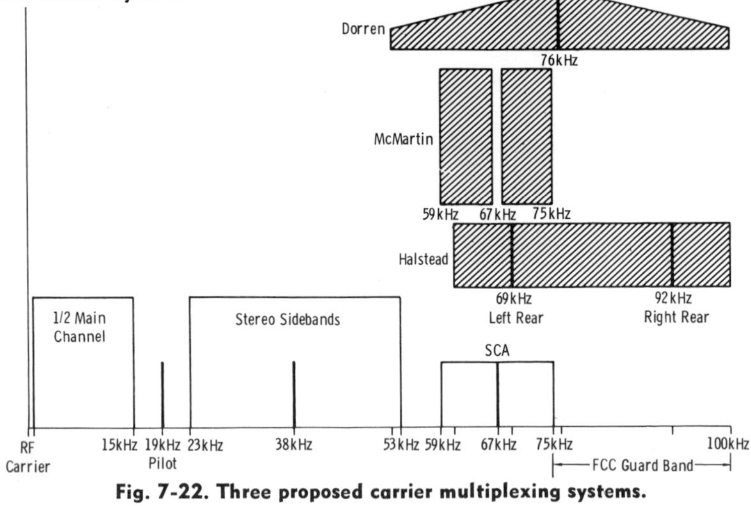

Fig. 7-22. Three proposed carrier multiplexing systems.

The Halstead system also provides 8-kHz channels for the left-rear and right-rear speakers. The left-rear subcarrier is at 69 kHz, and the right-rear subcarrier is at 92 kHz. Thus the upper sidebands of the 92-kHz subcarrier extend to the limit of the assigned guard band at 100 kHz. However, the fm subcarriers are at low level and produce only first-order sidebands, and the stereo modulation is reduced below the maximum deviation. Proponents of the system claim that tests with a panoramic spectrum analyzer reveal that signals produced by this technique extend no further at the limit extreme than do modulation products of the conventional stereo transmitter using 100-percent modulation. The degradation in signal-to-noise ratio is approximately 1.5 dB for each 10-percent reduction of modulation in an fm system.

The Dorren system employs a 76-kHz subcarrier which supplies decoding information to direct the four discrete channels to their proper speakers. Full 50-15,000 Hz range for all speakers is claimed for this technique, with channel separation of 40 to 42 dB over this range. The signal-to-noise ratio is said to be 60 to 62 dB per channel. Very little information has been released at the time of this writing, probably due to patent processing.

7-7. TRANSMITTER REMOTE-CONTROL FACILITIES

It is becoming almost standard practice to operate the transmitter remotely from the studio control room. The control units are normally interconnected by telephone lines, although microwave links have been used. Transmitter on-off, power adjustment, directional operation, and various monitoring circuits are provided.

The control panel of a typical remote-control system is shown in Fig. 7-23. The transmitter control unit contains the calibration controls. These are adjusted so that the remote meter readings at the studio (frequency,

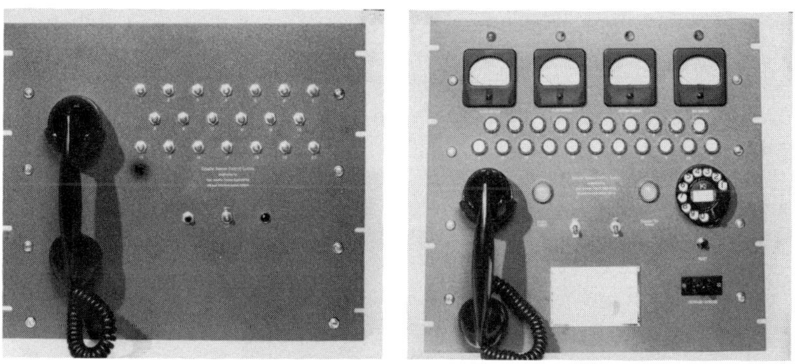

(*A*) *Studio unit.* (*B*) *Transmitter unit.*

Courtesy Schafer Electronics Corp.

Fig. 7-23. Remote control system panels.

modulation, etc.) are the same as those at the transmitter. Built-in telephone communication eliminates the need for an extra phone line. This is an all-dc system which uses no vacuum tubes and operates on any two ordinary metallic telephone circuits. There are 24 metering circuits and 40 control circuits included in this particular studio control unit. The telephone-type dial operates pulsing relays which connect the monitoring point desired.

The following controls are required:

Transmitter start on-off
Tower lights on-off
Final-stage tuning
Output-power control

The following metering circuits are normally provided:

Deviation of carrier from assigned frequency
Modulation percentage
Final-stage plate voltage
Final-stage plate current
Antenna current or currents
Antenna-current phase (for directional antennas)

Additional controls and metering points may be added.

Two telephone circuits are required, one for the control line from the studio to the transmitter and one for the metering-line return from the transmitter to the studio. A diagram of the RCA remote-control system operation is shown in Fig. 7-24. A description of this system follows.

When switch S is closed on the control unit, power supply A through voltage divider B and telephone dial C normally supplies 12 volts dc to the control line. Operation of the lower switch in unit B increases this voltage to 35 volts, and operation of the raise switch in unit B increases this voltage to 100 volts. Telephone dial C interrupts the line voltage with a number of impulses corresponding to the number dialed. Stepping system D with its power supply E controls the indicating lights showing the number dialed.

In the transmitter unit, relays F, G, and H are bridged across the control line. Relay F is energized when switch S is closed and follows the dial impulses. Relay F operates slow relay N which does not release during dial impulses. Relay N controls the transmitter power so that in the event the control-line voltage fails, the transmitter is taken off the air.

Relay G operates when the lower switch is operated, and relays G and H both are operated when the raise switch is operated. Relay F operates stepping system I, which controls the indicating lights showing the number dialed and also connects the raise and lower relays to any one of ten external operating units consisting of reversible motors or control relays. Stepping system I also connects the metering line to any one of nine external metering elements which convert any desired ac or dc potential or current being measured to a value suitable for remote measurement.

The entire metering system is calibrated against a standard source of voltage on position 0. This accurately compensates for any variation in the line resistance with temperature changes.

A metering circuit is required from the control unit at the remote-control point to the transmitter unit. This circuit is to be a low-grade telephone pair with a maximum total dc loop resistance of 4000 ohms. This

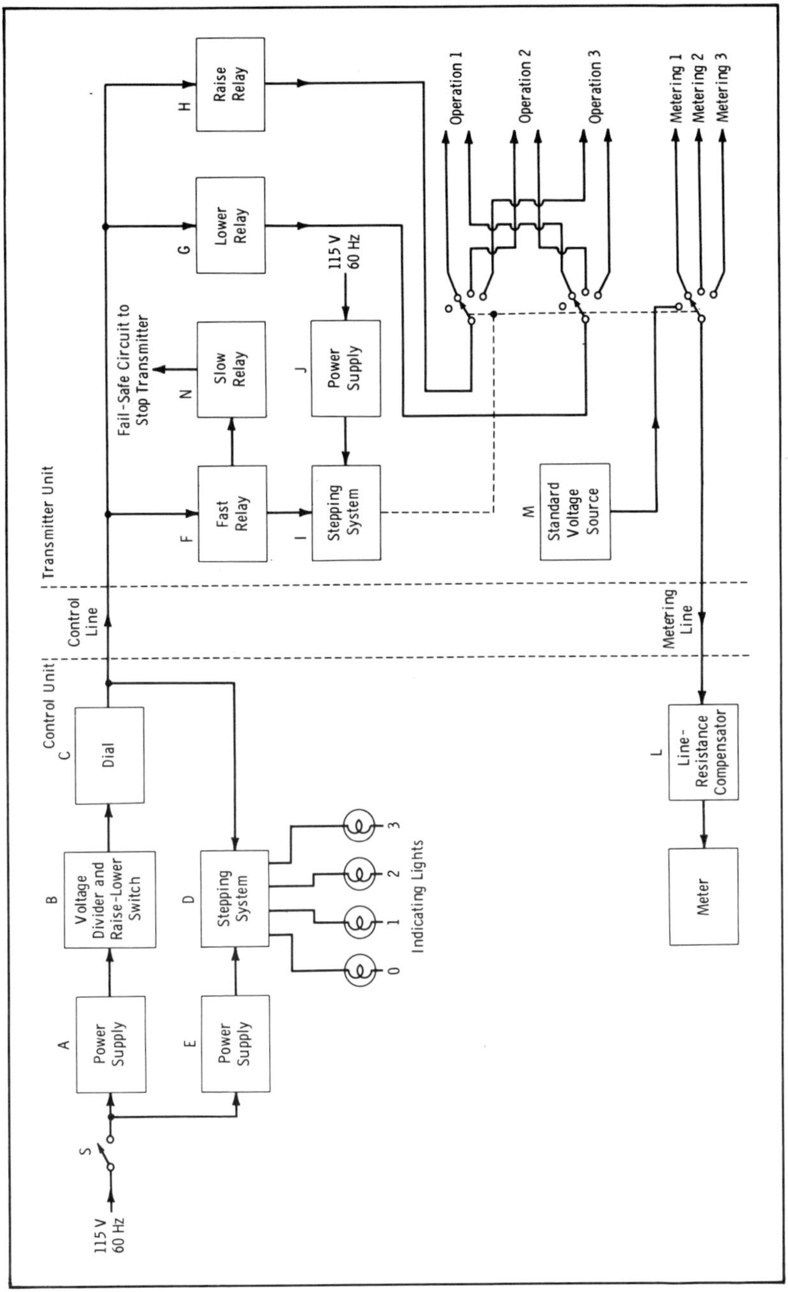

Fig. 7-24. Block diagram of remote-control system.

circuit must furnish a dc path, and no simplex or grounds of any type are permissible.

Maximum voltage line-to-line: 0.5 volt
Maximum current through circuit: 200 microamperes
Line termination protection: 200-mA fuses and Thyrite protectors to ground
Maximum voltage line-to-ground: 100 volts
Maximum current with line shorted: Approx 10 mA

A control circuit is required from the control unit at the remote control point to the transmitter unit. This circuit is to be a low-grade telephone circuit with a total dc loop resistance of 4000 ohms maximum. This circuit must furnish a dc path, and no simplex or grounds of any type are recommended. The circuit carries only pulsed dc (10 pulses per second).

Maximum voltage line-to-line: 100 volts
Maximum current through circuits: 15 mA

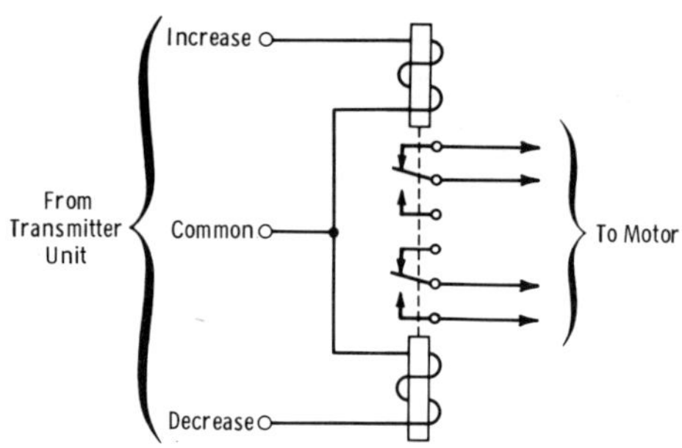

(*A*) *Circuit for control of quantities such as plate tuning or antenna loading.*

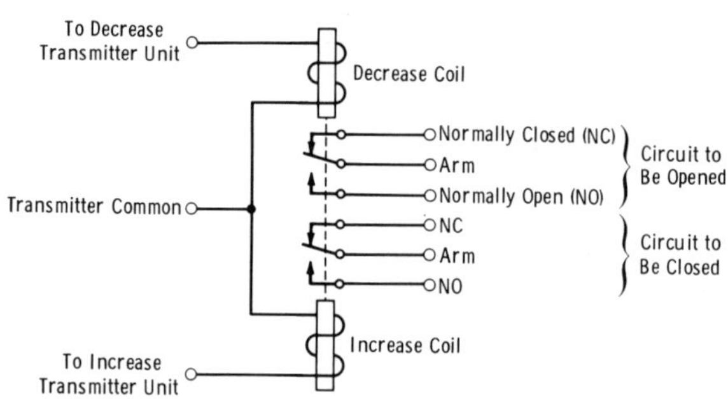

(*B*) *Details of latching-relay circuit.*

Fig. 7-25. Typical remote-control circuits.

Line termination protection: 250-mA fuses
Maximum current with line shorted: 82 mA

The transmitter unit provides the required number of calibrating potentiometers so that the remote meters at the studio read the same as those at the transmitter. Fig. 7-25 illustrates typical transmitter remote-control circuitry. Fig. 7-26 shows typical remote-metering circuitry. Table 7-1 shows typical a-m station dial functions.

The maintenance procedures for components of a remote-control system such as resistors, capacitors, relays, etc., are covered in other chapters. Procedures for the system as a working entity may be outlined as follows:

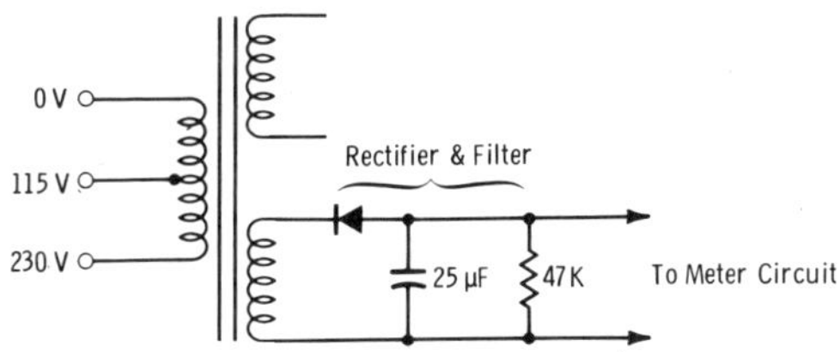

(A) Line-voltage metering.

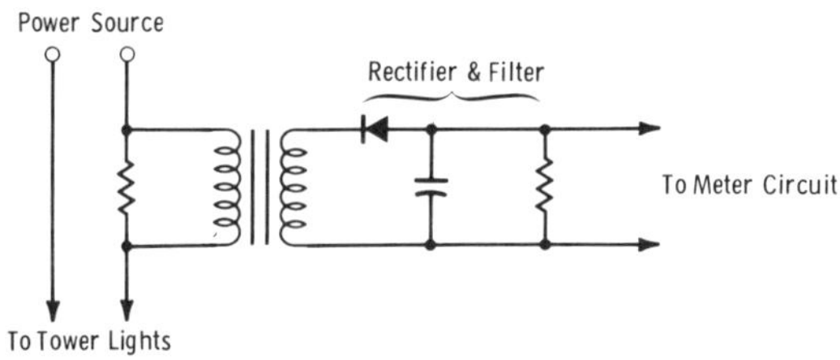

(B) Tower-light current metering.

Fig. 7-26. Typical remote-metering circuits.

Table 7-1. Control-Unit Dial Functions

Dial	Control	Meter
0	....	(Set meter full scale)
1	Tower lights on-off	Tower lighting current
2	Filaments on-off	Filament line voltage
3	Plates on-off	Final stage plate voltage
4	Final stage tuning	Final stage plate current
5	Output power control	Antenna current
6	Additional positions for future requirements	
7		
8		
9		
0		

1. Always check and record the dc resistance of the telephone pairs employed. This serves as a reference for future use.
2. Check the dc resistance of the telephone pair to ground with a sensitive vtvm and record it for future reference.
3. Periodically, or whenever conditions warrant, run a noise-level measurement on the line. Electrical storms or other conditions along the pair routing may produce a low-resistance ground or cross talk, which can result in erratic operation of the remote-control or metering circuits.
4. Check and calibrate all metering circuits at least as often as required by current FCC regulations or special conditions that may be contained in the station remote-control authorization.

NOTE: A transmitter remote control involving microwave and telemetry is described in Section 8-6.

7-8. AUTOMATION SYSTEMS

Automation in am-fm broadcasting is here to stay, and to grow. A basic block diagram of automation as applied to technical operations is shown in Fig. 7-27. The type and extent of program control varies considerably as evident in following descriptions, but the overall function is similar to that illustrated.

The CCA Mini-Automation System

Fig. 7-28 shows four lever switches, each with a four-position capacity. The event at which each program source occurs can quickly be arranged by setting the four switches to the desired positions. Normally, the cue detectors associated with the tape machines advance the system from one prearranged program source to another. However, upon request, the built-in priority clock can be used to interrupt the sequence. Located above each of the switches is a status light which, when registered, indicates that this event is next in the sequence.

An inexpensive but reliable one-hour clock is provided. It has facilities for presetting two switches in one-minute increments. Each of these switches is normally supplied prewired to the first two audio sources, which can thus be used on a real-time priority basis rather than on a sequential basis. For example, one source could be a combination of station identification, time announce, and prerecorded news which would occur every half hour. The second source could be one or several multiple cartridge machines which could be set up to play every 10 minutes. Please note that the system will not normally interrupt its programming and switch to another source in the middle of a message. The cue detector will normally restore the program line, and if a priority has been called, this source will then supersede the sequential event.

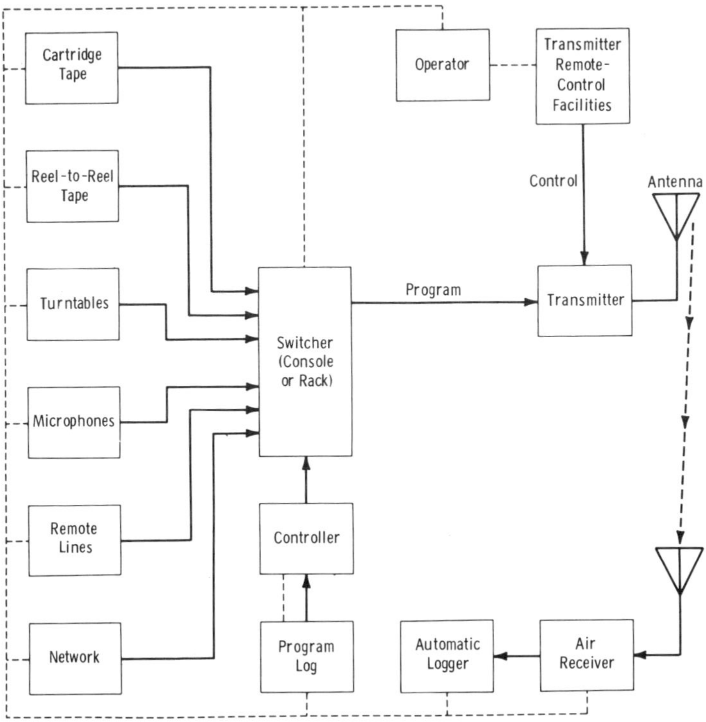

Fig. 7-27. Basic technical automation facility.

The system contains two plug-in 25-Hz cue detectors. These units are designed to detect the presence of a 25-Hz tone at a level as low as −30 dBm and will not overload or cause miscues at levels as high as +10 dBm. In a normal automation system, the 25-Hz end-of-message tone is generally recorded on the left channel at a level of approximately −5 dBm. The 40 dB dynamic range of the CCA Q-Tectors permits them to function over a wide range of errors in the production and playback of 25-Hz cue information. Each detector has controls for achieving an adjustable overlap.

The unit contains a built-in 25-Hz generator with dual 25-Hz filters. It is preset to operate for 2 seconds and can be controlled from either the front panel or a remote position. It is independent from the normal program lines, and thus production work can be done while the system is providing programs.

The system also contains solid-state circuitry which can sense the absence of signal on the program line. This circuitry can be adjusted to operate after any predetermined period of silence from 2 to 20 seconds. Thus, if one of the machines should become inoperative, the silence sensor will detect the problem and sequence the system to the next scheduled event.

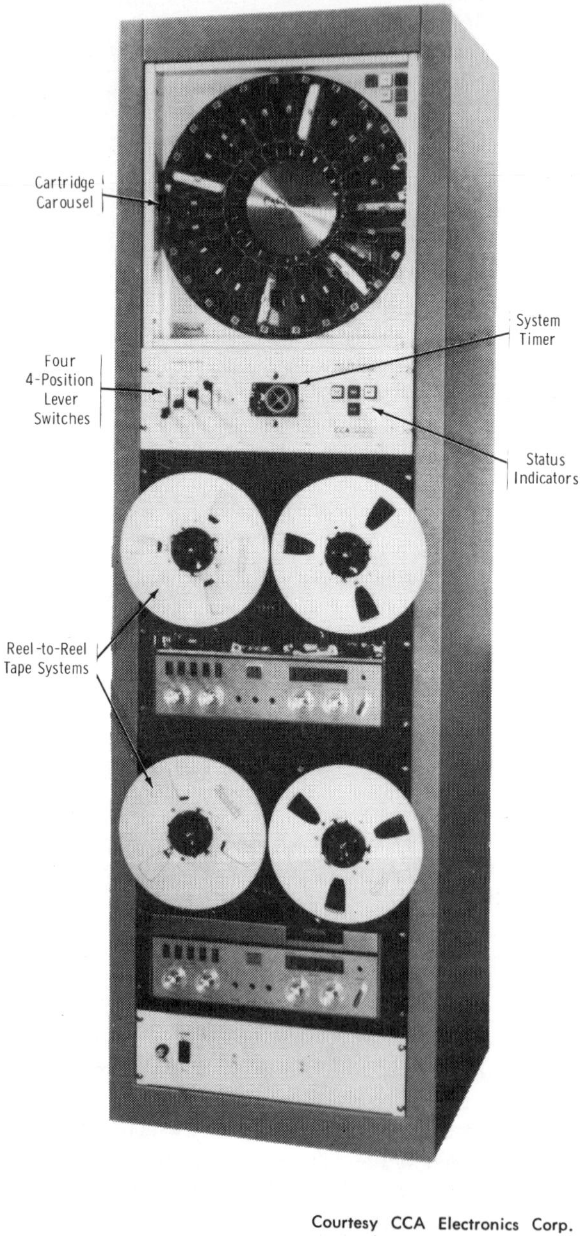

Courtesy CCA Electronics Corp.

Fig. 7-28. CCA Mini-Automation system.

The unit contains a front-panel switch which when depressed serves as a priority. Thus, once this button is operated, the next source of the automation system would be that associated with the remote. The system may then be used to serve as a program source, and periodically news or some other priority may be inserted in the system. The silence sensor is disabled during use of the remote source to allow flexibility of program content.

A series of panels can be used in conjunction with the CCA automation systems to expand a single channel from one to three more audio sources. The simplest version of the family of expansion panels is the Model EC3-C. It is designed to expand single-play tape cartridge machines or multiple-cartridge machines such as the Carousel. The additional three sources and the original source of the audio channel can be controlled from the front panel of the EC3-C. Each subsource has a switch associated with it. This switch programs the source or bypasses the subsource. There is also a provision for an external override, which when set to override the system will skip all remaining subsources to be played and will restore operation to the basic automation system at the end of the selection being played. This permits priority override of the expansion system.

On the rear of each panel which contains a Q-Tector are controls for adjusting the time after the end of the cue tone at which the reel-to-reel tape machine will stop. This adjustable delay permits achieving a very tight format as well as any practical overlay. A switch for each internal cue detector is available on the rear for overriding any preset overlay.

The CAA RA-50-1 (random access unit) is designed to permit rapid, reliable control on a random basis of the 24 cartridges stored in the Carousel multiple-cartridge playback machine. This panel permits the broadcaster rapidly to arrange the sequence in which the various cartridges will play without the necessity of removing the cartridges from their positions within the Carousel.

The random-access unit contains 50 levers, each of which refers to an individual event. Placing these levers in any one of 24 positions determines the cartridge which will operate when that event is sequentially selected. Each lever also has a skip position which, in the CCA automated system, will bypass to the next cartridge source. A front panel light indicates the next event to be played in the Carousel. To the right of the panel are three switches which provide for advancing the events rapidly, converting to a consecutive mode of operation of the system, and resetting the programming to event No. 1.

The RCA Basic Automation System

The RCA automation systems for radio cover a broad range of applications as fundamentally illustrated in Fig. 7-29. The System I shown in Fig. 7-29A consists of two reel playbacks, the programmer, a one-hour timer, and two cartridge units. For a simple program pattern, the operator

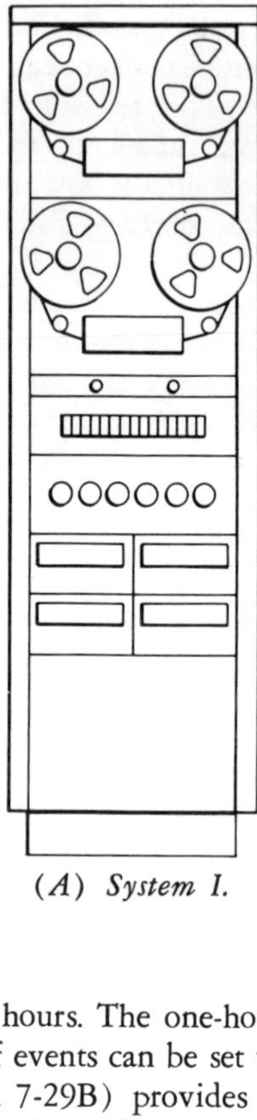

(*A*) *System I.*

Fig. 7-29. RCA radio

could program four or five hours. The one-hour timer is a cycling device so that the same sequence of events can be set up to happen as many times as required. System V (Fig. 7-29B) provides 24-hour programming with complex patterns. Systems II, III, and IV are in between these extremes so that individual tailoring of a system to specific requirements can be accomplished.

Many automation systems use some form of memory storage—paper tape, magnetic tape, punched cards, coded log, or other—which is programmed in advance for a certain sequence of events resulting in the automatic control of tape machines, turntables, and other program sources. There are advantages to memory storage, but as the complexity of the sequence of events increases, it sometimes becomes very rigid and difficult to change.

In the simplest concept, one program event starts the next, so a memory is not needed. What can be used instead is a method of routing events that permits quick and easy change of the program format, event, or sequences at any time. Live and automated operations are readily integrated. Uncertainty can be eliminated by continuous program status indication to the operator, who may be a nontechnical person with very little training. This is the operation of RCA's basic automation system. It borrows solid-state

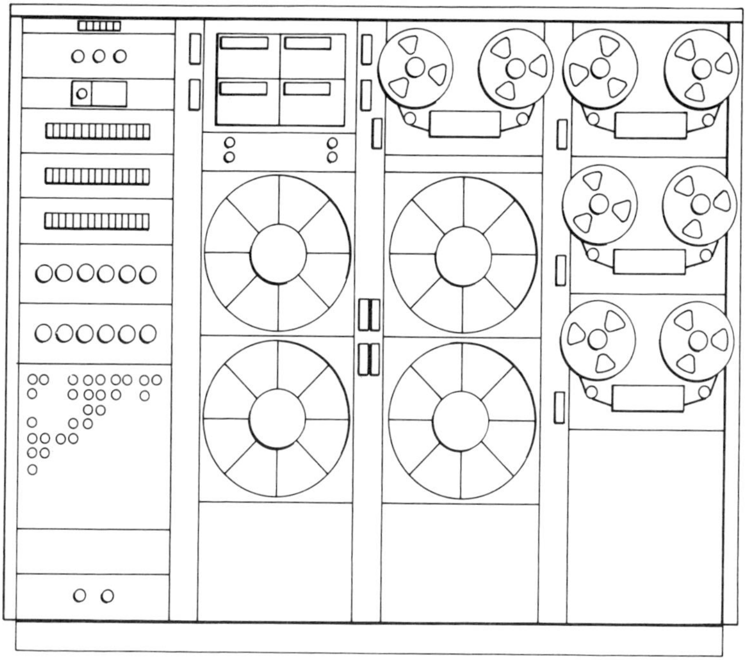

(*B*) *System V.*

automation systems.

computer techniques to permit flexibility of programming with simple setup requirements.

A major unit of RCA automation is the BCA-15B programmer (Fig. 7-30). It is designed to switch between as many as 18 preselected audio sources, such as reel and cartridge tape machines, turntables, and microphones, and continue to sequence them automatically in any preset pattern of events as long as required.

The sequence of events is programmed by means of thumbwheel switches (Fig. 7-31), which select events from any of the up to 18 program sources. Control is given in sequence by circuits in the unit, the end of one event (in the smallest systems) initiating the beginning of the next. The event being played is shown by a tally light above the switch. The capacity is 15 events. Modifications to the program may be made at any time either by (A) resetting the thumbwheel switches, (B) starting an event using the front-panel push buttons, or (C) manually starting the audio source itself.

Programmers may be cascaded to provide 30, 45, 60, or more events. Or, a second unit may be used to provide a subsequence to the first BCA-15 programmer, further expanding the programming capability of the combination. The recycling function increases the capability by repeating fre-

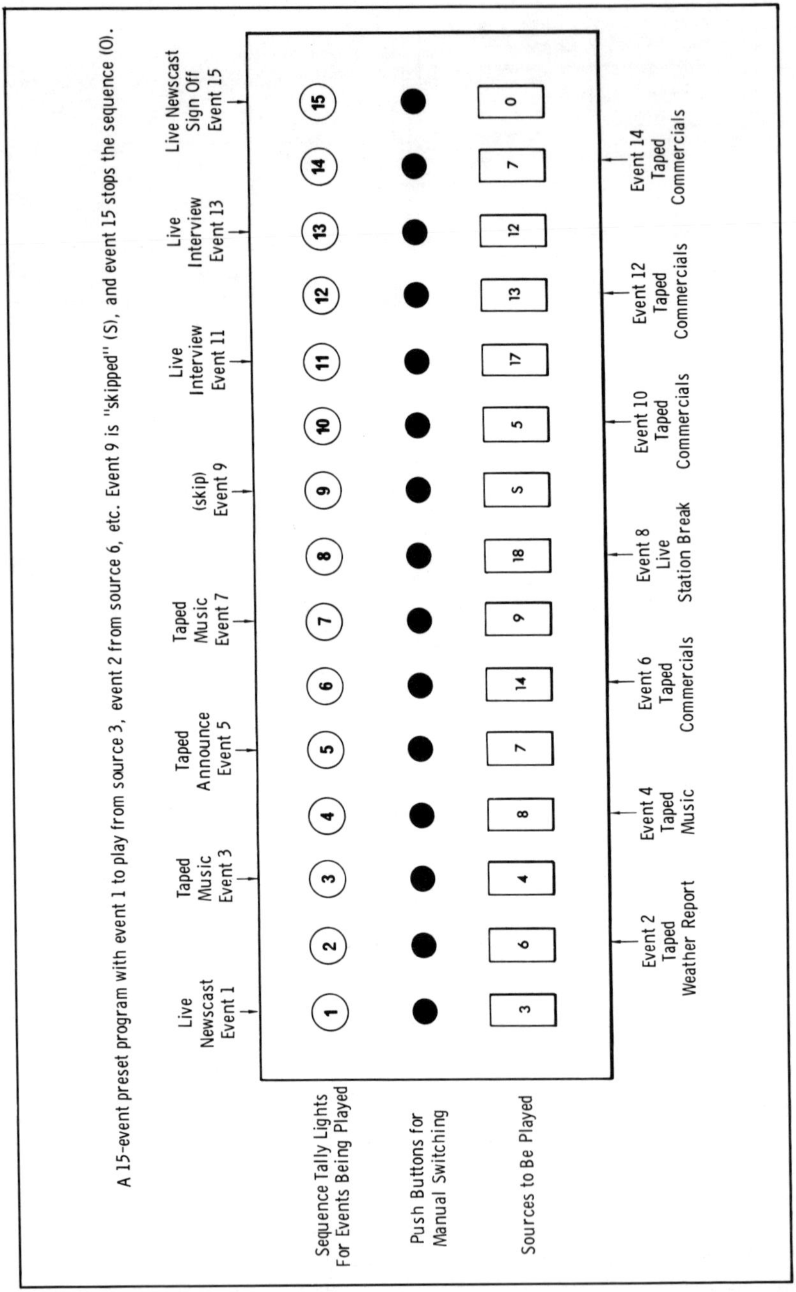

Fig. 7-30. RCA BCA-15B programmer.

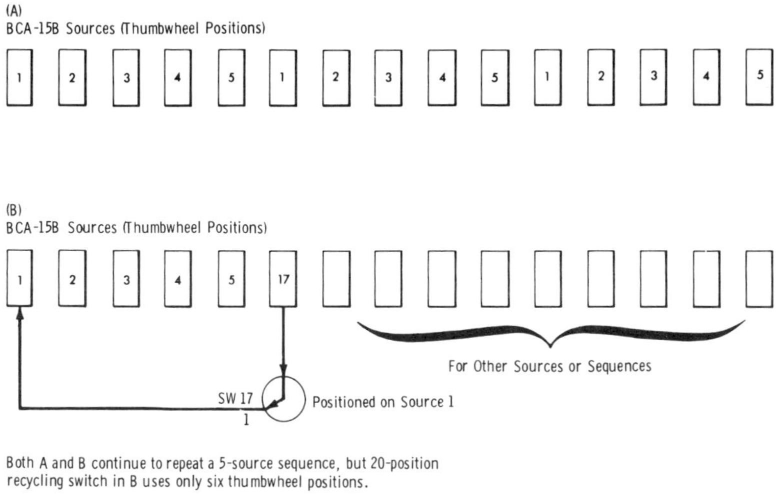

Fig. 7-31. Use of thumbwheel switches for automation programming.

quently used program segments. By use of cycling and subsequencing techniques, anywhere from a few hours to a full day's programming may be set up on two BCA-15B programmers.

Another major unit of the automation system is the program clock. This is actually a timer which synchronizes the program on an average time basis. The timer does not chop or fade program material; rather, for smoothest operation, it adds or deletes events from programmed musical fill at the end of a time segment to guarantee station identification within legal time limits. It then starts a new event after the ID segment is completed. The clock does not work on real time, but rather prearms the system so that the station break occurs in accordance with FCC requirements.

A typical clock function is shown by Fig. 7-32. For simplicity, each of sequences A, B, and C is repeated using identical sources. Starting on the hour, sources 1 to 9 (sequences A and B) play in consecutive order. After sequence B finishes, source 17 returns the program to source 6, through the program clock timers, and sequence B plays again. Sequence B will repeat until 15 minutes have elapsed; then the quarter-hour timer routes the program to source 11 in sequence C. By the time sequence C is finished playing, the quarter-hour timer will have returned to source 6. Sequence B will continue to repeat until 30 minutes have elapsed and the half-hour timer changes position. This routes the program back to sequence A, and the A-B-C-B cycle that occurred during the first half hour repeats in the second half hour under control of the three-quarter and one-hour timers.

For stations requiring rigid timing, as when joining and separating from network lines, the station break can be clocked precisely. Here the equipment, activated by network signaling tones or a digital clock, fades the local programming at the precise time and restarts the automation system at the finish of the network activity.

The event being played and the next to play are read out continuously on the faces of the BCA-15 programmer, the clock timer, and the status indicators on each source. Thus the portion of the program in play and the entire program sequence are always visible to the operator for reference or change.

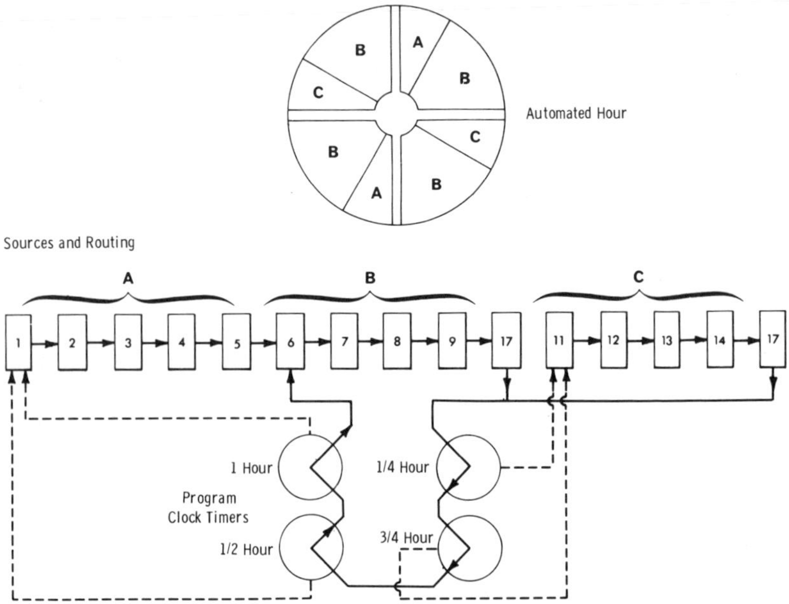

Fig. 7-32. Use of clock timers for breaks at quarter-hour intervals.

Even further identification is available in an accessory known as the source status readout system. This is a three-function illuminated display showing the status ("play," "next," and "pass") of each audio source, and it incorporates a pass switch to remove a source from the system for service, loading, cueing, or recording use. The play light shows that the source is operating on-air. When the pass indicator is lighted, the source is not available for use and is automatically skipped when called up by the programmer. The skip is achieved with no dropout or pause of the audio, just as it would be for a blown fuse, slack tape, or improperly seated cartridge. The pass light goes out when the source is properly loaded and ready to play. If it does not do so, the operator is immediately warned of a malfunction. The next indicator shows the next source programmed to play, and if the source is to be skipped automatically, the actual next source to play is indicated.

The studio override and manual control system allows a live microphone or other console-controlled source to be inserted into the automation programming sequence with considerable ease. These three insertion modes are provided:

1. The live source can be programmed into the BCA-15 programmer in the normal manner by dialing up the assigned source number. When this point in programming approaches, the operator will receive a "next" indication to which he must reply with a "ready" signal to indicate his presence in the studio. He then is supplied an "on-air" indication to start. At the end of the live segment, the operator presses a "pass" switch which returns control to the programmer. If it should happen that the studio is not manned when programmed, an automatic skip will occur, and the programmer will select the next event.
2. The announcer can interrupt the automation sequence and go live at the end of the on-air event by operating a "next" switch in a control center located at the console or announcer's booth. At the conclusion of a live insert, the program will automatically continue from the point of interruption.
3. The announcer can instantaneously override the on-air automation program and interrupt the program sequence for news flashes, etc., by operation of a play control. The program will automatically continue from the start of the next programmed event at the end of the interruption.

An additional operator aid is the automatic cue accessory. With this, any audio source placed in pass status will automatically feed its audio into an audition bus monitored by a cue amplifier-speaker unit. This system allows audio sources to be cued up, or played off line, while the automation system is on the air. For stereo, a lever switch allows selection of channel A, B, or A and B.

The RCA automation equipment is fail-safe in that no source malfunction will shut down the entire system, and most failures will be skipped without interruption of audio. Failures such as tape breakage during play or failure of the operator to return the system to programmer control at the end of a live segment will be picked up after a few seconds (adjustable from 2 to 20 seconds) by a silence sensor, and the next event will start. Of course, the operator can always override the programmer and play the system manually. The sensor is equipped with a balanced input and a bridging mixing network to combine stereo inputs while maintaining channel-to-channel isolation. A remote alarm may be triggered by the silence sensor.

The Schafer 900 Series Automation Systems

An early model of Schafer automation was shown in Fig. 5-1, Chapter 5. This has been replaced by the 900-series systems, Models 901, 902, and 903. The Model 902 shown in Fig. 7-33 is the switch memory system.

When this system is to be used with a network, it also needs the time gate unit for joining and leaving network and other auxiliary functions. Each Carousel in the system employs an M-52-C Carousel memory unit. This is a switch matrix device with 52 positions to select the plays for each Carousel. The Model 903 system utilizes an MOS memory and digital techniques. Programming the system is accomplished by use of the keyboard shown on the table. Each random-select Carousel is addressed directly by the keyboard. Both systems feature a solid-state digital clock. The 902 clock is synchronized with the power-line frequency. The 903 clock is crystal controlled and has a battery backup.

Both systems use the Model 901 control panel as the interface. Thus both systems have individual 25-Hz gates, individual time delays on each transport, with closed-loop control.

The Model 901 control unit is outlined below:

Program Sources

Basic

A. Up to 19 audio sources.

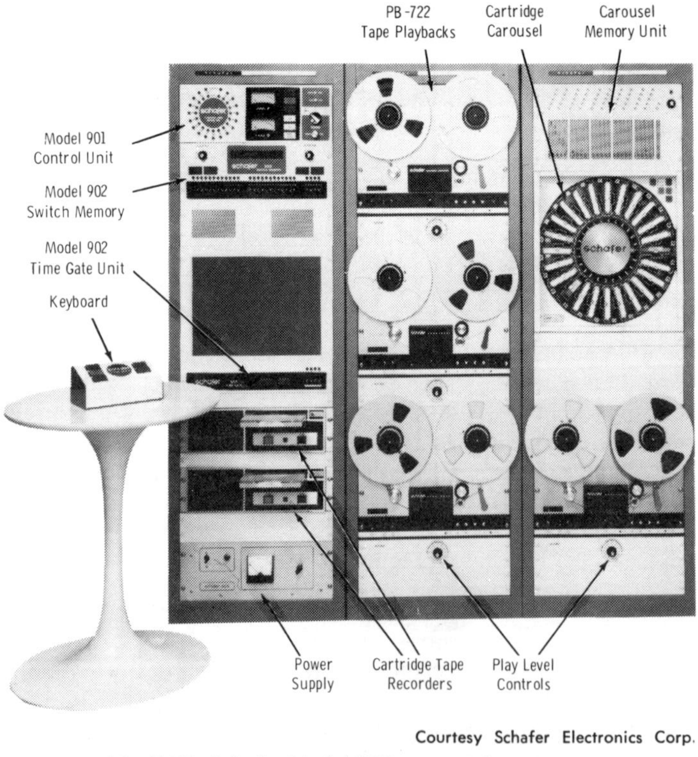

Courtesy Schafer Electronics Corp.

Fig. 7-33. Schafer Model 902 automation system.

B. May be reel-to-reel, cartridge, or cassette.
C. A separate interface card is used for each audio source.
D. One cable assembly from the interface card to each audio source.
E. Solid-state audio switching used throughout.
F. Manual control of each deck is provided.

Reel-to-Reel Interface Card

A. Has an active 25-Hz gate.
   1. To provide automation advance pulse if deck was on the air.
   2. To start adjustable cue delay after 25 Hz leaves.
   3. To hold audio on air (if it was on air) until 25 Hz stops.
B. Audio input single ended.
   1. −10 dB to +16 input level.
   2. 10K to 150 ohms impedance level.
C. 25-Hz lockout.
   1. Adjustable delay of 1 to 10 seconds from start of deck.
   2. Set at factory for specific deck.
D. Cue Delay.
   1. Adjustable delay of 1 to 10 seconds from end of 25 Hz.
   2. Adjustable from front by opening front panel.
E. Fade Control.
   1. Fade on 0.1 second minimum, 5 seconds maximum.
   2. Fade off 0.1 second minimum, 5 seconds maximum.
   3. Normal setting 0.5 second on, 2 seconds off.

Cartridge Interface Card

A. Accept contact closure from 150-Hz detector in cartridge deck.
   1. To provide automation advance pulse if deck was on the air.
   2. To hold audio on the air (if it was on the air) until the contacts open.
B. Audio input single ended.
   1. −10 dB to +16 dB input level.
   2. 10K to 150 ohms impedance level.
C. Contact closure or 150-Hz lockout.
   1. Adjustable delay of 1 to 10 seconds from start of deck.
   2. Set at factory for specific deck.
D. Fade Control
   1. Fade on 0.1 second minimum, 5 seconds maximum.
   2. Fade off 0.1 second minimum, 5 seconds maximum.
   3. Normal setting 0.5 second on, 2 seconds off.

Cassette Interface Card

A. With cassettes having own automation advance and cue sensors, use specifications for cartridge interface card.

B. With cassettes not having separate sensors, use specifications for reel-to-reel interface card.

System Controls & Indicators

A. *Start:* The Start function causes the audio source selected under Next to be started and played on the air if the system was in a stop condition. If a source is already playing, no action will occur except to cancel a stop request. The start light will come on and stay on as long as a source is on the air.
B. *Step Now:* The Step Now (sometimes called "panic") function operates only in the run condition by causing the audio source selected under Next to start now and the audio source playing to be removed from the air.
C. *Stop:* The Stop function places a request for the system to stop running at the end of the present selection. This stop request can be canceled by pressing the start button.
D. *Auto Pulse:* An auto-pulse lamp gives indication that an automation advance signal was received from the audio source now playing.
E. *Closed-Loop Control:* The automation advance pulse (auto pulse) generates a start command for the audio source selected to play next. If for some reason that source is unable to play (power off to the deck, broken tape, programming error selects nonexisting source, etc.), another auto pulse is generated within 300 milliseconds. This sequence is repeated until a selected source is able to play; this source is then monitored as long as it is on the air.
F. *Silence Sense:* The line outputs are monitored for continuous presence of signals above −20 dB (adjustable −40 to −10 dB). Falling below the set point initiates a time delay adjustable from 1 second to 10 seconds; if the time delay is not reset by a return of the audio level above the set point, an auto pulse is generated and advances the system to the next source. Silence sense inhibit capability is provided for studio or network sources.

Audio Monitoring

A. Monitor Switch
    1. Line
    2. Cue
    3. Clock Cue
    4. Network Cue
B. VU Meters
    1. Channel A and channel B indication of line output.
    2. Channel A and channel B indication of cue output.
C. Stereo Speakers for monitoring line or cue with level control.

D. Mono Check push button to give visual indication of monaural receiver reception level.

Audio Distribution

A. Line Output
   1. Balanced 600 ohms.
   2. Level adjustable from −10 to +8 dB.
   3. 25-Hz filter in each line.

B. Cue Monitor Output
   1. Single-ended 10K.
   2. Level at 0 dB.

Verified Encoded Logging (Optional)

A. Time-Code Generator
   1. Time is sampled from the digital clock and typed out in hour, minute, second, and a.m. or p.m. at the start of an audio source. Audio source number is typed after the time.
   3. Special symbols are typed around the audio device number if an rf monitor contact opens.

B. Clear-Text Log
   1. After the time and device number are typed, a clear-text message is read directly from the tape to produce a verified encoded log. This message does not appear if the rf monitor contacts are open.

EXERCISES

Q7-1. What must be checked when a new and different type of microphone is employed in a stereo application?

Q7-2. Quad sound adds what information to conventional two-channel stereo?

Q7-3. Is an encoder required at the studio for matrixed quad sound?

Q7-4. What is the basic action of an encoder for quad sound?

Q7-5. What does "front center" mean in stereo or quad sound?

Q7-6. May a transmitter be operated by remote control without FCC authorization?

Q7-7. Can studio automation be used without FCC authorization?

Q7-8. Can matrixed quad sound be broadcast without FCC authorization?

Q7-9. Can discrete quad sound be broadcast without FCC authorization?

CHAPTER 8

Mobile and Field Facilities and STLs

Broadcast programs are as varied as the interests of the more than twelve million people who make up the listening audience. It is inevitable that a great number of programs must originate at some point other than in a carefully designed studio that has a permanent and complex control console and amplifier racks.

Many stations today are investing a considerable sum in specialized remote facilities. A complete two-turnable control console which collapses into a suitcase and is easily transported by one man is available. Such a facility is adaptable to remotes staged in automobile salesrooms, supermarkets, shopping centers, etc.—wherever special promotional broadcasts are originated. Some stations own and operate complete mobile studios wherein a special truck or van houses the mobile equipment. Modern radio is witnessing a significant growth in remote and mobile service. The application of transistors to compact remote-control gear allows an adequate amount of facilities to be housed in quite small spaces.

Equipment used for remote broadcasts must provide the same means of mixing the outputs of the microphones and sufficient amplifier gain for the use of low-output, high-quality microphones as does the main studio equipment. It is obvious, however, that the equipment must be conveniently portable and, therefore, limited in size and weight. For this reason, many of the earlier types of remote amplifiers used low-level mixing circuits requiring only one preamplifier tube for all microphone inputs. Recent advances in circuit design, however, have permitted the use of high-level mixing circuits with all their advantages, without adding to the bulk of the equipment. Power is supplied by batteries or from the power line. In case of power-line failure, or in mobile applications, dynamotor supplies or batteries are used.

8-1. BASIC REMOTE MIXER-AMPLIFIERS

A typical transistor remote amplifier is the Collins 212Z-1, which provides for four microphones with high-level mixing. It weighs 22 pounds, including the batteries. This unit includes a power source for both 115 volts ac and batteries, with automatic changeover when the ac power fails and when it is restored; the self-contained batteries have a life of approximately 75 hours. The unit has a maximum gain of 90 dB, a tone oscillator for line-level setup, an auxiliary output for public-address feed, and step faders.

One or two headsets may be plugged into the monitor jacks. Where speaker monitoring or feed for local pa is desired, the pa terminals are used, and an individual gain control allows the operator to handle the program and simultaneously ride gain on the pa system. A multiple jack is located on the side of the unit, permitting two units to be used simultaneously while being controlled by one master gain control.

Fig. 8-1 shows a block diagram of this remote unit. The four preamplifiers, Q1 through Q4, use 2N106 heremetically sealed low-noise transistors. The input faders feed second preamplifier Q5 (also a 2N106) through the tone-oscillator switch. Booster Q6 feeds the master gain control, which is followed by driver Q7.

The booster and the driver both employ 2N64 transistors. The output amplifier has push-pull 2N44 transistors, Q8 and Q9, with transformer coupling on the input and output sides. The output transformer feeds the program monitor, VU meter, public-address line, and program switch. Provision is made for two program lines and telephones through the output switch. The power supply is a shielded, filtered full-wave circuit employing germanium diodes and multisection filtering. A relay connects the batteries to the amplifier whenever the ac line voltage fails.

The 400-Hz oscillator employs a Colpitts circuit and feeds a low-level signal to the second preamplifier through a selector switch. A power interlock switch insures that there is no battery drain when the unit is in its closed carrying case.

The four-channel mixing circuit incorporated in the amplifier is designed to work with all microphones from 30 to 600 ohms. The output circuit is designed to match a 600-ohm line. To work into 150 ohms, the use of an external repeat coil is recommended. Minor rework of the unit will also provide a 150-ohm output. When a telephone set is connected to the TEL posts, the line can be used for communication with the master control room.

Although simultaneous program feed and communication cannot take place over a single line, the output switch allows rapid interchange between the telephone set for communication and the amplifier output for program transmission. This facilitates operation where only one line is available to the control point or radio transmitter.

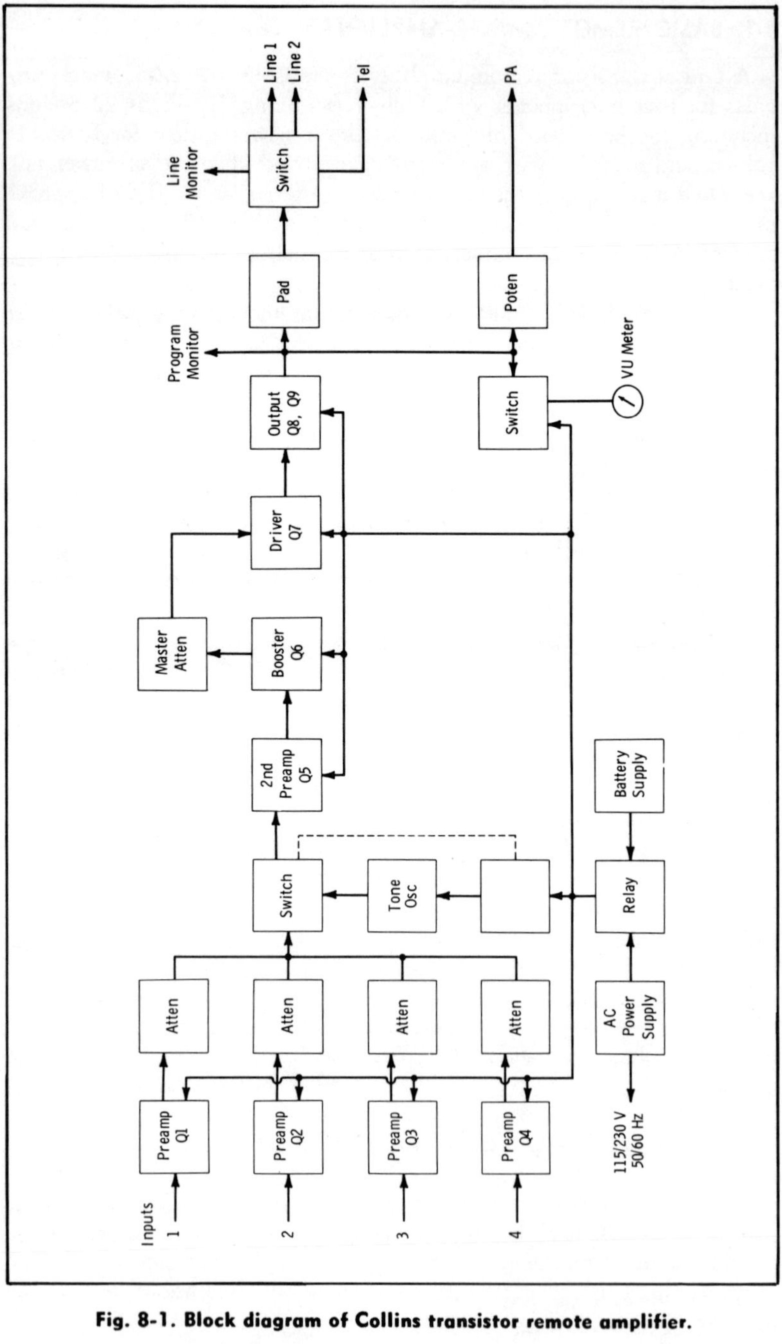

Fig. 8-1. Block diagram of Collins transistor remote amplifier.

When two lines to the master control are available, one can be used for program feed or receipt of cue preceding transmission, and the other can be used for simultaneous communication. With this arrangement, the communication line can be substituted for broadcast immediately by simply turning the output switch and making a corresponding switch in the master control room.

Fig. 8-2 shows the block diagram of another transistor portable remote amplifier, the RCA BN-6A. The VU meter monitors either the output level or battery current. The line-monitor output is a phone-type jack for headphones or speaker. Normally the signal from the studio is strong enough that the CUE-MIC 1 switch may be left in the normal Mic-1 operating position as shown. If the cue-back from the studio is weak due to long-line effects or other reasons, this switch may be turned to the Cue position. The signal will then be amplified through the amplifier to the headphones so that the cue may be heard easily. This is a spring-return switch so that the normal Mic-1 position is restored on release of the switch.

8-2. MOBILE EQUIPMENT

Remote-pickup mobile and base-station equipment is licensed under Part 74, Subpart D, of the FCC Rules and Regulations, for the express purpose of broadcasting program material from a remote point outside of the studio location. Such a system may be licensed by the Commission for this purpose whether telephone land lines are or are not available. Chart 8-1 gives the frequencies allocated to remote-pickup broadcast base and mobile stations. CAUTION: Always check current FCC Rules and Regulations.

Mobile Studio Facilities

The *Traveler,* complete mobile studio facilities of radio station WXLW, Indianapolis, Indiana, is shown in Fig. 8-3. It is housed in a custom-built trailer which is 47 feet long and 8 feet wide and weighs 9 tons net. It is equipped with electric brakes, oil heat, air conditioning, water tanks, and leveling jacks. It has double-paned windows, and each pane is 1 inch thick. Concealed fuel tanks supply the heating system and the stove, which is located in the combination lounge and kitchen (Fig. 8-4A) at the forward end of the trailer. The exterior skin of the trailer is aluminum. The interior is finished in acoustical tile on the ceilings and walnut plywood on the sidewalls. The floors are carpeted.

Fig. 8-4B shows the studio and its equipment. In the center is the operating console, with a remote-control unit for the Ampex tape recorder located in the control room. Also mounted in the console are controls for the two turntables, a VU meter, a control for the air monitor, and a potentiometer for the news "beeper." The two turntables are equipped with transistor preamplifiers and a cue amplifier. A cabinet for the storage of tapes and discs can be seen between the console desk and lounge chair.

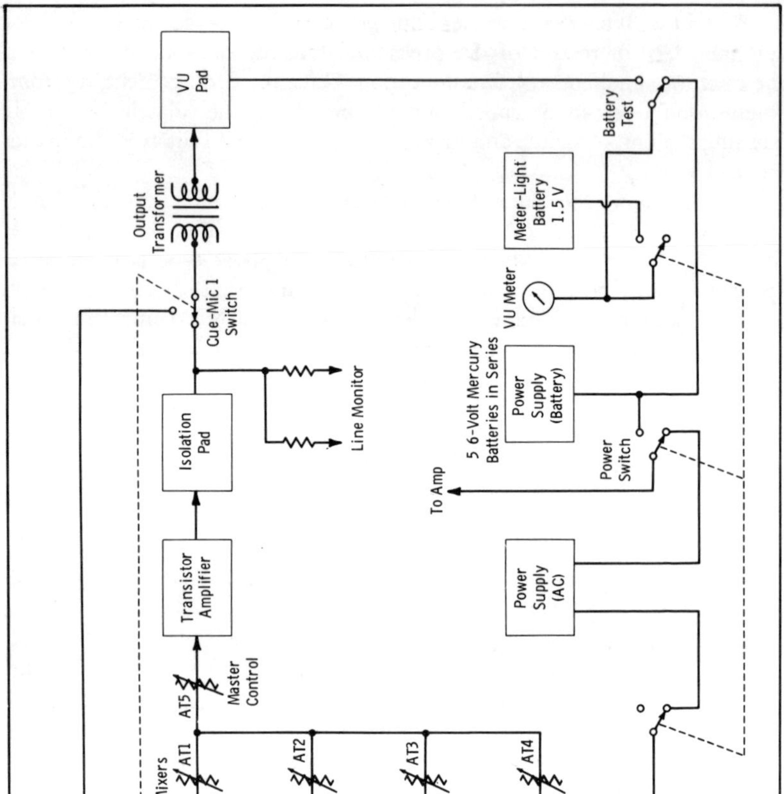

Fig. 8-2. Block diagram of RCA transistor remote amplifier.

Chart 8-1. Frequencies Allocated for Remote-Pickup Broadcast Base and Mobile Stations

Group K (MHz)

[3] 152.87	[3] 153.17	[5] 161.64*
[3] 152.93	[3] 153.23	[5] 161.67*
[3] 152.99	[3] 153.29	[5] 161.70*
[3] 153.05	[3] 153.35	[5] 161.73*
[3] 153.11		[5] 161.76*

Group L (MHz)	Group M (MHz)
[4] 166.25*	[4] 170.15*

Group N (MHz)

450.05*	450.55*	455.05*	455.55*
450.15*	450.65*	455.15*	455.65*
450.25*	450.75*	455.25*	455.75*
450.35*	450.85*	455.35*	455.85*
450.45*	450.95*	455.45*	455.95*

Puerto Rico and Virgin Islands Only:

(MHz)	(MHz)	(MHz)
160.89	161.07	161.25
160.95	161.13	161.31
161.01	161.19	161.37

NOTE: These frequencies are shared with the Land Transportation Radio Service.

*Frequencies indicated by asterisk are exclusive to radio broadcast except as noted:

[3]Subject to the condition that no harmful interference is caused to stations operating in accordance with the Table of Frequency Allocations.

[4]Operation on the frequencies 166.25 MHz and 170.15 MHz is not authorized (i) within the area bounded on the west by the Mississippi River, on the north by the parallel of latitude 37°30′ N, and on the east and south by that arc of the circle with center at Springfield, Ill., and radius equal to the airline distance between Springfield, Ill., and Montgomery, Alabama, subtended between the foregoing west and north boundries; (ii) within 150 miles of New York City; and (iii) in Alaska or outside the continental United States; and is subject to the condition that no harmful interference is caused to government radio stations in the band 162-174 MHz.

[5]These frequencies may not be used by remote pickup stations in Puerto Rico or the Virgin Islands. In other areas, certain existing stations in the Public Safety and Land Transportation Radio Services have been permitted to continue operation on these frequencies on condition that no harmful interference is caused to remote pickup broadcast stations.

The control room is located approximately in the center of the trailer. Some of the control-room equipment is shown in Fig. 8-5A. A Gates console is mounted beneath the studio window. Beneath the desk in the farthest left rack are a patch panel, a-m air monitor, monitor amplifier, console power supply, and the switch and fuse panel. The second rack houses an RCA transistor tape recorder, a communications receiver for the 166-MHz link, and a 55-watt power amplifier for the outside speakers.

Courtesy Greater Indianapolis Broadcasting Co., Inc.

Fig. 8-3. Mobile studio facility designed to provide broadcasting of main-studio quality.

(A) Combination living room, lounge, and kitchen.

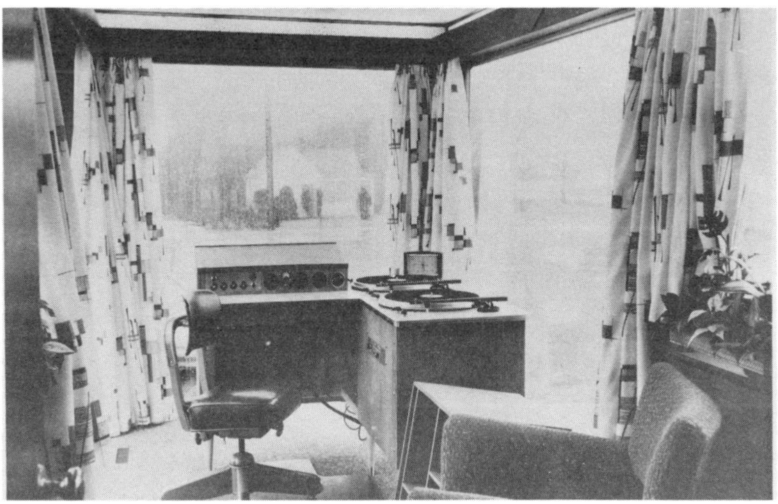

(B) Studio with operating console and turntables.

Courtesy Greater Indianapolis Broadcasting Co., Inc.

Fig. 8-4. Interior views of lounge and studio.

(*A*) *Operating position.*

(*B*) *Additional equipment.*

Courtesy Greater Indianapolis Broadcasting Co., Inc.

Fig. 8-5. Interior views of mobile control room.

Fig. 8-5B shows the remaining control-room equipment. The main tape transport and its record/playback amplifier are mounted in the third rack, beneath the control-room counter at the rear. A 15-watt transmitter provides the main 455.55-MHz link to the home studio; it can be seen in the right rack in Fig. 8-5B. This device, with the yagi transmitting antenna (12.2-dB gain) on its 50-foot crank-up tower and the receiving antenna at the main studio (7.5-dB gain), provides a range of 25 miles while maintaining a program-quality signal.

The tractor was custom modified from a two-ton truck. The frame of the truck was shortened, a shorter drive shaft was installed, and the drive wheels were moved forward so that the overall trailer-truck length could be within the 60-foot limit. A specially built housing on the truck accepts a 15-kW generator and also provides storage space for cables and power accessories. A waterproof nylon cover encloses the rear of the truck.

This elaborate setup is an example of the steps radio stations have taken to increase their program capabilities through remote broadcasting. This mobile unit provides full studio facilities and is completely independent of power and telephone lines.

Typical Mobile Equipment

Fig. 8-6A is a basic block diagram of the mobile broadcast technique. Either vertical or horizontal polarization may be used.

Recent revisions to the Rules and Regulations permit the use of unattended, automatic relay of remote-pickup signals, to extend the coverage of remote-pickup broadcasting. This is particularly advantageous to those stations in mountainous areas or any location where it is difficult to cover a primary service area because of terrain conditions. A block diagram of an automatic relay system appears in Fig. 8-6B.

More recent rules permit the use of low-power devices such as walkie-talkies, wireless microphones (described in Section 8-3), or any hand-held, portable transmitter to initiate the program material and broadcast it back to the mobile unit or remote transmitter for relay to the studio location. This is becoming particularly important in extensive news operations for adequate coverage of events such as disasters and certain sporting events where it is practical to use such devices.

Remote-pickup equipment can also serve the very important function of a back-up, emergency program link between studio and transmitter in the event of a loss of the program line, since the equipment is capable of broadcast sound quality and continuous-duty operation.

Station WIRE operates a fleet of mobile units employing Marti Electronics transmitter/receivers licensed for two frequencies 60 kHz apart in the 150-MHz region. A base transmitter and receiver are used for each channel, and the mobile units are each switchable to either channel. Separate horizontal and vertical antenna polarization is employed at the base station. Mobile units normally stay switched to the cue channel for cues

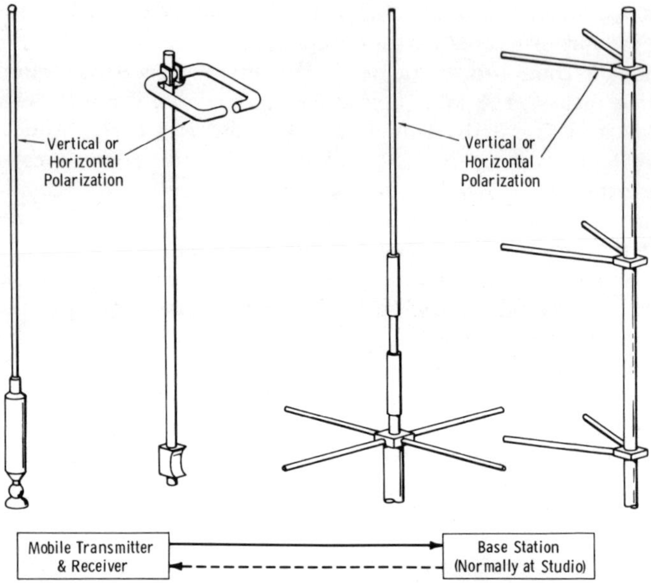

(*A*) *Block diagram, basic installation.*

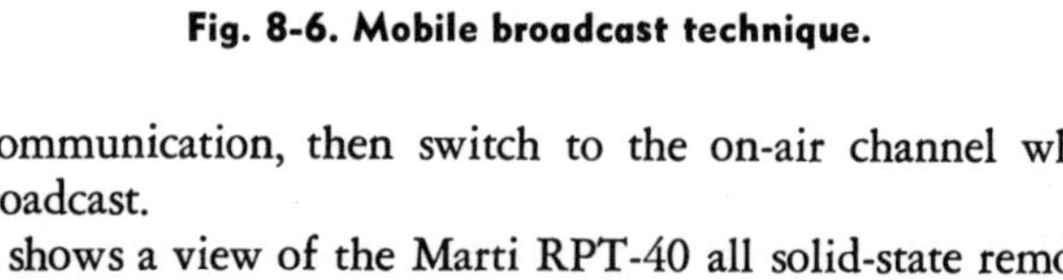

(*B*) *Use of automatic relay station.*

Fig. 8-6. Mobile broadcast technique.

and intercommunication, then switch to the on-air channel when called upon to broadcast.

Fig. 8-7 shows a view of the Marti RPT-40 all solid-state remote-pickup transmitter. This unit may be obtained to operate in either the 150- or 450-MHz band. The R-30/150 receiver is identical in appearance to the stl receiver described in Section 8-5. The RPT-40 has a 40-watt output in the 150 MHz band, whereas the RPT-25 has a 25-watt output in the 450-MHz band. These two transmitters are identical in appearance.

As shown by the block diagram of Fig. 8-8, two separate crystal oscillators generate the two operating frequencies, selectable by a switch. The switch is followed by the phase modulator, which accepts the audio input, normally from a compressor or limiter amplifier. This is followed by a series of frequency multipliers to raise the original crystal frequency to the desired operating frequency. In Marti equipment, this multiplication is 36 times for the 150-MHz band and 108 times for the 450-MHz band.

The frequency deviation accomplished by the phase modulator is also increased by the same magnitude as the operating carrier frequency. The final frequency swing is about plus and minus 7.5 kHz. A driver stage is then used to excite the final power amplifier, which drives the antenna through harmonic filters and contacts of the transmit-receive relay. It should be noted that the multiplier and remaining stages are sufficiently broad in response that no retuning is required for the two assigned operating frequencies. The maximum difference in frequencies is about 120 kHz.

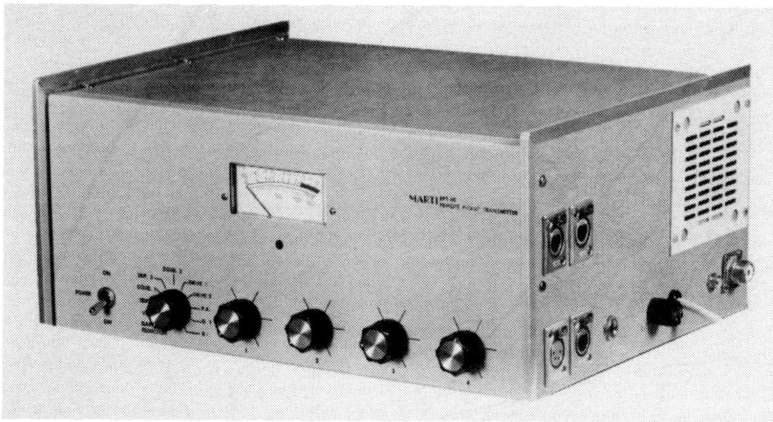

Courtesy Marti Electronics, Inc.

Fig. 8-7. Marti RPT-40 remote-pickup transmitter.

It is usual in remote-pickup transmitters that only one meter is used to measure various operating parameters including stage-by-stage tuning of the rf section. The meter is normally arranged to measure the input drive current resulting from tuning of the previous stage. Thus each stage is tuned to result in a maximum reading on the meter. However, in Fig. 8-8, note that for the final-stage tuning the meter is placed in series with the rf tank current feed so that this output tuning is adjusted for minimum meter reading. Then the antenna coupling capacitor is adjusted also to obtain a minimum final input-current reading, and the two tuning procedures are repeated until exact resonance is obtained. The final adjustment for the antenna coupling capacitor is to obtain the manufacturer's recommended power input to the final stage. This is normally specified as a given current reading on the meter when monitoring the input to the final output tank circuit.

Fig. 8-9 illustrates how a typical Marti transmitter is installed in an automobile. Table 8-1 gives expected coverage for the conditions and types of antennas listed. This assumes flat terrain and a 30-watt transmitter in the 150-MHz band. Mobile applications in practice will be discussed in Chapter 12.

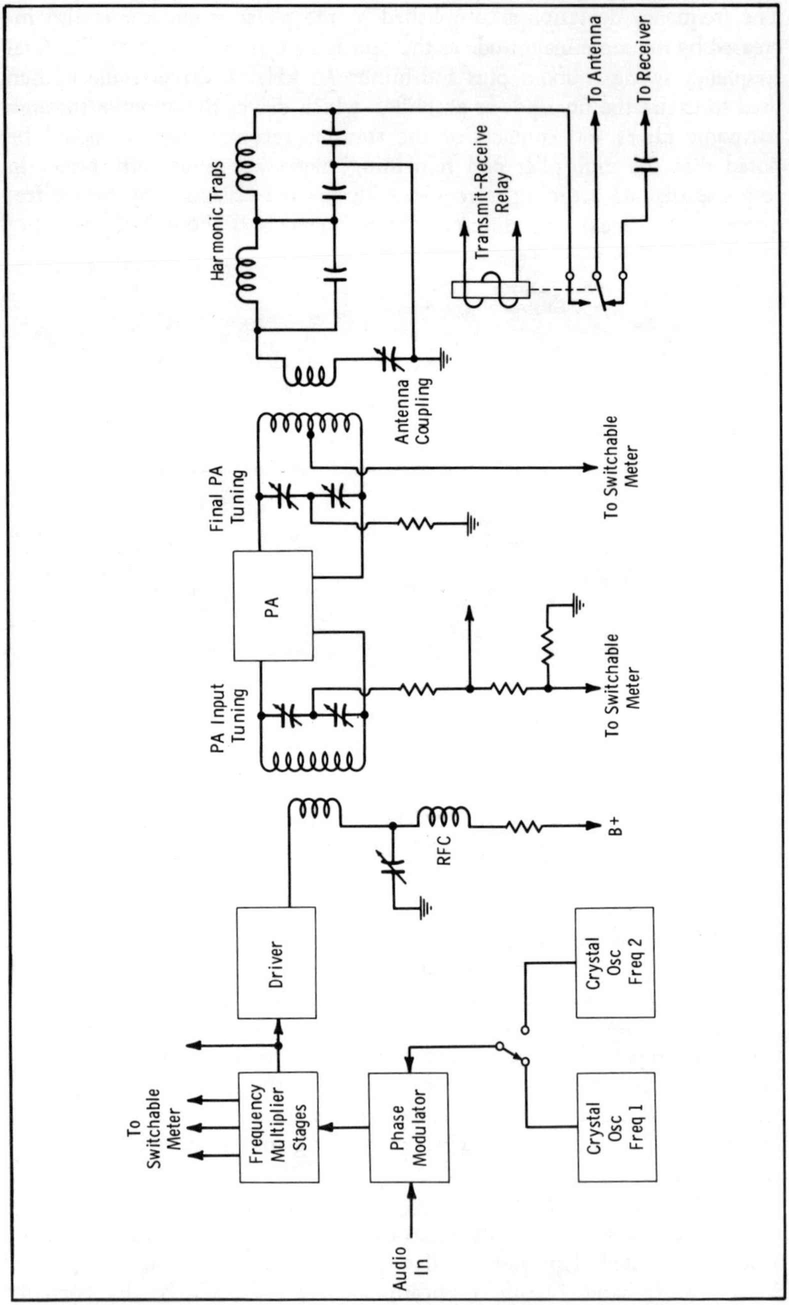

Fig. 8-8. Rf section of dual-frequency mobile or base station.

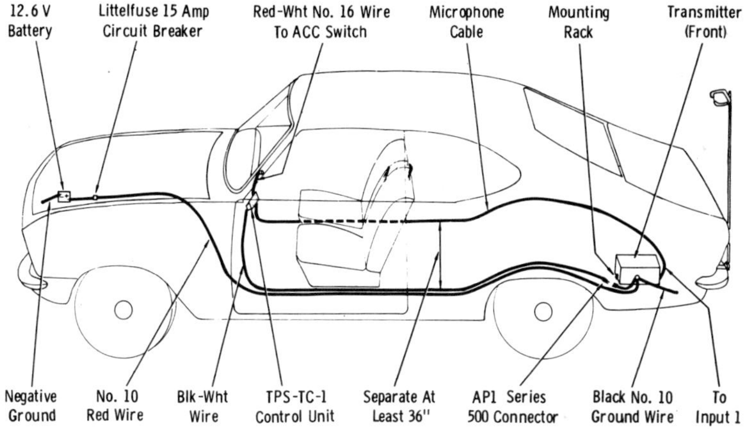

Courtesy Marti Electronics, Inc.

Fig. 8-9. Automotive installation of Marti remote-pickup transmitters.

8-3. PORTABLE TRANSMITTERS AND WIRELESS MICROPHONES

Small portable transmitters and wireless microphones are often employed in conjunction with the mobile relay transmitter. Fig. 8-10 shows the fundamental layout of equipment used to broadcast special events involving mobile relay service. Fig. 8-11 illustrates a transistor pocket fm transmitter with a push-to-talk switch and an internal microphone. The unit is battery-powered and operates in the 132-174 MHz range.

Miniature transmitters of this type are normally good for line of sight only. Therefore the signals are often picked up on a receiver in the mobile truck, and the receiver output modulates the main mobile transmitter. Mobile transmitters with their associated antennas are mounted in trucks or cars, variously known as "newsmobiles" or "mobile units." These units are also employed to keep in constant touch with the base station (often in the news room) so that when local news breaks it can be covered immediately.

Solid-state technology and integrated circuits have made possible the design of wireless microphones that are contained in very small cases such as the lavalier or tie-clasp type. The antenna may be a thin wire hanging from the bottom of the case, or it may be wired to an earpiece which contains both the transmitter and receiver for cues. In the latter instance, the antenna is usually a small whip mounted directly on the earpiece. Very small mercury cells often supply the necessary power. The wireless-microphone transmitter is normally a frequency-modulated device for best signal-to-noise ratio. As a general rule, the distance between the wireless-microphone transmitter and the receiver is kept to 300 feet or less.

Frequency-modulated oscillators vary greatly in design, but most are based on the circuit in Fig. 8-12. Transistor Q1 is sustained in oscillation

Table 8-1. Average Coverage Over Flat Terrain (30-Watt Transmitter, 152-172 MHz Frequency Band)

Receiving Antenna Height (Feet)	Antenna Combinations		Coverage (Miles)
	Receiving	Transmitting	
* 75	5-Element Yagi	Single Ring	9
** 150	5-Element Yagi	Single Ring	13
* 75	Stacked 5-Element Yagis	Single Ring	11
** 150	Stacked 5-Element Yagis	Single Ring	15
* 75	5-Element Yagi	5-Element Yagi	14
** 150	5-Element Yagi	5-Element Yagi	18
* 75	Stacked 5-Element Yagis	5-Element Yagi	16
** 150	Stacked 5-Element Yagis	5-Element Yagi	20
** 150	RA-4 Antenna	Single Ring	10
***300	RA-4 Antenna	Single Ring	14
** 150	RA-4 Antenna	5-Element Yagi	16
***300	RA-4 Antenna	5-Element Yagi	20

Code:

* Measurement based on length of RG-8U transmission line not to exceed 80 feet.

** Measurement based on length of ½" FHJ4 transmission line not to exceed 200 feet.

***Measurement based on length of ⅞" FHJ5 transmission line not to exceed 350 feet.

Notes:

The above measurements are based on a transmitting antenna height of 6 feet above surrounding objects. An increase in height of the transmitting antenna to 30 feet will increase the coverage by approximately 50%. Do not use RG-58U; there is too much loss.

The ring antennas are nondirectional within ±3 dB. The yagi antennas are unidirectional. The gain of the single-ring antenna is unity. The gain of a YC-Series yagi antenna is 9 dB. The gain of stacked YC-Series yagi antennas is 12 dB. When yagi antennas are used for receiving, a standard tv-type rotator is recommended.

Data Supplied by Marti Electronics

by a regenerative-feedback tap on L1. The tuned tank circuit (L1 and C1) establishes the frequency of the oscillator, and C1 is sometimes made variable to adjust the frequency over a narrow range.

The modulating signal from the amplified microphone output varies the emitter-base current, and therefore the collector voltage, of Q2 at the corresponding audio rate. When the Q2 collector voltage increases, the collector-emitter capacitance (C_{CE}) decreases. When the collector voltage decreases, C_{CE} increases. Thus an increase in the capacitance across a portion of the Q1 tank circuit causes the resonant frequency to decrease, and a decrease in capacitance causes the frequency to increase. In this manner, the oscillator frequency is varied at a rate corresponding to the signal applied. It may be observed that some amount of amplitude modulation also occurs due to changes in the collector-to-emitter voltage of Q1, but this is

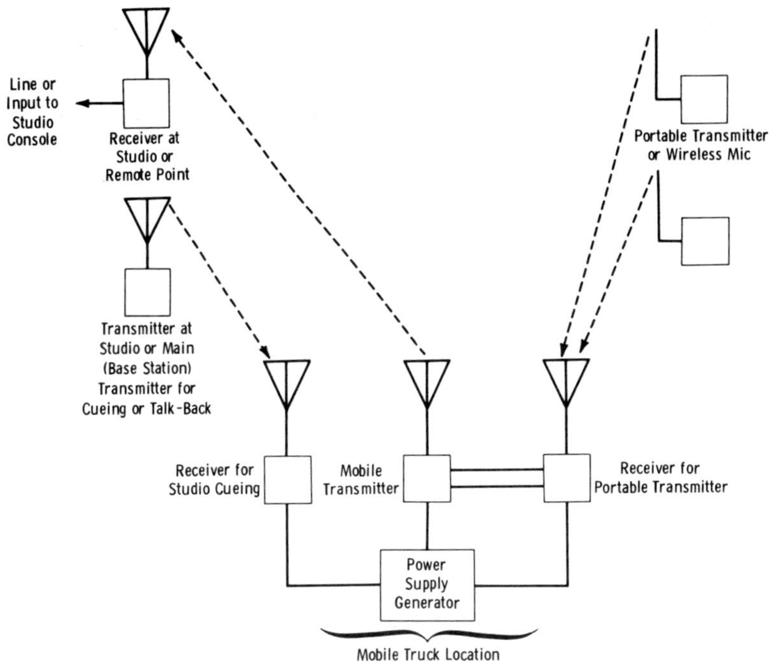

Fig. 8-10. Block diagram of remote-relay system.

Fig. 8-11. Pocket-size transmitter for remote use.

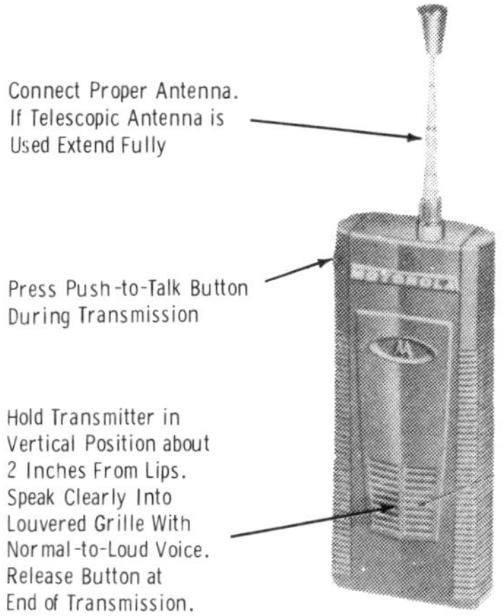

Courtesy Motorola

slight. Normally, the rf link feeds a frequency multiplier so that practically all a-m is removed; the multiplier also increases the available frequency swing of the oscillator. The final swing is usually between 10 and 20 kHz.

Many of the latest wireless microphones employ a single integrated circuit for the microphone amplifier, oscillator, modulator, and frequency multiplier or power stage. Power output ranges from 30 to 60 milliwatts. The frequency of operation for broadcast is normally 26.25 MHz.

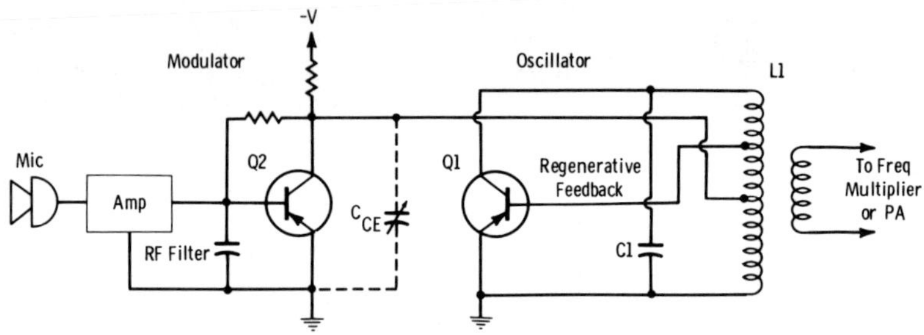

Fig. 8-12. Basic frequency-modulated oscillator.

8-4. USE OF TELEPHONE LINES

When a program is to originate at some point remote from the regular studios, a number of facilities besides the actual equipment must be considered. A remote broadcast point must check in to the main studio before each broadcast and be able to receive the cue for the start of the remote broadcast. Although some stations, particularly when feeding a network, use two separate lines for this purpose with a "PL" (private line) connected to the talk circuit, most stations now use a single line for broadcast and cue-back.

There are three ordinary positions of a remote key at the studio console which permit proper checking and coordinating functions to be performed. These are:

1. *Override*—The key is connected to the proper remote-line pair and is in a position so that when the remote operator calls in he will be heard over the monitor system. In some cases, a separate amplifier and speaker are used so that the check-in by the remote operator does not come in over the regular monitoring system which must, of course, be simultaneously monitoring the program signal.
2. *Cue and Talk*—In this position, a microphone in the control room (or a telephone handset) may be connected to the line so that the operators may talk to one another for test of the line, time checks, etc. It may also be terminated from a bridging connection of the monitor lines so that the program at the studio may be fed back over the line for purposes of cueing.

3. *Broadcast*—In this position, the line is terminated in the mixer control which feeds to the program bus. The remote point is thus fed through the control console in the normal manner.

Fig. 8-13 is a simplified schematic of one general type of remote check-in, cue-in, cue-back arrangement. Key A is a three-position remote key and is the advance selector circuit for keys B, C, and D. For example, when key A is in neutral (the monitor-phone position) and key C is in the override position, the remote operator will be heard over the monitor or an auxiliary amplifier and speaker. Key B is the headphone selector key, which may be placed in the remote position when it is desired to talk back to the remote point. To talk back, remote key A is left in neutral, phone-jack key B is placed in the remote position, and the phone or microphone circuit is actuated to talk to the remote point. Usually a talk-back button is used to connect the microphone and disconnect the speaker by relay action. When it is desired to feed the cue-back signal over the line, remote key A is placed in the upper, or cue, position, and the signal from the monitor amplifier is bridged through the adjustable pad into the remote line. When going on the air, key A is simply moved down, or to the mixer input position, and remote-channel key D is thrown to the program position. The remote signal may be fed into the monitor amplifier when key A is in the down position by throwing key D to the audition position.

This is a complete control-coordinating facility for remote controls. Although the variety of arrangements and components used in practice is great, the general principles are usually the same as those outlined.

Simplex Control of Remote Amplifiers

Many times it is desirable to be able to turn the remote amplifier on and off from the studio. This is desirable for a regularly scheduled broadcast from a point requiring only one microphone where no mixing adjustments are necessary.

Such switching may be accomplished by means of a *simplex* installation, as shown in Fig. 8-14. When the remote line is patched into the equipment through the simplex coil, the battery circuit will be completed to activate the relay at the remote point and energize the power-supply circuit of the remote amplifier. This procedure eliminates the necessity of sending an operator to this point for each broadcast.

Equalization of Remote Lines

Although the telephone company usually equalizes the incoming network lines and regular studio-transmitter broadcast lines, it is beneficial to equalize lines from remote pickup points where the cost of high-class line service is not practical to the station. Most stations have an equalizer of adjustable characteristics in the control room which may be used for this purpose.

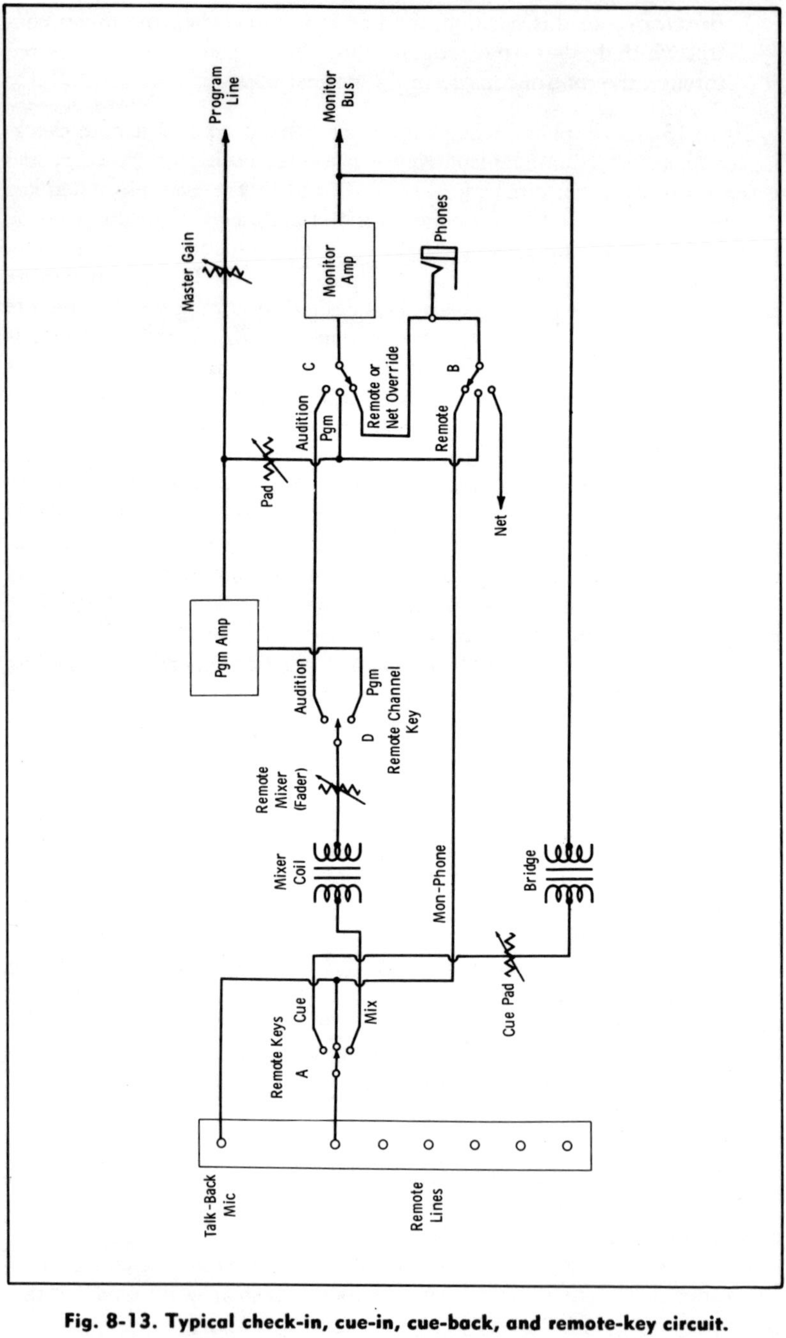

Fig. 8-13. Typical check-in, cue-in, cue-back, and remote-key circuit.

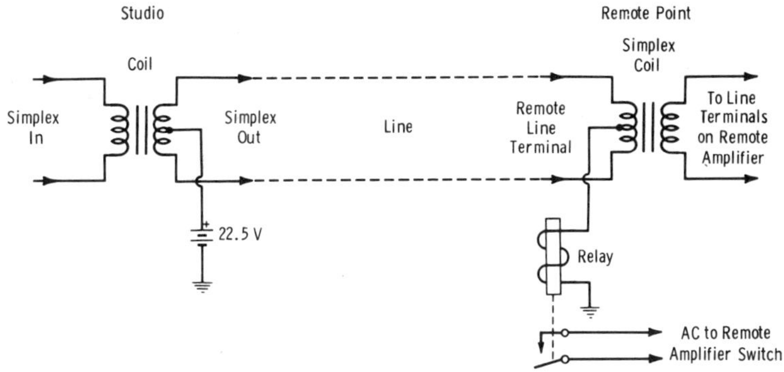

Fig. 8-14. Simplex circuit for control of remote amplifier.

Fig. 8-15 illustrates a typical setup for equalizing a broadcast line. The signal is a steady tone from an audio oscillator which is terminated in an isolation pad. The same load must be presented to the volume indicator at any frequency, and this instrument should, therefore, be bridged on the oscillator side of the pad. The repeat coil properly terminates the line to feed the line equalizer, which is on the load side of the coil and the line side of the line amplifier.

A 1000-Hz tone is usually used and fed to the line at a 0-VU level. The gain of the receiving amplifier is adjusted to give a 0-VU reading at that point. The oscillator is then adjusted to 100, 1000, 3000, and 5000 Hz with constant level maintained at each frequency, and the equalizer is adjusted at the receiving point to compensate as much as possible for the line characteristics at each frequency. This will determine the approximate setting of the equalizer, after which finer adjustments may be made over the entire frequency range.

Ordinary a-m services usually equalize to 5 or 10 kHz, but fm lines may be equalized to 15 kHz. The insertion loss of the equalizer is approximately

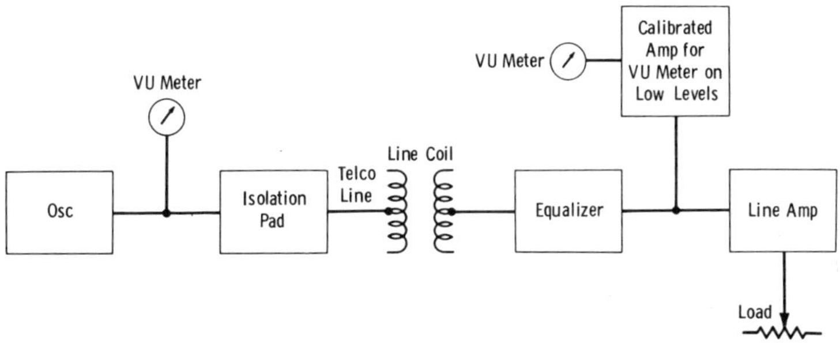

Fig. 8-15. Block diagram of line equalization.

equal to the amount of equalization. A calibrated attenuator selects the amount of equalization at the required frequency. The input and output impedances of the line equalizer are usually 600 ohms.

While considering the use of an equalizer, the reader should remember that some loss of signal energy will take place, and some form of extra amplification is necessary. A one- or two-stage isolation amplifier is usually used in conjunction with an equalizer to compensate for this loss; it should have at least a 30-dB gain throughout the frequency range considered.

Telephone Line "Gimmicks"

Telephone talk programs occupy a sizable amount of time at many stations today. Always talk to the local telephone company before initiating the use of the telephone in broadcasting. There are many options available, some of which the telephone company can accommodate with leased equipment at varying monthly charges; you can custom build others with the phone company's blessings. You must *always* first ascertain the permissible amount of activity in tampering inside the telephone circuitry. Such official activity seems to vary considerably between states and even between various localities within a given state.

The most common official telephone company equipment for broadcast talk programs is the recorder connector illustrated in Fig. 8-16. This unit injects a "beep" into the line every 15 seconds and provides both the send and receive sides of a conversation to the recorder. In the past, this warning signal was required by law so that the calling party was aware that the conversation would be recorded or broadcast. This signal is no longer required by law for regular telephone talk shows, such as a news report from a remote point, or any prearranged program where all participants are aware of the use intended.

The simplest device obtainable is the telephone pickup coil. One type consists of a coil loop which clips to the earpiece of the telephone to supply a medium-level input (about −20 dBm) to a recorder or console amplifier. A symmetrical hum-cancelling pickup coil is best when this simple method is used. It is available at very low cost from most electronic supply stores.

If a direct connection to the line terminals of the telephone set is to be made, use the arrangement of Fig. 8-17. A low-level input, such as the microphone input, may be required for sufficient voice level. When the telephone line is terminated in the set, the impedance is considerably lower than 600 ohms. A 1000- to 2000-ohm bridging impedance is satisfactory.

Another unit that can be built by station personnel is the *hybrid* circuit of Fig. 8-18. With this arrangement, telephone callers are able to hear the in-studio participants directly from their microphones. The main disadvantage is that the balance control can be nulled to eliminate feedback and minimize return echo distortion for a given line, but it usually needs readjustment for a different incoming line. This control can be quite touchy even with more elaborate bridge-balancing circuitry than that shown.

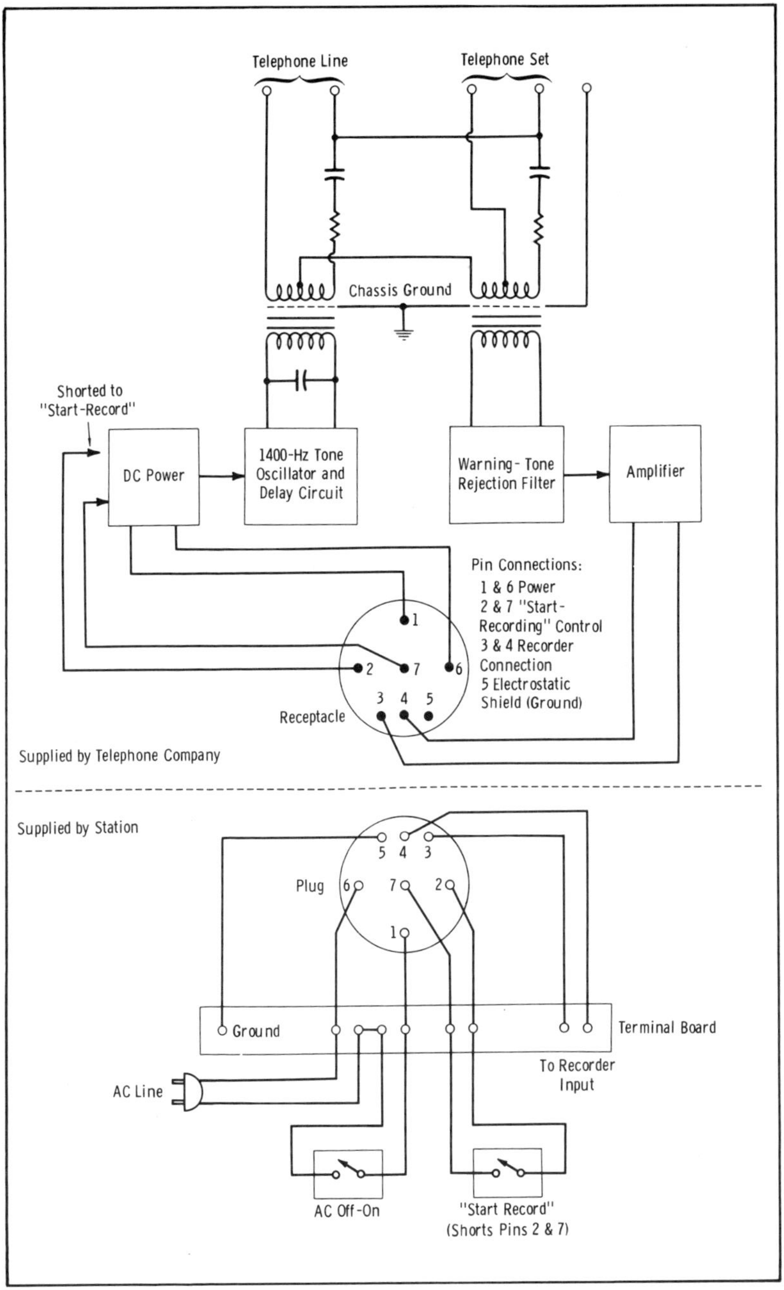

Fig. 8-16. Typical recorder-connector arrangement.

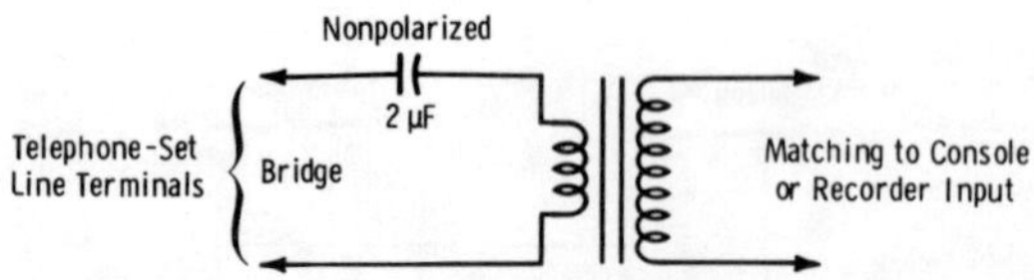

Fig. 8-17. Direct connection to telephone-set line terminals.

One of the most satisfactory methods of telephone-talk programming is presented in Fig. 8-19. This system employs the speaker phone often used for business conferences. The phone is simply removed from the standard cradle, the call made, and the phone placed on the amplifier cradle as shown. The caller's voice appears at the speaker so that all studio participants can hear, and a tap is made from the voice coil to supply the mixing console through an impedance-matching amplifier. The amplifier cradle is normally placed on a common table shared by the microphones of the studio participants. When people in the studio speak in a normal voice, this voice is picked up by the studio-quality microphone for broadcast, and simultaneously by the microphone for the phone amplifier, which enables the caller to hear. The speaker is automatically muted when this microphone is excited by a normal voice level. The mixing-console operator has complete control over each individual voice level.

8-5. STUDIO-TO-TRANSMITTER LINKS (STL'S)

Studio-to-transmitter links (stl's) are often employed rather than wire facilities to relay programs from studio to transmitter. Ten channels in the

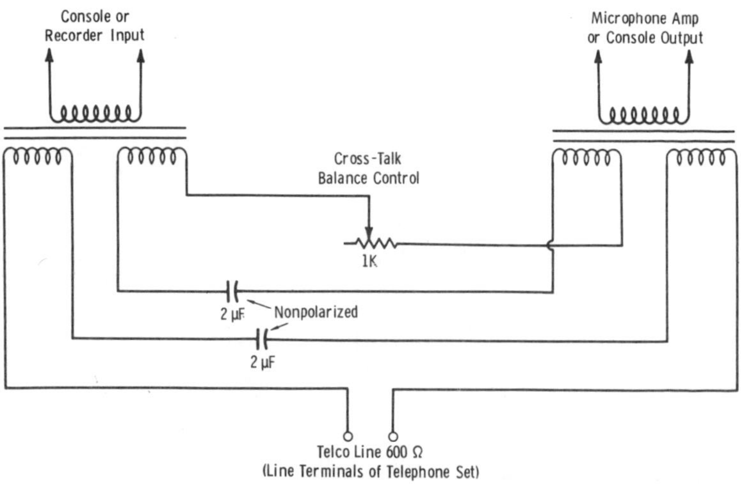

Fig. 8-18. Hybrid circuit for two-way communication.

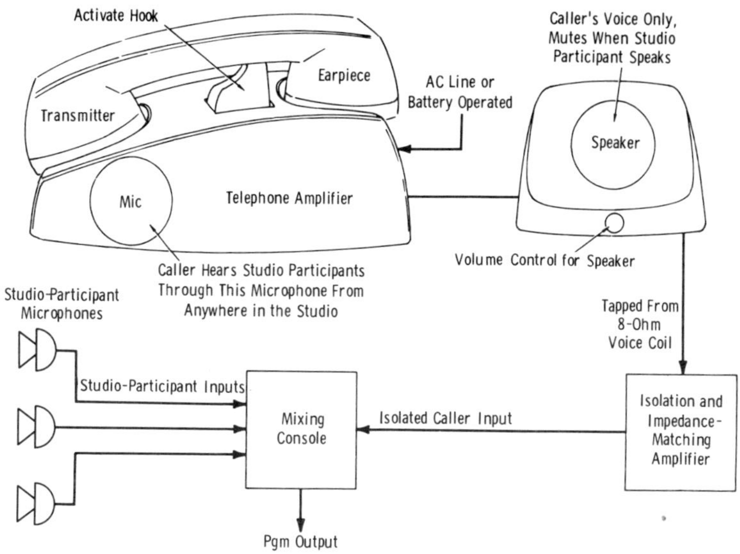

Fig. 8-19. Use of speaker phone in broadcasting.

band 947-952 MHz are allocated by use by stl's. CAUTION: Always check current FCC assignments.

Propagation-Path Calculations

Both the transmitting and receiving antennas are usually half-wave dipoles in parabolic reflectors from 4 feet to 15 feet in diameter. Table 8-2

Table 8-2. Parabolic-Antenna Gain in dB in 950-MHz Range

Diameter (Feet)	Gain (dB)
4	18.9
6	22.0
8	25.0
10	27.0
15	30.0

NOTE: Gain is relative to isotropic radiator at midband.

lists the gain in decibels (over an isotropic radiator at midband) for parabolic reflectors in the 950-MHz range. A strict line of sight must be maintained between antennas used in stl service. The distance to the horizon is based on the fundamental formula:

$$D = 1.23\sqrt{H}$$

where,

D is the distance to the horizon in miles,
H is the height of the transmitting antenna in feet.

Thus if the antenna may be placed 100 feet high, the theoretical distance to the horizon (line-of-sight) is 12.3 miles. However, if it is possible to raise the receiving antenna to 100 feet also, then the distance to the horizon may be found from the equation:

$$D = 1.23\,(\sqrt{H_t} + \sqrt{H_r})$$

where,

H_t is the height of the transmitting antenna,
H_r is the height of the receiving antenna.

In this case, the effective line-of-sight path becomes 24.6 miles.

Since the preceding basic formulas assume a flat surface, modifications for practical application consider the profile elevations along the projected route of signal propagation. Maps showing elevation contours (height above sea level) at as little as 10-foot elevation intervals are available from local Geodetic Survey offices. Unless the proposed pickup point allows distinct line-of-sight service as observed with the eye, the engineer cannot give a definite opinion until these additional factors are considered.

Microwave energy beamed between two points actually takes an infinite number of paths depending on atmospheric and weather conditions as well as terrain. Refraction may result in either upward or downward directions.

Energy arriving over the various paths is categorized into *Fresnel zones.* These zones are numbered to correspond with circles of different radii centered on the direct line between antennas. The zone with the smallest radius is the first Fresnel zone (Fig. 8-20). The first Fresnel zone contains the energy of a straight-line path plus energy arriving over a closely adjacent path (normally phase additive to the energy on the most direct path).

Many Fresnel zones exist. Energy in the second and all other even-numbered zones has a half-wavelength (180°) relationship to energy in the first Fresnel zone and therefore is a canceling signal. Energy in the third

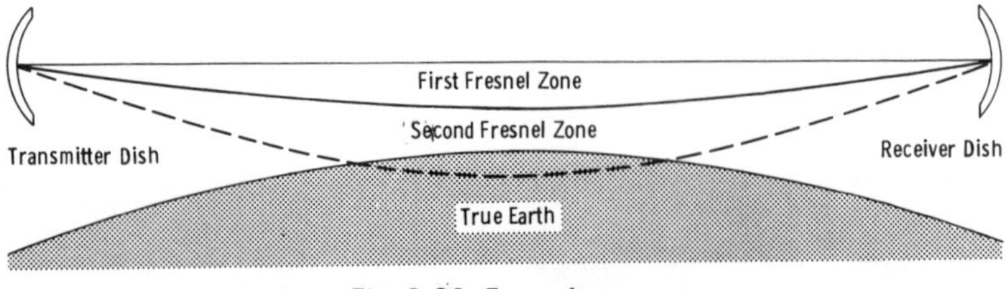

Fig. 8-20. Fresnel zones.

Fresnel zone and all other odd-numbered zones has a full-wavelength (or even multiples of a wavelength) relationship to energy in the first zone and therefore is phase additive.

The primary energy is contained in the first Fresnel zone, and energy contained in even-numbered zones is phase canceling; therefore, it is desirable to obstruct the energy contained in all but the first Fresnel zone. At the same time, the first zone must be provided ample clearance. A value of 0.6 times the radius of the first Fresnel zone normally is taken as the absolute minimum clearance.

The radius of the first Fresnel zone (at the point of a major obstruction in the path) may be calculated from the equation:

$$R = 72\sqrt{\frac{AB}{Pf}} \qquad \text{(Eq. 8-1.)}$$

where,

R is the radius in feet,
A is the distance from one end of the path to the point of obstruction in miles,
B is the distance from the other end of the path to the point of obstruction in miles,
P is the total path length in miles,
f is the frequency in GHz.

The following procedure will serve to give the reader a basic understanding of plotting stl paths:

1. Plot a profile of the transmission path. Graph paper which presents the curvature of the earth on a radius 4/3 times its true value may be used. Since ordinary linear graph paper is more convenient for limited use, you may use it and the data of Fig. 8-21. Paper with ten squares to the inch is ideal for this purpose.
2. The path profile and obstructions on the path may be charted from topographic maps. The topographic map gives the heights above sea level of the surface of the earth, to which are added the heights of major obstructions, such as heavily forested areas, water towers, grain elevators, and high-rise buildings. Maps for specific areas may be obtained from the U.S. Geological Survey, Washington, D.C. 20025. For maps of areas west of the Mississippi River, write to the U.S. Geological Survey, Denver, Colorado 80215.

Take the example illustrated by Fig. 8-22A. The bulge (h) of the earth in feet at distance d1 miles from the near end and d2 miles from the far end of the path is:

$$h = 0.5\, d1\, d2 \qquad \text{(Eq. 8-2.)}$$

Using equation 8-2 to find the correction for earth curvature at the point of maximum obstruction:

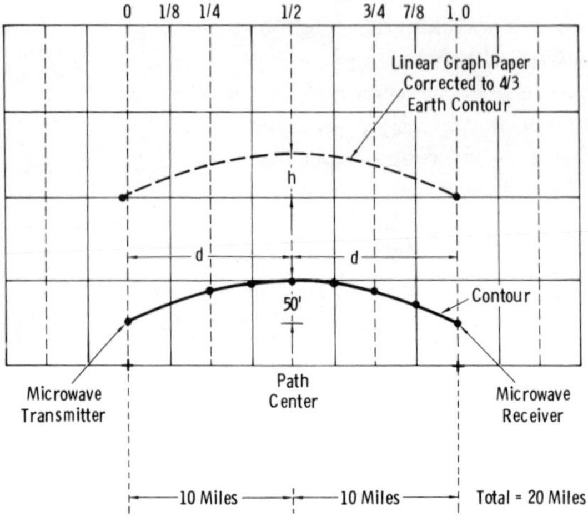

At center of path: $h = 0.5\,d^2$ (From Eq. 8-2)

Example for 20-Mile Path: $h = 0.5(10)^2 = 50$ feet

Fig. 8-21. Method of plotting profile of transmission path.

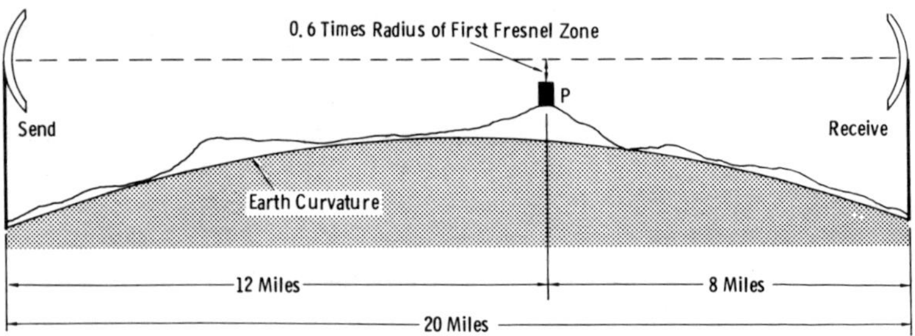

(A) Equal heights.

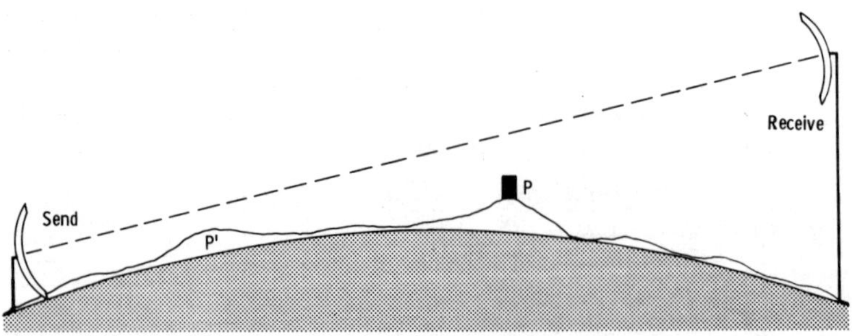

(B) Unequal heights.

Fig. 8-22. Determination of required antenna heights.

$$h = 0.5\ (12)\ (8) = 0.5\ (96) = 48 \text{ feet}$$

Using equation 8-1 to find the radius of the first Fresnel zone at the point in question (point P), for 950 MHz (0.95 GHz):

$$R = 72 \sqrt{\frac{(12)(8)}{(20)(0.95)}} = 72\ (2.23) = 160 \text{ feet (approx)}$$

Then the minimum clearance is (0.6) (160) = 96 feet.

Now assume that the height above ground of a structure in the path at point P is 100 feet. The earth curvature correction has been found to be 48 feet, and the clearance must be at least 96 feet. The sum of these quantities is 100 + 48 + 96 = 244 feet.

Now further assume that the topographic maps show the sending-end and receiving-end locations to be 1000 feet above sea level and point P to be 1100 feet above sea level. This adds another 100 feet to the "negative clearance" of the dishes. Thus 244 + 100 = 344 feet is the required height of the sending and receiving dishes for a straight-line path. In case the installation is an stl, the transmitter location normally includes a tall tower on which the receiving dish can be mounted, and the method of Fig. 8-22B might be used. Note that now point P′ limits how low the sending dish can be mounted for adequate clearance.

Usually, energy in the second and higher-order Fresnel zones is severely attenuated as a result of normal terrain differences. If the path contains variations of at least 75 percent of the radius of the first Fresnel zone at the center of the path, very little problem is expected from other than first-zone clearance. If the path is very smooth (as over water), the problems can multiply. This is so because reflections are not dispersed and can become quite strong.

When accurate topographic maps are not available, accurately calibrated, sensitive altimeters (properly adjusted in accordance with the barometric pressure) can be referred to a known point or bench mark (American Paulin system). Or, aircraft-mounted absolute altimeters may be used. In the absence of these devices, conventional civil-engineering techniques must be employed. This is always recommended when there is doubt about a proposed stl path.

Attention to proper first-Fresnel-zone clearance is all that is necessary for the average path, so long as the path length and antenna system are such that the proper *fade margin* is provided. In evaluating microwave relay performance, the following fundamentals serve as a basic guide:

1. What field strength in dBW is required at the stl receiver to meet s/n ratio specifications? This is 50 dB for a-m use and 60 dB for fm use.
2. Since a microwave beam is sometimes bent and scattered by atmospheric conditions, not only adequate clearance, but also adequate *fading margin* must be provided.

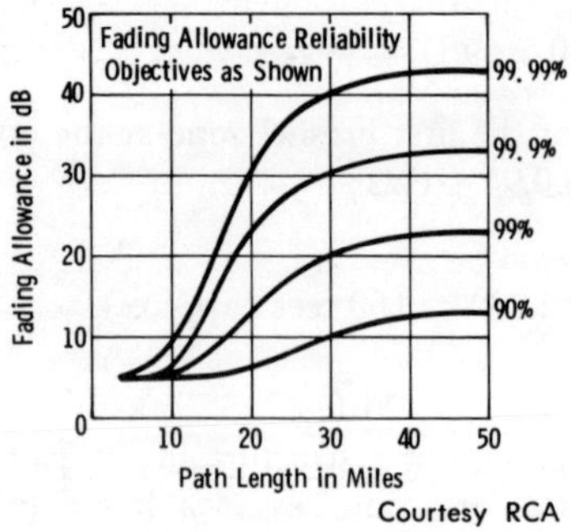

Fig. 8-23. Fading allowance.

3. Fig. 8-23 is a graph of fading allowance in decibels versus path length in miles for various reliability factors. For an stl system, the reliability factor must be 99.99 percent. This means that out of a 24-hour program day (86,400 seconds) the signal *could* deteriorate to something just less than normal program quality for a period of 9 seconds. This 9-second period would not necessarily occur all at one time, but could be accumulative over the 24-hour period. A 99.9 percent reliability factor would allow some 87 seconds in the same program day to deteriorate to less than standard, which is entirely unsatisfactory for stl use.

Table 8-3. Free-Space Path Loss at 950 MHz

Path Length (Miles)	Loss (dB)
5	109
10	115
15	118
20	121
30	124

4. The field strength at the stl receiver is dependent on the free-space path loss (Table 8-3), the total line loss (Table 8-4), the total antenna gain (Table 8-2), and the transmitter output in dBW (decibels above 1 watt).

Table 8-4. Transmission Line Losses per 100 Feet at 950 MHz

Line	Loss (dB)
½" Foam	3.2
½" Air	2.7
⅞" Foam	1.9
⅞" Air	1.2
1⅝" Air	0.65

The values of free-space loss in Table 8-3 are based on the following relationship:

$$A' = 37 + 20 \log f + 20 \log D$$

where,

A′ is the free-space loss in dB,
f is the operating frequency in MHz,
D is the distance in miles

For example, the free space attenuation at 950 MHz for a 20-mile path is:

$$\begin{aligned} A' &= 37 + (20)(2.9) + (20)(1.3) \\ &= 37 + 58 + 26 \\ &= 121 \text{ dB} \end{aligned}$$

Suppose you desire to evaluate an stl system that has the following characteristics:

1. Transmitter power output: 6 watts (8 dBW) at 950 MHz
2. Type of transmission line: ½-inch air
3. Length of transmission line: 330 ft
4. Diameter of reflector at transmitter and receiver: 4 ft
5. Path length: 15 miles
6. Minimum 0.6 Fresnel-zone clearance assumed

The transmission-line loss is 9 dB (Table 8-4). The antenna-system gain at each end is 19 dB, or a total of 38 dB (Table 8-2). The combination of transmission-line loss and antenna gain is $38 - 9 = 29$ dB. The *net* path loss is:

$$A = A' - G_t$$

where,

A is the net path loss,
A′ is the free-space loss (Table 8-3),
G_t is the combined transmission-line loss and antenna gain.

In this example:

$$\begin{aligned} A &= 118 - 29 \\ &= 89 \text{ dB} \end{aligned}$$

The receiver power input is:

$$P_r = P_t - A$$

where,

P_r is the receiver power input,
P_t is the transmitter power output in dBW,
A is the net path attenuation.

Thus in this example:

$$
\begin{aligned}
P_r &= 8 - 89 \\
&= -81 \text{ dBW}
\end{aligned}
$$

The fade allowance for 99.99-percent reliability (15-mile path) is approximately 16 dB (Fig. 8-23). Therefore the signal at the receiver with fade is −97 dBW, or 100 microvolts (Table 8-5).

Table 8-5. Signal Level in dBW Across 50 Ohms and Signal Strength in μV

dBW	μV
−78	1000.0
−83	500.0
−87	320.0
−97	100.0
−107	32.0
−117	10.0
−127	3.2
−131	2.0

In this example, if the receiver requires 34 μV in 50 ohms for a 60-dB signal-to-noise ratio, there is a safety factor of 100/34 = 2.94, or about 9 dB. This is over and above the required fade margin, and the installation normally would be considered adequate.

The Marti STL System

The Marti STL-8 transmitter is shown in Fig. 8-24A, and the companion R-200/950 stl receiver is shown in Fig. 8-24B. The basic block diagram of the transmitter appears in Fig. 8-25 and that of the receiver in Fig. 8-26.

The maximum licensed power of the transmitter is 8 watts, and this unit is normally operated at around 6 watts under average safety-factor applications. Direct fm is employed with plus and minus 52.5 kHz constituting 100-percent modulation. The standard 75-μs pre-emphasis network is used at the transmitter, with corresponding de-emphasis in the receiver. Remote-control and/or subcarrier inputs are provided.

The fm modulator (Fig. 8-25) employs a phase-shift network in the transistor oscillator circuit, with varactors (voltage-variable-capacitance diodes) used to determine the oscillator frequency. The varactor bias is initially adjusted to 6 volts, and audio is fed to this network to vary the oscillation frequency at the corresponding audio rate. The oscillator frequency is 1/216 of the final operating frequency.

Both the frequency and power are multiplied in the following stages as shown in Fig. 8-25. The actual rf power-amplifier output is 23 watts, which results in 8 watts at the output of the final varactor multiplier and filter

(*A*) *Transmitter.*

(*B*) *Receiver.*

Courtesy Marti Electronics, Inc.

Fig. 8-24. Equipment for stl.

network. This final stage employs varactors in tuned transmission lines, which function as highly stable frequency multipliers, with an output containing spurious emissions more than 60 dB below the carrier. The overall frequency stability of this unit is at least 0.0005 percent between temperature extremes of $-30°$ and $+60°C$, with $+25°C$ as the reference.

The R-200/950 receiver requires only 32 μV for a 60-dB or better signal-to-noise ratio. A crystal controlled, double-conversion system is used (Fig. 8-26).

The Marti SCG-8 subcarrier generator and the SCR-8 subcarrier receiver are intended to be used in conjunction with the 950-MHz aural stl, to transmit any type of auxiliary program material from the studio to the transmitter location over one of the subchannels of the link. Used primarily

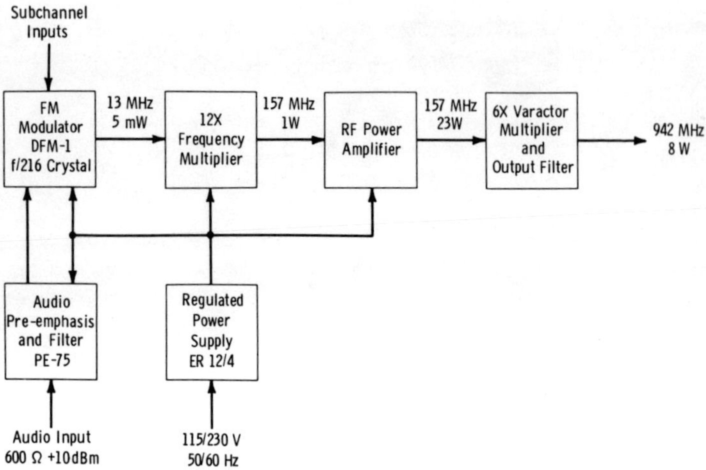

Fig. 8-25. Block diagram of stl transmitter.

to transmit background music material, they can be used equally as well (with the proper auxiliary equipment) to provide a telephone communications circuit to the transmitter.

The SCG-8 subcarrier generator can also be used with most fm broadcast transmitters to inject background music onto the 41- or 67-kHz fm subcarrier. Many stations which use radio remote pickup equipment prefer to move their base receiving antenna and remote-pickup receiver to the fm transmitter site and feed the rpu audio back to the studio over the 67-kHz

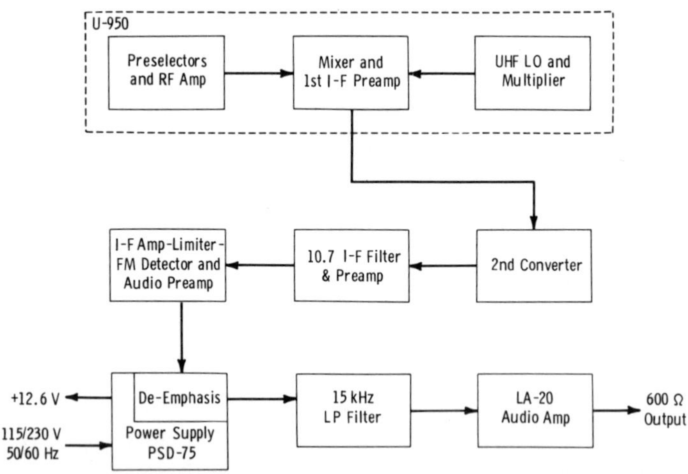

Fig. 8-26. Block diagram of stl receiver.

subcarrier, in order to achieve maximum tower height for the rpu antenna. When the SCG-8 generator is used to feed the rpu audio to the studio, the audio is recovered on a subchannel frequency-modulation monitor for reprocessing.

Both the generator and receiver are supplied with an extremely sharp 6-kHz low-pass filter to prevent cross talk between the subchannel and main channel. Care has been exercised to insure extensive shielding and filtering for operation in strong rf fields.

For fm stereo, maximum reliability is reflected in the dual- or split-channel approach, in which two transmitters and two receivers are used, combined into a common antenna system. Fig. 8-27 is a diagram of a typical system employing stereo, SCA, and transmitter remote control. (Transmitter remote control using stl's is described in the next section.) The transmitter combiner is a device commonly used for operating two stl transmitters into a common antenna and transmission-line system. While used primarily for fm stereo application in the split- or dual-channel system, it can also be used in "hot-standby" applications. The combiner presents a 3 dB loss to the system, but in most cases is less expensive than using separate antenna and transmission-line systems. More than adequate iolation is maintained between transmitters, and the combiner has a 20-watt resistive load built in. A special precision-cut T matching section is available for receivers; it also presents a 3-dB loss. Thus the total loss of the combiner technique is 6 dB, and this must be accounted for in path computations.

The dual-channel concept allows the stereo generator to remain at the fm transmitter site, preventing continual adjustment by unauthorized and uninstructed personnel, and eliminating the need for expensive interface equipment between the stl system and the fm broadcast transmitter. In addition, such problems as inadequate separation can be diagnosed more easily when the system is arranged so that the stereo generator is not combined with the stl system.

8-6. TRANSMITTER REMOTE CONTROL VIA STL

Fig. 8-28A is a view of the Marti RMC-2AXS studio unit, and Fig. 8-28B is a view of the companion RMC2-AXT transmitter unit. Designed for a-m and fm subaudible telemetry, this remote-control system requires no interface equipment to meet requirements of the FCC Rules and Regulations pertaining to the mixing of the subaudible telemetry, filtering, and prevention of overmodulation. These circuits and components are an integral part of the basic studio and transmitter units.

Of all solid-state design and employing modular construction techniques, the RMC-2AX is available as a 22-function system having 10 metering positions, or, for the more complex installation, an optional 50-function system with 24 metering positions is available.

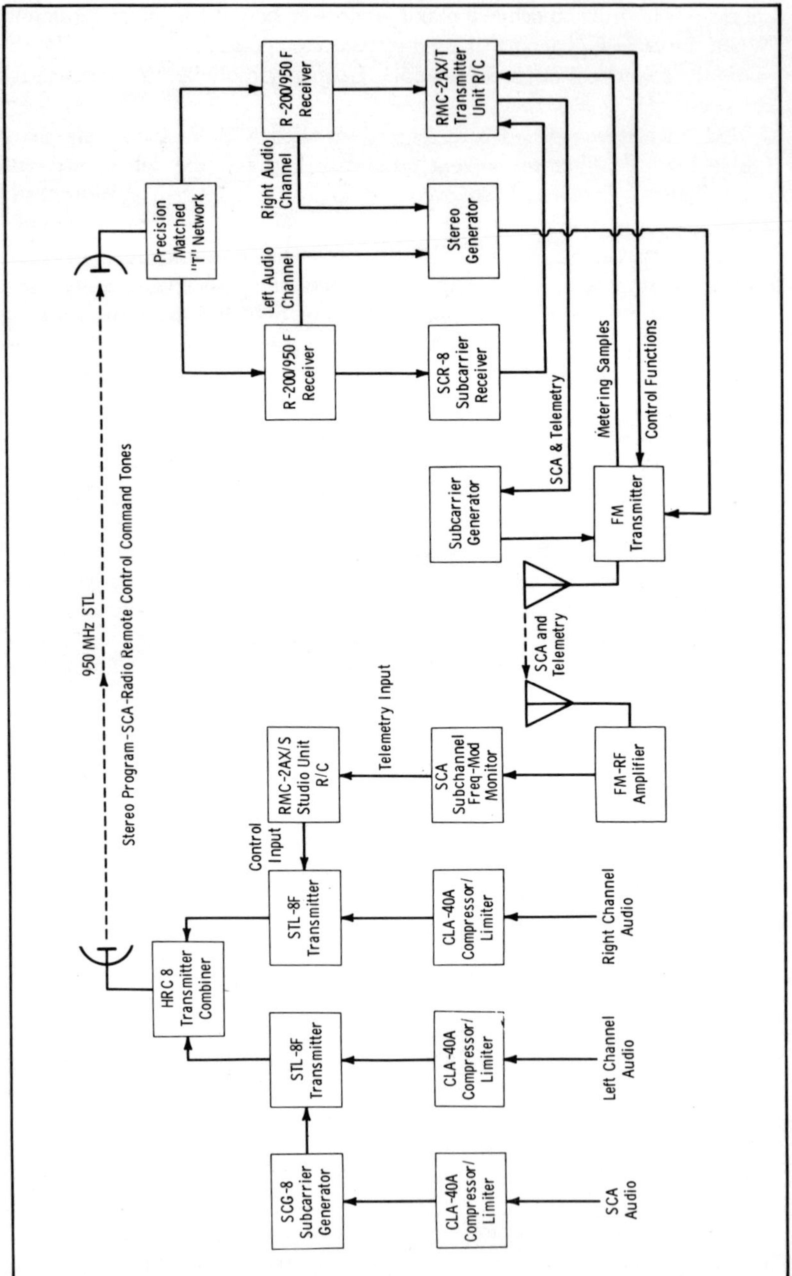

Fig. 8-27. System for fm stereo, SCA, remote control, and telemetry.

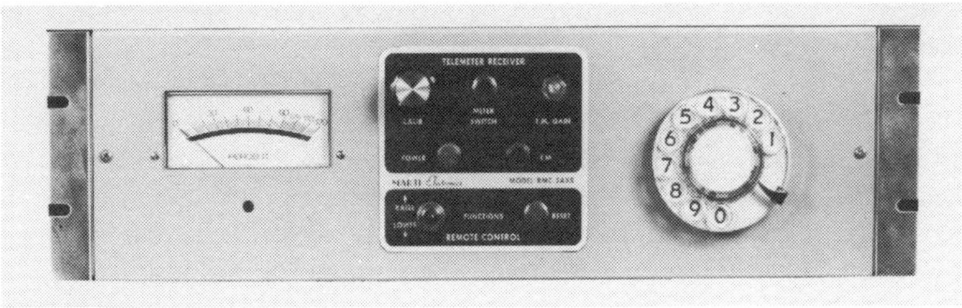

(*A*) *Studio unit.*

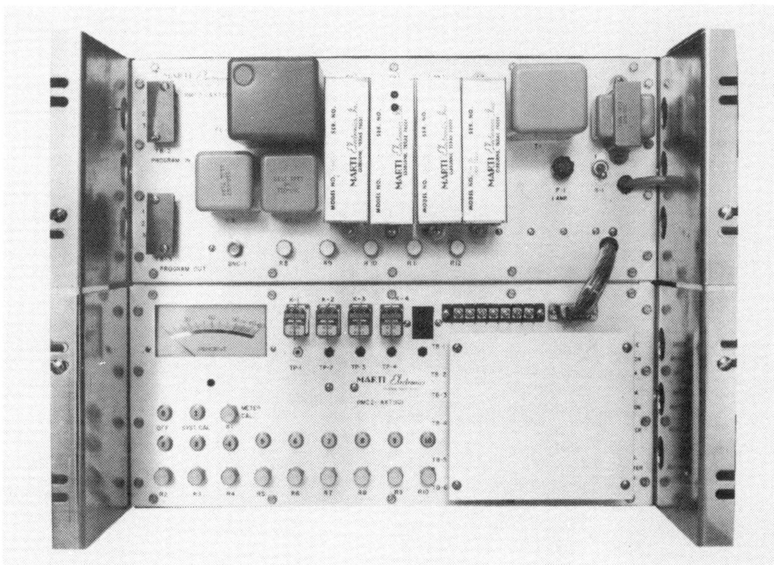

(*B*) *Transmitter unit.*

Courtesy Marti Electronics, Inc.

Fig. 8-28. Remote control system.

Subaudible telemetry is accomplished through the use of a voltage-controlled oscillator with a frequency shift of from 22 to 28 Hz at a low percentage of modulation. A high-pass filter is used to prevent the program audio component from modulating the metering channel. Automatic compensation is provided to limit the total modulation to 100 percent modulation while telemetering.

The same system can be used to meter an fm transmitter, except the fm 67-kHz subcarrier is modulated with the metering information through a subcarrier generator. The information is usually recovered at the studio on a subchannel frequency-modulation monitor or on a multiplex receiver. The subcarrier may also be used for background music.

Remote control is accomplished by means of four supersonic tones injected into J1 of the Marti STL-8 transmitter (Fig. 8-29). These tones are recovered from the output of the microwave receiver and connected into the RMC-2AXT(A) unit (Fig. 8-30) at the transmitter through a short RG-58A/U cable with BNC connectors. These four tones are applied to four tuned amplifiers which select one tone each, amplify it, and rectify it to operate a function relay. The function relays perform the dial and reset operation of the stepping switch and the raise and lower adjustment (of the transmitter under control) in each position of the stepping switch.

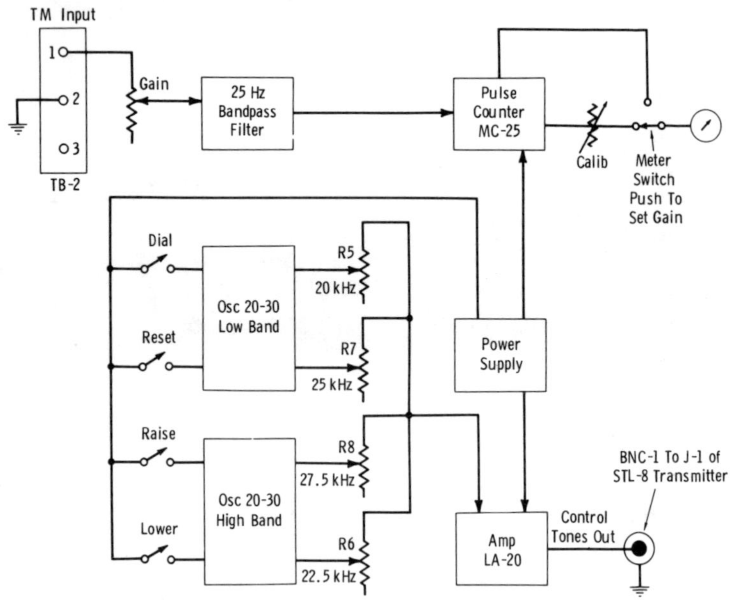

Fig. 8-29. Block diagram of remote-control studio unit.

The telemetering of the data concerning transmitter operation back to the studio is accomplished as follows. Each circuit to be telemetered is equipped with a sampling resistor to provide a dc voltage drop of 1.6 to 2.5 volts. (Modern broadcast transmitters are supplied with voltage and current sampling circuits.) Each sampling circuit is connected to a pair of RMC-2AXT metering terminals. A 10-turn potentiometer is provided for each metering circuit to provide a precise adjustment to 100 percent on the RMC-2AXT meter. Thus, when, for example, the transmitter plate current is at its normal (100-percent) value, the sampled voltage drop developed by this current is adjusted to read 100 percent on the RMC-2AXT meter. A decrease in transmitter plate current of 5 percent would then produce a 5 percent reduction in this sampled voltage drop and be telemetered back

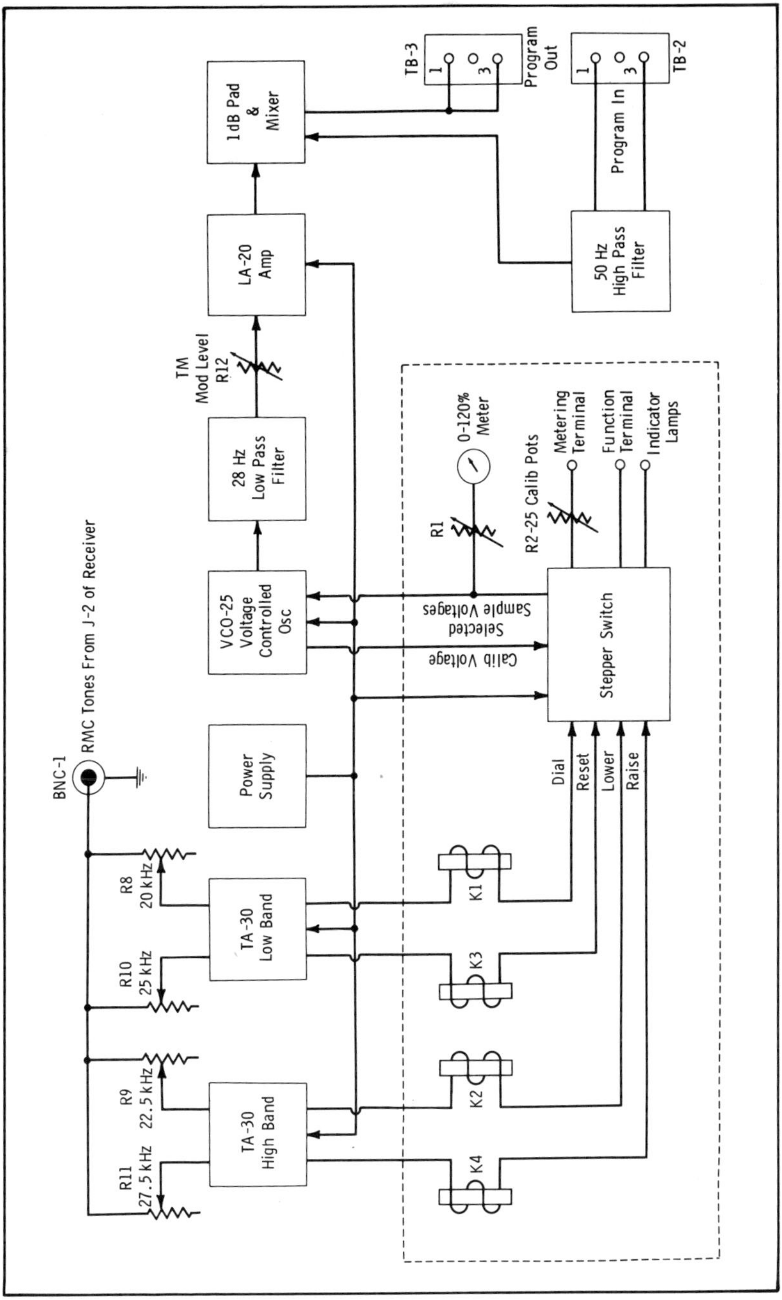

Fig. 8-30. Block diagram of remote-control transmitter unit.

to the studio as such. At the studio, a chart must be prepared to allow fast conversion of percentage readings to voltage, current, and power quantities for entry into the station log. Details of this procedure as well as remote adjustment and control are covered in Chapter 11.

In order for the sampled dc voltage to be transmitted by radio, a converter must be employed to produce an ac signal which is a linear function of the dc voltage sample. This converter is a voltage-controlled oscillator which operates over a frequency range of 22 to 28 Hz (corresponding to dc sample voltages of zero to 120 percent). These subaudible frequencies are selected to permit the transmission of the telemetry information at low level on the audio program channel (a-m or fm SCA) without noticeable degradation of the program quality. In order to prevent extraneous low-frequency program audio components from interfering with the telemetry tone, a high-pass filter is used in the program line before the transmitter modulator to attenuate frequencies below 50 Hz. A bandpass filter is also required at the studio unit to separate the 22-28 Hz telemetry tone from the program material. Reference to the system block diagrams (Figs. 8-29 and 8-30) will aid in the understanding of the signal processing described above.

Another converter must now be used to convert the recovered telemetry tone linearly to an analog display (meter reading). This conversion is performed by the MC-25 pulse-counter module in the studio unit. The system is accurately calibrated before the readings are taken by dialing position 1 at the studio. This switches in a standard reference voltage at the transmitter unit and allows the studio meter to be adjusted to "Cal" (100 percent) by means of the CALIB knob.

The Marti STL remote-control system has the following specifications:

Remote Control Functions

Model RMC-2AXT(10)	10 Raise commands 10 Lower commands
Model RMC-2AXT(25)	25 Raise commands 25 Lower commands

Metering

Model RMC-2AXT(10)	10 Telemetering channels including calibration
Model RMC-2AXT(25)	25 Telemetering channels including calibration

Metering Input Requirements

1.6 to 2.5 volts dc, either polarity to ground or referenced above ground not to exceed ±200 volts dc. Metering input resistance greater than 12,000 ohms.

Metering Calibration

Calibration voltage derived from double regulated zener diode, oven temperature controlled. Ten-turn vernier potentiometers provided for all metering adjustments. Four-inch, mirror-scale, taut-band meters with one-percent accuracy provided.

Remote Control Frequencies

For STL:	20 kHz	Dial function
	22.5 kHz	Lower function
	25 kHz	Reset function
	27.5 kHz	Raise function
For Line:	600 Hz	Dial function
	980 Hz	Lower function
	1450 Hz	Reset function
	2250 Hz	Raise function

Telemetry Tone Frequencies

22 Hz to 28 Hz corresponding to metering sample voltages of 0 to 120 percent of normal (100 percent)

Telemetry Tone Purity

Less than 2 percent thd at 27 Hz (10 dBm)

Telemetry Accuracy

Plus or minus 2 percent of full scale

Telemetry Tone Output Level, RMC-2AXT(A)

Adjustable from −35 dBm to +10 dBm (measured at "program out" terminals)

Impedances

Program in, program out, and telemetry input (TB-2 of RMC-2AXS) are all 600 ohms

Operating Environment Air Temperature Range

−10° C to +40° C

Power Requirement

110 to 125 volts, 50/60 Hz

EXERCISES

Q8-1. Give the frequency band for stl microwave systems.

Q8-2. Do the frequency bands differ between stl service and mobile service?

Q8-3. What is the absolute minimum first-Fresnel-zone clearance recommended?

Q8-4. Define "first Fresnel zone."

Q8-5. Why should strictly over-water paths be avoided where possible?

Q8-6. Is the signal modulation for microwave systems normally a-m or fm?

Q8-7. What type of line couples the energy from the output stage to the antenna?

Q8-8. What is the signal-to-noise ratio required in an stl system for:
(A) Use with a-m broadcast transmitters
(B) Use with fm broadcast transmitters

CHAPTER 9

AM Transmitters and Antenna Systems

Although a knowledge of basic electronics is assumed in this text, a brief review of some important principles is presented here.

9-1. REVIEW OF ANGULAR VELOCITY

When a sine wave is plotted on paper, its instantaneous values are related to the radius of a circle (Fig. 9-1). The progressive points on the curve are in relation to the radius (A) as it rotates counterclockwise from the starting point at zero degrees. There are 360° in the entire circle, and the corresponding degrees of the curve are shown in the illustration. If the angle is measured in terms of the path along the circumference with the length of the radius as the unit of measurement, the angle is determined in terms of *radians.* When the length of the arc (B) is equal to the length of the radius (A), the angle is one radian. From geometry, the circumference of a circle is 2π times the radius. (The circumference of a circle divided by its diameter is always π, which is equal to 3.1416. Since the radius is one-half the diameter, there are 2π, or 6.28, radii in the circumference.) The radius divides the circumference 2π times. In degrees, then, one radian is 360° divided by 2π, or 57.3°.

As can be noted from the illustration, an angular rotation of 180° is equivalent to π radians, and 90° is one-half this amount, or $\pi/2$ radians; 270° is $3\pi/2$ radians, and 360° is 2π radians (Table 9-1).

The sine wave goes through one complete cycle for each rotation of the radius. The *angular velocity* (sometimes termed *angular frequency* or *radian frequency*) is the rate of rotation of the radius expressed in radians per second. The number of radians in one cycle, which is 2π (6.28), is one of the most common numerical values in basic radio formulas. For example,

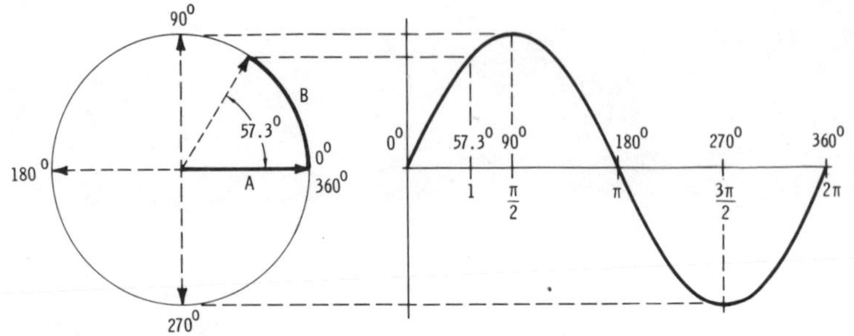

Fig. 9-1. Relationship between a sine wave and the radius of a circle.

inductive reactance is found by multiplying the value of inductance in henrys by 6.28 times the frequency. The value of capacitive reactance is 6.28(f)(C), divided into 1. In any of these formulas (with which the reader should already be familiar), the value $2\pi f$ should be understood to be the number of radians per cycle times the number of cycles occurring per second. The primary definition is:

$$\omega = 2\pi f$$

where,

ω (Greek letter omega) is the angular velocity in radians per second,
f is the frequency in hertz (cycles per second),
2π is 6.28.

Table 9-1. Relation of Radians to Degrees

Degrees	Radians	In Terms of π	In Terms of λ
90	1.57	$\pi/2$	1/4
180	3.14	π	1/2
270	4.71	$3\pi/2$	3/4
360	6.28	2π	1

9-2. REVIEW OF FREQUENCY-WAVELENGTH RELATIONSHIP

To understand the wave theory of propagation, it is necessary to correlate the *speed* of travel, the *angular velocity,* and the *shape* of the radiation pattern. The speed of travel, or *velocity of propagation* (do not confuse it with angular velocity), of a radio wave in space is the same as the speed of light, which has been accurately measured to be very nearly 186,000 miles per second. Since the metric scale is often used in scientific studies, the speed of propagation is taken as 300 million meters per second. There is an immediate relationship that occurs between the number of cycles per second (frequency) of the radio wave and its "physical" length, or wavelength.

NOTE: The unit of frequency is the *hertz,* abbreviated hz. One hertz is equal to one cycle per second.

One wavelength is the measurement of the length in space (or any given medium) occupied by one cycle of the wave. It is obvious that if the frequency is doubled, one cycle, or one wavelength, will occupy one-half as much space. In other words, the higher the frequency, the shorter is the wavelength; the lower the frequency, the longer is the wavelength.

The symbol used by engineers for wavelength is the Greek letter *lambda* (λ). The relationship of wavelength and frequency is expressed mathematically as follows:

$$\lambda = \frac{300{,}000{,}000}{f}$$

where,

λ is the wavelength in meters,
f is the frequency in hertz.

This formula states that to find the wavelength in meters, divide the velocity of propagation in meters per second by the frequency in hertz. Since it is much easier to use kilohertz (kHz) and megahertz (MHz) for radio frequencies, the formula may be expressed in the following alternate forms:

$$\lambda = \frac{300{,}000}{f\ (\mathrm{kHz})}$$

or,

$$\lambda = \frac{300}{f\ (\mathrm{MHz})}$$

Although the wavelength is always given in meters in technical discussions, it is useful to be able to convert meters to the more familiar English measurement in feet or inches. The following conversion factors may be used:

$$1 \text{ meter} = 3.28 \text{ feet} = 39.37 \text{ inches}$$

For example, the wavelength at the approximate center of the a-m broadcast band (1000 kHz) is calculated as follows:

$$\lambda = \frac{300{,}000}{1000} = 300 \text{ meters}$$

Since 1 meter is equal to 3.28 feet:

$$(300)(3.28) = 984 \text{ feet}$$

For comparison, find the wavelength at the approximate center of the regular fm broadcast band (88 to 108 MHz). This would be at approximately 100 MHz, so that:

$$\lambda = \frac{300}{100} = 3 \text{ meters, or } 9.84 \text{ feet}$$

As a further comparison, find the wavelength of a 950-MHz stl signal. This is found to be 0.316 meter, or approximately 1.03 feet. Wavelengths at microwave frequencies are measured in centimeters and inches (1 centimeter equals 0.01 meter, or 0.3937 inch).

9-3. REVIEW OF ELEMENTARY WAVE PROPAGATION

The transmitting antenna may be considered to be a device used for coupling radio-frequency energy to space. Fig. 9-2 illustrates a vertical radiator fed at the base, as is standard for a-m broadcasting. The ground sets up a "mirror," or "image," antenna (as shown) as if there exists an actual antenna of the same length as the one above the ground. Fig. 9-2A shows the action at an instant when the first positive alternation is applied, making the upper wire terminate in a positive pole and the lower (ground) terminate in a negative pole (surplus of electrons). Since each electron has minute lines of force associated with it (as does any charged particle), an electric field of force exists between upper and lower sections of the antenna. The similarity to a capacitor, which is simply two conductors separated by a nonconductor, can be seen. The two sides of the antenna itself are the conductors, and the air is the dielectric (nonconductor) between the conductors.

In Fig. 9-2B, the first polarity reversal (second alternation of the first cycle) is taking place. The top of the antenna is now charged negatively, and the image is charged positively. The lines of force produced during this half-cycle are in the opposite direction from those produced in the preceding half-cycle. When the current again reverses, the lines of force also reverse (Fig. 9-2C). The lines of force are propagated outward from the antenna at the speed of light. This moving energy field is termed the *electric field.*

In addition to the electric field, the antenna also radiates a *magnetic field,* which is at right angles to the electric field. Thus the radiation field includes two components of force. The *electromagnetic* field, by definition, includes both the electric and magnetic fields at right angles to each other.

To complete the basic picture, the receiving antenna is a conductor, fixed in space (or within the receiver cabinet itself), across which the electromagnetic lines of force pass in their travel from the transmitting antenna. A corresponding emf is set up in the conductor (in this case, the receiver antenna-ground system); thus, the receiver is able to reproduce signals from the electromagnetic field that passes the receiving antenna.

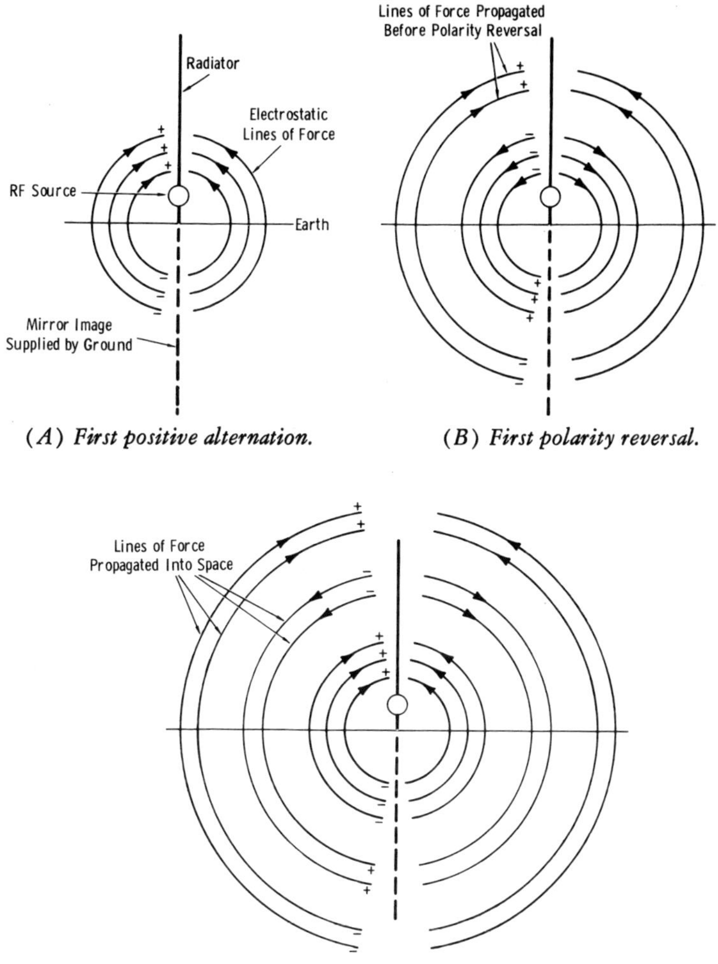

(*A*) *First positive alternation.*

(*B*) *First polarity reversal.*

(*C*) *Second polarity reversal.*

Fig. 9-2. Wave propagation.

The radiation field, which is the actual radio wave, should not be confused with the *induction field,* which exists very near the transmitting antenna. The behavior of the field is different in this region, and the electric and magnetic components are out of phase in time as well as in direction.

In general, it may be stated that a vertical radiator propagates *vertically polarized* waves and a horizontal radiator propagates *horizontally polarized* waves. Since the radiated wave is composed of two components at right angles to each other, it is desirable to examine what is meant by a "direction of polarization."

Fig. 9-3, which illustrates one cycle of the radiation at a distance from a vertical antenna, should help visualize the two components of the electromagnetic wave. One cycle (one wavelength) of the electric component is shown by the arrows in Fig. 9-3A. The shaded waveform indicates the magnetic field at right angles to the moving electric field. The term "at a distance" from the radiating antenna means that the induction field in no way influences the two components. Note that the electric and magnetic cycles

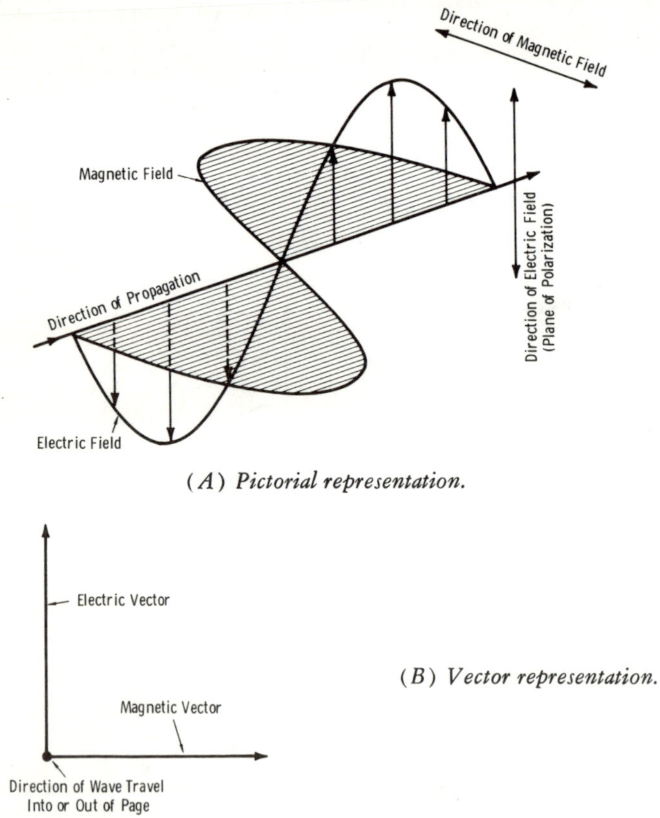

(*A*) *Pictorial representation.*

(*B*) *Vector representation.*

Fig. 9-3. Components of an electromagnetic wave.

pass through the zero reference axis at the same point in space, although at right angles in relative directions of change. Fig. 9-3B shows the instantaneous electric and magnetic vectors, assuming the direction of travel to be into or out of the page. It should be obvious that during the following alternation, the electric vector would extend downward and the magnetic vector to the left. *The direction of the electric field is taken as the direction of polarization.*

The direction of polarization for various broadcast services is specified by the Federal Communications Commission. Standard a-m broadcast stations must radiate vertically polarized waves, and fm and tv broadcast stations must radiate horizontally polarized waves. (Vertically polarized waves may be radiated by an fm station in addition to the required horizontally polarized waves.)

The chief reasons for selection of vertical polarization for standard a-m broadcast service are as follows: (1) The primary purpose of the broadcast station is to provide the maximum possible ground-wave signal (signal unaffected by sky waves) to override the noise level at the receiver location. The vertical antenna has superior ground-wave characteristics in comparison to a horizontal antenna. (2) Due to the length of the wave between 540 and 1600 kHz, the vertical antenna is best from the standpoint of design and adjustment simplicity. (3) When directional arrays must be used (very common in practice), vertical arrays are much more practical than horizontal arrays at these frequencies.

At the high frequencies utilized by fm and tv, the shorter wavelengths and line-of-sight propagation factors result in a situation opposite to that existing at a-m frequencies. Consequently, horizontal polarization is used at these frequencies.

Little difficulty is experienced in reception on the standard a-m broadcast band regardless of the type or orientation of the receiving antenna. This is true because of the strong ground-wave signal and the fact that such vertically polarized waves do not retain an exact vertical polarization. In field-intensity surveys, however, polarization may become important. At other frequencies, such as those assigned to fm and tv, the effects of polarization are highly important.

The next characteristic to be considered regarding the radiation field is the shape, or configuration, in space from the transmitting antenna. This shape is determined by the *wave angle* and has two components. Fig. 9-4 shows the *azimuth* (horizontal) angle (ϕ), which is measured in degrees clockwise from north, and the *elevation* (vertical) angle (θ), which is measured in degrees upward from the horizontal axis. This figure also shows that a vertical radiator gives a strong nondirectional ground-wave signal, while the vertical radiation is minimized. Radiation in the vertical direction is useless to the primary area to be served by a standard a-m broadcast station.

Note that the signal strength (in volts per meter) at some point represented by elevation angle θ is equal to the maximum value (represented by E) times the cosine of angle θ. Assume E to be 100 millivolts per meter at the indicated point of the horizontal axis. The value of e at any angle from this axis is obtained by multiplying 100 by the cosine of the angle involved. Assume it is desired to obtain the value of e at an elevation angle of 30°. From a table of trigonometric functions, the cosine value is found to be 0.866. Therefore:

$$E \cos 30° = 100\ (0.866) = 86.6 \text{ millivolts per meter}$$

It should be observed that the cosine of 90° is zero, and the value of e in the vertical direction is zero. At standard broadcast frequencies, the angle of elevation is of no consequence when considering the primary service area, but it does influence effective field strength at a distance.

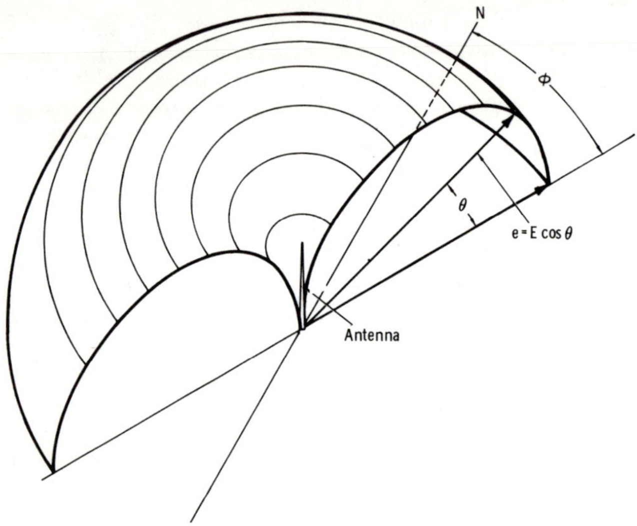

Fig. 9-4. Cross-section of radiation from a vertical antenna.

9-4. REVIEW OF PROPAGATION UNITS OF MEASUREMENT

Field strength is measured in millivolts per meter or microvolts per meter. The measurement is taken in the plane of maximum electric-field polarization. This measurement is in terms of carrier-wave stress, which is equivalent to the voltage induced in a 1-meter-long conductor as the wave cuts across the conductor at the velocity of light.

Since the carrier wave has two components, electric and magnetic, the two most important properties of the transmission medium are the *dielectric constant* and the *magnetic conductivity.* Since the main transmission medium for standard broadcast frequencies is the ground, these two properties of the soil, rocks, sand, or water over which the wave is carried must be considered.

The dielectric constant is a value given to the dielectric, taking the value for air as unity, or 1. Basic theory states that if two large metal plates separated by air are connected across a battery, the plates are given a charge that depends on the battery voltage and the area of the plates. If a sheet of mica, for example, is inserted between the plates, the "capacitor" will immediately acquire a much greater charge. This is because mica has a dielectric constant, or storage capability, much greater than air. Thus, if the dielectric constant of the ground is relatively high, giving it a high inductivity, the efficiency of transfer of the electric component of the wave will be correspondingly high. If the conductivity is also high, the ground forms an efficient medium for electromagnetic wave propagation.

As stated previously, the dielectric constant is simply a numerical value given to a substance to denote its storage capabilities relative to air (unity). The conductivity is measured in *electromagnetic units,* abbreviated emu. Table 9-2 indicates the values of inductivity and conductivity used for various types of terrain in the absence of specific surveys. They should therefore be taken as approximate for any particular and limited area. It should be noted that ground with a high inductivity invariably has high conductivity also. Sea water is notably the most efficient carrier of radio waves. The best locations inland are low areas of marshy, or "crawfishy," soil which is damp the maximum percentage of time. For standard a-m

Table 9-2. Inductivity and Conductivity for Various Types of Terrain

Types of Terrain	Inductivity	Conductivity	Absorption Factor at 50 Miles 1000 kHz[1]
Sea water, minimum attenuation	81	4.64×10^{-11}	1.0
Pastoral, low hills, rich soil, typical of Dallas, Tex., Lincoln, Nebr., and Wolf Point, Mont., areas	20	3×10^{-13}	0.50
Pastoral, low hills, rich soil, typical of Ohio and Illinois	14	10^{-13}	0.17
Flat country, marshy, densely wooded, typical of Louisiana near Mississippi River	12	7.5×10^{-14}	0.13
Pastoral, medium hills, and forestation, typical of Maryland, Pennsylvania, New York, exclusive of mountainous territory and sea coasts	13	6×10^{-14}	0.09
Pastoral, medium hills, and forestation, heavy clay soil, typical of central Virginia	13	4×10^{-14}	0.05
Rocky soil, steep hills, typical of New England	14	2×10^{-14}	0.025
Sandy, dry, flat, typical of coastal country	10	2×10^{-14}	0.024
City, industrial areas, average attenuation	5	10^{-14}	0.011
City, industrial areas, maximum attenuation	3	10^{-15}	0.003

[1]This figure is stated for comparison purposes in order to indicate at a glance which values of conductivity and inductivity represent the higher absorption. This figure is the ratio between field intensity obtained with the soil constants given and with no absorption.

broadcast frequencies high locations are not at all important, unless the conductivity happens to be better in that area than in surrounding areas. It is entirely possible at these frequencies for a transmitter in a low area to have greater coverage than one located on a mountain top.

It is apparent from the preceding that the dielectric constant and permeability of free space (or air) will be less than that of the ground. It might be deduced from this statement that the attenuation in air would be much greater than attenuation along the ground. In one sense this is true, but not in the way that might be most expected by the casual reader.

It is necessary to bear in mind that "free space" is entirely theoretical and is used to establish a basis for comparison. The definition of free space implies that all energy radiated into such a medium is done so freely with no reflections such as occur in practice from the ground and the ionosphere. From this it is possible to deduce that free space may be given a theoretical *characteristic resistance* which is of such value that no energy is reflected and consequently none ever returns. The value of this characteristic resistance of free space has been determined to be about 377 ohms, and through such a medium the waves are propagated at the speed of light, or 3×10^8 meters/second.

If one watt of power is radiated from a point in free space, one watt of power still exists at a distance of, say, 100 miles from the radiator. However, this one watt of power has now expanded into a sphere extremely large in comparison to the point from which it was propagated, and the watts per square meter is considerably less. That is, a receiving antenna covers only a small area of space, and when a watt-sphere spreads its area considerably, much less power per unit area exists. The area of the sphere of a wave varies as the square of the distance. If the distance is doubled, the total area increases 4 times, etc. Therefore the strength of the wave in units per square meter varies inversely as the square of the distance. If the distance is doubled, the strength in units per square meter decreases by a factor of 4. Thus at 100 miles from the antenna in free space (and remember the existence of free space is theoretical), the signal strength per unit area will be only $1/100^2$, or 1/10,000, as strong as at one mile.

This characteristic of free-space transmission may now be compared with the ground-wave signal that is being carried by a medium having a much higher dielectric constant than air. The characteristic resistance (R_o) of any material in relation to free space may be taken as:

$$R_o = 377 \sqrt{\frac{M_x}{E_x}}$$

where,

R_o is the characteristic resistance in ohms,
M_x is the relative permeability with respect to air,
E_x is the relative dielectric constant with respect to air.

For all practical purposes, it may be considered that the permeability (M_x)

will be the same as for free space, whereas the dielectric constant will be high, such as 10. From this it can be seen that other mediums, such as the ground, will have less characteristic resistance than the 377 ohms of free space. For a dielectric constant of 10, the value of R_0 would be 377 times the square root of 1/10, or about 116 ohms. With this change in R_0, the velocity of propagation in free space is changed according to the relationship:

$$c = \frac{R_0}{M_0}$$

where,

c is the velocity of propagation in meters per second,
R_0 is the characteristic resistance in ohms,
M_0 is the permeability in henrys per meter.

This relationship for free space becomes:

$$c = \frac{377}{4\pi \times 10^{-7}} = \frac{377}{12.56 \times 10^{-7}} = \frac{377}{12.56} \times 10^{7} = 3 \times 10^{8}$$

The answer is approximate because the numbers used in the computation were rounded off to a convenient number of significant figures. The error is negligible in practice.

Another example illustrates what happens when a wave enters a medium, such as the ground, with a lower R_0:

$$c = \frac{116}{12.56} \times 10^{7} = 9.28 \times 10^{7} \text{ or } 0.928 \times 10^{8}$$

As the wave is slowed down, the spherical surface becomes less in a given time, and the strength per unit is therefore greater than in free space or air.

9-5. PRIMARY AND SECONDARY COVERAGE AREAS (A-M)

In its allocation of frequencies, the FCC specifies a signal-intensity contour of the station coverage area to be protected from objectionable interference. This varies with the class of station (clear, regional, or local, and their existing subclassifications). No attempt is made here to list such protected contours and station classifications. The chief engineer, in particular, should subscribe to current FCC Rules and Regulations to keep abreast of all changes that occur. The cost of such a subscription is normally borne by the station as part of the overhead expenses.

The term *primary service area* (Fig. 9-5) designates the area in which the ground wave is not subject to objectionable interference or objectionable fading. The term *secondary service area* designates the area served by the sky wave and not subject to objectionable interference (Fig. 9-5). The signal in this area is subject to intermittent variations in intensity. The sky wave (at standard a-m frequencies) is almost completely absorbed in the

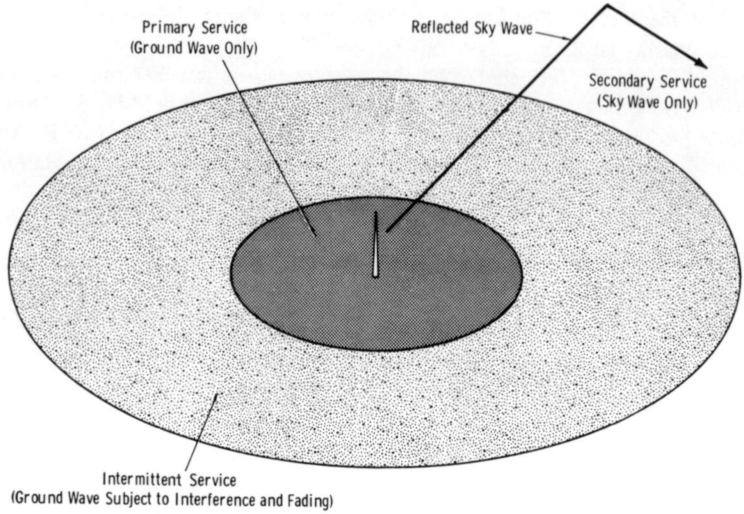

Fig. 9-5. Service areas of an a-m broadcast station.

daytime; thus, a secondary service area of any appreciable extent appears only at night. Fig. 9-6 shows how the attenuation of the sky wave varies during the sunset period. The term *intermittent service area* designates the area that receives service from the ground wave but which is beyond the primary service area and subject to some interference and fading.

9-6. GROUND-WAVE FIELD-INTENSITY CHARTS (540-1640 kHz)

The primary service area of a transmitter having a given frequency and power depends on the conductivity of the earth and the directivity of the antenna system. The graph in Fig. 9-7 illustrates the effect of soil conductivity on signal attenuation. This type of graph is published by the FCC for blocks of frequencies; some 20 graphs are required to cover the broadcast-band assignments. They show the ground-wave field-intensity curve plotted against distance for various conductivity values.

The reference assumes an antenna power and efficiency such that the inverse-distance field is 100 mV/m at 1 mile. Note the upper group of curves, which apply to the top row of miles from the antenna. The topmost curve intersects the 100 mV/m line at 1 mile. This line is transposed to the top of the lower set of curves, where 10 mV/m occurs at 10 miles from the antenna.

The inverse-distance field (100 mV/m divided by the distance in miles) corresponds to the ground-wave field intensity expected from an antenna with the same radiation efficiency and located over a perfectly conducting earth. To determine the value of the ground-wave field intensity corresponding to a value of inverse-distance field other than 100 mV/m at 1 mile, simply multiply the field intensity as given on these charts by the desired value of inverse-distance field at 1 mile divided by 100; for example, to determine the ground-wave field intensity for a station with an inverse-distance field of 1700 mV/m at 1 mile, simply multiply the values given on the charts by 17. The value of the inverse-distance field to be used for a particular antenna depends on the power input to the antenna, the nature of the ground in the neighborhood of the antenna, and the geometry of the antenna.

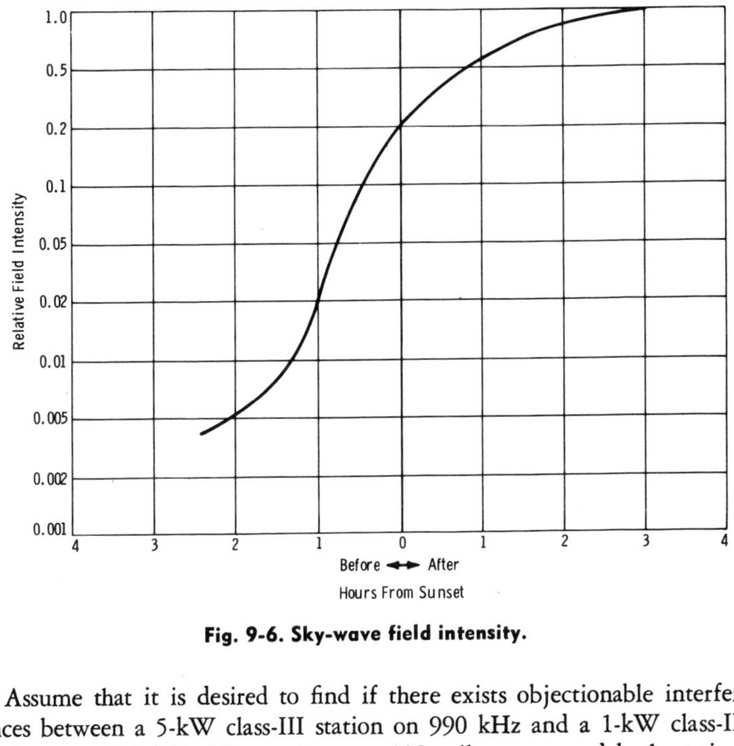

Fig. 9-6. Sky-wave field intensity.

Assume that it is desired to find if there exists objectionable interferences between a 5-kW class-III station on 990 kHz and a 1-kW class-III station on 1000 kHz. The stations are 130 miles apart, and both stations use nondirectional antennas having such height as to produce an effective field for 1 kW of 175 mV/m at 1 mile. The conductivity at each station and of the intervening terrain is 6 mmhos/m. The protection to class-III stations during daytime is to the 500-μV/m contour. The distance to the 500-μV/m ground-wave contour of the 1-kW station is determined by the use of the appropriate curve (Fig. 9-7). Since the curve is plotted for 100 mV/m at 1 mile, to find the distance to the 500 μV/m contour of the 1-kW station it is necessary to determine the distance for 285 μV/m on the graph, because (100)(500)/175 = 285. The estimated radius of the

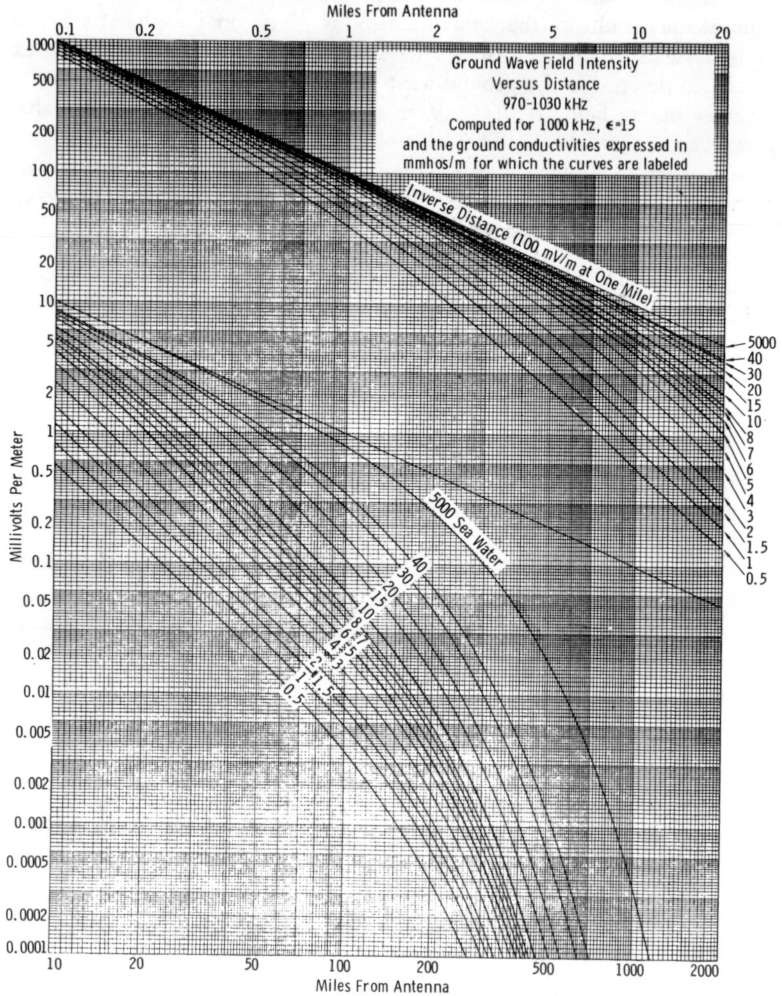

Fig. 9-7. Ground-wave field-intensity chart.

service area for the desired station is found from the appropriate curve to be 39.5 miles (using the curve designated as "6," for 6 mmhos/m). By subtracting this distance from the distance between the two stations, the interfering signal is found to travel 90.5 miles. From the curve it is found that the signal from the 5-kW station would be 158 μV/m at this distance. Since a one-to-one signal ratio applies for stations separated by 10 kHz, the undesired signal at that point can have a value up to 500 μV/m without causing objectionable interference. If the undesired signal had been found to be greater than 500 μV/m, then objectionable interference would exist. For other channel separations, the appropriate ratio of the desired signal to the undesired signal should be used.

When a signal traverses a path over which different conductivities exist, the distance to a particular ground-wave field-intensity contour is determined by the use of the equivalent-distance method. When the unattenuated field of the antenna, the various ground conductivities, and the locations of discontinuities are known, reasonably accurate results may be expected in determining field intensities at a distance from the antenna by an application of this method. The method assumes a wave to be propagated across a given conductivity according to the curve for a homogeneous earth of that conductivity. When the wave crosses from a region of one conductivity into a region of a second conductivity, the equivalent distance of the receiving point from the transmitter changes abruptly, but the field intensity does not. From a point just inside the second region, the transmitter appears to be at that distance where, on the curve for a homogeneous earth of the second conductivity, the field intensity equals the value that occurred just across the boundary in the first region. Thus the equivalent distance from the receiving point to the transmitter may be either greater or less than the actual distance. An imaginary transmitter is considered to exist at that equivalent distance. This technique is not intended to be used as a means of evaluating an unattenuated field or the ground conductivity by the analysis of measured data.

An example using the equivalent-distance method follows. It is desired to determine the distance to the 0.5-mV/m and 0.025-mV/m contours of a station on a frequency of 1000 kHz. The inverse-distance field is 100 mV/m at one mile, and the path has a conductivity of 10 mmhos/m for a distance of 15 miles, 5 mmhos/m for the next 20 miles, and 15 mmhos/m thereafter. By the use of the appropriate curves (Fig. 9-7), one can observe that at a distance of 15 miles on the curve for 10 mmhos/m the field is 3.45 mV/m. The equivalent distance to this field intensity for a conductivity of 5 mmhos/m is 11 miles. Continuing along on the propagation curve for the second conductivity, the 0.5 mV/m contour is encountered at a distance of 27.9 miles from the imaginary transmitter. Since the imaginary transmitter was 4 miles nearer (15 − 11 miles) to the 0.5-mV/m contour, the distance from the contour to the actual transmitter is 31.9 miles (27.9 + 4 miles). The distance to the 0.025 mV/m contour is determined by continuing on the propagation curve for the second conductivity to a distance of 31 miles (11 + 20 miles), at which point the field is read to be 0.39 mV/m. At this point the conductivity changes to 15 mmhos/m, and from the curve relating to that conductivity, the equivalent distance is determined to be 58 miles—27 miles more distant than if a conductivity of 5 mmhos/m had prevailed. By using the curve representing the conductivity of 15 mmhos/m, the 0.025 mV/m contour is determined to be at an equivalent distance of 172 miles. Since the imaginary transmitter was considered to be 4 miles closer at the first boundary and 27 miles farther at the second boundary, the net effect is to consider the imaginary transmitter 23 miles (27 − 4 miles) more distant than the actual transmitter; thus the actual distance to the 0.025-mV/m contour is determined to be 149 miles (172 − 23 miles).

Considerable detail has been presented on the use of the FCC propagation curves even though the percentage of readers involved in the location of transmitter sites is relatively small. However, the understanding of such data is very important in maintaining broadcast antenna systems (particularly directional systems). The proper techniques in field-intensity measurements for directional systems are assuming an ever-increasing importance to the FCC and therefore to your responsibilities in maintenance.

Notice that the preceding discussion involves the determination of the interference between ground waves only. When sky-wave interference is present, more complex computations are necessary.

NOTE: An amendment of Part 73 of the FCC Rules and Regulations modifies the calculation of radiation patterns for new directional-antenna installations. Essentially, the theoretical pattern is modified by adding a quadrature component and then multiplying the result by 1.05. The quadrature component is the larger of (1) 0.025 of the root-sum-square (rss) value of the inverse fields of the elements in the array, or (2) a value of 6 times the square root of the power in kilowatts. (Both of these values are modified by the vertical field distribution factor for the shortest tower in the array.) For details, refer to a current copy of Part 73 of the FCC Rules and Regulations.

To find rss: (1) Take the square of the individual signals from each tower; (2) add the individual squares (sum of step 1); (3) take the square root of the total in step 2.

9-7. BROADCAST ANTENNAS (STANDARD A-M SERVICE)

The efficiency of service depends principally on five factors: the frequency of operation, operating power, ground conductivity, orientation of the transmitter with respect to the distribution of population, and design of the radiation system.

When an application is made to the FCC for new, additional, or modified broadcast facilities (such as changing transmitter location), the applicant must specify the nature of the radiating system to be employed. This system must comply with efficiency standards adopted by the FCC to meet the requirements of good engineering practice for the particular class of service concerned.

Fig. 9-8 shows the FCC standards which specify a minimum effective field intensity for particular classes of stations. (*Caution:* Always check current FCC Rules and Regulations, since modifications may be made at any time.) An observation of Fig. 9-8 shows these requirements:

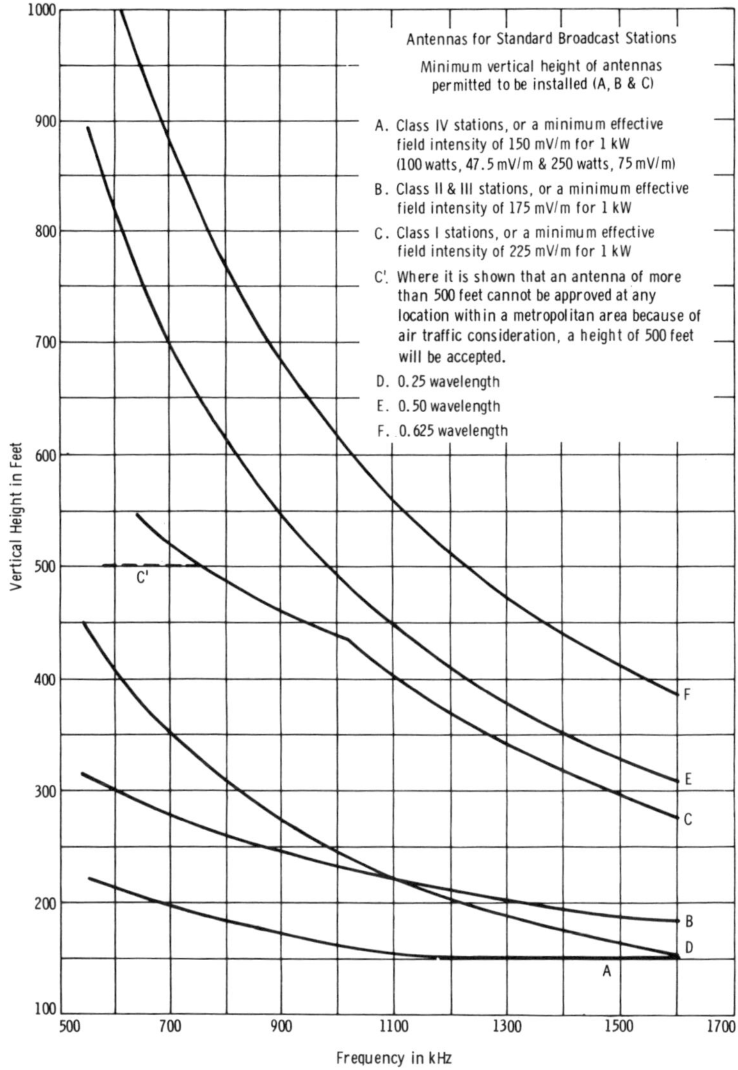

Fig. 9-8. FCC antenna requirements for a-m stations.

Curve A. Class IV stations, a minimum height of 150 feet (for frequencies 1200 kHz and higher), or a minimum effective field intensity of 150 mV/m for 1 kW (47.5 mV/m for 100 watts and 75 mV/m for 250 watts).

Curve B. Class II and III stations, minimum effective field intensity of 175 mV/m for 1 kW.

Curve C. Class I stations, minimum effective field intensity of 225 mV/m for 1 kW.

Curves D, E, and F: The physical heights of the antenna for 0.25, 0.5, and 0.625 wavelength for any frequency from 550 to 1600 kHz.

Some interesting points are involved in the design of radiating systems for broadcast frequencies. Fig. 9-9 illustrates the comparative vertical radiation patterns for antennas of 0.25, 0.311, 0.5, and 0.625 wavelength. Observation of this figure reveals that although an antenna of 0.625 wavelength has a large low-angle lobe, a secondary lobe exists at a higher angle and decreases the effective fade-free area. Fading occurs when the sky wave reflected from the ionosphere meets the ground wave and tends to cancel out the signal due to phase reversal.

It has been found in practice, for example, that the strength of the ground wave at a given distance is increased only a few decibels by increasing the height of the antenna from 0.125 to 0.5 wavelength, but the *effective* fade-free area is greatly increased due to the reduction in strength of the high-angle radiation that produces a sky wave that returns to ground close to the transmitting location. Increased intensity in the horizontal plane is the main purpose of using antennas higher than a quarter wavelength. It has been found that an antenna height of 190°, or 0.53 wavelength, is the most efficient to use where the cost of such an installation is warranted by the conditions involved.

An adequate ground system must be employed with the broadcast antenna in order to obtain maximum efficiency. The FCC specifies that where the vertical radiator is used with the base on the ground, a ground system must be employed consisting of buried radial wires at least one-quarter wavelength long. They require at least 90 such radials, and recommend 120 radials of 0.35 to 0.4 wavelength spaced every 3°. In case of high base voltage (as in antennas approaching 0.5 wavelength), a base screen of adequate dimensions should be employed to prevent high dielectric losses.

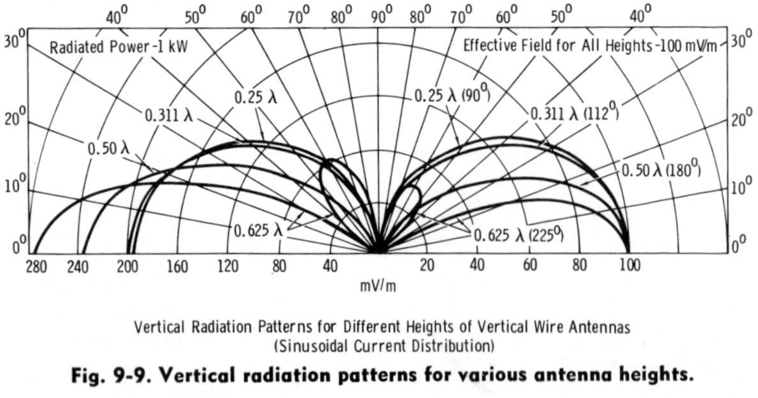

Vertical Radiation Patterns for Different Heights of Vertical Wire Antennas (Sinusoidal Current Distribution)

Fig. 9-9. Vertical radiation patterns for various antenna heights.

Note on Fig. 9-9 that the curves to the right on the polar graph are plotted on the basis of a 100-mV/m effective field for all antenna heights. The curves to the left are plotted for a specified radiated power of 1 kW. This gives the comparative effective field in millivolts per meter at one mile. For example, a 0.5-wavelength antenna with a radiated power of 1 kW produces a field intensity of 237.5 mV/m at one mile.

This information is duplicated by curve A of Fig. 9-10. Observe that the increase in field strength when going from a quarter-wave antenna (190 mV/m) to a half-wave antenna (237 mV/m) is a voltage ratio of 1.25 to 1. This is an increase of approximately 2 dB. However, as mentioned before, the effective fade-free area is significantly increased with the half-wave antenna due to the reduction of high-angle radiation.

Curve B of Fig. 9-10 indicates a complete wavelength (in feet) at any frequency between 500 and 1600 kHz. For example, a wavelength at 1400 kHz is 700 feet; at 700 kHz it is 1400 feet. Thus a half wavelength at 1400 kHz is 350 feet, and at 700 kHz it is 700 feet.

> NOTE: When guyed towers are used, the guy wires are broken with insulators spaced a fraction of a wavelength at the operating frequency to avoid radiation interference to the tower.

9-8. DIRECTIONAL ANTENNA ARRAYS

Directional arrays must be used in a great majority of present-day broadcast installations. In most instances, they must be used to protect the service areas of other stations on the same or adjacent channels. Such a system also can be used to cause the carrier wave to be reinforced in the direction of the densely populated area intended to be served. Still another application is the elimination of multiple-ownership problems by using directional antenna patterns to prevent overlap of the coverage from transmitters owned by a common licensee.

The design of directional antenna systems is a specialized field requiring considerable training, the use of complex mathematics, and experience. Just as in the case of transmitter-site preliminary tests, this work is undertaken by licensed consulting engineers who specialize in such work. The discussion here is to eliminate the usual elements of mystery surrounding directional systems for the average operator and technician.

Conditions Governing Number of Towers

Directional arrays commonly have from two to six towers, although as many as 12 have been used. The radiation pattern must have a shape which results in adequate coverage of intended service areas from the necessary location of the transmitter, and at the same time provide the required protection to other stations. Economics obviously dictate a need to accomplish the results with a minimum of towers and associated phasing equipment. Under the more severe requirements, the number of towers must be in-

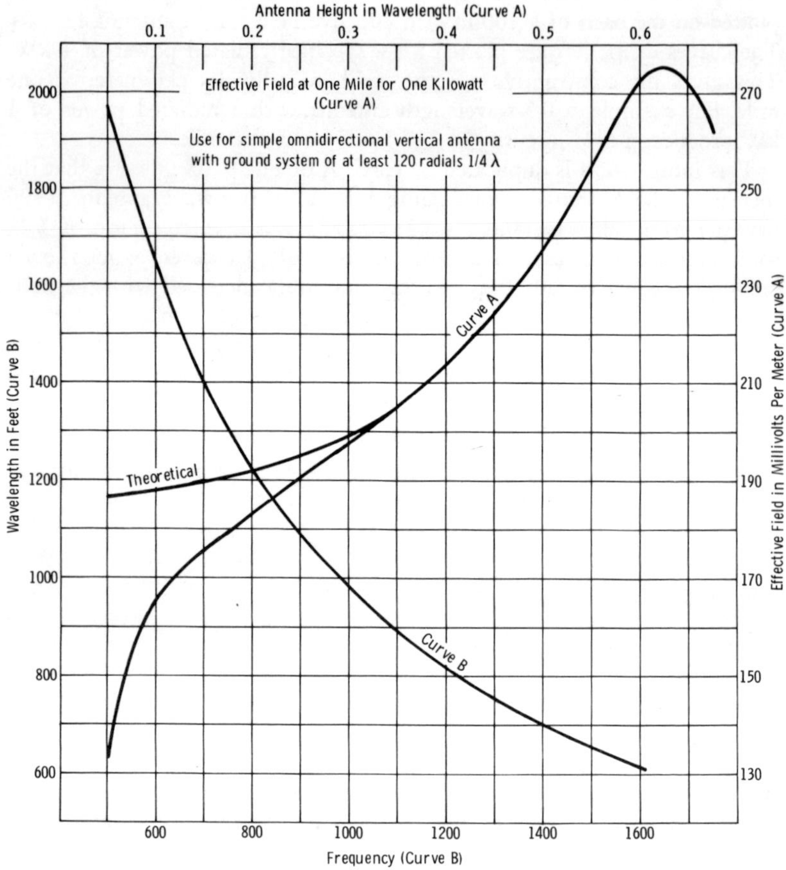

Fig. 9-10. Chart of wavelengths and effective field intensity.

creased until sufficient control of the radiated energy is achieved to result in the critical shape.

A broadly flexible rule, but one that is fundamental and serves as a starting point, states that two stations may be given the required protection with two towers. Three stations may be adequately protected with three towers, or in some cases of less severe requirements, only two towers will serve. Similarly, four towers will provide control of four nulls for four stations. Again, however, the required results often can be accomplished with three towers. Additional requirements entering into the overall problem, such as the need to concentrate energy into one or two directions to provide adequate service in addition to protecting stations in other directions, may complicate the control problem and call for additional towers.

Radiation Pattern Control

The simplest directional array consists of two half-wave towers spaced a half-wavelength apart with no associated phasing equipment. This combination serves as a good starting point to illustrate the fundamentals of directional broadcast arrays.

In this example, the two towers are fed with currents of equal amplitude, in step (in phase) with each other. The currents in both towers reach their maximums and minimums at the same instant, assuming the transmission-line lengths are equal. Fig. 9-11 shows the resulting wave interactions which control the final radiation pattern.

In Fig. 9-11A, the towers are a half wavelength apart, and the wave from antenna A (solid line) will have traveled a half wavelength by the time it reaches antenna B. The current in antenna B (shown by dashed line) will then be 180° out of phase in this direction with that of antenna A, and the field will tend to cancel out. Conditions in the opposite direction in line with the towers will result in the same canceling effect, the wave of antenna A now being 180° behind that of antenna B.

Fig. 9-11B illustrates the wave reinforcement that occurs in the direction perpendicular to the line of the towers. The waves from the two towers, being in phase, add to one another, and the energy field is the total of the radiated energy from both. The same condition holds true in the opposite direction.

Fig. 9-11C shows the field interaction at an arbitrary point, P, at some direction between those of the previous examples. The radiated waves are now slightly, but not 180°, out of phase; therefore, the total energy at this point is the difference of the two energies. The resulting directional pattern in the horizontal plane is shown in Fig. 9-11D. When this two-element array is fed with currents 180° out of phase, the conditions are reversed.

In actual practice, there are many variables involved, such as differences in the amplitudes of the currents, phasing of the currents, spacing of the towers, etc., dictated by the requirements to be met and the convenience of controlling the current magnitudes and phases. The design mathematics are very involved and complex, and only the basic principles concern the average operating and maintenance engineer.

As a practical example, consider a typical two-tower directional array. Nondirectional (one-tower) operation is specified for daytime, with a directional pattern (approximately a figure eight) necessary from local sunset to local sunrise. This dual operating function requires the relay switching system illustrated in Fig. 9-12A, which shows the relay in the position for directional operation. Fig. 9-12B shows the equivalent circuit for nondirectional operation, where the relay contacts disconnect the north tower, and only the south tower is excited. Fig. 9-12C shows the equivalent circuit when the relay is thrown to contacts B and D for directional operation.

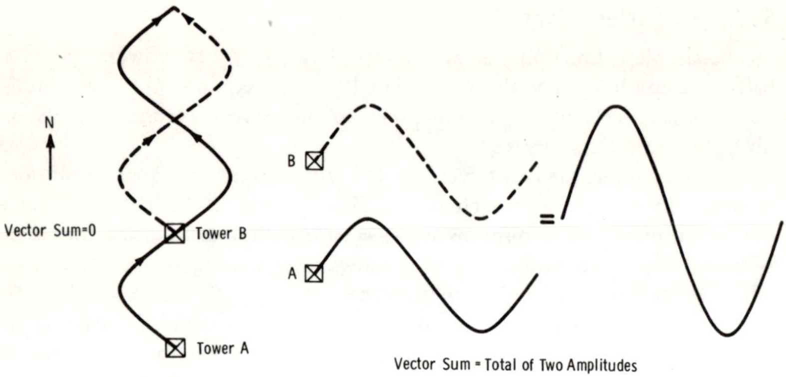

(*A*) *Towers one-half wavelength apart.*

(*B*) *Sum of radiation from towers shown in A.*

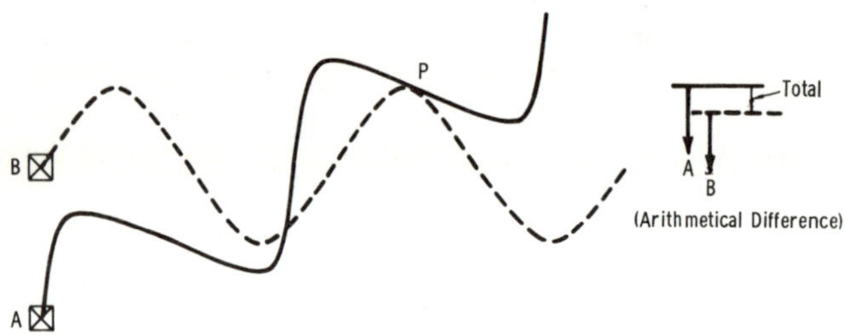

(*C*) *Field interaction between towers.*

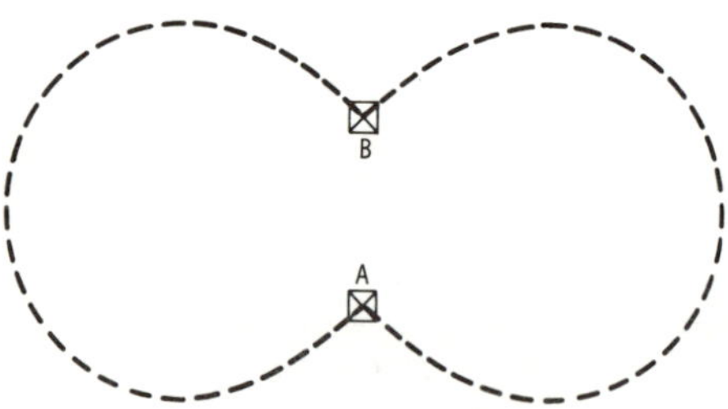

(*D*) *Pattern in horizontal plane.*

Fig. 9-11. Simple directional array.

This particular directional array is defined as follows: The spacing between the towers is 351 feet (183.9°) on a line bearing 165° true. The current in the south tower leads the current in the north tower by 160.6°. The current ratio is: south tower 1, north tower 0.91.

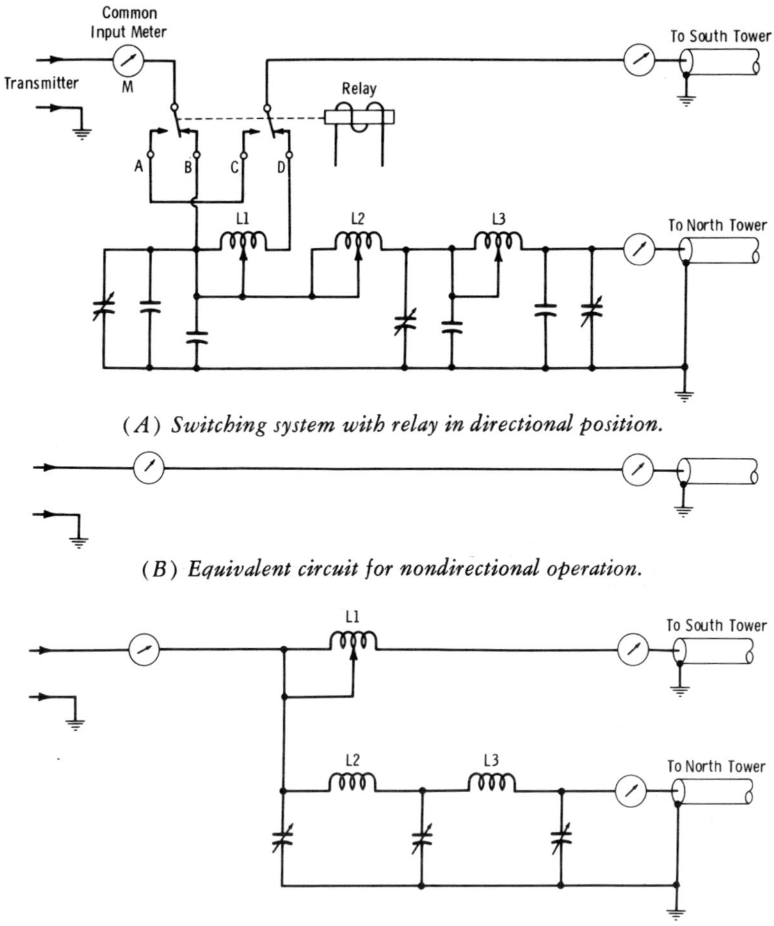

(A) Switching system with relay in directional position.

(B) Equivalent circuit for nondirectional operation.

(C) Equivalent circuit for directional operation.

Fig. 9-12. Two-tower directional-array switching system.

A transmission line introduces a delay (in degrees) of:

$$\theta = 360^\circ \frac{l}{\lambda}$$

where,

l is the length of the line,
λ is the wavelength.

The phasing unit must divide the transmitter currents into two parts with the required amplitudes (power ratio) and phase angles, taking into consideration the delay occurring in the transmission lines.

In order to illustrate the function of a phasing unit, consider the tower array and transmission lines without the phasing control. The transmission line to the south tower is 144 feet, 5 inches long, or 74.1 electrical degrees. The line to the north tower is 306 feet, or 160.2 electrical degrees. The operating frequency of the station is 1430 kHz. Hence 360°, or one wavelength is:

$$\frac{300{,}000}{1430} = 209 \text{ meters}$$

or 685.5 feet, approximately.

Therefore the current in the south tower is delayed by 74.1° from the transmitter, and the current in the north tower is delayed by 160.2°. Under these conditions, the current in the south tower would lead that in the north tower by 160.2 − 74.1, or 86.1°.

Since the directional design factors to obtain the necessary pattern of radiation call for a current lead in the south tower of 160.6°, the phasing network must make up the difference by providing an additional delay in the north leg or a phase advance in the south leg of the circuit. The choice is usually a matter of convenience, and the configuration that is simplest to design and control for any specified circumstance is used. In the example, a phase-delay circuit was chosen for the north tower to delay the current by an additional 74.5° so that the total current lead in the south tower is 160.6°.

Variable capacitors are shunted across each fixed capacitor (Fig. 9-12A) for purposes of maintaining control after the array has been adjusted initially. These capacitors are always locked on the panel to avoid slight changes that might otherwise inadvertently occur as a result of brushing against the controls or in cleaning and maintenance routines. The FCC requires periodic checks on the directional radiation pattern. Any deviation caused by changing parameters over a period of time may be corrected by adjusting the variable capacitors and tapped coils.

It is necessary to maintain the correct phase and magnitude relationships between the currents in the different branches of directional-antenna systems. The phase monitor is an instrument designed for the remote indication of the relative amplitudes and phases of the antenna currents in the various elements of directive arrays.

9-9. ANTENNA TUNING UNITS

The circuit of a single-antenna tuning unit often consists of a single T-section, low-pass filter (Fig. 9-13). The two series inductors allow independent adjustment of their respective terminating impedances; L1 is for the transmission line and L2 is for the antenna circuit. The capacitive shunt leg, common to both branches, is given a fixed value determined by the operating frequency.

This T-section network has two primary functions, to match impedances and to tune the antenna to the exact frequency of the station. For tuning the antenna, coil L2 is adjusted to series resonate with the capacitive reactance of the antenna, or, if the reactive component is inductive, it is to be considered absorbed into the inductance of L2. Coil L1 is then used to match the resistance of the transmission line to the resistance component left in the antenna circuit.

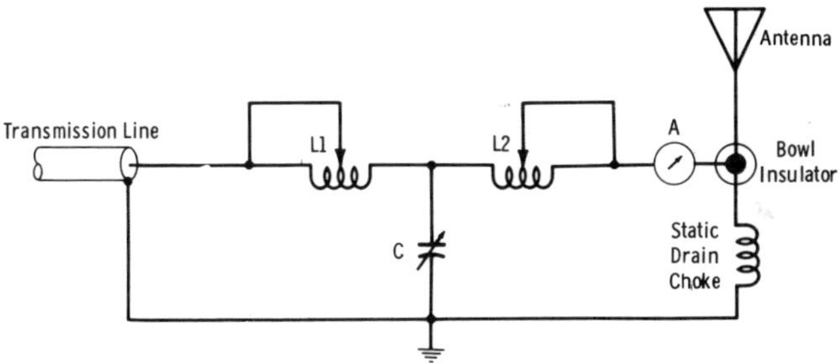

Fig. 9-13. Single-antenna tuning unit.

Table 9-3 presents the average resistance and reactance values for both self-supporting and guyed towers of various heights. These are not absolute values, which must be determined accurately by an rf bridge or other means, as described in Chapter 14.

Table 9-3. Antenna Resistance and Reactance Values

Antenna Height (Electrical Degrees)	Self-Supporting Towers		Guyed Towers	
	Resistance (Ohms)	Reactance (Ohms)	Resistance (Ohms)	Reactance (Ohms)
50	7	−j100	8	−j222
60	9	−j70	13	−j170
70	14	−j25	19	−j175
80	20	+j11	28	−j28
90	40	+j35	36	+j0
100	60	+j80	80	+j140
110	90	+j90	140	+j320
120	175	+j80	220	+j500
130	190	+j15	370	+j600
140	165	−j70	660	+j480
150	130	−j85	1100	+j0
160	82	−j55	550	−j250
170	60	−j25	280	−j450
180	40	−j5	180	−j500
190	28	+j25	120	−j430
200	23	+j50	80	−j400

This type of circuit forms a low-pass filter network with excellent harmonic attenuation. It should also be observed that the reactance values may cause a phase shift between zero and 180° (cutoff). In practice, a value of around 90° is chosen, in which case the following formulas hold true for each reactive element:

$$X_1 = +\sqrt{R_1R_2}$$
$$X_2 = +\sqrt{R_1R_2}$$
$$X_3 = -\sqrt{R_1R_2}$$

where,

R_1 is the line impedance,
R_2 is the antenna resistance,
X_1 is the reactance of L_1,
X_2 is the reactance of L_2,
X_3 is the reactance of C.

There are two accepted methods of tuning a broadcast antenna, the rf bridge method and the substitution method. Both methods are described in Chapter 14.

9-10. LIGHTNING PROTECTION

An a-m tower insulated from ground is subject to severe static build-up and high potential storage relative to ground upon the approach and actual duration of a storm. It can be seen in Fig. 9-13 that a static drain choke is employed; it has very high impedance at the operating frequency but low resistance to ground for dc.

Fig. 9-14 illustrates conventional methods of lightning protection used at most a-m stations, both within the tuning house (commonly termed the "dog house") and at the tower base. The antenna meters should be provided with heavy knife-blade switches which are kept closed (meter shorted) except when readings are to be made. A 10- to 14-inch loop is usually provided just below the bowl insulator to provide a certain amount of inductive reactance to that path. The radio-frequency choke (rfc) will provide a sufficient amount of static drain under approaching storm conditions, or for normal static build-up from rain, snow, icing, etc. However, under severe electrical storm conditions, the impedance of the choke is usually too high for adequate protection. The series ammeters, when not shorted, usually burn out and open the antenna path.

The ball gap across the antenna base insulator should be spaced an amount just beyond the distance at which arcing occurs on 120-percent (positive-peak) modulation of the rf carrier. Thus, a lightning "hit" in the vicinity will cause an arc across this gap which, in turn, causes the final-stage overload relay in the transmitter to trip, removing the carrier voltage so that the arc will not be sustained. Most transmitters employ automatic recycling such that the carrier is restored after about 1 second.

9-11. TOWER MARKING AND LIGHTING

The precise tower painting and lighting requirements for each station are specified in the station authorization. The general rules for tower lighting and painting that apply to most stations are given in the following paragraphs. (*Caution:* Always check the latest FCC regulations.)

Painting

In general, towers are required to be painted with alternate bands of aircraft-surface orange and white, with the top and bottom bands being air-

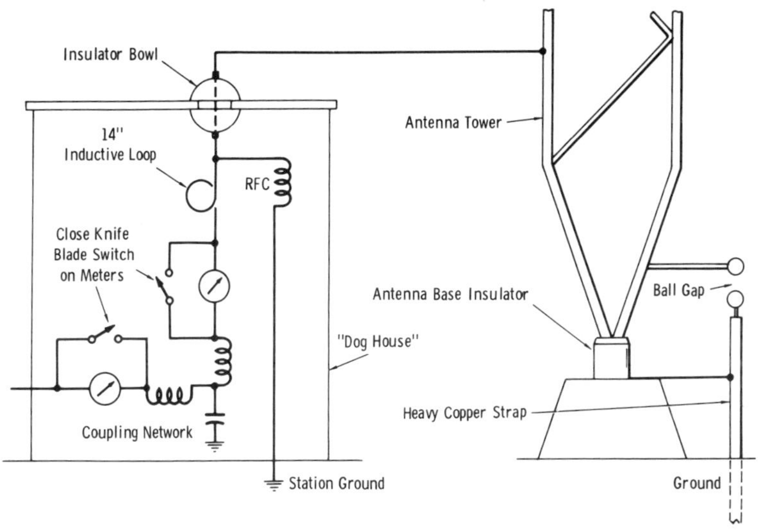

Fig. 9-14. Lightning protection for a-m tower.

craft surface orange. There are to be 7 bands, except that the bands must be no narrower than 1½ feet and no wider than 100 feet. All towers must be cleaned or repainted as often as necessary to maintain good visibility.

Lighting

The lighting requirement for a tower depends on the height of the tower (Table 9-4). The obstruction lights on towers 300 feet high or less consist of at least two 100-, 107-, or 116-watt lamps (No. 100 A21/TS, No. 107 A21/TS, or No. 116 A21/TS, respectively) enclosed in aviation-red obstruction globes. On taller towers, at least one such lamp is required at each corner of the tower at each level.

Table 9-4. Tower Lighting Requirements

Tower Height (Feet)	Flasher Levels (H = Tower Height)	Obstruction-Light Levels (H = Tower Height)
150 or less	None Required	H
Over 150 up to 300	H	1/2H
Over 300 up to 450	H	1/3H, 2/3H
Over 450 up to 600	1/2H, H	1/4H, 3/4H
Over 600 up to 750	2/5H, H	1/5H, 3/5H, 4/5H
Over 750 up to 900	1/3H, 2/3H, H	1/6H, 1/2H, 5/6H
Over 900 up to 1050	2/7H, 4/7H, H	1/7H, 3/7H, 5/7H, 6/7H
Over 1050 up to 1200	1/4H, 1/2H, 3/4H, H	1/8H, 3/8H, 5/8H, 7/8H
Over 1200 up to 1350	2/9H, 4/9H, 2/3H, H	1/9H, 1/3H, 5/9H, 7/9H, 8/9H
Over 1350 up to 1500	1/5H, 2/5H, 3/5H, 4/5H, H	1/10H, 3/10H, 1/2H, 7/10H, 9/10H

Each flasher is a 300 mm electric code beacon equipped with two 500-, 620- or 700-watt lamps (PS-40, code beacon type). On larger towers where an unobstructed view of the beacon at any level is not possible from all angles of approach, two such beacons must be installed at that level.

All lights at each level are required to burn simultaneously. On towers of 150 feet or less, the lights are required to burn from sunset to sunrise. The lights may be controlled manually, by a timer, or by a light-sensitive device. On larger towers the lights must either burn continuously or be controlled by a light-sensitive device adjusted so that the lights will be turned on at a north-sky light-intensity level of about 35 footcandles and turned off at a north-sky light-intensity level of about 58 footcandles.

During the construction of a tower, two obstruction lights of the type previously described must be installed at the top of the tower and at each level where lighting is required. (Alternately, the permanent lights may be installed at each level as the level is reached.) The temporary lights must be displayed from sunset to sunrise until the permanent lights have been placed in operation.

In special cases, a station may be required to mark its tower or towers in a manner different from that previously described. Many older installations were built when other requirements were in effect. In every case, the required type of marking is specified in the station authorization.

Fig. 9-15 shows a typical lighting plan for towers up to 150 feet high. The double obstruction light uses 100-watt A21/TS lamps. The wire-feeder size should be based on no more than a 5-percent voltage drop from the rated voltage of the lamp.

Maintenance Requirements

At least once every 24 hours, an observation of the tower lights must be made. This may be done either visually or by means of an automatic and properly maintained indicator designed to register any defect of the lights. As an alternative, an automatic alarm system may be used to indicate a failure in the tower-lighting system. Any observed failure of a top light or code or rotating beacon which is not corrected within 30 minutes must be reported by telephone or telegraph to the nearest Flight Service Station or office of the Federal Aviation Administration. Further notification by telephone or telegraph must be given immediately on resumption of the required illumination.

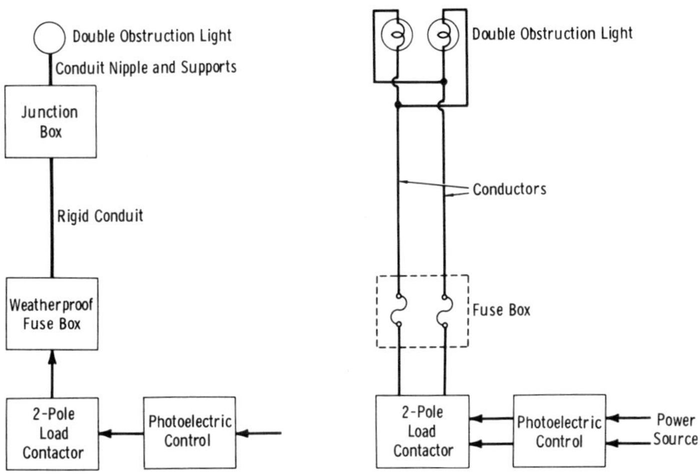

(*A*) *Physical layout.* (*B*) *Wiring diagram.*

Fig. 9-15. Typical lighting system for a 150-foot tower.

At least once every three months, there must be an inspection of all automatic or mechanical control devices, indicators, and alarm systems associated with the tower lighting. This is to insure that the apparatus is functioning properly.

Entries must be made in the station record of the following:

1. The time the tower lights are turned on and off each day, if manually controlled.
2. The time the daily check of proper operation of the tower lights was made, if automatic alarm system is not provided.
3. In the event of any observed or otherwise known failure of a tower light:
   A. Nature of such failure.
   B. Date and time the failure was observed or otherwise noted.
   C. Date, time, and nature of the adjustments, repairs, or replacements where made.
   D. Identification of Flight Service Station (Federal Aviation Administration) notified of the failure of any code or rotating

beacon light or top light not corrected within 30 minutes, and the date and time such notice was given.

E. Date and time notice was given to the Flight Service Station (Federal Aviation Administration) that the required illumination was resumed.

4. On completion of the periodic inspection required at least once each three months:

A. The date of the inspection and the condition of all tower lights and associated tower-lighting control devices, indicators, and alarm systems.

B. Any adjustments, replacements, or repairs made to insure compliance with the lighting requirements and the date such adjustments, replacements, or repairs were made.

The FCC lists specifications for tower-lighting equipment and paint. These specifications are given in Chart 9-1.

Chart 9-1. Specifications for Tower Paint and Lighting Equipment

Item	Specification	Number
Outside White	Federal Specifications	TT-P-102[1]
Aviation Surface Orange	Federal Specifications	TT-P-59[1] (Color No. 12197 of Federal Standard 595)
Aviation Surface Orange, Enamel	Federal Specifications	TT-E-489[1] (Color No. 12197 of Federal Standard 595)
Code Beacon	FAA Specifications	446 (Sec. II-d-Style 4)[3]
Obstruction Light Globe, Prismatic	Army-Navy Drawing	}
Obstruction Light Globe, Fresnel	Army-Navy Drawing	} AN-L-10A[2] or FAA Specification L-810[3]
Single Multiple Obstruction Light Fitting Assembly	Army-Navy Drawing	}
Obstruction Light Fitting Assembly	Army-Navy Drawing	}
100-Watt Lamp		100 A21/TS[4]
107-Watt Lamp		107 A21/TS (3000 Hours)
116-Watt Lamp		116 A21/TS (6000 Hours)
500-Watt Lamp		500 PS-40/0 (1000 Hours)[4]
620-Watt Lamp		620 PS-40 (3000 Hours)
700-Watt Lamp		700 PS-40/0 (6000 Hours)

[1]Copies of this specification can be obtained from the Specification Activity, Room 1643, Federal Supply Service Center, General Services Administration, 7th and D Sts SW., Washington, D.C. 20407.

[2]Copies of Army-Navy specifications or drawings can be obtained by contacting the Commanding General, Air Materiel Command, Wright Field, Dayton, Ohio 45433, or the Naval Air Systems Command, Navy Department, Washington, D.C. 20360. Information concerning Army-Navy specifications or drawings can also be obtained from the Federal Aviation Administration, Washington, D.C. 20553.

[3]Copies of this specification can be obtained from the Federal Aviation Administration, Washington, D.C. 20553.

[4]The 116-watt, 6000-hour lamp and the 700-watt, 6000-hour lamp may be used instead of the 100-watt and 500-watt lamps whenever possible in view of the extended life, lower maintenance cost, and greater safety which they provide.

RF Chokes for Lighting Systems

Broadcast installations in which the tower itself forms the antenna and is insulated from ground (all a-m towers except shunt-fed types) must employ a means of preventing the radiation energy from following the tower-lighting system into the ground, thus detuning the tower. This is accomplished by lighting filters consisting of inductors and capacitors.

The filter must present a high impedance to rf while presenting negligible impedance to the 60-Hz ac. Such a filter also serves to drain static charges on grounded ac lines. The coil must be able to carry the amperage of the lighting circuit. Fig. 9-16 illustrates typical lighting-filter circuits. The choke coils should have an inductance of at least 350 microhenrys at 1000 kHz. The value of each capacitor is 0.01 μF.

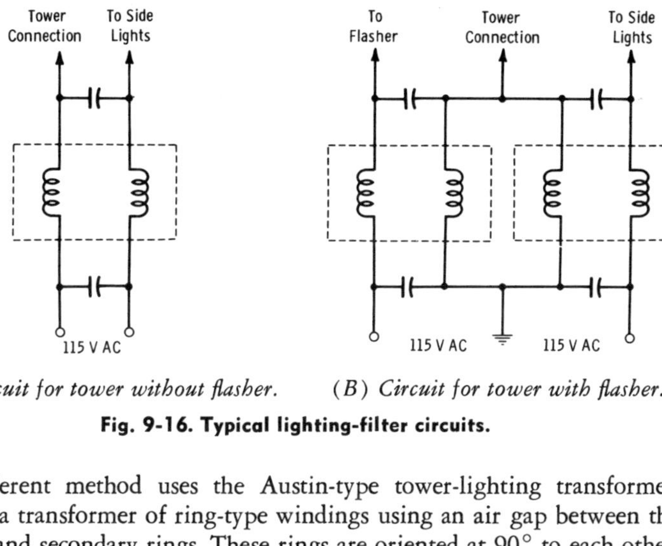

(*A*) *Circuit for tower without flasher.* (*B*) *Circuit for tower with flasher.*

Fig. 9-16. Typical lighting-filter circuits.

A different method uses the Austin-type tower-lighting transformer, which is a transformer of ring-type windings using an air gap between the primary and secondary rings. These rings are oriented at 90° to each other, adding negligible capacitance at the tower insulating zone. This transformer requires no housing, chokes, or filters. The primary is usually attached to the base of the tower insulator or to the pier supporting the insulator. The secondary may be supported by a conduit attached to the top of the insulator or to the tower.

9-12. ANTENNA REMOTE-INDICATING METERS

An indicator of antenna current is normally used inside the transmitter building (Fig. 9-17). This gives a relative indication of current in a single antenna, and one meter for each element in an array is used. In directional arrays, a means must be provided for insuring the correct relationships between elements, and hence, proper field patterns. When remote control of the transmitter from the studio is involved, additional circuitry is used inside the building. Two basic methods of sampling exist, tuned and nonresonant.

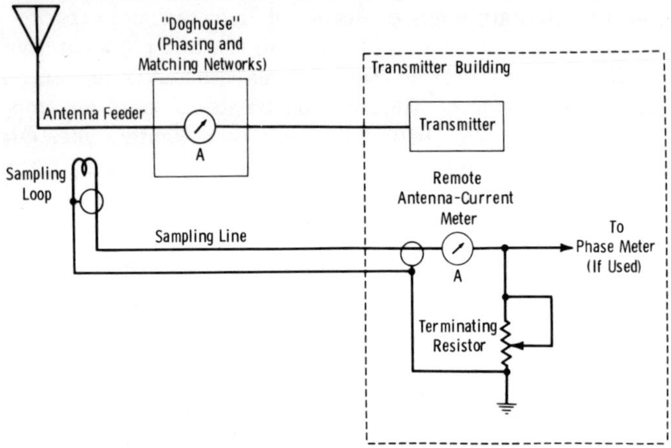

Fig. 9-17. Typical arrangement for remote indication of antenna current.

Use of Tuned Circuits

In many installations, the tuned sampling coil can be coupled at practically any point along the radiator or feed line and still pick up enough voltage to operate the current and phase indicators. The considerations involved in locating the coil are discussed later. The coil can be made small, and if a phase meter is not used, the pickup voltage can be conveniently varied over a limited range by slightly detuning the circuit, rather than by varying the degree of coupling, which presents a mechanical problem in some installations. The tuned coil has one disadvantage for use with a phase meter; unless the coil is kept precisely tuned to the transmitter frequency, errors in phase measurements due to the reactive component in the sampling unit will result.

Since the tuned circuit is more sensitive than a nonresonant loop, care must be exercised in deciding its location. Otherwise, misleading indications may be produced by pickup from adjacent towers or voltages induced by other inductors in the installation. Pickup will cause error in the phase indication, and it may also produce nonlinearity in current indication; therefore, current indications will be in error when a change of operating conditions occurs.

If the sampling equipment is to be located within a tuning house, a shielded compartment can be constructed of copper or copper-lined steel.

For outdoor locations, the enclosure can be weatherproofed by the use of 5⅜-inch ceramic bowl insulators for powers up to 5 kW and for radiators with operating impedances less than approximately 200 ohms at the sampling point. For higher power, larger insulators should be employed. For 50-kW installations where the radiator is sampled at a point of high impedance, the clearance of the sampling coil from the antenna bus should be increased an additional inch to prevent voltage breakdown.

For satisfactory results, care must be taken in the placement of the pickup coil. A position should be chosen that will eliminate magnetic coupling to all sources except the one which is to be measured. If more than one antenna element is to be monitored, all coils should be placed in the same relative physical location with respect to the antenna leads to which they are coupled; otherwise a 180° error may be introduced. Pickup may be reduced by slightly rotating the coil assembly in the horizontal plane. Ninety degrees of rotation reduces the induced voltage to essentially zero. In cases where the current in the antenna lead is too low to give sufficient output voltage from the coil, the spacing between the coil and the antenna lead may be decreased to increase the output, but in no case should the clearance be less than one inch, in order to minimize the danger from voltage surges. In extreme cases, the antenna lead can be formed into a single-turn loop parallel to the turns of the pickup coil, and the spacing reduced to approximately one inch.

Use of Nonresonant Loops

An untuned pickup loop is easy to build and usually can be fabricated on the site. Since it requires no tuning, an initial adjustment usually proves satisfactory for longer periods of time than can be expected with a tuned pickup loop. Also, there is little danger of phase shift being introduced in the nonresonant pickup circuit. As previously stated, phase shift in the sampling circuit can cause erroneous indications to be given by the phase meter.

The nonresonant loop must be rigidly mounted, and in cases where it must be coupled to points of low current, the loop might necessarily be very large to provide sufficient pickup. Since the size of the loop limits its rigidity, it is desirable to couple it to points of relatively high current if possible, particularly in high-power installations where high rf voltages may be present.

The shielded nonresonant loop in Fig. 9-18 is made of a single-conductor cable and can be made rigid and rather large to provide adequate pickup. The shielding must be cut back at the end and taped so that it does not short-circuit the single-turn loop formed by it and the inner conductor. This part of the loop can be effectively weatherproofed for outdoor installation by using a ¼-inch copper-tubing T section and insulating hose coupling, as shown in Fig. 9-18. This type of loop can be pressurized along with the transmission line, if air-dielectric lines are employed.

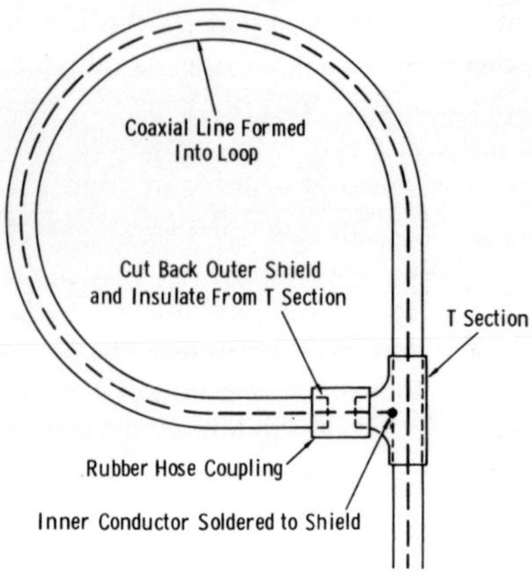

Fig. 9-18. A nonresonant pickup loop.

Location of Sampling Coils

The ideal location for the sampling coil varies with different installations and depends on such factors as the electrical design of the elements of the array and the feed lines, the construction of the towers or other supports, and the power used. In high-power installations, sufficient pickup is insured at many points along either the feed line or the antenna elements, but the problem of securing adequate high-voltage insulation between the sampling system and the antenna system requires consideration. In low-power installations, the desired type of sampling coil may not provide sufficient pickup unless it is coupled to a point near a current maximum, which in some installations might be near the center of the antenna elements, thus presenting a problem in mounting and subsequent adjustment of the coil.

In installations where the tower is fed at its base, sufficient current is usually present to induce ample voltage into the sampling system. A sampling unit which is installed at this point is easily weatherproofed and is accessible for frequent readjustment. However, a variety of stray currents usually are encountered at this point. In some installations, these stray base currents may be appreciable compared to the absolute antenna current of the system. For example, in a vertical half-wave antenna, the capacitance of the tower to ground across the base insulator may draw an rf current comparable in magnitude to that fed into the antenna (Fig. 9-19). Moreover, the stray currents at the base of the tower do not always provide accurate antenna-current indications under a change in operating conditions. A decrease in base-feeder current might be indicated, for example, with an increase in absolute antenna current. Before a sampling installation at the base of a tower or mast is made, it is advisable to determine both the magnitude of the base currents and their relationship to the absolute antenna current under different operating conditions.

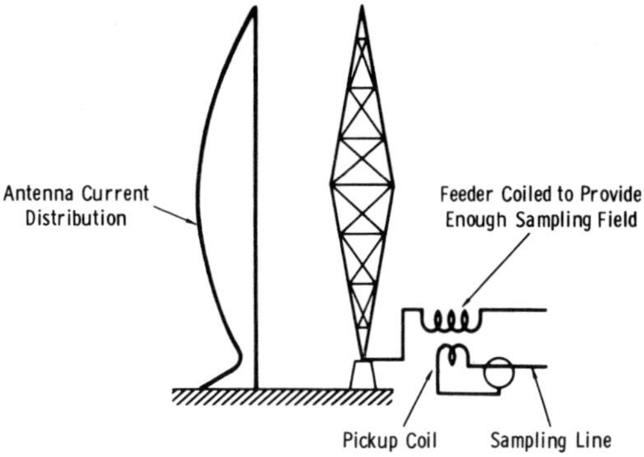

(*A*) *Half-wave antenna.*

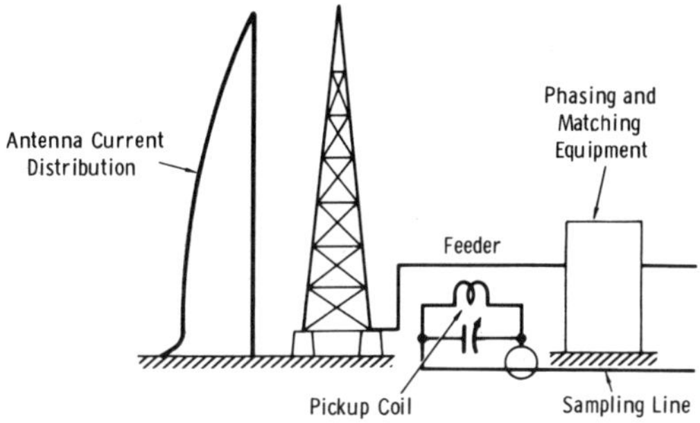

(*B*) *Quarter-wave antenna.*

Fig. 9-19. Antenna current distributions.

A sampling system installed on the tower structure will indicate the effective value of the antenna current. It is often difficult to determine the absolute current in the tower; therefore, direct correlation between the remote meter reading and the true antenna current is difficult to achieve. An installation on the tower structure is advantageous in that ample current is usually available for sampling, and indications are not influenced by base currents if the loop is located several feet above the base. But such a system usually is costly and difficult to install. The sampling line, which can be clamped to the tower, must be brought across the tower base insulators. This requires the introduction of a high-impedance circuit at the base of the tower, which is obtained in practice by forming a sufficient length of the sampling line into a coil that can be tuned by a shunt capacitor to the transmitter frequency (Fig. 9-20A). The tuned circuit must be kept accurately tuned to the transmitter frequency; otherwise the sampling system will disturb the electrical characteristics of the antenna and will produce inaccurate phase and current relationships. In some cases, the sampling line can be spaced from the tower by high-voltage insulators and brought across the base without an isolating network.

To overcome the difficulty of bringing the sampling line across the base of the tower, the sampling coil can be mounted at the top of a special mast erected adjacent to the antenna tower (Fig. 9-20B). If the height of the antenna tower is one-quarter wavelength at the frequency used, the sampling unit at the top of a short mast erected close to the tower will provide ample voltage for antenna current and phase indications. Moreover, these indications will not be influenced by stray currents existing at the base of the tower. The sampling line can be attached to the mast and its outer shield grounded at the base. Care must be exercised in the location of the mast so that stray fields from other adjacent towers do not induce appreciable current in the sampling system.

The principal disadvantages in this system are mechanical. The auxiliary mast must remain rigid under the most adverse weather conditions to pre-

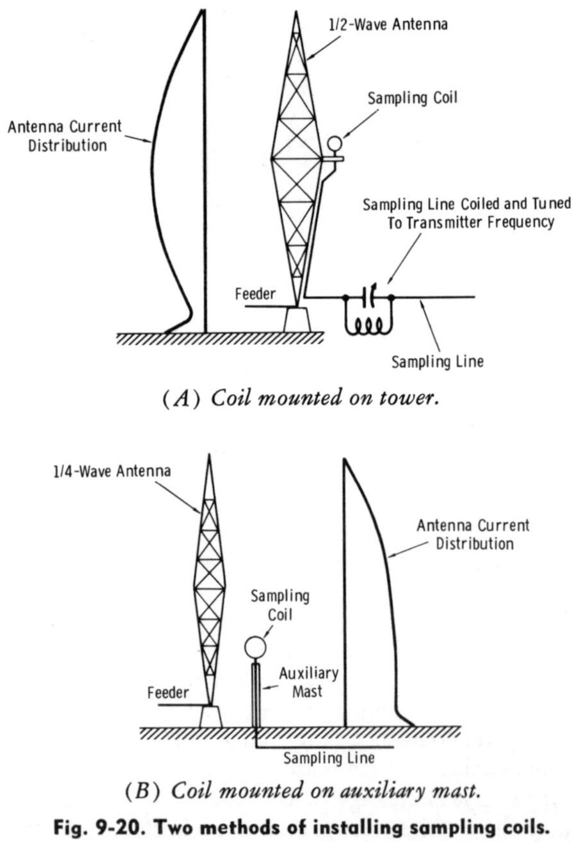

(*A*) *Coil mounted on tower.*

(*B*) *Coil mounted on auxiliary mast.*

Fig. 9-20. Two methods of installing sampling coils.

vent variation in the spacing between the pickup coil and the antenna tower. If the mast is laterally supported by the tower structure, the supports must be adequately insulated from the mast, which might prove too costly in some installations. In order to obtain the desired current in the sampling system, some provision must be made at the top of the mast for adjusting the position of the sampling coil and perhaps its distance from the antenna tower. In either case, where the sampling line is mounted near the antenna structure, the effect of the line on the electrical characteristics of the antenna system must be carefully considered.

Sampling Lines

The current induced in the sampling coil by the field surrounding the antenna or its feeder is fed by concentric transmission line to the remote antenna-current and phase indicators. In practice, the degree of coupling between the coil and antenna or feeder is usually adjusted so that the pointer deflection on the scale of the antenna-current indicator is identical to that of the corresponding antenna ammeter.

The sampling line can be any one of a number of types of concentric lines with surge impedances from approximately 50 to 100 ohms. In general, open-wire lines prove unsatisfactory; if they are used in the vicinity of the antenna, objectionable currents will be induced in the lines. Beaded coaxial line is an entirely satisfactory type for sampling. This type of line can be obtained with surge impedances ranging from 70 to 150 ohms. Its construction provides an efficient, low-loss transfer of energy and makes it suitable for long periods of outdoor use. An ideal sampling-line installation would consist of beaded coaxial line installed in gas-filled copper tubing. Such a line could be depended on to give reliable service for long periods of time. Solid-dielectric coaxial lines which should give long trouble-free service have been developed. They do not require pressurizing. So that a phase shift is not introduced in the sampling system, particularly if a phase meter is used, each sampling line must be terminated in its characteristic impedance.

9-13. COAXIAL TRANSMISSION LINES (AM-FM)

The transmission line connects the transmitter output to the antenna or antenna tuning unit. Details of electrical design characteristics are well covered in existing literature and are not important to this text. The important features to be considered are mechanical construction as it is related to installation problems and the proper choice of line to meet the requirements for power-handling capacity, allowable power loss, proper impedance match, and economic factors. The differences in lines suitable for a-m and fm service will be discussed.

The coaxial line uses inner and outer conductors made of rigid or semi-flexible copper tubing, with either air as the dielectric or a solid dielectric of polyethylene. Solid-dielectric cables, although they have the advantages of flexibility and comparative ease of installation, are less efficient in the transmission of rf energy and have less power-carrying capacity.

In the construction of a coaxial line, cross-pin insulators are used to support and space the inner conductor. Insulators of steatite are spaced throughout the length of the line at intervals depending on design. Close spacings help to make the line more uniform and are desirable for constant impedance characteristics. Longer spacings, however, minimize insulator loss and give a higher relative efficiency. In practice, the insulator loss is minimized by using insulators as small as possible for the required mechanical strength. At standard broadcast frequencies, the insulator spacing is not extremely important and is usually dictated by mechanical considerations alone. At the higher frequencies used by fm broadcasters, the spacing is usually made such that the impedance at the highest operating frequencies varies no more than 3 percent from the lower-frequency (standard broadcast band) values. This practice has resulted in an average spacing of 12 inches for fm lines.

Impedance Values of Coaxial Lines

Currently, 72-ohm coaxial lines are most common for standard broadcast transmitters, and 51.5-ohm lines are most common for fm transmitters. This holds true whether air-dielectric or solid-dielectric cables are used. By means of suitable connectors, it is possible and practical in many cases to connect the rigid type of line to a solid type of cable without a matching section. NOTE: Many of the latest a-m transmitters are designed to feed 51.5-ohm transmission lines.

Power-Handling Capacity

The maximum power that a coaxial cable can carry is limited primarily by the temperature rise in the line when the rf is applied. At standard broadcast frequencies, voltage breakdown is the limiting factor. Above approximately 50 MHz, a maximum temperature rise of 40°C in the outer conductor is considered safe. Aside from the actual rf power applied, various conditions may affect the temperature rise in the line. The amount of ventilation, the nature of any type of enclosure, and, most important, the standing-wave ratio are all contributing factors.

In practice, it is customary to consider the probable standing-wave ratio to be encountered (which is considered later) and then divide the maximum power rating of the line by this ratio. For example, if the standing-wave ratio is 1.5 and a line rated at 5000 watts is considered, the maximum power recommended to be fed into the line is 5000/1.5, or 3333 watts.

The four standard diameters of rigid (air-dielectric) coaxial lines are as follows: 7/8, 1 5/8, 3 1/8, and 6 1/8 inches. Most manufacturers design their equipment to accommodate these standard sizes only. In addition, there is a 3/8-inch line used in some very low-power transmitter installations.

Choosing a Coaxial Line

The choice of the impedance value of the line is, of course, dictated by the application. Choose the impedance value specified by the manufacturer.

The choice of the size of the line depends on the maximum power to be applied and the maximum allowable attenuation. Table 9-5 gives the average maximum power ratings, based on possible flashovers, for the standard broadcast band. It should be remembered that for a-m service the maximum power occurs at 100-percent modulation and is more than the rated power output of the transmitter. Except when lines are very short, they should not be operated at the maximum power rating.

Table 9-5. Typical Maximum-Power Ratings of Air-Dielectric Coaxial Lines

Size (Inches)	Average Maximum Power Ratings (Watts)
3/8	500
7/8	3000
1 5/8	12,000
3 1/8	50,000
6 1/8	150,000

The choice of a coaxial cable for fm broadcast services is somewhat more involved. The procedure may be outlined briefly as follows:

1. To find the power required at the antenna terminals, divide the erp (effective radiated power) by the power gain of the antenna to be used.
2. To find the minimum allowable transmission-line efficiency, divide the antenna input power by the maximum power output of the transmitter.
3. Choose a line from one of the standard sizes that has an efficiency equal to or greater than the minimum allowable value just obtained (Fig. 9-21).
4. To check the necessary transmitter output power, divide the antenna input power by the efficiency of the line.
5. Check the power rating of the cable. If this rating is less than the necessary transmitter output power, a larger line must be used.

As a practical illustration of this procedure, assume that an erp of 33 kW is required. A 10-kW transmitter will be installed with a 4-bay fm antenna that has a power gain of 3.7. A 200-foot line will be necessary. These values are used in the preceding steps as follows:

1. The erp of 33 kW divided by the antenna power gain of $3.7 = 8920$ watts required at the antenna terminals.

2. 8920 watts divided by 10,000 watts = 0.892, or 89 percent.
3. Inspection of Fig. 9-21 shows that the smallest line suitable for a length of 200 feet has a 1⅝-inch diameter.
4. 8920 watts divided by 0.89 = 10 kW.
5. Table 9-6 shows that the 1⅝-inch line is rated at 10 kW at unity power factor. It was previously shown that to be perfectly safe this power rating must be divided by the probable voltage standing-wave ratio. The antenna in this example has a vswr of less than 1.2, but it is always good practice to assume the maximum allowable vswr of 1.75 in the computations. The average power rating of the 1⅝-inch line with 1.75 vswr is only 5700 watts. Therefore, we must look to the next larger line, 3⅛ inches, which is observed to have a power

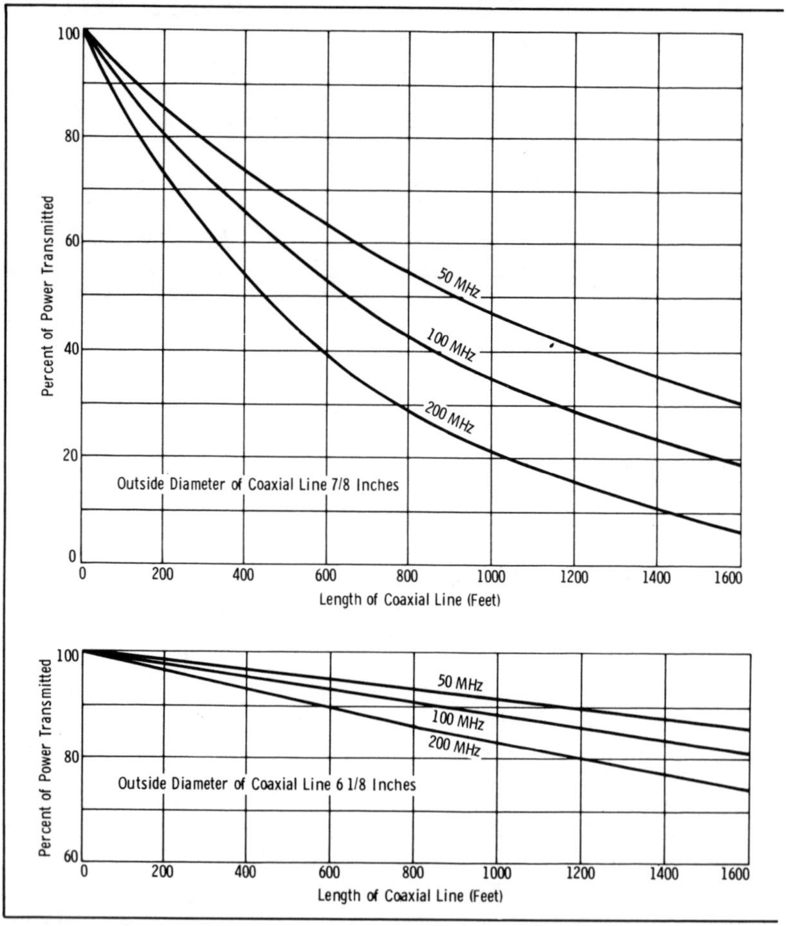

Fig. 9-21. Percent of power

capacity of 24 kW at a vswr of 1.75. The 3⅛-inch line, then, would be the correct choice in this example. The higher efficiency of this line must then be considered in the computations, and it will be found that a lower transmitter output will be required.

Installation of Coaxial Lines

Semiflexible, soft-copper lines of the ⅜-inch size usually are manufactured in 100-foot lengths which are spliced together by silver brazing to obtain a gas-tight and electrically secure connection. They are crated in coils and are usually cut to the exact length specified. This type of line is often used in larger stations for rf sampling lines such as those used with remote antenna-current meters and phase monitors.

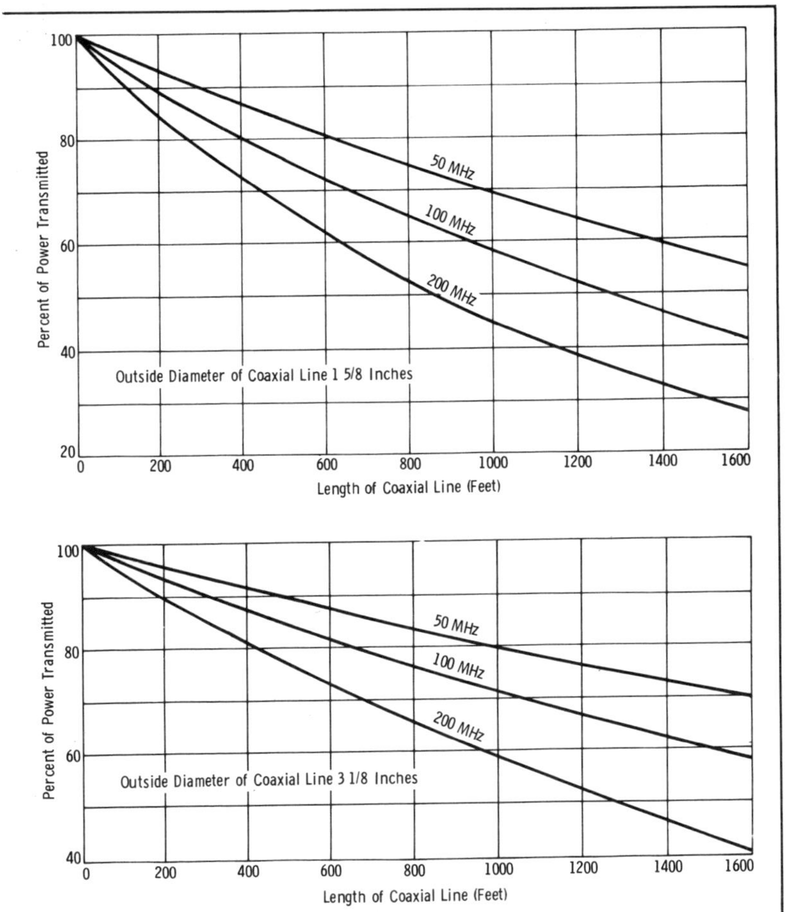

transmitted by coaxial cables.

Table 9-6. Typical Power Ratings of Standard-Size Lines

Size (Inches)	Maximum Power in FM Broadcast Band (Unity Power Factor) (Watts)	Maximum Power with 1.75 Voltage Standing-Wave Ratio (Watts)
3/8	1370	772
7/8	3000	1710
1 5/8	10,000	5700
3 1/8	42,000	24,000
6 1/8	166,000	95,000

Rigid lines of hard-temper copper tubing are normally shipped in 20-foot straight lengths. Flanges are usually silver-soldered on both ends of each length at the factory. The flanges have bolt holes so that the lengths of line may be bolted together. Grooves in the faces of the flanges accommodate a round gasket (known as an *O ring*) which makes the connection gas-tight and weatherproof. The inner conductors are connected by a slotted *bullet* spring, which provides a solderless connection necessary for field installation.

When it is necessary to change the direction of line runs, special elbows of 45° or 90° are used. The flange that provides the O-ring groove is silver-soldered to the bend in the usual manner, and the second ring containing the bolt holes is assembled loosely on the bend. This provides a means of rotating the flange so that proper hole alignment may be obtained to accommodate any orientation of the elbow without drilling bolt holes.

It is sometimes necessary to cut a 20-foot section of line in the field. When this must be done, a device known as a *clamp connector* may be used to provide a solderless fitting which substitutes for the flange that was cut away. When cutting is necessary, it is preferable to make the cut at the midpoint between the insulators, since cutting the line at any other point will disturb the impedance of the line. This is very important in fm-line installations where the vswr must be kept as low as possible. If a cut is necessary near the transmitter rather than near the antenna end, the effect of the cut is not quite as important. The midpoints between insulators usually are marked by bands on the outside of the line.

Providing for Expansion

Soft-temper coaxial lines in the 3/8- or 7/8-inch size rarely need any special provision for expansion due to temperature changes, since they are somewhat flexible by nature. Rigid lines, however, must have some means of allowing for expansion and contraction. It has been found from experience that in severe climates a temperature range from winter cold to summer heat will cause an effective variation in line length of about 1 1/4 inches per 100 feet. Thus, for example, on a 400-foot run, provisions must be made for a variation of about 5 inches in length.

There are two general types of expansion, one which must be considered in the design and construction of the line itself, and one for which the installation engineer must provide. These two types are known as *differential expansion* and *overall expansion.*

Differential expansion results from unequal temperatures of the inner and outer conductors. Under normal operating conditions, the inner conductor develops a temperature rise with respect to the temperature of the outer conductor. This causes an expansion of the inner conductor relative to the outer conductor. Another factor causing differential expansion is that it takes considerably longer for the inner conductor to follow the variations of temperature in the outer conductor due to sudden changes in weather conditions. For example, the outer conductor may be cooled suddenly from 100° to 70°F during a summer shower. It will take longer for the temperature of the inner conductor to change the same amount.

This type of expansion is provided for by making the inner conductor slightly shorter than the outer conductor. Thus when a spring bullet inner connector is used, one end will not fit up tightly against the shoulder, providing for relative changes in the lengths of the inner and outer conductors.

The overall expansion must be provided for when the line is installed. Long horizontal runs must be given some means of relief from such expansion, and when the line runs up a steel tower to feed an fm antenna, consideration must be given to the fact that the expansion coefficient of copper can be as much as 50 percent greater than that of steel. The two metals cannot, therefore, be connected inflexibly together. It is said that several tons of pressure may be developed when no provision for expansion is made, causing failure of transmission-line couplings, supports, or even the tower.

There are several alternative methods of providing for overall expansion. When the line is located so that it may change directions several times, expansion is allowed by the 45° or 90° elbows. Support standards using roller supports will allow a longitudinal motion of the line. A combination of these two provisions usually allows a sufficient safety factor for the expansion of the line and is widely used for long horizontal runs to the towers.

These methods, however, are not entirely practical on vertical runs up a tower such as those for an fm installation. For such applications, various types of expansion joints have been developed. One type consists of telescoping outer and inner conductors with a sliding gas-tight seal. This joint is capable of a 4-inch line displacement, which is entirely adequate for a 200-foot line used in an area having temperature variations over a range of −50°F to 150°F. Such expansion joints are often used on horizontal runs as well as vertical runs.

The entire line-support system consists of rigid mounting clamps, support brackets or railings allowing axial movement, and sometimes one or more expansion joints. On long vertical runs of more than 150 feet, expansion joints should be considered a prime requirement. On vertical runs of this type, the line is assembled starting at the bottom, the bottom section being rigidly attached to the tower. Additional line sections are supported at approximately 15-foot intervals, preferably with sliding support brackets up to a distance of 200 feet of line. At this point, an expansion joint should be installed, with rigid clamping used on the next 20-foot section. The process is then repeated for each 200 feet of line.

Support brackets may be obtained to fit practically any mounting problem encountered in the field. When it is necessary to insulate the transmission line from the tower, insulated support brackets are used.

A word of caution should be given. When it is necessary to insulate a transmission line from the tower, the line should be sufficiently separated from the climbing ladder so that a man climbing the tower cannot touch the line. Such a contact could cause severe rf burns. Towers not incorporating a climbing ladder are particularly hazardous to the climber under these circumstances, and he should be cautioned about the danger.

Coaxial-Line Procedures for AM-FM Towers

When an fm antenna must be mounted on an a-m tower, it is necessary to mount the transmission line so that the a-m electrical characteristics of the tower remain undisturbed. Two practical methods for accomplishing this are the insulated-line method and the *bazooka* construction.

In either of the methods to be described, it is necessary to provide isolation between the two rf voltages by causing a high impedance to exist between the base of the a-m tower and the outside surface of the fm transmission line. In the first method, the fm line is insulated from the a-m tower by means of insulating brackets for one-quarter wavelength (at the a-m frequency) starting at the exact base of the tower. The outer conductor is then shorted to the tower at a point about one-quarter wavelength (at the a-m frequency) up from the base. Along the insulated portion of the line, the tower itself is considered to form the outer conductor. In practice, the shorting point is determined by trial and is only approximate. The optimum point is the point at which minimum detuning of the a-m tower is obtained. The arrangement is equivalent to a transmission line that is one-quarter wavelength long with the end shorted, and a high impedance exists across the tower base as far as the a-m power is concerned. Above the shorting point, the fm line is attached directly to the tower.

In actual installation practice, it is advisable to short the line slightly less than one-quarter wavelength from the tower base and connect a variable capacitor across the open end at the base (from outer conductor of line to tower). An exact adjustment may then be made. The capacitor should have a reactance of 400 to 1000 ohms at the a-m frequency and should have an adequate voltage rating.

The bazooka method differs from the preceding method only in physical construction and retains the electrical theory of isolation by means of quarter-wave lines. A bazooka is a quarter-wave line (at the a-m frequency) assembled along the ground from the base of the tower. It actually consists of a one-quarter wavelength shield installed around a transmission line (Fig. 9-22). The line is insulated from the shield, and the shield is grounded to eliminate radiation at the a-m frequency. The fm line runs through the shield as an inner conductor. The line is connected to the tower throughout the entire vertical run, eliminating the need for insulating the line from the tower. The outer shield may be either a metal hood or a wire cage consisting of six wires supported around the line by a metal ring with the wires running through holes around it.

As in the first method, the bazooka section may be cut shorter than an actual quarter wavelength and a variable capacitor shunted across the open end to achieve exact resonance. The reactive and voltage values are the same as in the first case.

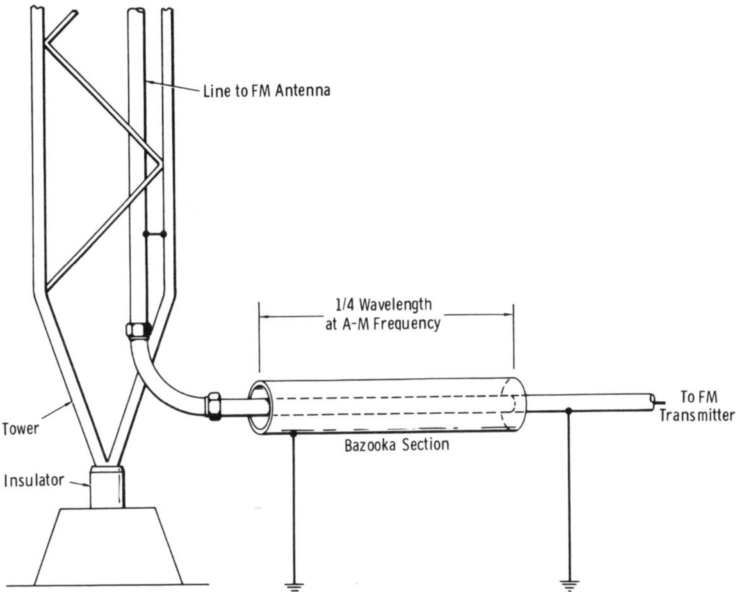

Fig. 9-22. Bazooka used in attaching fm line to a-m tower.

Gassing Provisions

In air-dielectric coaxial lines, it is advisable to introduce dry air or nitrogen into the line to prevent moisture condensation. This is done unless the line installed is very large for the applied rf power.

When gassing is used, it also is necessary to bleed the line by using gas-opening flanges at intervals on extremely long lines or at the antenna end. This means that any moisture which collects may be removed by allowing the dry air or nitrogen to blow through the line, evaporating the moisture. When lines are run to the top of a tower for an fm antenna, a line is run back down from the gas seal at the antenna to the tower base so that the gas-release valve may be operated there. The gas pressure is not critical, and in practice it may vary from about 1 to 20 pounds per square inch.

Solid-Dielectric Coaxial Line

The advantages of solid-dielectric coaxial line are maximum flexibility, comparatively low initial and installation costs, no gassing requirement, and fewer mounting problems since expansion is not a consideration. The disadvantages are greater attenuation per foot (due to the fact that up to a certain point no further increase in diameter will lower the insulation loss) and less permanency in use when the line is exposed for long periods of time to outdoor weather.

Solid-dielectric lines are composed of a highly resistant vinyl outer jacket and an inner conductor of stranded wire (size determined by type number) separated from an outer conductor of shielded braid by a polyethylene dielectric. The spacing is maintained within close tolerances. Such lines are often used in low-power a-m and fm installations where the run may be less than 100 feet and the line is adequately shielded from extreme weather conditions. They are also popularly used for monitoring lines, such as remote antenna-monitor lines.

The technical specifications of the most popular sizes of solid-dielectric coaxial lines used in a-m and low-power fm transmitter installations are listed in Table 9-7. With regard to the maximum power rating, always remember to divide this rating by the maximum allowable vswr (2 for a-m

Table 9-7. Solid-Dielectric Coaxial Cable Data

Type No.	Actual Z_0	Maximum Operating Volts rms	dB Loss per 100 ft Average for Operating Frequencies	Maximum Total Watts Input for Operating Frequencies
70-Ohm Cables (Usually Used for AM)				
RG 11 U	75	4000	0.115	1800
RG 34 U	71	5200	0.115	4500
RG 59 U	73	2300	0.260	860
Amphenol 21-125	72	10,000	0.020	8000
52-Ohm Cables (Usually Used for FM)				
RG 8 U	52	4000	2.10	800
RG 17 U	52	11,000	0.850	3150
RG 19 U	52	14,000	0.700	4700
RG 58 A	52	1900	4.10	210

and 1.75 for fm). Although the actual vswr should be lower than these values, this procedure allows a safety factor.

9-14. FUNDAMENTALS OF A-M TRANSMITTERS

A block diagram of a typical 1000-watt a-m transmitter is shown in Fig. 9-23. The functions of each block are as follows:

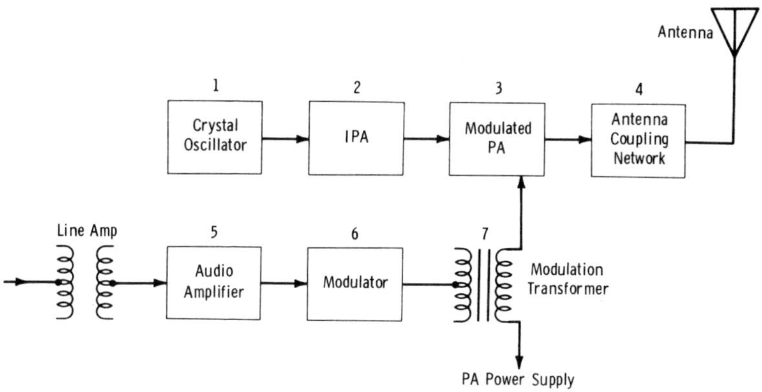

Fig. 9-23. Block diagram of typical 1-kW a-m transmitter.

1. The crystal oscillator generates the carrier frequency. Usually two separate but identical circuits are employed; the extra may be switched in if failure of the other occurs.
2. The intermediate power amplifier (IPA) amplifies the low output from the oscillator stage up to a level sufficient to drive the power amplifier.
3. The power amplifier provides the power to the antenna system. This stage is normally the one which is amplitude-modulated by the audio signal (described under 7).
4. The antenna coupling network provides the proper load for the power amplifier and the proper driving load to the antenna. In the case of directional antenna systems, the power is divided and phased.
5. In the audio amplifier, audio signals from the studio line output are raised to a level sufficient to drive the modulator stage.
6. The modulator further amplifies the signal to a level sufficient for driving the power amplifier to 100-percent modulation.
7. The modulation-transformer secondary is in series with the power-amplifier plate-voltage feed so that the power-amplifier output amplitude is varied in accordance with the program content.

A typical 1000/250-watt a-m transmitter is shown in the normal operating condition with the front door closed in Fig. 9-24A. Behind the front door is a full-length perforated grille, interlocked for personnel protection but affording full view of the components with the transmitter in operation. This shield may be removed in seconds by means of snap locks (Fig. 9-24B).

Many of the most recent a-m transmitters employ all solid-state circuitry in the low-power stages. In transmitters with greater than 5-kW final power output, the intermediate stages, final power amplifier, and modulator still employ vacuum-tube circuitry to meet power requirements.

The block diagram of Fig. 9-23 shows the final amplifier as the modulated stage. When the final rf stage is the one modulated, the process is termed *high-level modulation.* When an intermediate stage is modulated, followed by linear rf amplifiers, the process is termed *intermediate modulation. Low-level modulation* is the term applied when a very low-level stage just following the oscillator is modulated.

Fig. 9-25 illustrates the appearance of the modulator and final tube stages of a 50-kW transmitter. The large radiating fins on the tubes are physically

(*A*) *Operating condition.*

Fig. 9-24. A 1000/250-watt

connected to the tube anode, and forced air is ducted through these fins to dissipate the heat.

Oscillator and RF Intermediate Stages

Amplitude-modulated transmitters depend on a crystal-controlled oscillator for frequency stability. The crystal is normally operated within a constant-temperature oven to avoid any possible drift caused by changes in ambient temperature.

Intermediate stages prior to modulation are tuned, narrow-band class-C stages for maximum efficiency and minimum harmonic content. Following modulation, all stages are linear rf power amplifiers.

Plate Modulation

Fig. 9-26 shows the basic principle of plate modulation. The rf stage is assumed to have 3000 volts applied to the plate and to have sufficient drive that conduction occurs on the tips of the applied rf signal. Thus plate-cur-

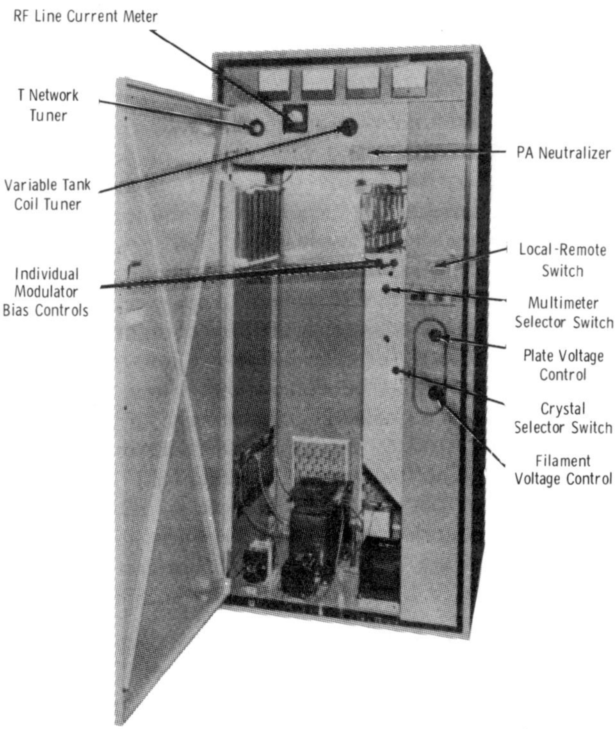

(B) Screen removed.

Courtesy Gates Division Harris-Intertype Corp.

a-m broadcast transmitter.

rent pulses occur as shown in the diagram; this is typical class-C amplifier operation.

From time T1 to time T2, no audio signal is applied, and the plate current depends on the plate impedance determined by the loading of the output tank to the antenna transmission line or another stage. The tank circuit, resonant at the operating frequency, converts the plate pulses to a signal that swings on both sides of zero (due to the flywheel effect of charging and discharging the LC combination). The voltage developed in the tank is dependent on the LC ratio and the Q (ratio of reactance to resistance) of the resonant circuit.

At time T2, a positive-going sine-wave voltage is applied from the modulator. This audio-frequency voltage is added to the dc voltage of 3 kV. If the audio voltage peaks at 3 kV, the total voltage applied is 6 kV, doubling the amplitude of the rf plate-current pulse. At the positive peak of the audio signal, the carrier has twice the amplitude of the unmodulated value.

Courtesy RCA

Fig. 9-25. Modulator and final stages of RCA BTA-50F (50 kW) a-m broadcast transmitter.

When the applied audio reaches the negative peak (T3), the 3 kV of audio subtracts from the 3-kV dc voltage applied, resulting in zero plate voltage and zero current. Thus for 100-percent modulation, the carrier swings from zero to twice the unmodulated amplitude.

The plate-current meter measures the average current being drawn by the modulated tube. This reading remains constant for linear modulation (equal positive and negative peaks). At resonance, the tank circuit presents a purely resistive load. Thus, if the unmodulated current in V1 is 500 mA, this meter will indicate the same 500 mA under modulation. If an ammeter is inserted in any of the coupled antenna circuitry, the variation of rf energy with modulation will be indicated.

This is illustrated by Fig. 9-27, which shows the relation between antenna current and percent modulation. Thus if 100-percent modulation occurs, the increase in antenna current is 22.5 percent over the unmodulated value.

To make this relationship clear, bear in mind that when a carrier is modulated 100 percent the total radiated power is 1.5 times the power in the

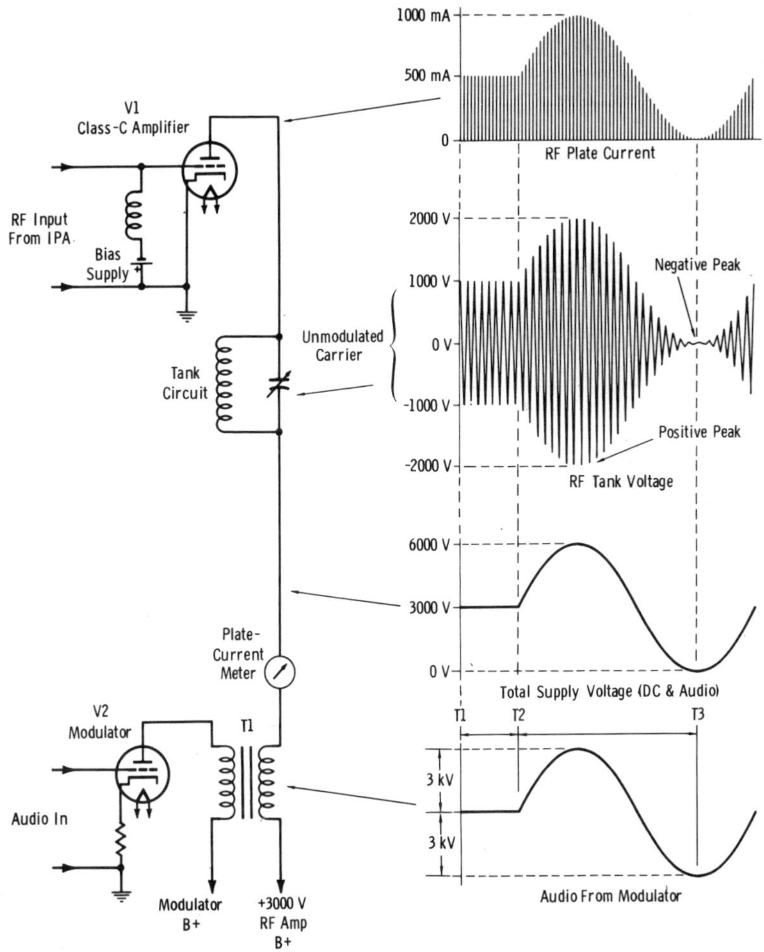

Fig. 9-26. Principle of plate modulation.

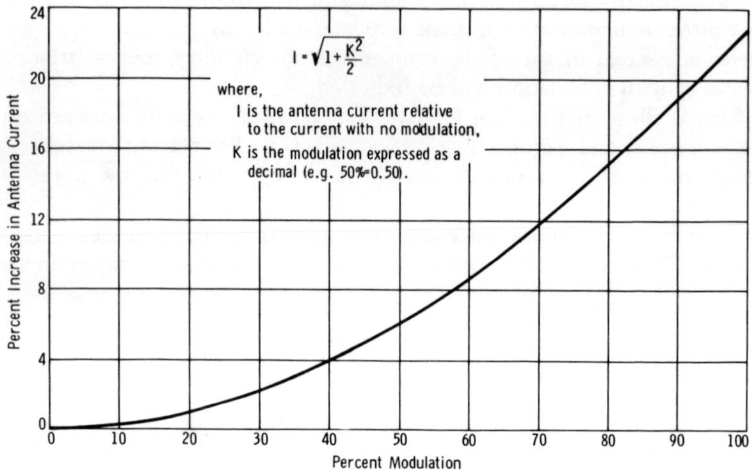

Fig. 9-27. Relation between antenna current and percentage modulation.

unmodulated carrier. (This is developed in section 9-15.) Now assume an ammeter in the antenna circuit is measuring the rf current in a 50-ohm line with a 100-watt unmodulated carrier.

$$\begin{aligned} I &= \sqrt{\frac{W}{R}} \\ &= \sqrt{\frac{100}{50}} \\ &= \sqrt{2} = 1.414 \text{ amperes} \end{aligned}$$

When this 100-watt carrier is modulated 100 percent, the power is 150 watts, and the antenna current becomes:

$$\begin{aligned} I &= \sqrt{\frac{150}{50}} \\ &= \sqrt{3} = 1.732 \text{ amperes} \end{aligned}$$

Thus a 0.318-ampere increase in antenna current has occurred, and this is equal to a 22.5-percent increase over the unmodulated value of 1.414 amperes.

Sometimes the rf modulated stage employs a pentode tube. Since the screen voltage also affects the plate current, modulation of the plate alone does not normally allow linear modulation. Therefore, the screen voltage is usually varied in ratio to the plate voltage, resulting in a combination of plate and screen modulation (Fig. 9-28). Resistor R is the screen dropping resistor, which establishes the proper screen-to-plate voltage ratio. Capaci-

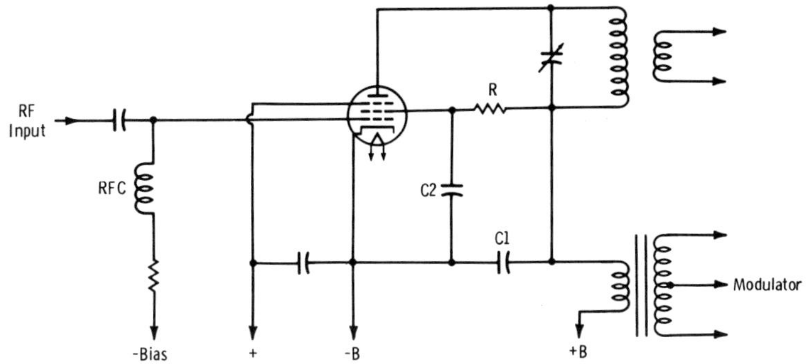

Fig. 9-28. Plate-screen modulation of rf pentode.

tor C1 presents a high reactance at the highest audio frequency but very low reactance at the radio frequency. The screen bypass capacitor (C2) is normally smaller than 0.002 μF.

Control-Grid Modulation

When modulation occurs in a low-level or intermediate-level stage, grid modulation is sometimes employed. A conventional class-C amplifier is used with the secondary of the modulation transformer in series with the grid-bias supply rather than the plate supply. The carrier amplitude applied to a grid-modulated stage is less than that normally used for plate modulation, to allow the positive peaks of the carrier wave to shift upward on the transfer curve during the positive alternation of the modulating signal. Thus, when the carrier is not modulated, the peak amplitude of the resulting plate-current pulses is much lower than for corresponding plate-modulation techniques, resulting in less efficiency of operation. This usually is not important, since the necessary number of linear rf stages must be used to obtain the required output power. Grid modulation requires very little power from the modulator.

Diode Modulation

A diode modulator is often used in solid-state exciters where modulation occurs at a low level. See Fig. 9-29. A sufficient amplitude of rf is applied to diode X1 to cause conduction on the positive peaks. The modulating audio voltage is applied across R1 and therefore to the same terminal of X1. Capacitor C3 serves as an rf bypass for the audio input.

Due to the high impedance presented by the tank circuit, the diode is cut off until the applied rf reaches a given amplitude. Thus the output of the diode appears as a series of pulses just as in the class-C modulated plate circuit of Fig. 9-26. The applied audio signal then adds to and subtracts from the rf carrier, and the rf pulses are converted to the normal envelope by the tank circuit. For good linear modulation, the ratio of rf to audio

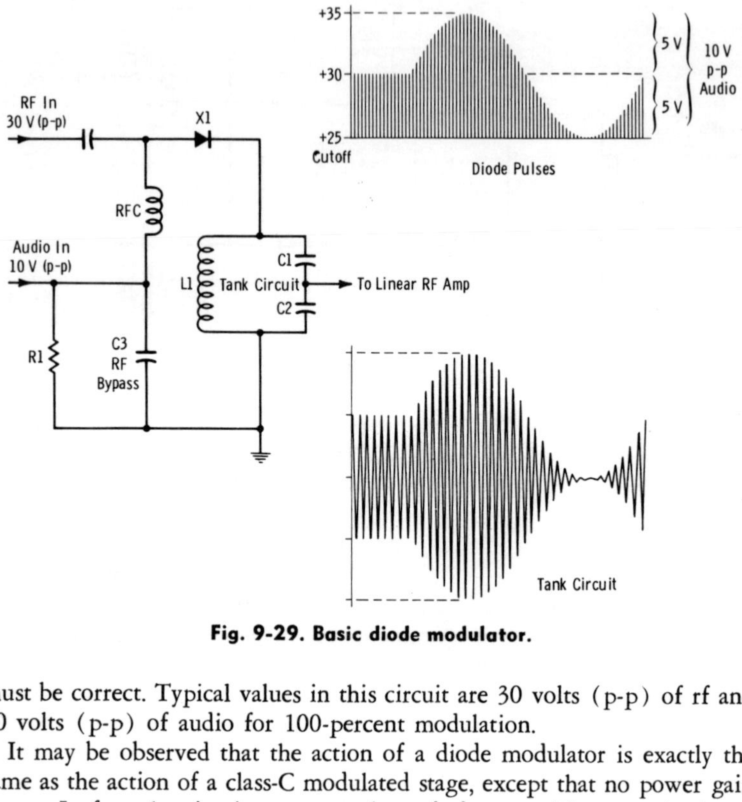

Fig. 9-29. Basic diode modulator.

must be correct. Typical values in this circuit are 30 volts (p-p) of rf and 10 volts (p-p) of audio for 100-percent modulation.

It may be observed that the action of a diode modulator is exactly the same as the action of a class-C modulated stage, except that no power gain occurs. In fact, the circuit presents a loss of rf power. The stage is always followed by a series of class-AB or -B linear rf amplifiers to achieve the desired power output.

Phase-to-Amplitude Modulation

Phase-to-amplitude modulation, sometimes termed "outphasing modulation," provides high transmitter efficiency and stability, minimum effect from changes in the characteristics of high-power tubes, and elimination of modulation transformers and reactors. The latest RCA *Ampliphase* transmitter (RCA's terminology for phase-to-amplitude modulation) uses no iron-core transformers anywhere in the audio system.

Fig. 9-30 is a simplified diagram of one model of RCA Ampliphase exciter unit. The crystal oscillator is followed by a buffer stage which feeds a center-tapped tank circuit (L1). Since the center tap is grounded, the feeds to the identical rf channels are 180° out of phase. If these signals were recombined, zero output would result. For this reason, "dc modulators" (V3 and V4) provide an appropriate phase relationship and an output of some desired value of carrier voltage.

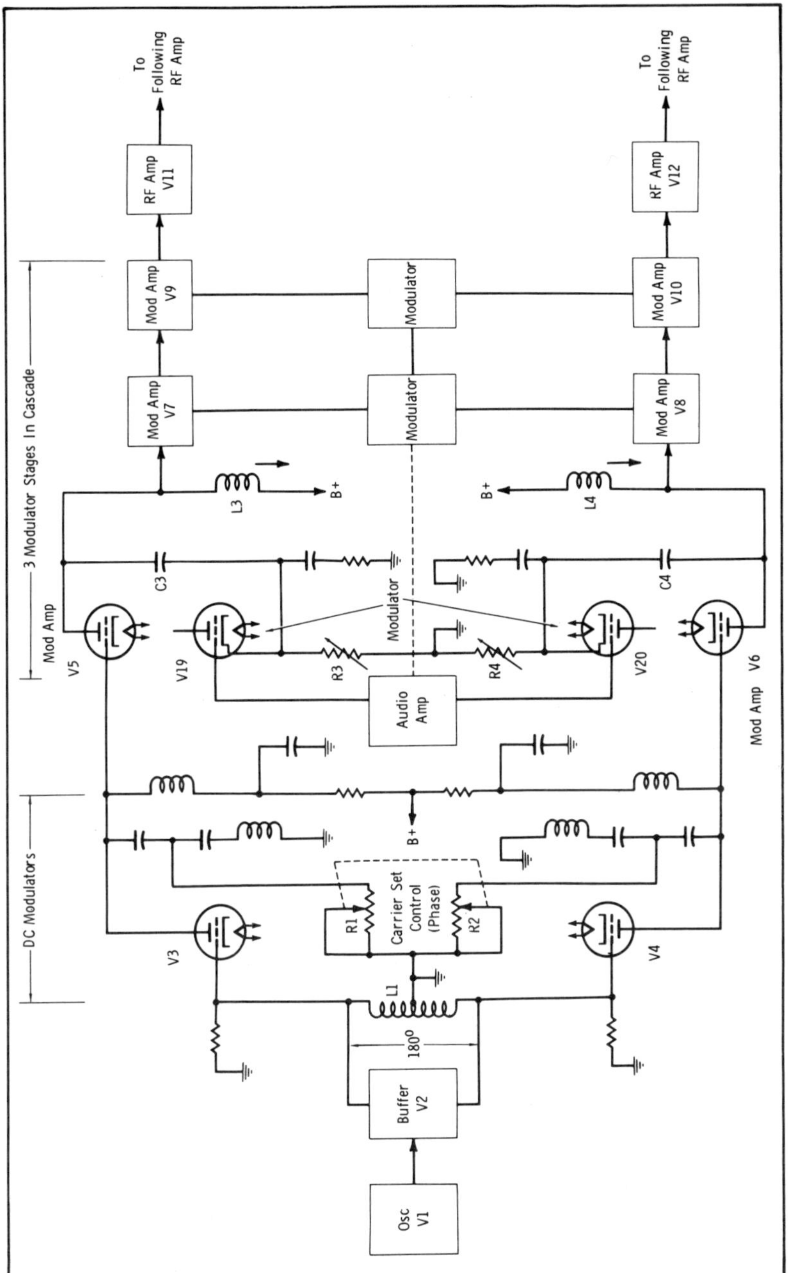

Fig. 9-30. Simplified diagram of exciter unit in RCA Ampliphase transmitter.

This initial phase relationship is obtained by adjusting R1 and R2 (Figs. 9-30 and 9-31) for a 135° phase relationship. These variable resistors in conjunction with the capacitive reactances of the plate circuits allow precise adjustment of the phase between the two identical rf transmitter paths. In this manner, the 180° signals (vectors A and B of Fig. 9-31) are converted to a fixed (unmodulated) carrier voltage (vector C). Low-Q plate tank circuits are used, and the values of L and C are selected so that variation of R1 or R2 affects only the phase and not the tuning or impedance.

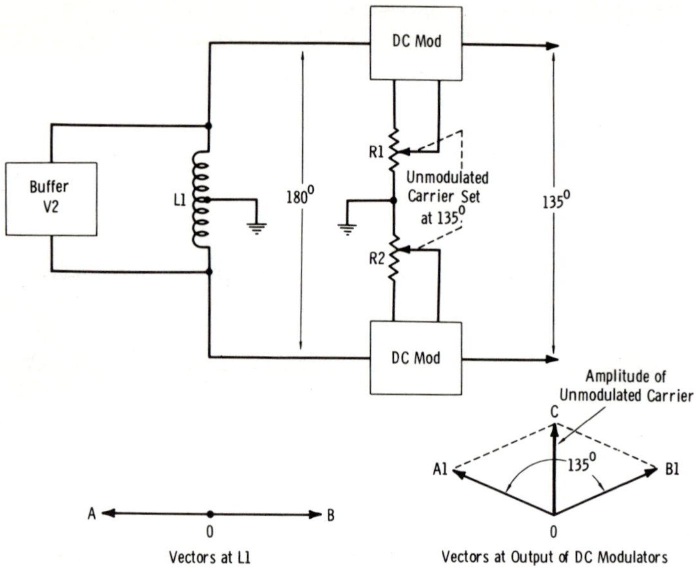

Fig. 9-31. Principle of carrier phase set.

The outputs of the dc modulators feed the first of three cascaded modulator stages. These circuits act in a manner similar to the dc modulators, except that the variable resistance (R3 and R4, Fig. 9-30) is the cathode impedance of an audio cathode follower (V19 and V20). This variable resistance is in series with a capacitive reactance (C3 and C4) in a parallel-resonant circuit, in which L3 and L4 have twice the reactance of the corresponding capacitor. The resulting phase modulation at an audio rate is highly linear because of the linear change in cathode-follower transconductance, with no effect on the magnitude of the impedance seen by the plate of the rf tube.

It will be shown below that a magnitude of phase modulation of ±22.5° for each carrier is required at the output network to represent 100 percent modulation. This is achieved by allowing ±7.5° variation in each of the three modulator stages to result in the desired ±22.5° phase shift.

NOTE: In the earliest RCA Ampliphase transmitters, the carrier frequency generated by the crystal oscillator was divided by three, and a frequency-tripler stage was used following a single phase-modulated stage.

Fig. 9-32A presents a simplified overall block diagram of a typical 50-kW RCA Ampliphase transmitter. The outputs of the modulated stages are fed to a series of conventional broadband amplifiers to achieve 25 kW on each side; the 25-kW outputs combine to result in 50 kW in the combining network. Remember that all rf stages following a modulated stage must present the same impedance from the carrier frequency to the maximum sideband frequency. For example, a response to 10 kHz means the amplifier bandwidth must be $\pm$10 kHz, or 20 kHz total.

Each output has its own tank circuit, which is a conventional 90° pi type as shown. The output of the pi network employs a common capacitor for both sides. The input capacitor of each network is adjusted to provide a nonreactive load for the tube, with the characteristic impedance to convert the load resistance to the value required for the tube.

Note the vectors for 100-percent modulation in Fig. 9-32B. The unmodulated carrier is represented as OC1, resulting from the vector addition of A1 and B1 (135° phased). If the two components are changed to A2 and B2 (90° phased), the carrier is OC2, or twice the amplitude of the unmodulated carrier. This represents the carrier amplitude at 100-percent positive modulation. If the components are changed to A and B, the carrier amplitude reduces to zero, corresponding to 100-percent negative modulation. Thus a total phase excursion of $\pm$22.5° from the unmodulated carrier phase (on each of the two carriers) results in full modulation.

9-15. CHARACTERISTICS OF THE MODULATED ENVELOPE

Fig. 9-33 shows oscilloscope presentations of an amplitude-modulated rf envelope. Considerable important information for the engineer is contained in a detailed analysis of this envelope. A carrier wave modulated by voice or music signals contains its information in the resulting upper and lower sidebands, which consist of continuously fluctuating individual frequencies. Thus speech or music modulation results in an envelope that is extremely difficult to analyze. In this section, only single-frequency sine-wave modulation (as shown by Fig. 9-33) will be considered for the purpose of steady-state analysis. Chapters 11, 13, and 14 will treat the special and practical problems of operations and maintenance associated with actual program transmission.

Modulation Factor and Percent Modulation

To understand degrees of modulation less than 100 percent, we use a *modulation factor* (M) to indicate relative magnitudes of rf carrier and audio modulating signal. This modulation factor is:

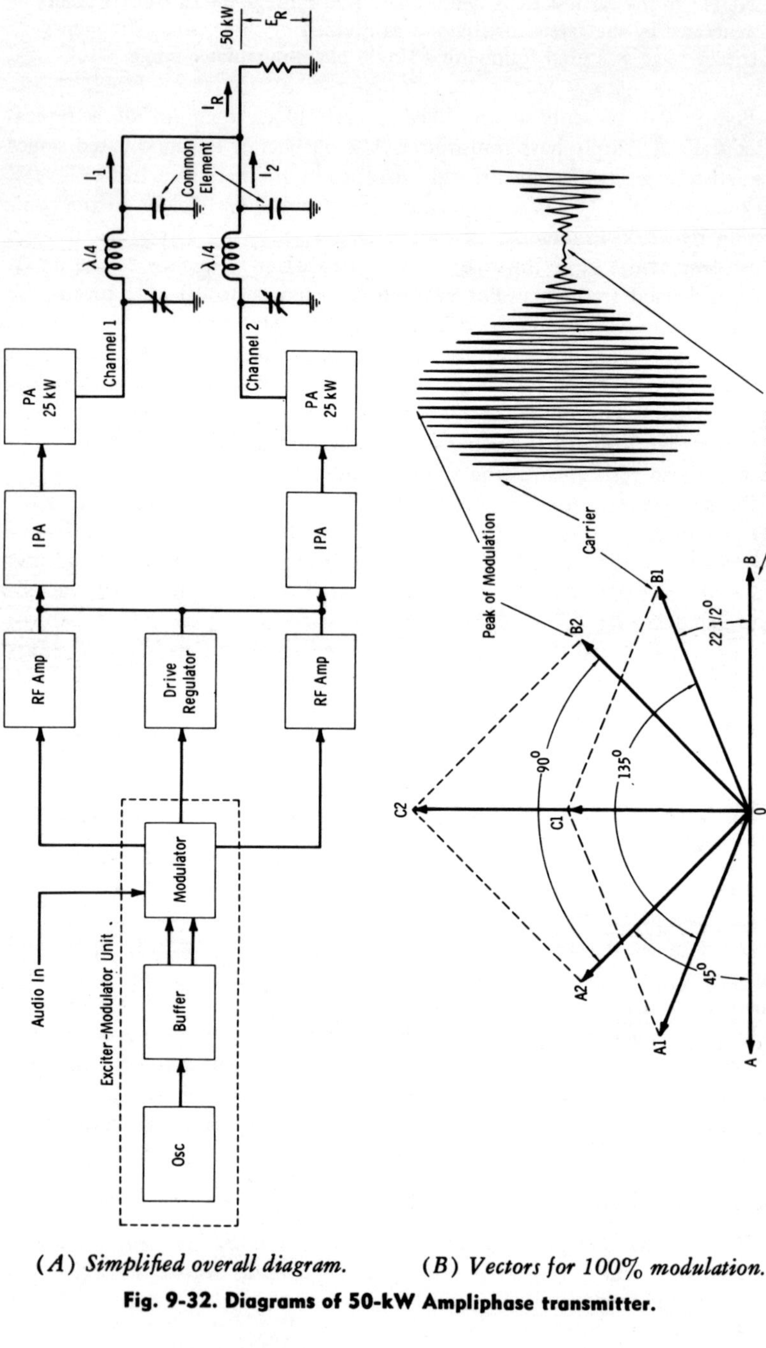

(*A*) *Simplified overall diagram.* (*B*) *Vectors for 100% modulation.*

Fig. 9-32. Diagrams of 50-kW Ampliphase transmitter.

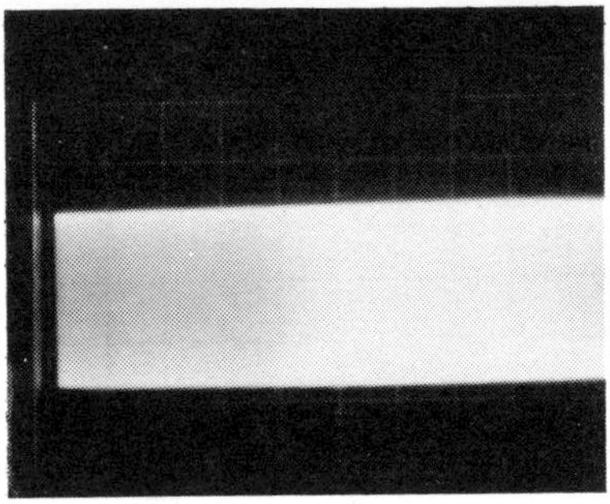

(A) Unmodulated carrier.

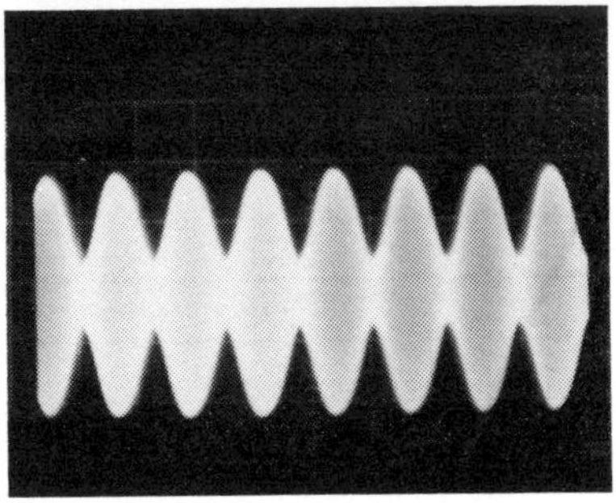

(B) 50% modulation.

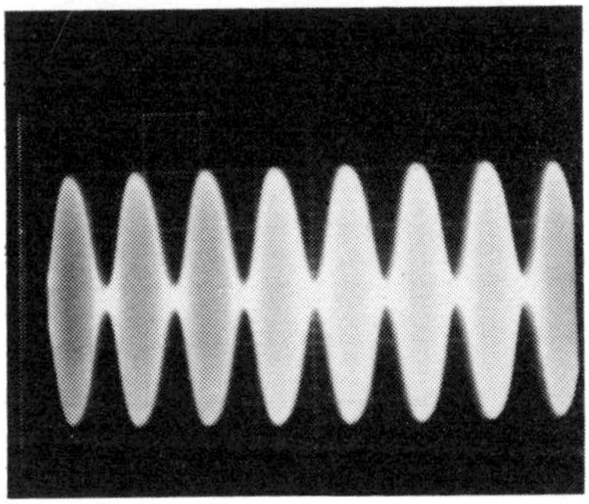

(C) 100% modulation.

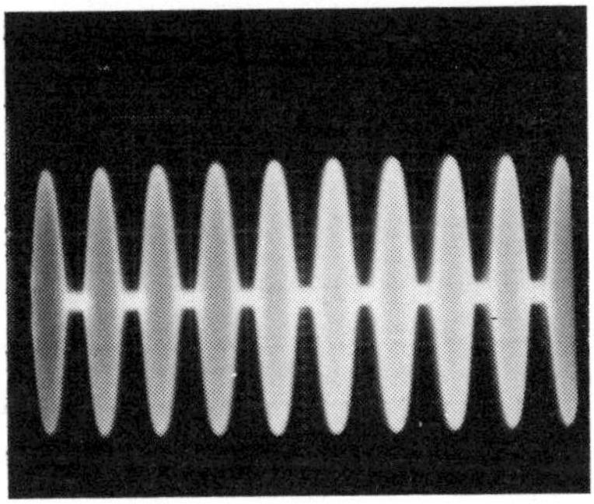

(D) Overmodulation.

Fig. 9-33. Oscilloscope waveforms of amplitude-modulated envelope.

$$M = \frac{E_m}{E_c} \qquad \text{(Eq. 9-1.)}$$

where,

E_m is the peak-to-peak or rms value of modulating voltage,
E_c is the carrier voltage in the same units as E_m.

For example, if a carrier wave with a peak-to-peak amplitude of 1000 volts is modulated with an audio sine wave of 500 volts, the modulation factor is:

$$M = \frac{500}{1000}$$
$$= 0.5$$

Multiplying the modulation factor by 100 gives the percent of modulation (%M):

$$\%M = \frac{E_m}{E_c}(100) \qquad \text{(Eq. 9-2.)}$$

Thus the percent modulation in the above example is:

$$\%M = \frac{500}{1000}(100)$$
$$= (0.5)(100)$$
$$= 50\%$$

Since percent modulation is a ratio, actual voltages need not be measured to find this value. If E_m and E_c of Equation 9-2 are considered to be in peak-to-peak values, the measuring oscilloscope can be set to any convenient gain (Fig. 9-34). The peak-to-peak value of the modulating voltage (E_m) is observed to be equal to $e_{max} - e_{min}$. Similarly the peak-to-peak carrier voltage (E_c) is seen to be equal to $e_{max} + e_{min}$. Therefore, substituting these values in Equation 9-2:

$$\%M = \frac{e_{max} - e_{min}}{e_{max} + e_{min}}(100) \qquad \text{(Eq. 9-3.)}$$

In Fig. 9-34, the maximum peak-to-peak amplitude on the scope presentation has been arbitrarily adjusted for 6 graticule divisions. The peak-to-peak minimum carrier amplitude is 2 divisions. Therefore the percent modulation is:

$$\%M = \frac{6-2}{6+2}(100)$$
$$= \frac{4}{8}(100)$$
$$= (0.5)(100)$$
$$= 50\%$$

Amplitude-Modulation Sideband Power

The total intelligence of the amplitude-modulated envelope is contained in the sideband frequencies. The greater the sideband energy, the greater the demodulated output in the receiver. The relationship of carrier to modulating voltage was given by equation 9-1. Solving this equation for E_m:

$$E_m = ME_c \qquad \text{(Eq. 9-4.)}$$

Thus for a fixed carrier magnitude, the sideband voltage developed across a common impedance varies directly with the modulation factor (M). Although the audio modulating voltage (E_m) is divided between two sidebands (upper and lower), we will first look at the voltage of a single sideband (E_{ss}), which is one-half the voltage of both sidebands:

$$E_{ss} = \frac{ME_c}{2} \qquad \text{(Eq. 9-5.)}$$

Since power varies as the square of the voltage across a given load resistance (R_L), the power in the single sideband is:

$$P_{ss} = \frac{(E_{ss})^2}{R_L} \qquad \text{(Eq. 9-6.)}$$

Substituting Equation 9-5 into Equation 9-6:

$$P_{ss} = \frac{\left(\frac{M\,E_c}{2}\right)^2}{R_L} \qquad \text{(Eq. 9-7.)}$$

and simplifying Equation 9-7:

$$P_{ss} = \frac{M^2\,E_c^{\,2}}{4R_L} \qquad \text{(Eq. 9-8.)}$$

which can now be written:

$$P_{ss} = \left(\frac{M^2}{4}\right)\left(\frac{E_c^{\,2}}{R_L}\right) \qquad \text{(Eq. 9-9.)}$$

We know that the carrier power (P_c) is:

$$P_c = \frac{E_c^{\,2}}{R_L} \qquad \text{(Eq. 9-10.)}$$

Substituting Equation 9-10 in Equation 9-9:

$$P_{ss} = \frac{M^2}{4}P_c \qquad \text{(Eq. 9-11.)}$$

where,

M is the modulation factor,
P_{ss} is the average power in one of two equal-amplitude sidebands,
P_c is the average carrier power using rms values of voltage and current.

For single-frequency sine-wave modulation, two equal-amplitude side frequencies exist: carrier plus the modulating frequency, and carrier minus the modulating frequency. Thus the *total* sideband power (P_{sb}) is twice the power given by Equation 9-11. Therefore:

$$P_{sb} = 2\frac{M^2}{4}P_c \qquad \text{(Eq. 9-12.)}$$

which can be simplified to:

$$P_{sb} = \frac{M^2}{2}P_c \qquad \text{(Eq. 9-13.)}$$

where,

P_{sb} is the total sideband power,
M is the modulation factor,
P_c is the average carrier power.

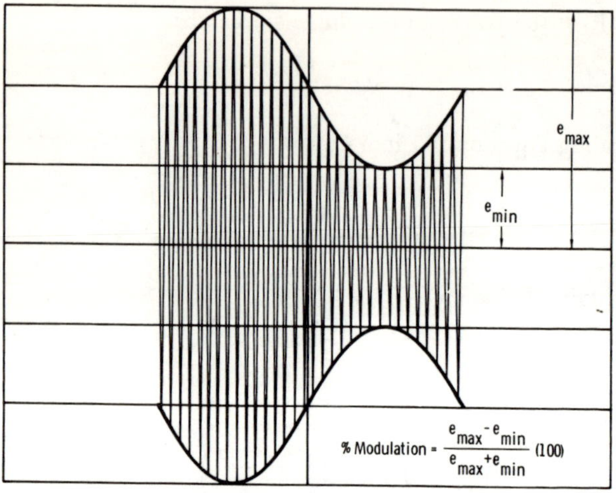

Fig. 9-34. Measurement of percent modulation.

Assume for example that a 5-kW carrier is modulated 100 percent. The sideband power (Eq. 9-13) is:

$$P_{sb} = \frac{1^2}{2}(5000)$$
$$= (0.5)(5000)$$
$$= 2500 \text{ Watts}$$

This illustrates that 100-percent modulation results in sidebands that contain one-half as much power as the unmodulated carrier.

If the same 5-kW carrier is modulated only 50 percent:

$$P_{sb} = \frac{0.5^2}{2}(5000)$$
$$= \frac{0.25}{2}(5000)$$
$$= (0.125)(5000)$$
$$= 625 \text{ Watts}$$

This illustrates that halving the modulation from a given original value reduces the sideband power to one-fourth the original amount.

The total power contained in a modulated wave is equal to the carrier power plus the sideband power. The total power (P_T) is:

$$P_T = P_c + P_{sb} \qquad \text{(Eq. 9-14.)}$$

To put this more conveniently in terms of the modulation index (M), substitute for P_{sb}:

$$P_T = P_c + \frac{M^2}{2} P_c \qquad \text{(Eq. 9-15.)}$$

and factor out P_c:

$$P_T = P_c\left(1 + \frac{M^2}{2}\right) \qquad \text{(Eq. 9-16.)}$$

For example, the total power of a 5-kW carrier modulated 100 percent is:

$$\begin{aligned} P_T &= 5000\left(1 + \frac{1^2}{2}\right) \\ &= (5000)(1.5) \\ &= 7500 \text{ Watts} \end{aligned}$$

This illustrates that for 100-percent modulation, the total radiated power is 1.5 times the unmodulated carrier power.

As has been stated previously, when a carrier is modulated with a sine wave, sideband energy exists at the carrier frequency plus and minus the modulating frequency. For a 100-percent modulated carrier (Fig. 9-35A), the *combined* sideband voltage is equal to the carrier voltage and appears as a single set of sidebands above and below the carrier frequency. Since the sideband voltage is divided equally between the two sideband frequencies, each side frequency has one-half the carrier amplitude.

When the modulating voltage exceeds that required for 100-percent modulation, the negative peak of the modulating signal drives the plate voltage of the modulated amplifier negative for a period of time that depends on the degree of overmodulation. For the duration of this negative plate voltage, the carrier is cut off (Fig. 9-35B). The resultant clipping of the negative peaks of the modulating signal causes spurious frequencies at both odd and even harmonics of the fundamental frequency, above and below the carrier frequency. Depending on the degree of overmodulation, the sideband "splatter" can result in spurious sideband frequency content beyond the fifth or sixth harmonic. Obviously, this not only causes audio distortion in the receiver, but also is a direct violation of FCC rules and results in severe interference to adjacent-channel stations because the added side frequencies overlap the adjacent channels.

At the time of this writing, the FCC allows a maximum of 120-percent modulation on positive peaks, provided the carrier never reaches cutoff on negative program peaks. With speech and program modulation, this is a practical possibility, and it is treated further in the operations sections of this text.

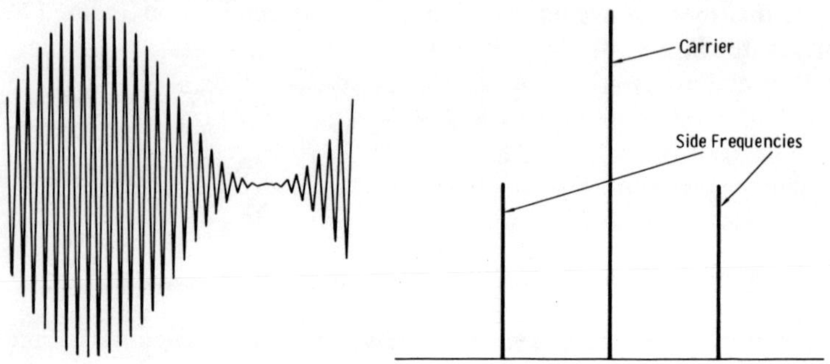

(*A*) *Modulated 100%*.

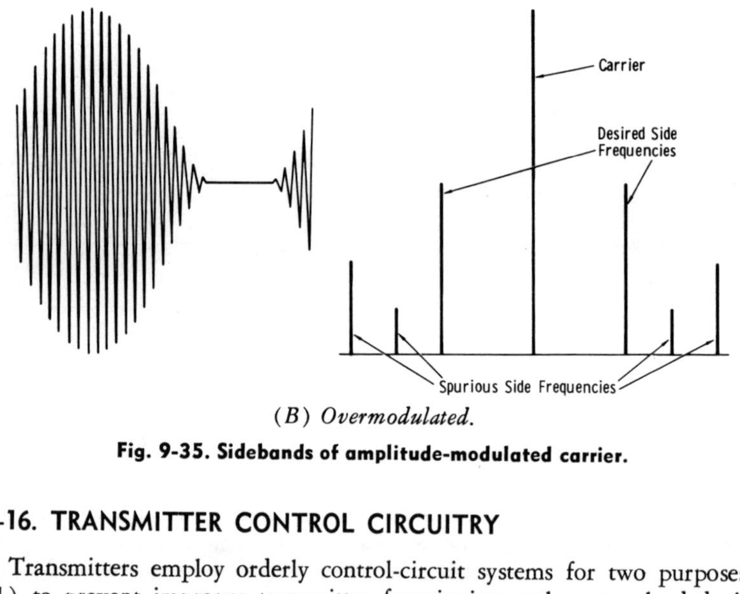

(*B*) *Overmodulated.*

Fig. 9-35. Sidebands of amplitude-modulated carrier.

9-16. TRANSMITTER CONTROL CIRCUITRY

Transmitters employ orderly control-circuit systems for two purposes: (1) to prevent improper transmitter functioning such as overloaded circuits or inadequate time delays for application of high voltages, and (2) to protect the operating personnel from contacting high-voltage terminals. The first function also serves to prevent application of voltages to certain elements if they are not receiving a normal flow of cooling medium such as forced air.

A typical sequence of operation is as follows: When the start button is pressed, this action applies filament voltages to the tubes and initiates action in a number of other relay circuits. Blower motors are started, and, until the air stream is of sufficient strength to actuate mercury switches on the air vanes, full filament voltage is not applied (in higher-power transmitters). At the same instant, time-delay relay motors or electrical networks are started. These do not close the high-voltage circuit until a specified time, such as 30 seconds or one minute, has elapsed. Some transmitters employ relays that automatically apply high voltages upon timing out of the delay relay. When the transmitter is first placed on the air, however, switches are normally set to the manual position rather than the automatic position. This allows an inspection of all operating parameters up to that point before high-voltage application. Door interlock switches prevent application of high voltage if any door is open in a cubicle containing high voltage. An open door also sometimes actuates grounding switches that short the high-voltage supplies to ground so that large capacitors cannot discharge through an operator's body should he accidentally come in contact with a high voltage conductor.

After all filament meters have been checked and normal operation is obtained, the low voltages usually are applied to the rf exciter stages. This permits checking the operation of these stages before application of high voltages to the final stage or series of high-level linear amplifiers.

A description of a typical transmitter control circuit (in this case, for a General Electric 250-watt transmitter) will serve to introduce the student to practical applications. Refer to Fig. 9-36 during the following analysis.

The entire control circuit operates from the secondary of T33, which is energized upon the closing of S3. Assume S3 to be closed. Motor-driven timing relay K1 is energized, and after approximately 30 seconds K1 times out and its normally open contact closes. Its normally closed contact opens and removes power from the relay motor.

Transformer T31, for the green indicating light (IS1), now is energized through the normally closed contact of K2, provided rear-door interlocks S1 and S2 are closed. The green light indicates that the transmitter is ready for the application of plate voltage. The timing relay, K1, prevents application of plate voltage before the tube filaments are warm.

Plate voltage is applied by pressing momentary switch IS1. This energizes relay K3, closing its normally open contact. Relay K2 now energizes through the closed contact of K3, the normally closed contact of K4, and the normally closed contacts of overload relays K7 and K8. As K2 operates, power is removed from green light IS1 and applied to red light IS2; power is applied to the low- and high-voltage rectifier plate transformers; a shunt is placed across the contact of K3 which locks K2 in, and K3 may be de-energized by release of IS1; the plate-contactor interlock between the 250-watt set and an additional power amplifier (if used) closes (terminals 11-6 and 11-7).

Momentary-contact switch IS2 serves to remove the plate voltage. Depressing S2 energizes K4. The normally closed contact of this relay opens and de-energizes K2, which falls out; the normally open contact of K4 closes and removes plate voltage from an additional amplifier (if used).

One set of normally closed contacts of low-voltage overload relay K7 and of high-voltage overload relay K8 are in series with K2. Another set of

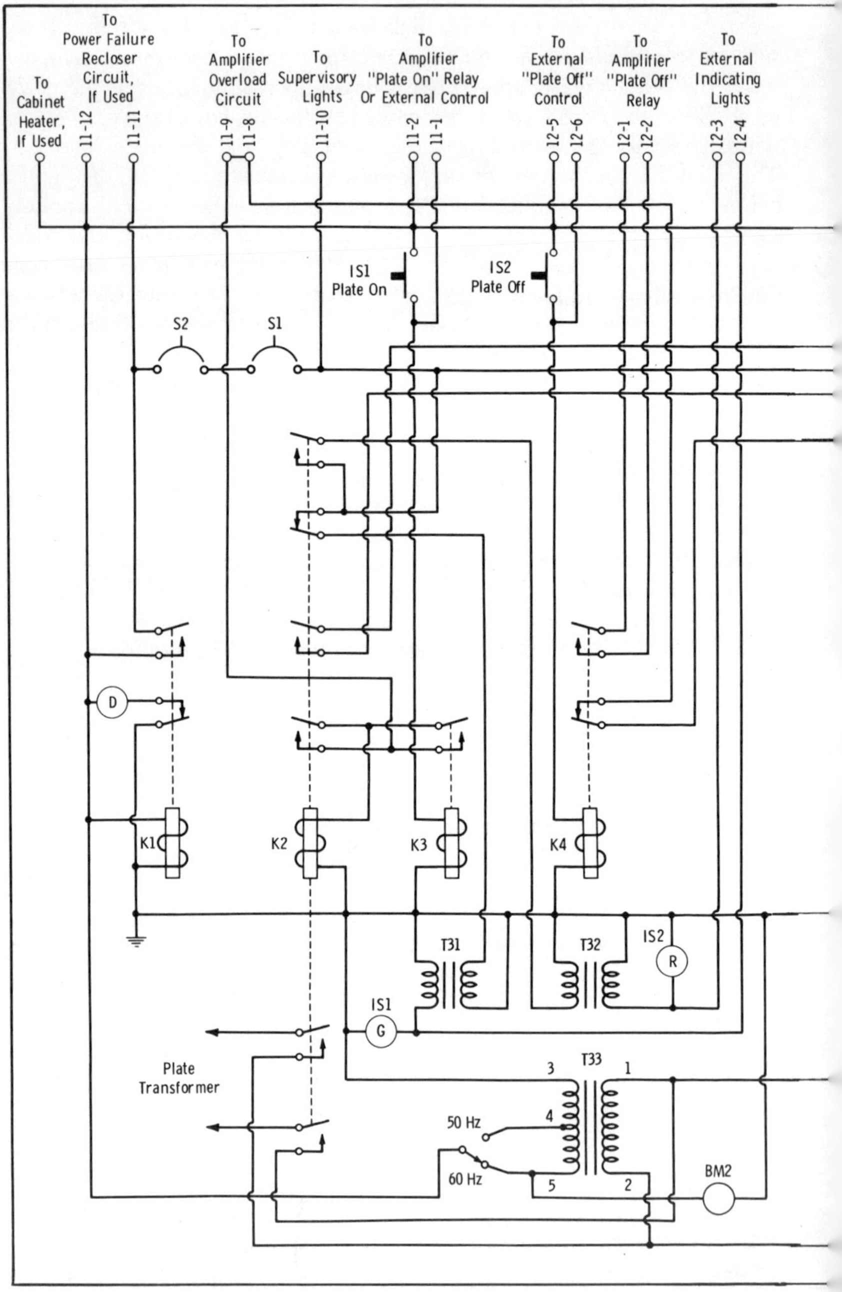

Fig. 9-36. Control

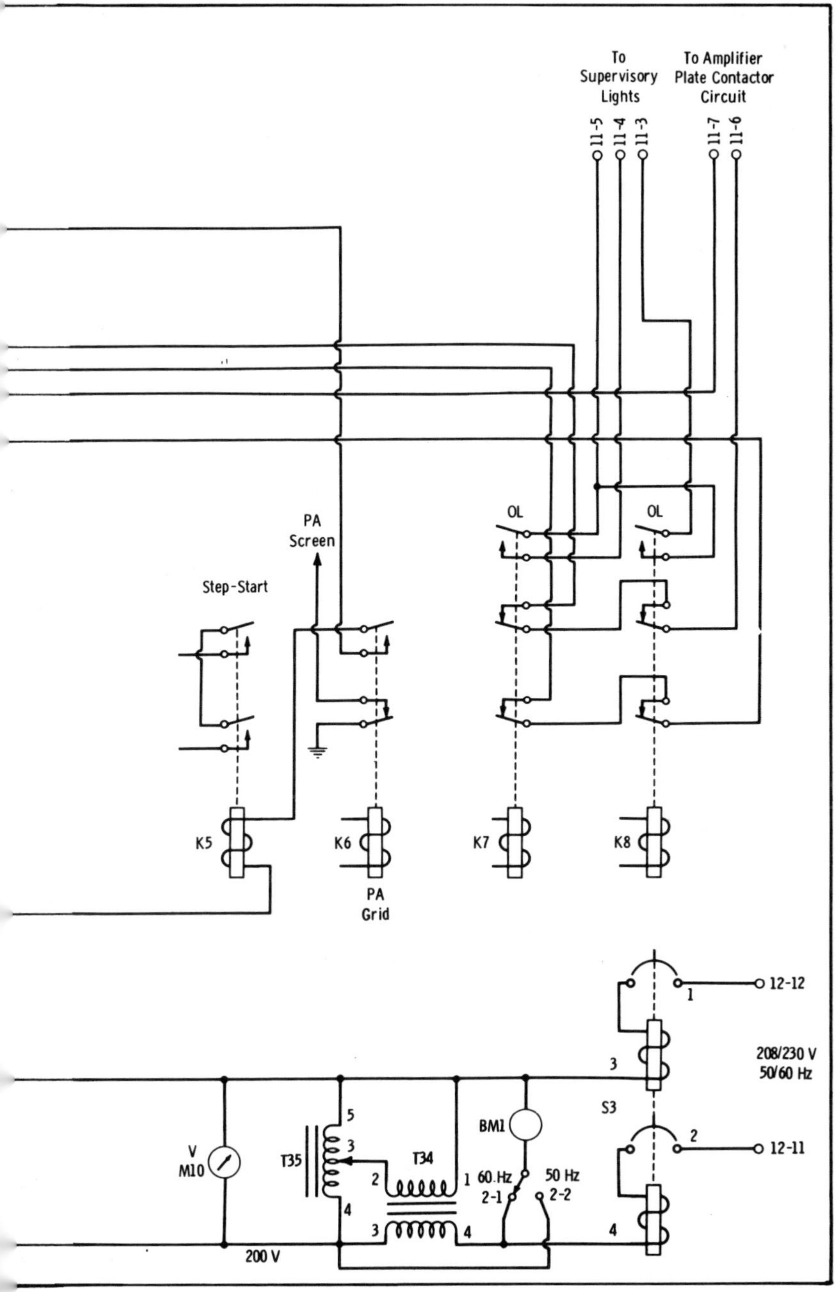

circuits in a transmitter.

contacts is extended for use in conjunction with a higher-powered amplifier. A set of normally open contacts is extended to a supervisory light circuit, if used. Hence, an overload in the low- or high-voltage supply will cause K2 to drop out, will cause an amplifier plate contactor to drop out, and will cause a supervisory light circuit to operate and indicate in which circuit the overload occurred. The supervisory lights must be operated by auxiliary relays which will lock themselves in when the normally open contacts of K7 or K8 close for an instant. As soon as K2 falls out, K7 or K8 will return to the unenergized position.

Relay K5, when energized, shorts out a surge-limiting resistor (not shown). This resistor limits the peak charging current passed by the high-voltage rectifier tubes to a safe value when plate voltage is applied. Relay K5 is energized through a contact of K6 after K6 is energized. Relay K6 closes when the power-amplifier grid current exceeds approximately 8 milliamperes. Previous to this, a normally closed contact of K6 shorts the power-amplifier screens to ground, thus protecting the tubes from excessive dissipation before the grid excitation is sufficient to produce safe grid bias. By the time the power-amplifier grid current builds up to 8 milliamperes and closes K6, the high-voltage rectifier capacitors are almost fully charged. The subsequent energizing of K5 and shorting of the surge-limiting resistor then results in negligible surge current through the high-voltage rectifier tubes.

Leads may be extended in shunt with IS1 from terminals 11-1 and 11-2 for an external plate "on" control, and in shunt with IS2 from terminals 12-5 and 12-6 for an external plate "off" control. External indicating lights (6.3 volts) may be wired between terminal 12-4 and ground (green) and between terminal 12-3 and ground (red).

The cabinet heater, if used, has a relay coil connected to terminal 11-12 and ground. The normally closed contacts carry power to the heater when power is removed from the transmitter.

An automatic power-failure recloser may be employed. It must be of a nature to place a shunt across terminals 11-11 and 11-12, and also a shunt across terminals 11-1 and 11-2, in case of a brief power failure. These terminals are extended to the amplifier (when used), which contains an automatic power-failure recloser.

9-17. PARALLEL TRANSMITTER OPERATION

Some a-m transmitters employ *redundant* circuitry to avoid loss of air time or the necessity of purchasing a complete standby transmitter installation, usually of lower power than the main transmitter. Redundancy takes the form either of actual parallel operation of two transmitters, or of a special means of diplexing that allows part of the power amplifier to be used (at reduced output) while the other part (feeding a dummy load) can be repaired.

Fig. 9-37 is a block diagram of a CCA *Dual Reliable* a-m transmitter. In this system, two transmitters operating at one-half the authorized power are combined at the output, with an automatic monitoring and switching system. In case of failure in one of the two transmitters, it is automatically disconnected and fed to a test dummy load. The other transmitter can then be fed directly to the antenna transmission line, bypassing the loss incurred in the reject load of the combiner network. This allows servicing of the faulty stage while continuing to operate (at reduced power), avoiding loss of actual air time. In this case, since the two transmitters are identical, repair of the faulty unit is facilitated since the electrical characteristics can be compared step-by-step with the correctly operating unit. Obviously, this also reduces the stock of spare tubes and critical components required.

Not shown in the block diagram of Fig. 9-37 is the fact that there are actually two electrically independent oscillators. Both transmitters are fed from the same oscillator through a relay operated by an rf level detector. In case of failure of the No. 1 oscillator (or the detector system itself), the relay connects the No. 2 oscillator to feed both transmitters. A manually operated switch allows this type of automatic operation, or the switch may be set for operation of oscillator No. 2. A third position of this selector switch maintains the automatic system but turns oscillator No. 2 on to allow testing and servicing when necessary. Each crystal oscillator employs an independent power supply.

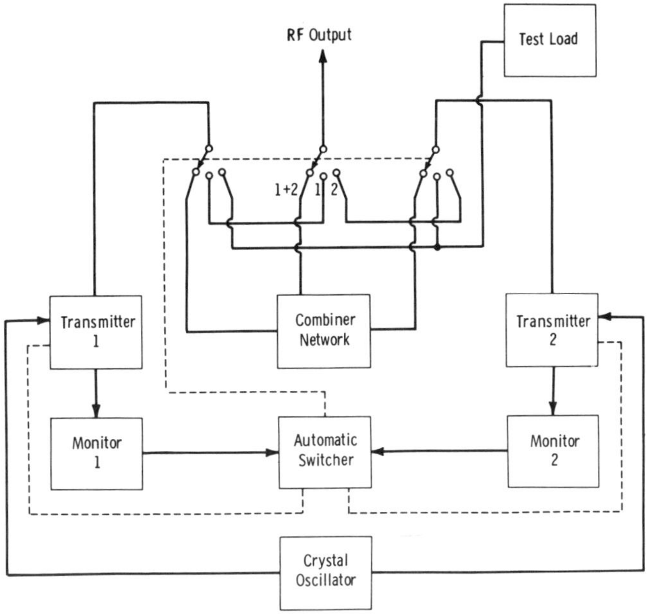

Fig. 9-37. Block diagram of CCA Dual Reliable transmitter.

The rf output of each transmitter is fed to its individual monitoring system which employs a tuned circuit feeding a diode detector. A dc component (proportional to the carrier amplitude) is separated from the demodulated audio envelope. Departure of the carrier level from a preset amount operates relays to remove plate voltage from *both transmitters* so that no rf arcing can occur in the output switches that are activated. Motor-operated rf switches connect the faulty transmitter to the test load and the operating transmitter to the antenna. Voltages are then automatically reapplied to the transmitter feeding the antenna system.

The audio component of the demodulated rf signal in each monitor is also amplified and fed to a peak detector with a very slow decay time constant of about 10 seconds. If this peak rectified signal falls below a preset value, an audio relay will close at the start of this 10-second interval to preset decision-making relays for action. If the 10-second-delayed output of the peak detector shows the output is still missing or low, control relays are enabled to remove the faulty transmitter from the air. The same action occurs in each of the two monitoring systems, which are interconnected in such a way that if one closes when the other does not, the transmitter with the faulty modulator or audio system is removed in 10 seconds, in the same way as for an rf fault. If a similar reduction of modulation occurs in both transmitters, corrective action is not taken, since this condition is interpreted as a lack of modulation due to the signal source.

When the rf switches are in the position shown in Fig. 9-37, the outputs are combined by means of the diplexer circuit shown in Fig. 9-38. The power combination occurs in the two inductive sections, L1 and L2, with common loading capacitor C2. Across this combining network are an RC network (R1C1) and reject power meter M2. When the transmitters are delivering equal power to the combiner, each one is loaded by an impedance equal to that connected at the output terminal, which, with the arms in the position shown in Fig. 9-38, is the antenna system. If one transmitter fails or is reduced in power, the opposite transmitter is still loaded by the same impedance, with part of its power shunted through the reject branch, R1C1. If the transmitter fails completely, the output power of the properly operating transmitter would be split into two equal portions, one-half in the antenna system and one-half in reject load R1. Because of this constant loading characteristic of the diplexer network, the operating transmitter selected can be connected directly to the antenna line without a need for retuning. Thus it is possible to bypass the combining network and eliminate the dissipation of half of the power in the reject load resistor.

9-18. LIMITER AMPLIFIERS

The limiting amplifier, or compression amplifier, is a very important link in a broadcast transmitter installation. The limiting amplifier employed at the transmitter should not be considered as a "volume limiter," but as a peak limiter intended to prevent overloading of transmitter components and carrier cutoff on negative peaks.

Fig. 9-39A illustrates the Marti CA-40 all solid-state compressor/limiter. Fig. 9-39B is a simplified block diagram of this unit. The unit operates on the variable-attenuator principle. A full-wave audio detector provides a voltage proportional to the audio peak level, and this voltage controls a

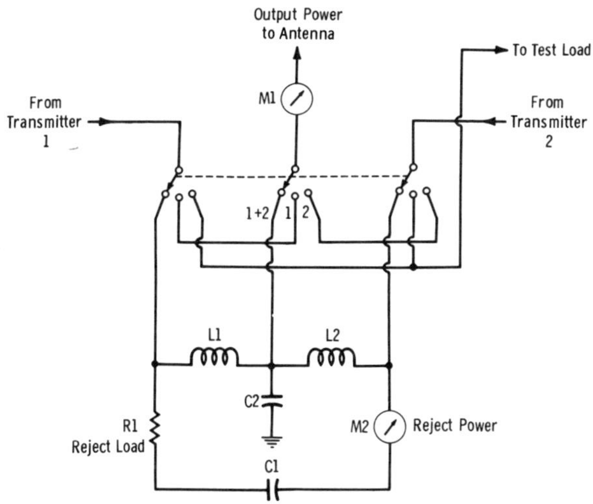

Fig. 9-38. Simplified diagram of combiner network.

voltage-variable resistor (VVR) that acts as an average-level attenuator. This VVR device has a large dynamic operating range, adds no audio distortion, but has a comparatively slow attack time. For instantaneous control of audio waveforms with short rise times, the control voltage is differentiated and applied to the gate of a field-effect transistor. This device is connected in parallel with the VVR and serves as an extremely fast attenuator, operating only on waveforms with short rise times, thus complementing the slower VVR device. The result is a compressor with a 40-dB dynamic range and a limiter with microseconds attack time.

The release time is adjustable to 800 milliseconds, 2 seconds, or 5 seconds. Control systems allow symmetrical peak limiting for fm and selectable asymmetrical or symmetrical peak limiting for a-m. The peak limiting level is adjustable.

9-19. MONITORING FACILITIES FOR A-M TRANSMITTERS

In addition to conventional audio monitors for checking transmitter input and demodulated output, two additional monitoring facilities are required. These are the frequency monitor and the modulation monitor.

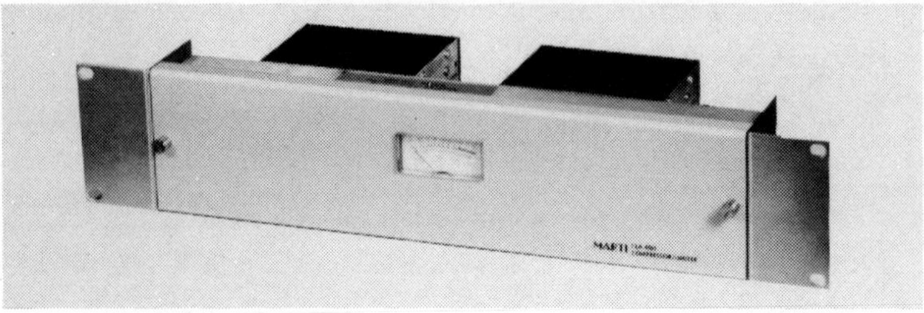

Courtesy Marti Electronics, Inc.

(*A*) *Front view.*

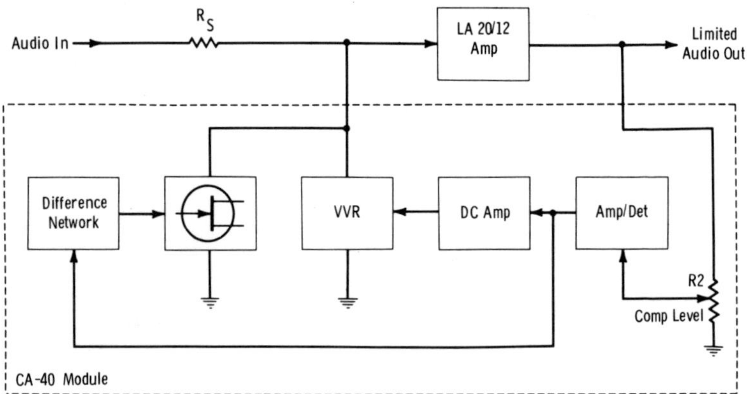

(*B*) *Block diagram.*

Fig. 9-39. Marti CA-40 compressor/limiter.

The Frequency Monitor

A frequency monitor must be included at all transmitter installations to aid in maintaining the frequency within prescribed limits. This device is known as a secondary standard and is normally checked against a primary standard (by a frequency-measuring service) at intervals of 30 to 90 days in order to insure its accuracy.

A station frequency monitor consists essentially of a crystal oscillator which operates at a frequency differing from the transmitter frequency by a given amount. The output of this oscillator is beat with an rf signal picked up from the transmitter, and the amplified beat is applied across a diode circuit which drives the indicating instrument.

The operator at a standard broadcast station must ascertain whether his transmitter is within ± 20 Hz of the assigned frequency, the exact assigned frequency being indicated by zero (center position) on the meter. All oscillator units in transmitters have a small trimmer capacitor which makes exact adjustment of the frequency possible.

A block diagram of the Metron Model 510 analog frequency deviation monitor for a-m use is shown in Fig. 9-40. It operates as follows: The transmitter signal (0.1 volt to 3 volts rms) is fed to the rf amplifier and then to the mixer. The signal from the local oscillator, which is set 100 Hz below the transmitter frequency, mixes with the input signal to produce

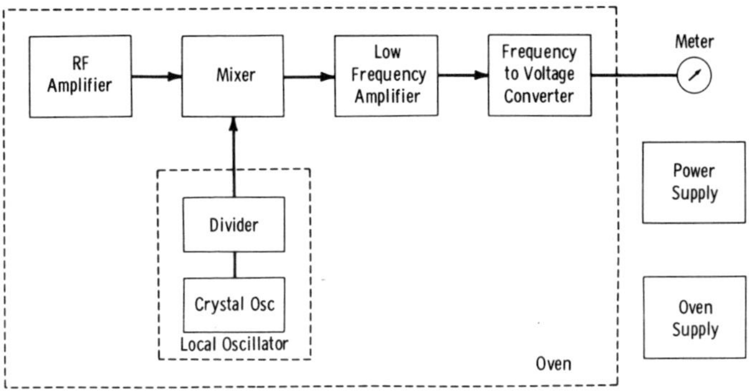

Fig. 9-40. Metron Model 510 analog a-m frequency monitor.

100 Hz when the transmitter frequency is at zero deviation. The 100-Hz signal is amplified and fed to a frequency-to-dc converter which produces a dc current. At zero deviation, this current is at the half-scale value of the indicating meter. If the transmitter frequency decreases, the current to the meter decreases, and if the transmitter frequency increases, the current to the meter increases. All circuits except the power supply are located in the oven to give good stability.

The latest a-m frequency monitors employ digital techniques to provide electronic numeric readouts directly in plus or minus hertz deviation from the assigned carrier frequency. Fig. 9-41 shows a block diagram of the Metron Model 520 digital a-m frequency deviation monitor. This unit has a two-digit electronic numeric readout plus a polarity indicating display. It incorporates provisions for tuning the transmitter over a frequency range of plus or minus 500 Hz. This is accomplished by the use of a numeral 1 in the polarity-indicating display on the front panel. This 1 flashes once for every 100 Hz deviation in excess of that read on the digital readout. An image-check push button located on the front panel provides a convenient method of insuring that the frequency read is not an image of the true transmitter frequency.

The monitor in Fig. 9-41 operates as follows: The modulated transmitted signal (0.1 volt to 3 volts rms) is fed to the rf amplifier and then to the mixer. The signal from the local oscillator, which is set 100 Hz below the transmitted frequency, mixes with the input signal to produce 100 Hz when the transmitter frequency is at zero deviation. This 100-Hz signal is amplified and shaped to produce pulses for the digital logic cir-

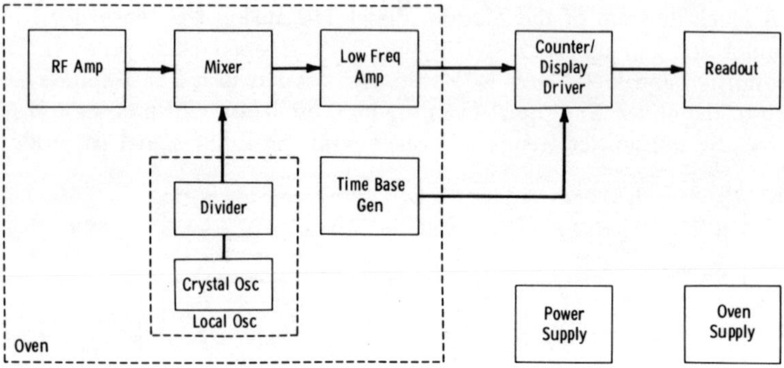

Fig. 9-41. Metron Model 520 digital a-m frequency monitor.

cuitry. A low-frequency time-base generator produces a stable gate period which is also fed to the digital logic circuitry. During the first half of this gate period, the logic circuit presets to 99, and during the last half of the period, the logic circuit counts down for one second. If the transmitted frequency is correct, the frequency fed to the logic circuitry is 100 Hz, and the countdown is 100, from 99 to 0. The frequency deviation of the transmitted signal is read every two seconds. The indication on the digital readout tubes remains on constantly except for instantaneously registering any change which occurs during the previous second.

Modulation Monitors

The modulation monitor is designed to give continuous direct reading of percentage modulation of the a-m carrier. The unit also normally provides a carrier-level meter to provide constant indication of any carrier shift under modulation (Chapter 14).

The CCA Model AMM-1T a-m modulation monitor is shown in Fig. 9-42A. This unit provides for the determination of modulation asymmetry through the use of two individually settable peak flashers, one for positive peaks (calibrated from 50 to 130 percent) and one for negative peaks (calibrated from 50 to 100 percent). (In addition, the negative-peak lamp functions as a carrier-off alarm.) Modulation is displayed on a 4½-inch meter that is switchable to positive or negative modulation as desired. Carrier level is displayed on an expanded scale (80 to 120 percent) for accurate determination of carrier shift.

The instrument has built-in test functions which allow it to be calibrated periodically with only an accurate dc voltmeter used as an external reference. Built-in meter drivers can drive an external line of 5000 ohms impedance, and shorting or opening the remote lines will not affect the meters on the monitor.

Refer to the block diagram in Fig. 9-42B. The amplitude-modulated rf is fed to an input level potentiometer and through a wide-band rf matching

Courtesy CCA Electronics Corp.

(*A*) *Front view.*

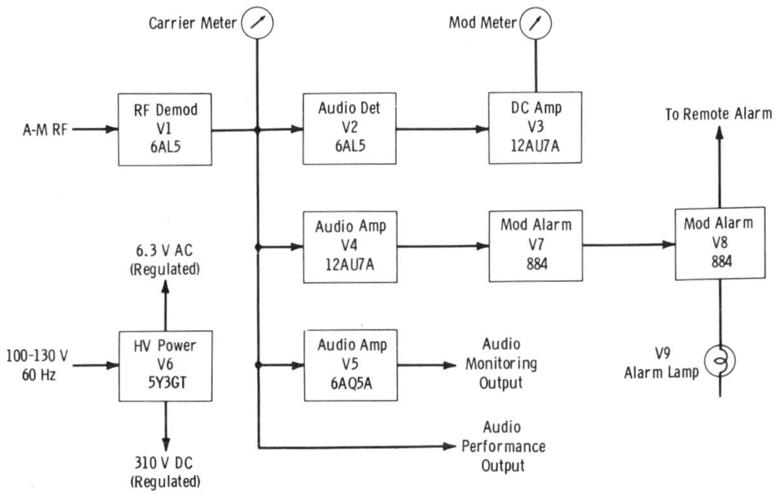

(*B*) *Block diagram.*

Fig. 9-42. CCA Model AMM-1T a-m modulation monitor.

transformer to the demodulator stage, V1. One-half watt of rf power is sufficient to provide full-scale meter readings. A maximum of four watts of rf can be attenuated by the input potentiometer.

The demodulator consists of two diodes connected in opposite polarity such that two demodulated audio signals of opposite polarity are available. By means of the front-panel polarity switch, either of these audio signals can be fed to the metering and alarm circuits; the remaining signal is used for monitoring purposes. When the switch is in the positive position, diode 1 of V1 supplies the rectified carrier to the carrier-level meter and demodulated audio to a low-pass filter. This filter eliminates the rf but not the audio frequencies up to 50 kHz.

The output of the filter feeds the audio peak rectifier, V2, which constitutes a peak-reading voltmeter that follows the envelope of the modulation.

Between this diode and the dc amplifier, V3, there is a network that overshoots during the rise of the pulse and provides a slow decay during the fall. The time constants of this network, together with the ballistic characteristics of the modulation meter, provide the timing characteristics for the modulation metering specified by the FCC.

Tube V3 is a balanced dc amplifier, which drives the modulation meter. A zero-set potentiometer is used to balance the two sections of V3 so that the meter will rest at zero when there is no signal applied to the rf input. It is set at the factory and it will not require resetting until V3 is replaced or has aged considerably. A potentiometer called "audio set" is also set at the factory to provide a 100-percent modulation reading when the transmitter is being modulated 100 percent and the rf input is such that the carrier-level meter reads 100 (set mark).

The alarm circuit is preceded by a cathode follower, V4, that isolates the thyratron tubes from the audio circuits. It is fed from the same audio source that actuates the carrier meter. The polarity is the same as that set for the modulation meter by means of the polarity switch.

The overmodulation-alarm control thyratron, V7, fires at a voltage determined by the bias, which is set by means of the peak-level potentiometer on the front panel. An additional potentiometer determines the audio level fed to V7; it is set at the factory, in conjunction with a resistive network, in order to match the overmodulation flashings with the position of the peak-level control.

Thyratron V7 controls front-panel flashing light V9 and drives switching thyratron V8, which operates the alarm relay. The operation is such that even a brief overmodulation peak will hold the relay closed for a definite length of time. A set of contacts is connected to operate a remote overmodulation indicator or counter. These contacts are rated at 1 ampere, 115 volts ac.

Two indicating meters are provided. The carrier-level meter is graduated from 0 to 120. The 100 mark corresponds to the set position and is near the end of the scale; this allows a better reading of carrier shift.

The modulation-percentage meter is calibrated from 0 to 120-percent modulation and is also calibrated in decibels referred to 100-percent modulation. This meter meets the ballistic characteristics specified by the FCC.

> NOTE: The modulation-percentage meter is not a VU meter, nor an instantaneous peak-reading device like an oscilloscope or the peak-level flasher. These devices will not read the same under actual program modulation, but they will coincide under steady modulation.

The frequency response of the modulation metering and alarm circuits is 30 to 15,000 Hz $\pm$ 0.5 dB. External meters can be connected.

A second low-pass filter, with a cutoff frequency of 70 kHz, feeds the demodulated audio from diode 2 of V1 to the distortion-meter terminals. This filter reduces any trace of carrier to below the noise level and allows up to the third harmonic of 15 kHz to reach an external distortion analyzer connected to a terminal board provided. The output level at these terminals is more than 4 volts rms for 100-percent modulation; it is constant ±0.5 dB from 30 to 25,000 Hz; distortion is less than 0.2 percent; and hum and noise is lower than 70 dB below 100-percent modulation.

The same audio signal is fed to the monitoring cathode follower, V5; this cathode follower can drive a 600-ohm line with more than 2 volts rms at 100-percent modulation. Hum and noise is lower than 60 dB below this level; distortion is less than 1 percent. Frequency response is 30 to 25,000 Hz ±0.5 dB.

EXERCISES

Q9-1. Define an angle of one radian.

Q9-2. How would you convert degrees to radians?

Q9-3. For a single cycle, how many radians are there between the maximum positive and maximum negative peaks?

Q9-4. What is the approximate wavelength for (A) 600 kHz, (B) 1600 kHz?

Q9-5. Does the primary service area of a standard broadcast station depend primarily on ground-wave signals or sky-wave signals?

Q9-6. Assuming the same ground conductivity throughout, if the field strength is 100 millivolts/meter at 1 mile, what inverse-distance field strength would you expect at 10 miles?

Q9-7. What is the basic difference between a conventional class-C amplifier and a linear rf amplifier?

Q9-8. What percentage increase occurs in antenna current when the carrier is modulated 100 percent by a pure sine wave?

Q9-9. See Fig. 9-34. If the maximum peak-to-peak value is 6 divisions and the minimum peak-to-peak value is 0.5 division, what is the percent modulation?

Q9-10. What is the total radiated power for a 10-kW carrier when modulated with a sine wave at (A) 100 percent, (B) 50 percent, (C) 25 percent?

CHAPTER **10**

FM Transmitters and Antenna Systems

Authorizations for fm stations involve certain FCC specifications as to power and antenna-height requirements. We will look briefly into these FCC regulations and explore the general characteristics of fm propagation before examining the details of conventional fm and stereo fm transmitters.

10-1. FUNDAMENTALS OF FM TRANSMISSION

Except as provided below, the fm licensee is required to supply a minimum effective radiated power (erp) as follows:

Class A Station: 100 watts (−10 dBk)
Class B Station: 5 kW (7 dBk)
Class C Station: 25 kW (14 dBk)

NOTE: If you are not familiar with the term dBk, please review Figs. 2-17, 2-18, and 2-19 (and associated text) in Chapter 2.

No minimum antenna height above average terrain is specified by the FCC. However, maximum antenna height is correlated with maximum erp as listed in Table 10-1.

Antenna heights exceeding those specified in Table 10-1 may be used provided the erp is reduced an amount determined by use of the appropriate curve in Fig. 10-1. If use of this curve requires an erp reduction to an amount less than the normal minimum specified for the class of station involved, then the erp determined from Fig. 10-1 becomes the minimum for the station involved. That is, when such conditions exist, the station is authorized to operate at an erp lower than the minimum value normally specified by the FCC.

Table 10-1. Maximum ERP vs Maximum Antenna Height

Class	Maximum Power	Maximum Antenna Height (Feet) Above Average Terrain
A	3 kW (4.8 dBk)	300
B	50 kW (17 dBk)	500
C	100 kW (20 dBk)	2000

Determination of applicable rules depends on the zone in which an fm station is located, and the procedures are quite varied and complex. Current issues of the FCC Rules and Regulations should always be readily available to responsible station personnel.

FM Coverage

Signal propagation in the 88-108 MHz region differs radically from propagation at the standard broadcast frequencies studied in the preceding

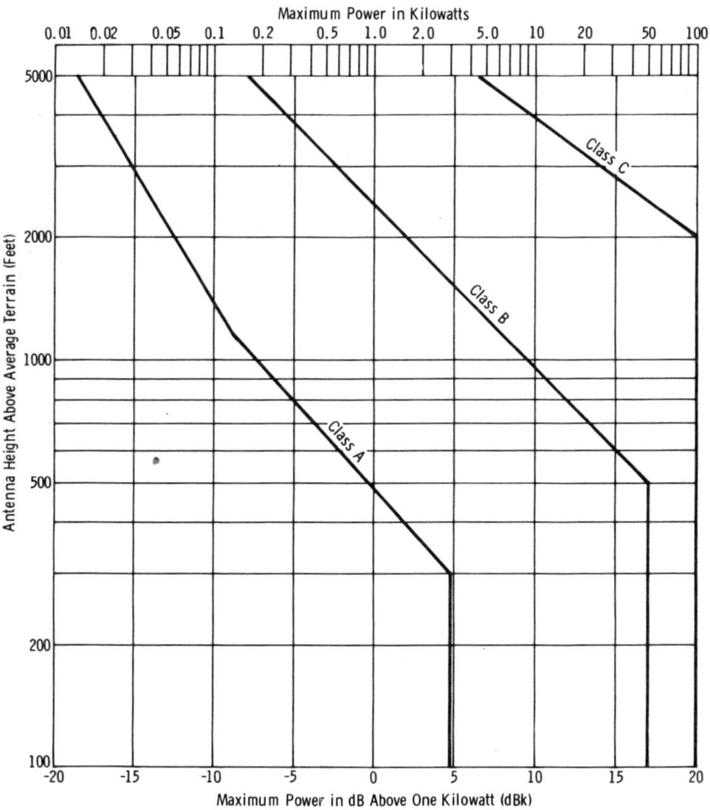

Fig. 10-1. Maximum power versus antenna height for fm stations.

chapter. A high power gain may be achieved by concentration of radiation in the horizontal plane at the expense of radiation in the vertical plane. In the past, horizontal polarization was used exclusively by fm broadcasters. With the rapid increase in the number of automobile fm receivers, the vertical component of a radiated wave assumed increasing importance, and combined horizontal and vertical polarization through the use of two individual sets of antennas became popular. The most recent development is the circularly polarized antenna. All systems are covered in Section 10-2.

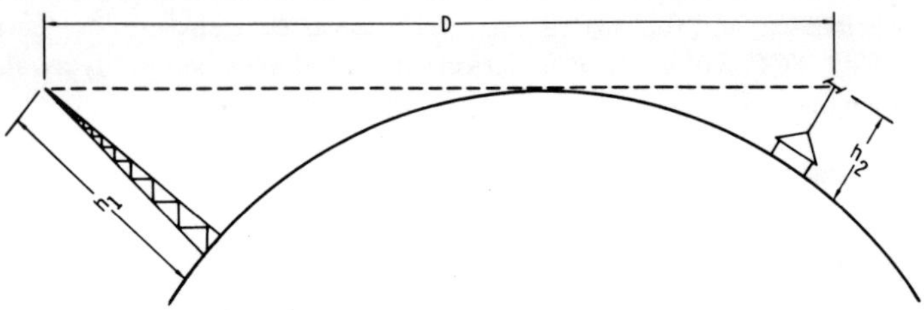

Fig. 10-2. Calculation of line-of-sight distance.

The propagation of fm power, since it occurs in the vhf band, is based primarily on line-of-sight reception. Fig. 10-2 shows the optical path. The line-of-sight distance is:

$$D = 1.22\ (\sqrt{h_1} + \sqrt{h_2})$$

where,

D is the line-of-sight distance in miles,
h_1 is the height of the transmitting antenna in feet,
h_2 is the height of the receiving antenna in feet.

It should be remembered that in practice tall steel structures must be considered when predicting field strengths of received signals. The height of the transmitting antenna is measured from the ground to the *effective radiating center* of the antenna elements.

When a high erp is used, the usable range of the fm signal is extended past the theoretical calculation of Fig. 10-2. As a practical example, the line-of-sight distance for a 500-foot transmitting antenna and a 50-foot receiving antenna is found as follows:

$$D = 1.22\ (\sqrt{500} + \sqrt{50}) = 1.22\ (29.4) = 35.9 \text{ miles}$$

Therefore, the optical path is 35.9 miles. In one case, a combination of a 10-kW transmitter and an antenna with a power gain of 12 at a height of 500 feet placed a 20-microvolt per meter signal at a distance of 88 miles. This is sufficient signal strength for most modern fm receivers in rural or small-community locations.

The free-space field intensity is:

$$e = \frac{7\sqrt{P_R}}{d}$$

where,

e is the free-space field intensity in volts/meter,
P_R is the effective radiated power in watts,
d is the distance in meters (1 mile = 1610 meters).

Assume that a half-wave dipole is radiating 1 kW of power. To find the free-space field intensity at 1 mile (1610 meters), perform the following calculation:

$$\begin{aligned} e &= \frac{7\sqrt{1000}}{1610} \\ &= \frac{(7)(31.6)}{1610} \\ &= 0.1374 \text{ volt/meter} \\ &= 137.4 \text{ mV/meter} \end{aligned}$$

All antenna gains (or losses) are related to this standard half-wave dipole which gives 137.4 millivolts/meter at one mile with 1000 watts of power. In free space, e is independent of the frequency (wavelength) of operation.

In practice, ground effects must be considered, and e becomes a function of frequency or wavelength. In this case, the approximate formula for e becomes:

$$e = \frac{3.2\, ah\sqrt{P_R}}{d^2\lambda}.$$

where,

e is the field intensity in microvolts per meter,
a is the height of the transmitting antenna in feet,
h is the height of the receiving antenna in feet,
P_R is the effective radiated power in watts,
d is the distance in miles,
λ is the wavelength in meters.

Note that since λ is in the denominator, the shorter the wavelength (higher the frequency), the greater is the received field strength. This is directly offset by the *effective antenna length* which, for a half-wave dipole, varies inversely with frequency.

Fig. 10-3 is used to determine either a given field-strength contour or the required effective radiated power to produce a given field strength at a certain distance, for a given antenna height. The chart is based on a receiving-antenna height of 30 feet. As an example of its use to estimate the approximate radius of the area within a given field-strength contour, consider the following problem.

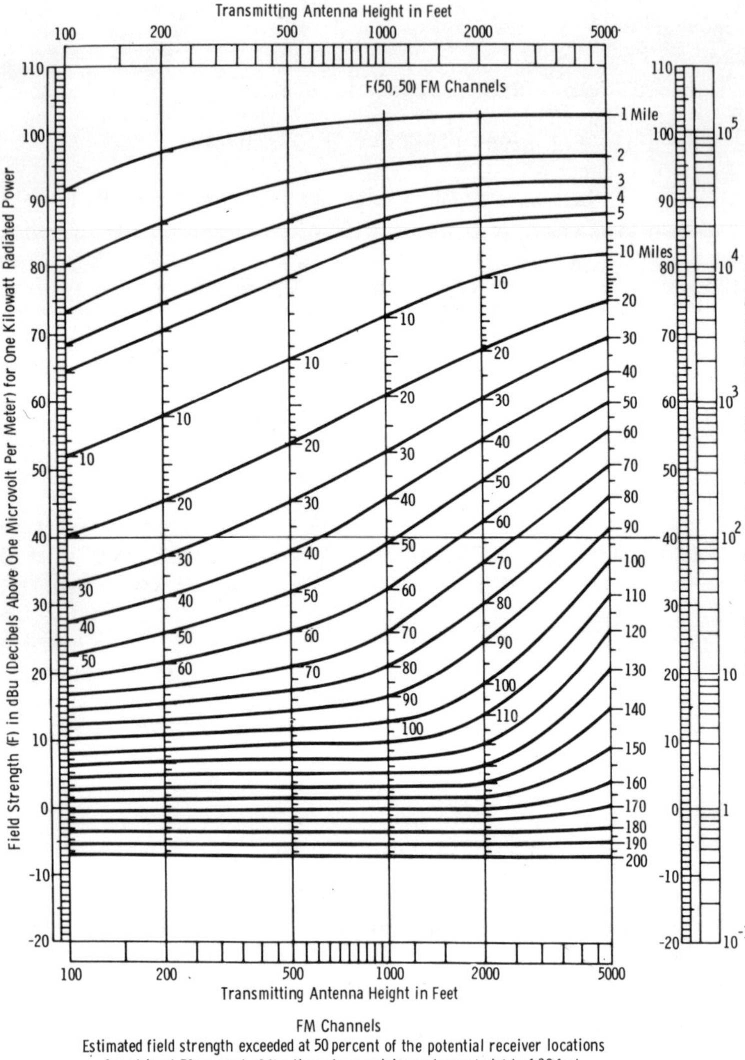

FM Channels
Estimated field strength exceeded at 50 percent of the potential receiver locations for at least 50 percent of the time at a receiving antenna height of 30 feet

Fig. 10-3. Signal-intensity chart for fm stations.

Assume that it is desired to determine the distance from the antenna to the 1000-microvolt per meter (designated $\mu V/m$) contour when the antenna height above average terrain is 500 feet and the effective radiated power is 1 kW. This distance would probably be different in each direction unless the land is very flat. Note that 1000 $\mu V/m$ is +60 dBu. Place a straightedge from 60 on the right edge to 60 on the left edge of the chart. Mark the point at which the vertical line representing 500 feet (noted along the top and bottom of the graph) intersects the straightedge. In this example, the distance to the 1000 μV/m contour is approximately 15 miles.

The graph is drawn on the assumption of 1-kW power. To use the chart for any other power, subtract the number of dBk for the power used from the number of dBu corresponding to the desired signal strength; then use the resulting number in the manner just described. The charts issued by the FCC are supplied with a sliding scale that permits making this conversion graphically.

To find the value of effective radiated power necessary to produce a field intensity of 50 μV/m (34 dBu) at 50 miles from an antenna 500 feet high, locate the 50-mile point on the 500-foot antenna-height line. Place a straightedge (preferably transparent) across the chart through this point, being sure it is parallel with the bottom line. The field strength resulting from a 1-kW erp can then be read on the left edge of the chart. In this case, it is 32 dBu. The required field strength is 34 dBu, or a difference of 2 dB. Adding this same number of dB to the 1-kW erp gives a required erp of 2 dBk, or 1585 watts.

10-2. FM ANTENNA SYSTEMS

Gain is achieved in an fm broadcast antenna by stacking *bays* of radiators (normally about one wavelength apart) in the vertical direction, as shown by Fig. 10-4. This method nulls radiation in the vertical plane and concentrates radiation in the useful horizontal direction. Do not confuse this statement with direction of polarization; the same type of stacking of antenna bays results in maximizing radiation in the horizontal direction for either vertically or horizontally *polarized* radiating elements. Table 10-2 gives the gain (relative to a half-wave dipole) for various numbers of bays used. Antenna field gain is the square root of the power gain. Of particular interest is the correlation of power gain, decibels, and field gain (field strength) given in the table, which is typical for all fm antennas when horizontal polarization (only) is used. This does not apply to recent circularly polarized antennas (vertical *and* horizontal polarization), which are discussed later in this section.

The RCA Type BFA broadband fm antenna (Fig. 10-4) is of sectionalized construction; each section consists basically of four radiating rings attached to a supporting frame. An insulated feed assembly and a section of 3⅛-inch transmission line is provided with flanges to fit 3⅛-inch coaxial feed lines. Adapters are available for other sizes of lines. Standard antennas have power gains from 0.9 to 16.5. Special designs are available on application. All BFA antennas are factory tuned to any channel in the frequency range of 88 to 108 megahertz. In standard and multiplexing operations, a voltage standing-wave ratio of 1.1 to 1 can be achieved with a minimum of field trimming. A transformer section is located near the input fitting.

The horizontal radiation pattern of the BFA antenna is essentially omnidirectional for top mounting. The horizontal pattern in free space is circular within 1 dB. The extent of deviation from a circular pattern for a side-mounted array is dependent on the type and size of the tower. It is recommended that the array be mounted, if possible, above the top set of guys on a guyed tower. Where this is not possible, the guys in the immediate area of the antenna should be broken by insulators every 3½ feet for a distance of at least 14 feet. In addition, each guy in the vicinity of the antenna should be insulated at the point where it connects to the tower.

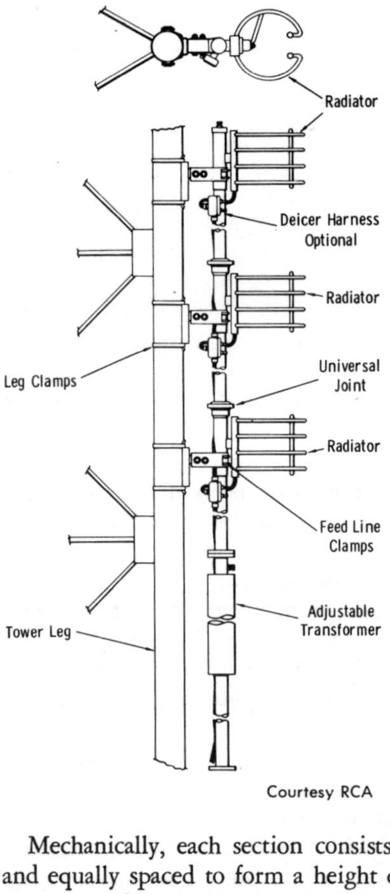

Courtesy RCA

Fig. 10-4. Three-bay fm antenna side-mounted on tower.

Mechanically, each section consists of four stainless-steel rings stacked and equally spaced to form a height dimension of 12 inches. The sections are mounted on 3⅛-inch coaxial line with an insulated feed stud energizing each radiating section. Only one coaxial transmission line is used to feed all sections of the antenna, and the individual radiating sections are

Table 10-2. Typical FM-Antenna Gains for Horizontally Polarized Elements

No. of Bays	Power	dB	Field
1	0.9	−0.5	0.95
2	2.0	3.0	1.41
3	3.0	4.8	1.73
4	4.1	6.1	2.02
5	5.2	7.15	2.28
6	6.3	8.0	2.51
7	7.3	8.63	2.70
8	8.4	9.25	2.90
9	9.4	9.72	3.07
10	10.5	10.2	3.25
11	11.5	10.6	3.39
12	12.5	11.0	3.55
13	13.6	11.33	3.69
14	14.6	11.65	3.83
15	15.6	11.93	3.95
16	16.6	12.20	4.07

identical mechanically and electrically. The radiators are both shunt fed and mechanically supported by this interconnecting feedline, which consists of modified lengths of 3⅛-inch rigid coaxial transmission line. The BFA-1B through BFA-8B antennas terminate mechanically in a pressurized top cap with bleed valve and a bottom input flange for coupling the antenna to the desired type of transmission line. Type BFA-10B through BFA-16B antennas are center fed through a matching tee, and the lower and top radiators terminate mechanically in pressurized caps.

In some installations, it may be desirable to mount the antenna inside of a large tower to protect it from falling ice. Where tower widths are over six feet, this is quite feasible. In this type of installation, the horizontal pattern is the result of several factors, including the tower width, tower-member size, and the operating frequency of the station.

Tower-leg mounting or side mounting to the tower has several advantages. Guyed towers and self-supporting towers of modest strength can be used, resulting in economy. Nearly all standard a-m towers will support a side-mounted broadband antenna.

Pole, or top-mounting, antennas are available. They may be mounted on top of existing towers, both guyed and self-supporting. Pole mounting is extremely useful in situations where the antenna is to be mounted on top of a building the esthetic qualities of which are to be maintained.

In areas where a quarter inch or more of ice is routine, deicing is recommended. Long strip heaters are inserted into the element arms, providing excellent heat transfer to ice-sensitive members of the radiating element. Depending on the type of connection, either 250 watts or 1000 watts of heat is available from a 110-volt ac source; 220 volts may also be used. An automatic control unit that operates the deicers when temperatures fall below preset levels and turns them off when temperatures rise can be used with each deicer system. Interbay deicing cable and conduit and electrical junction boxes are provided by the manufacturer to meet local electrical codes.

Combined Horizontal and Vertical Polarization

Fig. 10-5A shows half of the "doughnut" pattern of a half-wave dipole oriented horizontally. Maximum radiation occurs at right angles to the axis of the dipole. At increasing angles from the perpendicular, the radiation becomes weaker until, at the ends of the dipole, practically no radiation occurs at all. This is the familiar figure-8 transmitting or receiving antenna pattern; hence the basic half-wave dipole is considered to be a directional antenna.

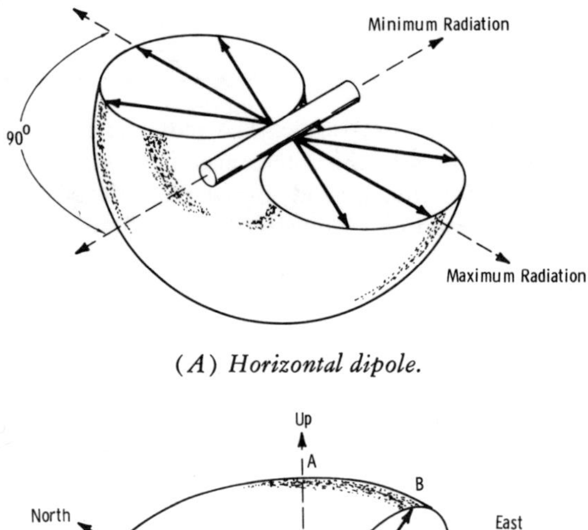

(*A*) *Horizontal dipole.*

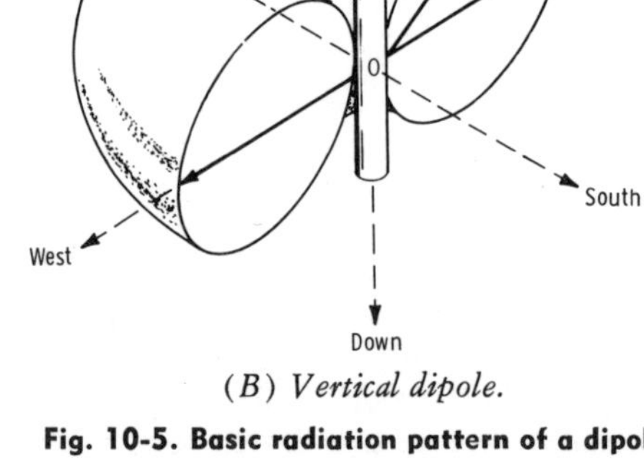

(*B*) *Vertical dipole.*

Fig. 10-5. Basic radiation pattern of a dipole.

The ring-type antenna discussed above is based on the half-wave dipole principle, but it is bent physically into a circle to obtain a 360° radiation pattern. Bear in mind, however, that the propagated wave is still horizontally polarized.

Fig. 10-5B illustrates the resultant radiation pattern when the dipole is oriented vertically. The electric field is now vertical with respect to ground, and the wave is vertically polarized. Note that although no radiation occurs (theoretically, in free space) from the ends of the dipole, a considerable amount can occur in other directions. Note that although radiation along the OA axis is zero, radiation along the OB axis is considerable, but reduced in magnitude from the value along the east-west axis. Thus field-strength readings in a horizontal plane from a vertical radiator must specify the vertical angle of radiation for which the readings apply.

It was pointed out in the previous chapter that the wavefront from an antenna consists of electric and magnetic fields which are in phase but at right angles to each other. The direction of the electric field is taken as the plane of polarization.

Now assume two antenna systems are fed from the same final rf amplifier; one of them consists of dipoles oriented vertically, and the other consists of dipoles oriented horizontally. Further assume that extra transmission line is employed in the horizontal feed to cause the rf current to lag the vertical current by 90° (one-quarter wavelength). The result is shown by Fig. 10-6. This assumes that the currents are of identical magnitudes.

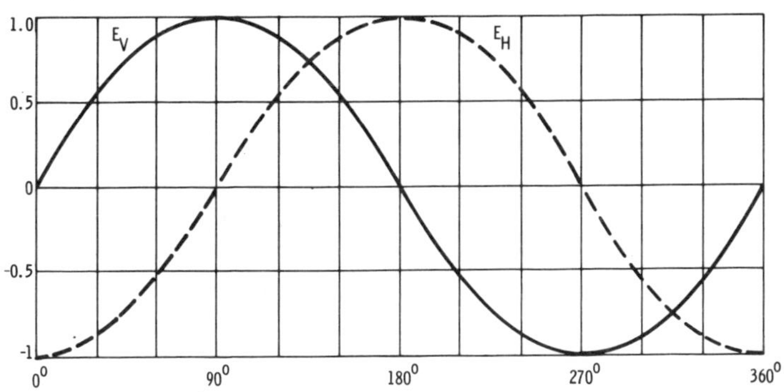

Fig. 10-6. Horizontal component of radiation delayed from vertical component by 90 electrical degrees.

Let E_V be the vertically polarized component, and let E_H represent the horizontally polarized component. Table 10-3 lists values for every 30° of the graph in Fig. 10-6.

At zero degrees, $E_V = 0$ and $E_H = -1$. The magnitude of energy is the vector sum, or:

Table 10-3. Instantaneous Values of E_V and E_H at 30° Intervals

Degrees	E_V	E_H
0	0.0	−1.0
30	+0.5	−0.866
60	+0.866	−0.5
90	+1.0	0.0
120	+0.866	+0.5
150	+0.5	+0.866
180	0.0	+1.0
210	−0.5	+0.866
240	−0.866	+0.5
270	−1.0	0.0
300	−0.866	−0.5
330	−0.5	−0.866
360	0.0	−1.0

$$E = \sqrt{E_V{}^2 + E_H{}^2}$$

Then if the vector sum at zero degrees is designated E1:

$$\begin{aligned} E1 &= \sqrt{(0.0)^2 + (-1.0)^2} \\ &= \sqrt{0+1} \\ &= 1 \end{aligned}$$

At 30°:

$$\begin{aligned} E2 &= \sqrt{(0.5)^2 + (-0.866)^2} \\ &= \sqrt{0.25 + 0.75} \\ &= \sqrt{1} \\ &= 1 \end{aligned}$$

If similar computations are made for the other angles in Table 10-3, the result in each case is $\sqrt{1} = 1$. Thus the intensity of the electric field remains constant, but the angle of polarization changes throughout the entire 360° of the wave, resulting in *circular polarization.* Fig. 10-7 illustrates this fact for the condition where E_H lags E_V by 90° as in the graph of Fig. 10-6. Note that here the direction of polarization rotates clockwise. Rotation may be either clockwise or counterclockwise, depending on the component shifted and the direction of the phase shift. This fact is not important for fm but is used in radar applications.

If the magnitudes of the rf currents are not equal in the two antenna systems, elliptical polarization results (Fig. 10-8). Power dividers can be used to separate the antenna systems to achieve greater amplitude in one plane of polarization than in the other.

For circularly polarized fm antenna systems, power gain in the horizontally polarized plane or in the vertically polarized plane is approximately

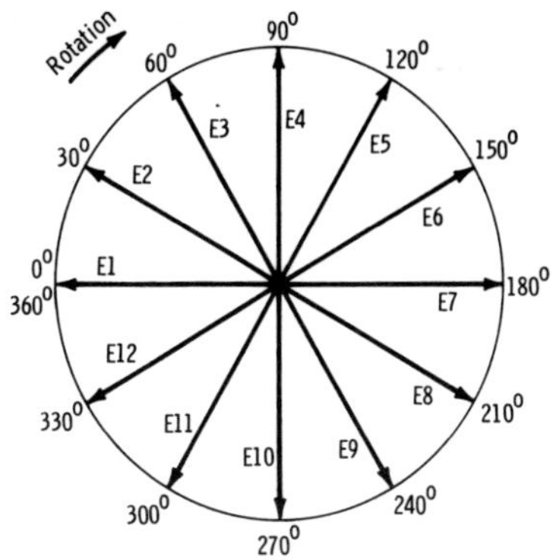

Fig. 10-7. Circular polarization, E_H lagging E_V by 90°.

equal to the number of stacked bays divided by two (the number of equivalent planes of polarization). When circular polarization is used, the transmitter power can be doubled without exceeding the licensed horizontal erp, since the additional power radiated is in other planes of polarization. Conversely, for a given transmitter power, the antenna gain can be doubled for the same reason.

Most recent circularly polarized fm antenna systems do not employ two separately polarized units, but each bay or element gives circular polarization by design. Fig. 10-9 illustrates one example, a single element of the CCA FMC antenna. Complete systems incorporate from 1 to 16 elements. Normally, systems with eight elements or fewer are fed at the bottom, but systems with more than eight elements are fed at the center. The element is essentially a 1½-turn helix constructed of 2-inch diameter, thick-walled copper with rounded surfaces to minimize corona problems. Individual elements are placed one wavelength apart. The power gains listed in Table 10-4 are for equal horizontally and vertically polarized components; other ratios are available. The radiation centers of both the vertical and horizontal

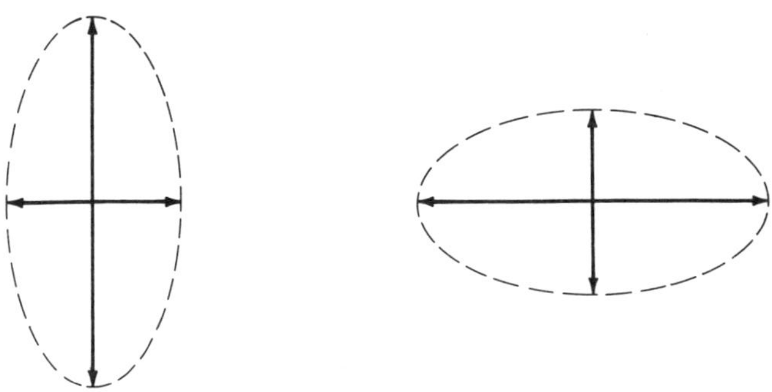

(*A*) E_V *greater than* E_H. (*B*) E_H *greater than* E_V.

Fig. 10-8. Elliptical polarization.

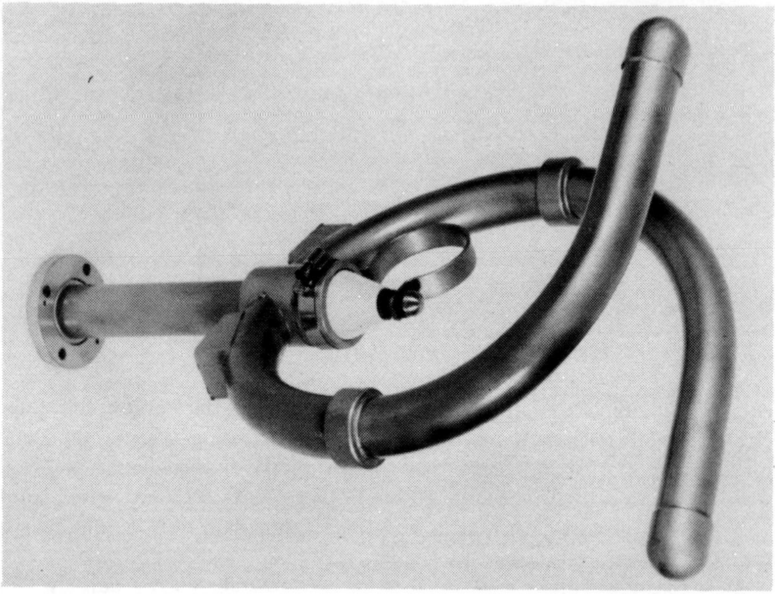

Courtesy CCA Electronics Corp.

Fig. 10-9. Single element of CCA FMC circularly polarized fm antenna.

components are identical, thus achieving true phase coincidence, which is essential for true circular polarization.

The basic element can serve with a minimum of retuning as a radiator at any frequency in the standard fm broadcast band. Each of the elements in the antenna system represents an impedance of 50 ohms times the number of elements. For example, an 8-bay antenna would have 400-ohm elements. Thus, since all elements are spaced one wavelength apart, all elements are fed in parallel with equal power and represent a nominal 50 ohms to the feed line.

Because of the substantial diameter of the basic elements, they are fairly low-Q broadband elements. Thus it is possible, with associated combiners, to have one CCA Type FMC antenna radiate several fm channels.

The free-space circularity patterns (Fig. 10-10) of the FMC antenna are within ±1 dB of perfect circularity. However, when side mounted on a wide tower, the FMC, like all antennas which radiate a vertical component, produces a pattern in the vertical field with nulls and some directivity. However, in practice the antenna will provide the urban area with considerably better coverage than a horizontal-only radiating system.

The antenna is supplied with a triple stub tuner which provides adjustable capacitors at discrete positions in the feed line. These adjustments are such that they compensate for impedance changes due to the mounting environment and still achieve a matched condition.

Table 10-4. CCA FMC Antenna Gain Specifications

Type No. & Bays	Power Gain	Gain in dB	Field Gain	Field Strength at 1 Mile for 1 kW (mV/m)
FMC-LP-1	0.45	−3.24	0.67	93
FMC-HP-1	0.475	−3.21	0.69	95
FMC-LP-2	0.95	−0.2	0.98	136
FMC-HP-2	1.00	0.0	1.0	138
FMC-LP-3	1.5	1.8	1.23	170
FMC-HP-3	1.55	1.9	1.25	172
FMC-LP-4	2.05	3.1	1.44	199
FMC-HP-4	2.15	3.30	1.47	203
FMC-LP-5	2.55	4.1	1.6	221
FMC-HP-5	2.70	4.3	1.65	227
FMC-LP-6	3.15	5.0	1.78	246
FMC-HP-6	3.30	5.2	1.82	251
FMC-LP-7	3.65	5.6	1.92	265
FMC-HP-7	3.85	5.9	1.97	273
FMC-LP-8	4.2	6.2	2.05	283
FMC-HP-8	4.4	6.4	2.10	290
FMC-LP-10	5.2	7.2	2.28	315
FMC-HP-10	5.5	7.4	2.35	325
FMC-LP-12	6.25	8.0	2.5	345
FMC-HP-12	6.60	8.2	2.57	355
FMC-LP-14	7.3	8.6	2.7	373
FMC-HP-14	7.7	8.9	2.8	386
FMC-LP-16	8.4	9.2	2.9	400
FMC-HP-16	8.8	9.5	2.97	410

Low power (LP) uses 1⅝-inch feed line.

High power (HP) uses 3⅛-inch feed line (still higher powers available).

Antenna polarization is circular clockwise for all directions of azimuth.

Length of antenna in feet is 984 divided by frequency in MHz times one less than number of bays, plus 6½ feet for transformer.

The deicers of the FMC series are stainless steel rods which require approximately 400 watts of power consumption per bay. They are capable of achieving ice-free operation at ambient temperatures of 0°C and in 50-mph winds. They require 230 volts, single phase, balanced to ground.

The Harmonic Filter

A harmonic filter is normally used in series with the antenna feed to provide a high attenuation to harmonics and negligible attenuation to the fm band. A description of the Jampro JF-50 harmonic filter follows.

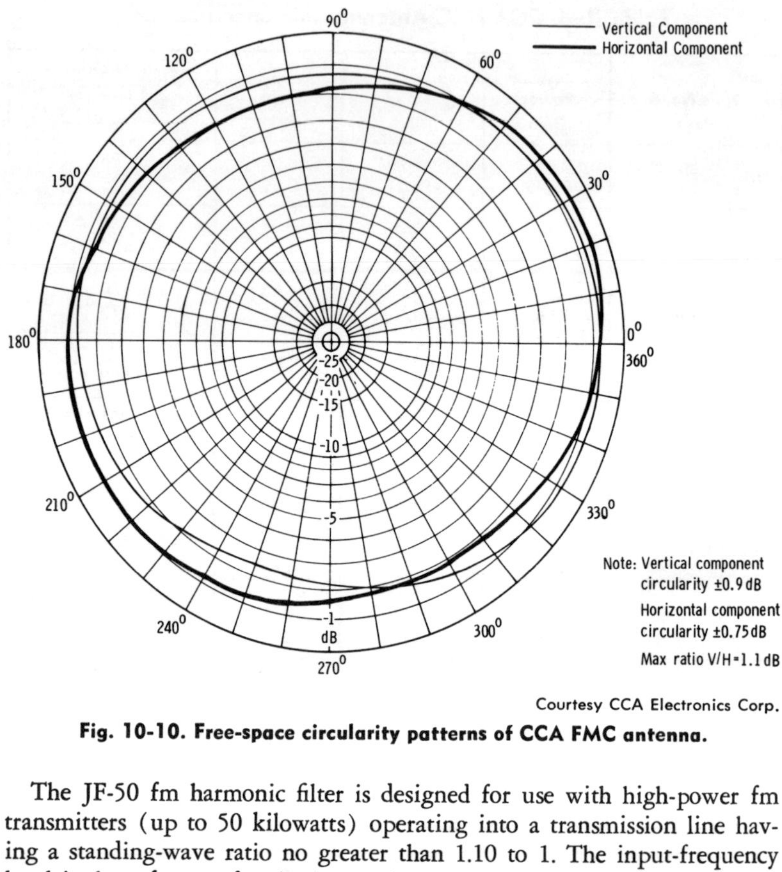

Courtesy CCA Electronics Corp.

Fig. 10-10. Free-space circularity patterns of CCA FMC antenna.

The JF-50 fm harmonic filter is designed for use with high-power fm transmitters (up to 50 kilowatts) operating into a transmission line having a standing-wave ratio no greater than 1.10 to 1. The input-frequency level is the reference for the harmonic attenuation. The filter is tuned for best response for a given frequency in the fm band and is fixed-tuned to a given channel prior to shipment. The design passband of the filter is two megahertz wide. There are no spurious passbands in the range from the fundamental to the tenth harmonic that are not attenuated at least 10 dB. The attenuation in the desired passband is negligible due to the basic design of the filter.

The filter is designed to be inserted into a standard 3⅛-inch, 50-ohm transmission line between the transmitter and the antenna. The filter is airtight but will pass air as required in a pressurized system. To install the unit, simply replace a section of transmission line with the filter and make connections in the standard manner. In the installation, observe the input and output markings on the filter. The filter is provided with 3⅛-inch EIA flanges and fixed bullets.

The filter consists of a series of constant-k midsections with a constant-k half-section at the input and output to provide an impedance match to the

50-ohm transmission line. The constant-k sections have been designed according to conventional filter theory. They are assembled as shown in Fig. 10-11A. The equivalent electrical circuit is shown in Fig. 10-11B. The midsections have a cutoff frequency of 110 MHz, and the half-sections are designed with a slightly higher cutoff frequency to provide a good impedance match over the fm band.

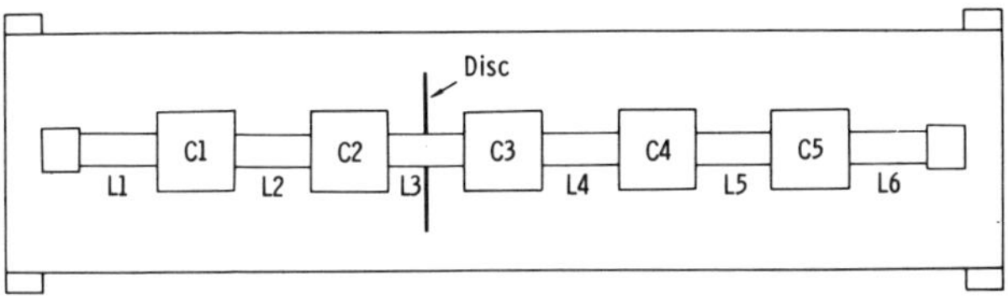

(A) Physical arrangement.

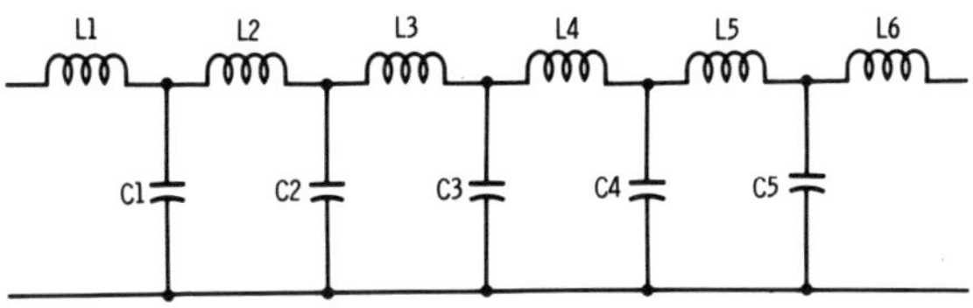

(B) Schematic diagram.

Fig. 10-11. Typical harmonic filter.

The L and C circuit elements are made up of lengths of coaxial transmission line. A section of low-impedance line is used for C, and a section of high-impedance line is used for L, as is illustrated in Fig. 10-11A. In a filter of this type using a series of coaxial elements, there will be several frequencies in the attenuation band at which compensation between elements occurs, causing transmission through the filter. The small disc between C2 and C3 is used to eliminate or reduce this effect in the harmonic ranges of the filter out to the tenth harmonic.

The filter has a very low vswr (under 1.06 to 1 for the frequency of operation). During tests of the unit, the positions of the C sections are adjusted slightly to obtain the best vswr.

The extremely low insertion loss makes it possible to use this filter under high-power conditions without heat damage. All of the internal insulation is Teflon.

In normal operation, the filter will show a very poor vswr to harmonic energy. The harmonics, depending on their amplitude with respect to the fundamental, may cause excessively high reflectometer readings when the reflectometer is inserted between the transmitter and the harmonic filter. In this event, it is recommended that the reflectometer probe be inserted in the transmission line between the filter and the antenna.

10-3. FREQUENCY MODULATION (FM) AND PHASE MODULATION (PM)

The two basic methods employed to obtain a frequency-modulated carrier wave are phase modulation (pm) and direct frequency modulation (fm). Basic block diagrams of the two types of transmitters appear in Fig. 10-12.

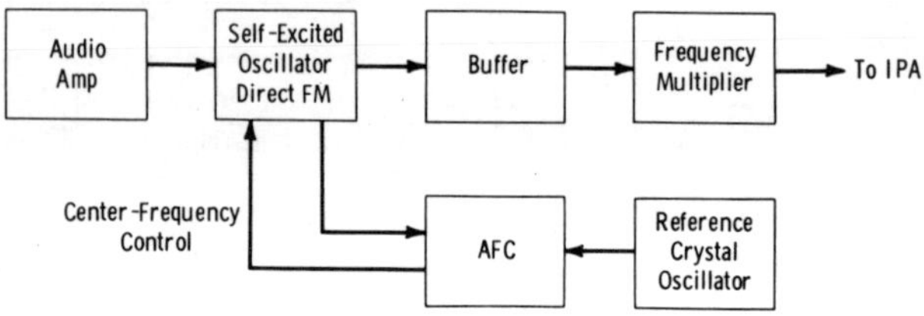

(*A*) *Direct fm.*

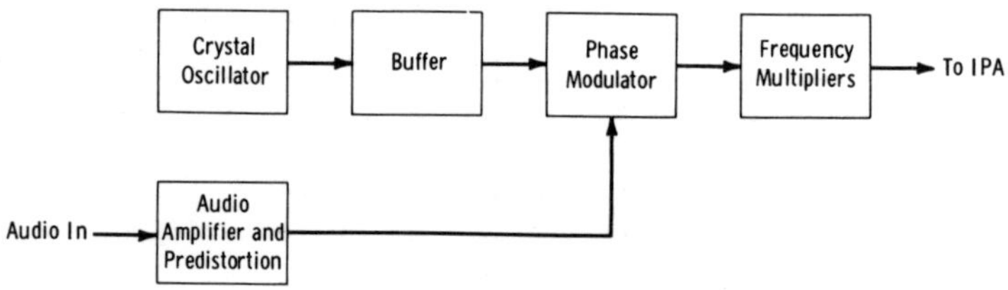

(*B*) *Phase modulation.*

Fig. 10-12. Block diagrams of pm and fm transmitters.

Modulation Methods

In direct fm (Fig. 10-12A), a self-excited, voltage-controlled oscillator is varied in frequency at the audio rate applied. The *frequency deviation* is proportional to the instantaneous *amplitude* of the modulating signal. The frequency of the modulating signal determines the rapidity of the swing through the center carrier frequency.

In pm (Fig. 10-12B), frequency deviation is also proportional to the instantaneous amplitude of the modulating signal, but the rapidity (frequency) of *phase shift* is determined by the modulating signal frequency. Therefore, the frequency deviation in pm is proportional to both the amplitude and frequency of the signal applied.

Fig. 10-13 illustrates the difference between pm and fm carriers modulated by identical sine-wave signals. Note that for pm, maximum carrier frequency occurs during the maximum rate of change of the modulating sine wave (intervals between peaks). For fm, maximum carrier frequency occurs on the negative peaks of the modulating signal, and minimum carrier frequency occurs on the positive peaks (for the particular system

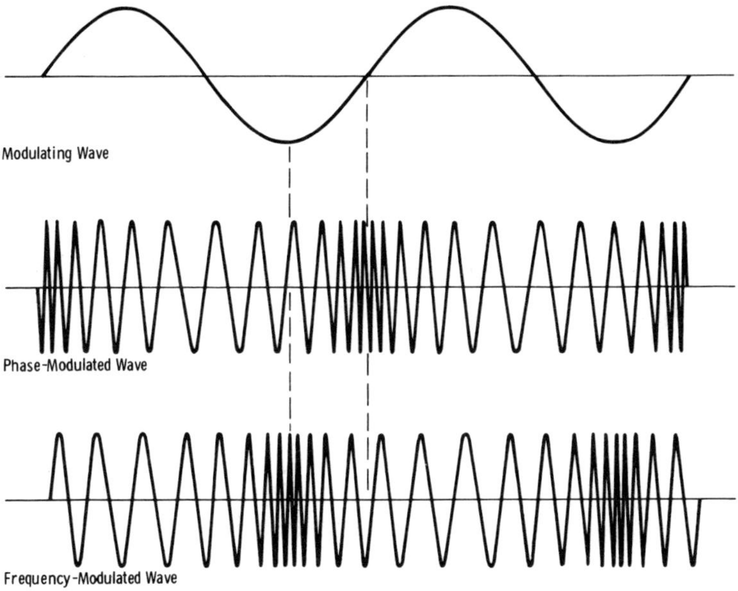

Fig. 10-13. Comparison of phase and frequency modulation.

illustrated). An fm receiver is designed to demodulate a pure fm signal. For this reason, the audio prior to phase modulation is made to have a response inversely proportional to frequency ($1/f_m$). This is shown in Fig. 10-12B as *predistortion* in the audio block. In this manner, pm is made equivalent to fm.

In the pm transmitter, the audio signal is applied in an amplifier stage, which allows a highly stable crystal oscillator to be used without further frequency control. In direct fm, a self-controlled oscillator must be modulated, and a separate highly stable crystal oscillator must be used to control the center operating frequency of the modulated oscillator. This is done by counting down from the modulated oscillator frequency, comparing a difference between this frequency and the reference crystal frequency, and deriving a dc voltage proportional to any amount of correction required. This correction voltage is fed to the voltage-controlled modulated oscillator (vco) to hold the center carrier frequency at the assigned value. Both pm and fm transmitters are covered further in Section 10-4.

FM Sidebands

In angular modulation of a carrier wave (pm or fm), the modulating signal adds no power to the carrier as it does with a-m. The amplitude of the modulated carrier envelope remains constant, regardless of the amplitude or frequency of the modulating signal. Note that it is the modulation *envelope* that remains constant; the individual carrier cycles, of course, cross (positive and negative) through zero upward and downward to form this carrier envelope. Individual cycles of the carrier component and the various sidebands go to zero amplitude at specific values of the *modulation index* (m). This characteristic can be used to measure fm deviation under sine-wave modulation, as covered in Chapter 14.

In contrast to a-m, angular modulation of a carrier with a single sine wave yields an infinite number (theoretically) of sidebands spaced by the modulation frequency, f_m. In practice, the required bandwidth is determined by considering the number of *significant* sidebands. The term "significant" means those sidebands which have a voltage at least 1 percent (−40 dB) of the unmodulated carrier voltage. Sideband amplitudes become negligibly small beyond a certain frequency range from the center carrier frequency, depending on the magnitude of the modulation index.

The modulation index (m) is:

$$m = \frac{\Delta F}{f}$$

where,

m is the modulation index,
ΔF is the peak deviation in hertz,
f is the modulating frequency in hertz.

For fm, the ratio of the peak frequency deviation of the carrier to the frequency of the modulating signal is m, the modulation index. The modulation index is equal to the peak phase deviation of the carrier in radians.

For pm, the peak phase deviation of the carrier in radians is the modulation index. As in fm, the modulation index is equal to the ratio of the peak frequency deviation of the carrier to the frequency of the modulating signal. Note therefore that the expression for the modulation index is the same for either fm or pm.

In fm broadcasting, 100-percent modulation is related to ±75 kHz frequency deviation. Thus for 100-percent modulation ($\Delta f = 75$ kHz) and assuming a 15-kHz modulating sine wave (f_m):

$$m = \frac{75}{15} = 5$$

For $m = 5$, actual spectrum analysis shows that the resulting signal has eight significant sideband pairs spaced f_m apart (A in Fig. 10-14). This actually exceeds the 150-kHz signal bandwidth alloted to fm broadcasting (150 kHz plus 25-kHz guard band at each side of the channel). The actual significant bandwidth required for 100-percent modulation at 15 kHz may be seen to be:

$$(2)(8)(15 \text{ kHz}) = 240 \text{ kHz}$$

The justification for this characteristic is that program content does not

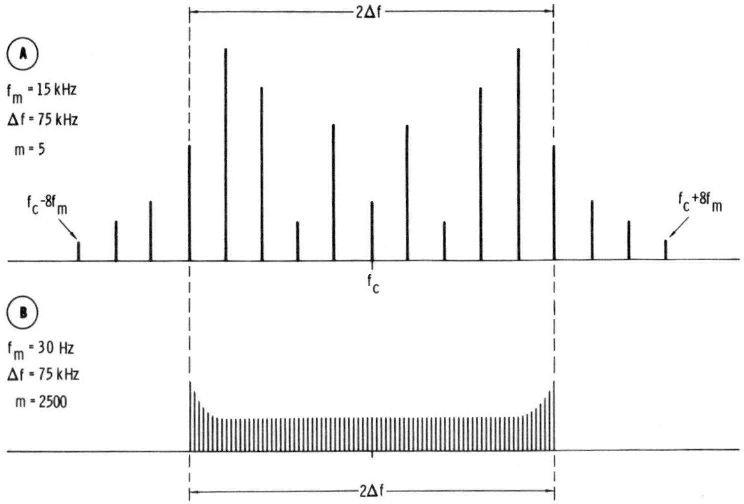

Fig. 10-14. Spectrum analysis of fm signals.

have 15-kHz components that approach 100-percent modulation when the average program content itself is held to this amplitude.

The required audio response for fm broadcast transmission is 30 Hz to 15 kHz. The modulation index is 2500 for a 30-Hz signal at 100-percent modulation. As m approaches and exceeds a value of 100, the bandwidth becomes simply $2\ \Delta f = 150$ kHz at the lower frequencies (B in Fig. 10-14).

10-4. FUNDAMENTAL FM TRANSMITTER CIRCUITRY

An fm transmitter has essentially the same outward appearance as an a-m transmitter. However, most of the circuitry is significantly different.

Basic Method of Phase Modulation

A block diagram of a typical pm exciter unit is shown in Fig. 10-15. The basic function of each numbered block is as follows:

1. The crystal oscillator generates a stable sine-wave signal at a relatively low submultiple of the final assigned carrier frequency.
2. Pulse shaping is accomplished by overdriving a tube so that plate current flows in short pulses and is cut off between pulses by the automatic bias developed by grid current. This shaping action is necessary for the following sawtooth generator.
3. The sawtooth is generated by charging a capacitor over a small linear time duration, then rapidly discharging it as the tube is driven into

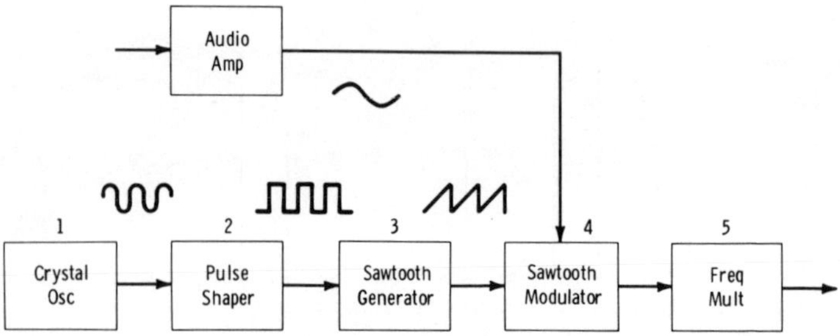

Fig. 10-15. Block diagram of pm exciter.

conduction by the plate pulses from the preceding stage. Thus a sawtooth at the crystal-oscillator frequency is formed.

4. One of the tube elements, such as the control grid, receives the sawtooth from the generator, while another element, such as the cathode, receives the amplified program audio. This modulator stage is usually operated with low plate-to-cathode voltage. The audio voltage superimposed on the static cathode-bias voltage shifts the point on the saw slope at which the tube begins to conduct so that the phase of the conduction point is advanced or retarded at the audio rate (Fig. 10-16). The actual phase shift, which is quite low for good linearity, is about 2 radians (roughly 120°).
5. A string of frequency multipliers increases the crystal frequency to the assigned carrier frequency. As the frequency is multiplied, so is the phase shift; thus, the full 100-percent modulation (which represents ±75 kHz for standard fm broadcast) can be achieved.

The method just described is termed *phase-shift modulation employing pulse techniques,* or sometimes *serrasoid modulation.* The average carrier frequency is tightly controlled by the crystal and is independent of the modulation process. The carrier is deviated only during the time the phase is being changed by modulation. Since the oscillator and modulator stages are fully isolated, the modulation process cannot affect the average carrier frequency. The pulses are converted to a phase-modulated sine-wave carrier by the following tuned amplifiers. There are other methods of pm, but this form is the most popular.

The audio signal is pre-emphasized at the transmitter before modulation. A complementary de-emphasis circuit is employed in the receiver and in transmitter monitoring equipment. This arrangement improves the signal-to-noise ratio so that extended dynamic range is a practical reality. Also in a pm transmitter, the response of the audio circuits driving the modulator must be inversely proportional to frequency. This is necessary to make pm equivalent to fm, as developed below.

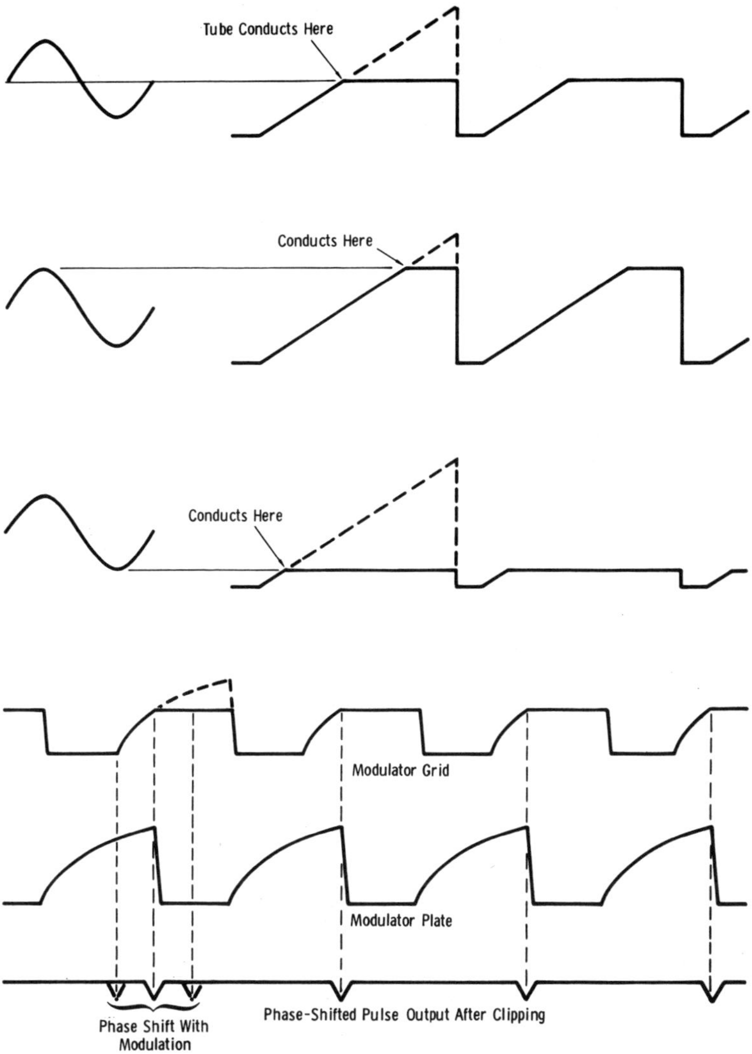

Fig. 10-16. Modulated sawtooth waveforms.

The relationship between the phase shift ($\Delta\theta$) in radians and the equivalent frequency deviation (ΔF) from the mean carrier frequency is:

$$\Delta\theta = \frac{\Delta F}{f_m} = m$$

It should be noted that since the required phase shift in radians is directly related to the modulation index (m), the maximum shift in radians occurs at the lowest modulating frequency to be considered. This is 30 Hz for fm broadcast. If the oscillator frequency is 100 kHz and the assigned carrier frequency is 100 MHz, a frequency multiplication of 1000 is required to obtain the proper carrier frequency. It was shown above that $m = 2500$ for a 30-Hz modulating frequency at 100-percent modulation ($\Delta F = 75$ kHz). Then the required phase shift in radians of the 100-kHz signal is $2500/1000 = 2.5$ radians. Since the phase shift is multiplied in direct ratio to the frequency multiplication, the required 2500-radian phase shift occurs at the multiplier output. Note also that at $f_m = 15$ kHz ($m = 5$) only a 0.005-radian phase shift should occur in the modulator. Thus the audio input to the modulated stage must be made to have a response proportional to $1/f_m$.

The Direct FM Oscillator

Many recent fm transmitters employ *direct fm,* which means that a self-controlled oscillator is directly varied in frequency by the modulating signal. In earlier tube-type fm oscillators (many of which are still in use), the *reactance-tube* method was the most popular form of circuitry, and we will review this method of modulation briefly before going to modern solid-state techniques.

A typical reactance-tube circuit is illustrated in Fig. 10-17A. The circuit uses a pentode tube with the plate connected across the tank circuit of a self-excited oscillator. The two blocking capacitors on each side of the pentode plate lead have negligible reactance at the oscillator frequency and become essentially short circuits. The plate current and the grid voltage of the reactance tube are placed in quadrature (90°) with the plate-voltage swing by R and C. Since the plate current is in quadrature with the plate voltage, the tube appears as a voltage-controlled reactance across the oscillator tank circuit and therefore controls the oscillator frequency. With R much greater than the reciprocal of ωC (where $\omega = 2\pi f$), the reactance tube acts as a variable inductance. (The current lags the oscillator rf voltage by 90°.)

The audio voltage on the grid of the reactance tube varies the transconductance (g_m) of the tube and therefore the magnitude of the simulated inductive reactance. The relationship is:

$$X = \frac{R\omega C}{g_m}$$

An increase on the positive alternation of the applied signal increases g_m. Since g_m is in the denominator of the equation, an increase in value lowers the resultant magnitude of the inductance in the oscillator tank circuit and hence increases the frequency. A negative signal swing decreases g_m and hence increases the inductive reactance, lowering the oscillator frequency. The center operating frequency is held constant by a dc voltage from the afc circuit.

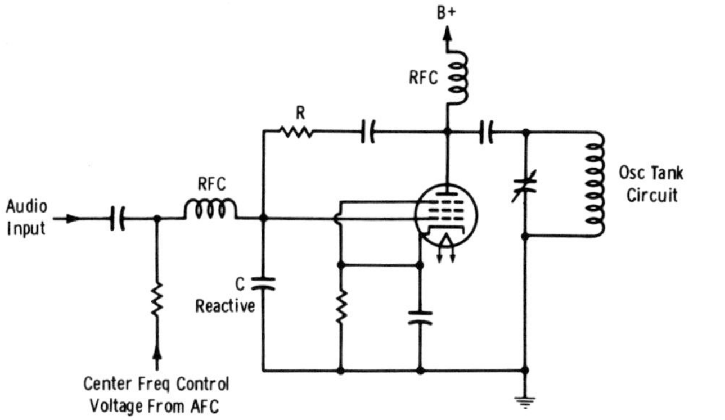

(A) Simulated variable inductance.

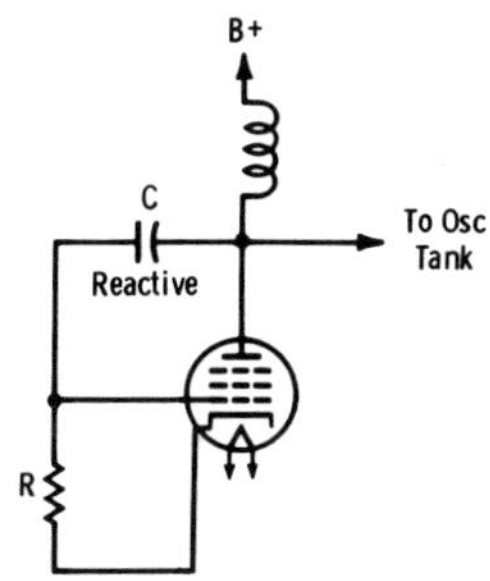

(B) Simulated variable capacitance.

Fig. 10-17. Fundamental reactance-tube circuits.

If R and C are interchanged and R is made much less than $1/\omega C$, the circuit of Fig. 10-17B applies. In this case, the reactance tube appears as a voltage-variable capacitance across the oscillator tank circuit. The reactance is:

$$X = -\frac{1}{g_m R \omega C}$$

A positive signal swing increases g_m, lowers the magnitude of the capacitive reactance across the oscillator tank, and decreases the frequency. A negative signal swing decreases g_m, raises the value of the reactance, and increases the oscillator frequency.

The signal plate current of the reactance tube must be kept small relative to the oscillator tank current; otherwise, amplitude modulation may result. Also, the transconductance of the reactance tube must vary linearly with grid voltage over the audio operating range for good modulation linearity.

Some direct fm transmitters employ push-pull reactance-tube modulators (employing a split oscillator tank circuit) to achieve very low harmonic distortion. Since push-pull stages are fed in opposite phase, one reactance tube acts as an inductance, and the other acts as a capacitance. A greater linear frequency deviation can be achieved in the direct fm oscillator by this method.

Solid-State Direct FM

In order to understand solid-state circuits for direct fm, it is necessary to be familiar with the characteristics of the voltage-variable-capacitance diode. The width of the barrier in a pn junction is influenced by the voltage across the junction. Since an electrostatic field is thus produced, the junction is the equivalent of two plates of a capacitor in which the spacing between the plates is governed by the voltage (and current) of the junction.

Figs. 10-18A and 10-18B illustrate this effect. Anything that increases the width of the pn junction barrier is equivalent to spreading apart the plates of the capacitor, resulting in less capacitance. As the reverse bias is decreased, the junction capacitance increases.

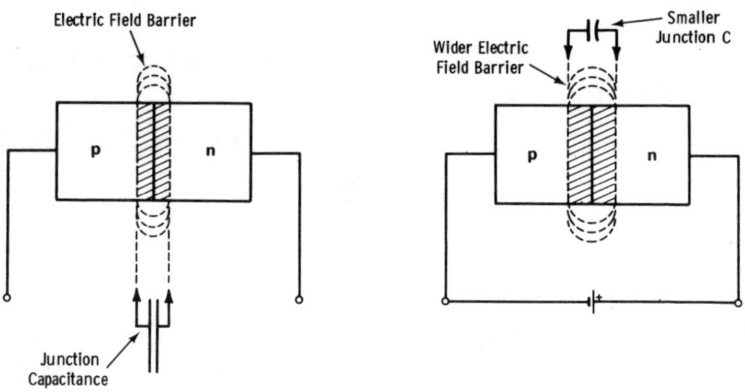

(*A*) *Zero bias.* (*B*) *Reverse bias.*

Fig. 10-18. Voltage-variable-capacitance diode.

Fig. 10-19A illustrates the junction-capacitance curve typical of nearly all signal diodes. Note that, in the reverse-voltage region of operation, the capacitance change is quite small over a 100-volt excursion. By comparison, Fig. 10-19B is an example (only) of the curve for one type of variable-capacitance diode. There are many different types of variable-capacitance diodes with different sensitivities in capacitance change per volt of change. All of these diodes are typically nonlinear in capacitance-to-voltage change, as shown by this graph.

Some common circuits employ a voltage-variable diode capacitor which may be known as a *variable-capacitance diode, varicap,* or *varactor.* (The term "varactor" is usually applied to microwave devices.) This is a diode

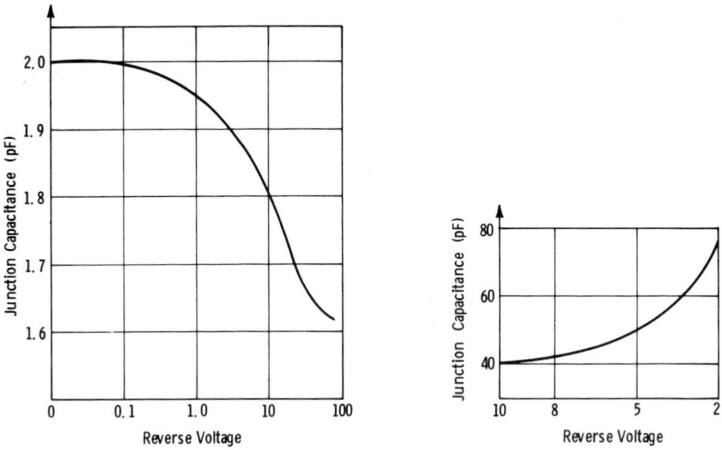

(*A*) *Typical small-signal diode.* (*B*) *Variable-capacitance diode.*

Fig. 10-19. Variation of junction capacitance with voltage.

which is manufactured to maximize certain physical and chemical features affecting junction capacitance. Fig. 10-20 shows three symbols for this type of diode that may be found in practice on schematic diagrams.

Like the zener diode, the variable-capacitance diode is always operated in the reverse-biased condition. The capacitance decreases with increased reverse bias; that is, the capacitance varies inversely with the reverse bias voltage. The capacitance-versus-voltage relationship is nonlinear. The diode normally has a greater change in capacitance for a given change in applied voltage than the effective capacitance change in a conventional reactance-tube circuit.

When the total capacitance used to resonate a tuned circuit is split and used to form a feedback voltage divider, a Colpitts oscillator results. Fig. 10-21A illustrates a transistor Colpitts oscillator, in which C1 and C2 together set the effective capacitance with which L resonates. Remember that part of this capacitance is "hidden" in junction capacitance of the transistor. The important point at this time is to understand the effect of a capacitance diode on an LC tuned circuit.

Fig. 10-20. Symbols for voltage-variable-capacitance diodes.

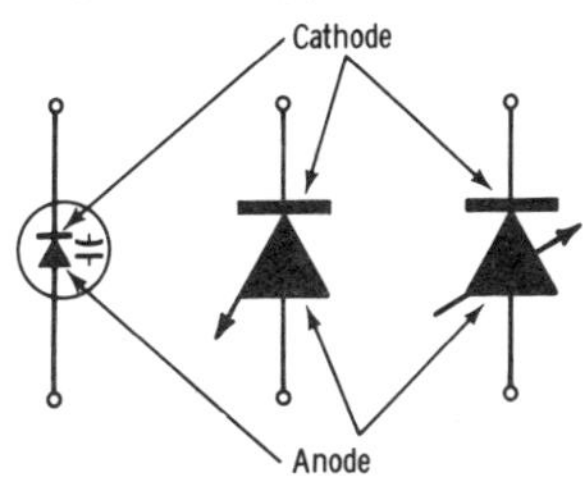

Now look at Fig. 10-21B. This is the same circuit as the one in Fig. 10-21A, but with added components. Variable-capacitance diodes X1 and X2 must remain reverse biased at all times; this bias is supplied through resistors R1 and R2. Capacitors C3 and C4 are large and have negligible reactance at the oscillator frequency. Now assume that a signal is applied with a swing in the negative polarity. The effective capacitance of X1 and X2 varies with the instantaneous audio voltage. Since the diodes are effectively connected to the tank circuit, the resonant frequency is varied in step with the instantaneous audio voltage variation. For a negative excursion of the modulating signal, the reverse bias is increased. Since capacitance varies inversely with the reverse bias, the capacitance decreases, *increasing* the output frequency of the oscillator. Conversely, as the signal goes positive, the frequency *decreases.* The result is a frequency-modulated (but nonlinear) output.

Note that with the same type of transistor and the same polarity of audio input signal, X1 and X2 could be reversed in connection of polarity, with a reverse bias now of negative rather than positive polarity. In this case, the oscillator would be caused to *decrease* in frequency for negative swings and increase in frequency as the signal swings positive. If both of the foregoing oscillator circuits are used and the same audio signal is applied to both inputs, one oscillator increases in frequency as the other decreases. This is a push-pull type of frequency modulation which tends to cancel the nonlinearity of modulation resulting from the nonlinear voltage-capacitance relationship of the diodes. When this type of modulation is not used, a nonlinear amplifier with characteristics inverse to those of the modulator must be used to restore linearity.

Audio Pre-Emphasis

The relatively low energy of the higher-frequency components in speech and music makes it desirable to pre-emphasize the high-frequency compo-

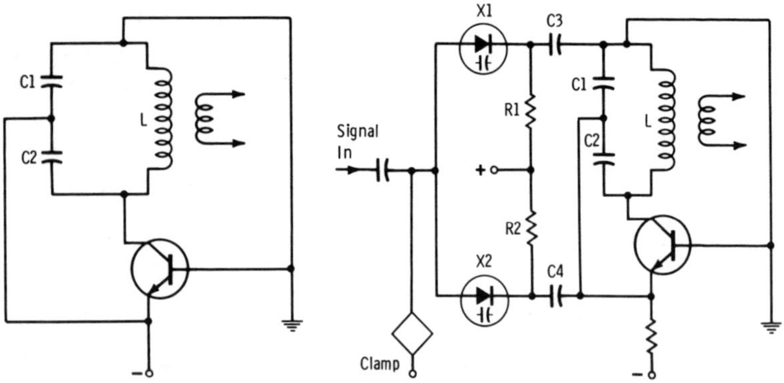

(*A*) *Colpitts oscillator.* (*B*) *Fm Colpitts oscillator.*

Fig. 10-21. Use of voltage-variable-capacitance diodes for direct fm.

nents prior to modulation in the transmitter, and to employ a corresponding de-emphasis subsequent to demodulation in the receiver. This raises the resulting signal-to-noise ratio for a wideband receiver. The pre-emphasis curve employed at the transmitter is shown in Fig. 10-22 with the FCC limits prescribed.

The FCC standard is stated in terms of the L/R time constant of a series combination of inductance and resistance. The complementary de-emphasis constant is the RC time constant of a series combination of resistance and capacitance. The specified time constant is 75 microseconds.

The pre-emphasis network is normally part of, or immediately follows, the first amplifying stages for the audio signal within the transmitter. In a pm transmitter, this network preceds the $1/f_m$ network required to convert pm to fm. Therefore, the $1/f_m$ network works from the standard pre-emphasized signal, and the pm transmitter produces the same pre-emphasized signal as the fm transmitter. The receiver de-emphasis network receives the same signal characteristic whether the transmitter is pm or fm.

AFC for Direct FM

Since a crystal oscillator cannot be employed for the direct fm process, some form of automatic frequency control (afc) must be used. In general, this depends on a highly stabilized crystal oscillator unit operating at some frequency below the assigned carrier frequency. This oscillator is used to obtain a beat frequency with the carrier, and the beat frequency is then counted down to a very low frequency to derive a dc with a magnitude and polarity that correspond to any drift of the center carrier frequency. This dc is applied to the modulated oscillator for center-frequency control.

Fig. 10-23A is a basic block diagram of the CCA FM-10DS exciter unit, which employs direct varicap fm with afc control. The reference crystal oscillator is a transistor stage. The crystal is a third-overtone type that operates at ½ of the required frequency. The oscillation and frequency multiplication are accomplished in the same stage by means of resonant circuits for the oscillator frequency and a frequency doubler. The output frequency of the circuit is 250 kHz below the desired carrier frequency of the fm exciter. The output of the oscillator is fed to a transistor mixer stage.

A sample rf signal obtained from the output stage of the exciter is fed through a resistor to the base of the mixer transistor. The i-f (difference frequency) is obtained at the collector of the mixer after filtering of the high frequencies to ground by means of a capacitor. The i-f is amplified by a transistor stage and is further amplified and limited by an additional stage. The i-f thus obtained is centered at approximately 250 kHz, and it includes the modulation content of the fm exciter output.

A 250-kHz square-wave output of the limiter stage is fed to the flip-flop (integrated circuit), which divides this frequency by two and provides a consistent square-wave output. The output of the flip-flop is fed to the sec-

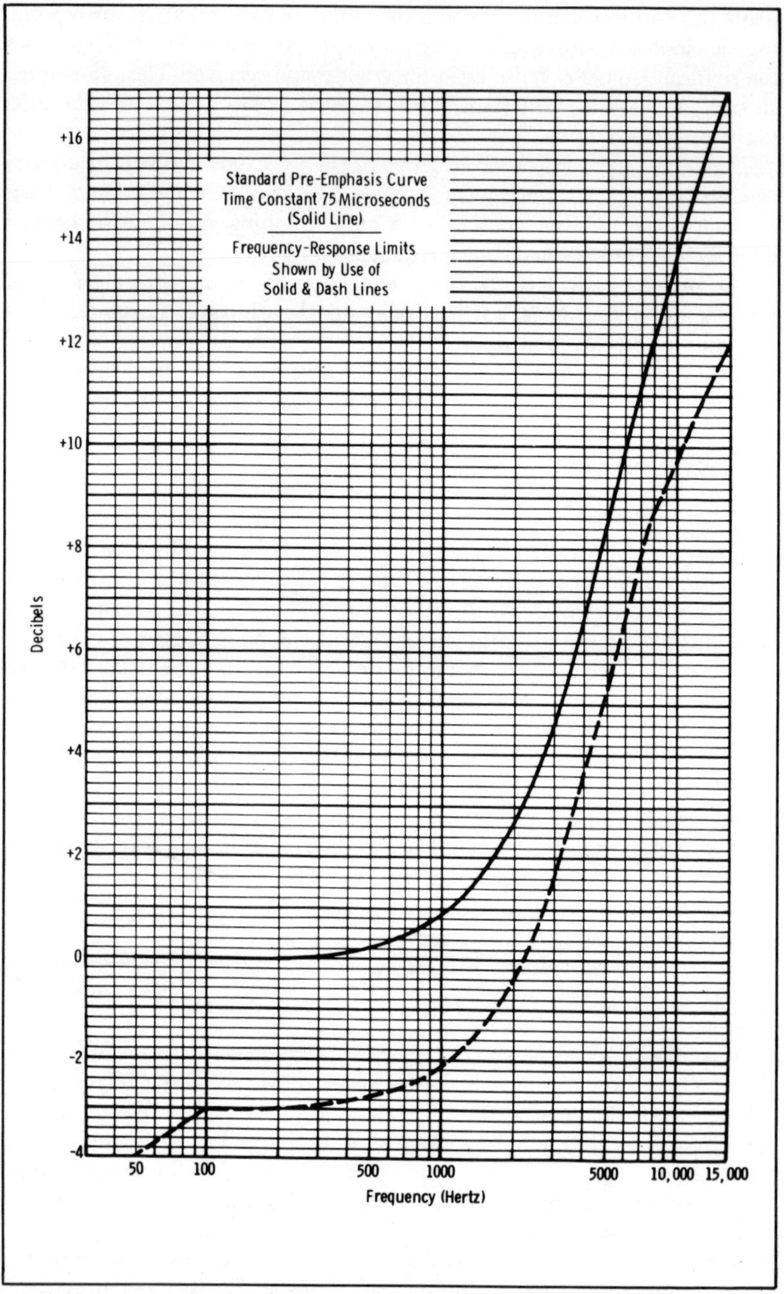

Fig. 10-22. FCC pre-emphasis requirements for fm stations.

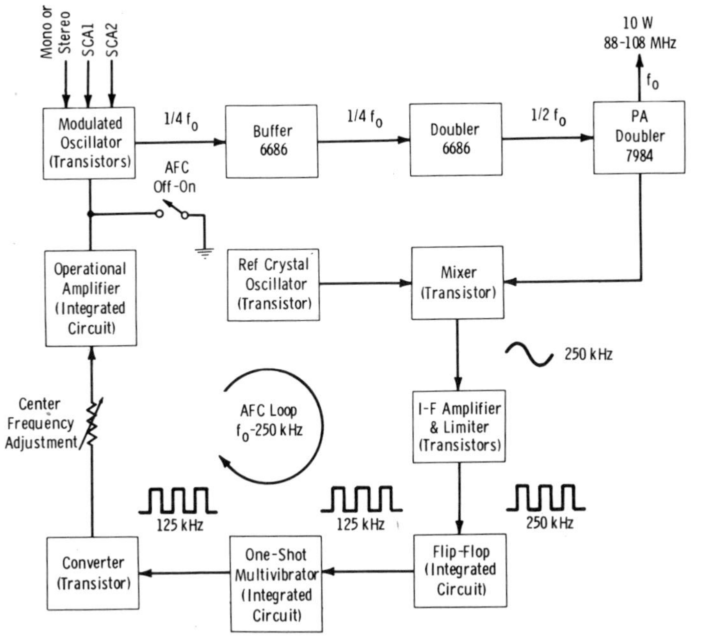

(A) Block diagram of exciter.

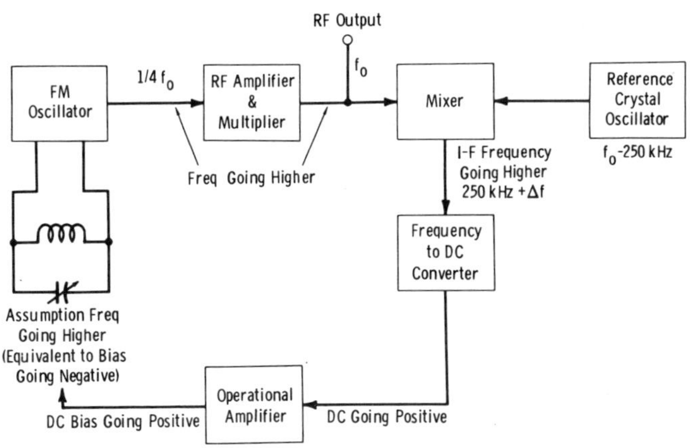

(B) Block diagram of afc loop.

Fig. 10-23. CCA FM-10DS exciter.

ond integrated circuit, which is connected as a one-shot multivibrator. This circuit provides a square pulse of constant width and amplitude, regardless of the shape of the input pulse. The square pulse is obtained every time that the input signal goes in the positive direction.

The output of this IC consists of positive pulses of constant width and amplitude and a repetition rate that depends on the input frequency. The average dc content of the square pulses varies according to the frequency. These pulses are applied to the converter stage, which consists of a transistor and a filter circuit. The output consists of a dc level proportional to the repetition frequency. The filter circuit is so designed that the modulation content of the intermediate frequency does not change the average dc output; that is, the filter is unable to follow any frequency variation of more than 5 Hz. The dc circuit is completed through resistors to the −15-volt dc power supply.

The afc adjustment potentiometer (center-frequency adjustment) determines the amount of the dc voltage that is fed to the operational amplifier. The position of the arm of this control can be selected such that its voltage is about zero volts dc when the intermediate frequency centers around 250 kHz. Any deviation from that frequency provides a dc voltage which will be amplified by the operational amplifier.

The gain of this amplifier is about 100 times, and its output is clamped through two gate diodes in order to hold the excursions to not more than plus or minus 4 volts. This afc correction voltage is applied to the anodes of the voltage-variable-capacitance diodes to hold the center frequency of the fm oscillator at the assigned value. When the afc action is to be disabled, a front-panel afc switch simply shorts the afc correction voltage to ground.

Fig. 10-23B is a simplified block diagram of the afc loop only; it shows the action required when an increase in center carrier frequency occurs. In this transmitter, an increased negative potential on the varicap diodes increases the carrier frequency. Therefore, a corresponding shift of dc bias in the positive direction must be obtained from the afc loop.

It should be noted that a direct-fm transmitter requires very little frequency multiplication to obtain 100 percent modulation. Whereas pm transmitters require frequency multiplication of from 800 to 1000 times or more, the particular transmitter shown in Fig. 10-23 requires only a 4-times multiplication. Some direct-fm transmitters employ a frequency-modulated oscillator which operates at one-half the carrier frequency, and only one doubler stage is required.

The RCA 20E fm transmitter uses direct fm in which the full deviation occurs in the direct-fm oscillator, and no frequency multiplication is required. A basic block diagram is shown in Fig. 10-24. The diagram shows one of two units that are combined to make a 40-kW transmitter.

Automatic frequency control of the on-frequency basic oscillator is achieved by taking a sample of the buffer output frequency and dividing it by two, 14 times. A low-frequency reference crystal operating at 1/1024th of the desired output frequency is divided by two, 4 times. Integrated circuits operating in the saturated mode are used in both binary dividing chains. The outputs from the reference and basic-oscillator binary dividers are phase compared in a time-sharing IC comparator. The output of the circuit, which represents the afc error voltage, is filtered and applied to another pair of varicap diodes coupled to the tuned circuit of the basic oscillator. Thus, the basic oscillator is phase locked to the 1024th harmonic of the oven-controlled reference crystal.

An off-frequency detector is incorporated in the design of this fm exciter. When the basic-oscillator frequency is not phase locked to the reference crystal, an ac component appears at the afc output. This voltage is rectified to operate a relay the contacts of which can be used to turn off the transmitter.

Parallel Operation of FM Transmitters

The parallel operation of a-m transmitters was covered in the previous chapter. This practice is becoming quite popular in fm installations as well. Fig. 10-25 is a block diagram of the RCA BTF-40E1 fm transmitter, which employs parallel operation. The function of parallel fm transmitters operating through a hybrid coupler is basically the same as that described previously for a-m transmitters. Fig. 10-26 shows the layout and dimensions of the BTF-40E1.

The BTF-40E1 is a diplexed transmitter consisting of two 20E transmitters united by a combining panel. The diplexed output combiner (a 3-dB hybrid coupler) and one 6⅛-inch harmonic filter are external to the cabinets of the transmitter.

The combining panel consists of five sections, the first of which contains the following control and metering functions: Six illuminated push buttons are provided for operating the combined-unit on, off, plate on, and plate off, and to indicate transmitter overload and output-line vswr overload. Also located here are three meters essential for operation. A reject-power meter is used to indicate power into the air-cooled reject load. A reflected output power meter serves to indicate reflected power and to protect the transmitter from an unusually high vswr in case of a line or antenna fault. The control can be adjusted to any value of vswr. The third meter indicates power output.

The second section in the combining panel is used to mount the set-point module of the meter control and to mount the reflectometer adjustments and controls. Below this, a third section contains illuminated switches for exciter 1 or 2, a control-circuit line breaker, and a meter indicating reject power in the input balun. The fourth section has a control for adjustment of the line stretcher used for phasing the input circuits of the combined 20E transmitter units. The last section in the combining panel is a blank panel.

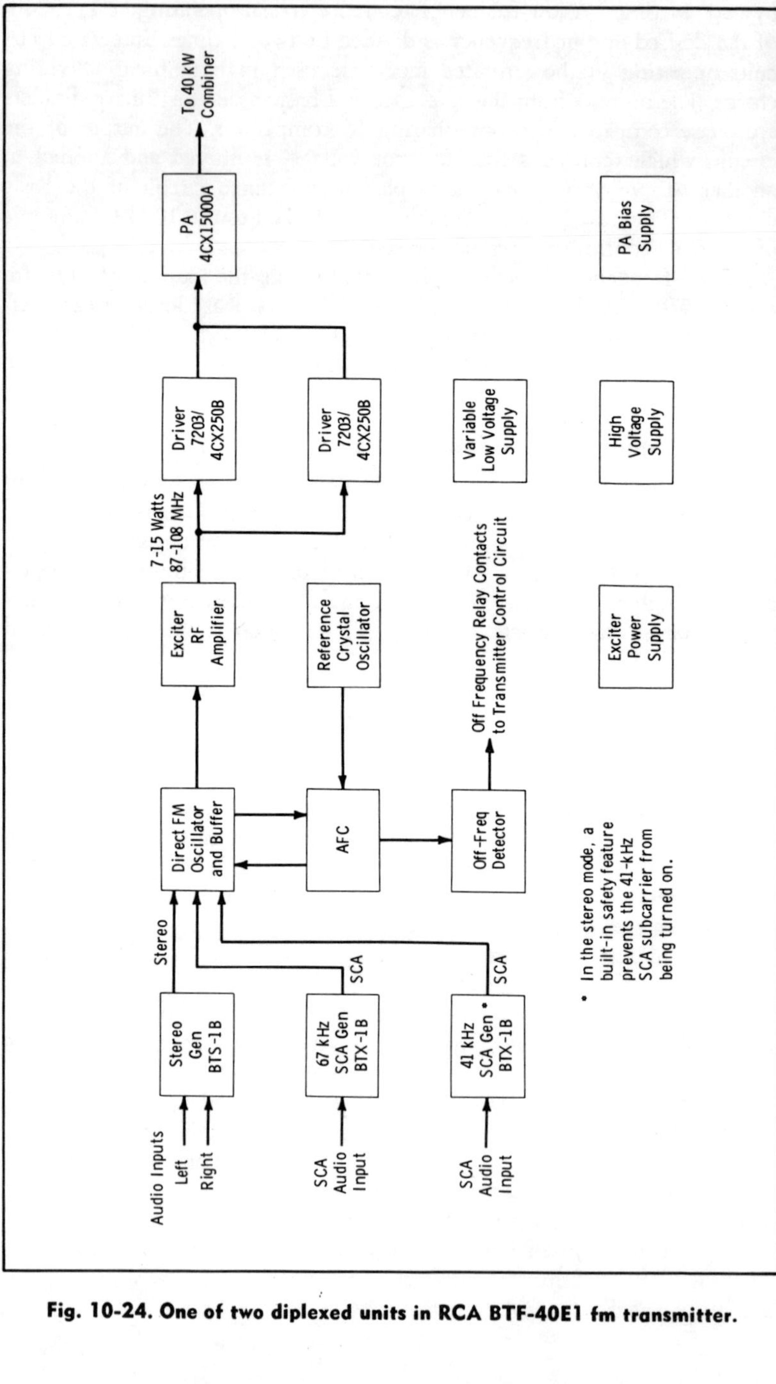

Fig. 10-24. One of two diplexed units in RCA BTF-40E1 fm transmitter.

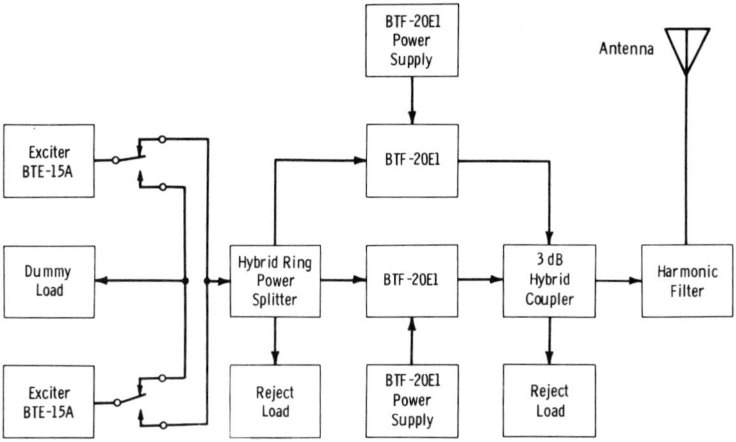

Fig. 10-25. Block diagram of RCA BTF-40E1 fm transmitter.

In the top of the combiner unit is a thermostatically controlled blower so that in case of failure of either of the 20-kW transmitters, the heat from the 10-kW reject loads will be removed from the cabinet. Each air-cooled 5-kW load is also equipped with a thermostatically controlled fan for cooling.

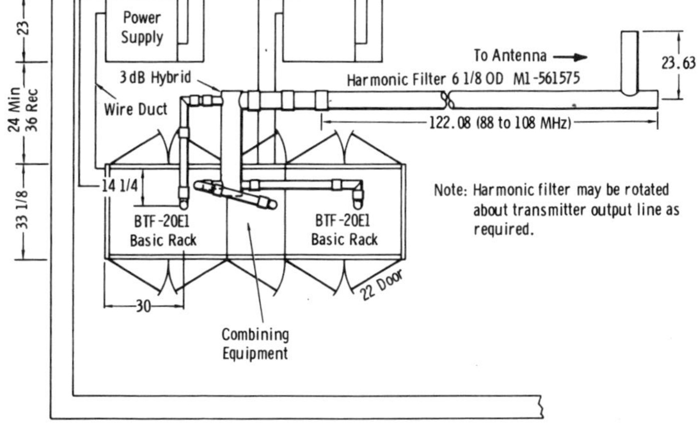

Fig. 10-26. Physical layout of BTF-40E1 transmitter.

The transmitter can be controlled as a 40-kW unit from the combining-cabinet panel, or as individual 20-kW transmitters from the individual control positions. The BTF-40E1 transmitter or the individual 20-kW sections can be operated by a remote-control system.

Relays automatically switch all inputs (stereo and two SCA channels) from one exciter to the other. The primary power to each exciter is always fed from the transmitter that is not shut down.

Two simplified, single-ended amplifiers (operating class C) follow the exciter in each 20-kW unit. The ipa stages consist of two ceramic 7203/CX250B tetrodes operating in parallel, and both final power amplifiers contain Type 4CX15000A tubes. Variable vacuum capacitors are used to tune the interstage network between driver and power amplifier.

The power amplifiers also use pi-network circuitry; however, the tuning of these stages is accomplished by variable inductors operating at ground potential. The output tubes are designed for very high power gain with little drive. The power output is controlled by means of motor-driven variable transformers connected in the primary of the low-voltage power supply for the driver amplifiers. This controls the ipa-plate and the pa-screen voltages simultaneously.

For increased transmitter stability and reliability, separate grid-bias supplies are incorporated. These supplies are semiconductor rectifiers.

To keep spurious emissions at a minimum, the transmitter is furnished with a 6⅛-inch harmonic filter. The filter consists of a series of transmission-line elements with a uniform-diameter outer conductor, a stepped inner conductor, and a shunt stub. The conductors are fabricated of a high-grade copper alloy. Attenuation of all harmonic radiation above channel limits is accomplished in an m-derived section and a series of constant-k T-sections. This design provides a broad passband with a sharp high-frequency cutoff and excellent attenuation of frequencies above the passband.

The high-voltage power supplies are installed at any convenient place in the station. The cabinets house the high-voltage plate transformers, a bank of plug-in semiconductor rectifiers, a line circuit breaker, a low-power circuit breaker, and the plate contactor. Personnel are fully protected from shock through interlock and grounding switches.

The rectifier section contains silicon-junction diodes (with equalizing resistors and capacitors) in a three-phase, full-wave bridge circuit. Circuit breakers are used instead of fuses in the transmitter to add to the dependability, particularly for remote-control operation.

The VSWR Meter

The power indicator and voltage-standing-wave-ratio (vswr) meter usually takes the form of a *reflectometer*. This is a combination of a directional coupling device and a diode detector circuit. This indicator provides a constant check on power output as well as showing the condition of the transmission line and antenna system as they affect standing waves on the line.

A basic diagram of a directional coupler is shown in Fig. 10-27. The loop might have both magnetic and capacitive coupling to the transmission line. The capacitive coupling is small, with a large reactance at the carrier frequency. Therefore the current through the resistor is in quadrature (90°) with the line current. The loop and resistor voltage drops are in series. For a wave traveling in one direction, the voltage across the transmission line and the current in the line are in phase at a magnitude set by the characteristic impedance of the line (normally 50 ohms). Since both the coupled voltage and the resistor voltage are in quadrature with the line current, the loop voltage is in phase with the resistor voltage, and the sum represents the directional-coupler output voltage.

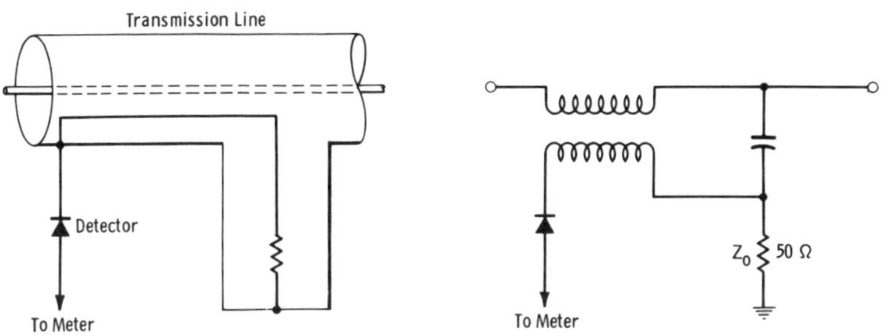

(*A*) *Coupling arrangement.* (*B*) *Equivalent circuit.*

Fig. 10-27. Principle of directional coupler.

When any degree of impedance mismatch occurs at the sending or receiving terminals of the transmission line, the line voltage and current maxima and minima are no longer in phase, resulting in standing waves. In the event of a standing wave (wave on transmission line from opposite direction), the loop induced voltage is out of phase with the resistor voltage drop. Now, if the loop is adjusted so that these voltage drops are made equal, the coupler output voltage will be zero. In this way, the directional coupler can distinguish between waves traveling in opposite directions. It can be calibrated to measure power output, or, by comparing voltages of opposite wave direction, it can measure load mismatch or voltage standing wave ratio.

It is necessary to calibrate the reflectometer and then ensure maintenance of correct calibration in arbitrary terms of power output. Calibration is done with a dummy load and rf wattmeter.

10-5. SUBSIDIARY COMMUNICATIONS (SCA)

Paragraph 73.268 of the FCC Rules and Regulations pertains to fm modulation:

The percentage of modulation shall be maintained as high as possible consistent with good quality of transmission and good broadcast practice. In no case is it to exceed 100 percent on peaks of frequent recurrence. Generally, it should not be less than 85 percent on peaks of frequency recurrence; but where necessary to avoid objectionable loudness, modulation may be reduced to whatever level is necessary, even if the resulting modulation is substantially less than 85 percent on peaks of frequent recurrence.

NOTE: The "loudness" problem will be considered in greater detail in Chapter 11.

Paragraph 73.319 of the Rules and Regulations concerns the engineering standards for subsidiary communications multiplex operation:

A. Frequency modulation of SCA subcarriers shall be used.

B. The instantaneous frequency of SCA subcarriers shall at all times be within the range 20 to 75 kHz, *provided, however,* that when the station is engaged in stereophonic broadcasting the instantaneous frequency of SCA subcarriers shall at all times be within the range 53 to 75 kHz.

C. The arithmetic sum of the modulation of the main carrier by SCA subcarriers shall not exceed 30 percent, *provided, however,* that when the station is engaged in stereophonic broadcasting, the arithmetic sum of the modulation of the main carrier by the SCA subcarriers shall not exceed 10 percent.

D. The total modulation of the main carrier, including SCA subcarriers, shall meet the requirements of paragraph 73.268 (presented above).

E. Frequency modulation of the main carrier caused by the SCA subcarrier operation shall, in the frequency range 50 to 15,000 Hz, be at least 60 dB below 100 percent modulation, *provided, however,* that when the station is engaged in stereophonic broadcasting, frequency modulation of the main carrier by the SCA subcarrier operation shall, in the frequency range 50 to 53,000 Hz, be at least 60 dB below 100 percent modulation.

Modern SCA operations are usually on 41 kHz and/or 67 kHz. The latter frequency is the most popular, since it can be used with either mono or stereo transmission, and interference traps in fm stereo receivers are usually tuned to this frequency.

The effect of the FCC rulings for SCA and/or stereo transmission on modulation percentages (when maximum SCA deviation is used) is as follows:

1. For mono transmission and one SCA:

   Maximum mono modulation: 70 percent (±52.5 kHz)
   Maximum SCA modulation: 30 percent (±22.5 kHz)

2. For mono transmission and two SCAs:

   Maximum mono modulation: 70 percent (±52.5 kHz)

Maximum SCA modulation (each): 15 percent (±11.25 kHz each)

3. Stereo transmission is covered in Section 10-6. At this time, we will consider only the modulation percentages involved in stereo with one SCA (two SCAs cannot be used with stereo, and the SCA subcarrier must be between 53 and 75 kHz, usually 67 kHz):

   Maximum Main Channel (L + R) Modulation: 40 percent (±30 kHz)
   Maximum Stereo Sideband (L + R or L − R only): 40 percent (±30 kHz)
   Maximum SCA Modulation: 10 percent (±7.5 kHz)

It should be noted that the above modulation percentages are in terms of the *maximum allowed.* Many mono stations have 100-percent SCA modulation referred to ±7.5 kHz with frequency response limited to 30 Hz to 7.5 kHz (sometimes 30 Hz to 5 kHz) to allow greater modulation to be used on the main carrier. Similarly, some stereo stations employing simultaneous SCA have 100-percent SCA modulation referred to ±3.5 kHz.

The SCA subcarrier injection on the main carrier is of constant amplitude, but is frequency-modulated by the desired audio in the SCA generator. The system therefore can be termed fm on fm by the modulation process shown basically in Fig. 10-28. The subcarrier (modulated) deviates the main carrier at the SCA frequency. Since only a limited deviation relative to main-channel deviation is required, the SCA reactance circuit normally is connected only across a fixed portion of the master-oscillator tank circuit.

10-6. STEREO TRANSMITTERS

As was stated in Chapter 7, the number one requirement in stereo is compatibility, which means that (1) a standard receiver can reproduce a monophonic output from a stereocast and (2) a stereo receiver can reproduce a monophonic program. The left and right channels in the studio

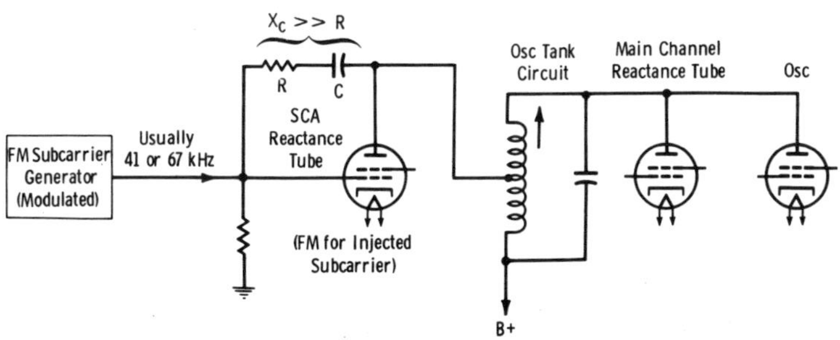

Fig. 10-28. Main-channel and SCA modulation of oscillator.

(which may originate at microphones, a stereo record pickup, or a stereo magnetic-tape head) are used to form sum and difference signals:

$$\text{Sum} = L + R$$
$$\text{Difference} = L - R$$

The L + R signal is fed into the fm modulator in the conventional manner and is the 30-15,000 Hz signal used in monophonic receivers. The L − R signal provides one input to a balanced modulator; a subcarrier of 38 kHz is the other input. The subcarrier is balanced out, or suppressed, and only the double sidebands of the L − R amplitude-modulated signal remain. These are added to the main-channel signal and are used by a stereophonic receiver to supply the right speaker. The right signal is subtracted from the L + R signal to supply the left speaker.

The respective modulation levels are shown in Fig. 10-29A. The sum and difference signals are allowed 45 percent modulation each, and the pilot signal is allowed 10 percent, for a total of 100 percent. (The pilot frequency is one-half the subcarrier frequency and is used by a stereophonic receiver to reinsert the subcarrier frequency.) Since an additional subcarrier is used in SCA, the respective modulation levels are reduced to allow for the 10-percent SCA subcarrier modulation (Fig. 10-29B).

Fig. 10-30 is a block diagram of a basic stereo transmitter. Each numbered block functions as follows:

1. Both left and right audio inputs pass through filters which are essentially flat to 15 kHz and roll off sharply beyond that frequency. This prevents any spurious signals in the 19-kHz pilot region or the region of the L − R sidebands.
2. Both channels are pre-emphasized by the standard 75-microsecond characteristic specified by the FCC.
3. The two channels feed a matrix which derives the sum and difference signals. This circuit is simply a cross-connected voltage divider (Fig. 10-31).
4. The sum signal feeds the normal audio input of the fm modulator. This signal occupies the conventional 30-15,000 Hz bandwidth.
5. An adjustable time delay is used to meet the FCC requirement that any phase difference between the two signals as received at the receiver matrix shall not exceed ±3° at any frequency between 30 Hz and 15 kHz. This is necessary to assure proper separation between left and right channels. (In practice, the delay adjustment may be found in either the left or right channel.)
6. A crystal oscillator generates the 19-kHz pilot.
7. The times-2 multiplier generates the 38-kHz subcarrier frequency.
8. If the subcarrier is fed to a single-ended input of a stage which is connected push-pull at the plates, the signal is cancelled, or suppressed. This is one form of obtaining the suppressed-carrier stereo

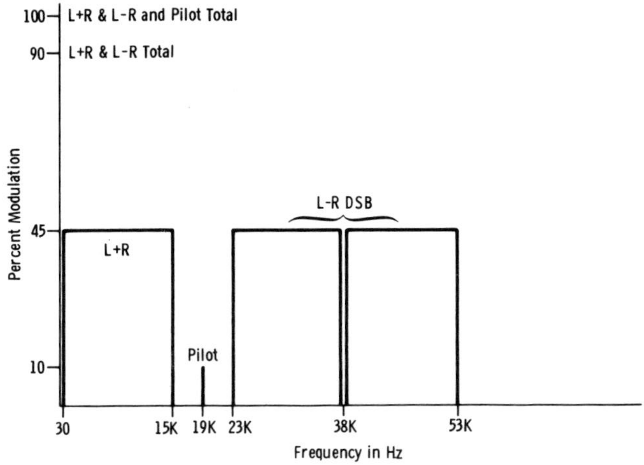

(A) Without SCA.

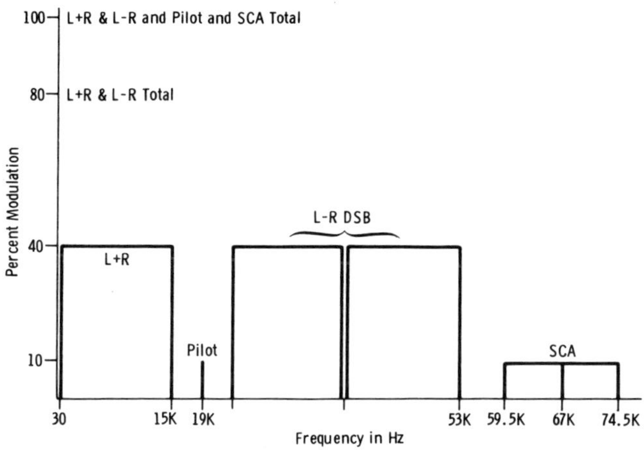

(B) With SCA.

Fig. 10-29. Frequency distribution of fm stereo signal.

signal. The L − R signal amplitude-modulates the subcarrier rf; therefore the output consists of a double-sideband, suppressed-carrier signal. Since the sidebands extend to a maximum value at the highest modulating frequency, the total spectrum is:

$$38 \text{ kHz} + 15 \text{ kHz} = 53 \text{ kHz}$$
$$38 \text{ kHz} - 15 \text{ kHz} = 23 \text{ kHz}$$

Thus, the sidebands extend from 23 kHz to 53 kHz, and the total

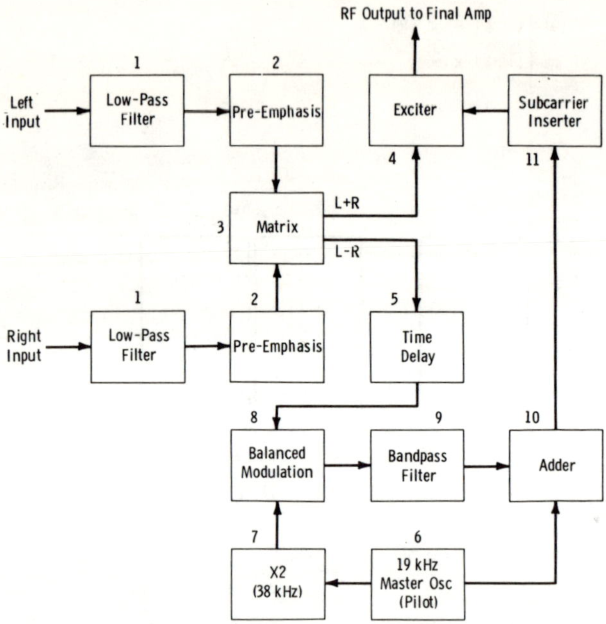

Fig. 10-30. Block diagram of basic stereo transmitter.

spectrum is 30 Hz to 53 kHz for stereo only or 30 Hz to 74.5 kHz when SCA is added.

9. The bandpass filter removes residual L − R audio components in the frequency range of 30-15,000 Hz and harmonics outside the range of 23-53 kHz.
10. The adder accepts the 19-kHz pilot and the output of the bandpass filter. At the output of this stage are the double-sideband suppressed-carrier difference signal and the proper amplitude of the 19-kHz pilot.
11. The subcarrier inserter is essentially a phase modulator which places, or frequency modulates, the subcarrier signal onto the main signal. The 19-kHz pilot also frequency modulates the main carrier.

Principles of the Stereo Receiver

The stereo receiver (Fig. 10-32) is conventional up to the discriminator output. Here, the L + R signal in a monophonic receiver would produce a compatible monophonic program.

The amplifier-filter arrangement in a stereo receiver separates the L + R signal, the L − R sidebands, and the 19-kHz pilot. The pilot is doubled in frequency to recover the 38-kHz subcarrier which in turn is mixed with the L − R sidebands to obtain a conventional amplitude-modulated signal.

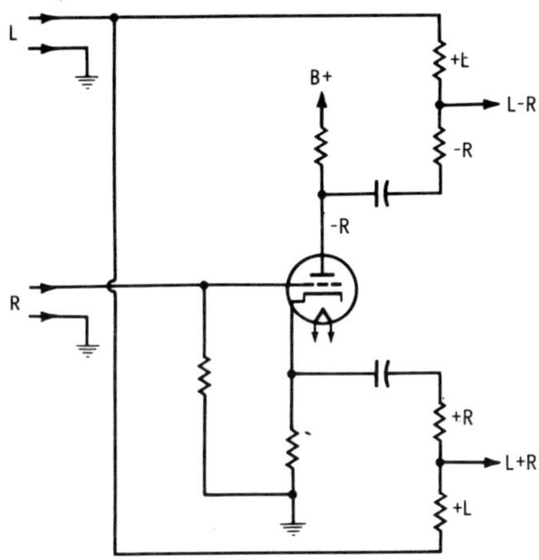

Fig. 10-31. Inverter-tube stereo matrix.

This is detected to produce the L − R signal for the matrix. The outputs of the matrix, after de-emphasis in the normal manner, produce the original left and right stereophonic signals.

Why are a suppressed subcarrier and subharmonic pilot used? This is done for two important reasons:

1. The 19-kHz pilot falls in a clear-channel portion of the discriminator output. Thus, a simple filter in the receiver can isolate the pilot to recover the subcarrier.
2. The suppressed subcarrier is used for the resulting interleaving which would permit 90 percent maximum deviation on the main channel as well as 90 percent on the subchannel. (The remaining 10 percent is used for the pilot.) This is possible because one channel is deviating the main carrier at a peak while the other is zero, and vice-versa. (The sum of two variables, L + R, is maximum when their difference, L − R, is minimum.) Thus, the monophonic listener experiences a signal-to-noise loss of less than 1 dB.

The subcarrier is also suppressed (to less than 1 percent) so that the main and subchannel fm deviations are limited only by the necessity to provide for the 10-percent pilot-subcarrier modulation.

Stereophonic Transmission Standards

The following standards are from paragraph 73.322 of the FCC Rules and Regulations:

A. The modulating signal for the main channel shall consist of the sum of the left and right signals.

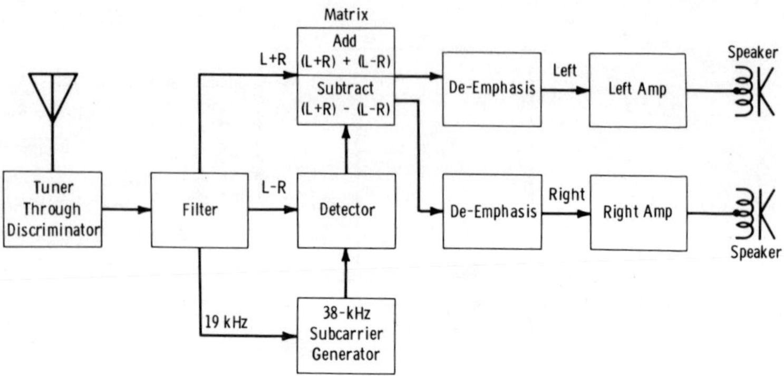

Fig. 10-32. Block diagram of stereo receiver.

B. A pilot subcarrier at 19,000 Hz plus or minus 2 Hz shall be transmitted that shall frequency modulate the main carrier between the limits of 8 and 10 percent.
C. The stereophonic subcarrier shall be the second harmonic of the pilot subcarrier and shall cross the time axis with a positive slope simultaneously with each crossing of the time axis by the pilot subcarrier.
D. Amplitude modulation of the stereophonic subcarrier shall be used.
E. The stereophonic subcarrier shall be suppressed to a level less than one percent modulation of the main carrier.
F. The stereophonic subcarrier shall be capable of accepting audio frequencies from 50 to 15,000 Hz.
G. The modulating signal for the stereophonic subcarrier shall be equal to the difference of the left and right signals.
H. The pre-emphasis characteristics of the stereophonic subchannel shall be identical with those of the main channel with respect to phase and amplitude at all frequencies.
I. The sum of the sidebands resulting from amplitude modulation of the stereophonic subcarrier shall not cause a peak deviation of the main carrier in excess of 45 percent of total modulation (excluding SCA subcarriers) when only a left (or right) signal exists; simultaneously in the main channel, the deviation when only a left (or right) signal exists shall not exceed 45 percent of total modulation (excluding SCA subcarriers).
J. Total modulation of the main carrier including pilot subcarrier and SCA subcarriers shall meet the requirements of Section 73.268 with maximum modulation of the main carrier by all SCA subcarriers limited to 10 percent.
K. At the instant when only a positive left signal is applied, the main channel modulation shall cause an upward deviation of the main carrier frequency, and the stereophonic subcarrier and its sidebands

signal shall cross the time axis simultaneously and in the same direction.

L. The ratio of peak main-channel deviation to peak stereophonic subchannel deviation when only a steady-state left (or right) signal exists shall be within ±3.5 percent of unity for all levels of this signal and all frequencies from 50 to 15,000 Hz.

M. The phase difference between the zero points of the main-channel signal and the stereophonic subcarrier-sidebands envelope, when only a steady-state left (or right) signal exists, shall not exceed ±3° for audio modulating frequencies from 50 to 15,000 Hz.

NOTE: If the stereophonic separation between left and right stereophonic channels is better than 29.7 dB at audio modulating frequencies between 50 and 15,000 Hz, it will be assumed that paragraphs L and M of this section have been complied with.

N. Cross talk into the main channel caused by a signal in the stereophonic subchannel shall be attenuated at least 40 dB below 90 percent modulation.

O. Cross talk into the stereophonic subchannel caused by a signal in the main channel shall be attenuated at least 40 dB below 90 percent modulation.

P. For required transmitter performance, all of the requirements of Section 73.254 shall apply with the exception that the maximum modulation to be employed is 90 percent (excluding pilot subcarrier) rather than 100 percent.

Q. For electrical-performance standards of the transmitter and associated equipment, the requirements of Section 73.317 (a) (2), (3), (4), and (5) shall apply to the main channel and stereophonic subchannel alike, except that where 100 percent modulation is referred to, this figure shall include the pilot subcarrier.

Additional Discussion of Standards

The following paragraphs contain clarifications of certain aspects of the FCC rules most important in maintenance.

FCC Paragraph 73.322(c)—Standardization of the polarity of the subcarrier relative to the pilot subcarrier is necessary to provide identification of the left and right channels. If the second harmonic should cross the time axis with a negative slope at the transmitter and a positive slope at the receiver, the left channel as transmitted would be reproduced in the right channel of the receiver, and conversely.

FCC Paragraph 73.322(k)—Essentially the same reasoning applies as for paragraph 73.322 (c). In this case, the relationship of the left-plus-right signal relative to the left-minus-right signal is maintained. If the transmission standards varied from station to station (deviation upward in some transmitters and downward in others), the left and right channels would be reversed in the receiver, depending on the particular fm station being received.

FCC Paragraphs 73.322(l) and (m)—These paragraphs give the FCC requirement for channel separation. Paragraph l specifies the required amplitude response of the two channels, and paragraph m specifies the phase response. With a steady-state signal in the left channel only, signals of the same frequency and equal amplitudes (within 3.5 percent) exist in the left-plus-right and left-minus-right channels. The phase difference between the left-plus-right and left-minus-right signals must not exceed ±3° between 50 and 15,000 Hz.

The preceding may be summarized by stating that for proper channel separation, the left-plus-right and left-minus-right channels must have the same frequency response, and all frequencies in both channels must arrive at the receiving matrix at the same time (in the proper phase relationship).

The FCC note following paragraph m specifies when further measurements in checking channel separation are required. For a given signal in the left channel only, adjustment is satisfactory if the signal recovered at the receiver (or receiver-type monitor) in the right channel is attenuated at least 29.7 dB. This must be true throughout the frequency range of 30 to 15,000 Hz. If this check is negative, further measurements of the frequency and phase response must be made.

FCC Paragraphs 73.322(n) and (o)—These paragraphs specify transmitter requirements with respect to cross talk from the left-minus-right channel into the left-plus-right channel (stereo subchannel into main channel) and cross talk from the left-plus-right channel into the left-minus-right channel (main channel into stereo subchannel). In either case, the ratio must be at least 40 dB relative to 90-percent modulation of the other channel.

Stereo Subcarrier Modulation

The conventional tube-type balanced modulator for modulating the 38-kHz subcarrier and suppressing the carrier frequency was mentioned in the preceding description. One of the earlier circuits using a diode bridge as a balanced modulator is shown in Fig. 10-33. The L − R audio signal across T1 is applied to terminals 1 and 2 of the bridge. The subcarrier input is applied to the center tap of the secondary of T1. The resulting amplitude-modulated output of the bridge from terminals 3 and 4 is applied across the primary of T2. Balance control R2 effectively grounds the center tap of this transformer so that, with proper bridge balance, the net value of the 38-kHz frequency can be made to be essentially zero. Balance control R1 can be adjusted so that the sideband components across T2 are of equal value on both sides of the carrier frequency.

In practice, stereo signals can be derived either by the matrix and balanced modulator technique covered thus far, or by a time-division or sampler technique. Both techniques are mathematically equivalent and per-

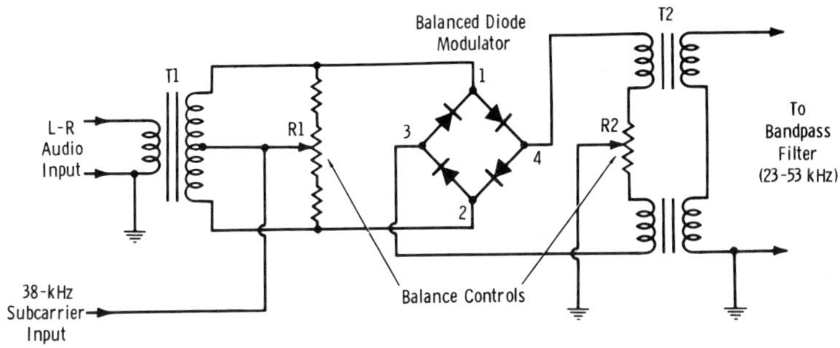

Fig. 10-33. Balanced modulator using diode bridge.

fectly valid. The sampler (time-division) technique gives excellent results with greater simplicity.

An example of a commercial fm stereo generator employing time division is the CCA Model SG-1D. A simplified block diagram of this system is shown in Fig. 10-34.

The most obvious advantage of the time-division method is the fact that mono and stereo signals follow the same path. When the mono and stereo channels are developed, they appear at a common point and do not go through different circuits, which could introduce undesirable delays. In Fig. 10-34, the left and right signals are fed through matching transformers into the left and right amplifiers. These are two identical amplifiers with flat frequency response; however, when the pre-emphasis switch is on, the frequency response of the amplifier is modified through a frequency-sensitive negative-feedback network, and the response of the amplifiers follows very closely that of the 75-microsecond pre-emphasis required for fm broadcasting.

The 38-kHz subcarrier frequency and the 19-kHz pilot are generated in the following manner. A crystal oscillator operates at 76 kHz. This oscillator drives a flip-flop that divides the frequency by 2, providing a 38-kHz square wave on each one of the two transistor collectors. At the same time, this flip-flop drives another flip-flop, which in turn divides the frequency to 19 kHz. This 19-kHz square wave is filtered by a low-pass filter, and only the fundamental 19-kHz sine wave is obtained at the output. This is followed by a pilot level control and a phase control. The 19-kHz signal is then added to the modulator output to form the composite signal.

The left and right signals are fed to transistor switchers that are turned on and off at a 38-kHz rate. This switching alternately shorts to ground the left and right signals.

The switching process creates high-order harmonics of the 38-kHz signal and its sidebands; thus a low-pass filter is included in the circuit in order to eliminate frequencies above 53 kHz. This low-pass filter has flat frequency response up to 53 kHz, and its phase linearity is such that the total

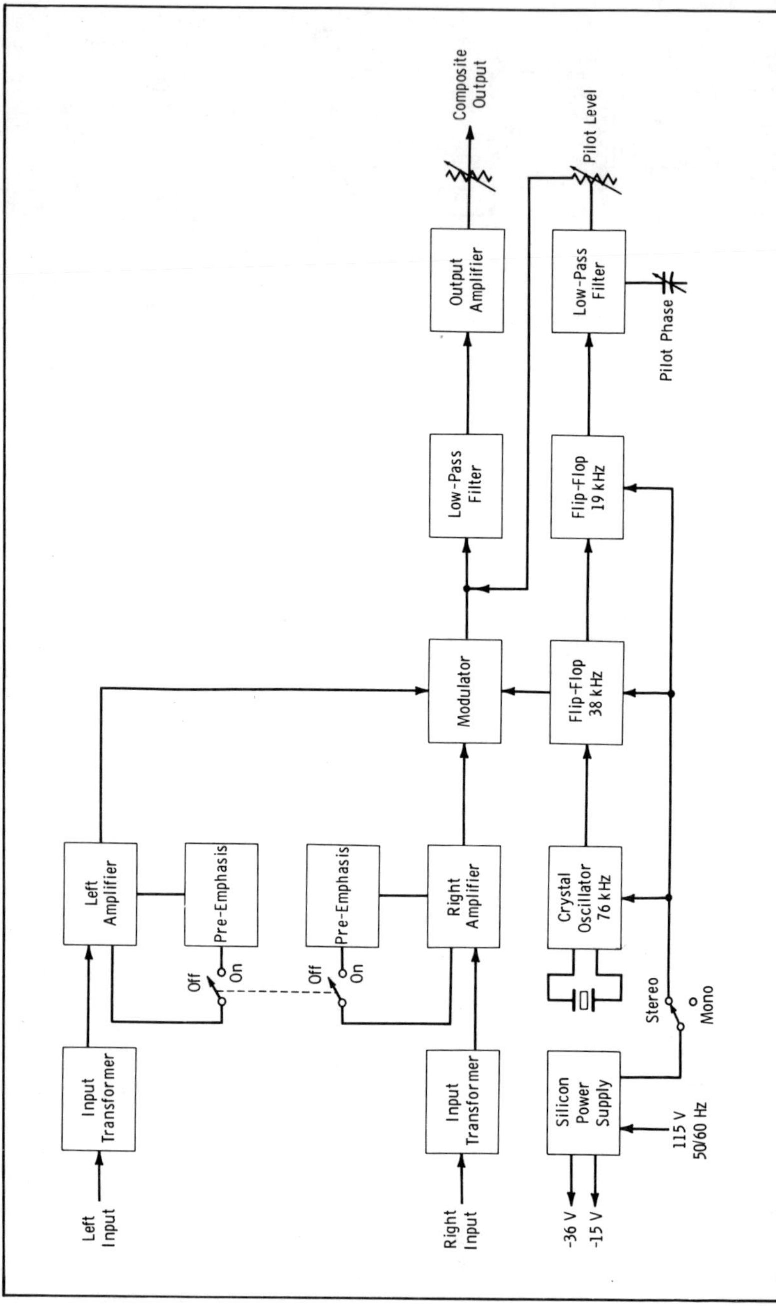

Fig. 10-34. Block diagram of CCA Model SG-1D stereo generator.

time delay is the same at any frequency from 30 Hz to 53 kHz. The output of the low-pass filter is fed to the output amplifier.

The power supply includes a full-wave bridge rectifier which provides voltages of −15 and −36 volts. These voltages are regulated with zener diodes so that the output signal level of the generator is independent of power-line variations.

Circuit Analysis of CCA Stereo Generator

See Fig. 10-35 for the following circuit analysis. The audio signal is connected to a balanced input transformer which provides proper matching with the input line. Potentiometer R22 is provided in the right channel to adjust the signal level slightly for compensation purposes.

The left and right amplifiers are identical. There is a negative-feedback loop the frequency response of which is flat when the pre-emphasis switch is in the off position; the frequency response can be changed to match a 75-microsecond pre-emphasis when the switch is on. This is done by means of capacitors C3 and C11.

Two switching transistors are utilized in the modulator. They operate as grounded-emitter stages with the bases connected to the 38-kHz square-wave generator.

During one half period, transistor Q4 is saturated and Q5 is cut off; Q4 is cut off and Q5 is saturated during the other half period. The saturated transistor shorts the output of the corresponding amplifier to ground; the cut-off transistor allows the other signal to be amplified normally. This is done at the rate of 38 kHz, and it can be shown mathematically that this action generates the monaural left-plus-right signal as well as the required sidebands for left-minus-right signals centered on 38 kHz. At the same time, the 38-kHz subcarrier is eliminated to a high degree. The switching system produces third-order harmonics that are eliminated by means of the low-pass filter.

A highly stable crystal oscillator operates at 76 kHz. This frequency was chosen in order to facilitate the generation of the 38-kHz square wave. In this circuit, the crystal operates as a series resonant circuit. The output of the 76-kHz crystal oscillator is differentiated and utilized to trigger a flip-flop which generates the 38-kHz square wave. This signal is used to switch the modulator transistors.

Another output of the 38-kHz flip-flop is differentiated and utilized to trigger a second flip-flop, the output of which is a 19-kHz square wave. The output of this flip-flop is filtered by means of a low-pass filter to eliminate all frequencies above the fundamental 19 kHz. Inductor L4 of the second section of the filter is utilized as a pilot-phase control.

The output of the 19-kHz low-pass filter is connected to a potentiometer labeled PILOT LEVEL CONTROL. This potentiometer is set to inject the proper amount of 19-kHz pilot signal directly at the output of the modulator.

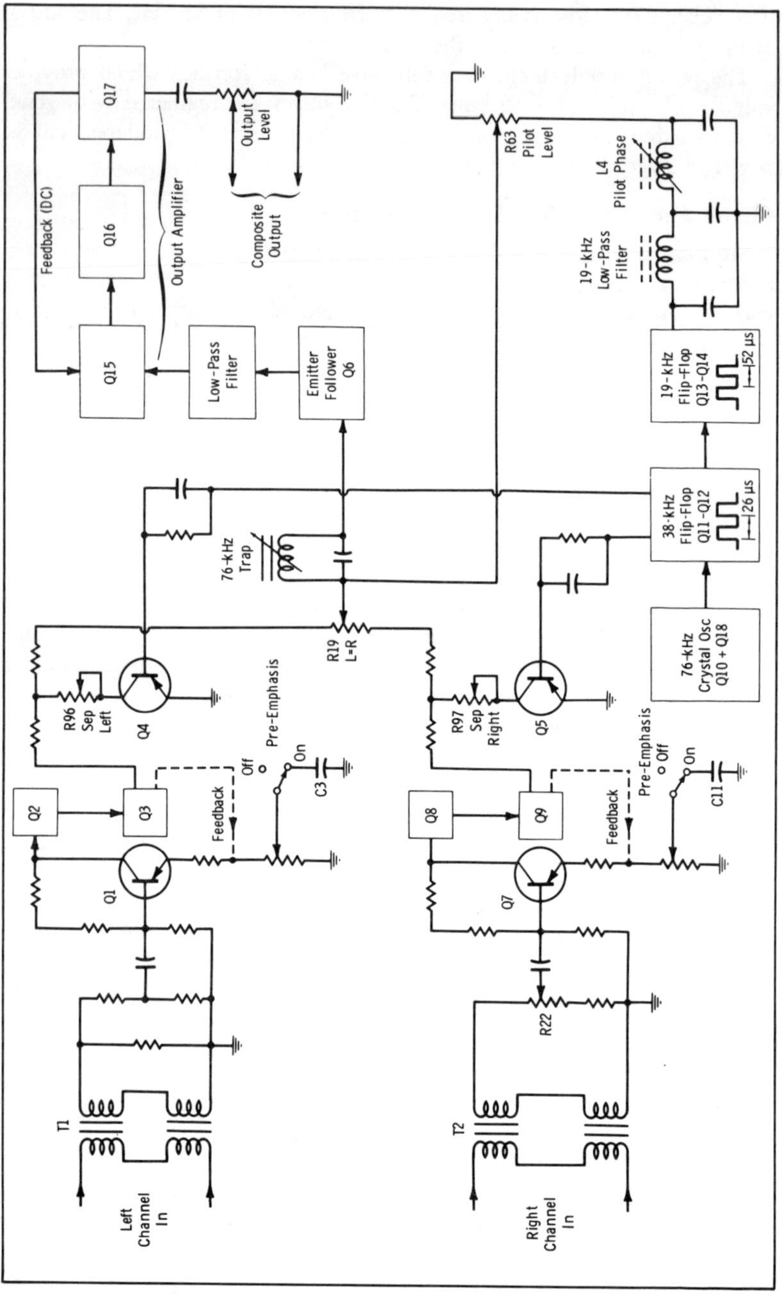

Fig. 10-35. Simplified schematic of CCA Model SG-1D stereo generator.

After the pilot subcarrier and the modulator output signal are combined, the composite signal passes through a parallel tuned circuit which eliminates the 76-kHz spurious components that remain after the modulating process. This trap allows a reduction of spurious content well below the interference level of an SCA system, if used. The signal is fed next to the base of an emitter-follower transistor, which provides impedance matching with the low-pass filter.

The low-pass filter provides a flat frequency response and linear phase characteristics up to 53 kHz. This filter reduces the second group of harmonics centered at 76 kHz, but the most important function is to attenuate the third-order harmonics centered at 3 times 38 kHz, or 114 kHz. It should be noted that the SG-1D would meet FCC spurious-signal specifications without the low-pass filter and the 76-kHz trap, but they are considered a worthwhile addition.

The composite stereo signal is amplified by means of a 3-transistor output amplifier. The circuitry is similar to the two input amplifiers, and it provides a signal level capable of modulating most of the existing fm exciters for broadcast transmission. The output-level control selects the proper level to modulate the fm exciter fully.

Not shown in the simplified schematic is the means of mono-stereo operation control. The transition from mono to stereo operation is achieved by a multiple-contact relay. This relay can be operated by means of the front-panel stereo switch, or it can be operated by a remote stereo switch connected to the proper terminals of the generator.

When the relay is energized, all the circuits are completed for proper stereo operation as described above. When the relay is de-energized, the following changes are accomplished:

A. The two transistor switches are disconnected from the circuit.
B. The two flip-flops are de-energized. In addition, a set of contacts connects a resistor, which has the same power consumption as the previously disconnected stages, in order to keep a constant load on the power supply.
C. Another function accomplished by the relay contacts is to introduce an attenuator pad so that the total modulation of the transmitter will be kept constant when the change is made from stereo to monaural operation. This compensates not only for the inherent change of signal level of the stereo system, but also for the 10-percent modulation utilized by the pilot.

NOTE: The proper adjustments of separation, pilot phase, etc., are covered in Chapter 14.

Main Carrier Modulation

The generation of the composite stereo signal, ready to frequency-modulate the main-carrier oscillator of the exciter unit, has been described. Fig.

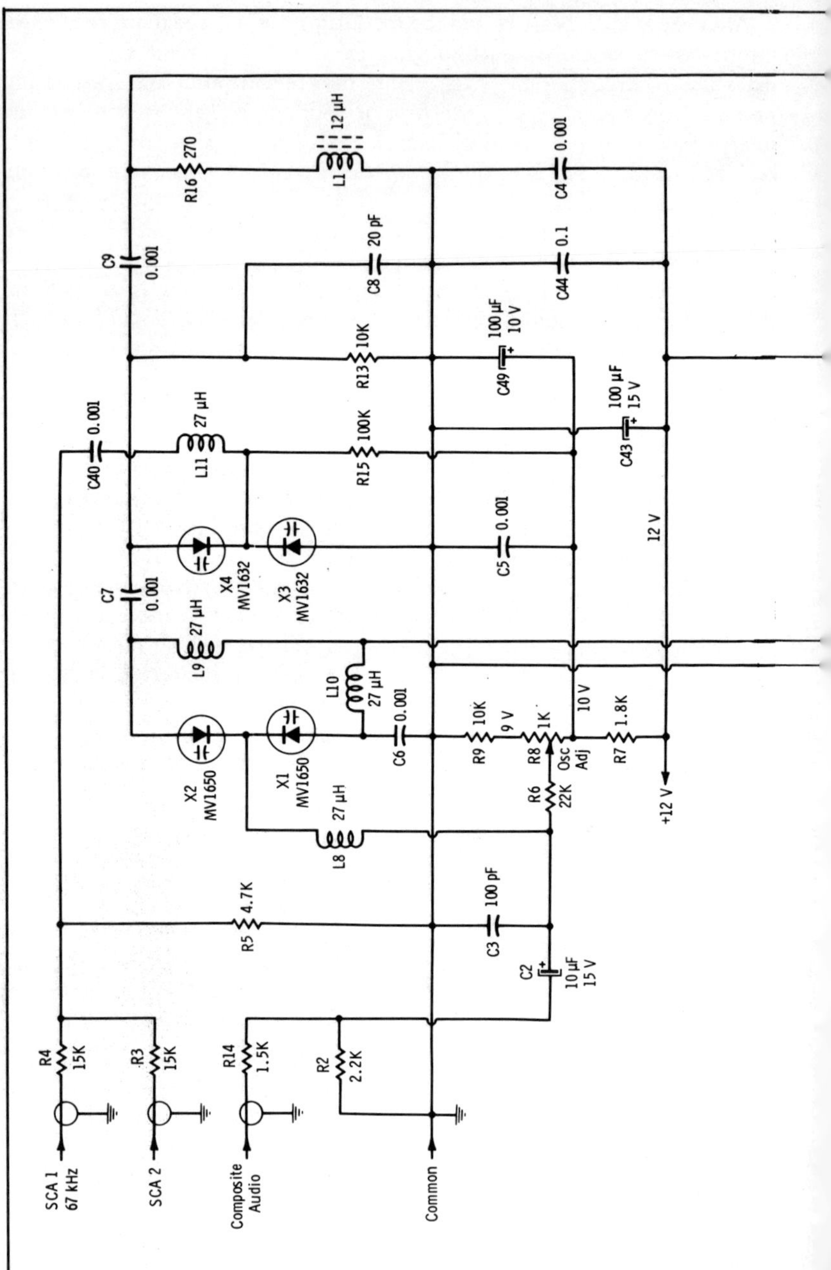

Fig. 10-36. Modulated oscillator

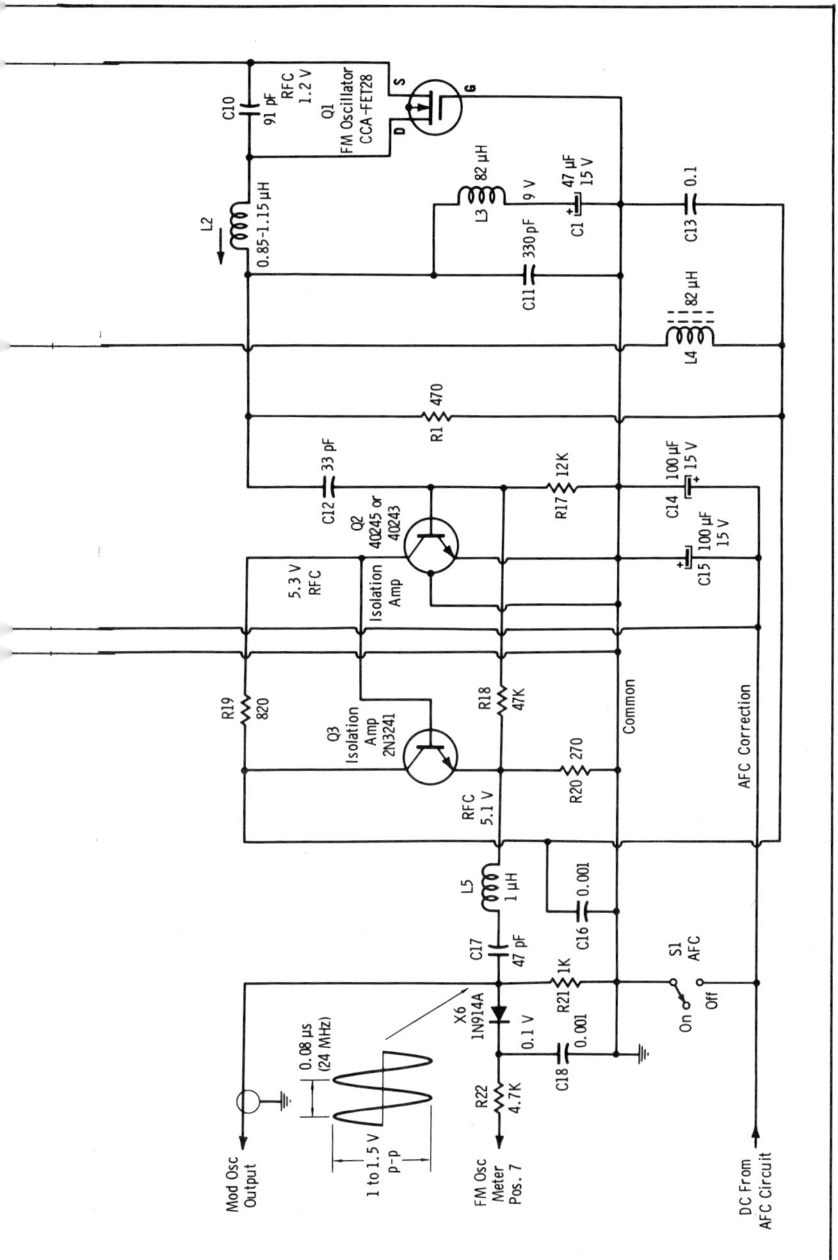

section of CCA fm exciter.

10-36 shows the schematic of the frequency-modulated oscillator (fmo) in the CCA FM-10DS exciter.

The frequency-modulated oscillator is built around Q1, an insulated-gate field-effect transistor (IGFET). The operating frequency of the oscillator corresponds to the resonance of inductor L2 and the capacitance connected from each end of the inductor to ground. The capacitance on one side of L2 is composed of C10 in series with a combination of several capacitors, including fixed capacitor C8 in parallel with two sets of voltage-variable capacitors (diodes X1 and X2 and diodes X3 and X4).

The voltage-variable-capacitor (vvc) diodes are connected as back-to-back pairs. These diodes are reverse-biased at about 10 volts dc so that they operate as capacitors with capacitance that varies inversely with the bias voltage. If the variation of voltage is small compared with the fixed bias, the capacitance change will be in a linear region.

From the point of view of the external bias applied to the pair of vvc diodes, the increase or decrease occurs in the same direction, so that even when they are connected in series in the rf circuit, the total capacitance still goes up or down depending on the change of bias. (The modulating signal can be considered a fast variation of dc bias.)

Any voltage applied across the two diodes in series will change the bias of the two diodes in opposite directions, thus tending to cancel the overall change of capacitance, making the pair quite insensitive to the application of voltage across the two diodes in series. This is the case for rf voltage applied to the diodes by the oscillator circuit; thus it can be seen that this approach minimizes frequency drift caused by changes in amplitude of the oscillator rf voltage.

A second pair of vvc diodes is biased with a fixed voltage and, in addition, receives the SCA subcarrier modulation.

The dc bias for X1 and X2 is obtained from a zener diode, and it goes through a trimming adjustment consisting of adjustable resistor R8, which is labeled OSCILLATOR ADJUSTMENT. A small change in the bias voltage is utilized as a fine adjustment to set the center frequency of the modulated oscillator. The coarse adjustment is obtained by tuning L2.

The dc path for the anodes of X1 and X2 is returned to ground through the output resistor of the afc circuit; thus if the afc system (previously described) applies a correcting voltage to the fmo, it will be applied to the anodes of X1 and X2. The composite audio modulation signal is applied to the cathodes of the two diodes through a combining circuit that adds it to the fixed bias previously described.

The output signal of the oscillator is obtained across C11, which constitutes a low-impedance output. This reduces any interaction on the oscillator from the following stages. The isolation stage that follows the oscillator consists of a pair of transistors that provide an increase in signal level as well as a high-impedance input and a low-impedance output. The amplified signal is finally applied to the external amplifier-multiplier in order to multiply the frequency by four and increase the output power to a nominal 10 watts (2-to-12 watt range).

The rf output signal level is detected by diode X6. The resultant dc component is utilized to measure the activity of the oscillator circuit when the multimeter switch is in the proper monitoring position.

10-7. FM FREQUENCY AND MODULATION MONITORS

For fm broadcasting, the modulation and frequency monitors are normally combined in a single unit. For stereo and SCA broadcasting in addi-

(A) Modulation and SCA frequency monitor.

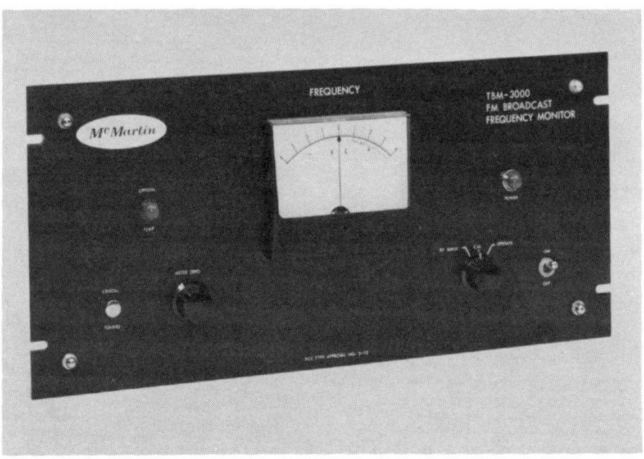

(B) Main-carrier frequency monitor.

Courtesy McMartin Industries, Inc.

Fig. 10-37. Fm broadcast monitors.

tion to regular service, two units may be required. Fig. 10-37A shows a monitor for the main and subchannel modulation and the subchannel frequency. This unit, combined with a standard frequency monitor (Fig. 10-37B) provides complete main-carrier and SCA-multiplex monitoring.

A block diagram of the RCA BW-73A fm multiplex monitor is shown in Fig. 10-38.This unit indicates total modulation on the main carrier, the subcarrier(s) modulation of the main carrier, percent program modulation on each subcarrier, and the swing on any external subcarrier. It also measures the rf input level to assure proper operating conditions in the monitor. A front-panel flasher lamp with adjustable threshold indicates modulation peaks in any of the above services. In addition, outputs are available for aural monitoring through phones or station lines. The moni-

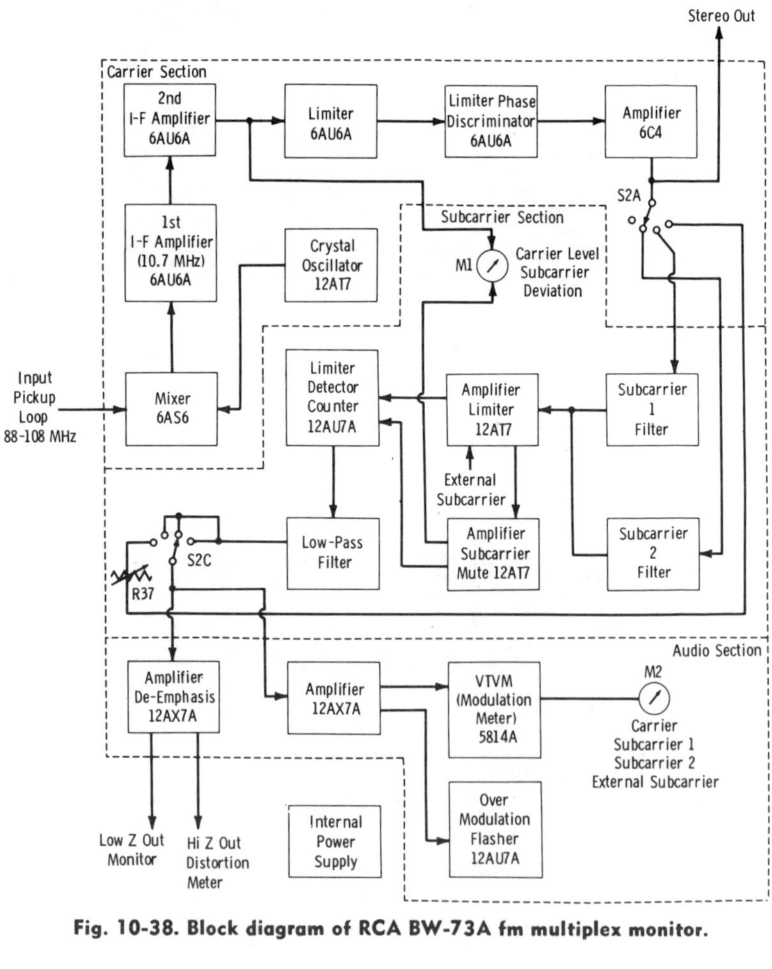

Fig. 10-38. Block diagram of RCA BW-73A fm multiplex monitor.

tor gives continuous indication directly in percent of main carrier deviation by the subcarrier.

Terminals are provided for the connection of external meters for remote monitoring of all metered functions. A separate output is provided for interconnection of the station's distortion and noise meter. With this arrangement, accurate measurements of signal-to-noise ratio, distortion, and frequency response may be made quickly. Similarly, the unit can be used to measure main-to-subcarrier and subcarrier-to-subcarrier cross talk.

The rf input signal is injected on a 50-ohm line, and the level is adjusted according to meter indication. The input level chosen was such that the monitor would be capable of making overall transmitter measurements and yet be relatively insensitive to spurious or unwanted signals. The oscillator is crystal controlled and operates at 10.7 MHz above the station channel. The mixer stage is followed by broadband i-f and discriminator circuitry. The succeeding amplifier output is switch selected for choice of monitoring mode. The subcarrier filters are plug-in units allowing fast, convenient changes or additions of subcarriers should the need arise.

The subcarrier chain employs two double-anode zener diodes in cascade to provide stability of the limiting level. At this point, the fm wave is applied to a pulse-counter detector where it is demodulated. Residual subcarrier is removed in a low-pass filter, and the remaining audio signal drives two separate stages. The first of these de-emphasizes the modulation, and it is this output which may be used for aural monitoring and distortion measurements in conjunction with an auxiliary distortion and noise meter.

The second stage is an amplifier for driving the peak-modulation indicator stage and the vacuum-tube-voltmeter modulation meter. The vtvm stage is compensated against zero drift. The ballistic characteristics of the meter meet the requirements of the FCC. The electronically regulated power supply is self-contained.

When used with the McMartin Model TBM-300 frequency monitor, this unit provides a complete station monitor.

EXERCISES

Q10-1. As a rule of thumb for horizontally polarized antenna systems, what is the power gain relative to the number of stacked bays?

Q10-2. If the power gain is 6, what is the gain in (A) dB, (B) field strength?

Q10-3. As a rule of thumb for circularly polarized antenna systems, what is the power gain relative to the number of stacked bays?

Q10-4. In terms of the modulating signal, give the characteristics determining the amount of frequency deviation for (A) fm, (B) pm.

Q10-5. What is done in a pm transmitter to make the output equivalent to an fm signal?

Q10-6. What is the instantaneous phase deviation in radians of an fm or pm carrier at 100-percent modulation for the modulating frequencies (A) 15 kHz, (B) 20 Hz?

Q10-7. What is the instantaneous shift in degrees for (A) and (B) of Q10-6?

Q10-8. What is the required frequency deviation of the oscillator stage in Fig. 10-23 for 100 percent modulation?

Q10-9. Is the SCA subcarrier amplitude modulated or frequency modulated?

Q10-10. Is the stereo subcarrier amplitude modulated or frequency modulated?

Q10-11. What information does the stereo subcarrier contain?

Q10-12. Give the frequency range of the stereo subcarrier sidebands.

CHAPTER **11**

Studio Operations

It is recognized that most studio operations are concerned with turntables and magnetic tape or tape cartridges. The major part of an operator's duties are concerned with prerecorded shows. "Live" studio programs normally are recorded on tape for playback at a later date. Whether a program is recorded on tape or aired live, the operator must be familiar with the proper operational techniques involved.

Different techniques are used for recording or airing stereo pickups than for monaural programs, even with compatible stereo. It should be understood that the term "compatible" applies to monaural-stereo receivers, not to pickup and recording techniques.

11-1. BASIC DUTIES OF THE OPERATOR

A good studio operator (whether an announcer/operator or operating engineer) must be very sensitive to art as well as science in broadcasting. It might appear that the art is overemphasized in the following pages; in reality, however, it cannot be overemphasized.

Broadly speaking, there are three kinds of program sources which concern the control-room operator:

1. Studio—This may be turntables, tape, or an actual studio performance to be aired or taped. The operator may be required to announce in addition to operating the equipment.
2. Remotes—These are pickups originating somewhere other than the studio, such as at a sporting arena, night club, or other location.
3. Incoming network program.

In nearly every case, the operator must also maintain the station program log. Where remote control of the transmitter is authorized, the transmitter readings specified by the FCC are logged at the specified time intervals.

For studio programs, microphones must be set up, or "spotted," in the studio in such a manner that all musical instruments and performers will be adequately covered. Microphone output is very weak and must be amplified sufficiently to be carried by wire lines to the transmitting plant. (This is true in all cases except where the transmitter and control rooms are installed together. Even in this case, the signal must be amplified considerably to drive the speech input stages of the transmitter.) Control of the various microphones is provided by the control console. A volume indicator must be used to indicate the relative magnitude of the program signals; it is mounted on the control console in an area convenient for viewing by the operator.

Fig. 11-1 is a view of the ABC production control room at Pittsburgh station KQV. The window looks into the studio for disc-jockey operations.

Courtesy American Broadcasting Co.

Fig. 11-1. Central production area at KQV, Pittsburgh, Pa.

One duty of the control operator is to place the microphones in the studio where the production director wants them for a particular program. If no production director is employed, the control operator must determine the positions of the microphones, or perhaps more correctly stated, he must determine the positioning of the performers in the microphone pickup area. The best positions are usually determined only by rehearsing the show before air time and changing the respective positions until the proper pickup is achieved.

During the progress of a studio show, the studio operator's position is at the control console. It is his function to operate the various controls so that the respective microphone outputs blend properly to give the effect desired. When a production director is employed at a station, he will assist the opator by telling him which sound or sections of sound to bring up or lower. Since, in any transmission system, definite limits exist as to maximum volume that can be handled and minimum volume that is adequate for transmission, the overall volume must be monitored and controlled by the operator. This is the purpose of the volume indicator. The operator must also operate the switching system to choose the proper studio or incoming program lines.

11-2. YOU'RE UP AGAINST SOUND

The control operator must be familiar with one of the most complex and perhaps one of the most controversial scientific fields, the field of sound. Since he must correlate the known facts of sound with operational techniques of equipment (even though only recordings may be involved for a local announcer), it is important here to review some fundamental principles.

Whenever the air is set in motion at a vibratory rate that happens to fall within the audible range of the human ear, we hear sound. This sound, in the strict physical sense, is a setting up of vibrations that consist of alternate condensations and rarefactions of the air. These variations in the air density constitute sound waves. The engineer or technician concerned with the transmission and reproduction of sound finds himself faced with problems of unwanted noise, loudness, tone color, and the more complex timbre. He may read somewhere that it takes about 5 seconds for a sound wave to travel a mile in air, that it travels four times this fast in water and almost fifteen times as fast through iron, but this picture of sound does not help him in any practical manner. He is taught in his physics courses that loudness depends on the amplitude of the air disturbance and that pitch depends on the vibratory rate of the sound wave, but in practice he finds many more variables than these fundamental truths of loudness and pitch. He soon becomes aware of a decided difference between loudness and volume.

Essentially, the reader is interested in the best possible methods of transmitting or reproducing sound, and the proper care, maintenance, and repair of the equipment involved. He knows that the sounds he hears in nature are very definite and distinct in themselves, that the rustle of dead leaves sounds different from the rumble of thunder. He is able to tell the difference between one person's voice and another, and even to interpret the very spirit that lies behind the inflections of a spoken word. The amplifier with which he is concerned must be fed with electrical impulses which may again be reconverted to all these original characteristics that are so much a part of the original sound in nature. This section, therefore, deals with the conversion and reconversion of sound.

Fundamentals

We usually think of sound as something we hear. Since this is the most practical approach to a highly complex phenomenon, we will approach the subject in relation to the effect of vibrations in the air on our hearing sense. The physicist and the laboratory technician consider the word "sound" to mean vibrations that are set up in the air or a solid by vibrations of the original sounding body, even though there is no ear to hear that sound. However, a broadcast technician or recording operator wants to be able to visualize the nature of his equipment as it affects the ear of the listener. For example, what could be the various causes of amplitude distortion in a microphone? What would affect the frequency response? If the characteristics of a particular microphone in normal operating condition are known, how can these particular characteristics best be used to achieve the utmost possibilities within the natural limitations of the microphone? In order to gain the feeling and the "know-how" with sound-converting apparatus, we will start with the fundamental nature of sound and the problems associated with the conversion of sound to electrical impulses in such a manner that the ear will be able to hear the reconverted sound as an almost exact duplicate of the original impulses. Sound, of course, may be transmitted in liquids and solids equally as well as in gaseous matter, but this text will be concerned with the nature of sound waves in air, which is the primary concern of the reader.

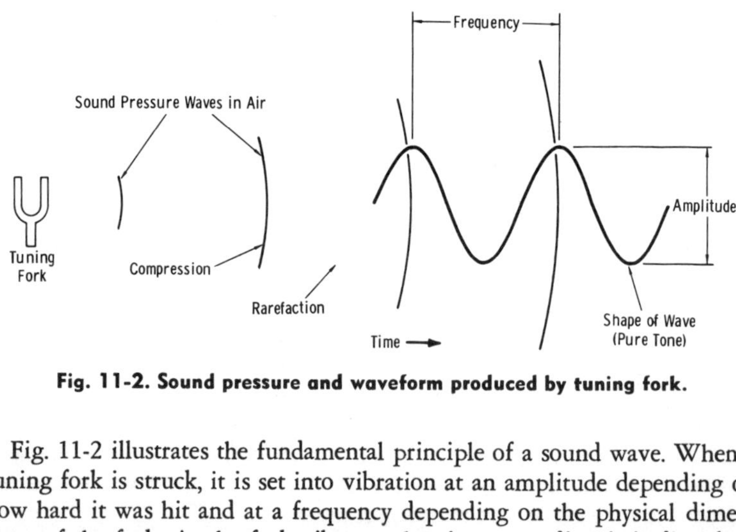

Fig. 11-2. Sound pressure and waveform produced by tuning fork.

Fig. 11-2 illustrates the fundamental principle of a sound wave. When a tuning fork is struck, it is set into vibration at an amplitude depending on how hard it was hit and at a frequency depending on the physical dimensions of the fork. As the fork vibrates, the air surrounding it is disturbed, and alternate condensations and rarefactions of the air occur. The waves travel in all directions from the fork and decrease in amplitude in proportion to the square of the distance from the vibrating fork. If these variations of the air pressure are charted with respect to time, a curve of the sound wave is obtained. The amplitude of the sound wave is represented as the overall height of the curve, and the frequency is represented by the distance between successive peaks or troughs of the curve plotted against a definite time axis. For example, if the time between successive peaks is one-thousandth of a second, the curve represents a tone of 1000 Hz.

Insofar as pure tones such as those from a tuning work or audio oscillator are concerned, it may be stated that the *intensity* of the sound is dependent on the height or amplitude of the wave, and the *pitch* of the sound is dependent on the number of wavelengths per unit of time or frequency of the wave. Note the amplitude of the sound wave has been defined as intensity, not loudness; the loudness of any sound wave is a more complex phenomenon.

Now consider what happens when two pure tones of different frequency are sounded. Fig. 11-3 shows two tuning forks, the individual waveshapes, and the resultant waveshape when the two are combined in the air. When the two waves are in phase (both rising and falling together), the resultant

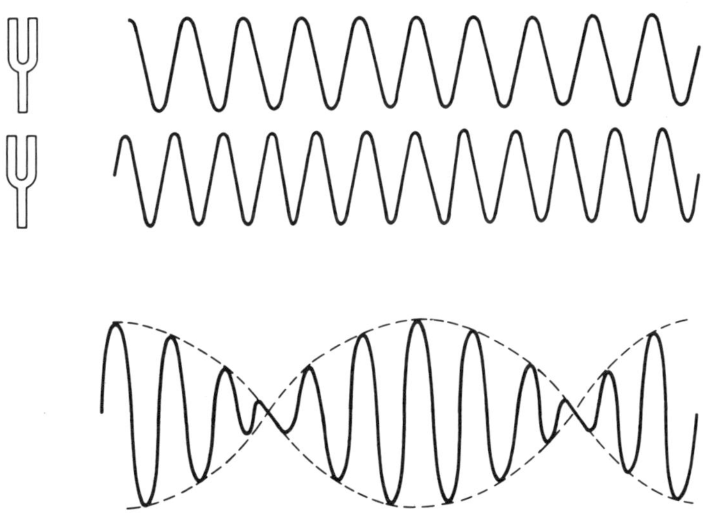

Fig. 11-3. Beat note produced from two fundamental frequencies.

wave is additive and a large amplitude results. When the waves differ in phase (one rising, the other falling), they tend to cancel out and the resultant amplitude is small. This physical growing and dying out of sound at regular intervals constitutes a *beat note,* which is equal in frequency to the difference between the two original frequencies. For example, if $f_1 = 2093$ Hz and $f_2 = 2349$ Hz, the beat frequency would be 256 Hz. In this case, the ear would hear not only the two real physical frequencies, but also the beat of 256 Hz. If a physical analysis of this tone is made, it will *not* reveal any wave component of 256 Hz, yet this sound will be just as real to the ear as are the two fundamental tones. Such tones have a great influence on quality as interpreted by the ear.

This tone quality, or *timbre,* of sound is related to the shape of the wave. For instance, a flute sounding the tone of middle C (261.63 Hz American Standard Pitch) will not sound the same as a clarinet sounding the same frequency. We still recognize one sound as being distinctly that of a flute and the other sound as being distinctly that of a clarinet. Fig. 11-4 is a graphical representation of these two sound waves, showing why the timbre, or tone quality, differs as interpreted by the ear, even though both tones are of the same frequency.

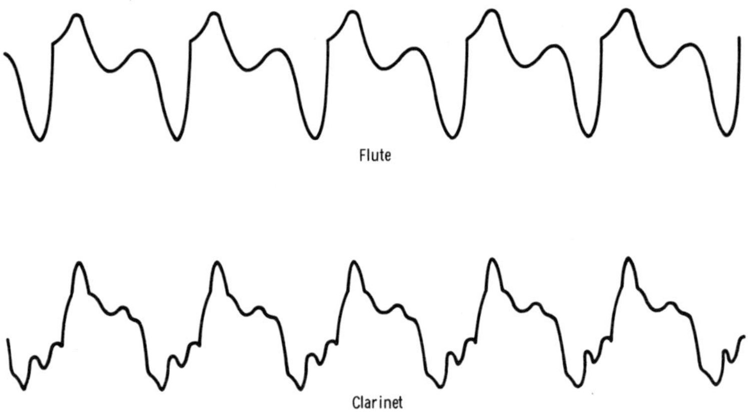

Fig. 11-4. Waveforms of two wind instruments.

In practical use, a microphone is subjected to a wide variety of tones that make up the character of speech or musical sound. These waves become very complex in character and are termed complex waves. When a word is spoken, the sound is given a certain timbre, or tone color, that is determined by the vocal chords, mouth, and nasal cavities of the speaker. This characteristic of speech is like the fingerprint of a person; no two people have exactly the same voice timbre, and this is the reason why we can recognize a person's voice even when we are blindfolded. For fidelity reproduction, as in broadcasting, any gateway to sound must be able to convey the different shades of expression of voice or music. In the case of the disc recorder head, for example, it must convert electrical vibrations into corresponding mechanical energy sufficient to move the cutting needle. The pickup head must convert the energy of the needle in the grooves of the record to corresponding electrical vibrations to excite the amplifier.

All of this must be accomplished so that the original shapes of the sound waves which determine timbre or tone quality are maintained as accurately as possible.

The Phenomenon of Hearing

The understanding of any apparatus connected with sound transmission necessitates a knowledge of human hearing. The previous paragraphs have dealt with the mechanics of sound in air, and how these waves must be handled to be used for transmission or reproduction. In the final analysis, however, the ear, or rather the interpretation by the ear, is the basis of all practice and theory in the practical field of sound.

The frequency response of an excellent ear is from 20 to 20,000 Hz. The range of the average normal ear is generally taken as 20 to 15,000 Hz. There is, however, a noted falling off of high-frequency response with advancing age.

When you think of response, consider the effect of differing frequencies on the hearing sense. When it is said that an ear responds to a frequency range of 20 to 15,000 Hz, it is *not* implied that the ear responds to a frequency of 40 to 50 Hz equally as well as one of, say, 1000 Hz.

Review for a moment the graph of Fig. 2-16, Chapter 2, which shows the frequency and volume range of speech and music. The lower curve, showing the threshold of hearing, is a picture of the deviation of response of the ear with frequency. That is, if the 1000-Hz tone is taken as zero dB and is of such intensity that it is barely audible, a 100-Hz tone must be approximately 40 dB to be barely audible. It is seen here that there is a notable falling off of both the low-frequency and high-frequency response when the sound intensity is low. When the intensity is so high that it approaches the threshold of feeling (upper curve), the frequency response of the ear is seen to constitute a comparatively flat line. It can also be observed from this illustration that the frequency range of speech is 100 to 8000 Hz, with a dynamic range (ratio of lowest to highest intensity) of 40 dB. The frequency range of music (in this case, an orchestra) is about 40 to 14,000 Hz with a volume range of 70 dB. There is a definite effect of frequency range on the intelligibility of speech and the quality of speech and music.

Loudness

It should be realized that loudness depends not only on intensity but also on frequency, due to the peculiar properties of the human ear. The radio service technician is acquainted with the bass-boost circuit associated with the volume control of a receiver, which boosts the bass response at low volume settings to help offset this characteristic of hearing. In actual practice, it may be noticed by the broadcast operator that two voices producing the same indications on the volume indicator will sound different in loudness. The loudness of a complex wave depends on the number of harmonics present in the sound wave, and on the phase relationships of these harmonics. Since the content of voice waves differs so radically from person to person, the technician should realize that this effect is not an indication of a faulty component, but is a natural phenomenon of the hearing sense. It may also be understood from this discussion why a microphone, pickup, or recorder head must introduce negligible phase or frequency distortion, since the timbre of sound depends so much on the waveshape.

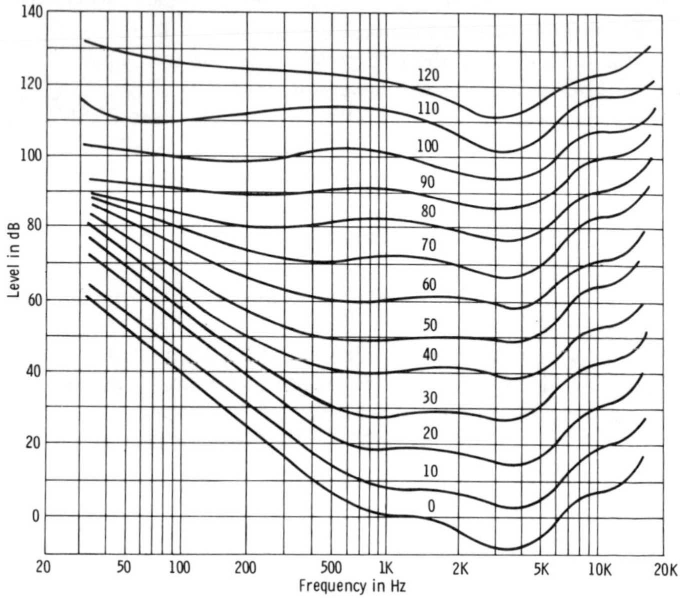

Fig. 11-5. Family of loudness-level curves.

Fig. 11-5 is a graph of loudness level curves adopted by the American Standards Association. The derivation of these curves is explained in most standard textbooks on sound and will not be duplicated here. An example will suffice to enable the reader to use this graph correctly. It will be noted, for example, that a 300-Hz tone 40 dB above the reference level (0 dB) corresponds to a point on the curve marked 30. This is the loudness level. It means that the intensity level of a 1000-Hz tone (reference frequency) would be only 30 dB in order to sound equally as loud as the 40-dB 300-Hz tone.

Now assume a fundamental of 392 Hz (G string of a violin) with an actual intensity of 40 dB above reference level. From the curve, the loudness level is approximately 36 dB. It was determined by Bell Laboratories that the addition of the overtones or harmonics of the fundamental raised the intensity from 40 to 40.9 dB, whereas the loudness level was raised from 36 to 44 dB. In other words, the addition of the harmonics raises the actual meter reading only 0.9 dB, while the loudness level increases 8 dB. The reference level of 1000 Hz would have to be 44 dB to sound equally as loud as the complex tone.

When it is realized that the vocal organs of human beings are all exceedingly different, and are associated with a particular resonating apparatus that gives to the voice its individual timbre, it becomes clear why two voices peaked at a given meter reading will often sound far different in loudness. Certain harmonics of the voice are emphasized while others are suppressed in an infinite variety of degrees. Fig. 11-6 illustrates waveforms of two male voices intoning the same vowel. A decided difference in peak factor (ratio of peak to average content) depends to a large extent on the harmonic content and the phase relationship of the harmonics to the fundamental.

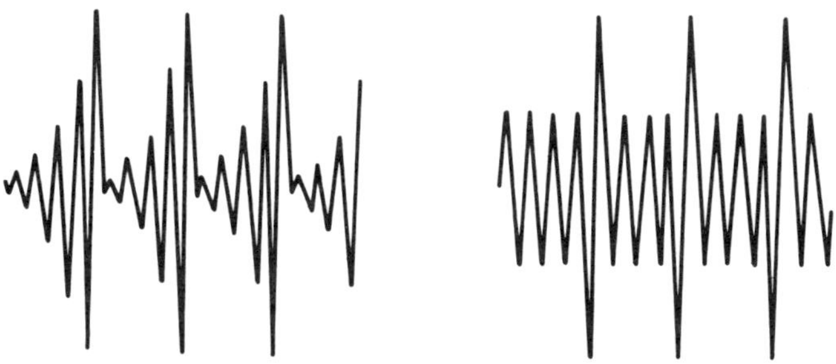

Fig. 11-6. Waveforms of two male voices intoning the same vowel.

At the present time, only one solution suggests itself. When it becomes necessary to transmit two voices so different in timbre as to be decidedly unequal in loudness for a given reference level, the good taste and judgment of the operator at the control panel must govern their respective levels. The author fully realizes the many and varied complications that arise from this condition, since loudness is not only a physical, but also a psychological phenomenon. The level at which the receiver in the home is operated will determine the extent to which changes in loudness intensity are noticeable; at low volumes, a greater change of intensity is required to be noticeable to the ear than at high volumes. Acoustics of the studio and the room in which the home receiver is operated will influence the effect of intensity changes on the ear. However, the control operator with good taste, a critical ear, and keen appreciation of values can find the "happy medium" between aesthetics and conventional transmission operations. This is of prime importance when voice and music are to be blended.

If some kind of guideline is desirable when "gain riding" by a meter, the compression meter (which indicates the amount of compression occurring in the agc amplifier) should be observed rather than the output VU meter. When experience dictates that a following sound source is louder than normal programming for a given VU meter reading, ride this source for zero compression. For example, if normal programming is causing 6 dB of compression, adjust the gain on the "loud" source for no indication of compression. The output VU meter will still indicate about normal level, but the *average* sound level will be 6 dB lower. In many cases, a difference of 10 dB is required between various sound originations for the loudness to seem the same to the ear.

NOTE: The subject of loud commercials has now become a source of FCC rulings, which are covered in the last section of this chapter. There are many variables involved, such as the amount of reverberation used and the type of filtering. Due to the wide gamut of factors involved, and the importance of this subject, we will cover these many variables before undertaking a study of the FCC rulings.

11-3. VOLUME-INDICATOR INTERPRETATIONS

For the purpose of discussing problems relating to the use and interpretation of volume-indicator readings, use will be made of the fact that any wave, no matter how complex, may be reproduced exactly by a number of sources of pure tones. Fig. 11-7A shows two simple harmonic motions of the same frequency, differing slightly in phase; tone B lags tone A by a certain number of degrees. The vector addition shows how the total amplitude is influenced by the reinforcement of the two tones. Fig. 11-7B illustrates the same tones of similar frequency differing in phase by exactly 180°. The vector diagram shows complete cancellation of the total energy, since the tones are now opposing each other. Keeping this clearly in mind, notice that the total amplitude will be larger for smaller angles of phase difference, and if the two tones are exactly in phase, the parallelogram collapses and becomes a straight line (that is, the sum of the individual amplitudes). As the angle of phase displacement becomes larger, the total amplitude becomes less, until at 180° it becomes zero.

Under program conditions, a large number of different frequencies with varying phase displacements are encountered, and the loudness sensation produced in the ear for a given meter reading is dependent on the number of harmonics present and the phase relationships of these harmonics. It then becomes obvious that the acoustical treatment of the studio and the type of program content will influence the correct interpretation of a volume-indicator reading. It is, of course, apparent that the volume indicator shows the magnitude of the waveform, whether it be distortion peaks, noise, or musical sound, that must be kept within the dynamic range of the transmission system. But when the reading of the volume indicator is correlated with the effect produced on the hearing sense of the listener, these problems must be met and analyzed. The loudness sensation for a given meter indication has already been discussed with regard to voices of differ-

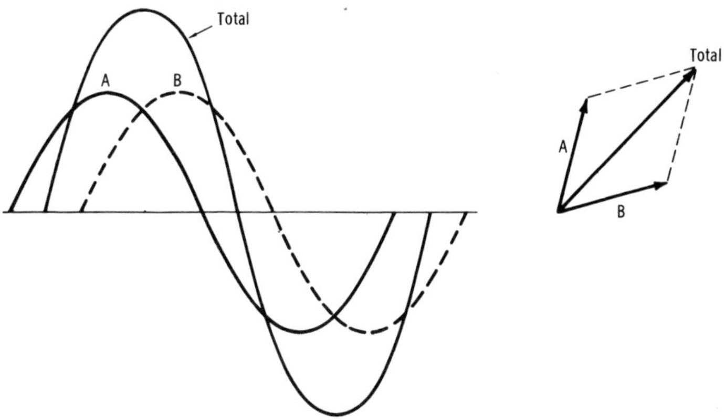

(*A*) *Small phase difference.*

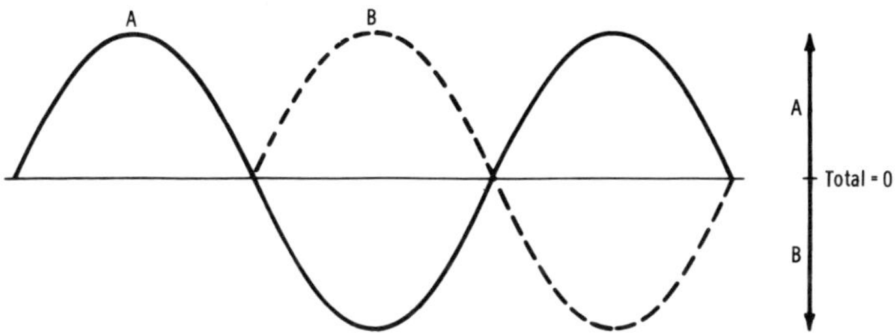

(*B*) *180° phase difference.*

Fig. 11-7. Two harmonic motions of the same frequency.

ent persons. The same characteristic is noted between individual musical instruments where the number of harmonics and their phase relationships may vary widely.

The volume indicator is used as a means of visually monitoring the magnitude of program waves for two primary reasons:

1. To compress the original wide dynamic range to an amount consistent with good engineering practice of the broadcast transmission system.
2. To locate the upper part of the dynamic range below the overload point of associated equipment.

For the latter purpose, a scale of 10 dB would be adequate; for the former, a much wider decibel range is desirable. Since the instrument is used for both applications, it was decided that a compromise on a scale length of 20 dB would be desirable. It appears that as the art and appreciation of high-fidelity service advances, not only the frequency range but also the dynamic range of transmission will be extended, particularly for certain types of program material. This feature becomes very important in fm service. With present-day meters, it is left to the experience and judgment of the operator as to how far the volume should be allowed to drop below the visual indication of the VU meter.

In practice, pure sine waves are not broadcast, except for special testing purposes. Program waves, whether speech or music, consists of relatively short bursts of energy with variously spaced peaks and valleys. The standard VU meter is characteristically damped such that the relatively sharp peaks in program audio cannot be indicated. Essentially, this meter shows a *kind* of *average* of program audio, and actual program voltage peaks exceed the resulting meter deflection.

Fig. 11-8A shows an oscilloscope display of a sine wave (arbitrarily adjusted for 2 divisions on the graticule) and the superimposed program audio signal when the VU meter indicates 0 VU (100-percent reference)

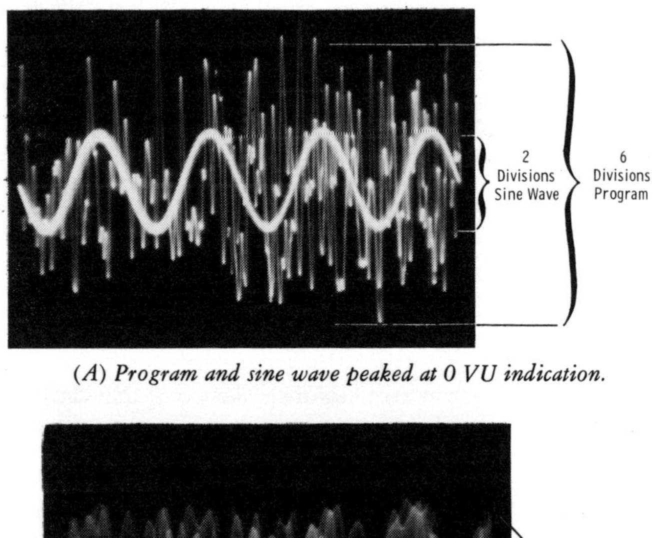

(*A*) *Program and sine wave peaked at 0 VU indication.*

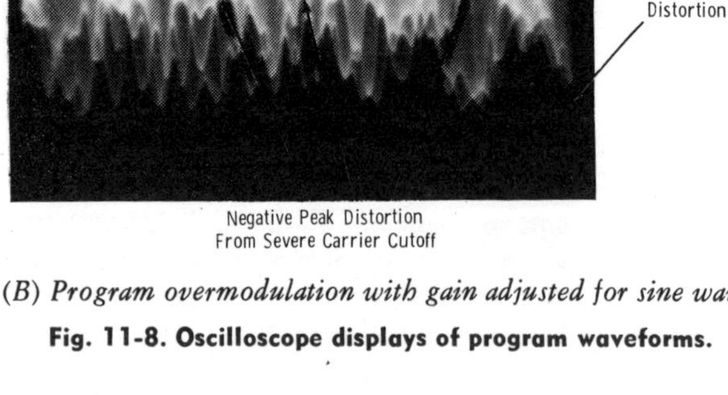

(*B*) *Program overmodulation with gain adjusted for sine wave.*

Fig. 11-8. Oscilloscope displays of program waveforms.

in both instances. An oscilloscope displays actual voltage peaks; the VU meter (even any so-called peak-reading meter) does not. The actual peak-to-peak excursion of program waves in Fig. 11-8A covers 6 divisions, for a three-to-one voltage ratio, or about 9.6 dB. In practice, depending on the program content, this ratio of program peaks to sine-wave voltage at the same reference VU indication can be anywhere between 8 and 14 dB.

This characteristic is very important to the studio engineer and transmitter engineer when the procedure of setting levels is involved. For example, if you operate the transmitter by remote control and you feed a sine wave at 100 percent reference on the VU meter, do not be concerned that the transmitter modulation monitor indicates only around 32 percent modulation when the input gain was previously set for 100-percent modulation on program.

If you make the mistake of adjusting the transmitter input gain to result in 100-percent modulation when the VU meter indicates 100-percent sine wave reference, you will have severe overmodulation of the transmitter when you apply program. Fig. 11-8B shows the result of this type of error.

The subject of correlation of meter readings between transmitter and studio will be covered more fully in Chapter 14. This is done primarily to keep the subject matter closely linked to the type of operation at hand, even though many installations use combined studio-transmitter operators because of transmitter remote-control operation.

11-4. MECHANICAL OPERATIONS

In the realm of the physical, mental, and psychological faculties of the control-room operator lies the success or failure of the broadcaster's daily schedule. A script-writer's masterpiece or a composer's dream can amount to no more than the original worth of his work, plus the ability of the control operator to interpret that work on the technical equipment at his command. Yet, perhaps paradoxically, the best qualified operators are the least conspicuous to the listener at home. This statement is just as applicable to operations in a so-called "automated" installation.

An ideal job of turntable and tape operation or switching and blending of microphones for the various performers of a given show is such that the listener is entirely unconscious of mechanical operations necessary to their performance. The operator who cuts his program at its conclusion, instead of fading out (even though because of time limitations it must be a quick fade), not only makes mechanical operations apparent to the listener, but marks himself as a man not entirely a master of his equipement. Exceptions to this rule exist, such as stunts of a technical nature that are sometimes aired to impart a technical flavor to the layman. In such programs, operations should be accentuated, of course, rather than subdued, but it is well to remember that, as a general rule, the test of the operating technique should be, "Are mechanical operations apparent?"

Music, speech, or background accompaniment that is too high or too low in level should be gradually adjusted to normal in a manner cognizant of musical and dramatic values. It might appear, at first, that this prime requisite for good operating practice would conflict seriously with good engineering practice. When levels are too high, overloading of associated equipment at the transmitter occurs. When compression amplifiers are used, as is commonly the case, distortion arises from excessive compression rather than from overmodulation of the transmitter. It has been proven from extensive tests, however, that distortion caused by momentary overloads simply is not noticeable even to highly trained ears. This appears to be due to physiological and psychological factors that determine the response of the ear to aural distortion, resulting in a lack of response to overload distortion of short duration and occurring at rare intervals.

The level of the speaking voice can be least obviously adjusted by correcting the fader setting between words or sentences where slight pauses occur, rather than increasing or decreasing the volume during actual excitation of the microphone. A comparison of these two methods by the operator on his audition channel will reveal the striking obviousness of level control; the former technique allows less conspicuous adjustment.

Anticipation can play a major part in smooth level control when circumstances permit. The operator soon becomes familiar with the approximate fader setting for each announcer, or for his own voice relative to a preceding one if he is a combination announcer-operator. It is obvious here, of course, that ample opportunity to adjust the mixer gain is needed before actual air time. This is also possible in some instances with transcribed and recorded shows, when the operator is aware of the brand of recording to be played next. This will also be influenced to a great extent by the type of filter used on the turntable for various recordings, since a different filter is used in many instances for different brands or conditions of recordings and transciptions. The gain settings for a given brand will usually be fairly consistent. Thus, when the operator has become familiar with the necessary fader adjustment for each brand of transcription or recording, he will be able to use the art of anticipation to good advantage. However, if uncertain of the level, it is well to remember that from an aesthetic as well as a technical point of view, it is far better to fade in the speech or music rather than to experience the shock of excessive volume which must be quickly lowered to normal values.

The foregoing discussion is likely to lead to an erroneous point of view on the part of a newcomer in a control room. One of the most common errors of new men in this field is to ride gain to the point of exasperation to a critical listener. The operator should endeavor at all times to give musical and dramatic values a free rein insofar as is practically possible. Remember that from the listener's point of view, the business and purpose of broadcasting is to provide entertainment by bringing music and dramatics into the home. The technical setup necessary for this purpose has been engineered to a point of perfection; it is only necessary that this equipment be operated in a manner that will promote musical and dramatic values in their original intent.

The fundamental rule of good operating technique is probably the most abused by innocent operators during the transmission of symphony broadcasts, live or recorded. Suppose that an orchestra of 40 to 80 members has just finished a number which for the past few minutes has been *pianissimo* (soft), say −15 to −20 VU. It is safe to say that the average listener to a symphony program will have his receiver volume adjusted so that comparatively high power exists in the speaker when the studio level reaches 0 VU. It is obvious then what will occur if the announcer suddently starts speaking at a 0-VU level. The listener may not be actually raised from his chair by this sudden roar, but the experience is quite unpleasant. Sudden crescendos in music are expected, welcomed, and appreciated, but a single announcer at an apparently greater volume than an 80-piece symphony orchestra simply is not only unwelcome, but extremely obnoxious.

It is a safe rule to remember that after such musical numbers as this, the announcer should be held down to about −6 VU maximum. The difference to be maintained between levels of voice and music will depend not only upon the type of program aired, but also upon the acoustical treatment of the studio where the program, live or recorded, originates.

The inadequacy of present-day broadcasting in the field of symphonic music transmission is quite apparent to most engineers. The discrepancy between the usual 70-dB dynamic range of a full orchestra and the actual 30 to 35 dB allowed by broadcast equipment is all too obvious to the control operator handling such pickups. It has been the practice of some operators who do not appreciate the symphonic form to bring all low passages up to around −4 VU, and then reduce the gain as the orchestra increases its power according to the continuity of the musical score. The fault in this technique should be apparent. If the very lowest passages are brought up to just "jiggle" the meter, and care is taken to use good taste in suppression of the crescendos, a satisfactory dynamic range may be experienced, since even a range of 25 dB will vary the output at the receiving point from 25 milliwatts to nearly 16 watts *on peaks.*

The technician on a symphony, or any musical program, should possess a good ear for music. Rules and regulations will never help a man with a pair of "tin ears" to handle a musical show properly. It is, nevertheless, important that the transmitter technician (if separate from the studio) understand that a great amount of modulation during classical music will be below 20 percent even with compression line amplifiers.

Recordings and transcriptions of symphonic music have already been compressed into the broadcast dynamic range, since the recording engineer has essentially the same problem to contend with in relation to this difficulty. Usually all that is necessary for the control technician to do is to set the peaks of the music to 0 VU or 100 on the scale, and "let it ride."

Don't Be Hypnotized

The foregoing discussion should serve to warn both newcomers and old-timers against one of the most common occurrences in control rooms—the hypnotized operator. A volume indicator is apt to exert a strong hypnotic effect on the person responsible for "riding gain" on a program. So long as the meter peaks at 0 VU, or 100 percent, on the scale the suggestion to the subconscious mind that all is well is almost overpowering, in spite of the aural reaction that ordinarily would warn the operator. Competition in the field of modern operating technique demands more and more coordination of aural effects and volume-indicator interpretations.

The rule here is to train the aural response to command attention over and above the VU-meter indication. If a certain voice sounds low in volume in relation to the music, peak the music just enough lower than the voice to compensate for this effect. If two or more voices on the same program sound different in loudness, balance them aurally, *then* note the respective VU-meter readings. It is quite often possible to alter the loudness sensation of a particular voice by changing the distance from the microphone or the angular relationship to the axis of the microphone.

Table 11-1. Suggested Program Transmission Standards

Program	VU Peaks	Gain Reduction
Speech	1. Normal passages. Peaks of 100.	6 to 10 dB
Music (Live or Recorded)	1. Normal passages. Peaks of 100. 2. Low-level passages. Not less than 40.	6 dB max 0-2 dB
Commercials (Recorded)	Prescreen for proper match in loudness to adjacent program material.	
Symphonic Music (Live)	Highest passages. Peaks of 100. Pianissimo. Just move VU.	Adj agc for 3 dB max. None.
Symphonic Music (Recorded)	Highest passages. Peaks of 100. Pianissimo. Just move VU.	Remove compression completely. None

The amount of compression of the agc or compression amplifier affects the loudness to the extent of raising the average volume level of transmission. Table 11-1 lists suggested audio program transmission standards and serves as a guide for radio operators. Note that the gain reduction (decibels of compression) is considered along with VU-meter peaks. In the individual station, where management permits, the operator is under a far more flexible set of rules which allows consideration of "loudness sensation" in gain adjustments.

Keeping Sound "Out of the Mud"

The problem of correlating volume levels with comparative loudness of speech and music has appeared as an item of major importance and should not be ignored by broadcast station personnel. As was mentioned before, because of the nature of classical music, the symphony listener at home will operate his receiver a great deal higher in level than he would for ordinary programs. Five minutes of symphonic music will have perhaps 3 to 4 minutes of low to very low levels; the average intensity level over a period of time is far lower than the average intensity level of a dance orchestra in the same time interval. It should then be obvious that a greater difference should exist in the ratio of music to speech levels for symphonic programs than for those of dance music.

The acoustical treatment of the studio in which the program originates affects to a great degree the loudness of voice and music, and in a different ratio. A studio that is overtreated with absorbent material deadens the sound because of high-frequency absorption and is an outstanding enemy of musical programs. Music from "dead" studios is "down in the mud," lacking in brilliance, and generally dull to hear. The effect on speech, however, is not so pronounced as that on music. Speech originates within a few feet of the microphone and requires much less reverberation to assure naturalness, whereas the space between the source of the music and the microphone is greater, and many things happen to the musical waveforms that must eventually be translated into perceptions of loudness.

So-called "optimum reverberation time" really is an expression of what constitutes pleasing sound, and it changes with experience. It may well be that standards of optimum reverberation time in the near future will decidely alter the preceding discussion of ratio in peaking voice and music. The point that is important to keep in mind is that a great majority of studios throughout the country are below today's standards of correct reverberation characteristics.

The newer "live-end, dead-end" studios, with musical instruments placed in the live end and microphones spotted in the dead end, present one solution for properly controlled reverberation. In these studios, voice and music peaked at the same level will appear the same in loudness sensation. In fact, the advancing state of studio development indicates a transitional era during which the brilliance of music is so great in some of the most modern studios using reflecting panels for musical pickups, that when the music is peaked the same as voice, the voices sound much lower in loudness than the music. This brings to mind again the importance of using judgment in aural perspective when riding gain on productions with the intent of achieving a properly balanced effect when the program is reproduced in the listener's home.

11-5. TECHNICAL PRODUCTION TECHNIQUE

This text is concerned with the type of productions normally encountered in the average a-m or fm station. More complex productions, such as those requiring many tracks of tape channels which are "mixed down" to a one- or two-track format (as is done in professional recording studios), are not considered.

Principle of Microphone Patterns

In determining the placement of microphones for any given setup, it is important that the operator become familiar with the pickup patterns of the microphones used. These patterns illustrate the amplitude response for varying directions about the face of the microphone. Fig. 11-9A shows the pattern of the RCA 44-BX velocity microphone, and Fig. 11-9B is the pattern of an RCA 77-B combination ribbon- and pressure-type instrument. There are several important points of interest relating to these patterns which show great differences in characteristics aside from the most apparent one, that of bidirectional pickup by the velocity type and unidirectional pickup by the combination type.

An analysis of the patterns reveals a much wider range of amplitude response for the combination pressure-gradient (ribbon) and pressure microphone than for the ribbon type alone. It is noted from Fig. 11-9A that at an angle of 70° the amplitude response is down about 10 dB with respect to the response at a given distance at 0°. Now note from the response curve of Fig. 11-9B (combination type microphone) that the amplitude response at 70° is down only approximately 3 dB from the 0° reference. These patterns are useful for determining the setups necessary for discriminating against unwanted sources of sound, and for obtaining a particular relation between sounds from different sources. It can be seen that as a performer is moved around the microphone, loss of sensitivity may be compensated for by moving closer to the instrument.

Microphone Placement

"The microphone is a mechanical extension of the human ear." This statement is purely a fancy. The fault in this definition lies not so much in the literal meaning as in the implications that are involved. Were it true, even if we had only one ear, we could walk into a studio or on a stage and place the microphone at the spot where the one good ear could hear the orchestra, the soloist, the chorus, and the announcer. It won't work. Yet this conception is probably the contributing factor in the reasoning of an operator who places a microphone and proceeds to handle the show by "riding gain." Of course, there is the more ambitious type who spots a microphone for each section of the orchestra, then one for the soloist and announcer, and several for the chorus. Granted that there are some cases (particularly encountered in remote pickups) where this is necessary, it will be made

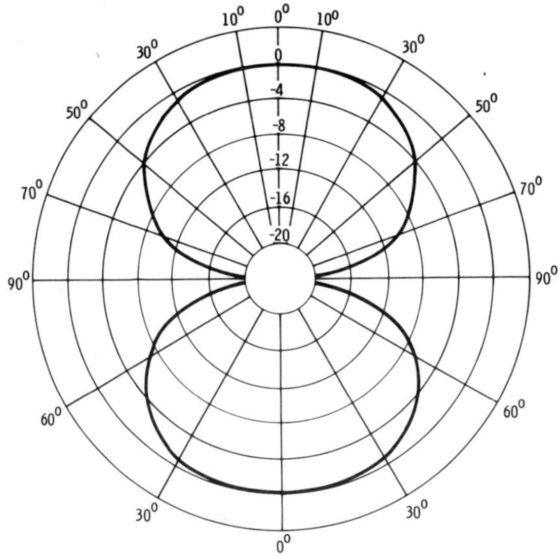

(*A*) *Bidirectional.*

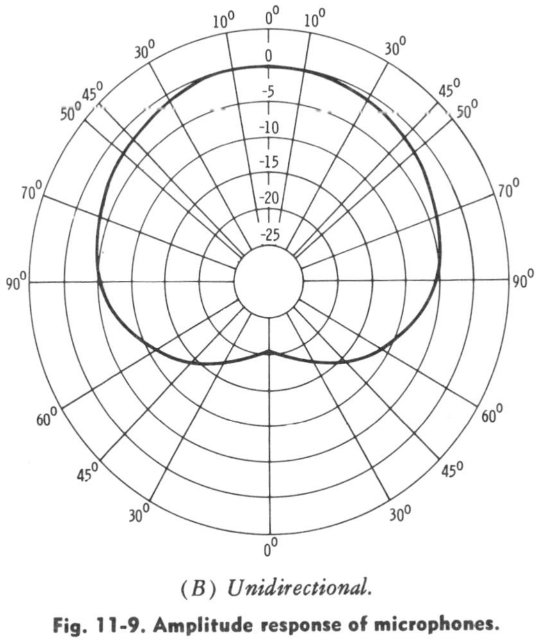

(*B*) *Unidirectional.*

Fig. 11-9. Amplitude response of microphones.

clear in discussions to follow that this condition is the exception rather than the rule for regular broadcast studios.

Why is it that a microphone cannot be treated as an extension of the human ear? To understand this it is necessary to review a fundamental theory that is an old story, but absolutely essential to understanding the microphone from an operational point of view.

Unlike the human hearing system, which is binaural (two-eared), the microphone is monaural (one-eared); this is true regardless of the number of microphones used to feed one channel. Whether one microphone or ten is used, the sound is collected into one channel and received through one speaker. Physically, the difference is that in a monaural system the sense of *direction* is lost, while reverberation is somewhat more noticeable; this makes the apparent distance of the sound source seem greater than it does when the sound is heard binaurally. Any operator who has ever set up a microphone in a particularly live hall has experienced this phenomenon. He is able to hear quite clearly the conversation of two people somewhere in the hall when listening with his two ears, but when he listens with the headphones through the amplifier, the sound seems much more distant, and extraneous noise is very high. This brings up an all-important psychological factor closely linked with the above physical one.

This psychological factor is the unconscious ability of *focusing power* when listening binaurally, or even with one ear plugged as much as is physically possible. The ear, associated as it is with the nervous system and the brain, tends to exclude extraneous noise that is present and focus attention on particular sounds. The microphone is a mechanical device not associated with any means of concentration except as it may be deliberately used to achieve a desired result in the listening ear at home.

A practical example is familiar to anyone who has dined in a public place where, for instance, a small orchestra renders dinner music. The table may have been out in the center of the room, and although the usual clatter of dishes, hum of conversation, and all the various noises were quite noticeable, the music may have been found very enjoyable when the diner cared to listen. Yet it takes little imagination to realize what the result would be if a microphone were placed at this same table. Very little of the music would be heard, and the few strains coming through would sound far, far away. The pickup would be nothing but a hopeless hodepodge of voice, noise, and confusion.

First, do away with the misleading idea that the microphone is a mechanical extension of the human ear. Think rather of the microphone as an instrument that must be used according to its characteristics and within its limitations to achieve an exact replica of the original program content in the listener's ear. Another statement heard occasionally is, "Microphone technique is different for fm that it is for a-m." This is still a somewhat controversial subject; however, the following analysis should be carefully considered before taking the statement literally.

It is obvious that the significant difference between a-m and fm is the means of rf modulation, transmission, reception, and demodulation. Many a-m stations today have an audio system that is as good as that of the best fm station. Now consider only the job of the control-room operator or production man responsible for microphone setups. He will monitor the program from the speaker in the control or production room, whether or not the audio signal is feeding an a-m or fm transmitter. The program will sound no better in the receiver than it does in the monitor regardless of the means of transmission. It is, of course, true that the fm listeners will hear more nearly what is heard in the control room than will the a-m listener. It appears then that the rule should be, "Microphone technique is somewhat more important for fm than for a-m."

Pattern/Studio Effects

Insofar as program setups are concerned, the foremost characteristic of a microphone is its pattern of response. There are several important factors in using a microphone pattern to obtain a desired result.

The basic point to keep in mind is that a response pattern as illustrated for any particular microphone is plotted in an acoustically dead room to avoid any practical amount of reflection of sound waves. In other words, a pattern of response will hold true when not influenced by any enclosure, such as when the microphone is used outdoors. The manner in which a microphone response pattern is affected by the acoustical nature of the surroundings becomes an interesting and highly important item for the technician to consider.

In general, the pattern will hold more nearly true in a dead studio, and it will be altered more and more as the surroundings are made more live. Take a hypothetical case of a setup involving two sound sources and a bidirectional microphone. Fig. 11-10A illustrates the theoretically dead room and ideal pattern of a bidirectional microphone. Sound sources A and B are of the same intensity and are the same distance from the microphone, but B is directly on the zero-response axis. Since sound source B will excite both sides of the ribbon equally, no movement of the ribbon will result from this source, and no output voltage will be created in the microphone. Therefore, the total output voltage will consist only of the impulses received from A. Thus, the pattern holds true.

Fig. 11-10B shows the same setup in a live room. It is still true that B will excite both sides of the ribbon equally, resulting in no response to the direct wave, but now there are reflections. Sound will be reflected from the walls back into the sensitive side of the microphone, and although reduced in intensity, the reflected sound will add to the sound from source A. The difference now between the two sound sources is one not only of intensity, but also of the ratio of reflected to direct sound. Naturally, this ratio is greater for sound source B than for source A. It is, of course, obvious that there is no such thing as a perfectly dead studio or hall. This example, therefore, illustrates the basic idea of how acoustical treatment influences the polar response pattern of a microphone.

From this and the previous discussion can be derived three fundamental operating rules for the microphone:

1. As a sound source is moved about the microphone, loss of sensitivity may be compensated for (if desired) by moving closer to the instrument.
2. The ratio of reflected to direct sound may be raised in sufficiently live surroundings by using greater angles from the zero axis of the microphone (especially with bidirectional microphones). The more live the acoustics, the greater will be the effect.
3. High-frequency sound sources must be more nearly "on beam" for a given distance to achieve the same intensity as lower-frequency sound sources.

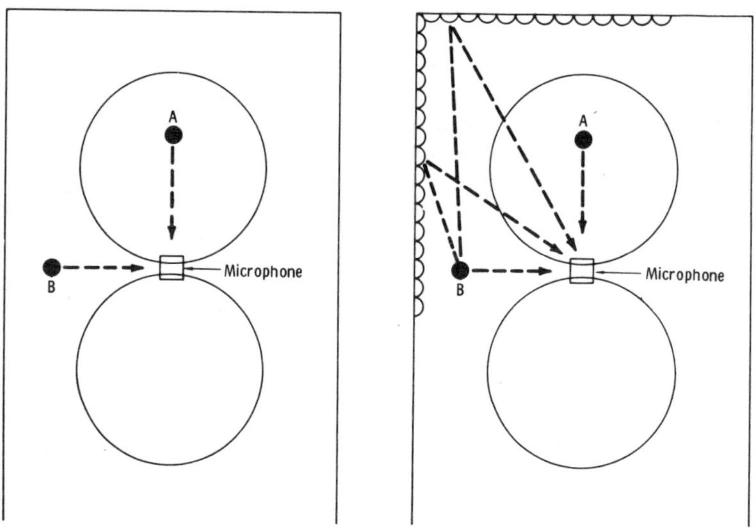

(*A*) *In dead studio.* (*B*) *In live studio.*

Fig. 11-10. Response patterns for a bidirectional microphone.

The nature of the acoustical enclosure about the microphone is the one factor that has most prevented any establishment of definite standards in microphone setups. If there were such a thing as a standard studio, designed to approach as nearly as practicable the ideal condition of sound dispersion, the setup for any particular musical organization would be a simple matter anywhere when once worked out. Naturally, such is not the case. It is probably safe to say that there are no two studios or auditoriums anywhere in the world that are acoustically alike.

There are, and will be, a great number of operators and producers who do not have adequately designed and acoustically controlled studios in which to work. This is an undesirable situation, but it is outside the influence of the average operator. Even some of the most recent fm broadcast installations, built in completely new studios independent of any a-m company, have neglected this feature. Most of the good articles appearing thus far on microphone setups have been concerned with well-designed, musically live studios of the network centers or more production conscious independent owners.

The so-called "modern studio" may not yet be the ultimate in design, but it is certainly a far cry from the previous rectangular-shaped, acoustically deadened rooms. The walls remain live to musical sounds but are broken up acoustically in some manner to avoid standing waves, while still achieving a maximum response to diffused, polyphased, high-frequency sound. Good tonal brilliance is obtained with a minimum of microphone and control-board manipulation. Recording studios have kept more abreast of recent acoustical development than have broadcasters.

Summary of Production Fundamentals

First, there is the monaural response of audio transmission. Sense of direction of the various sound sources is lost until stereophonic transmission is used. Sound perspective, then, must come from relative distances of sound sources. A sound will still appear close or distant in a monaural system. Also due to this type of response, "focusing power" must be achieved by orientation rather than by the subconscious and automatic means of the human hearing sense.

Second, strive in any microphone setup to achieve the best results possible as determined by the sound from the monitor speaker. Do not be confused by supposed differences in a-m, fm, recording, etc.

Third, bear in mind the microphone response pattern, and select one suitable for a particular purpose as explained later. Remember the influence of the acoustical treatment surrounding the pickup area on the basic pattern of the microphone.

And fourth, it is realized that many do not have available the latest in studio design so necessary for brilliance in musical tones. An attempt will be made to help overcome this difficulty by showing what operational modifications are necessary in the older-type general-purpose studios to achieve the best results possible.

11-6. REVERBERATION

When adequate sound reinforcement is not provided by the studio, several acoustical or electronic possibilities exist which may be exploited to achieve added reverberation. (Due to the special problems involved, stereo reverberation is covered in a later section.)

Fig. 11-11 illustrates one of the acoustical methods which has been used with some success and is self-explanatory. Note that the echo chamber is much smaller than that previously used for special cavern sound effects. Since sound travels roughly 1100 feet per second in air, the 12-foot delay provides a reverberant sound of approximately 10 milliseconds. Note particularly that the lower frequencies are attenuated through the delay or reverberant channel to add reinforcement of liveness without the boominess that would occur with lower-frequency reverberation.

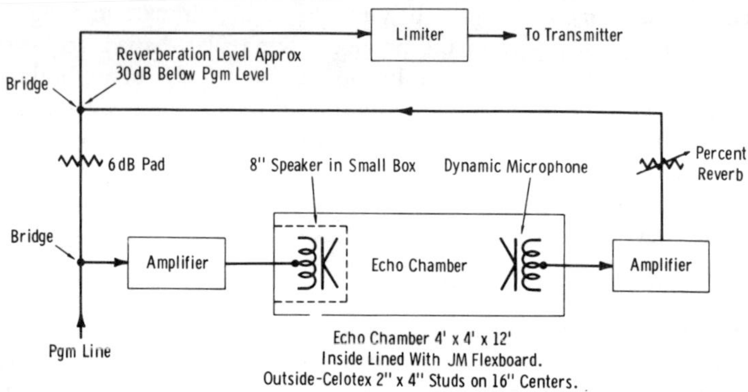

(*A*) *Block diagram.*

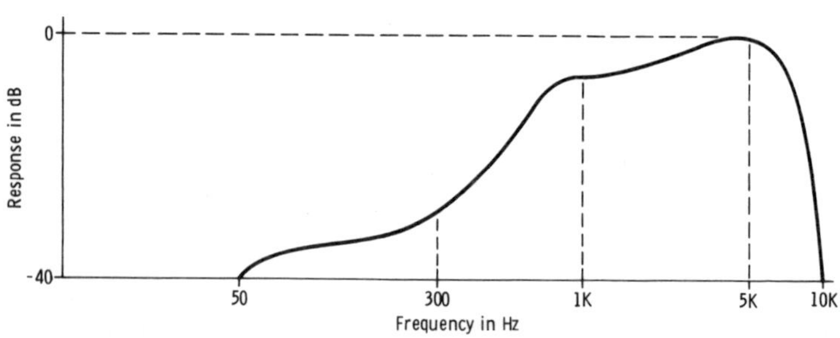

(*B*) *Relative response.*

Fig. 11-11. Acoustical reverberation system.

Electronic Method

An electronic device for adding reverberation is shown in Fig. 11-12. This reverberation unit is a tape recording system especially designed to replace an echo chamber in broadcasting and sound recording. The sound is allowed to circulate (as an electrical signal) in a closed loop formed by the multiple heads of a nine-head unit in the same way that sound circulates with multiple reflections in a room. The reverberant decay of this

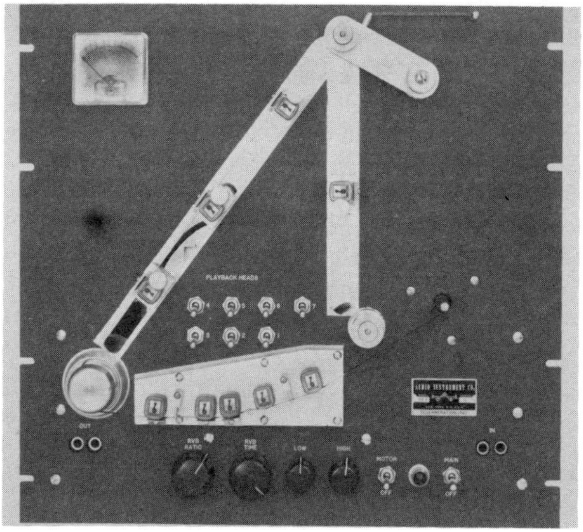

Courtesy Audio Instrument Co.

Fig. 11-12. Magnetic-tape reverberation unit.

circulating signal is precisely analogous to that of sound in a concert hall. The large number of heads provided allows long reverberation time with a natural overall effect.

The system is arranged to be patched between a microphone preamplifier and a mixer input, between a console output and a tape recorder, or between a tape reproducer and a disc recording system. It may also be wired permanently into a circuit and operated with or without reverberation.

When a note is played on a musical instrument, a fraction of the sound goes directly to the listener's ear, while some strikes the nearest surface and is reflected. The reflected sound and more of the original sound spread out, strike the next nearest surface, and are again reflected. The remainder of the original sound and the two reflection products spread out still further and are yet again reflected. The sound is attenuated by each reflection, and after a sufficient number of reflections, the sound becomes so weak as to be inaudible. Note that the sound circulates about the concert hall repeatedly, being reflected many times before it dies out. The tape reverberation device uses a process analogous to this, except that the signal is handled electronically and magnetically instead of acoustically.

The incoming signal is divided between two channels for its trip through the reverberation unit (Fig. 11-13). One portion goes through the direct-channel isolation amplifier to the output, just as some sound travels directly from the instrument to the ear. The balance of the incoming signal passes into the reverberation channel and undergoes multiple reflection in circulating about a tape-loop system. A controlled amount of this reverberant signal is fed to the output and mixed with the direct signal. By regulating the proportion of direct and reverberant signal, various effects may be produced.

Reverberation is generated as follows: A signal entering the reverberation channel passes first through high- and low-frequency equalizers, then to the recording head, and so on to the tape loop. The recorded signals are reproduced from the tape a fraction of a second later, and a controlled portion is fed back to the recording head without passing through the high- and low-frequency equalizers, to be recorded and fed through again. More than one playback head may be used so that the signal may be reproduced and rerecorded several times with different degrees of delay, just as sound experiences multiple reflection (with different time intervals) from the several walls of the concert hall.

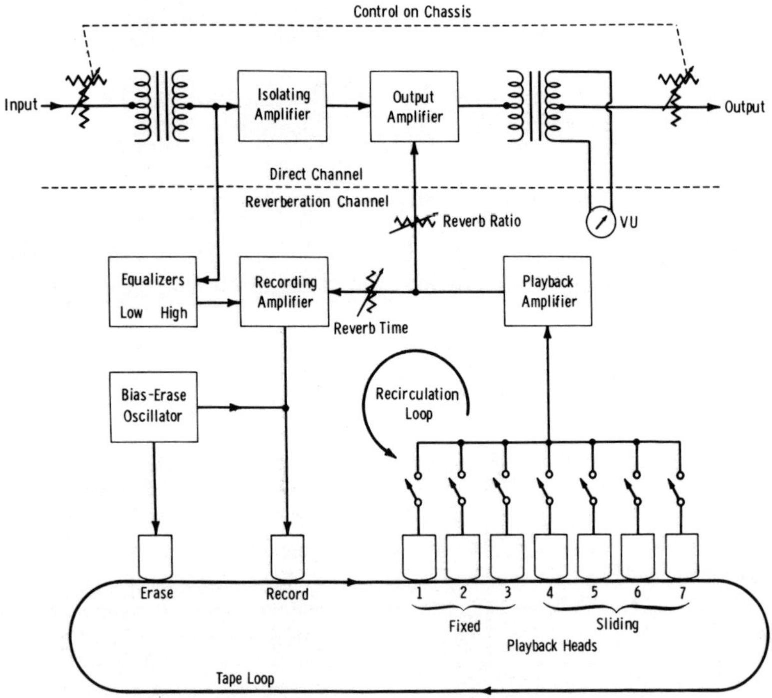

Fig. 11-13. Block diagram of tape reverberation unit.

The degree of amplification during rerecording is adjustable so that the signal may circulate many times, or only a few, around this reverberation loop before it decreases to inaudibility. By selecting the appropriate playback heads, a smooth natural reverberation decrease may be produced. Alternatively, a single reproducing head may be used to achieve sharp echoes for special effects, or to repeat words. High- and low-frequency equalizers modify the signal which feeds the reverberation loop so that a wide variety of other effects may be obtained. The unit shown in Figs. 11-12 and 11-13 has four auxiliary heads, adjustable in position, for producing long delayed echoes, for repeating words, and for very long, natural-style reverberation.

The signal passes through an input attenuator (optional) and an input transformer and then feeds the direct channel and the reverberation channel. Both channels are then combined and fed through an output amplifier, output transformer, and attenuator (optional). The input and output attenuators are controlled by ganged switches, providing the following combinations:

	1	*2*	*3*
Maximum input level (VU):	+4	−30	−30
Maximum output level (VU):	+4	−30	+4

The signal in the direct channel goes through an isolating amplifier and then into the output amplifier.

The reverberation-channel signal passes through low- and high-frequency equalizers to the recording amplifier, and on to the freshly erased tape by way of the record head. The signal is reproduced from the tape with a delay that depends on which of the playback heads are in use (determined by panel toggle switches).

After amplification in the playback amplifier, the delayed signal is split into two parts. One part is attenuated in the time potentiometer and fed back to the recording amplifier for rerecording. The remainder is fed through the ratio potentiometer to the output amplifier.

It is clear that the reverberating signal recirculates through a loop consisting of the recording and playback amplifiers, the time control, and the tape-loop system, decreasing in intensity by a controlled amount with each recirculation. Because the signal makes many trips around the recirculation loop, the playback amplifier must be equalized to compensate for both head and tape response, making the frequency response of the loop very accurately flat over the required frequency range.

In general, a combination of heads 1 and 2 affords moderately short-time reverberation and is useful for adding resonance to a single voice or to a small ensemble. A combination of heads 1, 2, and 3 develops a somewhat longer reverberation time and is useful with larger ensembles. For maximum natural reverberation time, heads 1 through 6 are used.

The ratio control regulates the amount of reverberant signal added to the direct signal in the output; the time control regulates the rate of decay of the signal in the reverberation loop. By analogy, we can say that the latter control regulates the sound absorption of the walls of the imaginary room and its reverberation time, while the former control regulates the ratio of direct to reverberant sound (in effect, the distance between the imaginary microphone and the imaginary orchestra).

A hasty guess may give one the idea that the controls interact. This is not correct; although both controls alter the quality of the signal, they change it in different ways. Proper setting of both the ratio and time controls is essential for good results.

Typical operation of the tape reverberation unit is as follows:

1. Set panel controls:
   Head Select—Switches 1 and 2 on
   Time—Fully counterclockwise
   Ratio—Straight up
   Low—Straight up (approximately flat response)
   High—Straight up (slight rolling off of highs entering reverberation path)
2. With signal being fed into the circuit, the direct channel should deliver signal to the console fader. Advance the fader until the system output level is as desired (the VU meter should be reading 0 VU).
3. Advance the ratio control until the reverberant signal is heard.
4. Advance the time control until the reverberation becomes sufficiently apparent.
5. Further variation of the ratio, time, and equalizer controls will now produce a variety of desirable reverberation effects. The output level will increase slightly as the amount of reverberation is increased.
6. Additional variation may be secured by using other head combinations. For example, a combination of heads 1 through 6 provides the longest and smoothest average decay and is often preferred for special effects and for large instrumental ensembles. (Be sure that at least one head switch is on; otherwise hum will result.)
7. If the time control is advanced too far, the circuit will either ring or be on the edge of ringing.
8. Be absolutely certain that the input level to the unit is not over the rated value. Complaints of distortion may usually be traced to excessive input level. If in doubt, compare the panel VU meter and system VU meter readings for zero with the same input signal. On the other hand, insufficient input level will lead to failure of adequate reverberation.
9. An exactly repeated word or phrase has been found to lend force to a spot announcement. To repeat words:
   A. Turn on head 5 only. Slide head 4 to bottom of track, slide head 6 to top. Turn ratio control fully clockwise.
   B. Turn time knob clockwise slowly, while someone repeats the desired word into the microphone, until the word repeats steadily (and gradually dies away).
   C. Stop the action by lifting the tape off the record head (next to right position on 5 head block) for 2 or 3 seconds.

D. Change the repeat tempo by sliding the head carriage up or down in the slot, or by switching to another playback head. After each change of head or head position, clear the tape by holding it off the record head as above.

10. In setting up playback-head combinations, two philosophies, natural and theatrical, may be used.
    A. For a theatrical effect, use one or a few heads, widely spaced. This produces a reverberation decrease which is stepwise and prominent. The result has acquired a wide vogue in the popular recording field.
    B. For a natural effect, always use heads 1, 2, and 3, perhaps plus 4, or 4 and 5, or 4, 5, and 6. The reverberation decrease is smooth, without stepwise character. Use as few heads as possible. The longer the reverb time desired, the more heads must be used to eliminate the stepwise character. Start with 1, 2, and 3 or 1, 2, 3, and 4. Head 4 at the start should be at the bottom of its slot, the head-5 carriage should be about ½″ away from carriage 4, and carriage 6 should be similarly far from carriage 5.

11-7. MICROPHONE TECHNIQUE

The announcer is the logical starting point for a discussion of microphone technique. It must be understood that we are not concerned at present with announcing over a background of music or any form of dramatic presentation.

Single-Voice Pickup

Announcing alone occupies a considerable portion of the broadcaster's schedule. Correct voice transmission is very important to radio because there is no sense of sight to aid impressions; the voice is the complete medium of expression. The intake of the breath, the most subtle inflections, the style of delivery, the original voice timbre, all are factors in naturalness. Any or all of them may be severely affected by the announcer's relation to the microphone. All of us in the technical end of radio must at some time in our career realize that our engineering training may not have encouraged a feeling of artistic values or sense of showmanship. We may be apt to lose sight of the real reasons behind the keys, faders, and perfectly matched impedance of the control system. They are designed this way so that the electrical impulses may correspond to the thunderous crescendos of Wagner's "Die Walkure" or the light, delicate strings of Debussy's "Festivals." They must reproduce the moods of music and the moods of voice. The wires and switches and dials and knobs of the technical department are concerned with these intangible qualities of human experience. If we grow ever more conscious of this in the course of our work, we will find whole new vistas of sound and sound control opening at our fingertips.

The first thing to do concerning the announcer is to get the microphone away from his face. "Mugging" a microphone is a terribly deep rut and a difficult habit to break. You will likely meet resistance from the announcer; in many cases you literally cannot do it. Some announcing desks have microphones permanently mounted on them, and the announcer cannot move back without holding his script uncomfortably in his hand with no arm rest. This is plainly and simply an engineering error. Like fingerprints, there are no two voices exactly alike. For every voice there is a definite relationship with the microphone which allows the most natural and pleasing reproduction for the combination of that particular voice and that particular microphone.

For voice work alone, *there is no distance less than 2 feet from the microphone that will assure naturalness.* Furthermore, a distance of 2 feet should be used only for the softest voices. Compare this with your own observations of studio operating practice. What is the distance most announcers use? Probably somewhere between four inches and a foot. Voice waves at this distance from the mouth and throat cavities do not create the electrical impulses that correspond to the natural character of that particular voice. Setting the microphone on the "voice" adjustment, if such an adjustment is incorporated, helps in cases where it is absolutely necessary to work very close (to cover up background for example), but it is not natural transmission.

Perform this experiment if possible: On rehearsal, set up microphones (in addition to the regular announce microphone) 3, 6, and 9 feet away. Stagger them enough so that no one microphone will be in front of another. If the regular microphone is immediately in front of the announcer's face, move it lower or to one side. Don't tell him what is occurring; you are just "testing microphones." If he is told to start reading at a great distance from one microphone and then to move closer, he will subconsciously alter his volume and you will not get a natural check.

Now turn on only one microphone at a time. Start with the farthest one. Unless you are so conditioned to hearing the "mugging voice" that you can accept no other, you will find a new experience in naturalness of voice transmission. A distance will be found at which the voice begins to sound "hollow" in a live studio, or "thin" in a dead one. Use a distance just a shade closer than this point.

In small announce booths where space is at a premium and where distance would create the "barrel effect" due to the proximity of the studio wall to the microphone, the microphone should be suspended over the announcer's head at a distance of several feet. This technique is borrowed from television, where the microphone must be kept out of the range of the camera and is often as much as 15 feet from a performer. It should be borne in mind, however, that the sound in television is only a secondary expression to the picture. In other words, all of the intent and meaning need not be embraced in sound alone as in radio broadcasting.

This same technique is most convenient when two persons are seated at a table and using a single microphone. The table top should be deadened to sound by acoustical treatment or a heavy cloth cover to prevent reflected sounds from entering the microphone. This is very important when the suspended-microphone technique is used.

Two or More Microphones

In studying specific program setups, the number of microphones that should be used will invariably be questioned. Always keep in mind the previous discussion on the monaural character of the system and the lack of focusing power unless deliberately used to concentrate the attention.

The greatest weakness in microphone setups for large shows has always been the use of too many. There is bound to be some distortion (however slight) in multiple setups because the time lag of sound waves creates phase additions and subtractions at the various positions of pickup. This source of distortion, however, is only minor compared to the other faults of this technique. Aside from the operational difficulties of handling a large number of channels on the mixing panel with greatly increased chances of error, the sound man and his board now share control with the conductor. All the dramatic interest, the emotional pattern written into the original score, plus even the conductor's interpretation, are now placed in the ratio of fader adjustments and depend on the reaction and psychological temperament of the operator; in other words, there are too many variables. This is just the opposite of professional recording techniques, where as many as 12 to 14 channels are later "mixed down" to achieve the proper balance and perspective in the final recording.

Microphone Phasing

It is very important in any sound application in which two or more microphones are used fairly close together to have the outputs properly phased. If this is not done, the outputs will oppose each other, reducing the total signal and introducing some distortion. This is the same thing that happens in multiple loudspeaker systems when the voice coils are out of phase, causing a "pumping" or bucking of sound at points in the overlapping fields of the speakers.

The best way to phase two or more microphones is to hold two in the hand at the same time while talking. Turn one microphone on and note the volume-indicator reading. Be sure not to alter the volume of the voice when turning on the second microphone with the same volume-control setting. If the volume now decreases, the connection of one of the microphone cables at the amplifier input terminals should be reversed. Each additional microphone should be checked in this manner against the first microphone. Phasing may be accomplished by using a patch cord between any two terminations of the circuit on the jack panel, and reversing *one* end at a time during the test.

How Many Microphones

Let us establish a foundation on which to build a workable structure for determining the correct number of microphones for a given show. Let us also be practical and realize that a great number of operators do not have the very latest studios and hours of rehearsal time to spend on one program.

1. Whenever it is physically possible in the alloted setup time, arrange the performers about a single microphone so that the overall balance is correct. If you have the time to do this, you achieve proper balance by positioning rather than mixing various sources on the control board.
2. Assume now that time is running out and you are still having trouble with a particular section in obtaining balance. (This is more apt to occur in a dead studio than in an acoustically correct live studio.) The use of one microphone is not achieving maximum fidelity, and another microphone for the troublesome section will have to be used. Many times in practice it will be found that the second pickup need be used only at times and not all through the show (such as rhythm-section accentuation of an orchestra at a few spots in the score.)
3. Some of the more complex shows will positively require more than one microphone for proper pickup. Take, for example, a variety show consisting of a drama cast, a chorus, and an orchestra. Remember that the microphone does not focus attention as your ears would do in the studio. Of course, it is likely that a program of this type will originate in a larger station having modern studios and plenty of rehearsal time with the operator.
4. The rule, then, is simply this: Use the *absolute minimum number of microphones* for a given program and set of conditions surrounding rehearsal time.

Piano Pickup

The single-piano pickup is the simplest setup in any kind of studio since it is perhaps the least affected by the acoustical nature of the room. But there are "tricks of the trade" even in the simplest programs.

First, regardless of how softly the pianist plays (so long as he is unaccompanied by other instruments), the microphone should *not* be placed under the lid. Operators who have had considerable experience with using the microphone have known the amount of distortion arising from the close proximity of large physical objects to the microphone. It should then be obvious that a pickup under the lid and close to the sounding board will not allow natural transmission of the piano tones. "Close micing" of this kind to almost any instrument will cause many "peaks" on the VU indicator that are inaudible, with a consequent necessity of holding the volume down with great loss of musical brillance. It is true that many operators have become so accustomed to this type of piano setup that they are in the same position as those in the habit of hearing the close-talking announcer. It may be a familiar sound, but it is not natural reproduction.

Here is the best method of determining the setup for a single piano. Start with a distance of 20 feet, head high. In a dead studio, this distance will probably result in a "thin" response, especially on low passages. In live studios, the sound may be too reverberant for clear-cut transmission. Move the microphone in on a line drawn through the center of the sounding board until the tones are full-bodied and just the right amount of reverberation exists. This distance is seldom less than 8 feet and will allow tonal brilliance and balance between lows and highs that is sacrificed in a close-up technique.

Now comes the final check for balance between bass and treble, in other words, the check of the player's left- and right-hand pressures. If the pianist happens to use approximately the same pressure with both hands, the above procedure is usually all that is necessary. But assume he has a heavy right hand and bass response is somewhat weak in relation to the highs. Keeping the microphone at the same distance, move it from the line drawn through the center of the sounding board toward the tail of the piano. This increases the response from the bass strings. If the pianist should have a heavy left hand causing loss of highs (this lack of highs is also apt to occur in dead studios), the microphone should be moved toward a line drawn through the hammer line of the piano, or even over to an imaginary extension of the keyboard. This will increase the response to the treble strings and decrease that from the bass strings.

The dual-piano team imposes only slight additional requirements. Fig. 11-14 illustrates the most satisfactory orientation of the two pianos with

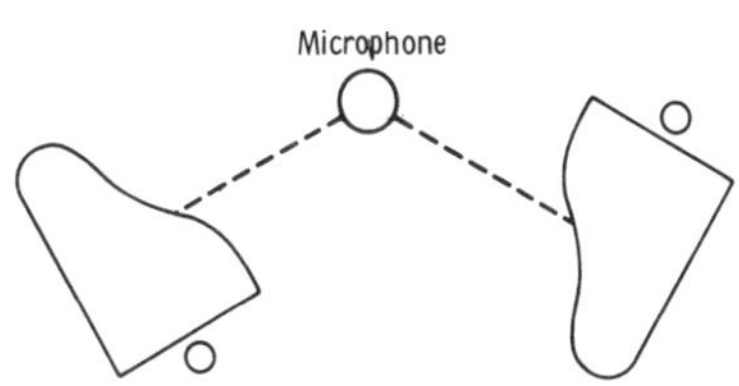

Fig. 11-14. Microphone placement for two-piano pickup.

the microphone. The temperament of the pianists must, of course, be considered and the lead piano given prominence by moving the microphone closer to that piano if the accompanist is "heavier fisted." The above procedure for bass and treble balance must be followed by moving not the microphone, but the piano itself.

Vocalist and Piano

There are three variables in the case of a voice and piano—the distance of the vocalist from the microphone, the distance of the microphone from the piano, and the distance of the vocalist from the piano. There is no reason to use more than one microphone unless no time is allowed for rehearsal.

Listen to the vocalist in the studio first. Does he or she "sing out" with the chest muscles, using a large volume and dynamic range? Or is the vocalist the "crooner" type, using only the larynx and throat for emphasis? Generally speaking, every vocalist fits in one category or the other.

Now excluding the piano accompaniment, it is known whether the vocalist must be near to or far from the microphone. One who sings with his full volume range will be placed anywhere from 6 to 12 feet from the microphone when a piano alone is used as background. On rehearsal, always start with the greatest distance. The goal is to be able to enjoy adequately the lowest volume and the highest volume *without riding gain on the fader control.* This part of the balance can be achieved by careful rehearsal checks.

The balance between vocalist and piano accompaniment is not always so simple. A good pianist or one familiar with a particular vocalist's style will automatically adjust his volume to the singer. The microphone may be placed about 8 feet from the piano with the vocalist on the opposite side (calling for a bidirectional microphone) at the distance determined by trial. It is well to point out here that a very common error of operators is taking a "vocal solo" too literally. The presence of the piano must always be considered, with only slight emphasis on the voice. It should be a blend, with, of course, the voice always a little predominant. This does not permit the so often noticeable weak background of piano tones.

When the pianist insists on playing so loudly that his accompaniment smothers the vocalist (some pianists cannot alter their volume and still play well), the dead side of the vocalist's microphone must be turned toward the piano. This is practically always the only adjustment necessary from the 8-foot distance between microphone and piano. If the studio is very live and the piano tones are still too predominant, turn the piano around so that the lid opens toward a wall of the studio, and deaden the wall with drapes. Do not at any time move the vocalist closer to the microphone, where riding gain is necessary for the natural dynamic range of the voice.

Small String Setups

The music of chamber groups is intimate in style and requires good instrumental definition. This calls for comparatively close microphone setups —but not too close.

Consider a small orchestra consisting (most popularly) of several violins, a viola, a cello, a string bass, and sometimes a piano. Due to their comparative volumes, the instruments are usually placed in that order from the microphone. Fig. 11-15 illustrates the general orientation of such a group with the microphone.

Now assume that the approximate distances are: violins 4 feet, viola 6 feet, cello 8 feet, string bass 10 feet, and piano somewhat off microphone at 10 feet. Also assume that the violins are too predominant for proper sectional balance and comparative "presence." What would you do first?

Here are the possibilities if you move the instruments: The violins could be moved farther back or to one side in a less sensitive zone, or the other instruments could be moved closer.

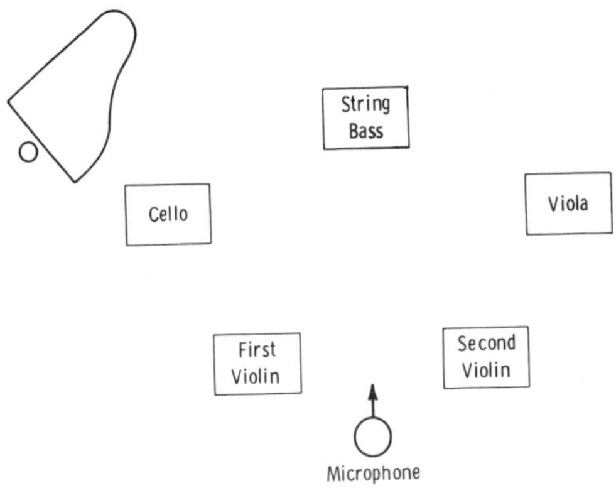

Fig. 11-15. Microphone placement for chamber orchestra.

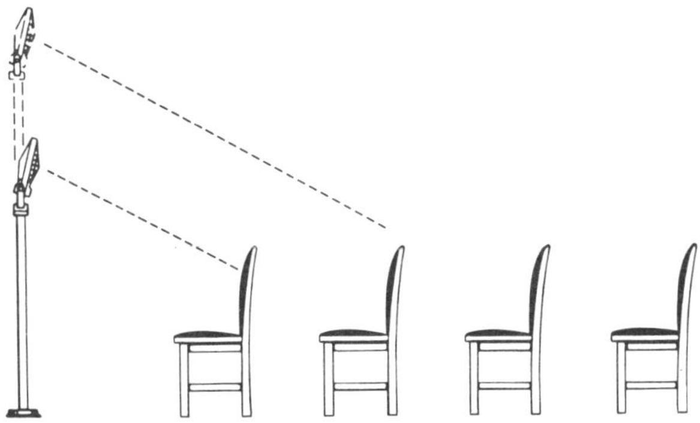

Fig. 11-16. Height and tilt adjustments.

But have you remembered the "focusing-power" principle? You can achieve quite a range of sectional balance by the simple expedient of height and tilt adjustment (Fig. 11-16). If the violins are too predominant, raise the microphone and focus on the other instruments. If the violins are weak, the microphone should be lowered and focused on the violins. This is a better method than moving the predominating instruments to one side in a less sensitive area of the microphone, since the higher frequencies containing the overtones are very important for wide-range pickups.

There is little difference in pickup for this type of musical organization between dead and live studios, since the setup must be intimate in character, with high direct-to-reflected-sound ratio. This minimizes the effect of the acoustical nature of the studio.

Small-Orchestra Setups

In the category of small orchestras are a large number of organizations presenting popular, serious, or variety music. They may consist of a combination of brass, strings, and reeds, and they may have from 4 to around 15 instruments. Again the first step is to visualize the instruments in terms of comparative power outputs, as follows (softest to loudest):

1. Violins, muted trumpets or trombones, and guitar
2. Clarinets, saxophones, xylophone, and vibraphone
3. String bass
4. Piano
5. Trombone (open belled) and trumpets (open belled)
6. Traps and bass drums; guitar (electrically amplified)

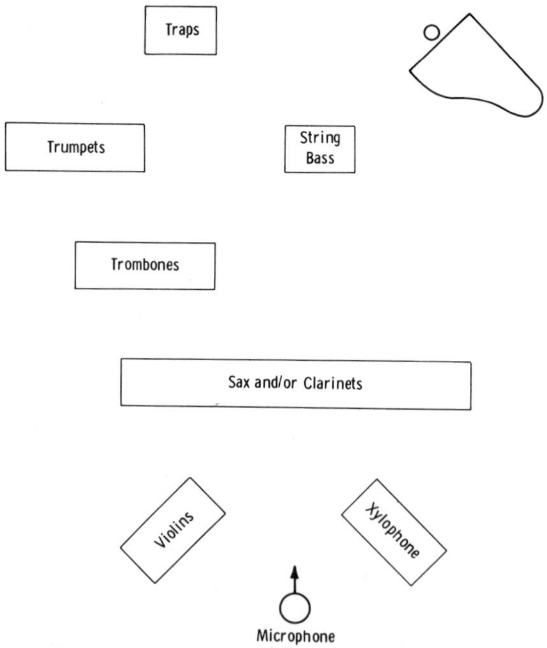

Fig. 11-17. Arrangement for small-orchestra pickup.

These are the most likely instruments to be encountered in such a setup. As before, the most likely approach is to arrange them in the listed order from a single microphone.

Fig. 11-17 shows a live studio layout, starting with the violins at about 8 feet. When initially checking balance of sections, remember the height and tilt adjustment for focusing power to obtain the proper blend. Then, and not until then, try moving any troublesome section.

When you think it becomes necessary to move instruments, keep these principles in mind: A predominating section might be too loud not because the relative distance between instruments is incorrect, but because the microphone is too close to the entire group. Move it back. A weak section might be too soft, not because it is too far from the microphone, but because all instruments are too close to the microphone. Move it back. This is to emphasize that the farther back a microphone is placed (within the limits of acoustically allowable distance), the better is the chance of a good balance among all sections.

Another very important item is the treatment afforded muted trumpets or trombones. When their bells are muted, the instruments must be very close to the microphone—about 2 feet. If the players cannot or will not step from their regular positions to one immediately in front of the microphone, a separate microphone must be spotted just in front of that section. Obviously, it need only be used during the time the instruments are muted.

Now consider a studio of older design without the liveness of a modern studio. When the musicians number around 12 to 15, it is often difficult to get good sectional balance on the above setup with only one microphone. Even though the farther instruments may contribute about the same number of volume units, their presence may be "thin" due to the lack of reinforcement of harmonics and overtones. Also, the microphone must be a little closer in a dead studio, emphasizing the discrepancy in sectional presence.

In nearly every case, however, one microphone can be used with the setup illustrated in Fig. 11-18. With a bidirectional ribbon microphone, the instruments must be more nearly "on beam" due to the narrower angle of response and the less lively acoustical nature of the studio. More time and movement of players will be necessary, but it is far better than the usual procedure of using additional microphones.

Adding a Vocalist

A vocalist added to a small orchestra imposes special problems unless the organization is trained thoroughly in broadcast production techniques. The ideal arrangement would be for the vocalist to be placed in front of the orchestra facing the microphone at a distance of several feet, depending on the singer's style. This arrangement, however, is often not practical when the microphone is raised and slanted so as to obtain proper balance of instrumental tones. It also imposes restrictions on the actual musical arrangement of the accompaniment since, if brass is used instead of just strings and possibly reeds, the voice will be drowned out. When such a situation occurs, there is no alternative but to use a second microphone for the vocalist. Preferably it should be a unidirectional type, with the dead side toward the orchestra. The announcer may also use this microphone.

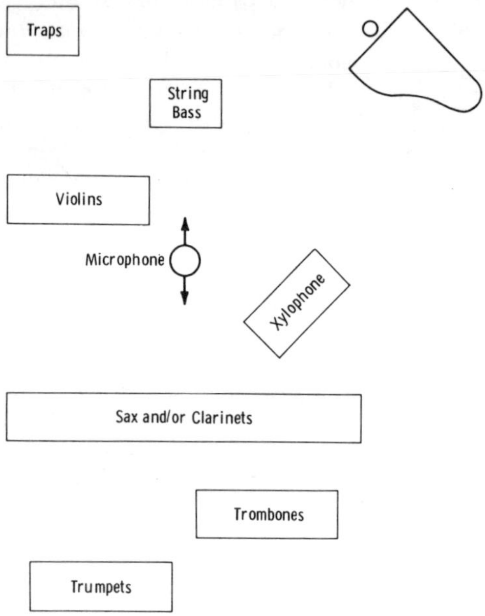

Fig. 11-18. Small-orchestra pickup with bidirectional microphone.

The Symphony

The large symphony orchestra pickup is based on exactly the same principles discussed above. The operator is, however, concerned with a combination of each type of orchestral setup into one grouping—strings, woodwinds, brasses, percussions, and quite often a choral accompaniment. The grouping of instruments and the type of musical score become much broader in scope. The most helpful characteristic here is the fact that such a program is less apt to be attempted in an inadequately designed studio than are the types of programs discussed previously.

The normal physical arrangement of the orchestra for regular audience listening will generally be satisfactory for broadcast purposes. On the initial trial of the main orchestra microphone, several microphones should be suspended at the most likely pickup positions in order to avoid the confusion of moving microphones, ladders, and personnel. The general area is 15 to 20 feet high and 15 to 25 feet in front of the violin section.

If a vocalist or instrumental solo is called for, it is almost always necessary to use a second microphone to achieve proper sound perspective. *The main orchestra microphone will not act like the human ear.* Remember, too, that vocal or instrumental solos are not to be entirely predominant; the orchestral accompaniment must be very much present. Always try, if possible, to get the conductor to listen to the control monitor for a final check on balance. If this is not possible, get some responsible member of the organization to pass on it. The very best control or production men do this simply because every symphony organization has an arrangement of score or possibly a distinct interpretation of the original score with which you cannot be expected to be familiar. This has, many times, necessitated a slight rearrangement of instruments in relation to the microphone.

Along this same line, it has become evident that in spite of the usual superior results obtained with a single well-oriented microphone, it is often necessary to deviate from this practice for true high-fidelity pickup. The main orchestra microphone, back far enough to obtain the proper blend of all instrumental tones, will faithfully pick up the delicate, distant tonal beauty of the violin passages in *Clair de Lune.* Most music lovers, however, criticize this same microphone setup for such numbers as the Strauss waltzes, where the tonal perspective of the strings should be closer and more strident in quality. It all boils down to the fact that the microphone is *not* a mechanical extension of the human ear. You will find many leading conductors and producers of symphony broadcasts insisting on an added microphone suspended directly over the strings or other sections of the setup to be turned up *only on cue.* This procedure supplies the missing psychological factor in obtaining microphone-to-sound perspective. When a choir is used with a symphony orchestra, it is also necessary to use a separate microphone due to the necessary distance between the total combination and the necessary focusing power to be obtained by the use of the microphone. A typical symphony and choir arrangement is shown in Fig. 11-19.

The major difference from the beginning in considering a setup for symphonies as compared to other broadcasts is the fact that the operator or producer has little to say about the physical arrangement of the orchestra since the interpretation of a score is dependent on the conductor and how he hears the individual and blended tones from the great number of instruments. It will be found in practice that little difference is made in physical arrangement of a large orchestra, since acoustical conditions are usually optimum wherever such organizations perform.

The first information to get when preparing a symphony setup is the program for the broadcast. This should include data on whether or not a choral group is combined with the orchestra and on the types of solos (piano, vocal, etc.). The next thing to do is to contact the conductor personally (in some cases the symphony business manager must be contacted instead of the conductor) and ascertain if any definite microphone placement is preferred or demanded as dictated by previous experiments or broadcasts. It is then possible to proceed with the preliminary microphone setup as outlined.

In the case of a piano solo with intermittent orchestral accompaniment, no change need be made in the location of the main orchestra microphone, since this nearly always proves ideal for the piano, which is placed in front of the first violin rows.

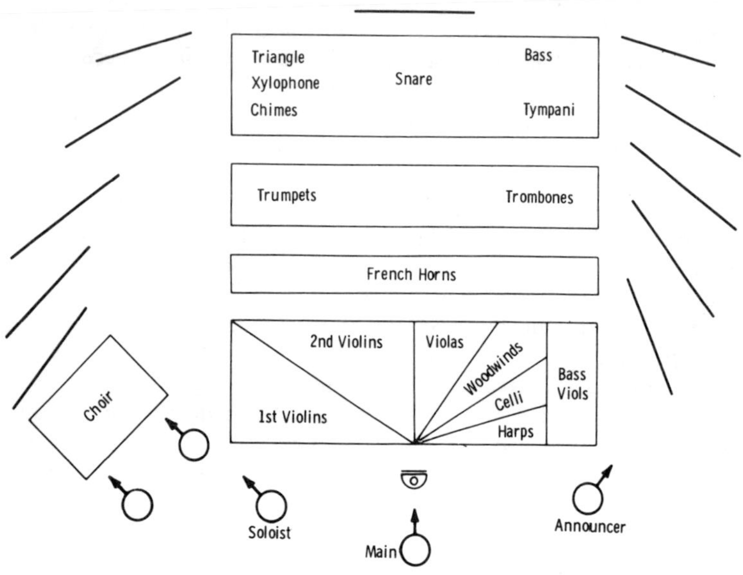

Fig. 11-19. Orchestra and choral arrangement for studio pickup.

Arrangements with a single voice accompaniment almost always require a separate microphone for the vocalist. Remember, however, that this voice must be given auditory presence while the auditory presence of the orchestral accompaniment is maintained.

Remember, too, the listening technique of the symphony listener and that music is the emphasized interest of the program. Announcers should be peaked lower than zero VU by an amount compatible with these conditions. Gain riding is to be held to an absolute minimum; any necessary adjustment of the fader control should be carried out in a gradual, undiscernible manner.

Organ Pickups

Pipe organs are seldom used in studios today, having been largely replaced by the electronic type. In a microphone setup for a pipe organ, sufficient distance must be maintained to get a good blend of tones, and this will naturally be governed by the particular acoustics involved. Cut and try is about the only approach that can be used.

Electronic organs are commonly used in studios. There are two general types of reproducer speakers: the older type with openings on both the sides and top (the major opening is on top), and the newer type that resembles closely the ordinary loudspeaker enclosure with an opening only at the front of the cabinet. Fig. 11-20 illustrates the best arrangements for the two distinct types of reproducers and is self-explanatory.

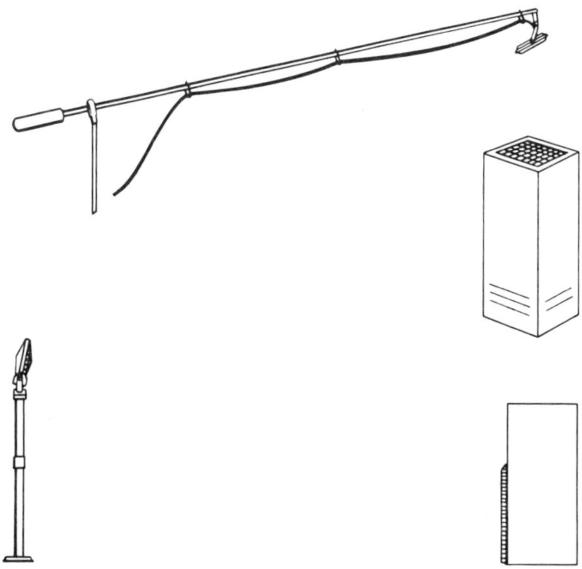

Fig. 11-20. Microphone placement for organ speakers.

Choral Pickups

The wide dynamic range as well as the size of large choral groups emphasize the importance of getting the microphone back as far as the acoustical conditions permit in most ordinary studios. In very large studios, the microphone should be back far enough that individual voices or sections are not emphasized. It is absolutely essential that risers be used for any large vocal group. When rows of singers are all on the same level, even a microphone raised high by means of a boom stand will not give satisfactory results.

Conventional choir arrangements are often not practical for broadcasting, whether in a regular studio or at a remote point. Fig. 11-21A illustrates the usual arrangement of a choir as used for auditorium or church presentation. On a broadcast, this arrangement nearly always results in a predominance of soprano voices with very little alto or bass. Fig. 11-21B shows an arrangement much more satisfactory for broadcast purposes, resulting in a better all-around balance of voices.

Importance of Rehearsals

The coordination of hand, ear, sight, and sound for the purpose of blending the component parts of a studio performance is best gained by the operator through rehearsals. Fading in or out of various microphones, turning them off and on, is the procedure which enables the engineer to play on the sound of voice and orchestra much as if the control panel itself were a musical instrument. The ratio of fader adjustments will determine the

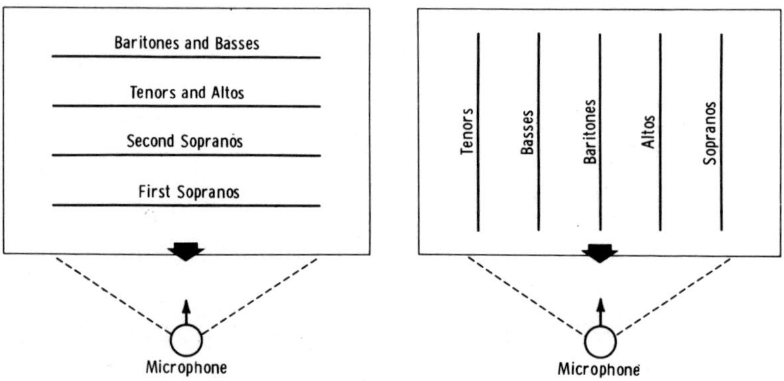

(*A*) *Wrong.* (*B*) *Correct.*

Fig. 11-21. Arrangement for choral pickups.

apparent distance of a singer from the audience; the voice may be smothered with music or may be made to stand alone with only a suggestion of background accompaniment. A proper blend of voice and music, or of dramatics and sound effects, can be created properly only through careful and detailed rehearsals. This is the one and only method of preventing a show from becoming only a caricature of the original idea.

Many experienced operators are familiar with the coloratura soprano who is nicely "adjusted" on rehearsal, then hits +20 VU on the air without batting an eyelash. This condition simply emphasizes one important point: The operator must be skilled at diplomacy as well as technically conscious. Talent *must* be made cognizant of the importance of treating rehearsals just the same as performances. If the performers are instructed in microphone technique from the point of view of making *their* performance sound just the way *they* desire to the listener, the operator will find ready and willing cooperation. Do not be shy of temperament. The more temperamental the performer, the more he likes to be fussed over at rehearsals to gain emphasis of his best talents.

The distance to be maintained between vocalist and microphone will depend on the type or style of vocal form used by the singer. While the crooner will employ a dynamic range of around 15 VU, the operatic-type singer will use a much wider range. For the former type, it is usually necessary to work close to very close into the microphone. The latter type must be placed a minimum of 4 feet, and preferably 6 to 8 feet, from the microphone. This may appear to be an excessive distance, but actually a much greater dynamic range and brilliance of voice may be realized by using this distance for singers who produce from extremely low to very high air pressures. There are, of course, a number of in-between singers, such as some of those who sing with dance orchestras, and they are usually placed from 2 to 3 feet from the microphone.

Microphone technique for actors in a dramatic program, or "skits" such as are commonly encountered for various charity fund-raising programs, spells success or failure in creating the desired illusion in the speaker. Usually one microphone is used for the entire cast, with a separate microphone for sound effects. As each actor plays his part, he steps up to the microphone, sometimes approaching from the fading zone into the announce zone to create the illusion of approaching the scene of action, sometimes leaving in the same manner. In some cases *board fades* are marked on the script. The operator fades the entire studio setup including sound effects by fading out with the master gain control. Shouts or screams must be performed in the shading area "off mic" to avoid excessive pressure on the microphone element which would require an excessive gain adjustment by the operator, which would cause loss of the effectiveness of the illusion.

While studio rehearsals are in progress, it is imperative that the engineer and production director be able to talk to the cast for the purpose of instruction in positions, microphone technique, etc. Switches on the control console and perhaps also on the production director's console (where such is used) are provided for the talk-back microphone. The control operator is sometimes provided with a foot switch to free his hands for the controls. When this microphone is turned on, the control-room speaker is cut off by a relay interlocked with the switch to prevent acoustic feedback. This microphone is also electrically interlocked with the on-air position of the output switch so that it may not be operated during the time a show is actually being broadcast.

11-8. SOUND EFFECTS AND SPECIAL EFFECTS

Sound effects include such things as rainfall, the burning of a building or forest, gunshots, etc. Special effects apply to special handling of a program, such as excessive reverberation to simulate a hall or cavern, or conversely, sound from cramped quarters such as a telephone booth.

Most sound effects are now on disc or tape. However, it sometimes occurs that individually styled sound effects are requested.

In previous years, major network practices in the art of sound effects developed into one of the most highly specialized fields of broadcasting. A sound-effects technician can take the lowly strawberry box and create illusions ranging from the squeak of a wooden gate or the creaks of a ship moored to a dock, to the terrible rending crashes and splintering of wood for collisions of any description. A bow of a bass viol is drawn in a particular manner over the edges of the box for the first effects, and the box is crumbled between the hands close to the microphone for the collision effect. Rainfall is simulated by the pouring of birdseed or buckshot onto a sheet of parchment.

Either a lazy palm-bordered beach or a veritable turmoil of angry waves in an ocean storm can be simulated with BB shot rolled back and forth with skillful timing over a copper screen. Cellophane crackled gently between the hands close to the microphone can create the illusion of the most terrible forest fire imaginable.

The sound technicians' collection includes contrivances such as hail and wind devices, boxes in which glass is shattered, thunder drums, hurricane machines, heavy doors on frames, keys, and a thousand items entirely beyond the scope of this book to describe. In addition, he has a console on which a number of turntables are mounted with their individual pickup arms and dials which automatically count the number of grooves set in from the edge for proper cueing. These turntables may be varied from 0 to about 150 rpm to make still more flexible the number of effects that can be obtained from recordings.

Ready-made recordings of sound effects are available in thousands of varieties and arrangements. A far from complete outline of general categories follows:

Airplanes: Amphibian, four engine, helicopter, etc.
Animals: Birds, crickets, all types of dogs, fowl, etc.
Automobiles: Ambulance, fire engine, auto in mud, auto with chains, etc.
Crowds: Airport background, applause, cheers, circus, etc.
Industrial: Air hammer, bulldozer, factory whistles, etc.
Marine: Foghorns, motor cruiser, outboard motor, etc.
Trains: Diesel, cable-car, freight, passenger, etc.
War: Air battle, gun shots, explosions, marching, etc.
Weather: Blizzards, rain, thunder, wind, etc.

The voice can be made to sound as though it were coming over a telephone by means of a *filter microphone,* which is simply a microphone run through a filter amplifier which clips the high and low frequencies so that the quality is similar to that heard in the telephone receiver. A good telephone effect can be achieved by feeding a microphone signal through a separate amplifier and exciting a pair of inexpensive headphones held immediately next to another microphone in the studio. Reverberation may be added by feeding the signal into a speaker at one end of a reverberation chamber. The farther away the microphone is placed, the larger becomes the hall or cavern meant to be simulated in the drama. This type of reverberation setup is being largely replaced by a continuous tape arrangement by which any amount of reverberation may be added synthetically to the original sound. The illusion of talking in close quarters, such as in a telephone booth, is created by placing a microphone in a sound-absorbent booth.

A slight reverberation effect may be obtained by placing a microphone immediately above the sounding board holes of a piano and directing the voice by means of a megaphone or tube over the strings. The sustain pedal is held down to allow the strings to vibrate freely when the voice waves impinge upon them.

Special effects, as distinguished from sound effects, are often called for in modern broadcast programming. Many such effects are achieved by special techniques in disc and tape recording, requiring complex "dubbing" procedures to unify the program.

Perhaps the most common example to illustrate this type of effect is the use of elfin voices and music on certain children's recordings. To get the elfin voice effect, the orchestral background is recorded at 15 inches per second. It is then played back at 7.5 in/s, and the voices of the singers are blended to this tempo on another 7.5-in/s recording. When this recording is played back at 15 in/s, the musical background returns to normal sound, while the voices are stepped up in pitch and tempo.

There are many variations of special recording techniques for special effects. Recording of musical backgrounds and vocal accompaniment on separate tapes allows adding reverberation to either vocal or background at points in the score where this is effective. Reverberation may be added to either one or both simultaneously, or varied at will. Many recordings have made use of such special effects.

11-9. OPERATING THE TURNTABLE AND TAPE EQUIPMENT

Many stations are on almost a complete "disc-jockey" schedule in which the announcer operates the control board, turntables, and tape equipment, in addition to his announcing duties. Fig. 11-22 is a view of the ABC radio studio for disc-jockey-type operation at KQV in Pittsburgh.

The practice of playing recordings and transcriptions varies considerably with different stations. In many stations, the control man operates the turntables as well as running the control console. In the majority of stations, either the announcer or a specially trained person runs the turntables, which may be in a separate room just for this purpose.

Recorded and transcribed shows constitute a most important part of a broadcaster's daily schedule. Transcriptions are recordings made for broadcasting purposes; they usually are 16 inches in diameter, use a turntable speed of 33⅓ rpm, and are recorded with standard grooves.

Turntables

A transcription "platter" may consist of a number of separate musical or voice selections on a single disc, in which case they are numbered on the label with the titles of each number listed. Also on this label will be the information as to type of cut (microgroove or regular), start on inside or outside groove, and reproduction speed (33⅓, 45, or 78 rpm). This is enough to keep any operator "on his toes," especially when a program consists of both recordings and transcriptions which may require changing the turntable speed, changing the stylus (1 mil or 3 mil for microgroove or regular), and noting whether the cut is started on the inside or outside groove.

Courtesy American Broadcasting Co.

Fig. 11-22. ABC radio studio for disc-jockey operation.

Also, a filter selector switch is employed to select a suitable frequency compensation for the particular disc used. For example, typical switch positions are as follows:

Lateral

1. Transcriptions, Orthacoustic, Columbia
2. Home records and worn transcriptions
3. Home records, World, Decca, and AMP
4. Test records and special recordings (wide open at highs)

In operating a turntable, it is necessary to be sure that the pickup selector switch is on the proper setting for the pickup arm used, that the turntable speed switch is on the correct speed adjustment for the particular recording used, and that the disc has been properly cued. This means that the pickup stylus must be at the spot on the groove where the announcement or music begins so that no time is lost in waiting for the arm to reach that point on the disc. This is usually accomplished by using headphones or a speaker on an auxiliary amplifier so that each disc may be cued preparatory to going on the air.

When the disc has been properly cued, most experienced operators find it advantageous to start the turntable moving while holding the disc (on the outside edge so as not to touch the grooves) to keep it from turning until start is desired. This practice (called *slip cueing*) eliminates wow that is likely to occur on starting due to the time taken for the turntable to gain proper running speed. When this trick of operation is not followed, the disc should be cued back at least one full revolution of the turntable so that the proper speed will be reached before the start of the signal. (Some modern turntables require only ½ revolution to reach proper speed.)

Fig. 11-23 shows the Gates CB-525 turntable. The cartridge may be one of three general types:

1. The *turn-around*—With this type, the knob on the top of the cartridge is depressed and turned so that the desired stylus, 1 mil for microgroove (0.7 mil for stereo) or 2.5 or 3 mils for standard groove, faces the front.
2. The *flip-over*—The handle at the front of the cartridge is simply turned to the desired position as indicated by the markings on the lever.
3. Some arms employ a complete plug-in head with the desired stylus installed.

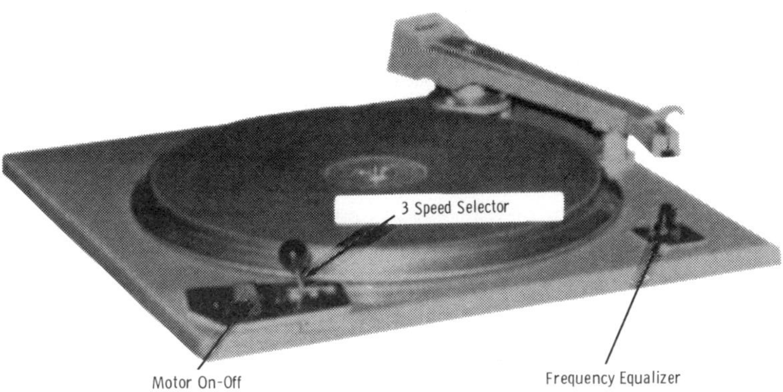

Courtesy Gates Division Harris-Intertype Corp.

Fig. 11-23. Broadcast transcription turntable.

The music library of a broadcast station may contain files of thousands of recordings and transcriptions, and their proper care in storing and handling is an important factor in on-the-air quality of reproduction. Excessive heat and dust in the air are major enemies to be considered in the storage room. The library should be well air-conditioned, with an efficient dust-filtering system. In any case, the disc should be cleaned with a soft dry cloth before playing. Static electricity causes dust to cling tightly to records, and all precautions such as linoleum floors in the library and turntable room to reduce static electricity should be taken. Finger marks cause noisy reproduction because the oil from the hands causes foreign matter to cling to the walls of the grooves. The discs should be handled by the edges only.

For the same reason, the permanent-type pickup stylus should not be wiped with the fingers in an attempt to clear it of dust. A small soft brush should be used instead.

Magnetic Tape Recording

When recording on magnetic tape, make certain the proper bias is applied to the recording head. This is normally monitored on the same meter used for volume indication by means of a selector switch. Never overload the tape with signal. The program amplitude should not exceed 0 VU or 100 percent (whichever type of scale is used) while recording.

Where short intervals occur between sections of the tape to be edited (portions cut out), the tape may be pulled by hand over the playback head at approximately the correct speed for intelligibility of the sound. A little practice should develop a fast and sure editing technique. As soon as the last sound of a particular sequence is heard, the tape should be stopped immediately and clipped at the spot where the tape has barely passed the playback head, unless some background sound is present that is desirable to hold a moment or two.

The correct splicing of tape is very important. An angular cut of about 45° should be made, then joined smoothly (do not overlap) with the succeeding portion of tape. Both ends should be cut simultaneously. Use splicing tape only, making sure that the trim is exactly even with the edges of the tape.

When it is known that the tape is to be edited, the technician should observe the following rules.

1. Be very careful of the level fed to the tape during recording. Although modern tape of good quality is hard to overload under usual operating conditions, changes of average level are heavily emphasized upon editing of the tape for playback. Try to avoid unnecessary rerecording for purposes of readjusting respective levels of different portions of the original take.
2. Watch levels of the background noise when tape is to be edited. An originally smooth program can be "choppy" on the air after editing

when background music or noise jumps noticeably due to an omitted portion of the tape. Remember that this ratio cannot be readjusted in dubbing.

3. For interviews, try to get some practice takes. The producer will often find these practice sessions more usable than the interview made for the final recording.
4. When musical numbers are to be part of a single show and must originate from different locations under different acoustical conditions, try to avoid even slight amounts of excessive reverberation. The needed reverberation may then be added for the show. This allows control over the separate musical originations, resulting in uniform characteristics. Too much original reverberation cannot be removed.
5. Check the relative microphone distances during all recording to appear on the same program. The reason should be obvious. If the announcer finishes one sentence that was spoken two feet from the microphone and starts the next sentence 6 inches from the microphone (due perhaps to a time lapse during which the announcer changed position several times, then this portion was edited out), the listener is apt to get the impression that another person is speaking.
6. Try to educate talent and producers alike that tape recording, especially when editing is required, is a critical technique and should be so handled.

The Automatic Tape Cartridge

Fig. 11-24 illustrates the Fidelipac tape cartridge which is widely used in the broadcast industry. Tape playback speed is normally 7½ inches per second for regular programming. The following operational procedures and drawings are by courtesy of TelePro Industries, Inc., manufacturers of the Fidelipac cartridge, and serve to introduce the reader not familiar with the unit to the proper handling, recording, and playback methods. To visualize the difference between the *continuous loop* and the *mobius loop,* study the following two steps:

1. Cut a piece of tape about 1 foot long, and mark one side with a colored crayon. Then turn the tape over and mark the other side with a different colored crayon. To assemble a continuous loop, splice both ends of the same color together forming a loop. Pull the tape between your fingers and observe that no matter how long you keep pulling the tape through your fingers, the same color remains visible at all times. This is termed a *continuous loop.*
2. Using the same piece of tape, cut the tape at the splice, and resplice the tape bringing the two different colors together. Again place the tape between your fingers and pull slowly. Note the tape does not turn over or buckle in your hand, but each time you pass the splice, the color changes. This indicates that on one revolution of the tape

one side appears, and when the end of the tape is reached at the splice, the other side appears. This is termed a *mobius loop.*

The tape when threaded on the automatic-tape-cartridge reel is no different than when threaded on a standard reel with the top side removed. The beginning of the tape is on the hub of the reel, and the end of the tape is on the outside (Fig. 11-25A). To make this reel of tape continuous in operation, simply splice the beginning to the end, forming a continuous loop (Fig. 11-25B).

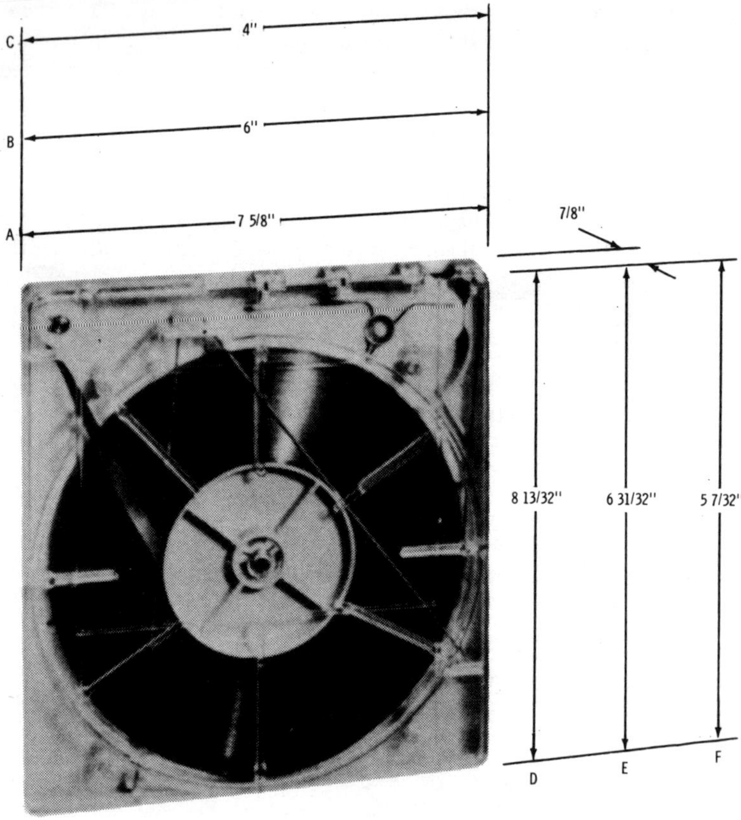

Courtesy TelePro Industries, Inc.

Fig. 11-24. Automatic tape cartridge.

If the reel is placed on a spindle and the tape is pulled from the beginning (inside) of the reel, the tape rewinds itself on the outside of the reel, and the free loop gets smaller as tape is pulled from the inside. This is due to the fact that the outside diameter of the tape is larger than the diameter at the hub. If this were a normal recording tape, the loop would become so

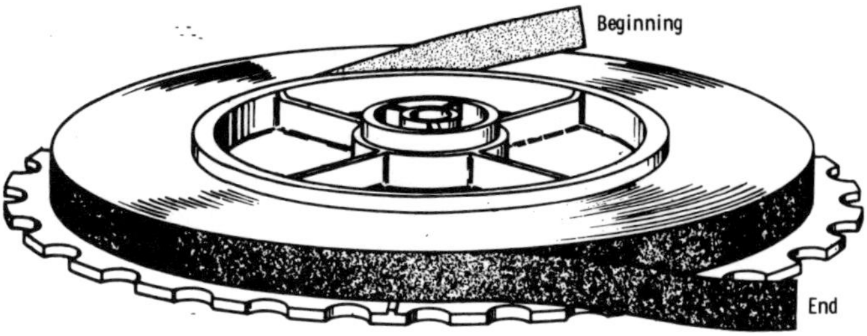

(*A*) *Beginning and end of tape.*

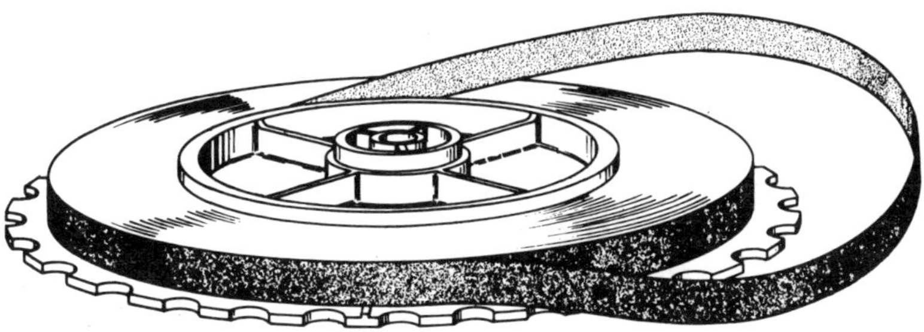

(*B*) *Tape spliced in continuous loop.*

Fig. 11-25. Reel for tape cartridge.

small that it would eventually bind and stop. For this reason, a specially lubricated tape is used to allow the tape to slide on itself at all times. Table 11-2 lists dimensions and recommended loading specifications for these tape cartridges.

NOTE: In the following recording procedures, it is assumed that the recordings are to be made on a conventional tape recorder, then loaded on the automatic tape cartridge. This is conventional practice at many stations where the cartridge machines are playback units only. Some stations also purchase units which are capable of both recording and playback, as described in Section 5-2, Chapter 5.

Table 11-2. Dimensions and Recommended Loading for Tape Cartridges

Cartridge	Width	Length	Continuous Loop		Mobius Loop	
			Min	Max	Min	Max
1200 (C)	A	D	25	1200	25	1052
600 (B)	B	E	25	600	25	500
300 (A)	C	F	25	394	25	285

NOTE: Dimensions are shown in Fig. 11-24; loop lengths are in feet.

Recording Continuous Loop—Place a reel of single-coated tape on the tape recorder, and follow the instructions for recording supplied by the manufacturer of the specific tape recorder used. If you have not used a complete reel of tape to record the information, cut the tape about 1 foot beyond the information recorded. Remove the supply reel. Take the reel from the take-up spindle on the recorder, and, without turning it over, place it on the supply spindle. Place an empty reel on the take-up spindle of the recorder, and wind the recorded tape fast forward in the normal manner. After the tape is completely wound, note that the recorded information now faces away from the recording head.

Place an empty cartridge reel on the supply spindle. Place a metal loading ring over the hub of the cartridge reel, and attach the end of the tape to the loading ring, making sure that the oxide (recorded information) faces outward. By means of fast rewind, wind the recorded information on the cartridge reel. (On the Fidelipac Model 300, simply wind a few turns around the hub; there is no loading ring for this model. The other models can also be loaded in this manner if a loading ring is not available.)

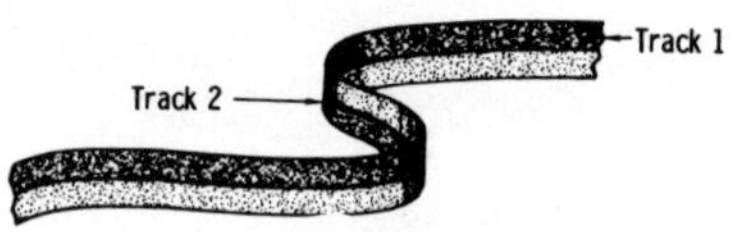

Fig. 11-26. Track positions for mobius-loop operation.

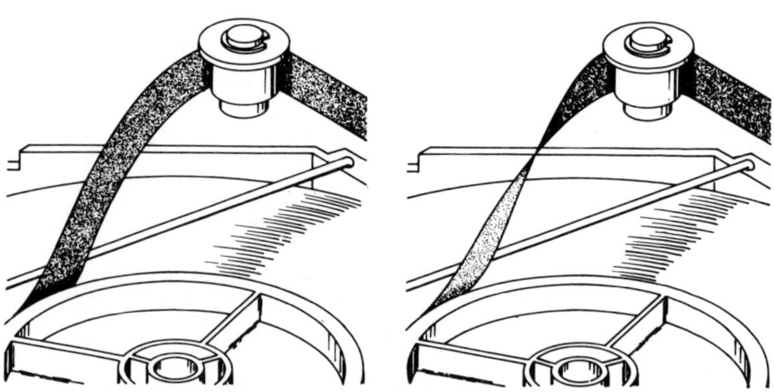

(*A*) *Continuous loop.* (*B*) *Mobius loop.*

Fig. 11-27. Tape configuration around guide.

Recording Mobius Loop—Place a reel of double-coated tape on the recorder, and record following the normal procedure. After the information has been recorded half-track on one side of the tape, rewind the tape to the supply reel in the normal manner. Take the beginning of the tape, and before threading the tape past the head, give it a half-twist. This makes it possible to record on the reverse side, half-track, but not back-to-back with the previous recording. Fig. 11-26 shows the position and the side on which each track has been recorded.

After recording on the reverse side, follow the instructions previously given for rewinding and loading on the magazine reel. When splicing for a mobius loop, care must be taken to twist the tape in the correct direction (Fig. 11-27). As the tape comes from the center hub, twist the tape so that it will slope toward the outside of the cartridge as it goes around the corner post. If the tape slopes toward the inside of the cartridge, it is a continuous loop.

The following points apply to any type of recording equipment:

1. The cartridge reel must revolve in a clockwise direction during loading of recorded information.
2. The tape must be loaded with the recorded information facing outward.
3. The beginning of the program must be at the center of the cartridge reel when starting to load.
4. The tape should be fed to the cartridge reel during loading with as little back tension as possible on the supply reel.

When the tape has been loaded on the cartridge reel, remove the reel from the machine, disengage the tape from the clip on the loading ring (if used), and carefully lift the loading ring off the hub.

Preparing to Splice—Place the loaded reel on a flat surface. Grasp the center of the reel with the right hand to hold it firmly while you pull approximately 18 inches of tape from the outside of the reel. Note that the tape will slip off, so the reel must not be permitted to turn. Remove about 9 inches of tape from the center of the reel by gently pulling on the center end of the tape.

Do this carefully so as not to spill the tape from the reel. Make sure there is no twist in either the beginning or end of the tape. For a continuous loop, place the two ends of the tape on the splicer and splice the tape in the normal manner, i.e., oxide to oxide, using Mylar-base splicing tape. (Acetate-base splicing tape generally will break after approximately 100 hours of operation on a fully loaded cartridge.)

For a mobius loop, follow the continuous-loop instructions with one exception. Take the beginning of the tape as it comes from the hub, give it one-half twist to the right, and then insert it in the splicer.

Loading the Cartridge—After the tape has been spliced, place the reel in the cartridge with the cartridge front facing forward. With the left hand, release the spring brake over the opening in the bottom left of the cartridge (Fig. 11-28A). This will allow the reel to turn. Pull the tape from the center with the right hand until the slack has been taken up. Place the tape around the guide post, through the front guide slots, and around the left guides as shown in Fig. 11-28B. Again, release the brake and take up all excess slack. Now turn the cartridge so the rear faces you, and pass the long straight wire guide under the tape coming from the center hub; insert the front and rear ends in the holes or slots provided at the front and rear of the cartridge.

On the Fidelipac Model 1200 (large size, now size C), place the wire with the V on the right side of the reel of tape with the shorter end of the wire guide facing the front of the magazine. The V guide wire should rest near, but not touching the hub, and the V portion of the wire should rest lightly on the tape. This is necessary to keep the tape down against the hub.

The reel must not be raised too high, since this would create excessive down pressure on the tape edge after the lid has been fastened in place.

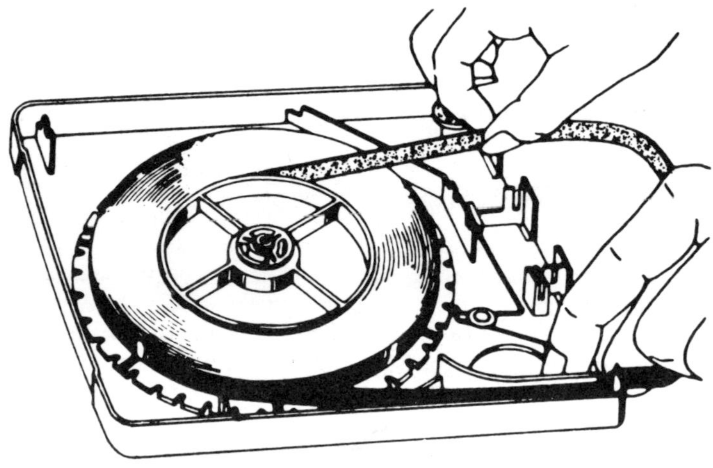

(*A*) *Take up slack.*

(*B*) *Place tape around guides.*

Courtesy TelePro Industries, Inc.

Fig. 11-28. Loading tape cartridge.

Before the cartridge is closed, place the cartridge in the tape player with head cover removed, which will allow the lid to be screwed down in place while the reel is in motion. Observe the reel motion carefully to make sure that while the lid is tightened, no binding is produced between the wire guides and the edge of the tape. After complete assembly, if there is too much space between the edge of the tape and the guide wires, remove the reel and add one or more Teflon washers around the center post until the edge of the tape has been raised sufficiently to just touch the wire guides without any pressure being exerted.

On the Fidelipac Model 600 (now Size B), the guide wire is in a single piece, but the same instructions apply. Only one guide wire is used on the Model 300 (now size A). Also on this model, use of the loading ring is not necessary. While holding the end of the tape, turn the reel three or four turns so as to give the tape a start; then proceed as instructed previously.

11-10. STEREO SETUPS

The stereo control console provides duplicate channels for feeding two output program lines, for the left and right channels. Any desired number (within design limitations of the console) of microphones may be used for either the left or right channel. An individual VU meter is provided for each channel. It is very important in stereo that the left and right channels be identical in frequency response and phase characteristics.

The console operation for stereo turntable and tape sources is the same as for monaural consoles, except that the disc pickup cartridge and the tape pickup heads each provide two outputs.

In a studio setup for air or recording, at least one microphone each is used for the left and right pickup areas. Directional microphones are normally employed so that each unit picks up first and most strongly the sound for its respective pickup area. When more than one microphone per channel is used, each is directed toward a given area of sound source for its respective (left or right) pickup area.

Important is the fact that no actual "curtain" exists between left and right. A certain amount of opposite-channel sound is picked up, subdued in amplitude and delayed by distance. The result in a properly adjusted receiver is to provide the illusion of dimension or realism found in natural hearing.

The center effect for the two-channel system must be provided by production techniques. With a properly microphoned and balanced production, a given sound source can be made to appear "front and center" even though no speaker exists in the center of the two-channel reproduction. Loss of this effect is termed a "hole in the center."

The term "compatibility," as defined previously, applies only to the technical aspects of broadcasting the sound to the listener, not to pickup technique. The problem of the relationship of background to foreground is the most troublesome aspect. As was emphasized earlier, it is necessary to use the microphone to obtain "attention focusing power" for the listener's mind in a monaural pickup. But in natural hearing, background sounds may be actually quite loud.

To illustrate the problem, consider the example of foreground monologue with musical background. The narrator in this case is at the center, which means the sound is equal in intensity and phase in both the left and right channels. The musical background is dimensioned by conventional stereo pickup. Assume now that a good stereo recording is made so that excellent reproduction results in a stereo system. Assume further that this recording is mixed electronically and presented through a monaural system. Chances are that the background music will be much too loud. Thus, some compromise is normally made, and this problem will persist until stereo becomes the rule rather than the exception.

Fig. 11-29 illustrates the basic rule of thumb for all stereo setups. Draw an equilateral triangle with the apex at the optimum listening point and the base on the first row of the pickup area. Place the two basic pickup microphones with the center axis (of the directional pattern) pointing along the sides of the triangle. The distance will depend on the dimensions of the pickup area, acoustic treatment of studio, and type of presentation,

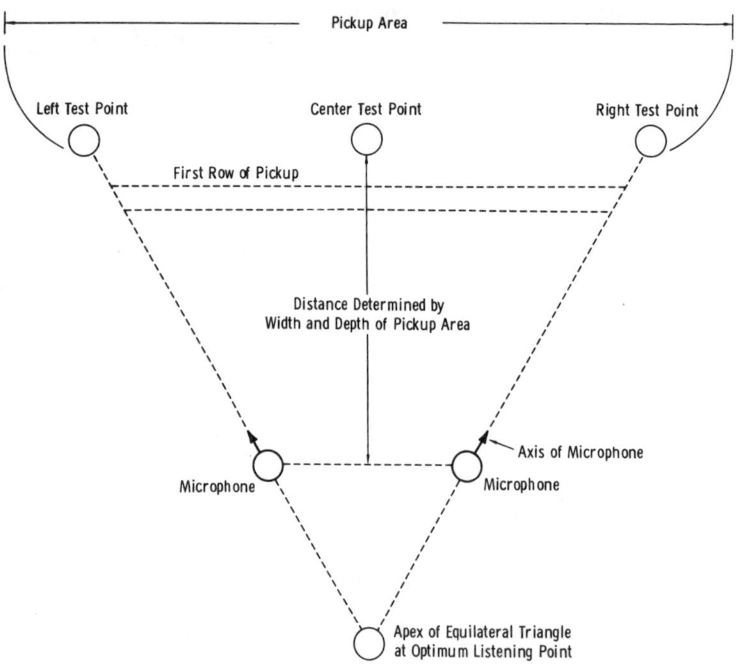

Fig. 11-29. Basic arrangement for stereo pickup.

and must be determined by trial. The basic characteristics to be obtained are:

1. A tone sounded at the right test point should result in an output 15 to 20 dB higher from the right microphone than the pickup from the left microphone.
2. A tone sounded at the left test point should result in an output 15 to 20 dB higher from the left microphone than the pickup from the right microphone.
3. A tone sounded from the center test point should result in equal amplitudes from both microphones.

It is sometimes necessary to obtain the center by experimentally moving the source along the base of the triangle until equal amplitudes result. This point may be displaced somewhat from the physical center and is influenced by the acoustics, the number and placement of participants, and the type of program.

Now take a specific example, broadcasting or recording a full orchestra with vocal soloist. Depending on the size of the overall pickup area, the two basic left and right microphones may need to be farther from the soloist than is optimum for good vocal pickup. The soloist may be "lacking in presence" on stereo reproduction, and this will be entirely unacceptable through monophonic systems. A third microphone may be added for the vocalist and bridged into both channels. The following precautions must be observed:

1. If the orchestral sound is too predominant in the center microphone, very little stereo will be evident in the reproduction. Thus the microphone must be turned off at all times except when required for the soloist.
2. The problem of (1) requires close pickup of the vocalist so that adequate isolation from the orchestra is obtained. (NOTE: This is overcome in multiple-track tape master recording by placing the vocalist in an isolated studio with headphones, and recording the solo on a separate track. The proper blend is then made in dubbing onto the stereo tape.)
3. When the soloist must be worked close to the microphone, natural liveness is sacrificed, and artificial reverberation is sometimes used on this one channel only (monophonic) to restore proper sound.
4. When an additional channel is bridged in, it must not introduce phase shift into the other two channels.

The use of artificial reverberation in monaural systems was covered in Section 11-6. It is just as desirable for stereo to correct the actual reverberation time to achieve a value which is suitable for the instruments used or the musical style. But the art of adding artificial reverberation to stereo is not as simple as it might first appear. If, for example, a reverberation device is used in each channel, not only is the phase-frequency problem a severe handicap, but a true "room effect" does not necessarily result.

One type of artificial reverberation unit which has proved satisfactory for both monaural and stereo applications employs a tinned steel plate in the arrangement shown by Fig. 11-30. (NOTE: This description appears here through the courtesy of Electronic Applications, Inc.) While the monaural type is illustrated here, the EMT 140st (stereo) unit provides an additional pickup to the left of the driver unit; this pickup is identical (in front-view appearance) to that shown on the right in the photo.

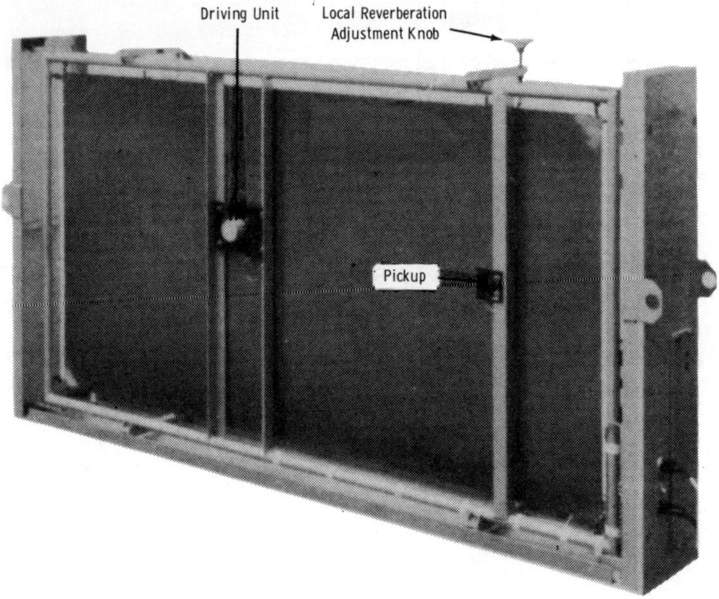

Courtesy EMT

Fig. 11-30. Electromechanical reverberation unit.

The plate, which is excited to transversal vibration, produces numerous reflections of a pulse. These have a linearly increasing repetition rate, which the ear cannot distinguish from a quadratically increasing repetition rate such as occurs in a three-dimensional room. With suitable dimensions and suitable material, the number of natural frequencies is adequate, and the magnitude and the frequency characteristic of the reverberation time are suitable for artificial reverberation. A plate represents the minimum, but nevertheless adequate, outlay for meeting the requirements of quality for artificial reverberation.

The plate is suspended in its own frame of tubular steel, which is mounted in the angle-iron frame of the outer casing by means of rubber springs. The casing is made of panels of Novopan, a pressed wood-fiber material with plastic binding. A porous damping plate is in a position parallel to the oscillating plate; the spacing between them may be adjusted approximately in the ratio of 1:30. This damping plate is used to vary the reverberation time of the device. The reverberant plate is excited electrodynamically to transversal vibrations. The vibrations are picked up by means of a piezoelectric contact microphone. The driving unit and the microphone are spot-welded to the steel plate. A power amplifier for the excitation and a voltage amplifier for amplifying the microphone output are contained in a built-in unit. The amplifier chassis is accommodated in one of the narrow sides of the casing, as is the additional remote-control device. By means of this remote control, it is possible to adjust the reverberation time to any desired value from the control desk. An indicating instrument on the control desk shows the reverberation-time setting of the moment. Alternatively, the reverberation time may be adjusted by means of a hand-operated wheel on the upper side of the casing, this wheel being connected with the swinging arms that support the damping plate. The shaft of the hand wheel carries a pointer which shows the reverberation-time setting on a scale on the top of the casing.

Due to the realism of stereophonic sound, special problems involved will become evident from the following review:

The listener will receive the sound from various directions at different intensities and with varying time delay.

1. The direct incidence: The sound waves traveling the shortest possible distance to the listener and which arrive with a minimum of delay.
2. The first reflection: Sound which is reflected from the nearest wall and reaches the listener with very little delay.
3. The second reflection.
4. The successive reflections: The successive reflections consist mostly of sound waves which have been reflected several times before reaching the listener and which are of decreasing intensity. From this it is quite obvious that the succession and intensity with which these reflections occur are strongly influenced by the dimensions of the room and the absorption factor of its boundaries, as well as by the locations of the speaker and the listener. There are many spots in a room at which the intensities of the direct and reflected sound are equal. This area of transition can be geometrically defined as a circle, the center of which is the location from which the sound is being radiated. Outside this imaginary circle, the intensity of the reflected sound predominates. The reflected waves are normally following in rapid succession, and they are decreasing in intensity in such a way that their decay curve can be shown as a function of ϵ(base of natural logarithms). From this curve the reverberation time can be defined. It is this time which is required for a sound pressure level to decrease 60 dB (a power ratio of one million to one) after the source is turned off.

While it is always possible for the listener to localize the sound source as long as direct radiation dominates, this will become impossible as soon as he finds himself in the reverberant field. Sound is coming at him from every conceivable direction, in many instances diffracted by various elements in the room, and it thus becomes impossible to single out any one directional impression.

From the foregoing it can be concluded that a signal in the reverberant field, while containing the information of the direct sound, will not provide information as to the direction from which it is originating because of the statistical pattern which it follows. An ideal reverberation device for stereophonic sound should therefore be an apparatus which transmits (without distortion) the information which is contained in the direct sound, but does not distinguish between signals which would contribute to giving a sound source an apparent location when the output of this device is reproduced through a stereophonic system.

The foremost requirement when stereo sound is processed through reverberation devices is to remove from the signal any information which makes it possible to determine the direction of the sound sources, without altering in any way the information which is contained in the program generally (direct sound). In other words, what is needed is a simple monophonic signal such as can be had by adding the left and right signals of stereo sound ($L + R = M$).

The signals are branched off the L and R channels and passed through isolation amplifiers. The amplifier outputs are added to each other, and the resulting sum signal is then fed to the driver system of the reverberation plate. To pick up the signal from the plate, two transmitters (contact microphones) are mounted on it at unequal distances from the driver. The signals delivered by these microphones are therefore not identical to each other, an important requirement for achieving random distribution of the directional information in the reverberated signal which is now to be added to the two stereo channels.

A reverberation plate has a very large number of points where resonances occur; in fact there are so many of these resonance points over the whole audio range that we can speak of a resonance spectrum of almost infinite density. This then results in a frequency-dependent phase shift between the two transmitters. If the plate is excited with a sinusoidal signal of any frequency, some form of phase relationship will exist between the outputs from the two microphone systems. Of these phase shifts, four are of special interest (Table 11-3).

The directional impressions indicated are those which would be created if two microphones were put in a room to pick up sound from a moving source. It is quite obvious that more than only the four distinct phase relationships will be formed; actually any number of intermediates will occur, and these will depend strongly on frequency. Therefore, if a complex signal such as music or voice is being applied to the plate, all imaginable

Table 11-3. Reverberation Microphone Phase Relations

Phase		
Microphone I	**Microphone II**	**Directional Impressions**
90°	0°	Reverberation (seemingly) from the right
0°	90°	Reverberation (seemingly) from the left
90°	90°	Reverberation (seemingly) from the center
90°	−90°	Reverberation (seemingly) from the sides

phase relationships will appear in the outputs of the microphones. Consequently, the reverberated signal will not indicate the original direction of the sound because of the statistical pattern which it follows, thus fulfilling the requirements.

11-11. TRANSMITTER REMOTE-CONTROL OPERATIONS

The majority of a-m and fm transmitters are remotely controlled from the studio control room. For a complete visualization of the circuitry involved, the studio operator should be familiar with the calibration and adjustment of such units. The following paragraphs describe the basic adjustment procedure for the Marti RMC-2AXS equipment as utilized with an stl transmitter remote-control system. (Review the description in Section 8-6, Chapter 8.)

Calibration and Adjustment

Remote Control Tone Level Adjustment (When Used With STL)—The tone-oscillator output levels at the studio unit, RMC-2AXS (Fig. 11-31), are set as follows:

1. With the system operating, hold the telephone dial on the studio-unit panel between any two numbers to key the 20-kHz dial oscillator. With an ac vtvm or sensitive ac vom, measure the tone-level output at BNC-1 of the studio unit. Set the dial level at 0.5 volt rms with R5. Measure with the unit connected to J1 of STL-8 or across a 560-ohm load resistor.
2. Depress the LOWER lever, and set the 25-kHz lower-oscillator level to 0.5 volt rms with R6.
3. Press the RESET button, and set the 22.5-kHz reset-oscillator level to 0.5 volt rms with R7.
4. Depress the RAISE lever, and set the 27.5-kHz raise-oscillator level to 0.5 volt rms with R8.
5. Reconnect the tone output cable to BNC-1.

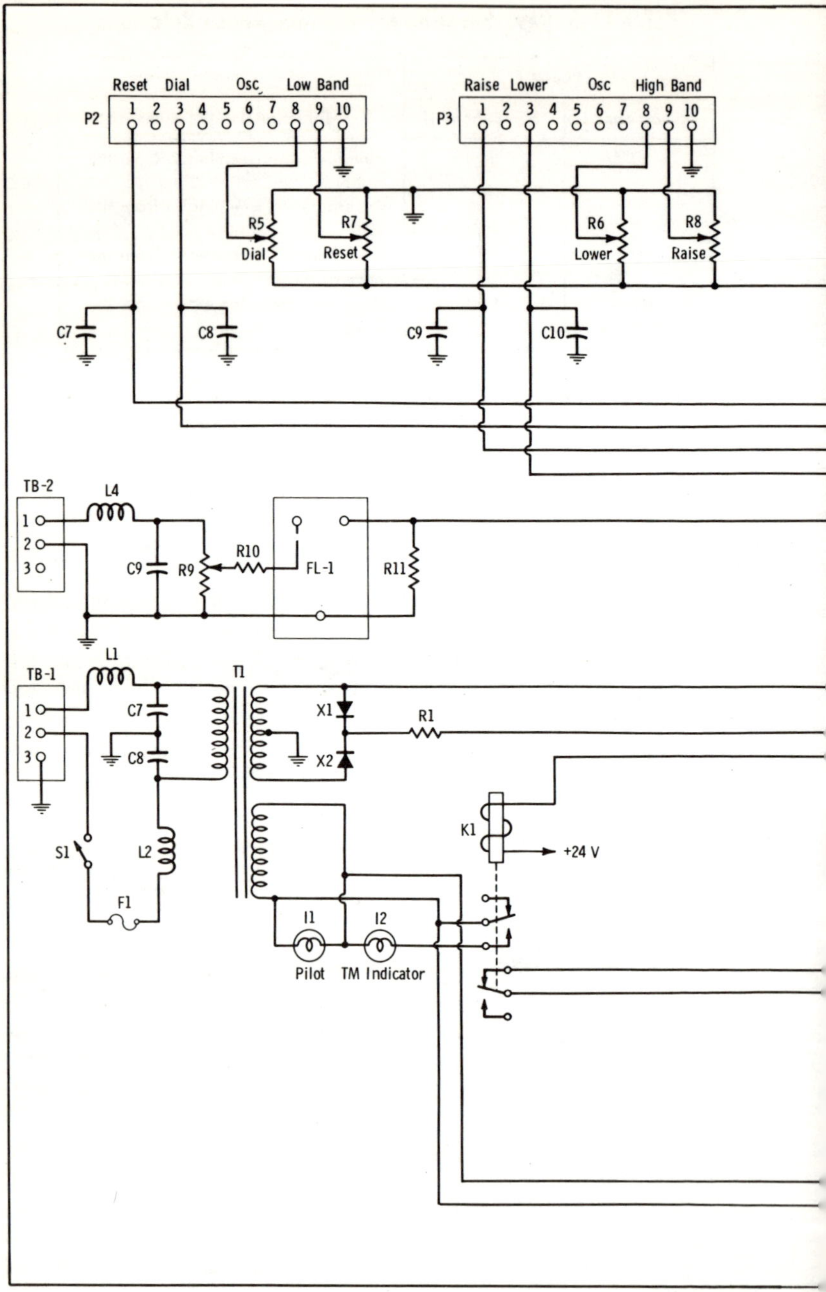

Fig. 11-31. Schematic diagram

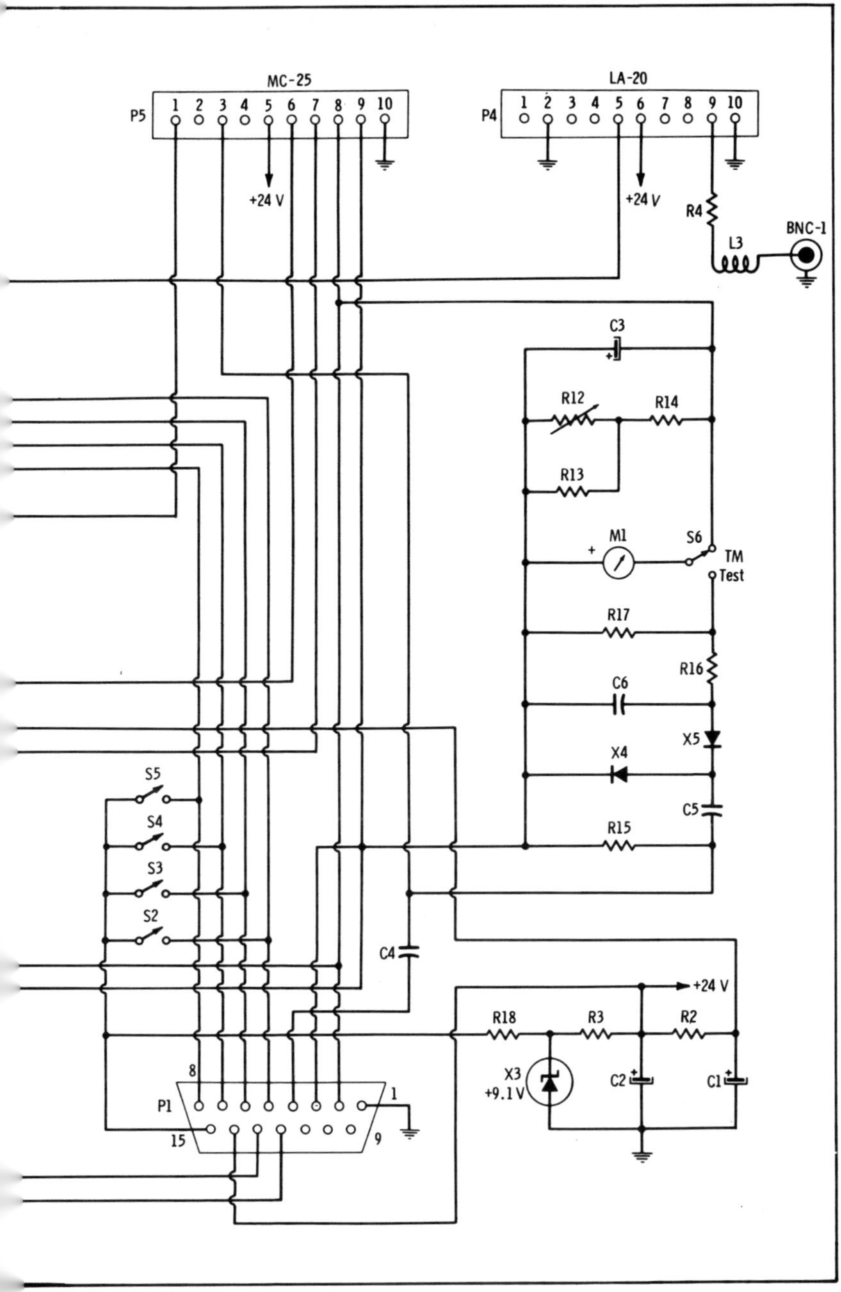

of RMC-2AXS studio unit.

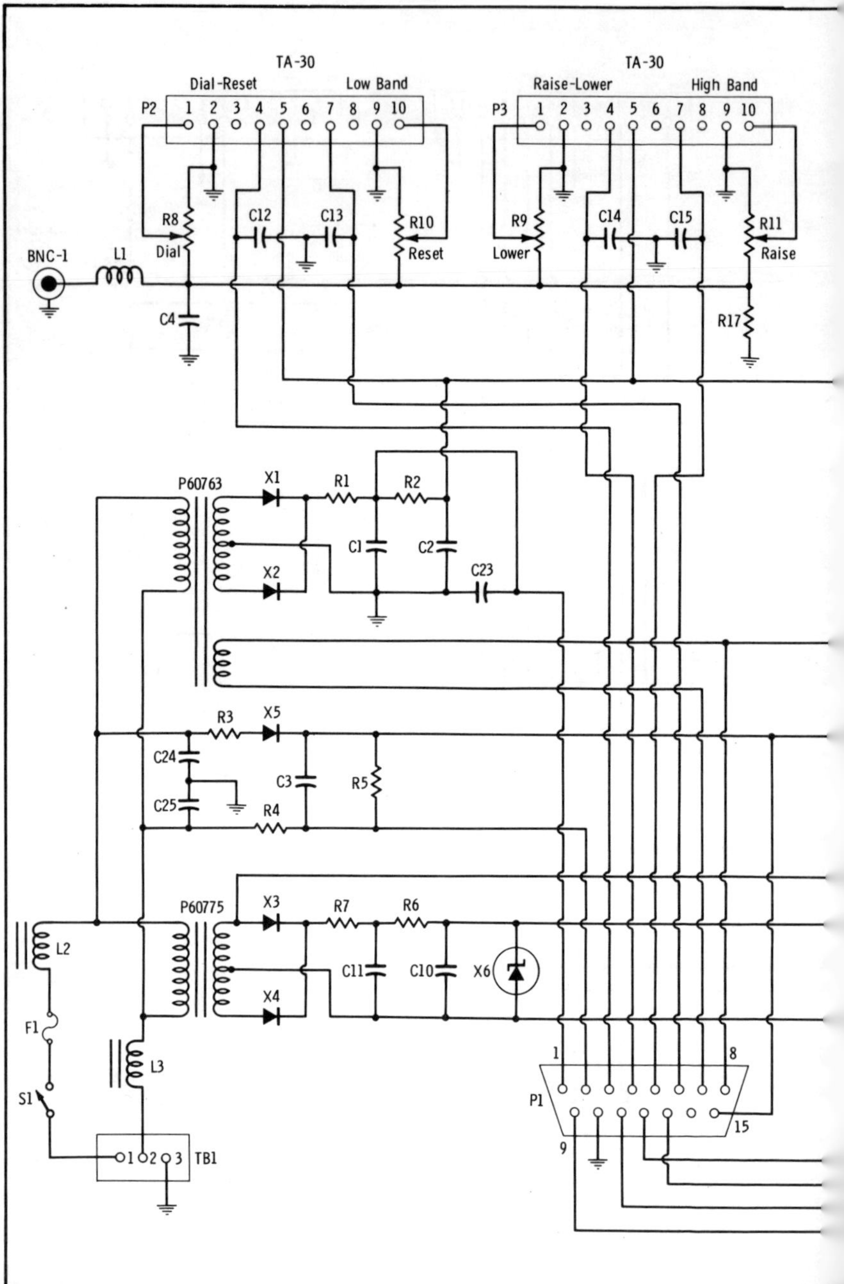

Fig. 11-32. Partial schematic of

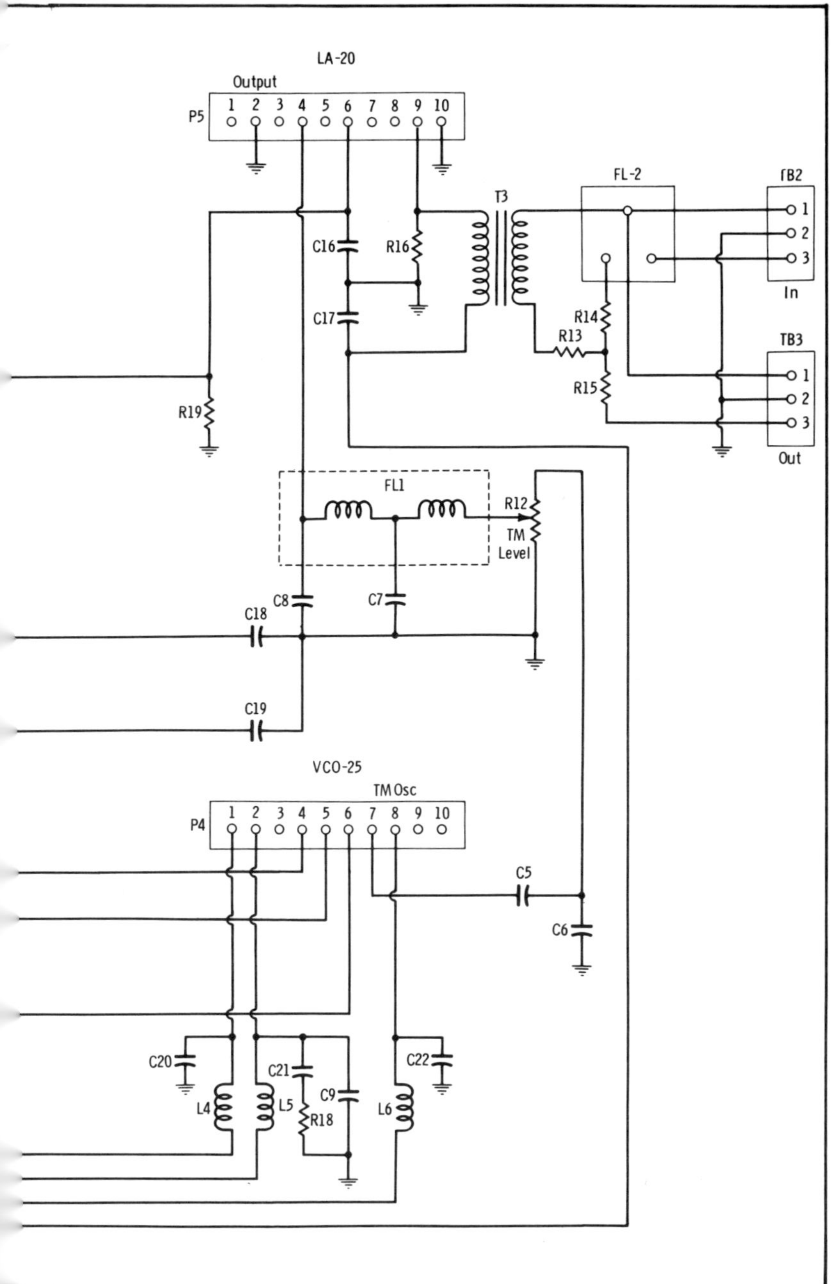

RMC-2AXT(A) transmitter unit.

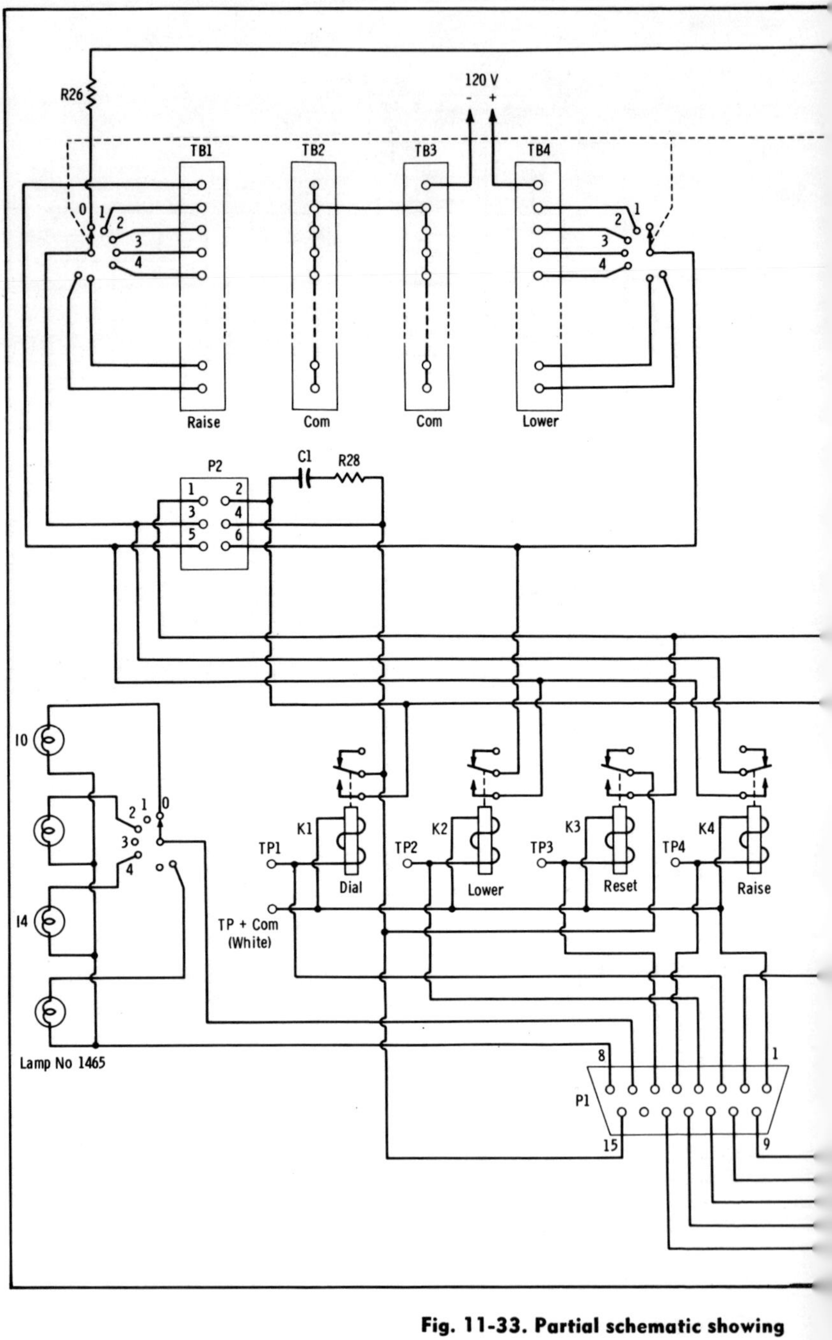

Fig. 11-33. Partial schematic showing

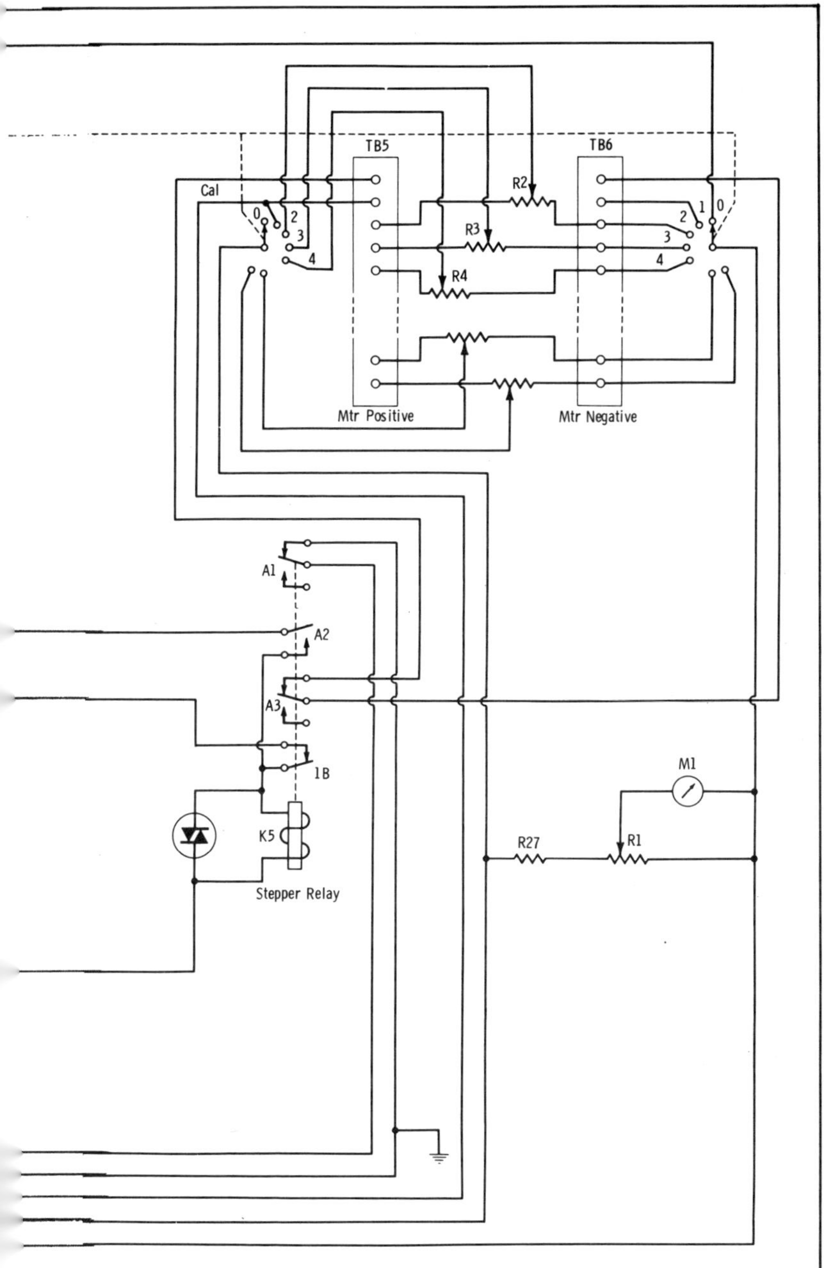

relay operations in transmitter unit.

The tone-amplifier levels at the transmitter unit, RMC-2AXT(A) (Figs. 11-32 and 11-33), are adjusted as follows:

1. With the system operating, hold the dial of the studio unit to key the dial oscillator as in step 1 above. At the transmitter unit, connect a dc meter (0-25 volt scale) with 20,000 ohms per volt sensitivity between TP-1 (negative) and the white common jack (positive). (See Fig. 11-33.) Set R8 (Fig. 11-32) for a reading of 11 volts dc at TP-1.
2. Depress the LOWER lever at the studio unit, and adjust R9 at the transmitter unit for 11 volts dc at TP-2. (Leave positive meter lead in white jack.)
3. Press the RESET button at the studio unit, and adjust R10 at the transmitter unit for 11 volts dc at TP-3.
4. Depress the RAISE lever at the studio unit, and adjust R11 at the transmitter unit for 11 volts dc at TP-4.

All four remote-control functions should now operate.

Telemeter Tone Modulation Level Adjustment for Telemetering to Studio Over SCA Channel

1. Reset the remote-control system to the off position by pushing the RESET button at the studio.
2. Connect a VU meter across terminals 1 and 3 of TB-3 of the transmitter unit (Fig. 11-32.)
3. Set R12 on the transmitter unit to zero (maximum counterclockwise position.)
4. Feed a 400-Hz tone through the entire SCA system, and set all levels necessary for proper operation at 100-percent modulation of the 67-kHz subcarrier. Record the level of the VU meter at 100-percent SCA modulation.
5. Remove the 400-Hz tone from the SCA-channel input. Dial position 1 on the telephone dial at the studio unit. Set the telemeter tone level with R12 on the transmitter unit for a level 10 VU below the level recorded in step 4 above.
6. With operation as in step 5 above, depress the METER SW on the studio unit, remove the cover knob from the GAIN potentiometer, and with a screwdriver slowly raise the gain control until the meter reaches maximum indication (limiting). This level should be between 90 and 120 percent on the meter. If this level is not obtained, check the SCA receiver gain setting. Replace the gain cover knob.
7. Release the METER SW and adjust the CALIB knob for Cal (100 percent) on the meter. The system is now calibrated in position 1.

Telemeter Tone Modulation Level Adjustment for Telemetry Over Telephone Line

1. Follow steps 1, 2, and 3 above.
2. Set the telemeter tone level with R12 on the transmitter unit for 2.5 volts rms (+10 VU) at terminals 1 and 3 of TB-3.
3. Follow steps 6 and 7 above.

Calibration of Meter Sample Inputs

1. With the system on calibrate (position 1 of the stepping switch), adjust R1 (meter calibration potentiometer) on the transmitter unit (Fig. 11-33) for Cal on the transmitter-unit meter. (The studio unit should also be reading 100 percent as adjusted in step 7 above.)
2. Dial in the next metering position, and adjust the potentiometer which bears the same number as that position, as indicated by the numbered light, for 100 percent on the meter. (The transmitter being controlled and telemetered must be operating at exactly 100 percent of normal in all readings.)
3. Repeat step 2 above for all metering positions.
4. Information other than transmitter voltage, current, and power can be telemetered if desired. Refer to Fig. 11-34A for a method of telemetering transmitter filament voltage. By means of current transformers, tower-light operation can be telemetered (Fig. 11-34C).
5. A conversion table (such as that in Fig. 11-35) must be prepared so that the operator taking a remote reading at the studio can quickly convert the percentage readings from the meter to actual voltage, current, and power values for entry into the station log.
6. For each required measurement, a graph must be prepared and posted at both the transmitter and studio showing the (linear) relationship between actual transmitter meter readings and remote readings.
7. Once weekly, the complete system must be calibrated to meet FCC regulations. (Check current FCC rules.)

Electrical-Zero Adjustment of Meter on Studio Unit RMC-2AXS

1. Push the RESET button to turn the system off.
2. Set the mechanical zero-adjustment screw located on the meter front.
3. Dial in a metering position which is not in use. If all metering positions are used, short-circuit one metering input terminal pair on TB-5 and TB-6 of the transmitter unit to provide a zero-voltage sample into the system.
4. With the system dialed into the zero voltage sample, adjust the trimmer potentiometer on the MC-25 module of the RMC-2AXS studio unit for a zero reading on the studio-unit meter. This trimmer is located on the back of the MC-25 module ¾ inch from the top edge and ⅞ inch from the left side. Use a small jeweler's screwdriver to punch through the paper label covering the adjustment hole; then make the zero adjustment.

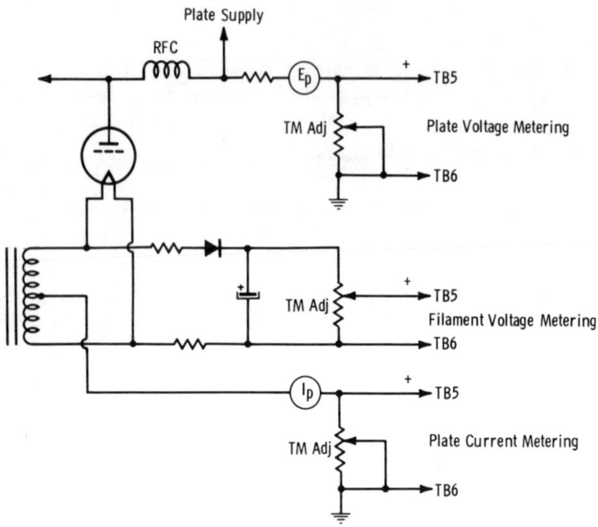

(A) Voltage and current.

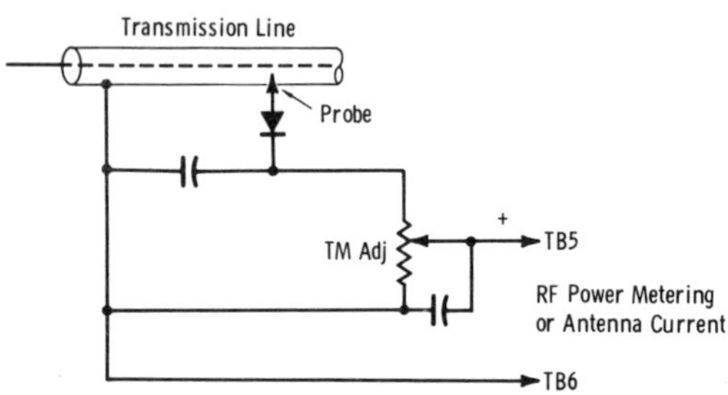

(B) Rf power or antenna current.

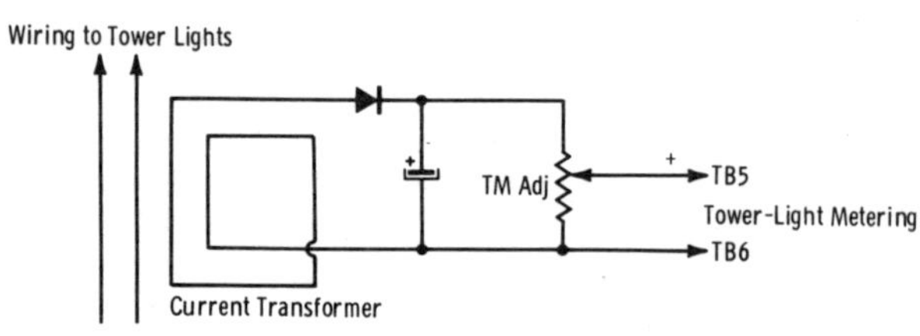

(C) Tower-light metering.

Fig. 11-34. Typical transmitter metering circuits.

MARTI Electronics MODEL RMC-2AXS REMOTE CONTROL / TELEMETER OPERATION CHART										
DIAL NO.										
MEASUREMENT	CAL.									
CONVERSION 90 %										
91 %										
92 %										
93 %										
94 %										
95 %										
96 %										
97 %										
98 %										
99 %										
100 %										
101 %										
102 %										
103 %										
104 %										
105 %										
106 %										
107 %										
108 %										
109 %										
110 %										
OPERATING ↑ FUNCTIONS ↓										
NOTES										RESET

Courtesy Marti Electronics, Inc.

Fig. 11-35. Remote control and telemetry operation chart.

Telemetering Linearity Test and Chart

1. Check for mechanical and electrical zero setting of the meter on the studio unit.
2. Turn off the transmitter unit (switch S-1), and set the zero-adjust screw on the meter of this unit.
3. Turn on the transmitter unit, dial position 1 (calibrate), and with potentiometer R1 (Fig. 11-33) set the meter of the transmitter unit to Cal (100 percent). Also adjust the studio-unit CALIB knob until that meter reads "Cal."
4. Connect a 3-volt battery (two flashlight cells in series) to an unused pair of metering terminals on TB-5 and TB-6 of the transmitter unit. Dial in this metering position.
5. Adjust the vernier potentiometer corresponding to the metering terminals to which the battery is connected for a meter reading of 120 percent. The meter on the studio unit should read the same value. Record the readings at the studio and transmitter units.
6. Repeat the above process in 10-percent steps down to zero percent.
7. On 8″ × 10″ linear graph paper (10 divisions per inch minimum), plot a graph of transmitter-unit meter readings versus studio-unit meter readings.

Operation of the RMC-2AX System

Control—This system provides remote transmitter off-on control as follows:

1. Turn on all microwave stl and associated studio equipment.
2. On the studio unit, dial the number to which "filaments on" is connected, and operate the RAISE lever to latch the filament-on relay at the transmitter.
3. Next, dial the number for "plates on," and operate the RAISE lever to latch the plate power-supply relay on at the transmitter.
4. To turn the broadcast transmitter off, repeat the above steps, but operate the LOWER lever to turn off the relays at the transmitter.

Measurements and Operating Adjustments

1. To make the required measurements for the log each half hour, first push the RESET button to make sure the stepping relay is in the zero (off) position.
2. Dial 1 on the studio unit, and allow four seconds for the TM (telemeter) light to come on, indicating the system is operating. The meter should now read "Cal," or 100 percent. If not, adjust for 100 percent using the CALIB KNOB. If the meter is not steady, push the the METER SW button, and note the telemeter tone level. If it is below about 90 percent, check the SCA fm receiver or make necessary adjustment of the GAIN control.
3. Proceed to dial each meter position, and record each reading in the log using the percentage-to-value conversion table for each reading. If the meter indicates an adjustment in the transmitter operation is necessary, operate the RAISE or LOWER lever as required until the correct reading is indicated on the meter.
4. When the last reading is dialed in, push RESET. The TM light will go off, and the system will be in standby.

11-12. MEETING EMERGENCIES

In spite of the most thorough maintenance schedules on broadcast equipment, trouble may occur at any time. A well rounded, consistent maintenance schedule will greatly reduce the probability of trouble during the broadcast day. When emergencies do develop, the experience of the operator and his familiarity with the equipment will determine the amount of time lost.

To this might be added a third factor—the "human factor." Perhaps this should be classed as a normal subpart of experience, yet many times the supposedly experienced operator will lose as much time under pressure as a comparatively new man. This human factor is the part of being *mentally prepared* at all times for emergency procedures.

It is obvious that experience and familiarity with specific equipment cannot be gained to a great degree by reading this text. Our purpose here is to make suggestions relative to a majority of troubles and illustrate specific examples of applications of emergency procedures. We hope to accomplish this in such a way as to make the reader mentally prepared.

In order that the following discussion be as applicable as possible for the particular type of installation with which the reader is concerned, it is necessary to divide the analysis into two general categories. These are the simple consolette class without associated patch panels, and the more complex installation with means of patching inputs and outputs of the equipment.

Meeting Emergencies Without Patch Panels

Contrary to what the student or new operator might think, the so-called "simple" type of control room, where only a control console is used without auxiliary amplifiers and patch boards, is the most difficult type to return to the air on short notice in case of major trouble. Some consolettes, such as the RCA 76-B5, have a means of switching the entire program circuit to the monitor or spare amplifier, and of feeding this amplifier to the regular program line. This type of installation will be discussed first.

Fig. 11-36 is a simplified block diagram of the 76-B5 consolette. Every operator on the job should be able to draw such a block diagram of his particular installation. We will assume that this equipment is installed as is, without any additional patch panels or auxiliary apparatus, and the microphones in Studio A are being used. (It makes no difference at present whether the system is on the air or on audition.)

One microphone channel goes dead. If this microphone has just been switched on, quickly double check for proper switching positions. If the channel fails while in use, there obviously is trouble.

The first check is to turn on the other microphone, if it is not already. If this microphone picks up sound, the program channel is operative, and the trouble lies in the microphone, the cable, or the associated preamplifier. If there is time, check the preamplifier tube, or simply instruct the persons to use the other microphone. If the program is being aired, the other microphone must be properly placed or the persons must be moved, as quickly as possible.

Suppose, however, that both microphone channels are dead. It is very unlikely that both microphones or cables would be at fault (provided they are plugged in—this illustrates the importance of rehearsals or pre-air microphone checks). This specific consolette uses a dual power supply; one supply is for the program amplifier and preamplifiers, and one is for the monitor amplifier. It is also unlikely that both preamplifiers would go bad at the same time. Therefore, the most likely points of trouble are the program amplifier (if on the air) or its power supply, or the monitor amplifier (if on audition).

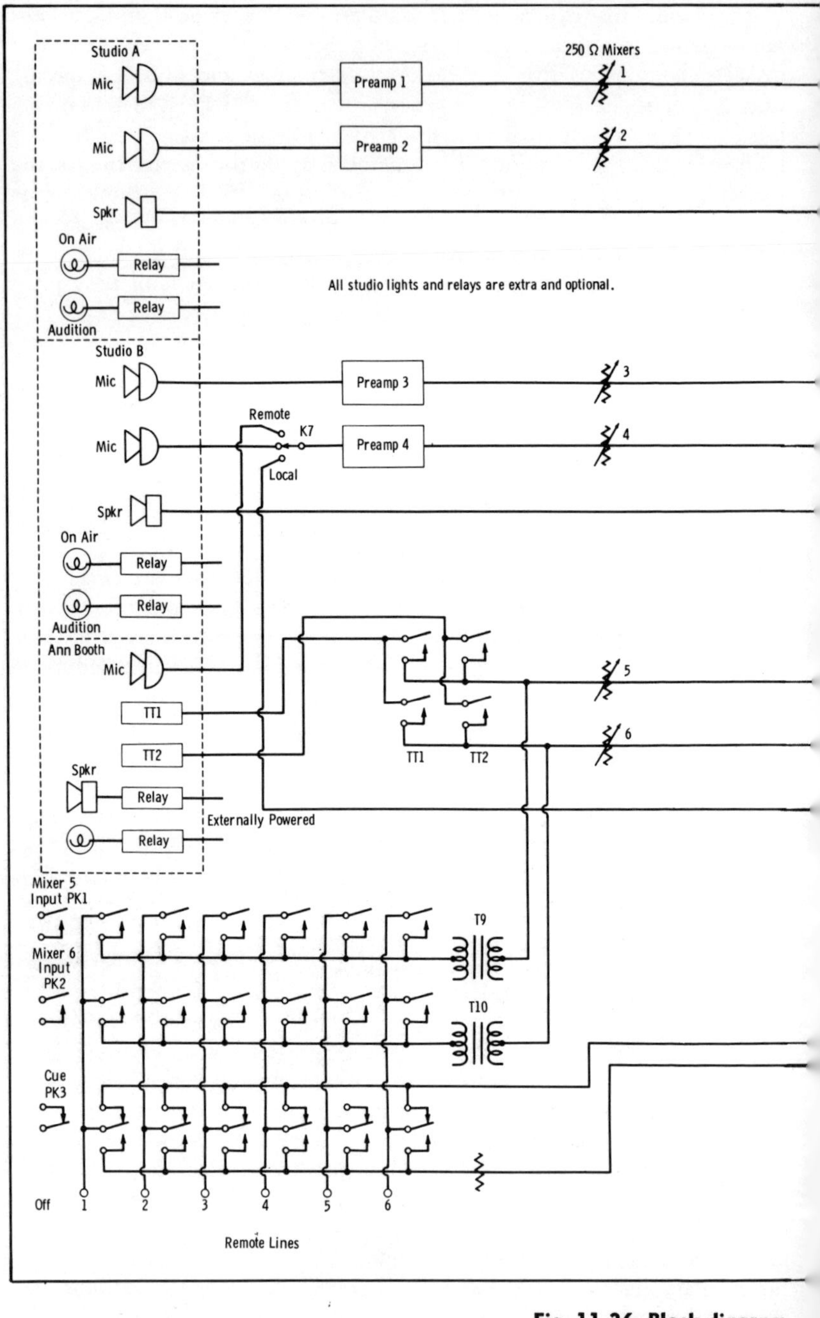

Fig. 11-36. Block diagram

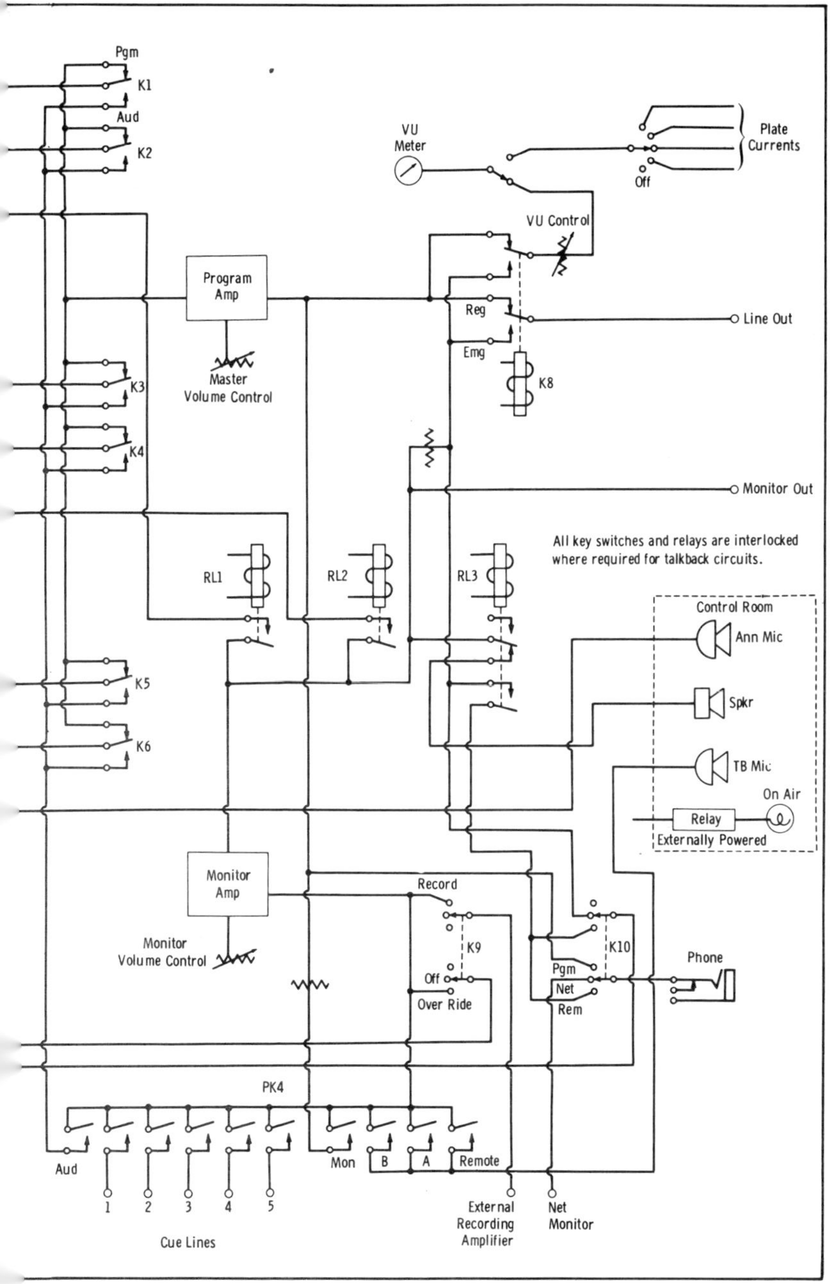

of control console.

If the program is being aired, the operator's duty is to get it back on the air in the shortest possible time. The 76-B5 has an emergency provision to supply the preamplifiers with power from the monitor-amplifier power supply by simply throwing a switch on the power-supply panel from the regular to the emergency position. The line out switch may then be thrown from the regular to the emergency position to connect the outgoing line to the monitor amplifier. By throwing the microphone keys to the audition position, the program can then be returned to the air. All operating personnel should be required to practice this procedure before or after broadcast hours when such a setup is provided.

As shown in Fig. 11-36, the turntables (TT1 and TT2) are connected through their respective keys and faders directly to the program amplifier (or monitor amplifier). This is also true of the remote and network lines. Turntables frequently have preamplifiers of their own installed in the turntable cabinets, and these should not be forgotten in case of a dead turntable. If both pickups are silent, the obvious procedure is to carry out the emergency measures. This bypasses any likely trouble in the power supply or regular program amplifier. Unless the preamplifiers in the turntable are supplied from the regular power supply, the emergency switch on the power supply need not be used, since the preamplifiers within the console are not in use.

In case of failure of an incoming remote or network line, use headphones to monitor the line as it comes in. If the program checks out at the line, it is again obvious that the same emergency procedure is necessary, without the power-supply emergency switching. Know your equipment, and plan every possible emergency check and procedure with the provisions at hand. This is the art of being mentally prepared. It is also the secret of avoiding panic and confusion, which cut down efficiency in meeting emergencies.

The Consolette

The simplest type of installation (from an initial cost standpoint) is the single-channel consolette with no emergency provision incorporated in the circuit and no associated patch board or auxiliary equipment. Obviously, emergency procedures may take somewhat longer, but as long as the operator meets an emergency with the highest efficiency possible, he is doing his best.

Fig. 11-37 is a block diagram of the Western Electric 23-C consolette when not used with an auxiliary patch panel. Microphone input circuits consist of switching keys K1, K2, K3, K4, and K9, with four single-stage preamplifiers terminating at the mixing potentiometers. Keys K1 through K4 select any of four microphones in either of two studios (up position for one studio, down for the corresponding microphone in the other studio). Key K9 connects an announcing or talk-back microphone in the control room to preamplifier 4 in place of the other microphones. Keys K5 through K8 connect any one of four incoming program lines (or high-level phono pickups or phono preamplifier outputs) to a mixing potentiometer or to the monitor amplifier for preliminary monitoring.

Now consider troubles with this type of installation similar to those discussed previously. Assume a microphone has gone dead. If the others in the studio are still operative, the trouble obviously is in the microphone, cord, or preamplifier associated with that particular channel. If all microphones are dead, there may be more serious trouble.

Naturally, if all the microphones are dead, it is necessary to look to a point common to all microphone channels. Where is the first common point in this example? It is the single-stage, low-level amplifier between the outputs of the microphone mixers and the master gain control. The operator should know where this tube is and be able to change it quickly. The Western Electric 23-C also has a meter selector switch and associated jack for the purpose of measuring the cathode current of each tube on an external meter. This tube is number 5 on the selector switch. Obviously, if all readings are zero, the trouble is in the power supply, and the 83V rectifier tube should be changed immediately. With consoles such as this that use all glass-type tubes, it is often quicker to lift the top cover and glance at the tubes, looking for a burned-out filament. If the trouble is not in the power supply or the low-level amplifier, it is in the program amplifier, which consists of a 1603 and a 42 output tube.

Assume now that all tubes are good, indicating some trouble in a component part of a circuit common to all microphone channels. Unless the trouble is a complete loss of power due to power-plant failure or a blown fuse on the console (indicated by all filaments being out), it is necessary to use the external meter and jack provision to measure all cathode currents. This generally (not always) provides a quick clue to the faulty stage. If there is simply a short or open in the signal circuit which would not influence dc meter readings, further emergency measures to return to the air are called for.

Since we are discussing, at present, the way of efficiently meeting emergencies rather than actual repair of the equipment, we will consider getting back on the air as quickly as possible without completely shutting down and starting a point-to-point service routine. For example, do the turntables feed into a key, K5 to K8, and is this channel operating? Such is possible, since in this specific instance a separate filter section of the common power supply is used to supply the preamplifiers. If trouble (other than a short) should develop here, it is still possible to use sources K5 through K8, since these bypass the preamplifiers. Recordings may then be played until the trouble is cleared or further emergency provisions can be made. If the station is affiliated with a network, the incoming network line may be jumpered to the transmitter line on the line terminal board.

If every source available on the consolette is dead and it is apparent that some time will be needed to clear the fault, there is still a last means available to nearly all stations to get back on the air quickly, even though under

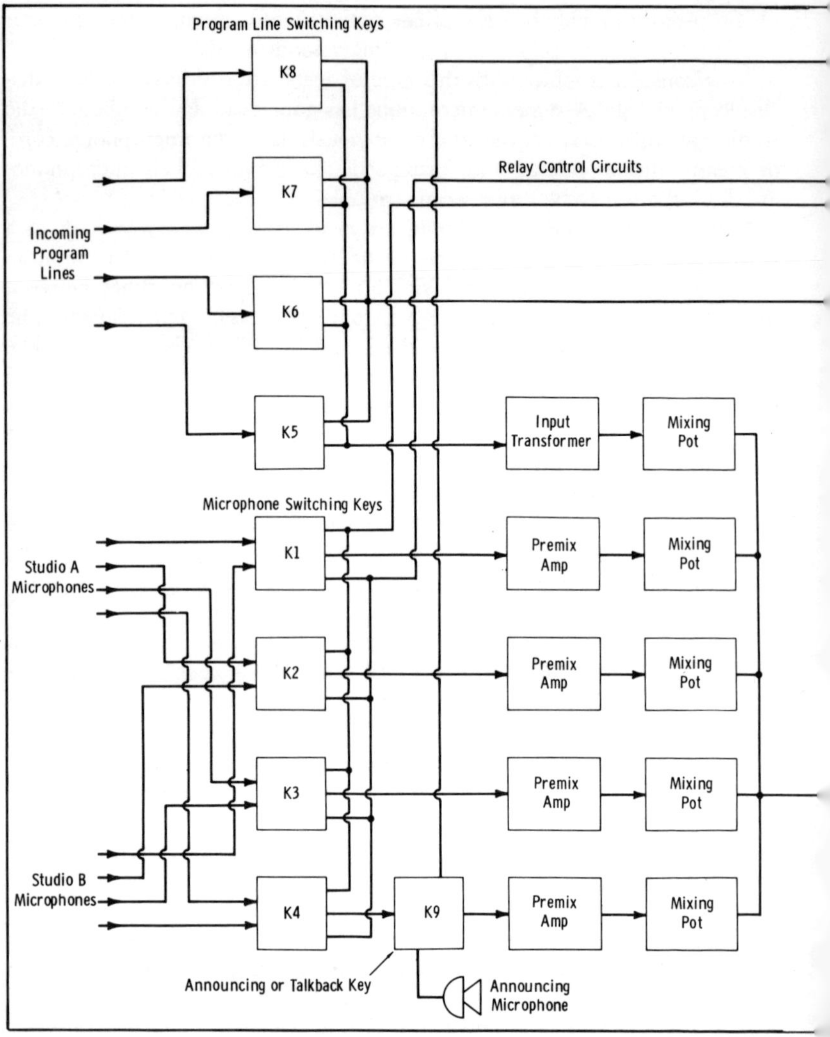

Fig. 11-37. Block

severe limitations. This is by use of a remote amplifier of the type taken to remote points for broadcast purposes. Studio microphones may be connected to this amplifier and the output fed to the transmitter line or to the spare line. Every operator should know where the line terminal board is and have special clip leads available to connect the output of a remote amplifier to the correct line. Certainly every broadcast station, no matter how small and economically limited, should provide this much of an emergency provision.

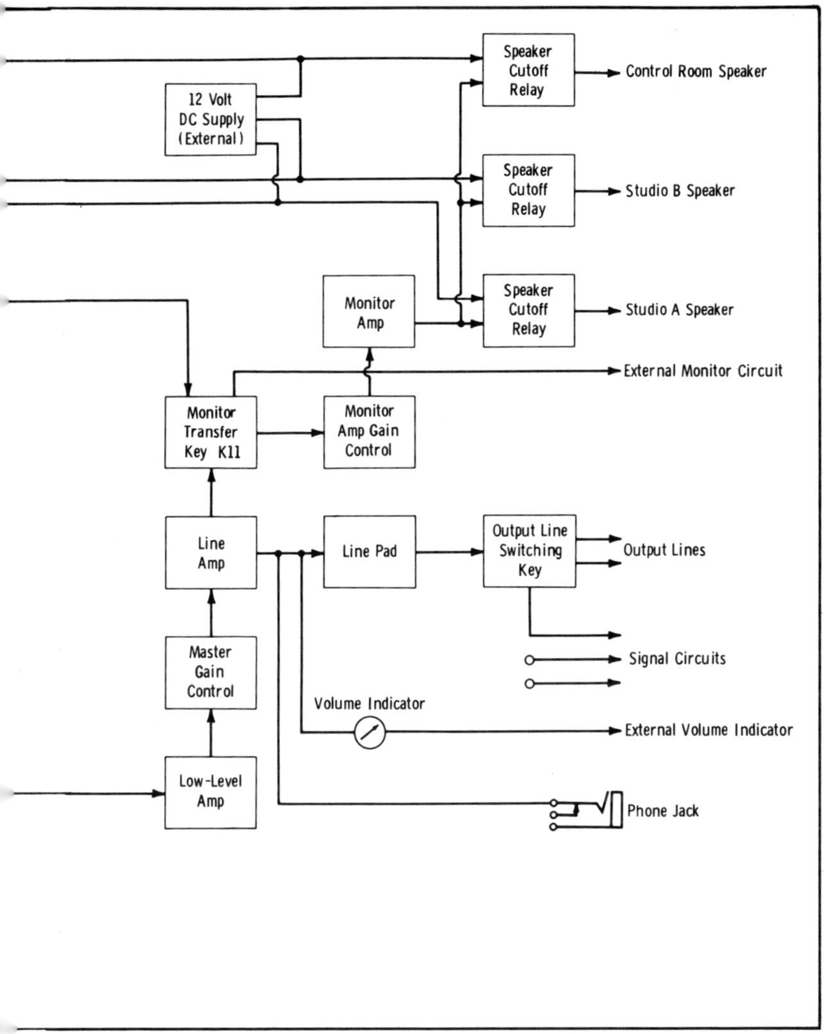

diagram of consolette.

Patching Around Troubles

Larger studios that have more auxiliary equipment and patching panels are much more efficiently arranged for emergency provisions than those types discussed above. Due to this very complexity, however, such an arrangement is apt to be very confusing, especially to the newcomer. The first rule is to become as familiar as possible with the general type of larger-station setup.

For purposes of discussion, a typical amplifier patch-panel installation may be arranged as follows:

Rack 1: All preamplifiers, low-level amplifiers, and high-level amplifiers used for studio A, with associated jack panel.
Rack 2: Same as rack 1, for studio B.
Rack 3: Same, except associated with transcription and tape sources and announce booth.
Rack 4: All amplifiers and jack panels associated with the master control panel. (Program amplifiers, monitor bus feeds, etc.)
Rack 5: Telephone-line terminations for remotes and incoming (and outgoing) network lines.

Fig. 11-38 illustrates the complete block diagram of studio A with corresponding jack-panel designations. Study this illustration closely to become familiar with the jacks in relation to the signal circuits. Microphone 1, for example, appears on the upper row of jacks as marked, and the input of BA-11A preamplifier 1 appears on the jacks just below. The following three blocks of jacks duplicate the same provision for microphones 2, 3, and 4. Then follow the outputs of preamplifiers 1, 2, 3, and 4, and the inputs to their respective faders. Then come the jacks associated with the input and output of the low-level isolation amplifier (BA-11A number 5), then the high-level amplifier (type BA-13A), the studio master fader, and the line-out switch connections. In this installation, the line-out switch may be thrown down for possible feed to the corresponding studio A key on the master control panel, or it may be placed in the up position, which connects the studio output to a monitoring amplifier for audition purposes without tying up one of the three main program channels. Also in this particular installation, the type BA-11A preamplifiers numbers 1 through 4, and the isolation amplifier, number 5, obtain their operating voltages from the high-level amplifier, type BA-13A.

It is obvious that such flexibility is more compatible with emergency measures than the arrangements previously discussed. Assume, for example, that the program stops as indicated by the monitor loudspeaker and the absence of movement of the master control panel VU meter. Since this type of installation uses a separate VU meter on the output of each studio panel, this provides a quick visual check as to whether the main program amplifier is at fault or if the trouble is in the studio circuit. Should the VU indicator of the individual studio still be indicating, it is simply necessary to push the corrsponding studio key on channel 2 or channel 3 on the master control panel, and feed the output of this new program channel to the transmitter line. If the studio VU meter is not indicating, the trouble obviously is in the studio high-level amplifier.

Except where the trouble is obvious from the start (such as when a quick visual examination of the rectifier indicates it to be out), it is advisable to reroute the signal by means of the patch panel. The studio B ampli-

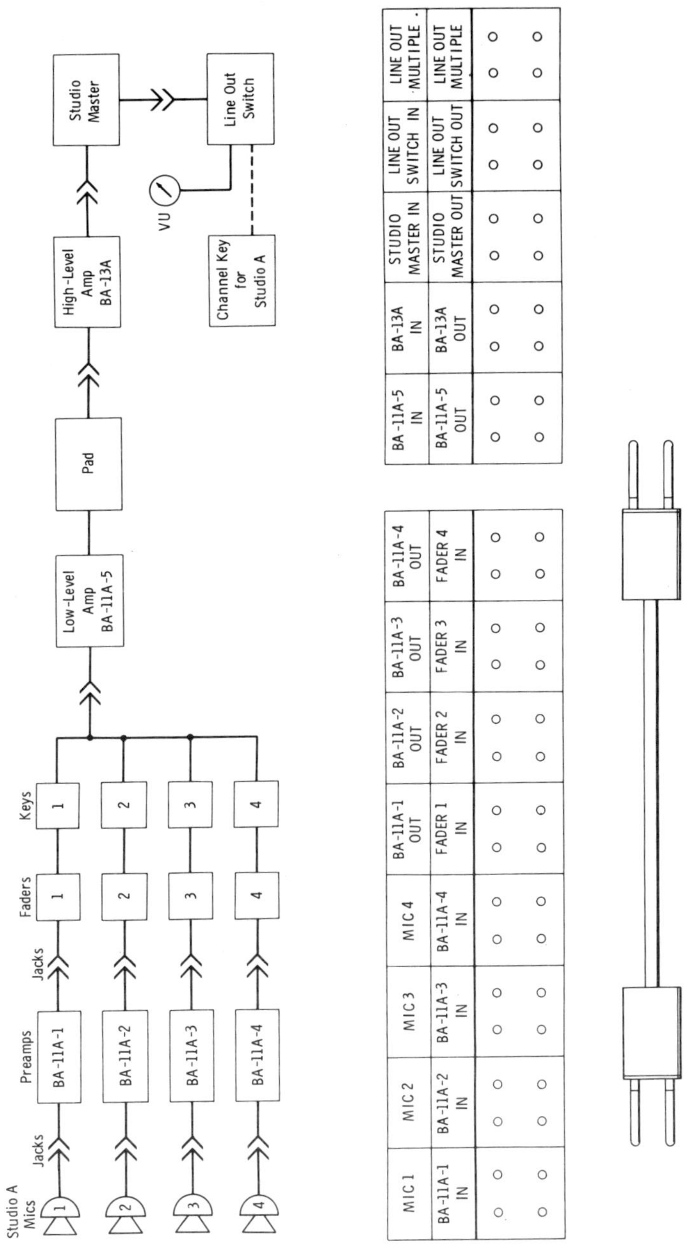

(*A*) *Block diagram.* (*B*) *Jack arrangement.*

Fig. 11-38. Jack-field layout and block diagram.

fier rack is an exact replica of the studio A rack and patch panel. If four microphones are used in studio A, it is necessary to patch the output of each microphone into the corresponding preamplifier of studio B. For example, microphone 1 of studio A would be patched to BA-11A-1 IN on the studio B rack, and so on. A patch cord may then be inserted from the studio B STUDIO MASTER OUT jacks to the studio A LINE OUT SWITCH jacks, and the studio B amplifiers are substituted for those of studio A; but now the faders on the studio B operating panel must be used.

It would be possible, of course, simply to patch the microphones of studio A over to the preamplifiers of studio B and then press the studio B key on the master control panel with the line-out switch on studio B down. Note, however, an important complication if this were done. When the line-out switch is in the down (program) position (as distinguished from up, or audition) and the studio key is pushed on one of the master-control amplifiers, the speaker is cut off in the studio to prevent feedback. If these microphones are patched into another studio and the corresponding key on master control for the substitute studio is pressed, the speaker in that studio will be cut but not the one in the studio where the microphones are. This is apt to cause feedback and result in confusion to the performers as well as the operator. If the operator must use this emergency procedure, he can throw the line-out switch of Studio A to the up (audition) position, which cuts the speaker in Studio A. It is best, however, to return to the studio A position, as in the example given, to avoid confusion in getting out of the emergency operating procedure when the trouble is cleared.

Assume now that only one microphone goes dead. As always, the first thing to do, if only one microphone is being used, is to turn on another microphone in the studio to ascertain if the entire circuit is dead. If only one channel has failed, it is a quick and simple matter to substitute another preamplifier by means of patch cords. For example, if microphone 1 in studio A fails, microphone 1 may be patched to any of the preamplifiers in studio B and from the output of that preamplifier back to FADER 1 IN on the studio-A patch panel. If the microphone is still dead, the operator is faced with a microphone failure and simply must substitute another one.

In patching installations of this type, it is also a simple matter to reroute network or remote signals when trouble develops in the control room. The network or remote line may be patched directly to the transmitter line-in jacks, bypassing all studio equipment.

All of the foregoing simply amounts to this—know the equipment and arrangement; then be *mentally prepared.*

When the station is affiliated with a network, either high-impedance headphones are provided across the incoming line or a bridging monitor selector switch is used to check network signals before application to the studio amplifiers. As quickly as it is ascertained that the network signal has been lost on the incoming line, the local telephone test board should be contacted. Normally a direct line is provided for this purpose. Trouble is usually quickly located and an alternate path provided by the operating center involved. Fig. 11-39 is a view of the control office for network broadcasting in AT&T Long Lines Headquarters, New York City.

Courtesy American Telephone and Telegraph Co.

Fig. 11-39. Control office for television and radio networks at the New York headquarters, Long Lines Dept., American Telephone and Telegraph Co.

11-13. FCC STATEMENT OF POLICY ON "LOUD COMMERCIALS"

The complex subject of "loudness" was discussed in Section 11-2. Because of a consistent barrage of complaints from the general public, the FCC has become involved in arguments between listeners and the broadcasting stations particularly as related to loudness of commercials as contrasted to adjacent program material. Although the FCC issued its Statement of Policy Concerning Loud Commercials in July of 1965, few stations have taken any practical steps in attempting a solution to the general problem.

Very recently, the FCC has taken disciplinary action against stations that have a large number of complaints about "loud commercials" filed with the Commission. Due to the extreme importance of the subject, we are presenting below extracts from the FCC report most important to the broadcast operator. In spite of special amplifiers termed "automatic loudness controllers," agc amplifiers, and special metering systems, the problem still exists.

Extracts from FCC Report, Public Notice B, July 12, 1965

By the Commission:

1. During the past two years, the Commission has studied intensively the problem of loud commercials in television and radio. We are told by industry engineers, broadcasters, and others that subjective loudness of commericals cannot be electronically measured, and therefore the Commission cannot act to prevent it. However, in hundreds of complaints from the public we are also told that some commercials are objectionably loud, often louder than adjacent programming—and often so objectionably loud that listeners are compelled to turn the volume down.

2. We conclude today in our Report and Order in the inquiry proceeding (Docket No. 14904) that objectionably loud commercials are a substantial problem, are contrary to the public interest, and that their presentation is to be avoided.

As industry parties point out, there is now no acoustic or electrical tool for determining precisely whether or not a given sound is objectionably loud. Nevertheless, it has been repeatedly held that objectionable or excessive loudness is both a proper subject for preventive government action, and a condition sufficiently definable for its existence to be established for legal purposes.

3. The purpose of this policy statement is threefold—to set forth our policy and the policy we expect licensees to follow in this respect, to detail some of the practices which are common causes of loud commercials, and to advise licensees not knowingly to broadcast commercials involving such practices. All licensees are expected to take appropriate measures to assure strict adherence to this policy. The Commission, through its complaint procedure or by spot checks at renewal time, will determine whether licensees are carrying out their obligations in this respect, and will take whatever action is appropriate on the basis of such review.

4. Among the practices which the Commission has identified as often causing loud commercials, and which licensees shall avoid, are the following:

(1) *Excessive modulation* on commercials as, for example, through inadequate control-room procedures. We are today amending our modulation rules to make it clear that minimum modulation on peaks of frequent recurrence need not be as much as 85 percent if a lesser level is required to avoid objectionable loudness.

(2) *Excessive volume compression* resulting from the use of automatic gain control, or similar devices—particularly in the broadcast of prerecorded commercial material which may have been prepared with extensive compression and other electrical processing. Excessive compression permits material to be broadcast at a higher than normal aver-

age level of modulation. At least on prerecorded commercial material, a maximum of 6 dB compression in broadcasting is recommended.

(3) *Excessive use of other electrical processing devices,* such as filters, attenuators, and reverberation units—again, particularly where prerecorded material is being presented.

(4) *The use of prerecorded commercials* which have been subjected to excessive compression, filtering, attenuation, "equalization," or reverberation (echo).

(5) *Voice commercials presented in a rapid-fire, loud, and strident manner.*

(6) *The presentation of commercial matter at modulation levels substantially higher than the immediately adjacent programs.* A maximum of 4 dB increase (40 percent to 60 percent to 100 percent modulation) is recommended.

To make sure that such practices are avoided, licensees are to adopt adequate control-room procedures to prevent them, and to take appropriate steps to provide for prescreening of recorded commercials for loudness.

5. Much of the loud-commercial problem arises in connection with the broadcast of prerecorded commercials. In fulfilling their obligations in this area, broadcasters are expected to take reasonable steps to get the cooperation of the recording industry so as to prevent the presentation of loud commercials.

6. We now turn to a brief discussion of the matters referred to above.

MINIMUM MODULATION REQUIREMENT

7. One argument advanced by some broadcasters is that they are prevented by our rules from avoiding loud commercials, because the rules require modulation on peaks of frequent recurrence to be at least 85 percent, thus prohibiting the operator from reducing the transmitter gain even if necessary to eliminate loudness. We do not so construe the rules. However, in order to make this matter completely clear, we are today amending the modulation rules to provide that, while in general modulation should not be less than 85 percent on peaks of frequent recurrence, it may be reduced to whatever level is necessary to avoid objectionable loudness in commercial and other material, even if this is substantially less than 85 percent on peaks. We expect television and radio broadcasters to observe this practice where necessary to avoid loud commercials.

CONTROL-ROOM PROCEDURES

8. Presentation of loud commercials is due partly to inadequate or lax control-room procedures. One cause is inaccurate reading of or inattention to the modulation monitor (required by our rules) or the widely used

volume unit (VU) meter. Another aspect is excessive reliance on automatic gain control (agc) or peak-limiting devices, which, unless properly regulated, are likely to result in loud commercials. Broadcasters are to adopt control-room procedures adequate to prevent the presentation of loud commercials which result from these deficiencies or practices. Attention is invited to a description of accepted procedures in the IRE (now IEEE) Standards on American Practice for Volume Measurement of Electrical Speech and Program Waves, 1953 (53 IRE 3.S2).

COMPRESSION AND OTHER PROCESSING

9. One contributing cause to the problem of loud commercials is the use of moderate amounts of volume compression, which permits material to be broadcast at a higher than normal average level of modulation without having peaks exceed 100 percent on the modulation meter. Compression in broadcasting, when used in moderation, appears to be desirable; but excessive use thereof, particularly in the broadcast of commercial material, is unquestionably undesirable and a major factor in causing objectionable loudness. Broadcasters are to exercise care in using devices causing compression. It is recommended that an appropriate maximum amount of compression is 6 dB, at least in broadcasting prerecorded commercials. Certainly, as a general rule, no more compression should be used in broadcasting a commercial than in presenting preceding material. Similar care should be used in connection with employment of other processing devices, such as attenuators, filters, or reverberation units. *Particular care is to be exercised when the commercial material has been prerecorded,* where substantial amounts of compression and other processing may have been used in the recording. The combination of such processing in recording and in broadcasting—e.g., what might be called "compression on compression"—may, when carelessly used, produce what one broadcaster has termed "a rather overwhelming effect" in terms of loudness. Therefore, the use of further compression or other electrical processing in broadcasting such commercials is to be avoided to the extent necessary to prevent objectionable loudness, and the amount thereof which may properly be used may well be substantially less (e.g., 6 dB of compression less) than that which is appropriate for other types of material.

USE OF RECORDED COMMERCIAL MATERIAL: PRESCREENING

10. Compression, filtering, "equalization," reverberation, and other processing are extensively used in recording commercial material, along with a generally high-volume level of recording. Again, these techniques when used in moderation serve desirable purposes, such as protecting equipment, producing a recording of good technical quality, and producing distinctive

effects other than loudness. But it appears that sometimes they are used extensively for no other purpose than to produce loud commercials. *Broadcasters are to exercise care in the presentation of recorded material in which such processing has been used resulting in an effect of excessive loudness.* Under the revised modulation rules, where a commercial has been prescreened and found too loud, a licensee should reduce modulation below 85 percent where necessary to avoid objectionable loudness. Also as mentioned above, care is to be exercised in the use of any further electrical processing in broadcasting recorded commercial material.

11. We note in "A Guide for Advertising Agencies and Television Stations in Handling Materials for Spot Television Commercials," a joint recommendation of the Station Representatives Association and the American Association of Advertising Agencies, that film, tape, and slides should be in the hands of licensees 48 hours in advance of use. The Guide further suggests that materials should be examined by the station on receipt for "damage, defects, and completeness." Clearly this contemplates delivery in adequate time to permit prescreening.

12. We are aware that in actual practice these guidelines are not alwavs observed. However, we expect broadcasters to adopt appropriate practices and procedures to provide time for prescreening, not only for damage, defects, and completeness, but for loudness.

13. We recognize that to require each station, large or small, to prescreen all commercials for loudness may impose some burden. The small radio licensee can engage in extensive spot prescreening, and if a loud commercial escapes prior detection through this process, he can be alert to the need for prescreening further commercials from the same source. Further, we suggest that the organizations, state or national, which represent advertisers, station representatives, agencies and licensees, should consider the establishment of a group to prescreen and label commercial material as to loudness for the industry.

STRIDENT DELIVERY

14. One common source of complaint is commercials which are delivered in a loud, rapid, and strident manner, with the maximum number of words crammed into the time period and all delivered at or close to maximum peak modulation. Presentation of such material is to be avoided.

CONTRAST WITH PRECEDING PROGRAM MATERIAL

15. Aside from differences resulting from varying degrees of electrical processing used in different types of material, another common source of complaint is the contrast between loudness of commercials as compared to the volume of preceding program material—e.g., soft music or dialogue

immediately followed by a rapid-fire, strident commercial. Such contrasts are to be avoided. For guidance, it is recommended that a maximum of 4 dB increase over the immediately preceding program segment (40 percent to 60 percent to 100 percent modulation) is appropriate for general observance.

CONCLUSION

16. We conclude that the presentation of objectionably loud commercials is contrary to the public interest. Therefore, to the extent it is within their control, broadcasters have an affirmative obligation to see that such material is not presented. In today's Report and Order we recognize that loudness—the impression created in the listener—is to a degree the result of factors beyond the broadcaster's control and varies as between individual listeners (for example, a reaction to a particular product or a particular sound effect other than volume). But we conclude that objectionably loud commercials result in large measure from factors of a technical or partly technical nature which are within the broadcaster's control, and which are not adequately controlled simply by adherence to our rules in the various broadcast services limiting modulation to 100 percent on peaks of frequent recurrence. While there is no evidence that broadcasters in substantial numbers deliberately "boost the power" in presenting commercials, neither is there indication of any concerted, industry-wide effort to deal with the problem. While most complaints of loud commercials are directed to television rather than radio (particularly at prerecorded commercials), the problem is by no means confined to television.

17. We have set forth above the broadcaster's general affirmative obligation, and specific practices and policies to which we expect strict adherence. The list of specifics is not intended to be all-inclusive, and there may well be other steps that can be taken. What is called for is a good-faith effort on the part of licensees to prevent the presentation of commercials which are too loud. In setting forth this policy statement, we recognize the underlying importance of advertising to the American system of broadcasting, and the legitimate interest of the advertiser in presenting his message attractively and understandably, and in drawing attention to what he has to say. But these are not irreconcilable alternatives.

18. We note with pleasure a recent suggestion by the American Association of Advertising Agencies that its Subcommittee on Commercial Production might assist in dealing with this problem, by screening commercials referred to it by the Commission about which loudness complaints have been received. We appreciate this offer of assistance, and if the circumstances appear appropriate, we will take advantage of the suggested procedure.

19. We also appreciate the consideration and attention being given this

problem by the NAB Engineering Advisory Committee. It is understood that investigations and studies are to be made by this committee, regarding the technical considerations that may be involved in the matter of "loudness," and also as to the possibility of developing a new volume measuring meter. The technical staff of the Commission is, of course, ready to cooperate in this endeavor as may be requested.

EXERCISES

Q11-1. For the human hearing sense, is a "flat frequency response" desirable all the way from the microphone to the loudspeaker?

Q11-2. What meter other than the VU meter is better in "riding gain" for handling the "loudness factor?"

Q11-3. When a transmitter input gain control has been properly set for 100 percent modulation on program peaks, what will the modulation monitor read on a pure sine-wave tone at 100 percent indication on the studio VU meter?

Q11-4. How many types of artificial-reverberation systems exist?

Q11-5. In Fig. 11-14, if the pianist on the left has a "heavy" left hand (more bass than treble), what would you do?

Q11-6. What is the basic rule in determining the number of microphones for a given pickup in radio broadcasting?

Q11-7. In the final analysis, who is responsible for public complaints about "loud commercials"?

CHAPTER 12

Remote Pickup Operations

The history of the development of radio since its earliest days, when the mere broadcasting of actual sound was miracle enough to create unbounded interest, has witnessed an almost fantastic evolution of technical equipment and technique of operation. Even during the earlier period, when amplifier response and the response of magnetic speakers so limited fidelity capabilities, broadcast engineers recognized the troublesome problems associated with the room or studio in which the program originated. Ordinary architectural construction did not satisfy the requirements for smooth control and faithful reproduction. This led to a detailed study and development of both architectural design and acoustical treatment to suit the needs of broadcasting. Although many experts believe that the final answer to this problem has not yet been found, they all concede that modern broadcast studios have spelled the difference between success and the utter uselessness of high-fidelity amplifiers, microphones, and line or relay links.

There are certain exacting requirements for remote equipment. (Review Chapter 8.) The remote operator encounters conditions far from favorable for the type of program to be broadcast. If the specific location produces very decided effects, he must either use them to his advantage or avoid them. It is the purpose of this chapter to discuss comprehensively the problems encountered in the best utilization of this equipment to achieve the desired results.

12-1. REMOTE OPERATING PROBLEMS

The remote operator is faced with conditions so varied and complex that any discussion of a specific type of pickup must necessarily present only the general principles involved.

A singer's voice is given a certain timbre by the breath which carries the sound from vibrating vocal chords into the modifying air cavities of the head. As these sound waves emerge, they disturb the air in the place of origin in all directions, but principally in the direction in which the singer is facing. The microphone will pick up the sound anywhere in the room. Good transmission depends on the position of the performer relative to the microphone and also on the position relative to the walls, floor, and ceiling of the room. The air cavities and acoustical condition of the air boundaries will affect the character or timbre of the sound just as the air cavities of the singer's head determine the original voice quality.

Thus, it becomes apparent that the varied acoustical conditions encountered place considerable importance on the type of microphone to be used and the technique of microphone placement. For example, the operator may find the surfaces bounding the point of pickup to be highly reflective to sound waves, causing distinct "slaps" and echoes to be prevalent. This condition is caused by deflecting surfaces parallel to each other and is the reason why "live-end" broadcast studios are constructed without parallel surfaces. Under this kind of handicap, the operator must use the directional characteristics of the microphone to the best advantage. He could not, for example, use a bidirectional microphone with one live side toward the pickup and the other live side toward a highly reflecting wall.

Due to the nature of remote-control pickups, the microphones used are nearly always of the unidirectional type. This type permits much better discrimination between wanted and unwanted sound, since the noise level at any remote point is quite high compared with a broadcast studio. The unidirectional characteristic is a convenient aid in preventing large amounts of reflected sound-wave energy from actuating the microphone elements. Since the intensity of a sound wave decreases as the square of the distance, increasing the distance between the sound source and the reflecting surfaces (where this is possible) will decrease the amount of reflected sound-wave energy at the microphone.

By experimenting with the distance between sound source and microphone, it may be observed that the relationship between original and reflected sound will vary over a considerable range. By decreasing this distance a greater proportion of original sound is obtained, and by increasing the distance (between wanted sound source and microphone) a greater proportion of reflected sound is obtained. Music in particular needs a certain amount of reflected sound for brilliance and color. Too much reflected sound will cause a "hollow" tone and uncomfortable overlapping of succeeding musical passages. If the amount of reflected sound is too small, such as in many studios overtreated with sound-absorbent material, the music will be lifeless.

Some of the older-type microphones and remote amplifiers using an unbalanced input (one side grounded) cause a considerable amount of extra work in certain locations where stray ac fields exist near the microphone or cable. Fluorescent-light circuits in particular are hazards for the remote operator.

When the noise level at a remote setup is noticeable, and it increases with an increase in microphone gain and stops when the microphone is disconnected, the operator must try reorientation and/or changing of the cable runs. Interference from fluorescent lights shows up as a raspy frying noise in the headphones.

At permanent installations using fluorescent lights, where microphones may be placed in proximity, the transformers of the lighting fixtures must be well shielded and placed some distance from the pickup area if possible. Shielded wires should then connect the fixture to the transformer.

At remote locations where the program is apt to be a one-shot affair, the lighting fixtures must be changed, moved, or turned off during the show. If this is not immediately possible, the microphone and performers can be moved until interference is eliminated.

The latest remote equipment gives negligible trouble in this respect due to improved shielding of microphones and input transformers, as well as balanced-to-ground circuits.

As stated previously, the problem of broadcasting concerns the transmission and reception of voice and music with the preservation of all the original values. In radio, the degree to which the sound of a performance can play on the emotions of the listener is affected by the transmitter and receiver equipment, studio conditions, and the skill of the engineers. Microphones and amplifiers today are of such good quality that no practical limitations to true fidelity exist because of mechanical or electrical characteristics. Modern broadcast studios impose only slight limitations on faithful transmission of sound. This emphasizes, insofar as remote broadcasting is concerned, that the skill of the engineer or producer responsible for microphone setups and operating technique is of utmost importance. This becomes doubly important when it is realized that each orchestra of any type has its own identifying qualities resulting from instrumentation, musical technique, and conductor's interpretations, all under the influencing factor of microphone placement and acoustical conditions of the point of origin.

The effects desired by the orchestra conductor may be achieved only by proper relationship of the microphones to the musical instruments. This proper relationship is directly influenced by the acoustical condition of the pickup area. For transmission of pure musical tones of a violin, the microphone must be far enough away from the sound holes of the violin that the reflected sounds may be caught fully developed in all their harmonic content. Conversely, when special effects are desired, the microphone might be placed near the violin to bring out the harshness of the resined bow drawn across the strings of the instrument.

Perhaps the most striking difference between studio and field pickups is the complete lack of permanent facilities of any kind in the field. The telephone company will install a broadcast loop on order from the program or traffic department of the station. Sometimes two loops are installed, one to be used as a "talk" line direct to the control room at the studio, or for emergency broadcast service in case of trouble with the regular broadcast line. These lines must be installed as conveniently as possible to the source of the broadcast, yet as inconspicuously as can be arranged. For this reason, it is often a matter of a "line hunt" for the field engineer. This is one reason why he should arrive at the remote point long in advance of broadcast time. The line, or lines, may be found under a table, behind a chair, piano, or organ console and possibly may be in a room away from the main area where the broadcast is to take place. The line is usually tagged with an identifying card.

Since good transmission of talks or speeches from remote points is not as difficult as good transmission of music, the major problems concern the latter. Musical programs may originate at such places as ballrooms, restaurants, night clubs, and cafes featuring dinner music and music for dancing and floor shows. The situation calls for a decided difference in technique of technical production between studio and remote broadcasts.

In the ideal studio musical setup, only one microphone is used at sufficient distance, with the musical instruments grouped and positioned so as to blend into the proper balance at the microphone position. This procedure not only simplifies the problem of control, which always makes for a better effect, but also leaves the problem of orchestral balance in the conductor's hands where it rightfully belongs. Multiple-microphone arrangements place the maximum responsibility for balance of the various sections in the hands of the operator who mixes the outputs of the various microphones.

At remote points, however, where so much activity such as dining or dancing occurs, microphones must be placed close to the musicians. This is inevitable, since otherwise the background noise would result in confusion. This close arrangement calls for the use of more than one microphone to achieve the desired balance; otherwise the instruments closest to the microphone would obscure the rest of the orchestra. Then the setup is divided into units of like instruments or combinations of instruments, each unit being covered by a separate microphone so that the volume from each unit may be adjusted at the mixing panel to achieve the desired balance in the combined program signal.

The practice has some advantages for remote pickups other than avoiding background noise. Acoustical conditions that might severely affect the broadcast are minimized to the fullest extent, since the ratio of any reflected sound to the original sound is small. Then too, although some loss of tonal brilliance results from close microphone arrangement, good instrumental definition is gained, which is an important factor for dance broadcasts.

Symphonic music and church broadcasts are different in this respect in that the audience is comparatively quiet, and the pickup area may be treated more as a studio by studying the acoustical conditions existing at the point of origin.

12-2. REMOTE MUSICAL PICKUPS

An observation of Fig. 12-1 will reveal the principles involved in a typical dance-orchestra broadcast. Insofar as the operator is concerned, this setup divides the orchestra into three separate units. One microphone is for the saxophones and clarinets, another is for the trumpets, trombones, and soloist, and a third is for string bass and piano. Microphone number 3 is handy for special emphasis on the rhythm section or for solo passages by the piano or string bass. Note that when the trumpets are open, they are behind the trombones and caught on the second microphone; when muted, they are placed ahead of the trombones and immediately in front of the microphone. Muted trumpets or trombones must be played with the muted bells very close to the face of the microphone. The same is true of any wind instrument on which the player is producing subtones. The subtones of any wind instrument are just as low in volume, even though open-belled, as the softest muted instrument. This, then, calls for close cooperation between the conductor and his musicians and the engineer responsible for proper pickup.

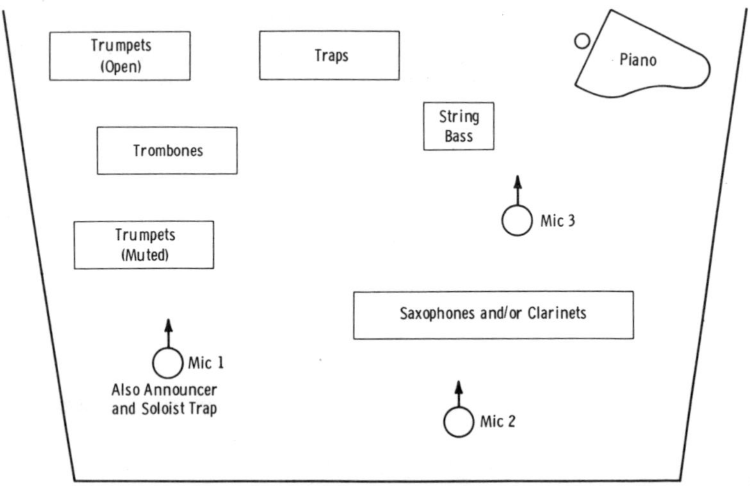

Fig. 12-1. Dance-orchestra arrangement for remote pickup.

Brass Bands

Although brass bands account for a comparatively small amount of radio time, their particular pecularities pose special problems in pickup. A number of community organizations, fraternal societies, and, of course, the armed services participate in radio through presentation of brass bands. These pickups very often must be made outdoors, the least favorable spot for broadcasting. With no outdoor shell or walls of any kind, no reflection of sound can occur to create the ideal polyphased sound dispersion so important to broadcasting technique. Under these conditions, it is necessary to use multiple-microphone pickups, grouping the units by means of spotting separate microphones where needed as determined by trial.

For a medium-sized band organization, the units are usually as follows: one microphone for the clarinets, piccolos, and flutes; one for the English horns, bassoons, bass clarinets, saxophones, and tubas; and one for the French horns, trombones, and trumpets. The tympani, traps, and chimes are usually placed in the lower-sensitivity zone of one of the microphones, which prevents the use of excessive distance for proper balance. Indeed, the sensitivity pattern characteristics of the particular microphones used must be thoroughly understood for any kind of musical pickup. Tympani, when used with brass, are predominant in character when placed in a zone where the sensitivity is the same as it is to the rest of the instruments. Just the opposite is true when they are used with strings, since the masking effect due to the characteristics of the musical instruments themselves tends to subordinate the tympani sound.

When well designed outdoor shells are used, the ideal condition exists for brass-band broadcasts. Usually only one microphone is used, suspended some 15 feet out and above the front-line musicians. As before, predominant instruments, such as tympani, traps, and chimes, are placed at the side in a lower-sensitivity area of the microphone.

Salon-Orchestra Remotes

Some dining places have salon or chamber music organizations which are picked up for broadcasting during the noon or early evening hours. Since a salon orchestra's library contains the more serious type of music, with many low passages, precautions must be taken to subdue the noise of the patrons as much as possible. An intimate microphone placement is therefore indicated under such circumstances.

Usually the chamber group is small, ranging from string trios and quartets to about ten members. For the smaller groups, one microphone raised quite high and slanted down at an angle of about 35 to 45° to the floor is adequate. A hard floor with no covering will aid in obtaining just the amount of brilliance necessary for this type of pickup.

Symphonic Pickups

Symphony-orchestra programs have become a regular feature on the air each season and quite often must be broadcast from a remote point rather than from a regular broadcast studio. Thus far, the musical setups discussed have involved a comparatively small number of musicians and a specific type of instrumental structure. The symphony orchestra, however, is many orchestras in one. The engineer is concerned with the proper grouping of four distinct instrumental sections:

1. Strings: violins, violas, cellos, string basses
2. Woodwinds: clarinets, bassoons, English horns, flutes
3. Brasses: trumpets, trombones, French horns, tubas, euphoniums
4. Percussions: snare drums, bass drums, tambourines, triangles, cymbals, piano, harp, xylophone, marimbas, tympani, etc.

As a general rule, the arrangement of the symphony orchestra for broadcast is the same as for a regular audience performance. The instruments vary in volume of sound produced and therefore in penetrative quality. Strings produce the least volume, then flutes, clarinets, horns, trumpets, and percussion instruments, in that order.

The acoustical situation for symphony broadcasts is generally better than for most other remote pickups, since the auditorium is usually designed for such large groups and made compatible with good listening for the audience, although not always ideal for broadcasting. It is easier from a good transmission standpoint to encounter an auditorium that is too "live" and reverberant so that wall, ceiling, and floor treatments may be added, than to start from one that is too "dead" to produce reflection.

The correct setup for a symphony orchestra is always arrived at on the first rehearsal by trial and error. A number of microphones are spotted at the most likely points so that each may be tried without the commotion of continually moving one microphone. The most likely setup is one microphone suspended at a height of about 15 feet and about 20 feet in front of the violins. A separate microphone must be used for vocal solos, since a closer relationship of vocalist to microphone must prevail in order to achieve proper balance.

Church Remotes

Programs from churches usually involve both music and a sermon. This ordinarily requires only one microphone when the pulpit is directly in front of the choir, as is the most common church arrangement. When vocal solos occur during the choral rendition, a separate microphone is necessary for proper pickup and balance. It will be noted in nearly all instances that, during solos being picked up by a microphone very close to the choir loft, the organ accompaniment must be brought up to the proper background level by use of the pulpit microphone or a microphone farther out in the congregation. This is due to the acoustical properties which are evident in nearly all churches, causing the organ tones to be much more predominant out in the congregation than up near the choir.

Although it would be impossible to cover all the details and complexities of remote pickups in a single discussion, it is hoped that the picture presented here sets forth the fundamental procedures that would help to approach a remote problem properly. To present an absolutely complete picture would be impossible, since acoustical conditions and orchestral intent vary as the number of places from which a broadcast can originate and the number of different music combinations existing. A good understanding of equipment and acoustical variations, however, will enable any engineer to achieve good results with this type of broadcast.

12-3. SPORTS AND SPECIAL EVENTS

There are many types of events with wide public appeal that cannot be covered adequately by the usual methods of remote pickup using wired communication. Among these are various kinds of sports, such as boat racing, cross-country events, and golf matches. Aside from these events, there are the inevitable times of disaster, such as floods, fires, earthquakes, and the myriad catastrophes that wreck ordinary communication services for many miles around the point of trouble. In order to be prepared for eyewitness accounts of these happenings, some stations are equipped with portable and mobile relay facilities that are independent of utility companies and any necessary wire lines for relaying the signal to the studio or main transmitter.

There is probably no other division of radio broadcasting that differs so radically from one station to another as the mobile-relay department. Fundamentally, however, the necessary inventory of equipment includes small portable transmitters, wireless microphones, mobile transmitter and antenna mounted in a car or truck, receivers, and power supplies. Typical equipment was described in Chapter 8.

12-4. LINE ISOLATION COILS AND PADS

Line-to-line coils, usually termed repeat or isolation coils, have numerous uses in the field department of a broadcast station. They are almost a necessity in most instances when an older-type or modified remote amplifier having an unbalanced output circuit (one side grounded) is used. Also, some dubious designs of remote cue-back facilities using an unbalanced circuit will not permit the use of a public-address installation at the remote point without the use of such a coil.

No electrical circuit can be called isolated, since moving electrons cause an electromagnetic field to exist about the conductors. In addition, all conductors have some capacitive effect existent to some point, since, by definition, a capacitor is two conductors separated by a nonconductor.

Consider now the electrical characteristics of the telephone line used to carry broadcast-program currents from studio to studio, studio to transmitter, and remote points to studio. Fig. 12-2A shows the electrical nature of a line operated with one side grounded. Electromagnetic voltages, such as might be set up in the line by adjacent power lines, may be represented by the generator shown in series with the line. Also, since some capacitance exists between the line and ground, any interference picked up by the capacitive coupling may be represented by the generator shown from line to ground. Program currents are shown by solid arrows, induced noise currents by dashed arrows. It is obvious that the two currents add together and will be transferred to the input of the transmitter equipment.

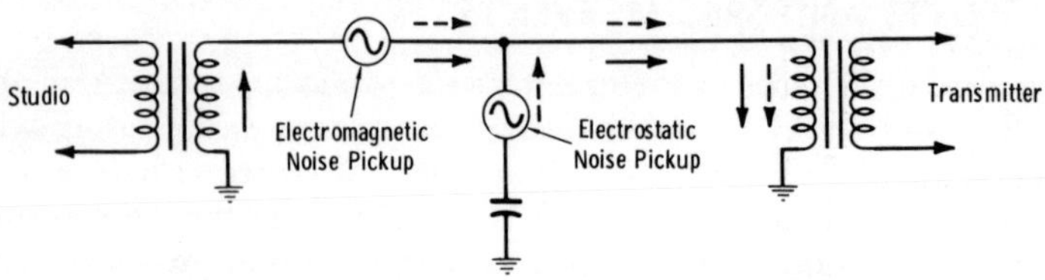

(*A*) *Unbalanced.*

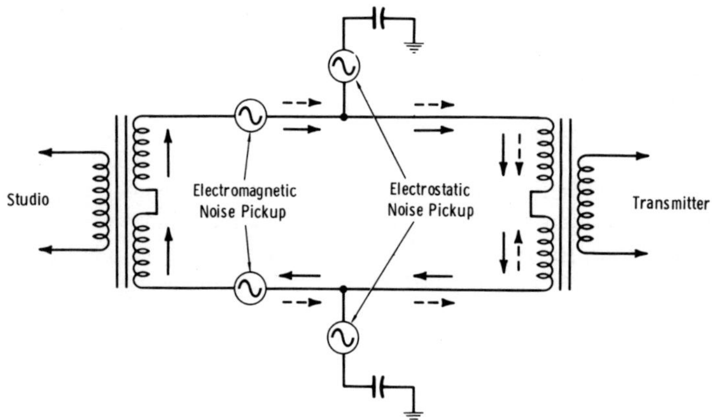

(*B*) *Balanced.*

Fig. 12-2. Remote telephone lines.

Such lines are not used for broadcast services. Many times, however, the operator is faced with adapting a spare public-address, monitor, or recording amplifier to remote broadcast service. Such amplifiers are almost always of unbalanced-output circuit design, which means that one side of the output is connected to the chassis or ground circuit of the amplifier. It is also often necessary for the broadcast operator to feed a supplementary public-address amplifier at the remote point from the output of his remote amplifier. Quite often this public-address amplifier will use an unbalanced input circuit which, if connected directly to the output of the remote amplifier, will ground one side of the circuit. Either of these conditions will result in effectively grounding one side of the program line.

Telephone lines leased for broadcast services are always strictly metallic circuits (no ground return), as shown in Fig. 12-2B, which shows the electrical characteristics of the line when balanced transformer windings are used at the terminating ends. It may be observed here that the noise currents from electromagnetic induction and electrostatic coupling are sent along the two wires in the same direction. Their respective directions of flow in the balanced transformer winding at the load end will be 180° out of phase and will cancel out, leaving only the program currents. In some circuits, the center tap of the windings is grounded; in the majority of circuits, however, the center tap is ungrounded.

Connections to Unbalanced Circuits

Whenever it is necessary to use an unbalanced output circuit to feed a program line, an isolation coil should be used. This coil may have an impedance ratio of 600 to 600 ohms, or 600 to 150 ohms if the line to be fed is quite long. Such a mismatch as this provides a beneficial equalizing effect to compensate for the usual high-frequency attenuation of the line.

At remote points where it is necessary to feed a public-address amplifier by direct connection, and the input circuit is unbalanced, the same type of coil should be used for the line connection. This permits the grounding of one side of the remote amplifier without disturbing the balanced line conditions.

As mentioned earlier, some remote cue-back circuits at the studio end are unbalanced. Such a circuit is not normally important, since it is used only as a cue device. Consider, however, a case in which it is necessary to feed the public-address amplifier with the cue-back signal from the studio. This often occurs during certain types of programs in which it is important for the audience at the remote point to hear signals from the studio. Obviously, the unbalanced line signals will not be of a quality suitable to feed the pa system due to hum and noise on the line. Here it is necessary to use the coil as shown in Fig. 12-3, both at the sending and receiving ends of the line, so that balanced line conditions prevail.

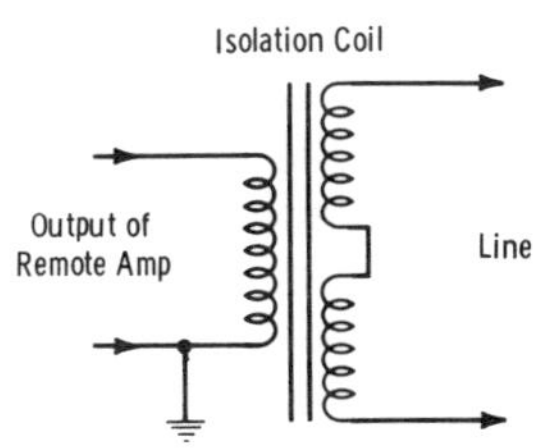

Fig. 12-3. Isolation-coil circuit.

Program Line Levels

Program signal magnitudes on broadcast lines must be high enough to override the noise level, but they must be limited to prevent cross-interference into other lines by electromagnetic and electrostatic coupling. The maximum level permitted in most states is +12 VU into the sending end of the line. The normal level is +8 VU. The multiplier switch is set to +8 so that 0 indication on the meter is referenced to +8.

The broadcast engineer faced with adapting spare amplifiers for remote amplifier purposes must keep this maximum level in mind. Public-address, monitoring, or recording amplifiers generally have an output far in excess of allowable broadcast line levels. For this purpose, pads must be designed as outlined in Chapter 6.

12-5. BELL TELEPHONE AND AT&T LINE SERVICES

The characteristics of various wire services available to the broadcaster are defined by the FCC tariff tables. Where customers have special or unusual requirements, the telephone company will work out services tailored to their needs.

Broadly speaking, services are classed into two main types:

1. Lines for continuous use, such as a studio-to-transmitter feed
2. Lines for occasional or one-time-only use, such as those ordered for remote pickups

Continuous-use lines are classified as follows:

Schedule AAA (frequency range of 50 to 15,000 Hz)
Schedule AA (frequency range of 50 to 8000 Hz)
Schedule A (frequency range of 100 to 5000 Hz)

Occasional-use lines are classified as follows:

Schedule BBB (same frequency range as AAA)
Schedule BB (same frequency range as AA)
Schedule B (same frequency range as A)

When a network of stations is interconnected, involving reversals of transmission direction at operating centers, special charges and rules apply.

The services listed are high-quality lines for full- and part-time use. They are amplitude and delay equalized for broadcast-program transmission over long distances from point of origin to studio. There are also several services of lower grades (C, D, and E) available to the broadcaster. Such services are designed for telephone communications and consequently make use of telephone-type amplifiers and repeat coils. In the case of these lines, accumulation of objectionable noise and delay distortion occurs over relatively short distances. Since remote-broadcast applications require that the line transmit in one direction only (thus eliminating the need for balanced configurations), small improvements in effective bandwidth may be made by the telephone company. Even with such adjustments, however, these lines are still restricted to short-distance use. This service is often used to save cost in the coverage of sporting events and similar programming for which wideband characteristics are not necessary.

EXERCISES

Q12-1. What is the basic difference in microphone setup for programs involving music at a remote location and in the studio?

Q12-2. You are feeding a balanced program line from a remote location, and it is necessary simultaneously to feed a public address amplifier that has an unbalanced input. What do you do?

Q12-3. You have a remote pickup at an airport, and you are picking up voice transmissions from the control tower. What can you do to minimize this?

Q12-4. What can be done at the studio to improve transmission from a remote pickup employing low-grade line service?

CHAPTER 13

Studio Maintenance

This chapter covers the testing and maintenance of all commonly employed studio units from the microphone to the line output terminals. Preventive maintenance and overall studio checks are then outlined.

13-1. MICROPHONES

The microphone is perhaps one of the most delicate pieces of equipment associated with broadcasting systems. Yet it is apparent to the experienced operator that with careful handling the microphone can outlast many sets of tubes and component electrical parts in the amplifier.

When it is necessary to transport microphones from one place to another, it is best to use a special box containing no other equipment. The box should contain sufficient padding not only to take up shocks of exterior bumps, but to prevent free movement of the microphone in the box. Sponge-rubber seat pads are excellent for lining the box, and heavy felt material is good to wrap around the instrument to prevent movement by filling the empty space.

Microphones having permanent magnets as component parts (moving-coil or ribbon types, etc.) should not be placed on a work bench or any place where there is a possibility that iron chips or filings might be attracted to the magnet.

The reader will find it helpful to review Section 6-9 (Audio System technology) and Fig. 6-50 with associated text for proper microphone grounding techniques. Also review Section 7-2 for stereo microphone input wiring and phasing, and Fig. 7-6 for examples of specific microphone wiring.

First Steps in Testing

There are a number of troubles in modern high-quality microphones which should be treated only at the factory of the manufacturer. It is the purpose of this section to acquaint the reader with test procedures that will determine what to do and what not to do regarding repairs. It is, however, obvious that the average technician is unable to check field-response patterns or run frequency-response curves which require laboratory apparatus and soundproof rooms.

First, of course, it is necessary to have a good audio amplifier of known characteristics and the proper input circuit and impedance to match the microphone under test. There are three general classifications of troubles: no response at all, high noise level with or without some signal, and no noise level but a weak and perhaps distorted signal.

As in all of the troubles of microphones, it is necessary to picture the relation of the input circuit to the schematic of the microphone. For example, consider a typical high-impedance input circuit with an open-circuit jack. The high-impedance microphone uses a two-conductor cable, the braided shield about the "hot" lead serving as the ground, or jack-sleeve, connection.

If the response from the microphone is zero when it is connected to an amplifier known to be good, several possibilities exist. Either the "hot" lead is open (if the grounded side were open noise would result), a short exists, or the internal element of the microphone is defective. The first places to check for a defect are in the plug or the point where the cable leaves the microphone housing.

In the case of the open-circuit jack, some noise usually exists before a microphone is plugged into the input. If this noise is lost on the insertion of the plug but no response is obtained, a short is indicated. If the noise level remains the same or is slightly raised in intensity, an open is likely. If a closed-circuit jack is used, a short will not result in lower noise level, but an open will probably raise the input noise, although very slightly in some cases.

Fig. 13-1 shows a typical low-impedance microphone circuit with three-conductor cable. In this case, an open in either the number 1 or number 2 wire will cause either a dead microphone or one with extremely "thin" frequency response. Obviously, a short in the pair would also cause no response. A break in the shielding (number 3) will usually result in a higher noise level than normal, or possible hum pickup.

Fig. 13-1. Typical low-impedance microphone input.

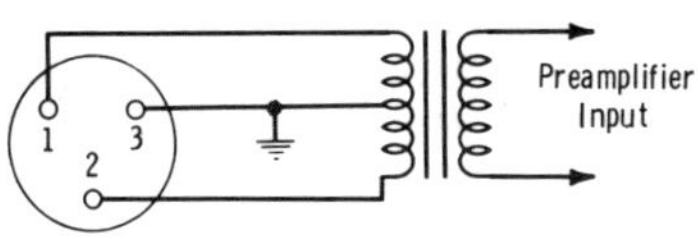

Sometimes the trouble is only intermittent and must be traced by jiggling the cable, starting at the plug and working back a foot or so at a time to the microphone housing. This is done by rapidly looping and straightening a small section of cable between the hands. The following is a good general procedure to check for microphone and cable troubles.

1. Check the plug and receptacle. All types are encountered in microphone input circuits. Some simple two-conductor microphones use the familiar jack and jack plug; some use a metal shell which is insulated from the outer conductor and which has a single pin in the center. For the latter kind, the receptacle is a matching female type with a spring connector to grip the center pin tightly and a knurled metal ring connector for the outer conductor. The three-conductor circuits vary considerably in design, but all are general as far as inspection is concerned. Some have on the receptacle shell a small lever which must be depressed in order to pull the plug from the receptacle. Other types have on the shell of the plug a small knurled knob which must be depressed. In connecting the plug to the receptacle, it is properly oriented in the receptacle, and when pressure is applied, a pin springs up through a hole in the receptacle and locks the two parts together.

   Plug connections that are made inside the shell require removal of the shell for inspection. In some cases the shell and plug body are both threaded and may simply be unscrewed. Others are held together by clamps and screws. Some cable conductors are soldered to the pins; some are held by screws on pin lugs.

   Check the connections to the pins for looseness, corrosion, dirt, faulty insulation, broken wires, or bent pins. Check the plug for damage, dirt, or corrosion. Check the shell for dents, cracks, dirt, or corrosion. While the assembly is taken apart, clean everything with a cloth and cleaning fluid. Corrosion may be removed with a small strip of crocus cloth.

   Check connectors of the spring type for proper contact and tension. In cases where the plug is difficult to connect or remove, coat the pins thinly with petroleum jelly or some other suitable lubricant.
2. Check the microphone cable in sections of about 10 inches to a foot. Loop and unloop this amount of cable between the hands while slowly twisting it. Listen to the output of the amplifier and continue to do each small section this way for at least a quarter of a minute. Broken insulation or wires will definitely show up in this test.

   If a break is found, it is far better to replace the entire cable than to remove and splice the faulty section of cable. Of course, splicing may be done in emergencies when new cable cannot be obtained.

   Some microphones have a switch for turning them off and on to allow greater flexibility in their applications. If this switch is a sealed type with nonaccessible contacts, it can simply be checked for proper working order and, if suspected to be faulty, replaced. If the contacts are accessible, inspect the terminal connections for tightness and cleanliness, and check the mounting for firmness.

   While operating the switch, observe all moving parts for freedom of movement, and look closely at the stationary spring contacts to ascertain their tension and if there is good or doubtful electrical contact. Contacts that have lost tension may be tightened with the fingers or pliers. Tighten all terminals. Any section of the switch that is dusty, corroded, or pitted should be cleaned with a dry cloth. For more serious conditions, the cloth can be moistened with cleaning fluid and the affected parts rubbed vigorously.

When the points of contact with the moving blade show signs of excessive wear, replace the entire switch. Crocus cloth dipped in cleaning fluid may be used to clean them. For severe corrosion, No. 0000 or No. 000 sandpaper should be used and the contacts polished clean.

If dryness and binding are noticed, apply a drop of instrument oil with a toothpick at the point of friction. *Do not allow the oil to flow into the electrical contacts.*

These steps are the preliminaries to checking a faulty microphone. A great majority of the common faults are found in the receptacles, plugs, or cables. If these items are not at fault, it is necessary to proceed according to the type of microphone used.

Ribbon and Combination Microphones

When cables and plugs have been definitely eliminated as sources of trouble, the transformer and terminal-block connections should be checked. Do *not* check transformer continuity with a battery-powered continuity checker without first removing the ribbon connections. Better yet, place a resistance of 47,000 ohms minimum in series with the checker test leads. Otherwise, permanent damage to the ribbon may result. Transformers may be replaced, but if the ribbon or ribbon assembly is damaged, the microphone must be returned to the manufacturer for factory repair. The only replacements normally made in the field are replacements of the cover, transformer, mounting parts, cables, and plugs.

Also remember that the microphone lines must not be checked with an ohmmeter without first disconnecting the ribbon microphone. The line then may be checked (unterminated) for high-resistance shorts or (terminated) for opens or high-resistance connections.

Hum and noise may occur in any part of the audio circuit. In the microphone circuit, it can result from ground loops or imbalance caused by faulty or improper cable connections to the bus or preamplifier board. Magnetic fields from power transformers or electrical machinery may induce hum into the microphone transformer or ribbon. This may sometimes be minimized by turning, tilting, or relocating the microphone relative to the magnetic field.

Another source of hum and noise is ground current between the microphone cable and the preamplifier. Fig. 6-50 (Chapter 6) shows the recommended practice for wiring. Fig. 13-2 shows the color coding and terminal-block connections for proper impedance matching of a typical microphone.

Microphone Phasing

It is well known that correct phasing may be important to the operation of any system employing more than one microphone simultaneously. This is especially true when two similar microphones are placed in a symmetrical relationship to a performer. In a lesser measure, correct phasing may be important for optimum operation of certain amplitude-modulated transmitters because of predominantly unsymmetrical aspects of speech waveforms.

Polarity of a microphone or a microphone transducer element refers to in-phase or out-of-phase conditions of voltage developed at its terminals with respect to the pressures of the sound wave causing the voltage. An exact in-phase relationship can be taken to mean that the phase of the voltage is coincident with the phase of the sound-pressure wave causing the voltage. In practical microphones, this perfect relationship may not always be attainable.

The in-phase terminal of a microphone is that terminal of the connector or conductor that is connected to the in-phase terminal of the transducer. On microphones using a connector as specified in EIA Standard TR-118, the in-phase terminal is No. 1, the out-of-phase terminal is No. 2, and the ground terminal is G. On a microphone with a cable but no connector, the out-of-phase terminal is black.

The polarity of a pressure (or omnidirectional) microphone does not vary with the direction of arrival of the sound wave. The polarity of a gradient microphone is reversed for sound waves directed toward the rear of the microphone. There may be a substantial phase shift in the microphone at the low- and high-frequency ends of the spectrum. Therefore, the definition of polarity is generally restricted to the midpoint of the useful transmission band.

When the outputs of two or more microphones are connected to a mixing circuit, it is necessary that the outputs of all the microphones have the

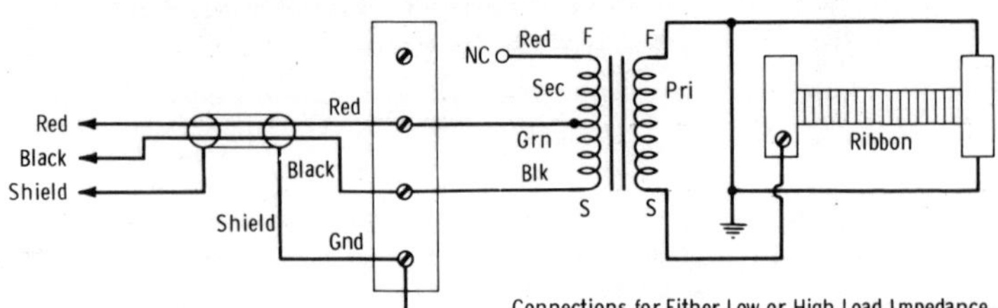

Connections for Either Low or High Load Impedance

Transformer Leads to Terminal Board	Output Impedance
Black and Green	200Ω
Black and Red	15,000Ω

Fig. 13-2. Schematic diagram of a typical microphone.

same phase relation. Otherwise, the output of one microphone will oppose the output of another, resulting in a reduction in output and the introduction of varying degrees of distortion.

To check the phasing of two or more microphones, connect one microphone to the associated amplifier input, and set the volume control to obtain the desired output when talking into the microphone. Then connect the second microphone in parallel with the first, and, without changing the volume-control setting, hold both microphones close together and speak into them. If the volume decreases from the previous level, reverse the connections of one of the microphone cables at the microphone plug. Similarly check each additional microphone for phasing. If it is necessary, reverse the cable connections to correct the phasing to match that of the microphone already connected.

In practice, polarity turnover may occur between microphone channels due to installation of different types of amplifiers. Turnover also may be due to the installation of the same type of amplifier-pad combinations when no attention was given to color-coded wiring with identical connections. This is best checked by employing a signal generator and oscilloscope. See Section 13-8.

For those interested in the exact standardization of microphone polarity, set up the circuit of Fig. 13-3A. The EIA suggests the following procedure:

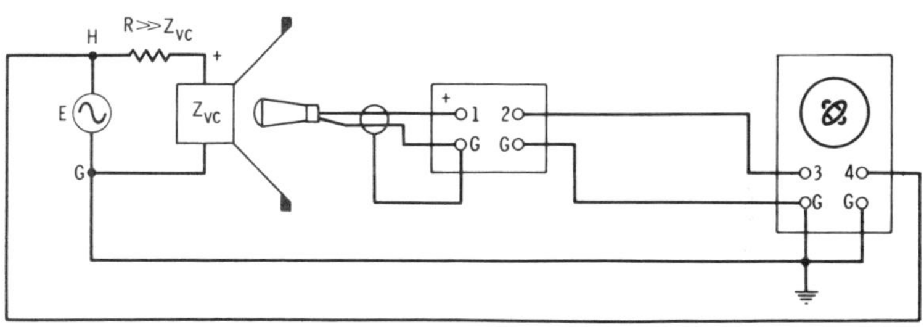

(A) For pressure microphones.

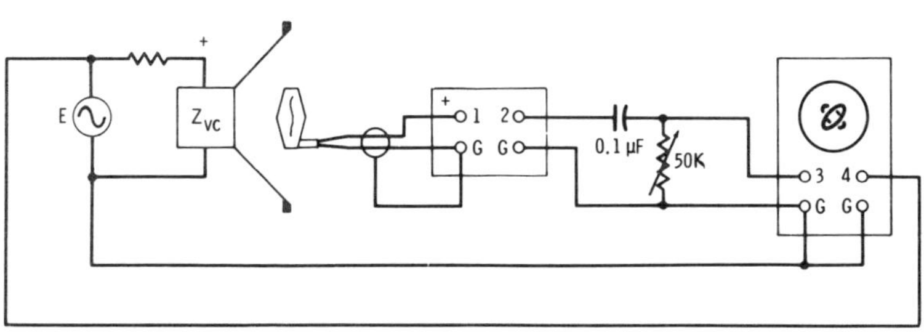

(B) For gradient microphones.

Fig. 13-3. Method for checking microphone phasing.

1. Check the proper phasing of the oscilloscope and amplifier. To do this, connect one terminal of the oscillator to ground; connect the other (H) terminal to terminal 1 of the amplifier and terminal 4 of the oscilloscope. The trace on the oscilloscope should be a line slanting from lower left to upper right.
2. Determine the in-phase terminal of the speaker. This is done by connecting a battery across the voice coil so that the cone moves toward the microphone; the terminal connected to the positive terminal of the battery is the in-phase terminal of the speaker. Connect the resistor (R) between the H terminal of the oscillator and the in-phase terminal of the speaker. The value of R should be at least 5 times the impedance of the voice coil.
3. Adjust the oscillator output for a suitable acoustic output from the speaker. Connect the microphone to the amplifier, and position the diaphragm of the microphone as close as possible to the surface of the vibrating cone. Check the orientation of the oscilloscope trace. If the trace is a slanted line or ellipse with its major axis oriented from lower left to upper right, then the in-phase terminal of the microphone is the terminal connected to the "high" terminal of the amplifier (terminal 1). The relationship should be constant throughout a range of frequencies (100 to 400 Hz).

When the outlined procedure is applied to a gradient (velocity) microphone, the trace is in the form of a circle. This is because of the out-of-phase relationship between pressure and velocity in a spreading wave. To remedy this, a phase-shift network consisting of, say, a 50,000-ohm resistor and a 0.1-μF capacitor may be connected as shown in Fig. 13-3B. Except for this one detail, the measurement can be performed exactly the same way as described before.

13-2. TURNTABLES

Preventive maintenance on turntables consists largely of cleaning, lubrication, and occasional adjustments in the speed-changing mechanisms to prevent chattering or binding. The mechanical details of the drive mechanism, along with detailed servicing procedures, are normally contained in the instruction manual for the particular turntable used. Always include the lubrication schedule or chart in the regular preventive-maintenance schedule.

The turntable alone (not considering the pickup and arm assembly with associated preamplifier) has three basic characteristics which concern the maintenance department: wow and flutter, rumble, and tolerance of operating speed.

The specifications of a typical broadcast-type turntable might be given as follows:

Wow and flutter: 33⅓ rpm, 0.25 percent half of peak-to-peak
45 rpm, 0.20 percent half of peak-to-peak
78 rpm, 0.20 percent half of peak-to-peak

Rumble: (All speeds) 35 dB below reference of 100 Hz test record with a 1.4 cm/s stylus velocity

Speed: 33⅓ rpm ±0.2 percent
45 rpm ±0.2 percent
78.26 rpm ±0.2 percent

Wow is a low-frequency modulation effect caused by variations in groove velocity. It may occur in recording, playback, or both. The NAB standards recommend that the maximum instantaneous deviation of the mean speed should not exceed ±0.1 percent for recording turntables or ±0.2 percent for playback turntables.

Flutter is similar to wow but is at a higher frequency (up to 200 Hz). This also may occur in both recording and playback. The ear is very sensitive to this type of distortion.

Rumble is a low-frequency, steady-state tone or series of random pulses generated at the pickup stylus. It is caused by vibrations of the turntable.

The variation in the groove of a lateral recording is termed *modulation.* The movement to one side of the mean path at a given time is termed the *amplitude of modulation.* When the recording is that of a sine wave, the maximum transverse velocity of the stylus tip occurs when the tip passes through the mean path, with zero velocity at the extremities of travel. The maximum transverse velocity can be determined by the formula:

$$V_T = 2\pi f A$$

where,

V_T is the maximum transverse velocity in centimeters per second,
f is the frequency in hertz,
A is the peak amplitude in centimeters.

The rms transverse velocity is then:

$$\text{rms transverse velocity} = 1.41\,\pi f A$$

The recorded level of a sine-wave test record is specified in terms of the rms velocity at 1 kHz, or in decibels with a reference level of 1 cm/s rms stylus velocity. Thus if a given test record specifies that a 1000-Hz reference level is 10 dB above 1 cm/s rms velocity, the resultant rms velocity is 3.16 cm/s. (Use the voltage-dB table for conversion.) Plus 16 dB becomes 6.31 cm/s, etc.

Whenever the manufacturer specifies rumble, he normally gives the conditions of measurement. For example, RCA specifies the rumble of the BQ-2B turntable as 35 dB below the reference of a 100-Hz test record with

1.4-cm/s stylus velocity, using an NBC Type ND-301 scratch and rumble meter (modified per NAB standards), with building rumble at −50 dB.

Turntable rumble can be checked with the station noise and distortion meter on an arbitrary basis as follows:

1. With the stylus in the 100-Hz test-record groove, set the reference level to the output of the turntable preamplifier. Use a low-pass filter which has a sharp cutoff above 300 Hz between the preamplifier output and the noise meter.
2. Place the stylus in the 1000-Hz (or higher) groove, and take the noise measurement on the noise meter. Use a groove as near the outer edge as possible, since rumble is most predominant with the stylus near the outer rim of the turntable.

NOTE: If the characteristics of the preamplifier are unknown, run a noise check on the preamplifier only, keeping the low-pass filter in place. This should be at least 58 to 60 dB below normal output. This measurement is made in the conventional manner as follows:

A. Feed sufficient 100-Hz tone input from the external test oscillator to obtain the rated output level of the preamplifier.
B. Remove the tone. Place the proper terminating resistor on the input and read the noise level.

When a turntable begins to develop rumble, the most common cause is lack of proper cleaning and lubrication of the drive mechanism. Adhere scrupulously to the manufacturer's instructions for cleaning and lubricating. As a general rule, it is *not* advisable to remove dust or dirt from the drive mechanism (or turntable platter) by air pressure. Use slightly oily, lint-free cloth to wipe the bushings and thrust balls before lubricating. The hub and spindle should also be treated in this manner.

Such items as rubber idlers and motor pulleys should be wiped with the same type of cloth dampened with naphtha or cleaning fluid. Wipe the inside surfaces of the platter rim (when rim driven) the same way. Be sure there is no oil on the motor pulley or rubber idlers. Do not use excessive amounts of cleaning fluid on rubber idlers since it may attack the rubber.

Check any shock mounts on the turntable motor or other parts. These may need to be replaced. Check all tensions specified in the instruction manual. Always keep the proper scales on hand for these measurements.

Wow and flutter are caused by small imperfections in the motor and/or the drive mechanism. They will be evident as a cyclic variation in the pitch when a steady tone such as the 1000-Hz band of a test record is reproduced. The term "wow" is applied to the very slow cyclic variation, the term "flutter" to the more rapid variation, up to around 200 Hz.

The actual measurement of wow and flutter requires special equipment. Such measurements are not normally made by station personnel. When the defect is noticeable, use the same maintenance procedures that are outlined above for rumble. An eccentric disc or turntable will also tend to cause wow. When all maintenance procedures have been tried and the condition persists, it is very likely that the motor needs to be replaced.

The turntable speed is checked by means of a stroboscope disc illuminated by a lamp supplied from the ac line. A neon bulb is best for this application. The stroboscope disc is simply placed on the turntable platter and the neon bulb held directly above it. There are 92 bars for 78 rpm and 216 bars for 33⅓ rpm. Not more than 21 dots per minute should drift past the visual reference point in either direction. If the dots drift in the direction of rotation, the speed is high. If the drift is opposite to the rotation, the speed is low. Most turntables employ a means of vernier control of the platter speed.

13-3. THE STYLUS, PICKUP HEAD, AND PREAMPLIFIER

Basic theory was covered in Chapter 4. Also review Figs. 6-30 and 6-31 with associated text for a visualization of constant-amplitude and constant-velocity recording methods.

The Pickup Unit

Pickup arms for broadcast use have an adjustable feature allowing control over the stylus pressure against the record groove. It is important that this amount of pressure be correctly adjusted for optimum system performance and minimum record and stylus wear. Several good makes of special-type scales for this purpose are on the market and should be a part of the equipment of every maintenance department. The pressure must be measured with the stylus in playing position. Thus when the scales are used, they must be placed off the turntable and adjusted so that the stylus is at playing height when the pressure is measured.

When the trouble has been traced to the pickup unit, the first step is to check the stylus. Excessive record noise and signal distortion are often caused by a defective stylus tip. Check the tip under a strong light and magnifying glass; chips or excessive wear may be readily detected in this manner. Of course, the easiest and best check is to replace the stylus with a new one if it is immediately available. In cases of permanent-type stylus assemblies (usually using diamond tips), the head must be removed and returned to the factory for stylus replacement if the tip has been damaged by dropping or other accidents.

Be certain that the stylus is properly secured in the holder. In the case of a bent-shank stylus, the bend must be properly aligned with the record grooves, not turned even slightly in either direction from the center of the groove.

Check all of the spaces around the stylus entrance for dust or lint that often clogs the free spaces around the stylus holder or the pole pieces in magnetic pickups. These spaces must be cleaned thoroughly. A *low-pressure* air stream aids considerably in this process.

Pickup heads must be examined closely for breaks in connections, wire, or shielding. Plugs and receptacles must be examined as previously suggested for microphone maintenance. When terminal boards or connecting panels are used, they should be carefully inspected for cracks, breaks, dirt, and loose connections or mountings. Each connection should be examined. Tighten all *clean* terminals, screws, lugs, and mounting bolts, being careful not to overtighten since this can cause cracks or breakage. Any connection that is dirty, rusty, or corroded should be disconnected. Clean each part individually and thoroughly with a clean cloth or crocus cloth moistened with cleaning fluid. Then replace and tighten the connection.

In the magnetic-type pickup, extreme care must be exercised in removing the cover to prevent damage to the delicate stylus and/or armature assembly. Check the centering of the armature between the pole pieces. Always check the air gaps for collections of dust and lint. In many instances, transformers are located inside the pickup arm. All leads and connections must be carefully checked for continuity and tightness. In some of the older-type magnetic pickups, the bushings and supports that center the armature become worn and deteriorated. Replacements for these parts must be ordered by model and part number from the manufacturer. It is always advisable and often cheaper in the long run to return the pickup head to the factory for repair. This is imperative in the case of the very high-quality magnetic pickups used in broadcasting and similar applications where a permanent-type precious-metal stylus tip is employed.

Checking Frequency Response

The broadcast engineer finds it interesting and helpful to know the actual frequency response of a given pickup unit. This information is necessary if complete individual equalization circuits are to be checked not only to meet the requirements of compensating a certain recording characteristic but also to compensate for the deficiencies of the pickup unit. Equalization, for example, would be unnecessary (indeed harmful) if the pickup unit has an inherent high-frequency rolloff.

Bear in mind that the *overall* combination of the stylus, pickup head, arm, and preamplifier with equalizer must provide the complementary reproduction characteristic given in Table 6-6 (Chapter 6). If the equalizer is external to the preamplifier, the preamplifier response curve is normally flat. However, most modern phono preamplifiers incorporate the equalizer as part of the preamplifier itself, and this must be considered by the maintenance technician running response tests.

Obviously, in order to check the frequency response of a reproducer unit, a frequency run must first be made on the amplifier. This is done in the orthodox manner using a variable-frequency audio oscillator (30 to 15,000 Hz) and a volume indicator on both the output of the oscillator and the output of the amplifier under test. Of course, a single volume indicator could be used with a suitable switching arrangement. The single tone is fed into the amplifier at the same level for each frequency, and the volume-indicator output-meter reading is noted. The reference frequency is usually 1000 Hz; that is, the gain is adjusted on the amplifier so that output meter reads 0 dB at 1000 Hz. Deviations at the other frequencies in the range to be tested are then observed and plotted as a graph.

For checking pickup units, major recording firms put out frequency test records that usually start with a 1000-Hz reference tone to set the level. They then go to 10,000 Hz and work down to 20 or 30 Hz, with a voice identification immediately before each tone. The output volume indicator on the amplifier is read for each frequency, the output level having been adjusted to zero at 1000 Hz as in the oscillator test. The curve thus obtained may then be plotted against the amplifier curve to obtain the pickup response curve.

If the amplifier response curve were perfectly flat, the pickup-curve run would be the actual response of the unit. Assume, however, that the amplifier curve is +4 dB at 2000 Hz and the pickup curve is +1 dB. The pickup response, therefore, is actually −3 dB at this frequency, since the 4-dB gain of the amplifier at this point on the curve must be subtracted from the pickup curve.

Test records made to EIA standards have two sides. Side A is intended for frequency-response measurements at normal levels, and side B tests the tracking ability of the pickup at different levels. Intermodulation test records are also available for checking distortion in terms of intermodulation—which is more meaningful than simple harmonic-distortion tests. The entire subject of intermodulation tests and measurements for studio equipment (including playback pickup heads and preamplifiers) is covered in Section 13-13.

Stereo Pickups

The dual-channel stereo system must be maintained tightly in the following characteristics: closely matched frequency response, closely matched phase-frequency response, and maximum separation (minimum cross talk).

The basic theory of operation of the variable-reluctance stereo pickup cartridge was covered in Chapter 4. Fig. 13-4 illustrates the General Electric Type VR-1000 stereo cartridge showing the stylus replacement feature. The broadcast-type VR-1000-5 employs a 0.5-mil diamond stylus and a tracking force of 1 to 3 grams.

Fig. 13-5 shows the frequency response and channel separation of the VR-1000-5 cartridge. The solid line represents the left channel, and the dashed line is a plot for the right channel. On the lower portion of the graph, the dashed line represents the signal induced in the right channel by modulation in the left channel, and the solid line represents the signal induced in the left channel by modulation in the right channel.

When a stereo test record is used, the left channel (solid line) is given a reference level of 0 dB (at 1000 Hz). With left-channel modulation

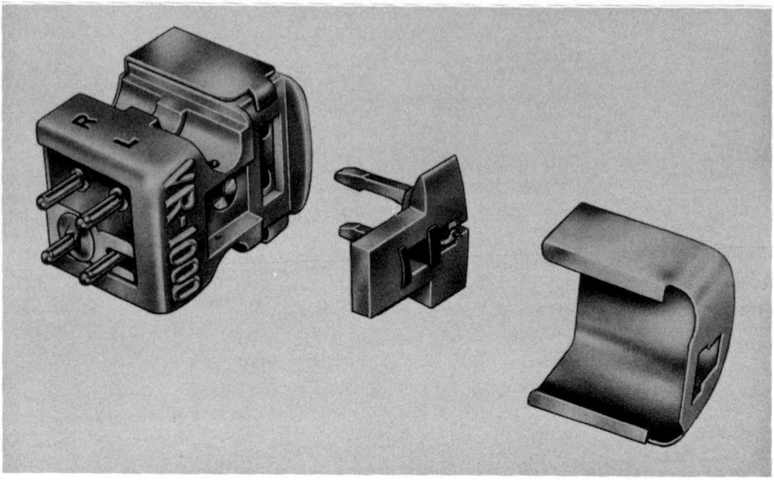

Courtesy General Electric Co.

Fig. 13-4. A variable-reluctance stereo cartridge.

only, the signal of the right channel is measured and plotted. In Fig. 13-5, the separation at 1000 Hz is 32 dB. The separation is then measured at frequencies up to 15,000 Hz. Without changing gains, the right-channel frequency response is measured. In Fig. 13-5, it is observed that this particular cartridge has a slightly higher output in the right channel. Although the responses of the two channels should be within a few decibels of each other for close frequency-response match, the actual output amplitude difference is of small significance since the respective channel faders are adjusted in practice for matched-level outputs.

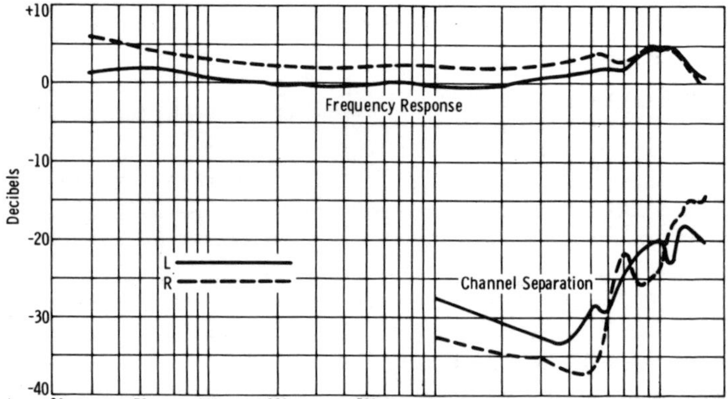

Fig. 13-5. Frequency response and channel separation, GE VR-1000-5 cartridge.

When frequency-response measurements are being made at the outputs of the pickup preamplifiers, it is obvious that the frequency response of the amplifier alone must first be plotted. This is then used for any correction factor necessary at a particular frequency. When channel separation is measured the noise level of each amplifier must be well below 60 to 70 dB down from the reference output level. This is to avoid measuring noise on the meter rather than actual signal.

Channel separation in a stereo pickup deteriorates at both the low- and high-frequency ends of the audio spectrum, as shown by Fig. 13-6. These curves, which are for a later version of GE stereo pickup (the VR-22 series), show the separation (lower curve) from 30 Hz to 20 kHz, in addition to the overall frequency response (upper curve). For completeness, the specifications of this pickup are listed in Chart 13-1.

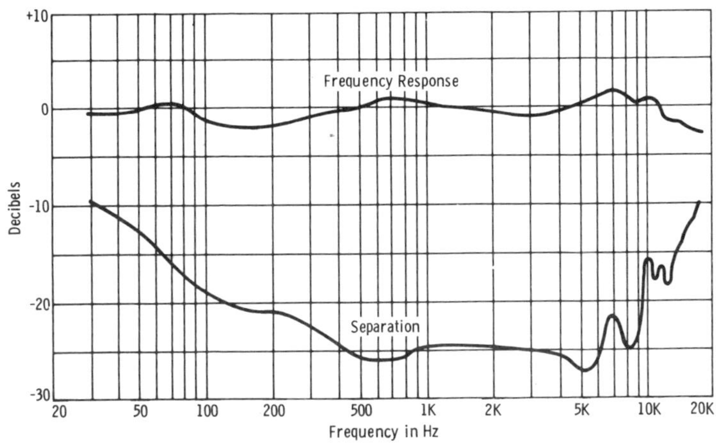

Fig. 13-6. Typical frequency response and channel separation of GE VR-22 series cartridge.

The RCA universal pickup cartridge has been designed especially for broadcast applications. This cartridge performs in both stereophonic and monophonic applications; the mode of operation is determined by external electrical connections to the cartridge. A replacement stylus is a feature of the cartridge design; the stylus plugs directly into the cartridge and is easily replaced. This plug-in feature also allows a user to buy one type of cartridge and several sizes of inexpensive styli in order to take care of all record-playing requirements. Fig. 13-7 illustrates the intermodulation distortion of the cartridge in both stereo and lateral applications.

Fig. 13-7 shows intermodulation distortion, which is a measure of the nonlinear distortion occurring at the lower test frequency. The RCA 12-5-39 test record on which the graph is based contains 400 and 4000 Hz mixed 4 to 1. Harmonic-distortion methods at 400 Hz could have been applied, but they have been found to be much less practical than the intermodulation method. Ten percent intermodulation distortion is not noticeable without a chance for direct comparison. This is comparable to a figure of 2 or 3 percent harmonic distortion. Intermodulation-distortion techniques are covered in Section 13-13.

In stereo broadcasting, it is of prime importance to be extremely careful in maintaining the proper polarities of the left and right channels. Particular attention must be given in the initial installation and, whenever new facilities are added, to the correct interconnections to maintain the proper phase of the two channels over the audio-frequency spectrum of 30 to 15,000 Hz.

Chart 13-1. Specifications of GE VR-22 Series Cartridge

(Specifications are similar for the 0.5-mil diamond stylus Model VR-225 and the 0.7-mil diamond stylus Model VR-227, except as noted.)

Frequency Response: 20 through 20,000 Hz, ±3 dB (VR-225); 20 through 17,000 Hz (VR-227)
Output: 5 millivolts, minimum, at stylus velocity of 3.8 cm/s
Separation Between Channels: Up to 25 dB nominal
Channel Balance, at 1000 Hz: One dB or better
Inductance: 307 millihenries, nominal
Resistance. 1386 ohms per channel, nominal
Load: 47K-100K, 47K optimum
Lateral Compliance: 4×10^{-6} cm/dyne (VR-225); 2.5×10^{-6} cm/dyne (VR-227), nominal
Vertical Compliance: 3×10^{-6} cm/dyne (VR-225); 2×10^{-6} cmdyne (VR-227), nominal
Tracking Force: Two grams minimum under optimum conditions—three grams for average use—four grams maximum (VR-225). Five to seven grams with VR-227
Dimensions: Width, 0.575 inch; height, 0.7 inch including stylus tip; length, 1.18 inches
Mounting: For 7/16-inch or 1/2-inch mounting centers
All measurements taken from RCA Victor stereophonic test record Number 12-5-71 and 12-5-73

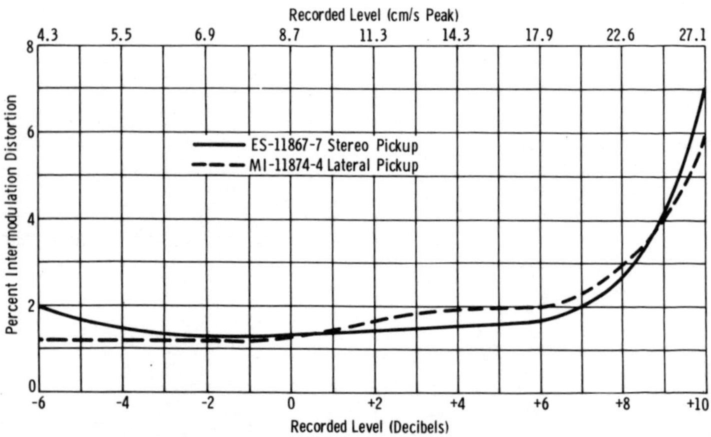

Fig. 13-7. Percent intermodulation distortion for one type of stereo cartridge.

The most prevalent form of trouble in pickups located near an fm transmitter (as when the transmitter is in the same building as the operating center) is rectification of the strong rf field at the phono preamplifier input. This rf interference can be eliminated by observing the precautions shown on Fig. 13-8. Many preamplifiers have built-in rf filtering networks, but some do not. Even when these are incorporated, the length of the shielded pickup cable is most usually left to the individual installation requirements. Be sure to cut this cable to an odd multiple (or submultiple) of a quarter wavelength at the operating frequency of the fm transmitter. For example, a quarter-wavelength at 100 MHz is 0.75 meter, or 29.5 inches. Be certain that the shield of the pickup cable is connected only at the preamplifier ground terminal, and that the preamplifier ground is properly connected to the station ground.

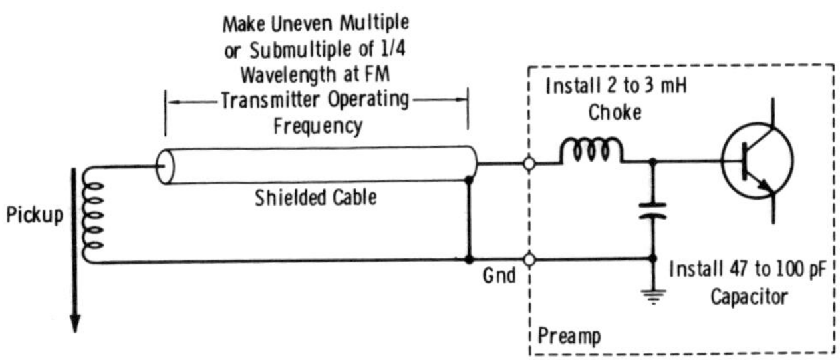

Fig. 13-8. Precautions to minimize fm-to-am rectification in vicinity of strong rf field from transmitter.

13-4. JACKS

Jacks, since they constitute either series or parallel connections in the signal path, must be kept free of dust and dirt and in perfect contact adjustment. They should be vacuum cleaned frequently, or preferably, a thin high-pressure stream of forced air should be blown through each sleeve from front and rear and be followed by a thorough vacuuming. Jack contacts may be kept clean by regular insertions and removals of patch-cord plugs (Fig. 13-9). Visual inspection of jacks and mounting structures should be made during this process to check for the proper operation and tightness of mounting. If jacks have not been maintained in this manner for a long period of time, a regular jack-burnishing tool may be used. When the burnishing tool is used, the handle is adjusted in such a way that about 2⅛ inches protrudes. This prevents the blade from touching the insulation in the spring pileup and prevents damage to the insulation. Insert the blade carefully in the jack until it just meets the normal contact. A slight turning movement of the tool will then tend to lift the normal spring from the normal contact, and a light forward pressure will allow the blade to slip easily between the contacts.

13-5. KEYS AND SWITCHES

Keys and switches vary considerably in their structure and functioning, but all types have a means of opening and closing single or multiple circuits by spring contacts and blade assemblies. Dirt or dust causes the majority of troubles here. Contacts usually may be cleaned with a clean, dry, lint-free cloth. A *clean* strip of silk is useful since it collects dust particles from the contacts by static electricity when it is drawn through the contacts while a light pressure is held on the blades or spring sheaves by the fingers. In the more serious situations, a toothpick immersed in carbon tetrachloride may be drawn through the contacts and then discarded to prevent its being reused.

Fig. 13-9. Method of "exercising" a jack field.

The mountings of keys or switches should be checked for tightness. Tighten every loose connection. Watch the mechanism as the key or switch is operated, and note the tension of the spring or stationary contacts. If the tension appears to be insufficient, adjust it with the fingers or long-nose pliers, or use a special switch and relay tool with a slotted end. Be very careful not to make it too tight.

13-6. FADERS (ATTENUATORS)

A regular cleaning schedule is necessary for continued, reliable, and noise-free operation of fader controls. Contacts should be cleaned with a clean, soft cloth and carbon tetrachloride. Also clean the leaves in this manner; be careful not to disturb the tension of the individual leaves. To help prevent wear, some engineers recommend a light lubrication of the contacts by applying a very slight amount of unmedicated petroleum jelly on a clean cloth.

Never disturb the tension of the leaves against the contacts unless absolutely necessary. If the attenuator must be dismantled, the leaves should be adjusted by means of the sliding mounting screw until there is just enough pressure to allow a reliable contact. Attenuators that turn hard against the fingers have excessive pressure between leaves and contacts, and their life is shortened.

13-7. AMPLIFIERS AND LOGIC CIRCUITRY

This text is not a detailed treatment of servicing techniques, and a fundamental radio background is assumed from the beginning. However, for the purpose of completeness in studio maintenance, helpful hints are given here for locating troubles in audio amplifiers. We will consider the tube-type amplifier first.

Filaments Do Not Light

When all filaments fail, it may generally be assumed either that the ac power line has failed or a fuse in the amplifier rack or amplifier has blown. First check all ac switches for the proper operating position. Then check the position of any circuit breakers involved. After that, check the fuses. Faulty switches may nearly always be detected by feel when one operates them. If a snap is not apparent, jumper the contacts until the switch can be replaced.

If only one tube fails to light, it is usually a bad tube. First wiggle the tube with a strong pressure against the socket. If the filament lights, a loose or rosin connection on the socket terminals may exist, or the contact springs may not have enough tension to maintain reliable contact with the tube prongs. See Chapter 14 for a detailed discussion of vacuum tubes and socket maintenance.

Filaments Light, No Response

First replace the rectifier tube with one known to be good. If trouble persists, check the plate voltages starting with the output of the filter system. Then check the output stage and so on back to the first stages. If all voltages are normal, check the continuity of the output transformer. If this test shows nothing wrong, the best procedure (assuming normal plate voltage) is to use a signal generator and headphones (with a series 0.1-μF capacitor) to trace the signal to the faulty stage. If no plate voltage exists or the voltage is very low, check the condition of the electrolytic capacitors and the continuity of the filter circuits. Check the wiring for continuity.

Check the condition of the power-supply switch and fuse holders. If voltage does not exist or is low in only one stage (in which case a small distorted signal is usually present), check the continuity and values of all resistors in the plate and cathode circuits, and the wiring to the component parts. In some cases, low plate voltages may be due to excessive plate currents which in turn may be caused by low or nonexistent grid bias or by a leaky coupling capacitor that places a positive voltage on the grid of the following stage. These cases require point-to-point analysis using any of the recognized methods of servicing, such as signal tracing, point-to-point resistance measurements, etc.

Amplifier Noise

Hum and noise are common troubles. First, try to locate the approximate origin of the noise by using the normal operating controls. For example, is the noise heard with all switches and attenuators off? If so, the noise could be in the monitor amplifier or high-level amplifier circuits. The possible sources of noise will, of course, vary depending on the layout existing at a particular installation.

It is always best to replace tubes first in case of noise. If this procedure does not clear up the trouble, look for loose connections or an apparent rosin connection. Move the wiring back and forth with the monitor-amplifier gain turned up. Wiggle the tubes in their sockets. If a faulty spring contact exists, see Chapter 14 for socket maintenance. Tap all coupling capacitors and resistors. If a distinct electrical noise (as distinguished from acoustic noise) exists, replace the part with an exact replacement. Examine all ground connections closely for looseness or bad soldering.

Steady noise (not intermittent) may be caused by some form of inductive interference. If input transformers are constructed in such a way that their windings may be oriented in any direction, try a reorientation for minimum noise pickup from any possible ac field. Some microphones are unusually sensitive to stray ac fields. If noise is coming from any particular microphone circuit, check the microphone and cable locations for interfering fields. This is particularly important at remote locations.

Servicing Printed Circuits

Care must be taken not to break or crack the board by undue stress or to damage the bonding adhesive by applying too much heat during soldering. The following tools are required: a small (25-watt) soldering iron, pair of small diagonal cutters, pair of small long-nose pliers, scribe or pick, and small knife.

A. If it is necessary, remove the hardware fastening the board to the chassis and remove the board, or tilt it up if it is hinged.
B. Isolate the defective component. If it is necessary to disconnect a component from the circuit for test, heat the junction of the compo-

nent lead and the printed wiring with the soldering iron. Concentrate the heat on the component lead rather than on the etched wiring pattern. Pry up and straighten the bent-over portion of the component lead with a knife blade, and then pull the lead through the hole with pliers.

C. To remove the defective component, snip off the leads at the component side of the board (Fig. 13-10A).

D. Using a small soldering iron (25 watts), heat the leads and remove them from the printed-wiring side of the board. To avoid damage to the conductors, be careful not to apply too much heat or force.

E. Clean and form the leads of the new component and insert them through the holes until the component body is tight against the board.

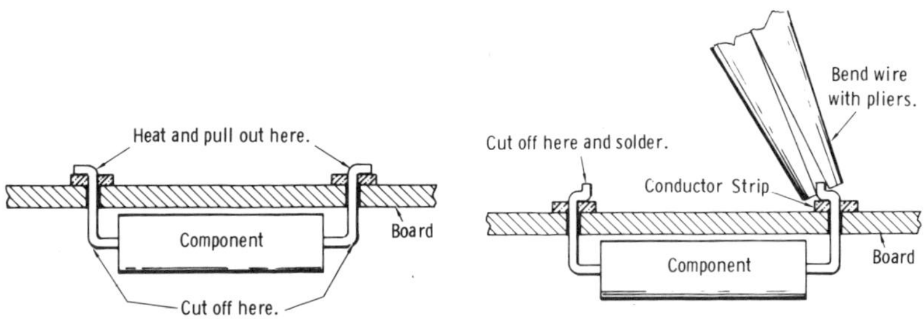

(*A*) *Removal of defective component.* (*B*) *Installation of new component.*

Fig. 13-10. Replacement of component in printed circuit.

F. On the circuit side, grasp the component lead and bend it over in the direction of the circuit pattern.

G. Crimp the wire tightly against the board (Fig. 13-10B), and cut off the excess component lead. Leave about $\frac{1}{16}$ inch of wire protruding from the edge of the hole.

H. Heat the lead and apply rosin-core solder. *Do not use paste or acid flux.* Remove the excess rosin from the joints with alcohol.

I. Replace the circuit board, using the original hardware.

Replacement of Transistor Socket

Heat each socket terminal and pry up and straighten it with a knife blade. Pull the socket out, applying heat to the terminal leads, if it is necessary. Clean the holes free of solder. Prepare the new socket for installation as follows. Using the old socket as a model, bend the terminal leads to fit the mounting holes provided in the board. Insert the socket terminals through the holes, making sure that the socket terminals are oriented as on the previous socket. Bend the socket terminals in the direction of the circuit pattern. If it is necessary, clip off the excess length to prevent short circuits with adjacent conductors. Solder the terminals to the etched wiring.

When it is necessary to remove a transistor that is not mounted in a socket, grasp the transistor lead with a pair of long-nose pliers while unsoldering. This provides a heat sink to avoid damage to the transistor from excessive heat. Use a 25-watt iron.

Logic Circuitry and Integrated Circuits

Most logic circuitry today is made up in integrated circuits (ICs). Therefore, it is normally necessary to know how to check inputs and outputs with an oscilloscope to isolate troubles in logic circuitry where the entire chip would need replacing. Let us go through the timing diagram of Fig. 13-11 to show how it is done.

> IMPORTANT NOTE: Pulses are rarely as perfectly shaped as indicated on such drawings as Fig. 13-11. In practice, there will usually be some rounding, and this may be normal. It is always desirable to scope such circuitry when it is working perfectly to ascertain the normal shape of pulses in any particular equipment.

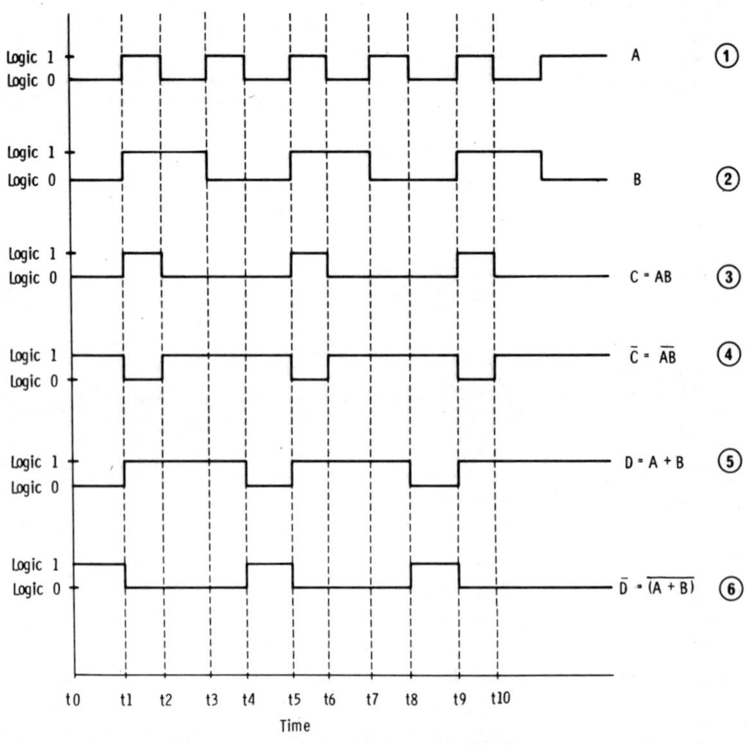

Fig. 13-11. Timing diagram.

In the timing diagram chosen, variable A (row 1) is a stream of data consisting of alternate 1's and 0's. Variable B (row 2) is composed of alternate *groups* of *two* 1's and *two* 0's. The remaining rows indicate the following:

Row 3: A and B combined in an AND gate. (C = AB reads "C equals A *and* B.")
Row 4: The complement of row 3 (NOT or NAND circuit).
Row 5: A and B combined in an OR gate. (D = A + B reads "D equals A *or* B.")
Row 6: The complement of row 5 (NOT or NOR circuit).

So here we have every combination fundamental to logic circuitry. If a technician understands the symbols for bistables and gate circuitry, he should know what to expect at the output of each chip for a given input or inputs. For example, he can quickly check inputs A and B to the AND gate. He must then determine the normal output as shown by row 3. For a NAND circuit, the output should be the inverted function of row 4.

As a matter of review, bear in mind that the bistable is a natural divide-by-two circuit. If the first pulse results in logic 1, a second pulse is required to result in logic 0. So to complete one output pulse, two input pulses are required. For example, if variable A (row 1 in Fig. 13-11) is the input to a bistable, variable B (row 2) would be the output. This is to say that the frequency of B is half that of A.

Digital logic circuitry becomes easy to troubleshoot with a little experience and familiarity with a particular system. Either an oscilloscope or any of the commercial IC pulsers and indicating lamps can be used. The latter permits using one probe as a pulser and another probe with an indicating lamp to show whether an output is obtained, either by half-brilliance, full brilliance, or flashing lamp for a train of pulses. Lack of lamp "turn-on" indicates low output or no output.

The important point to remember is that input conditions to a chip must be correct for the proper output condition to exist. An AND gate with four inputs must have simultaneous high levels (when high-level activation is called for) for an output to exist. One or more of these inputs might be a positive dc rather than a pulse. When a faulty chip is definitely found, the entire chip is replaced.

The three basic IC packages are shown in Fig. 13-12, along with their respective terminal arrangements. Figs. 13-12A and 13-12B are top views. Note that the pin numbers increase counterclockwise around the packages as seen from the top. There are also 16-lead packages which have 8 leads per side; pin 16 is opposite pin 1, and pin 9 is opposite pin 8.

Fig. 13-12C shows the TO-5 type of IC, for which it is conventional to show the bottom view. Note that the highest lead number is adjacent to the index tab. The numbers then run clockwise as viewed from the bottom. This type of package may have as many as 12 terminals.

The maintenance technician must be familiar with proper techniques for replacing ICs and other components on printed circuit boards. This must be done carefully and skillfully to prevent damage. There are special tools for desoldering all IC pins at the same time (Fig. 13-13). These heated elements are placed on the pins of the IC on the wiring side of the printed board, with a special extractor clamped to the IC body to exert a "pull-away" pressure as the solder is melted. Such tools must be used with extreme caution, since IC leads are sometimes folded against the board and must be bent up before extraction can be done without damage to the boards.

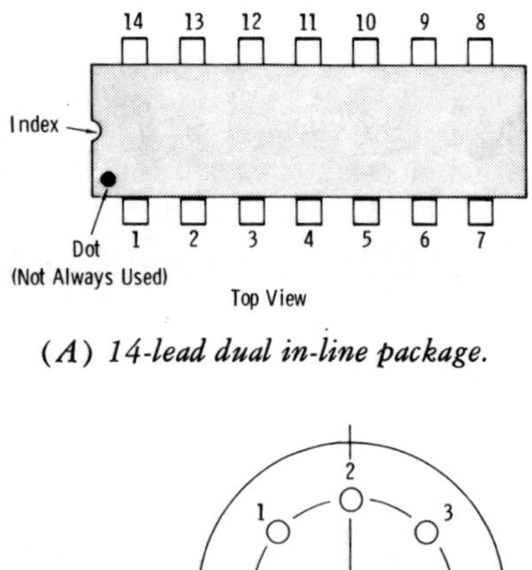

(*A*) *14-lead dual in-line package.*

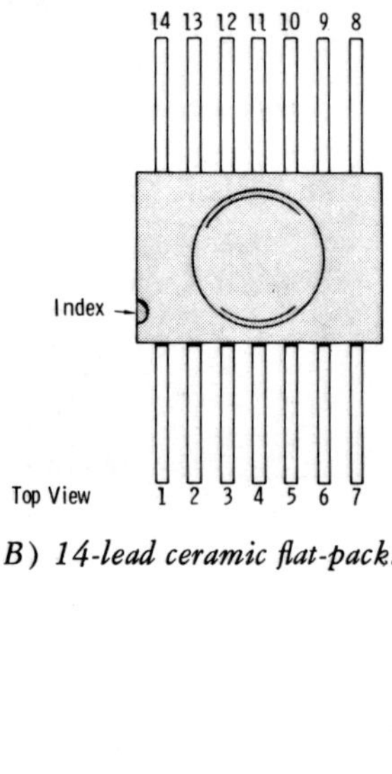

(*B*) *14-lead ceramic flat-pack.*

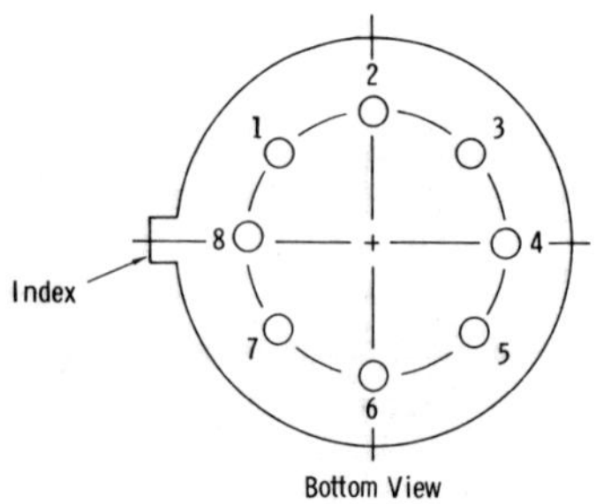

(*C*) *TO-5 style 8-terminal package.*

Fig. 13-12. Basic integrated-circuit packages and terminal arrangements.

The preferred method is shown in Fig. 13-14. The procedure is as follows:

1. With the soldering-iron tip applied, squeeze the desoldering bulb and hold its tip at an angle of approximately 45° against the solder fillet of the lead of the component to be removed.
2. As the solder melts, release the bulb quickly to draw solder away while continuing to hold the soldering-iron tip against the lead. All solder

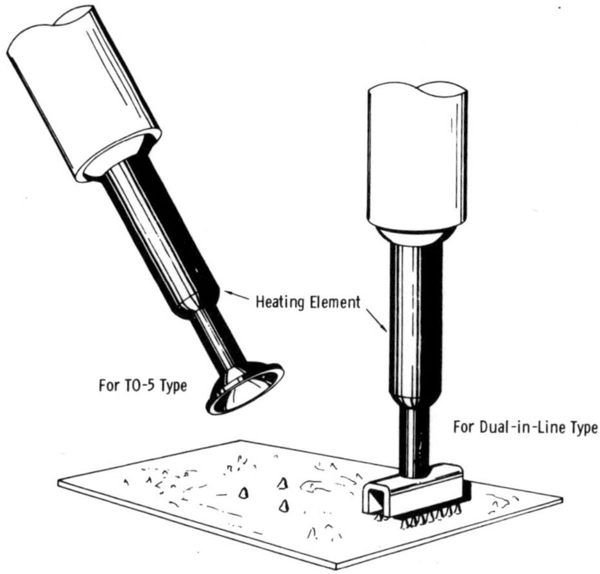

Fig. 13-13. Desoldering tools for integrated circuits.

must be removed from each lead. If the lead is bent against the board, pry it up and repeat the procedure until all solder is removed from the lead.

Caution: When the soldering bulb is released to draw in air (and solder), do not remove the soldering iron from the lead. If the iron is removed during this operation, air will cool the joint enough to prevent clean removal of the solder.

3. Repeat the procedure for each of the remaining leads. Note the position of the index (dot, notch, or tab) if the replaced component is an

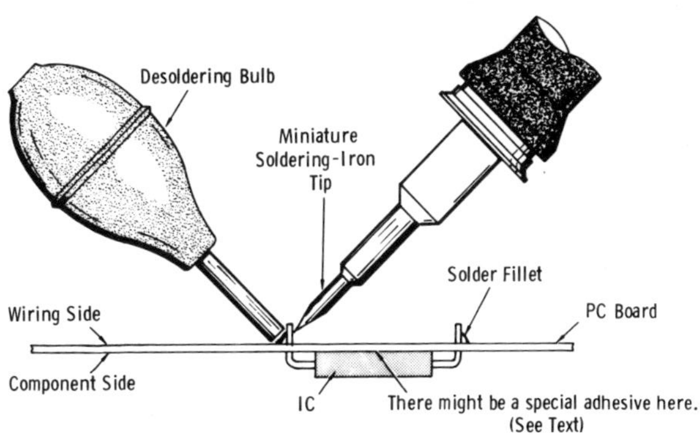

Fig. 13-14. Method for desoldering components.

IC. Make certain that all leads are free and clean of burrs; then remove the component. NOTE: Some manufacturers hold flat-packs in place by a small drop of special adhesive which remains flexible and will part with a small amount of pressure. The residue of adhesive from the old flat-pack should be sufficient to hold the new one in place before soldering.

4. Carefully inspect the holes in the board to insure freedom from excess solder or burrs which would prevent insertion of the replacement component. Remove burrs if necessary by gently reaming the holes with a sharp instrument such as a pick or soldering aid.
5. Bend the lead of the new component to correspond with those of the one removed. Place it on the component side of the board, making certain that the orienting dot, notch, or tap is in the same position as the one removed (step 3).
6. Insert the leads through the holes and press the component to the board. Do not trim the leads yet.
7. Solder the leads to the wiring eyelets or pads; use small-diameter solder to minimize the possibility of solder bridges between leads. Then clip the leads close to the board. Carefully inspect all work after the last lead is soldered to make certain that no solder bridges exist between any leads.

IMPORTANT NOTE: Examine the instruction book for any particular equipment for special instructions regarding component replacement.

A word of advice: This text can serve only to introduce the reader to logic circuitry. In broadcast systems, great numbers of such ciruits are being used in computer-type remote controls, signal generators and monitoring systems, station-break automatic switching systems, etc. The information presented in this book is written as a basic discussion that should remain valid with future advances in logic circuitry. Since a full-length book would be required to cover broadcast-system applications alone, the maintenance technician-engineer must undertake a training program on his own to remain competent in his field.

Amplifier Distortion Checks

It is good maintenance practice to run overall noise and distortion checks from the microphone input to the transmitter output at periodic intervals. If the maintenance personnel are divided between the studio and transmitter, the prevailing practice is to carry out this procedure separately for the studio and transmitter equipment. This is usually done in the intervals between the required proofs-of-performance in order to be ready for the official proof-of-performance run. The FCC requires a harmonic-distortion measurement. Detailed procedures are outlined in Chapter 14, since the transmitter is involved in the FCC requirement. When the run is made only at the studio, the noise-distortion meter is terminated by the line output of the studio. When excessive noise and/or distortion results, the audio oscillator is fed first to the line-amplifier input and then into preceding amplifier stages until the source of the defect is isolated. (Use the proper input level to the amplifier.)

Since harmonic-distortion measurements are required by the FCC, this type of measuring equipment must be on hand or available to the station. For studio use, however, more and more engineers are considering the use of intermodulation-distortion measurements to maintain a tighter and more definite control over studio amplifiers and transducers. This is covered in Section 13-13.

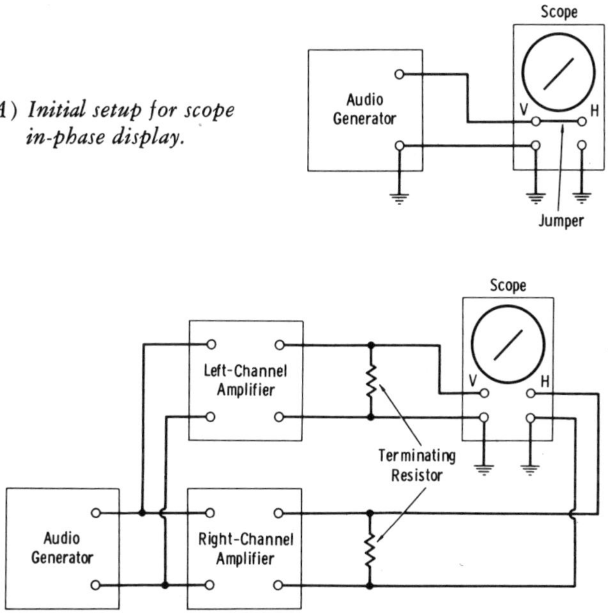

(A) Initial setup for scope in-phase display.

(B) Arrangement for phase-shift measurement.

Fig. 13-15. Method of measuring phase shift between stereo channels.

13-8. PHASING THE STEREO SYSTEM

A good audio oscillator of the type used in the station proof-of-performance runs and an oscilloscope provide the most convenient method of measuring the phase conditions. The technique is based primarily on that outlined previously for checking microphone phasing. The procedure is as follows:

1. Connect the audio generator (adjusted to 1000 Hz) to the vertical input of the scope. Run a jumper wire from the vertical to the horizontal input (Fig. 13-15A).

2. Adjust the horizontal and vertical gains of the scope so that a line slants from the lower left to the upper right of the screen. This is obviously the in-phase condition, since a common source is employed. If the vertical and horizontal amplifiers of the scope have a different number of phase reversals, the in-phase trace will be from the lower right to the upper left. A small difference in gains will not upset the accuracy of the reading. However, if an ellipse occurs (indicating phase distortion in the scope itself), it will be necessary to connect directly to the x- and y-deflection plates, through suitable (and identical-value) capacitors.

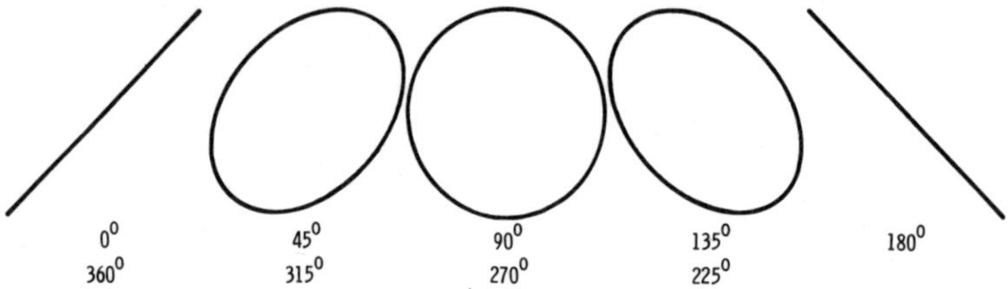

(*A*) *Typical patterns.*

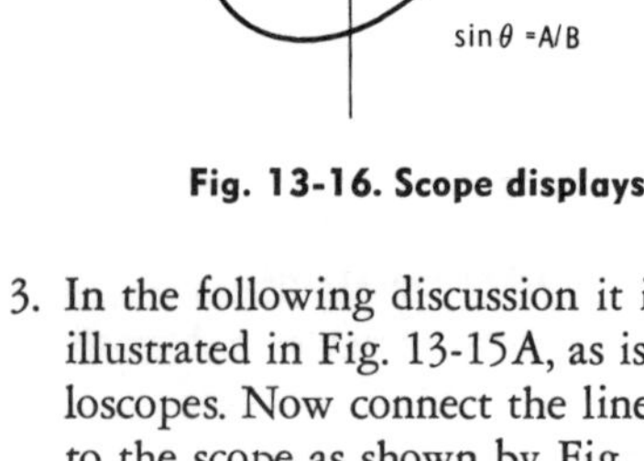

(*B*) *Determination of phase shift.*

Fig. 13-16. Scope displays in phase-shift measurement.

3. In the following discussion it is assumed that the in-phase trace is as illustrated in Fig. 13-15A, as is common with nearly all modern oscilloscopes. Now connect the line outputs of the left and right channels to the scope as shown by Fig. 13-15B. Adjust the respective levels to obtain the same sine-wave amplitudes used in step 1. Fig. 13-16A illustrates the basic phase-shift displays. If the 180° phase reversal is indicated, this means that a "turnover" in interconnecting wiring has occurred or that unlike amplifiers are employed. It is then necessary to proceed back toward the generator input until this discrepancy is isolated.
4. Fig. 13-16B indicates the method of measuring the phase angle. Adjust the centering so that the intersection of the x- and y-axis lines is at the exact center of the ellipse. For example, if A is 1.5 divisions

and B is 2 divisions, $1.5/2 = 0.75$, and $\sin \theta = 0.75$. From a table of sines, it is found that the angle whose sine is 0.75 is 48.6 degrees.

An alternate method is as follows. Measure the amplitudes of the two voltages individually. For example, adjust each channel for an output of 1 volt, a convenient value for a 600-ohm load. Then measure the sum of the two voltages. A zero-degree phase difference will result in 2 volts. A 180° phase difference will result in zero volts (complete cancellation). A 90° difference results in 1.4 volts, etc. This method is less accurate than the preceding one, but normally it is only important to ascertain that no phase turnover has occurred or that an excessive amount of phase distortion is not present.

Speakers must also be properly in phase. The most convenient and accurate method of checking speaker phasing is to connect a sensitive dc meter across the voice-coil transformer and note the polarity of meter deflection when the cone is pushed rapidly downward. All speaker connections are then made to achieve a like polarity of deflection. When the same make and model of speaker is employed throughout, it normally is only necessary to connect all color-coded leads in an identical fashion.

13-9. REEL TO REEL AND CARTRIDGE TAPE RECORDERS

There are a large number of different makes of commercial tape recorders. Although these machines embrace a wide variety of physical and electrical designs, the major troubles encountered by the operator have characteristics common to all makes. Some of the usual sources of trouble will be discussed here for the guidance of the user of commercial tape recorders.

Distortion

Distortion is the most commonly encountered trouble and probably has a larger number of causes than any other characteristic. There are three general types of distortion that are distinguished easily by the average listener. They are as follows:

1. Unintelligible speech and extreme lack of bass. This symptom is most likely caused (assuming that the audio amplifier and microphone are normal) by a complete loss of the supersonic bias to the recording head. Check all components in the bias oscillator-amplifier circuit. Check also for loose connections, and check the bias winding on the recording head.
2. Low output, muffled sound. Again assuming that the audio amplifier is normal, this trouble is usually traced to an accumulation of tape-coating residue, dirt, dust, or foreign matter on the pole pieces of the recording head. Clean them thoroughly. This should be a regular part of the maintenance schedule on high-quality magnetic recorders. Check the head alignment as outlined later.

3. Little or no output, mushy sound. This trouble can have the same causes just described; however, it is most likely an indication of a faulty component part in the audio amplifier or switching system. Check them in the orthodox manner. Keep heads clean by the application of isopropyl alcohol. Clean the capstan and the pinch roller with the same chemical. Follow the manufacturer's instructions on lubrication at the indicated intervals. *Be sure* to keep the heads demagnetized by using a small "wand" type demagnetizer.

Measurement of Harmonic Distortion

Manufacturers sometimes specify the percent of third-harmonic distortion in their system, since this is the most prevalent type of magnetic-tape distortion. It is desirable to measure the amount of third-harmonic distortion at normal signal input levels. In determining the maximum permissible recording level which can be used in the recorder, or in evaluating different tapes, this is an important consideration. It is desirable to operate the recorder at the highest possible recording level without exceeding the distortion limits.

The following method of measurement is suggested by the 3M Company, makers of Scotch brand recording tapes. The third harmonic is separated from the fundamental sine-wave frequency by a filter and directly measured with a vtvm (Fig. 13-17).

The equipment required is an audio oscillator that provides a good waveshape (normally the one used by the station for proof-of-performance runs), a vacuum-tube voltmeter, and a bandpass filter. Since it is customary to measure distortion at 400 Hz, a 1200-Hz filter is ideal. However, the more common 1000-Hz or 5000-Hz filters can easily be substituted if the test frequency is suitably adjusted. The filter should have a rejection of at least 60 dB at the fundamental test frequency if the highest accuracy is to be obtained.

Before the test is made, it is necessary to calibrate the system. This takes into account the insertion loss of the filter. Since the input termination affects this value, it is best to calibrate the filter from the actual recorder under test. To do this, the filter is disconnected, and the output level of the

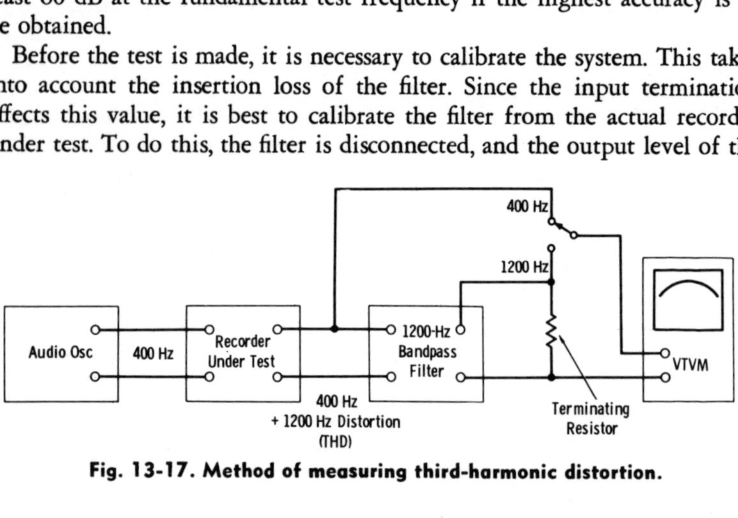

Fig. 13-17. Method of measuring third-harmonic distortion.

recorder is checked at 400 Hz and 1200 Hz to determine if it is the same at these frequencies. If it is not, the input to the recorder must be re-adjusted at one of the frequencies to compensate for the discrepancy. The filter is then connected to the recorder, and a level reading is taken at the input to the filter with a 400-Hz input to the recorder and at the output of the filter with a 1200-Hz input to the recorder. (If it is necessary, the input level to the recorder must be readjusted as previously explained.) The difference between these readings in decibels is the insertion loss of the filter.

In making the actual test, the 400-Hz signal is fed through the recorder, and level readings are taken at both the input and output of the filter. The difference between these readings in dB minus the insertion loss of the filter is the true ratio between the signal and the third-order harmonic component. This can be converted to percent by reference to the alignment chart (Fig. 13-18).

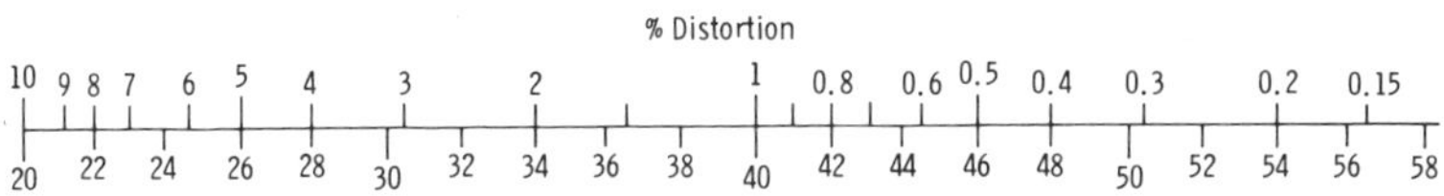

Fig. 13-18. Chart for converting decibels to percent distortion.

Once a particular system is calibrated and the insertion loss is known, this step need not be repeated in subsequent tests. With a little practice, distortion measurements can be made very quickly. All that is necessary is to patch in the oscillator, filter, and voltmeter and make two quick readings.

When distortion measurements are made not using the third-harmonic bandpass filter, the measurement is termed total harmonic distortion (thd). Normally, thd should be less than 2 percent over the frequency range of 50 Hz to 15 kHz.

Head Alignment and Loss of High Frequencies

If the reproducing gap is not parallel to the recorded poles on the tape, a serious loss of high frequencies may result. An attempt is made to align the head gaps on all recording machines exactly perpendicular to the tape so that tapes made on one machine can be played on any other machine in the station. This alignment is accomplished by rotating the head for the maximum output from an alignment tape on which a steady high-frequency signal has been recorded using a carefully aligned head. In making this adjustment, the extreme sensitivity of these high-frequency signals to alignment becomes immediately apparent. For example, in a full-width (¼″) recording of 1-mil wavelength, a misalignment of only 8 minutes of arc can reduce the output by about 6 dB.

An equally crucial problem in obtaining good frequency response is that of maintaining intimate contact between the tape and the head gap. In playing a low-frequency signal, a separation of one or two mils does not appreciably affect the level, but at a high frequency of, say, 1-mil wavelength, even a half-mil spacing results in a drop of more than 20 dB. It is apparent that any loss of contact between head and tape, however slight, has a profound effect on the high-frequency output. Therefore, always check the pressure pads and tensions to be certain that the tape is held in intimate contact with the heads. Be sure the heads are clean.

In playing a tape recording that was made on a different machine, head misalignment may be suspected, but one may not wish to disturb the head adjustment to verify this. In this case, the skew effect can be used to advantage because a deflection of the tape from its normal path across the head has the same effect as rotating the head. If, after carefully deflecting the tape a small amount in each direction, it is found that the greatest high-frequency output corresponds to the normal tape path, the head is correctly aligned. This test may be made on almost any kind of a recording, since the ear can readily distinguish the presence of high-frequency components in program material.

In case the highs can be improved by deflection of the tape (indicating misalignment), either the head used to make the recording or the one on the playback machine may be at fault. Regardless of the origin of the misalignment, optimum reproduction of this particular tape may be obtained by readjusting the playback head. After playing this tape, the machine should be rechecked with an alignment tape, and, if necessary, readjusted.

In summarizing these points, the following check list is suggested for locating the cause of poor high-frequency response due to head problems:

1. Check azimuth alignment with an alignment tape.
2. Check for tape skewing which seems to occur simultaneously with amplitude fluctuations.
3. Check to be sure the head meets the tape squarely.
4. Check for stability of the tape path in its guides.
5. Check for foreign deposits, nicks, or gouges on the head surface.
6. Check for "breakthrough" in the head gap. A magnifying glass or microscope is helpful.
7. Check for uneven wear.
8. Replace the head if necessary.

Fig. 13-19 illustrates the five basic adjustments in correctly positioning a recorder head. Note that two of these positioning adjustments, A and B, are concerned with the tape centerline. Ideally, the centerline of each component which is in direct contact with the tape should maintain an unvarying reference plane. Failure to do so results in bending of tape edges, tape skewing on a tangent from the normal path, azimuth error, and excessive friction.

Fig. 13-20 shows the details of the corresponding characteristics in Fig. 13-19. These may be outlined briefly as follows:

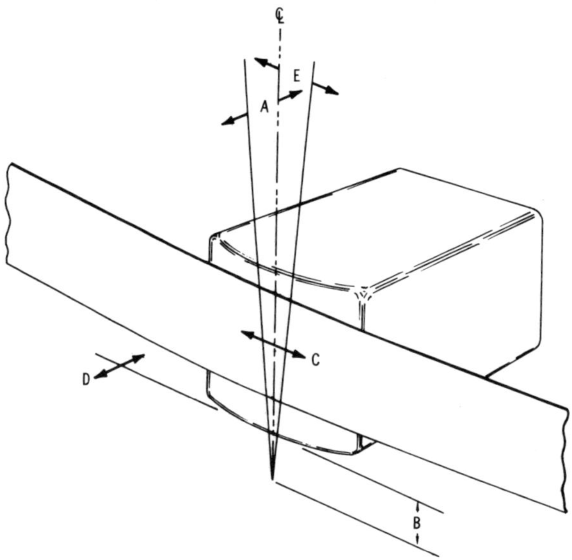

A. Tilt, in which the face of the head must be simultaneously tangent to the same degree with both edges of the tape and without distortion of either of the latter.
B. Height, in which the gap width dimension is centered on the standard track location.
C. Tangency assures that the tape contacts the portion of the head face containing the head gap.
D. Contact, head position into or away from the tape to assure proper contact pressure between head and tape ("wrap"). Not as critical with machines employing pressure pads at the heads.
E. Azimuth or skew, in which width dimension (corresponds to track width) of gap is exactly 90° with tape edge.

Courtesy 3M Co.

Fig. 13-19. Orientation in head adjustment planes.

A. *Tilt.* Designated as arc A in Fig. 13-19; must establish a true vertical position for the face of the head. Fig. 13-20A shows the exaggerated condition. If the head tilts even slightly, uneven tension results across the entire width of the tape. Intimate contact between tape and head is lost, and the tape can be caused to *skew* away from the centerline.
B. *Height.* Improper head height is particularly disturbing on multiple-track recordings due to noise, cross talk, and loss of signal.
C. *Tangency.* Characteristics A and B above have established the tape centerline path. Tangency (arc C of Fig. 13-19) squares the record and playback gaps to the tape surface. Even a slight error in tangency can cause high-frequency loss, and a high sensitivity to *dropouts* caused by slight contamination on the tape or head.

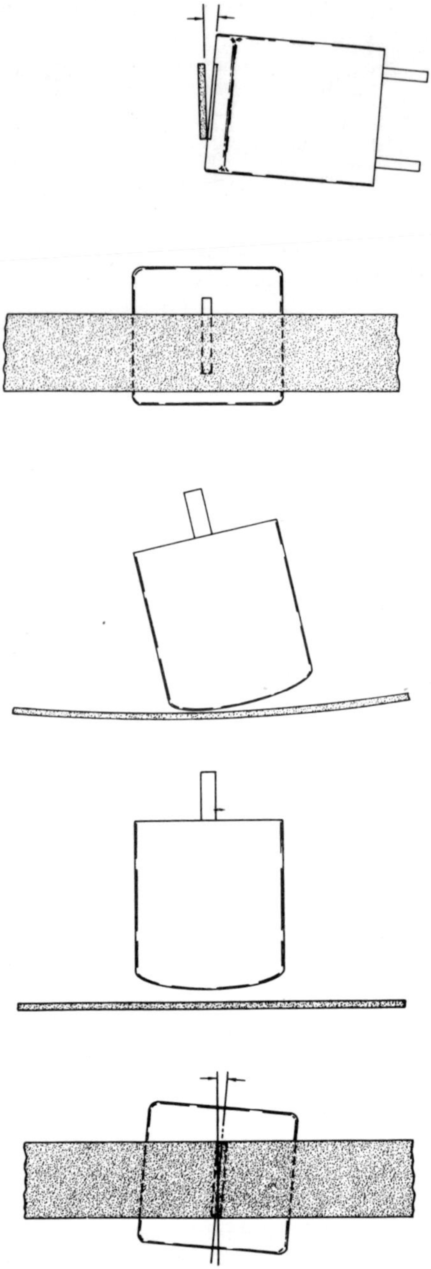

(*A*) *Tilt.*

(*B*) *Height.*

(*C*) *Tangency.*

(*D*) *Contact.*

(*E*) *Azimuth.*

Fig. 13-20. Details of head adjustment planes.

D. *Contact.* Proper contact is assured by a slight "wrap" in the path of the tape as it passes over the head. Contact is very important for good high-frequency response. When pressure pads are used, they should be inspected for signs of wear or damage.

E. *Azimuth.* The reproducing gap of the playback head must be exactly parallel to the recorded poles on the tape. Thus for compatibility and interchangeability, the record and playback heads must be adjusted so that the gaps are exactly perpendicular to the centerline of the tape path. An alignment test tape must be used for this adjustment. The 3M Company furnishes the following hints for this alignment procedure:

When an alignment tape is used to check azimuth, a variety of methods can be employed, the simplest being deliberately to skew the tape across the head while checking the output. If the output, as indicated by the signal level meter (or the playback volume), is highest with normal tape alignment across the head, it can be assumed that azimuth is correct. If the output signal level increases while the tape is intentionally skewed, the azimuth should be readjusted. The head should be realigned to yield maximum output. In the case of separate record and playback heads, the playback head should be peaked according to the output signal level determined while using the prerecorded alignment tape. The record-head azimuth should then be peaked by recording on a blank tape and playing back through the correctly positioned playback head. Only a studio-prepared prerecorded tape should be used for an azimuth test.

While checking head azimuth, it is also good practice to inspect the pressure pad (if used) for wear. If a pad which has become worn does not properly position itself against the tape, it will have a tendency to skew the tape out of alignment with the head gap, giving the same effect as incorrect head azimuth. If the pressure pad shows signs of a wear-created channel, it should be replaced.

Stereo Phase Problems in Tape Equipment

Proper head alignment is extremely criticial in stereo tape decks, because phase shifts between channels can result either from azimuth misalignment between the record and playback heads, or from a different skewing of the tape during the record and playback functions. This emphasizes that compatibility and interchangeability are much harder to obtain for stereo tapes than for mono tapes.

This type of distortion is most evident when a stereo tape which exhibits this problem is reproduced on a monophonic receiver. Bear in mind that a monophonic receiver essentially reproduces the *sum* of the two channels. Phase shifts between the two channels, which will differ with frequency, result in partial cancellation and reinforcement over the gamut of frequencies being reproduced. The net sum in this case can have a disastrous effect on the quality of the reproduced sound. This means that the stereo transmission is no longer compatible with monophonic reproduction when such errors exist.

For reel-to-reel stereo tape recorders, a very precise azimuth adjustment is required. Cartridge stereo systems have an additional problem in that all of the earlier cartridges did not provide for a tape-guide height adjustment, and the arrangement of the tape path allowed considerable "creeping" up the guide posts. More recent tape cartridges have provided adjustable guide height (as well as head adjustments), and some have an additional guide

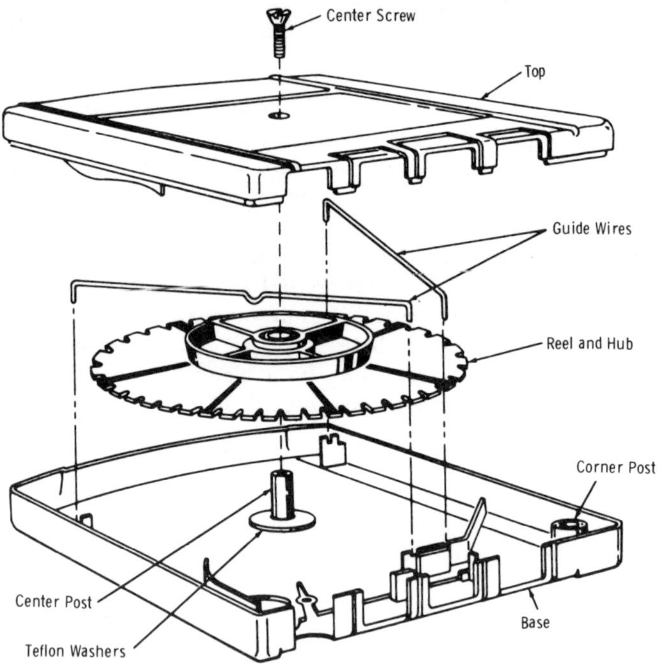

Courtesy TelePro Industries, Inc.

Fig. 13-21. Exploded view of tape cartridge.

post prior to the original input guide so that the tape enters its final path across the head in a more rigid and flutter-free state, with a standardized tension.

A good way to align the stereo head is to use a full-track monophonic alignment tape. This allows precise alignment of the two-head stereo assembly for equal amplitude and identical phase from the two tracks at the reference frequency on the tape. A scope with identical horizontal and vertical amplifiers can be used, as described in Section 13-8 for phasing the stereo system.

The Tape Cartridge

Servicing the automatic tape cartridge is necessary when the tape in the cartridge has become too tight, too loose, looped, or ridged. Telepro Industries, Inc., maker of the Fidelipac cartridge, recommends the following servicing techniques for each type of problem.

Tight Tape—One cause of tape tightening is the gradual wearing away of the graphite lubricant of the tape. When this occurs, the tape cannot slide on itself and will tighten to the extent that the reel will not turn and the tape may be torn or damaged. This problem can be corrected before the tape freezes or damages itself if periodic visual checks are made of the magazines. Just prior to the tightening up of the tape, it is possible to predict its failure by the very shiny appearance of the tape.

When this condition appears, it is possible for the tape to be relubricated temporarily with a flake graphite, such as Dixon's No. 635 flake graphite. To apply the graphite, remove the top cover of the magazine (Fig. 13-21), and with the top cover off, insert the magazine into a slow-speed player, preferably one operating at 1⅞ in/s. Place a small amount of graphite between the two guide wires near the front of the cartridge or next to the guide wire at the rear left of the cartridge. Note that the graphite tends to spread itself somewhat uniformly over the entire surface of the reel of tape.

If the tape is allowed to run long enough, the majority of the flake graphite will work itself between each turn of tape and eventually imbed itself either on the oxide-coated side or on the lubricated side of the tape. If the tape accepts the graphite lubricant, the eye, or large opening in the tape, will gradually move to the outside of the reel of tape and disappear. Then it will slowly reappear close to the hub if the magazine is permitted to run longer.

If the eye maintains a position near the hub, the tape will have a relatively long life. If the eye moves to the outside of the reel of tape, reappears near the hub and again moves to the outside, then no amount of graphite lubricating can salvage the tape. In these cases, it is best to scrap the tape and begin with a fresh reel of tape.

Tape tightening can also be caused by improper placement of the guide wires in the Fidelipac Model B and C cartridges. If the guide wires are allowed to rub against the hub of the reel, the reel will turn slower than normal. At the same time, the tape, which is being pulled from the center of the hub at a faster rate than the turning of the hub, will gradually tighten around the hub and stop the turning of the reel. This can be seen by watching the magazine in operation. If the guide wires are touching the hub, you will note that the hub does not turn freely, but turns with a slight jerky motion. Another way of noting this is to observe that portion of the hub which extends above the reel of tape. If the wire is touching, you will note a white mark drawn around the top edge of the hub by a wire guide.

To correct this, bend the guide wire away from the hub very slightly, just enough to clear.

The same difficulty may be encountered if the reel is warped enough to touch the base each time it makes a revolution. If either the lid or base is even slightly warped, tightening of the tape may again be introduced. Prior to assembly, visually check the cartridge for warping.

Failure of the brake to release the reel completely may cause the tape to tighten up also. When the pinch roller engages the brake, which in turn releases the reel, there should be a space of approximately $1/16$ inch between the edge of the reel and the brake. When the cartridge has been removed from the instrument, the brake should spring back and rest against the reel, thereby preventing it from turning during handling.

If a magazine is inserted in an instrument and the brake fails to release the reel, the tape will be pulled from the center of the cartridge. It will not rewind itself because the reel is not turning. In this case, the tape is spilled from the cartridge. Malfunctions of this type are rare and are immediately noticeable when a magazine is inserted in an instrument. When this occurs, remove the cartridge from the instrument. Check the tape for tears or twists. Holding the tape cartridge in your hand, insert your finger into the pinch-roller opening, and check for the release of the reel by pressing forward on the brake. Slowly pull the tape from the cartridge in the same direction the pinch roller would normally pull the tape, i.e., from the hub of the reel and not from the outside of the reel. You will note that the tape will pull itself back into the magazine and return to its normal position.

If the tape has been damaged, it will be necessary to disassemble the cartridge and cut out the damaged portion of the tape. After the damaged portion has been removed, resplice and reassemble the magazine. Check the magazine for proper release of the brake before inserting it in an instrument.

> NOTE: Be sure to become familiar with your particular type of cartridge. Some late-model types do not use this style of brake, but clamp the loose tape against the sidewall of the cartridge.

Loose Tape—When the tape is initially loaded on the reel, it is possible to have the tape too loose. This will show up only after a few hours of use in an instrument. It can be recognized by a very large eye, or opening, in the reel of tape itself. To correct this, remove the cover and guide wires. Unsplice the tape. Prevent the reel from turning, and pull the tape from the outside of the reel until the slack or eye of the tape has been reduced to normal. Then by turning the reel by hand, wind up all of the excess tape. Resplice and reassemble. The eye should stay approximately the same size during operation of the magazine.

Looped Tape—When the tape in the cartridge appears as shown in Fig. 13-22, the tape is beyond salvage. The cause of this condition is the lack of lubricant on the tape; consequently, the tape will not slide on itself. Note that the tape is shiny. As stated in a previous paragraph, this is a sign that the lubricant has worn off. Tape in this condition, however, can be used on a reel-to-reel machine; therefore, the information recorded on the tape may be re-recorded on fresh tape and thereby salvaged.

Ridged Tape—Under certain conditions, ridges of tape will rise above the normal height of the edge of the tape. If this condition is not corrected, the cartridge will bind up and stop. Severe damage also will result.

If, after assembly of the cartridge, there is too much space between the edge of the tape and the wire guides, ridging will result. To correct this condition, first inspect the wire guides to make sure they are perfectly straight and that they do not bind in the mounting holes or the mounting slots. Remove the reel, and add one or more Teflon washers until the edge of the tape has been raised sufficiently to just come into contact with the wire guides.

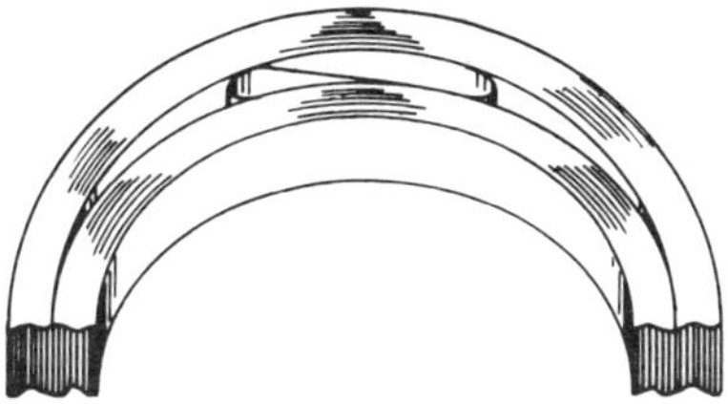

Fig. 13-22. When tape is looped as shown, it must be discarded.

Care must be taken not to raise the reel too high, because this creates too much downward pressure on the tape edge after the lid has been fastened in place. Before the lid is fastened in place, put the cartridge in a tape player that has its head cover removed to allow the lid to be secured in place while the reel is in motion. Observe the reel motion carefully to make sure that while the lid is being secured no binding is produced between the wire guides and the edge of the tape.

Checking Troubles From Tape Splicing

Splices in a wound reel of tape are subject to considerable pressure. Some of the pressure-sensitive splicing tapes currently available do not have satisfactory adhesives. Under pressure, adhesive "leakage" causes a "creep" or "bleed" around the splice. This not only weakens the splice, but also can contaminate the head and damage adjacent layers of tightly wound tape. The following information regarding tape splices is largely by courtesy of the 3M Company.

In the design of any pressure-sensitive tape, the two obvious components are the backing and the adhesive coating. In the development of a tape suitable for splicing magnetic recordings, both of these components were chosen with great care.

The backing had to be tough and durable while being as thin as possible. For this reason, paper was not suitable, and plastic was chosen. Both acetate and polyester are currently being used.

Developing an adhesive coating suitable for splicing tape was even more involved. Here, three basic qualities must be carefully evaluated. These are known as (1) *shear adhesion,* (2) *peel-back* or *ASTM adhesion,* and (3) *thumb appeal.*

Shear adhesion can be defined as the resistance of the adhesive to being parted from the surface to which it is adhered when pulled in what is commonly called the shear direction (Fig. 13-23A).

Peel-back or ASTM adhesion is, as its name implies, a measure of the resistance of the adhesive to being peeled away from the surface to which it is adhered. Fig. 13-23B shows how this test is performed.

The next property is "thumb appeal" or "quick stick." It is the quality of the adhesive to actually feel sticky. Oddly enough, it is not a particularly important quality as far as the strength of the bond is concerned, but it is a quality that is readily noticeable to the user. There seems to be an "old wives' tale" that has led some users to believe that "the stickier it feels, the better it will hold." This is not necessarily true when talking about splicing magnetic recording tape.

If the thumb appeal is high, the peel-back adhesion might be improved to some small degree, but this advantage must be paid for in two ways, neither of which can be tolerated. First of all, with a sticky adhesive the probability of its leaking out from around the bond is greatly increased. This *ooze,* as it is called, can be disastrous if it is permitted to exist in splicing tape. The adhesive oozing from under the splicing tape will tend to bond one layer of recording tape to the next layer in the roll. The result, when the recording tape is reused, would be possible removal of the oxide coating or complete blocking at that point in the reel. Secondly, with an increased thumb appeal, the shear strength of the splice is reduced. This is evidenced by a degree of parting of the once tightly butted ends of the recording tape, a condition referred to as *creep.* Not only will creep manifest itself as an absence of program material or a dropout; but now with the parted joint in the recording tape, the exposed portion of adhesive causes the additional problems cited above in the discussion of ooze. This, then, is why a properly designed splicing tape does not feel very sticky.

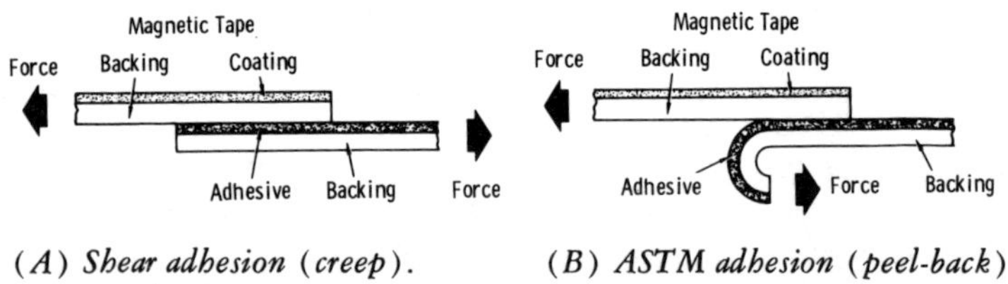

(*A*) *Shear adhesion* (*creep*). (*B*) *ASTM adhesion* (*peel-back*).

Fig. 13-23. Shear and peel-back adhesion.

Having defined some of the terms, we are now ready to examine the splice itself. There are several variations in splice geometry from which one can select the combination best suited to the conditions of use. These include the size of the spliced area and the angle at which the tape ends meet each other.

Initially, it would be well to discuss the length of a splice and the effect it will have on strength. The length of a splice is dictated, basically, by the amount of curvature it will have to sustain in its path from reel to reel (Fig. 13-24A). When the recording tape passes around the sharply curved surface of a guide as pictured in Fig. 13-24B, there is a tendency for the leading edge of the splicing tape to continue in its original direction. It is, in effect, attempting to peel itself away from the recording tape. In this case, the applicable adhesive parameter is peel-back or ASTM adhesion. With a given splicing tape, the amount of peel-back is decided in manufacture and, of course, is constant. The length of the splice has no effect on the tendency to peel but is important for another reason.

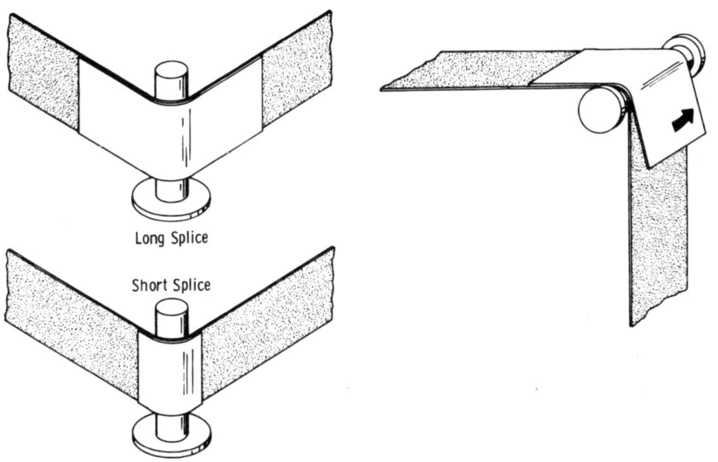

(*A*) *Splice length.* (*B*) *Bend radius.*

Fig. 13-24. Splice length and bend radius.

As shown in Fig. 13-25, a short splice may tend to loosen when subjected to a tight bend, because the area of peel may extend far enough into the tape bond to completely free one end of the recording tape. A longer splice will exhibit the same amount of peel, but the area of peel in this case does not extend all the way to the recording-tape junction. The bond at the junction is essentially undisturbed, and the splice passes the guide successfully. Of course, once the spliced area is wound on the take-up reel, the leading edge of the splicing tape that tended to peel is resecured to the recording tape by the pressure of the succeeding wraps as they they are wound onto the take-up reel. While it is impossible to assign a set of definite numerical values, generally speaking, use a long-length splice if small-radius bends or turns are expected.

As mentioned earlier, the tendency to creep is dependent on the shear strength of the splicing-tape adhesive. The force that opposes this shear strength is, of course, the amount of tension the tape encounters on the transport and while wound on the reel during storage. The amount of shear strength is constant for a given splicing tape. If the splice is subjected to a constant tension, the important variable affecting creep is then the area of the bond. The larger the bonded area, the better will be the creep resistance.

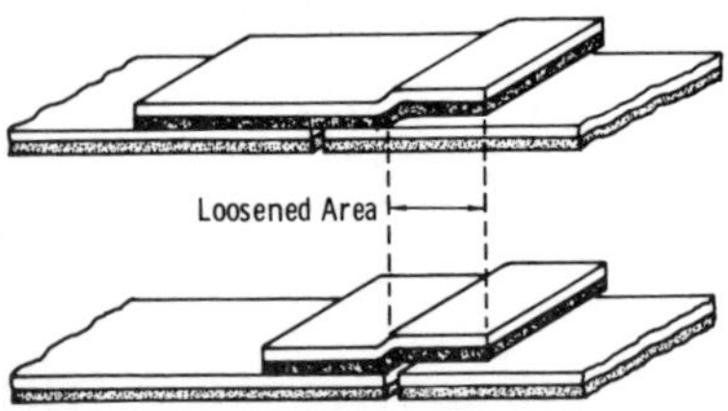

Fig. 13-25. Enlarged view of loosened splice.

A splicing tape with poor adhesive shear strength could be used if the area of the splice were greatly increased. Since the width dimension is limited by the recording tape, the area could only be increased by additional length. Since the program material may drop in level as much as 4 dB in the area of the bond because of the change in flexibility, the shorter the splice, the less disturbance there will be during playback. It is, therefore, important that the splicing tape chosen for use have high adhesive shear strength so that the spliced length can be kept short.

Much has been said and written about using splicing tape that is the same width as the recording tape and that which is somewhat narrower. It would be well to examine some of the variables and draw some conclusions.

When the full-width splicing method (Fig. 13-26A) is used, care must be taken to trim the splicing tape exactly at the edges of the recording tape. If the splicing tape is poorly trimmed (Fig. 13-26B), overhanging adhesive-coated splicing tape is apt to adhere to an adjacent layer on the reel, causing a problem similar to that encountered with ooze. Even though some splicing jigs are designed to cut an arc into each side of the splice (Fig. 13-26C) to insure against the possibility of overhang, this does not completely eliminate the chances of some adhesive oozing out of the edges.

Fig. 13-26D illustrates a splicing tape somewhat narrower than the tape to be spliced. This technique offers a number of advantages with no apparent disadvantages. Since the splicing tape does not extend to the edges of the recording tape, overlap is no longer a problem. A simple splicing jig

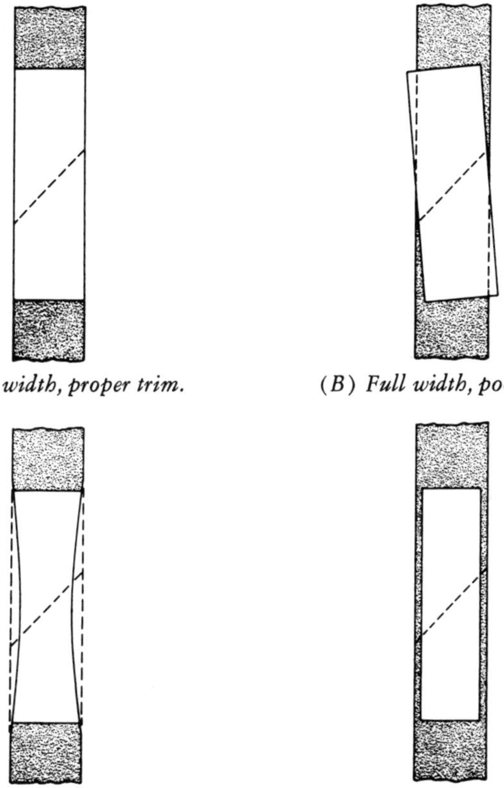

(*A*) *Full width, proper trim.* (*B*) *Full width, poor trim.*

(*C*) *Full width, undercut at edges.* (*D*) *Underwidth splicing tape.*

Fig. 13-26. Splicing widths.

can be used because there is no need to undercut the spliced area in an hour-glass configuration. Notice that the use of a somewhat narrower splicing tape does not appreciably sacrifice the overall bonding area when compared to full-width splicing tape that has been undercut.

In conclusion, let us examine the preparation of recording tape prior to the actual application of the splicing tape. The most desirable method is to cut the recording tape to be spliced at an angle of 45° to 60°, measured with respect to the tape edge. As the angle increases above 60° toward a perpendicular cut, the amount of electrical disturbance is increased because the head sees the discontinuity at the junction as an abrupt change. The shallower the angle, the less will be the amount of disturbance. But, as the angle is decreased below 45°, the pointed corners of the recording tape become vulnerable to being peeled back or debonded.

Regardless of the type of splice used, the first and possibly the most important consideration is cleanliness. The hands should be free of all dirt, dust, and oils, as one fingerprint on the oxide can drop the output several decibels. Also, contamination of the recording-tape backing or the adhesive of the splicing tape will usually reduce the strength of the bond between the two and can result in premature failure. After the recording tape is placed carefully in a splicing jig, it should be cut as carefully as possible, using a sharp, demagnetized razor blade. When handling pressure-sensitive splicing tape, take care not to handle the adhesive more than is necessary. After carefully laying the splicing tape down so as not to disturb the alignment of the splice, rub a finger over the tape to promote intimate contact between the two pieces. Then to remove the air pockets, use the flat of the fingernail. The selection of the proper splicing tape and the use of correct splicing techniques will insure a clean, long-lasting splice with no audible discontinuities.

13-10. FIELD EQUIPMENT

Preventive maintenance on field equipment is normally performed at the studio. Some of the larger stations that are heavily engaged in remotes employ a remote crew or field department.

The field department is concerned with a very important task, the job of properly carrying delicate broadcast equipment outside the studio and reaching the remote point without the necessity of any major maintenance work to be done on arriving. Microphones in particular must be firmly packed in such a manner that they do not roll around inside the case, and they must be protected against heavy physical shocks.

Field-department engineers almost unanimously prefer microphone cords that are detachable at the head of the microphone. This allows extremely firm packing of the individual microphone head between soft felt covers or rubber knee pads, without worrying about excessive bends of the cord where it enters the microphone housing.

Equipment-transporting facilities sometimes have an elaborate design for ease in the handling of a large amount of delicate equipment. A typical transportation trunk has several drawers with sponge-rubber-lined compartments for microphones. The other half of this trunk is arranged for the purpose of holding stand bases and shanks. A smaller trunk accommodates amplifiers, battery boxes, long ac and microphone extension cords, etc. Extra drawers accommodate tools and miscellaneous items.

Most modern remote-control amplifiers are conveniently packaged transistorized units that provide sufficient amplification to raise microphone outputs to a level ample for telephone-line transmission to the studio. They also include complete monitoring and cuing facilities. Usually either batteries or a self-contained ac power supply are employed; many units have an automatic switchover from ac to battery power in case of a power failure.

All field equipment should be inspected periodically for tightness of connections and general physical condition. Follow the precautions given previously when servicing transistor circuitry. Clean the faders and switches on a regular schedule depending on frequency of usage, and always before taking the unit to a remote location after long storage.

Motors and generators are often used in mobile units and as standby power sources at transmitters. The maintenance of these units is covered in Chapter 14.

13-11. TWO-WAY RADIO EQUIPMENT

Two-way radio is used for communication between the news room and mobile units for on-the-spot newsbreaks, and for the actual broadcast of news or special events. Each installation consists of a transmitter and a receiver. At the base station, the transmitter normally operates from ac power and has an rf power output of from 30 to 250 watts (approximately). The mobile transmitter averages 5 to 80 watts output and runs on 6- or 12-volt battery power. The receivers are usually tuned for a single frequency, use single- or double-conversion superheterodyne circuits with crystal control of the conversion oscillators, and have a selectivity of 20 to 40 kHz. The sensitivity is usually 1 μV or less, and a squelch circuit is used to quiet the loudspeaker in the absence of a signal.

The following is a summary of the FCC mobile-service maintenance requirements. *Caution:* Always check a current copy of the FCC Rules and Regulations.

A. *Frequency Check*—The frequency must be checked when the transmitter is initially installed, whenever a transmitter change is made which would affect carrier frequency or stability, and at intervals not to exceed one year when the transmitter is crystal controlled or at intervals not to exceed one month when it is not crystal controlled. This frequency check must determine if the frequency is within the percentage tolerance specified by the station license. Tolerances range from 0.02 to 0.0005 percent.

B. *Modulation Check*—A modulation check must be made when the transmitter is installed, whenever a change is made that might alter the modulation characteristics, and at intervals not to exceed one year. This must determine that the modulation does not exceed the FCC limits, which are: more than 70 percent but less than 100 percent (on negative peaks) for a-m transmitters; not to exceed from a 5-kHz to a 15-kHz swing, plus or minus, from the center frequency in fm transmitters.

C. *Plate Power-Input Check*—The plate power input of the output tube of the transmitter must be measured when the transmitter is installed, whenever a change is made that might change the power output, and at intervals not to exceed one year. The power must be within the tolerances permitted by the FCC.

The station normally uses tube testers, rf signal generators, vtvm's, and vom's in regular maintenance. Additional equipment required for two-way radio maintenance includes the following:

1. *Dummy RF Load*—With a dummy load, the transmitter can be operated at normal input and output without transmitting a signal beyond the shop. (The FCC requires that mobile units be measured under load conditions equivalent to actual operating conditions.) You can make this dummy load, or you can find several moderately priced ones (sometimes called dummy antennas) at most radio-parts distributors.
2. *Grid-Dip Meter*—The grid-dip meter is handy for checking the resonant frequencies of tuned circuits without applying power to the equipment. It is also convenient for determining the approximate output frequency of the various multiplier and amplifier stages in transmitters and for checking conversion oscillators in receivers.
3. *FM Modulation Meter*—The fm modulation meter must measure the fm deviation, plus or minus, from the center frequency of the transmitter. Frequency-modulated transmitters in these services operate on frequencies from 25 to 500 MHz. Since the FCC regulations say the *maximum* modulation deviation shall not exceed a certain limit, the meter should respond to actual *voice peaks,* not just to average or sine-wave modulation.
4. *Frequency Meter*—The frequency meter must show whether or not a transmitter carrier frequency is within the tolerance specified in the station license.
5. *A-M Modulation Meter*—An a-m modulation meter may not be needed, because the vast majority of transmitters in these services use frequency modulation.

As noted previously in the FCC requirements, the frequency, modulation, and plate power input must be measured at specified intervals. Many broadcast stations assign these measurements to independent mobile-service organizations, either on a time-plus-parts basis as the need arises, or on a contract basis. However, with the significant growth of mobile relay use in am-fm broadcasting, some stations prefer to invest in the equipment necessary to allow these measurements to be performed by the engineering department.

Examples of Equipment

Fig. 13-27A illustrates an fm modulation meter. The tuning range is 25 to 500 MHz (20-1 spread) without changing coils or switching. The placement of the monitor may be from 10 to 2000 feet from the transmitter, depending on the field strength and frequency. The relative field strengths of different transmitters, or of the same transmitter with different adjustments, are indicated. The operation is as follows:

1. Rotate the tuning control (with selector switch in "Tune Max" position) for maximum deflection of the "Peak kHz" meter on the right. Full scale may be set for either 12.5 kHz or 25 kHz.
2. Flip the selector switch to the "Tune Zero" position, and check for zero deflection of the meter. A slight adjustment of the vernier-tuning disc may be required to produce zero deflection.
3. Reset the switch to the "Modulation" position. Modulate the transmitter with the desired voice or tone, and the peak fm deviation in either the positive or negative direction (as determined by a polarity switch) is indicated on the meter.

A built-in speaker permits aural monitoring and can be used as an aid in tuning and for communication with the transmitter being monitored. A cro connection is also provided. Thus a three-way check is obtained on transmitter modulation: aural, visual display of waveshape, and meter indication of peak deviation.

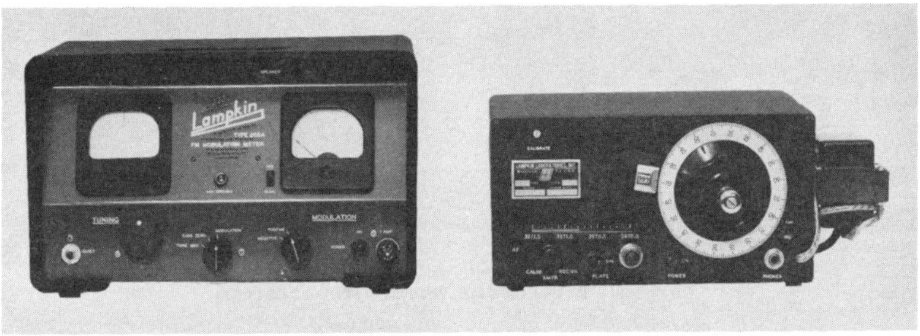

(*A*) *Fm modulation meter.* (*B*) *Frequency meter.*

Courtesy Lampkin Laboratories, Inc.

Fig. 13-27. Equipment for two-way radio transmitter maintenance.

The instrument just described indicates modulation swing only; it does not measure center frequency. A micrometer frequency meter (Fig. 13-27B) built by the same manufacturer is used for measuring carrier frequency. The procedure in the measurement is first to standardize the dial calibration either by means of WWV transmissions or against the internal crystal in the instrument. Then with headphones on, turn the dial for zero beat with the transmitter during a period of no modulation. Take the reading, and with the chart supplied convert this reading to the actual deviation from the assigned frequency. The result can be compared directly with the percentage tolerance allowed by the FCC.

The Motorola Motrac mobile fm two-way radio (Fig. 13-28) houses in one unit a transmitter of 25, 60, or 80 watts rf power output along with a fully transistorized receiver and power supply. The primary power is supplied by a 6-, 12-, or 24-volt battery.

The receivers include rf preselection tuned for maximum interference rejection. The carrier-squelch models include a noise-actuated squelch circuit consisting of a noise limiter, a noise detector, and a dc control stage (switching circuit) to cut off the audio amplifier. This eliminates noise otherwise heard in the speaker during the intervals between received messages.

The transmitters of the dual-squelch Private Line models are modulated by a continuous subaudio-frequency tone signal in addition to the voice modulation. The receivers accept only correctly tone-modulated signals and reject all others when the PL switch is in the on position. Several Private-Line networks can use the same rf carrier frequency in the same area when each network uses a different Private-Line tone frequency.

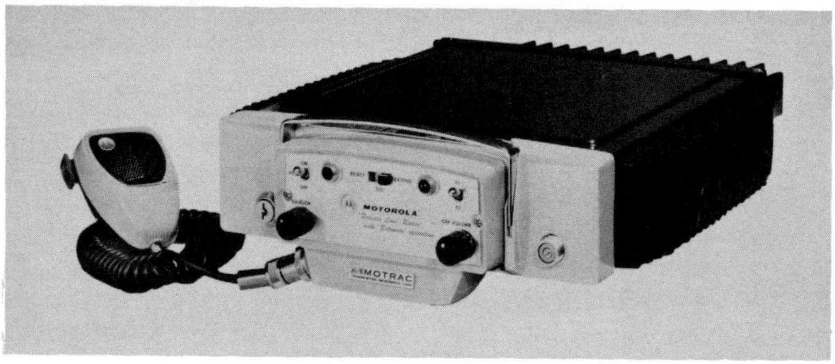

Courtesy Motorola, Inc.

Fig. 13-28. A mobile transmitter-receiver.

Dual-squelch Private-Line models also include noise-actuated squelch circuitry as previously described for carrier-squelch models. This enables the operator to monitor the channel before transmissions (PL switch in off position) and prevent interference with other users of the frequency.

The squelch control is not used in the Private-Line operation, and the setting of it does not affect the Private-Line circuit. In carrier (noise-actuated) squelch operation, however, it must be set for proper operation.

In dual-squelch Private-Line models, the Private-Line circuit is operative at all times. The PL-ON-OFF switch on the control head places the noise-actuated squelch circuit in or out of operation.

The fully transistorized power supplies use long-life transistors instead of vibrator units. Vibrator replacement costs are eliminated, and at the same time the overall operation is improved. Silicon-diode rectifiers provide the greatest efficiency. All operating voltages for the transmitter are provided by the power supply. Heat sinks along the sides of the housing surround the power-supply transistors.

A netting switch in the Motrac radio equipment allows a rapid check and adjustment of receiver and transmitter frequencies. Pressing this switch causes the receiver and transmitter to function simultaneously. If one of the two units is known to be on frequency, the oscillator of the other may then be adjusted with a minimum of test equipment.

The radio-set housing can be mounted either under the vehicle dashboard or in the trunk. The same drawer assembly, including the transmitter, receiver, and power supply, slides into the housing of either type of installation. In mobile-radio fleets where both dash-mount and trunk-mount installations are used, this design provides a full flexibility in exchanging radio sets between vehicles. Trunk-mounted radio sets are remotely controlled by means of a dash-mounted control head.

The microphone used with the Motrac radio set is a palm-type, dynamic microphone with a transistorized preamplifier. The microphone has a 12-inch tinsel coiled cord and a four-pin connector for plugging into the receptacle on the control head.

Two-Way-Radio Maintenance

Motorola, Inc. suggests the following outline as a basic guide in maintaining two-way radio systems.

The two phases of maintenance are preventive-maintenance measures and repair of the equipment when it breaks down. Preventive maintenance is performed on equipment to insure a good working condition so that breakdowns and needless interruptions in service will be kept at a minimum. Preventive measures include visual inspection of electrical and mechanical parts, periodic cleaning, and lubrication. These techniques can be performed with a minimum of equipment and effort. Potential failures can often be spotted and corrected before they have a chance to occur. Systematic recording of meter readings and the observation of performance characteristics also aid in the prevention of failures. A comparison of data taken on equipment at regular intervals may reveal slow, progressive drifts that may be too small to appear significantly in one test. Any marked variations in the regularly obtained data should be regarded as a sign of an approaching failure and investigated promptly.

Table 13-1 is part of the regular preventive maintenance procedure. These procedures should be performed at least as often as recommended. Under adverse conditions, such procedures as cleaning and inspection for signs of deterioration of exposed components should be performed more often. The following procedures should be followed for the specific items indicated.

Fuses—Fuses, sometimes even when new, may have too much resistance. Check them, and replace them if the voltage drop across the fuse exceeds 0.1 volt.

Fuse Holders—Fuse holders also lose their temper because of the heat resulting from a poor contact. Acid causes corrosion. An accumulation of oil and dirt increases the contact resistance. The remedial action is to clean the contacts, tighten, or replace them.

Table 13-1. Routine Maintenance Check Chart

What to Check	What to Look For	Remedy
	WEEKLY	
1. Cabinet and chassis	Dirt or moisture collected on equipment.	Remove loose dust and dirt with a soft brush or blower; remove dirt and grease which adheres to the chassis and parts with a brush or cloth moistened with a cleaning solution. Wipe gently with a clean soft cloth such as cheese cloth.
2. General condition of equipment	Rust and corrosion on parts and wiring, loose connections, frayed or burned insulation, loose screws, or charred resistors and coils.	Use No. 0000 sandpaper to remove rust and corrosion. Do not use steel wool or emery cloth, since minute particles frequently enter the equipment and cause harmful internal shorting or grounding of circuits. Use paint which is authorized and consistent with existing regulations to touch up bared surfaces.
3. Controls	Binding, scraping, excessive looseness, misalignment, or positive action.	Replace if necessary. Tighten any loose knobs or controls.
4. Plug-in items	Proper seating of readily accessible plug-in items such as plug-in type transformers.	Replace if necessary.
5. Cables	Cuts, breaks, fraying, deterioration, kinks, or strains.	Repair or replace if necessary.
6. Connectors	Loose connections and bent or broken contacts.	Straighten all bent contacts. Replace if contacts are broken.
7. Pilot-light assemblies	Looseness, broken lenses, or burned out bulbs.	Replace if necessary.
8. Dust covers	Check to see that all protective dust covers are in place.	Properly position and fasten securely.
9. Meters	Damaged glass, broken cases or bent needles.	Replace if necessary.
10. Antenna-tower guy wires (if used)	Check for tension.	Tighten all guy wires for equal tension on each corner.
11. Mountings and connections	Loose components and cases, rack mounts, antenna mounts, shock mounts, coaxial cables, and cable connections.	Tighten screws, bolts, and nuts carefully. Fittings tightened beyond the pressure for which they are designed will be damaged or broken.
12. Power-supply voltages	Proper power-supply voltages by observing indications on the appropriate meters.	Check each voltage on transmitter and receiver to determine that the power-supply voltages are correct.

Table 13-1. Routine Maintenance Check Chart—cont

What to Check	What to Look For	Remedy
	MONTHLY	
1. Tubes	Loose envelopes, cracked sockets, insufficient spring tension. Low emission of receiver-type tubes.	Replace all defective tubes.
2. Terminal boards	Loose connections, cracks, or breaks.	Replace if necessary.
3. Transformers, chokes, potentiometers	Signs of overheating.	Replace if necessary.
4. Electrical connections	Loose connections, terminals, etc.	Tighten all electrical connections, terminal strips, etc. Do not tighten beyond the pressure for which they are designed.
5. Capacitors	Leaks, bulges, or discoloration. Inspect variable capacitors for dirt, moisture, misalignment of plates, and loose mounting.	Replace if defective.
6. Resistors	Cracks, chipping, blistering, discoloration, or moisture.	Replace if defective.
7. Switches	Poor operation, arcing, or corroded contacts.	Repair or replace if necessary.

Courtesy Motorola, Inc.

Primary Power Sources—Primary voltages are always suspected in ac circuits, since they can often be too high as well as too low. In some cases, an application of voltage regulators may be necessary. Loose connections on fuses occur even in ac circuits. Check for hot fuses and connections. The primary power sources should be the first check point in preventive maintenance procedures and should be checked periodically.

Unit Assemblages—Unit assemblages include the transmitter, receiver, power supply, and associated units. Very few adjustments need to be made regularly in the transmitter and receiver. To catch a maximum number of impending troubles in unit assemblages, the technician must use his eyes, his ears, his sense of touch, and meter measurements, as well as his good judgement and past experience. Basically four types of observations are required:

1. Listening Tests—An experienced technician can often detect warnings of impending trouble by noting the audio quality, the noise background, and the squelch operation with respect to the settings of the volume and squelch controls. He will come to know the normal signal level he should obtain at given locations with selected stations in a particular system. Degradation of performance will be immedi-

ately apparent to the technician who checks the set periodically, whereas the steady operator may become accustomed to a gradual decline.

2. Visual Inspections—Check for loose bolts, screws, and clamps; worn microphone cords; burned-out pilot lights; arcing conditions at any point in the equipment; loose or broken components; burned resistors; or loose connections. A good visual inspection not only means quick location of the trouble source, but also may prevent more serious trouble from developing.
3. Touch Tests—The sense of touch is a good detector of improper temperature conditions. Fuses, fuse holders, crystal ovens, transformers, and relay coils that are too hot indicate possible trouble sources. Also, cold crystal ovens (except for nonheated crystals), tubes, or resistors that should normally be warm are probably in trouble. Such tests are quick and should be made often. WARNING: When making touch tests of the equipment, the technician should be extremely careful, since *high voltages are present at all times.* Extreme caution must be employed to avoid hazardous electrical shocks. Even after the ac power is disconnected from the equipment, short the high-voltage components to the chassis ground by means of a screwdriver with an insulated handle.
4. Meter Measurements—Actual measurements of voltages, grid and plate currents, power output, and frequency deviation have no substitute; their meaning is definite. Experience will indicate to what degree variations in these measurements may be tolerated.

Antenna Systems—Visual checks of antennas, tower lights, antenna-mounting devices, cables, and connectors should be made periodically, as should meter checks for open or shorted lines or high standing-wave ratios. Remember that the antenna must efficiently radiate the power developed in the transmitter and also transfer to the receiver the minute voltages received from the distant transmitter. Failure of the antenna line is a failure of the entire communications system.

Netting—Netting of receivers and transmitters in any system is more important than is usually realized. The term netting means that all receivers and transmitters in a given system are aligned to the same frequency. With modern selectivity requirements, it is necessary more than ever to tune every receiver exactly to the station it must receive. Careful netting of all units in an integrated system to the correct frequency is a first-order requirement of any communications system and will eliminate a prime cause of poor system performance.

Transmitter Deviation—Checks of transmitter deviation will also improve poor system operation. The technician should be sure that his transmitters are swinging through the entire deviation range, since underdeviation does not utilize all the advantages of frequency modulation.

Adjustments—A good rule to follow is to check performance frequently and make adjustments only when actual tests and measurements show improper operation. Do not make any unit adjustments without meters and signals to indicate the actual conditions.

Finishes—When the finish on the components has been badly scarred or damaged, rust and corrosion can be prevented by touching up the bared surfaces. Use No. 0000 sandpaper to clean the surface down to the bare metal; obtain a bright smooth finish. Use paint which is authorized and consistent with the existing regulations for touching up bared surfaces. CAUTION: Do not use steel wool or emery cloth. When these materials are used, minute particles frequently enter the equipment and cause harmful internal shorting or grounding of circuits.

Tagging of Leads—Tagging of leads is essential to insure that correct rewiring will be made when a part is replaced. Before unsoldering any leads, tie together the leads that are attached to each part. Use small tags or short pieces of adhesive tape to identify all wires in accordance with their actual connections. Identify every lead that is to be removed.

Parts Replacement—When damaged parts must be replaced, identical parts should be used. If the identical replacement is not available, check the parts list for the proper part number. If substitutions must be made immediately, it is recommended to replace the substitutions as soon as the proper replacement part is available. Even the substituted part must have identical electrical properties and must have equal or higher voltage and current ratings.

Parts Relocation—Relocation of a substituted part may cause hum or other interference and is not recommended.

Parts Mounting—Mount the new or repaired part in the same mounting as that formerly used for the defective part. Fasten all mountings securely.

Soldering—Before soldering any connections, carefully scrape all parts to be soldered until all traces of rust, corrosion, paint, or varnish are removed. Dust the scraped parts with a small, clean brush. Tin all surfaces to be soldered. To obtain sufficient mechanical support, wrap the wire around the lug to be soldered. Solder the connection, using a minimum amount of solder with sufficient heat to make the solder flow evenly around the tinned surface.

Maintenance Program—A good preventive maintenance program requires a complete system check *at least* every three months. In a general way, the following items should be part of any preventive maintenance routine.

A. Clean and dust thoroughly.
B. Check primary voltage circuits.
C. Measure all required B+ voltages.
D. Measure power output of transmitters.
E. Measure sensitivity of receivers.

F. Check meter readings at the metering positions of transmitters and receivers.
G. Check audio output of receivers and audio input to transmitters.
H. Check dc control voltages at control panels.
I. Check operation of all relays, and clean relay contacts.
J. Measure frequency and deviation of transmitters.
K. Check and adjust frequency netting of the entire system.
L. Check all accessories such as cables, microphones, pilot lights, etc.
M. Check the antenna system, and clean antenna of any dirt or corrosion.
N. Check telephone lines, ac line, and antenna coaxial lines.
O. Check and adjust remote-control levels.
P. Check power-supply output voltages.
Q. Actually make an operational listening test of transmission and reception.
R. Keep individual records for each unit.

13-12. PREVENTIVE-MAINTENANCE SCHEDULES

No specific maintenance schedule can be devised to apply to all stations, even for stations of the same size. Therefore the treatment given here is on a broad, general base from which any preventive-maintenance schedule involving specific and special considerations can be drawn.

Preventive-maintenance man-hours for the average am-fm installation total about 16 per week—about 8 for the studio and 8 for the transmitter(s). This obviously varies with the extent and nature of the installation.

In general, preventive maintenance consists of the following procedures: inspect (I), clean (C), tighten (T), lubricate (L), tube check (TC), spot-proof (SP), and FCC station proof (P). The latter involves measurement of overall performance from the studio microphone input to the transmitter output and is described in Chapter 14.

It is most convenient to draw up a form which can be easily reproduced, (such as that shown for a monthly sheet in Fig. 13-29). Five vertical spaces should be provided to allow for the months in which five week-starting dates occur. This is the master sheet which the chief engineer or maintenance supervisor fills in for schedule reference for each month. A schedule number refers to the specific checkoff sheet or card that lists the maintenance steps for a particular type of equipment.

In the case of turntables, the manufacturer's lubrication points should be noted, along with the type of lubricant. For special instructions, the instruction-book number and page should be listed.

For tape recorders, some additional items are head-alignment checks and thd tests. The heads should be cleaned and degaussed daily, and the capstan and pinch roller should be cleaned daily also.

Tube checks of each unit should be made about every 90 days. It is more convenient (and most rapid) simply to use the GOOD-?-BAD scale, since

MONTH: ______________________

Week of	Sched. No.	Items	Comments	Initial	Date

Fig. 13-29. Example of maintenance schedule form.

this normally automatically indicates a 75-percent reduction in transconductance. In addition, check for shorts and gas. Check voltage regulators (gas type) for firing voltage and regulation range.

In addition to the 90-day tube checks, it is good engineering practice to run spot proofs on studio sections with the noise and distortion meter at the same intervals. Some tubes will invariably be found to be unsatisfactory in their noise or distortion characteristics regardless of the tube-checker indications.

With the increasing use of transistorized equipment, spot proofing is assuming a greater importance than ever before. This will reveal any further maintenance necessary on the transistor amplifiers and is obviously much less time consuming than checking transistors.

The overall FCC proof of performance should be run at least once each year. Both the station and personnel will benefit if such a proof is run more often.

13-13. INTERMODULATION DISTORTION

As mentioned previously, the FCC specifies that harmonic distortion must be measured in broadcast proof-of-performance checks. With the strong interest in stereo and high-fidelity broadcasts in modern fm practice, some stations are more thoroughly investigating the use of the intermodulation-distortion meter in order to maintain a tighter control over quality as it is actually judged by the ear. It has already been mentioned that the distortion characteristics of a disc pickup cartridge are most accurately specified in terms of intermodulation-distortion characteristics.

Fig. 13-30 illustrates an intermodulation meter manufactured by the Audio Instrument Co. The following description appears through their courtesy.

Why Measure Intermodulation?

In comparing the two leading methods of measurement, they may be judged by their likeness to the actual conditions of system use, by comparative listening tests, or by a mathematical analysis of the theory. Conditions for the harmonic method of measurement bear small relation to actual practice. Program material is typically complex in nature, and the distortion effects parallel those produced by the two tones of the intermodulation method. If harmonic distortion is low and intermodulation high, then a solo instrument will sound all right, but a full orchestra at the same audio level will sound muddy and distorted.

It is sometimes stated that intermodulation and harmonic distortion are interrelated in a 4:1 ratio, but actually there is no simple mathematical relationship between the two types of distortion. Hence, it is not ordinarily practical to calculate the intermodulation distortion from the harmonic-distortion data of an actual system.

Listening tests show that whereas harmonic readings bear little or no relation to the opinion of the ear, intermodulation readings correlate very reliably. This is why the intermodulation method has been used by the motion-picture industry for many years. The high-fidelity industry has had similar experience.

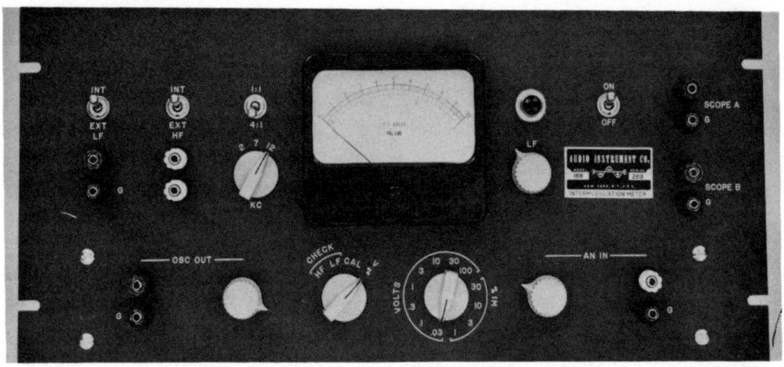

Courtesy Audio Instrument Co.

Fig. 13-30. An intermodulation-distortion meter.

Operation

The two-tone signal is generated by mixing the output of a low-distortion oscillator with a well-filtered 60-Hz tone from a transformer. The mixing is done in a bridge circuit that prevents interaction between the two sources and thus prevents false intermodulation. The output of the bridge passes through a constant-impedance attenuator for the adjustment of the intensity of the mixed tone, and the attenuated output appears at the OSC OUT posts (Fig. 13-31A).

By throwing the LF toggle switch to Ext and connecting the oscillator to the binding posts at the switch, the 60 Hz from the transformer may be replaced by a tone from an external audio oscillator. This is often desirable when a new design is being checked in the laboratory for the first time. One side of the external lf circuit is grounded.

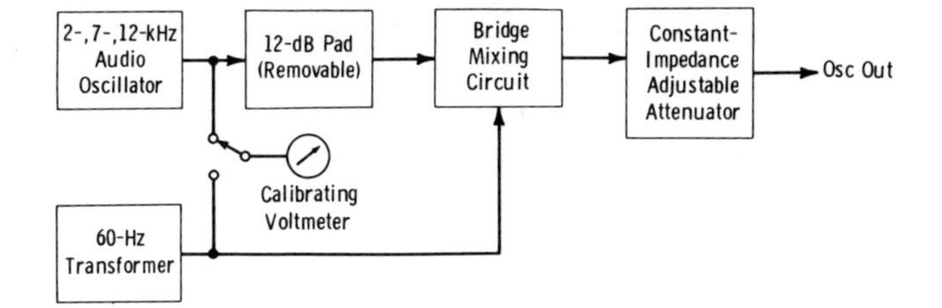

(*A*) *Signal-generating section.*

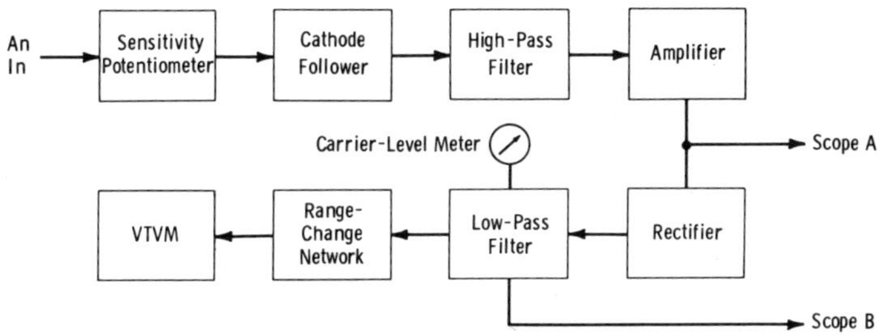

(*B*) *Analyzer section.*

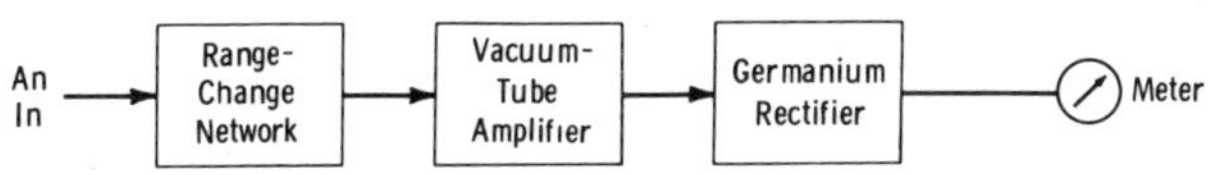

(*C*) *Vacuum-tube voltmeter.*

Fig. 13-31. Block diagrams of sections of Audio Instrument Co. intermodulation distortion meter.

The three fixed high frequencies from the high-frequency oscillator may be replaced by a tone from an external audio oscillator by throwing the HF toggle switch to Ext and connecting the oscillator to the binding posts at the switch. This is seldom necessary. The external oscillator for this specific circuit must be ungrounded. If the available oscillator has a ground on one side of its output circuit, it is possible to interpose an isolating transformer (usually 600:600 ohms) between the oscillator output and the mixing-circuit input.

The two signals which form the two-tone combination may be equal in intensity (1:1 ratio), or the low-frequency tone may have four times the intensity of the high-frequency tone (4:1 ratio). The change in the ratio is accomplished by inserting a 12-dB pad between the high-frequency oscillator and the mixing bridge following the circuit point at which the calibrating voltmeter is connected. Thus regardless of whether the ratio is 4:1 or 1:1, the calibrating voltmeter should read the same whether it is measuring the low-frequency or the high-frequency input to the bridge.

The two-tone signal passes from the OSC OUT terminals to the input of the amplifier or recorder under test. The intensity of the input is controlled by the adjustable attenuator at the output of the bridge. This attenuator is the constant-impedance type so that its adjustment will not unbalance the bridge and allow leakage of one tone into the other source. This would produce intermodulation.

If the audio system under test has distortion, the low-frequency tone will modulate the high-frequency tone. The degree of this modulation is measured in the analyzer section of the meter (Fig. 13-31B).

The output of the system under test is fed to the AN IN binding posts. The amount of the signal fed to the following circuits is regulated by the AN IN potentiometer, which feeds a cathode follower.

The output of the cathode follower consists of an amplified replica of the input, i.e., a low-frequency tone and a high-frequency tone. The low-frequency tone is removed, and the high-frequency tone is inspected for modulation. The first step is to pass the mixed-tone signal through a high-pass filter, which removes the low-frequency tone and passes the high-frequency tone without changing it. The latter passes through an amplifier and a full-wave rectifier, whose output consists of the rectified high-frequency tone and low frequencies corresponding to the modulation.

A low-pass filter removes the rectified high-frequency tone, and the low-frequency tones (which are a replica of the intermodulation distortion) are measured by a vacuum-tube voltmeter with its associated potentiometer network for range changing.

In order to standardize the voltage level at the rectifier, the rectified current is measured by a dc milliammeter, which reads the carrier level. To minimize cost, the panel meter is used in two roles. First, it is used to indicate carrier level as the latter is adjusted by means of the AN IN potentiometer. Second, it is shifted to the vacuum-tube voltmeter circuit, where it reads the percentage of distortion.

The vacuum-tube voltmeter (Fig. 13-31C) can be used separately to measure the voltage at the AN IN binding posts. The voltage passes from AN IN, via the switch, directly to the precision, adjustable, range-change potentiometer network; the output of the network is measured by a vacuum-tube voltmeter, which consists of an amplifier, a germanium rectifier, and a milliammeter. When the vtvm is used to measure distortion, the range-change network is switched to the output of the low-pass filter.

If the intermodulation meter feeds directly into the amplifier grid circuit, load the OSC OUT terminals with a 600-ohm, ½-watt resistor (Fig. 13-32A). If the amplifier has a transformer at the input, isolating and matching pads are most convenient, since they present 600 ohms to the meter and the appropriate output impedance to the amplifier under test (Fig. 13-32B). The resistance values in Fig. 13-32B are given in ohms; the first number is the calculated value, and the number in parentheses is the nearest standard EIA value.

With a 30- to 50-ohm amplifier input impedance, less loss will be introduced if a 600-ohm pad (the first pad in the table) is used followed by a suitable matching transformer.

Setting Up

Connect the signal-generator output terminals (OSC OUT) to the input of the amplifier or system under test. Be sure that the signal-generator output is terminated by 600 ohms; if it is to feed an unloaded transformer or a high-impedance grid circuit, connect 600 ohms across the OSC OUT posts, or use a pad. Note that one side of the signal-generator output circuit (the lower terminal) is grounded.

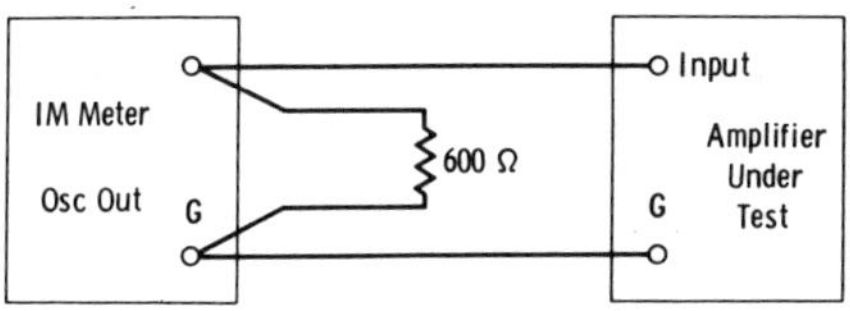

(A) Meter feeds directly into amplifier grid circuit.

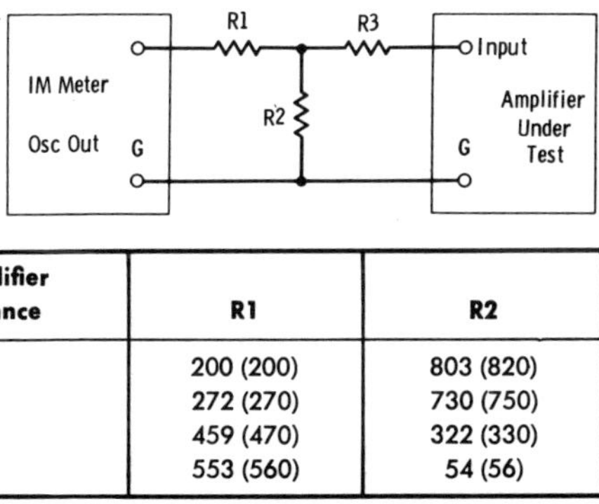

Nominal Amplifier Input Impedance	R1	R2	R3
600	200 (200)	803 (820)	200 (200)
500	272 (270)	730 (750)	105 (100)
250	459 (470)	322 (330)	0
50	553 (560)	54 (56)	0

(B) Meter connected through isolating and matching pad.

Fig. 13-32. Method of connecting intermodulation meter to amplifier.

Connect the analyzer-input terminals (AN IN) to the output of the amplifier or system under test. The analyzer-input impedance is about ½ megohm and may be bridged across most circuits without affecting their operation. One side of the input is grounded, and it is necessary to use an isolating transformer (usually 1:1 ratio) between the equipment and AN IN if the output circuit of the equipment under test has a grounded center tap. In any case, be sure to connect a load resistor of the proper value across the amplifier output. If the amplifier uses a great deal of negative feedback, it may occasionally be desirable to substitute a speaker for the load resistor in order to check the effect of an inductive load, or to bridge a capacitor across the load resistor to test the effect of a speaker dividing network.

If the equipment under test does not have one side of its input circuit grounded internally, it is desirable to ground the G post of the OSC OUT terminals. This avoids any tendency toward system oscillation. Allow the equipment to warm up for at least 5 minutes before use.

Although it has been common practice to test for low-frequency and high-frequency overload of an amplifier simultaneously, the result is often difficult to interpret. The following instructions have therefore been designed to test low- and high-frequency performance separately. Thus the cause of the trouble, if any, can be more clearly identified.

Measuring Low-Frequency Distortion

To measure low-frequency distortion, proceed as follows:

1. Set ratio switch to 4:1.
   Set HF switch to Int.
   Set LF switch to Int.
   Set frequency switch to 2 kHz.
   Set LF control (for internal 60 Hz voltage) fully counterclockwise.
   Set output and input controls fully counterclockwise.
   Set CHECK-CAL-% switch to Check HF.
2. Read high-frequency voltage on meter.
   Set same switch to Check LF, and adjust LF control until same voltage reading is produced as in previous step.
3. Set CHECK-CAL-% switch to V.
   Set meter range to the proper point on Volts arc of scale for desired system output voltage.
   Turn OSC OUT control until system under test produces desired output voltage.
4. Set CHECK-CAL-% switch to Cal.
   Set meter range to 30% IM, or a lower IM range if desired.
   Turn AN IN control until meter reads full scale.
   Set CHECK-CAL-% switch to %.
   Turn meter range switch to proper point on % IM arc of scale for a suitable meter reading.

Using External Low-Frequency Source

To use an external low-frequency source, use the following procedure:

1. Set LF control fully counterclockwise.
   Turn LF switch to Ext.
   Connect an oscillator (capable of delivering at least 10 volts across 600 ohms) to the EXT LF terminals. Any frequency from 20 to 200 Hz may be used with Model 168 or from 20 to 400 Hz with Model 168D.

2. Adjust the low-frequency voltage to the same value as the high-frequency voltage, as in step 2 in the previous section. However, adjust the low-frequency voltage by the output control of the external oscillator instead of by the meter-panel LF knob (which is now inoperative).

Measuring High-Frequency Distortion

To measure high-frequency distortion, proceed as follows:

1. Set ratio switch to 1:1.
   Set frequency switch to 7 kHz or 12 kHz.
   An internal 60-Hz source may be used, but an external oscillator providing 100 or 150 Hz is preferable.
2. Operate as in measuring low-frequency distortion.

Using External High-Frequency Source

To use an external high-frequency source, proceed as follows:

1. Set HF switch to Ext.
   Connect an oscillator capable of delivering at least 10 volts across 600 ohms to the EXT HF terminals. Any frequency from 2 to 20 kHz may be used. If the oscillator does not have both sides of its output circuit ungrounded, use a 600-ohm/600-ohm isolating transformer between the oscillator and the EXT HF terminals.
2. Operate as in the two previous sections.

Analyzing Distortion With Oscilloscope

An oscilloscope may be used to analyze distortion as follows:

1. Connext EXT SYNC post on oscilloscope to top EXT LF terminal, and turn oscilloscope sync selector to Ext Sync.
   Connect y-axis terminals of oscilloscope to EXT LF terminals temporarily.
   Set oscilloscope sweep and sync adjustments to show one low-frequency cycle on the screen.
2. Transfer y-axis terminal connections to SCOPE A terminals of meter. The oscilloscope will then show the high-frequency signal as it varies during one cycle of low frequency. If there is intermodulation, the high-frequency signal will be changed in amplitude during part of the low-frequency cycle. Analyze for distortion causes as covered in the next section. Alternately, the scope may be connected to the SCOPE B terminals.

Distortion Patterns

In the adjustment of an amplifier, every oscilloscope pattern has a meaning. A perfect unit would have a rectangular, notch-free pattern on the scope screen. The presence of one or more notches or bulges is a sign of intermodulation. The following remarks are based on using the SCOPE A terminals.

A typical pattern for a single-ended amplifier is shown in Fig. 13-33A. The notches may be at either the left or right side of the pattern, depending on the number of stages in the amplifier under test and in the oscilloscope deflection amplifier, and depending on the polarity of the oscilloscope sync circuit when it is in the external-sync position. The deeper the notches, the worse is the modulation.

The effect of insufficient bias in a single-ended stage is shown in Fig. 13-33B. Note the narrow additional notches at one side.

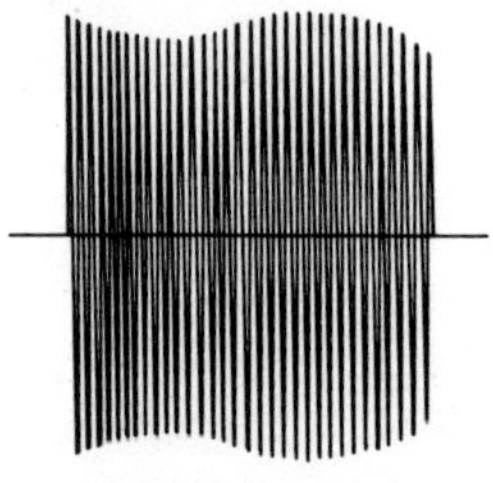

(A) Single-ended stage.

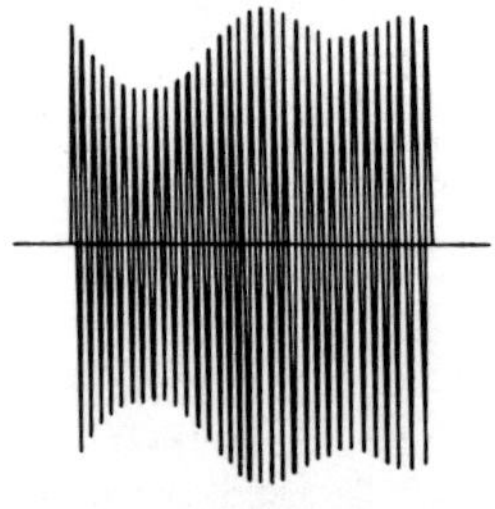

(B) Underbiased single-ended stage.

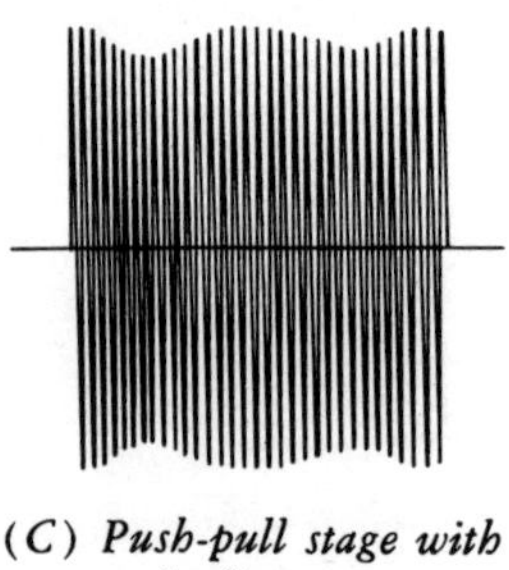

(C) Push-pull stage with good adjustment.

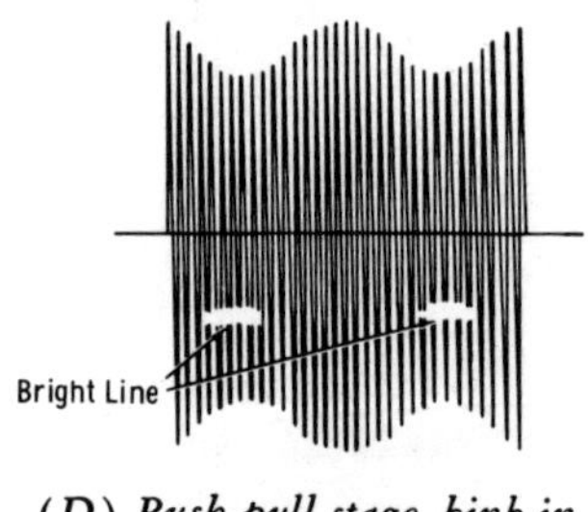

(D) Push-pull stage, kink in transfer characteristic.

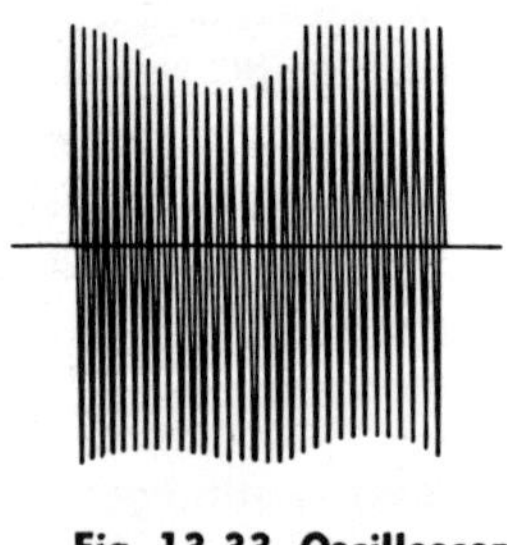

(E) Push-pull stage with overloaded single-ended driver stage.

Fig. 13-33. Oscilloscope patterns from Scope A terminals.

Three push-pull amplifier patterns are shown. Fig. 13-33C shows a good adjustment—all four notches are of equal size. Fig. 13-33D shows a highly undesirable condition, a push-pull stage which is overbiased and overdriven. A slight kink in the transfer characteristic of the tubes ensues; this appears as a bright line on the pattern. Depending on the amount of overdrive, as many as four bright lines can appear. An operating condition that shows a line on the pattern is highly fatiguing to the listener, regardless of the meter reading. The kink in the transfer characteristic can be made plain by readjusting the oscilloscope sweep. Shift the sync switch to internal sync, and set the sweep rate to the high frequency. This will put one cycle of the high frequency on the screen, and the kink will be clearly visible.

Sometimes a pattern shows bulges instead of notches. This is most likely to occur at low frequencies. It is a sign of highly undesirable regeneration and instability at the frequencies at which it appears—effects that lead to poor transient response. It can result from insufficient B+ supply isolation between the input and output stages, or from an excessive phase shift in a circuit using negative feedback. The effect is most likely to occur between 1 and 50 Hz.

If a push-pull system pattern shows three notches (Fig. 13-33E) instead of four, it is a sign that the driver stage (usually single ended) is overloading asymmetrically more rapidly than the output stage.

The preceding discussion is somewhat idealized, for it assumes that there is only one source of distortion (and hence only once cause of pattern notching). Actually, an audio system may distort at several points. For example, for many hearing aids having single-ended output stages it has been found desirable to cancel much of the output-stage second-harmonic distortion by driver-stage distortion in reverse phase. This in part resembles the action of a push-pull stage; a four-notch pattern may be expected.

A still more confusing result may ensue if there is a source of phase shift in the circuit between two distortion sources. The effect is to shift the position of the notches that correspond to the first source of distortion. Thus a six-notch pattern may be created.

Intermodulation Limits

Engineering observation has indicated that intermodulation distortion of less than 10 percent is not immediately noticeable without a chance for direct comparison. In the best commercial disc recording and reproducing systems, the intermodulation distortion is of the order of 2½ to 4 percent at nominal recording level. Radio transmitters average 6 or 7 percent at full modulation and 2 or 3 percent at 80 percent modulation; if conditions are worse than this, retuning often will be a cure.

In any audio system, the amplifier is usually the only portion that can be built to have very low distortion. It is necessary, therefore, to keep the amplifier distortion at a very small fraction of the preceding totals. Hence the amplifier should be used at well below its nominal overload point.

The desirable limit of distortion depends on how long a system is to be heard. For long-period listening, of hours at a time, lower intermodulation distortion (among other things) will reduce listener fatigue.

As was said previously, the relation between harmonic distortion and intermodulation distortion is extremely variable. For a single-ended amplifier, the relation seems fixed only if all the distortion occurs in a single stage and if it occurs in a specified type of tube. Ratios of less than 1 and more than 8 have been observed, as well as all sorts of intermediate values.

Recorder and Reproducer Testing

Both disc and tape recorders and reproducers can generate severe intermodulation. Disc recorders do so as a result of any of these causes: the armature not centered in the air gap; a worn or loose armature bearing; a defective cutting stylus; and amplifier-system distortion, often caused by a lack of sufficient electrical isolation between the amplifier and the cutting head.

Disc reproducers are subject to these faults: stylus tip excessively worn; armature not centered in the air gap; and poor tracking of the stylus in the groove, caused by insufficient pressure on the record. Poor tracking can be caused by the reproducer arm as well as by the reproducer head itself.

Magnetic recording systems may be affected by: incorrect bias current (usually too low), asymmetrical bias waveform, magnetized recording head, and amplifier distortion.

Distortion in a recording and reproducing system is a function of the recording level. Therefore, a series of tests with levels increased in 2-dB steps is often desirable.

Measuring a Disc Reproducer—A disc reproducer may be checked by feeding the output to an amplifier that has negligible intermodulation distortion under the conditions of the test. Play a test record and operate the analyzer section of the meter as usual.

Analyzing Disc Reproducer Distortion—To analyze the distortion from a disc reproducer, proceed as follows:

1. Connect as in the previous paragraph.
   Connect the ground side of the oscilloscope input to the G side of the SCOPE A terminals.
   Connect the EXT SYNC post of the oscilloscope to the high side of the amplifier output (that is, not to the G side).
   Temporarily connect the high side of the oscilloscope input to the high side of the amplifier output.
2. Start to play the intermodulation test record.
   Adjust the oscilloscope sweep and sync controls to show one cycle of the low-frequency wave.
3. Transfer the high side of the oscilloscope input to the high side of the SCOPE A terminals.

The notch pattern is then available for inspection. Meter readings are obtained in the usual way (calibrate and then read).

Measuring a Disc Recorder—Cut a lacquer disc, working near the outside of a 12-inch or 16-inch disc, using 60 and 2000 Hz or 60 and 7000 Hz, and recording at nominal recording levels of +10, +8, +6, +4, +2, 0, −2, −4, and −6 dB (0 dB = 7 cm/s). The highest level should be at the maximum diameter. At the higher recording levels, apply the tone for only a few seconds at a time to avoid heating and changing the compliance of damping materials; allow at least 10 seconds between tone applications for cooling.

The cut disc should then be played with a high-quality reproducer and amplifier that have been previously checked for low intermodulation in the range of stylus velocities used in these tests.

Measuring a Magnetic Recorder—If the recorder has separate record and playback heads, it may be measured in the same fashion as an amplifier. This is done by connecting its input to the OSC OUT terminals and its output to the AN IN terminals, and then operating the tape transport as in normal operation. Test frequencies should be 60 and 2000 Hz. Intermodulation should be less than 3 percent at the nominal recording level, and less than 8 percent at 6 dB above the nominal recording level. With good recorders, distortion is less than 1 percent at the nominal recording level, and less than 6 or 7 percent at 8 or 10 dB above the nominal recording level. If the recorder has a single record-reproduce head, the tone must be recorded first and played back for subsequent measurement.

Equivalent Single-Frequency Power

Manufacturers who rate their amplifiers on the basis of intermodulation tests have wanted a factor to use in converting the indicated power output to a basis comparable with that used by manufacturers who rate their product by harmonic distortion tests. This comparison can be made on a basis of equal peak power output, since distortion in any case is at a maximum at the peak of the output wave. Such a comparative figure is called *equivalent sine wave power;* it is calculated by multiplying the meter-indicated power output in the intermodulation (im) test by 1.47. The power output on an im basis should be measured using a ratio of 4:1.

EXERCISES

Q13-1. If you want to check the continuity of a microphone cable and transformer with a vom, what should you do first?

Q13-2. What is the most likely cause of hum and noise from a specific microphone in an otherwise acceptable pickup?

Q13-3. If you feed the output of a properly attenuated audio oscillator to the pickup leads of a turntable playback preamplifier (pickup disconnected), should you obtain a flat response at the output?

Q13-4. What must be checked first before using an IC extraction tool?

Q13-5. What are the two most important things to do for a magnetic tape system on a daily basis?

Q13-6. Name the five basic characteristics of a tape-recorder head to be considered in alignment.

Q13-7. Is a full-track monophonic alignment tape useful for stereo head alignment?

Q13-8. What does the term "netting" mean in two-way mobile radio?

Q13-9. Two basic means of distortion measurement, intermodulation distortion and total harmonic distortion, have been discussed. Which is required in FCC proof-of-performance measurements?

Q13-10. In any type of distortion measurement, what is the basic characteristic you must consider?

CHAPTER **14**

Transmitter Operations and Maintenance

In the majority of am-fm broadcast installations today, transmitter and studio operations are combined. However, in the event of transmitter problems as revealed by the remote-control monitoring devices, the duties of the personnel often become divided—those concerned with studio maintenance and those concerned with transmitter maintenance.

In any event, it is desirable from an orientation standpoint to treat transmitter installations as a separate, but correlated, study from studio installations. In the instance of remote control from the studio, it is still the responsibility of the operator to analyze any malfunction, to remove the transmitter from the air if it is out of tolerance (this is automatic in many instances), and to place the standby transmitter (or the properly operating portion of a parallel transmitter) on the air.

During the regular broadcast day, the transmitter operator keeps the circuits properly tuned, maintains correct power input to the final stage, logs meter readings in accordance with current FCC rules (which also aids in forestalling trouble), checks the frequency, and maintains the modulation at a level consistent with good engineering practice and the type of program in progress.

The first discussion to follow will pertain to the all-important operation of the broadcast transmitting installation in order to achieve the best results possible from the finely engineered equipment available and in use today. Operating practice at the transmitter is just as important to the final result of overall performance as it is at the broadcast studio. The operation of the transmitter and associated speech-input equipment may be shown to be a highly specialized art, and we have chosen the term "operational engineering" to define the content of the special study undertaken in this part of the chapter.

Following the operations sections, the testing and maintenance of transmitting equipment will be discussed. Proof-of-performance measurements are detailed, and a suggested preventive-maintenance routine is given.

14-1. OPERATOR'S DUTIES

It is true that the primary purpose of the transmitter operator is to *keep the station on the air.* But with the increased demands for higher-fidelity program transmission, the day when a typical "ship operator" possessing thorough technical understanding could step into a broadcast installation has passed. The operator of a broadcast transmitting plant has a specialized range of duties requiring a technical education as well as a thorough understanding and appreciation of the more intangible values of program material.

A number of his fundamental duties are, of course, strictly technical in nature. In brief, they consist of turning on the transmitter before the beginning of the daily program schedule, checking all meter readings to make proper adjustments, checking level with the studio, shutting down the transmitter after sign-off, repairing and maintaining equipment, and testing for noise and distortion levels. During the daily operating schedule, he consistently monitors the program with a monitoring amplifier and speaker, adjusts line-amplifier gain in accordance with good engineering practice pertaining to percentage modulation (the transmitter operator does not normally "ride gain" as does the studio operator), maintains correct transmitter tuning, logs all meter readings as required by the FCC, and corrects any trouble that develops in the shortest possible time in order to keep the station on the air.

The transmitter operator in all but the lowest-power local stations is usually scheduled to be on duty at least 30 minutes prior to air time for the purpose of getting the equipment ready for the broadcast day. The start of an operator's day (when on duty in person at the transmitter) may be outlined as follows:

1. Power is applied to the audio rack, including such measuring equipment as the frequency and modulation monitors. The audio line is used as a program loop and opened by inserting a patch cord into the line jacks. This removes the line from the input to the line amplifier and prevents any test program that might be on the line from the studio from being applied to the transmitter when it is first turned on.
2. A visual inspection of all relays in antenna-phasing cabinets (where used) and in coupling houses at the antenna is performed. Relay armatures are manually operated to ascertain freedom of movement. All rf meters are observed for bent hands or zero set.
3. An inspection of all safety gaps is carried out, including antenna and transmission-line lightning gaps for approximate correct spacings.

4. Air-cooling systems usually start the blower motors when the filament-on switches are operated. Transmitter filaments are next turned on and filament voltages checked. Minimum voltage should first be applied to large power tubes that have tungsten-type filaments; then the voltage should be run up to normal after about 3 to 5 minutes. This procedure is automatic in some transmitters and helps to lengthen the usable life of such power tubes. Tubes with thoriated-tungsten or oxide-coated filaments, such as those used in the low-power stages, are always operated at *normal* filament voltage for maximum tube life.
5. Plate voltage can then be applied to low-power units or exciter units (power installations of 1 kW or more) to check for proper excitation to the final stage.
6. Low power is then applied to the final stage. All meter readings are checked for normal low-power operation. If everything is normal, high power is applied and meter readings are checked.
7. Filament and line voltages are checked and adjusted for high-power operation. Final adjustment is made on the final stage for optimum meter readings regarding resonance and power input.
8. Since the control-room operator sometimes has circuits "hot" with his own testing procedure, the transmitter operator plugs a patch cord from the program line to a monitor amplifier to ascertain continuity of the program line. He then notifies the control operator to stand by for an overall circuit test. When this has been done, the transmitter operator removes the patch cord, automatically restoring the input line amplifier. A test tone may then be fed from the studio to check the overall continuity of the circuits from the studio to the transmitter modulators. In the event of remote-control operation, the studio operator has full control over the transmitter input.

Level checks with the studio are not required as a daily procedure after an initial installation has been made, tested, and operated for some time, since with properly operating equipment the level remains nearly the same over a period of time. At regular intervals, however, it is desirable to use a signal generator to check the frequency characteristics of the line and transmitting equipment. In this connection, it is advisable for the transmitter operator to understand the difference in modulator power requirements for sine waves and the complex waveforms of speech or music program content.

It will be remembered from circuit theory that for a class-C modulated amplifier the power requirement for complete *sinusoidal* modulation is 50 percent of the dc power input to the modulated tube or tubes. Fig. 11-8 (Chapter 11) shows how the peak factor of speech or music waves varies greatly from that of a pure sine wave, when "peaked" the same on a VU meter. This peak factor of program waves is 10 to 15 dB *more* than that of a sine wave. That is, the ratio of peak to rms voltage is far greater for complex waveforms than for sine waves. The *average* power for complete modulation of a transmitter over a period of time is far less than the average power required for complete modulation by means of a signal generator. It is a well known fact that for program signal waves the modulator power required may be 25 percent or less of the dc power input to a class-C stage. Therefore, if a signal generator is used at the studio for frequency runs or level checks, the transmitter operator must realize that if he adjusts the gain on the line amplifier to give 100 percent modulation on sine waves, the same adjustment will be 10 to 15 dB *high* for program signals. The gain adjustment must be lowered accordingly to the point that experience has dictated for program modulation before the actual program schedule starts. In the past, this has led to some confusion among transmitter operators.

This difference in peak factor between program and sine waves is also noticed when comparing the percent of antenna-current increase with 100-percent modulation. It is true that the antenna-current increase should be approximately 22.5 percent over no modulation when a sine wave is applied to the transmitter at 100 percent modulation. Antenna current increases for 100 percent program modulation, however, will be much less, due not only to the difference in peak factor, but also the sluggishness of the thermocouple rf meter action. This slowness of action is due to the heating effect of the two dissimilar metals on which the action of the meter depends.

14-2. CORRELATION OF METER READINGS

It is only natural that the program level being sent by wire from the studio be of utmost importance from a strictly operational point of view to the transmitter operator. With competent studio personnel, the line-amplifier gain adjustment may be set for 100-percent modulation on program peaks at the start of the day and left at that adjustment. Many times, however, a transmitter operator, who in some cases may not appreciate musical and dramatic values, will become piqued with the control operator when the program level is very low. He should realize, however, that broadcast stations are not strictly for "communications," but are intended to bring entertainment into the home with as much of the original content as possible consistent with the state of the art. Certain types of programs, symphony concerts in particular, are meant for those listeners in the primary service area and not intended to override the noise level at some secondary service point. If the monitor speaker is turned up in volume consistent with that of the interested listener at home for these types of programs, the transmitter operator will be able to use good judgment as to whether the signal is or is not entirely too low in level to be usable for proper modulation.

Program Levels

In relation to the study of program levels, it is of primary importance to understand the characteristics of the indicating meters used at both ends of the transmission system. These meters differ in characteristics because of the different functions which they are intended to perform. The standard VU meter, used in most broadcast studios today, is an rms-indicating full-wave rectifier device intended to give a close visual approximation of the sound waves emanating from the speaker. However, the concern at the transmitter is with modulating voltages, and a semipeak indicating device is necessary and required by the FCC. If peaks of the program signal content should be excessive and occur in rapid succession, danger of circuit component breakdowns would arise as well as severe adjacent-channel interference. Therefore, since the peak factor of program waves is high, the modulation meter is a peak-indicating device. It is also necessary that a phase-reverse switch be incorporated in the modulation-meter circuit to switch the polarity of the input to the metering circuit so that either the positive or negative side of the modulated envelope may be monitored separately. Thus, it is obvious that there are two distinct types of level meters, namely, a full-wave rms meter at the studio and a half-wave peak meter at the transmitter. In addition to these meters, there is a limiting-type amplifier (in most modern installations) which is used at the transmitter as a line amplifier. This has meters which measure the amount of compression (full-wave peak meter) and output level in VU (full-wave rms meter).

The number of different types of indicating meters should not confuse the operator as long as the proper interpretation is given to the readings. Fig. 14-1 shows the indication of a program peak at a given instant on the various meters involved. The studio VU meter has registered 100; the compression meter at the transmitter shows the normal 5-dB limiting; the line-amplifier output meter shows 100; and the modulation meter would show either 100-percent modulation on positive peaks or, if set to monitor negative peaks, might show only 60-percent modulation. This, of course, could be just reversed with a change in polarity of the microphone output or any connection in between.

It is a well known fact that speech waves are not equal in positive and negative peaks regardless of the type of microphone used. This may be observed from the graph of the speech wave shown in Fig. 14-2. Two speakers working from opposite sides of a *bidirectional* microphone and peaked the same on the studio VU meter will not give equal indications on the modulation meter when it is set to indicate a certain peak (either positive or negative) because of the negative-peak effect.

Assume, for example, that the modulation-monitor switch is set to monitor the negative peaks, and the indication of one voice is close to 100 percent. The indication of the voice on the other side of the microphone (therefore of opposite polarity at the microphone output transformer) may indicate only 40 to 50 percent, with the amplitude of the studio VU meter remaining the same. For this reason, it is obvious why misunderstandings sometimes arise between studio and transmitter personnel regarding the comparative levels of two or more voices.

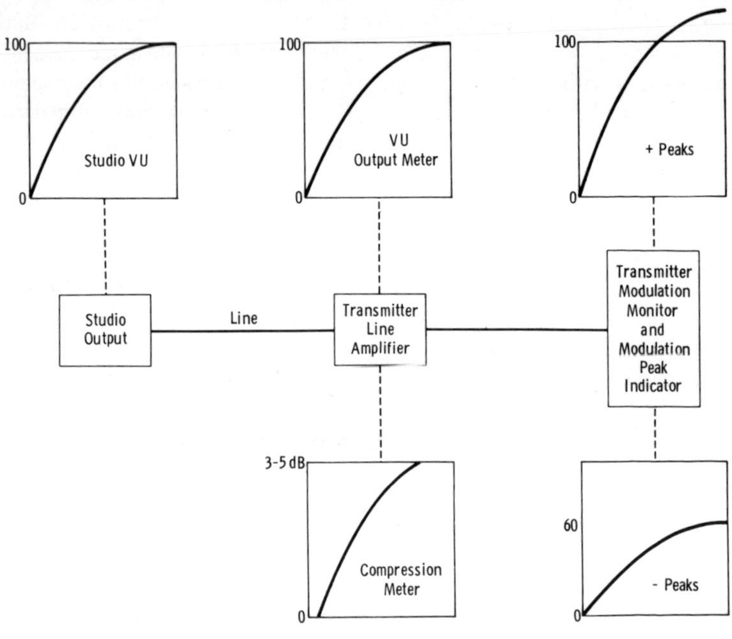

Fig. 14-1. Indications of a given program peak on several meters.

What indication exists at the transmitter plant to show a true indication of comparative levels from the studio? It has been shown that the half-wave reading of the modulation meter, which depends on the polarity of operation, is not a true indication of *comparative* levels from the studios. The VU meter at the output of a limiting amplifier would not be a true indication since the output level is limited by the compression taking place in the amplifier for signals over a predetermined level. The compression meter, although a full-wave indicating device, is a peak-reading instrument, and, since the peak factor of program waves varies considerably, it is not an absolutely accurate indication of comparative levels. It is, however, the most reliable indication (within limits) existing at the transmitter, since it is full-wave rectified and is limited by only wire-line characteristics. If two voices, for example, show about the same amount of compression, the comparative *levels* (not *loudness*) may be considered very nearly the same.

100-Percent Modulation (A-M)

Fig. 14-3A shows an oscillographic pattern of a carrier modulated 100 percent by a sine-wave tone. This illustration shows what constitutes positive and negative modulation of the carrier. It may be seen that negative, or trough, modulation cannot attain more than 100 percent of the available range, whereas positive, or peak, modulation may go over 100 percent. When a carrier is thus modulated with a pure tone, the degree of modulation is:

$$m = \frac{E_{av} - E_{min}}{E_{av}}$$

where,

m is the degree of modulation,
E_{av} is the average envelope amplitude,
E_{min} is the minimum envelope amplitude.

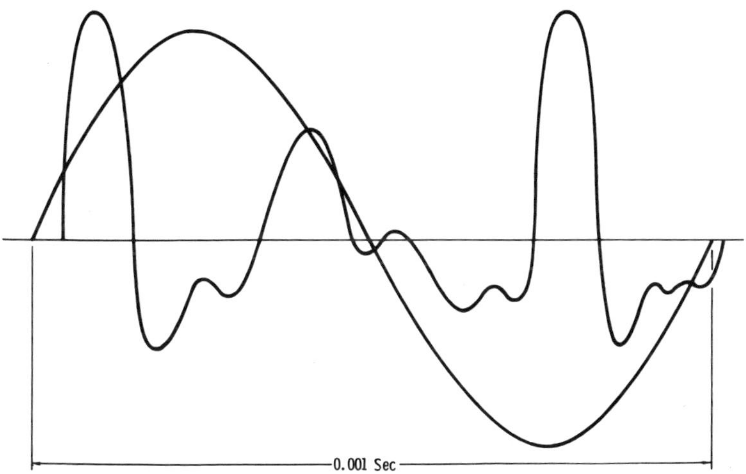

Fig. 14-2. Curves indicating peak factors of voice and sine waves.

The peaks and troughs of the envelope will be equal. When the minimum envelope amplitude (negative peak modulation) is zero in the foregoing equation, m is 1.0, and the degree of modulation is complete, or 100 percent expressed in percentage modulation.

When the envelope variation is not sinusoidal, such as is true for program signals, the positive and negative peaks will not be equal, and the percentage of modulation differs for the peaks and troughs of modulation as follows:

$$\text{Positive peak modulation} = \frac{E_{max} - E_{av}}{E_{av}} 100$$

$$\text{Negative peak modulation} = \frac{E_{av} - E_{min}}{E_{av}} 100$$

where,

E_{av} is the average envelope amplitude,
E_{min} is the minimum envelope amplitude,
E_{max} is the maximum envelope amplitude.

Thus, it is possible to understand why the trough modulation cannot exceed 100 percent, since the minimum voltage cannot be less than zero. It may be seen, however, that the positive peak voltage may be more than twice the average (or carrier) voltage, in which case the positive peak modulation will exceed 100 percent. What important information does this hold for the transmitter operator?

First, it should be clarified in the operator's mind that overmodulation *can* take place on the negative (trough) modulation as well as on the positive (peak) modulation. It is true that the degree of modulation can never exceed unity on the negative peaks, but it *can* exceed unity on the positive peaks. Complete modulation (of a class-C stage), however, requires that the peak values of the modulating voltage equal the dc plate voltage of the modulated stage. Fig. 14-3B shows an oscillographic pattern of a carrier wave with modulating voltage exceeding the dc plate voltage and causing overmodulation of the carrier. It is true that the positive modulation peaks exceed unity while the negative peaks are cut off by the excessive negative modulating voltage and cannot exceed unity. This excess energy, however, which allows the voltage applied to the rf-amplifier plate circuit to become negative with respect to ground, causes radiation in the form of spurious frequencies, resulting in "splatter" and adjacent-channel interference.

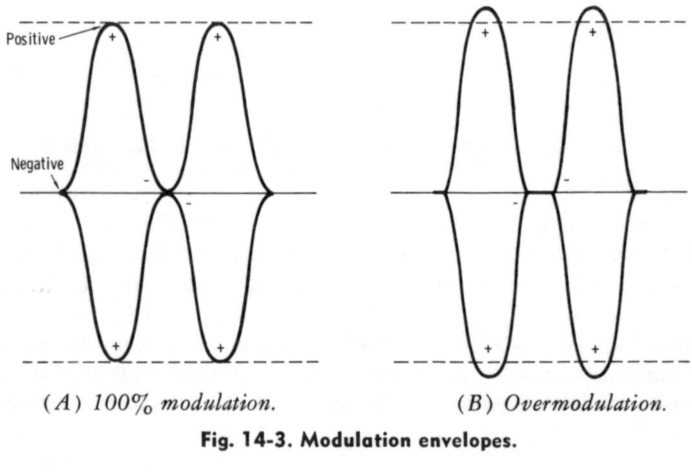

(*A*) *100% modulation.* (*B*) *Overmodulation.*

Fig. 14-3. Modulation envelopes.

This actually is overmodulation in its severest form, since positive peaks may extend beyond 100-percent modulation without amplitude distortion, whereas clipping of the negative peaks will cause severe amplitude distortion. It will be remembered that the bandwidth occupied by the carrier and sidebands depends (for amplitude modulation) not on the degree of modulation, but on the highest frequency being transmitted. Amplitude distortion resulting from negative-peak overmodulation generates a number of distortion harmonics that may extend high enough to spread the sidebands into adjacent channels.

This discussion has been presented in order to show the transmitter operator that the negative side of the modulation is the most important peak to monitor on the modulation meter. It should be held under 100 percent at all times. It is well to remember that a modulation meter of the vacuum-tube-voltmeter type will not be able to indicate over 100 percent (negative) on the meter because the peaks cannot attain more than this value. This is why an oscilloscope is sometimes used at a broadcast transmitter to show negative peak overmodulation, since the negative peak clipping shows up as white lines across the center of the modulated pattern. When the usual vacuum-tube-voltmeter type of modulation indicator is used, the flasher should be set for 95-percent modulation so that the warning is given usually before the meter ever swings to this value. A flasher will respond on a much faster increment of program peaks than will the indicating meter movement. NOTE: FCC rules at the time of this writing limit *positive* peak modulation to 125 percent. Always check current FCC rules.

FM Modulation-Monitor Interpretation

Peak amplitude variations in the positive and negative directions show up on an fm transmitter monitor as a decided difference in amplitude of the plus and minus excursions of the meter. In this case, however, we are not concerned with either peak in relation to distortion, since modern fm transmitters are able to overmodulate greatly either plus or minus without inherent distortion. Distortion does occur, however, in the receiver when overmodulation over noticeable periods of time occurs, since few receivers have the capability of providing faithful reproduction of a modulating signal that produces a frequency swing much in excess of the maximum of 150 kHz (±75 kHz, which is defined as the value of 100-percent modulation for fm broadcasting).

The fm transmitter operator, therefore, is concerned with preventing overmodulation on either peak. For this reason, it is of utmost value to have all studio microphones connected so that their maximum polarity occurs on a definite side, either plus or minus on the modulation monitor. Otherwise, the transmitter operator must check his peaks at each change of microphone before he is certain that the maximum peaks are not over 100 percent.

Automatic Correction of Nonsymmetrical Audio Peaks

The problem of peak polarity in proper modulating techniques, particularly important in a-m transmission, has been stressed. A special passive device designed to distribute equally nonsymmetrical audio peaks, particularly those produced by certain inherent characteristics of the human voice, is shown in Fig. 14-4. By removing asymmetrical energy, this unit permits higher average modulation, thus giving the effect of higher transmitter power with no change in the program signal level.

Courtesy Kahn Research Laboratories, Inc.

Fig. 14-4. A device for removing asymmetrical energy from audio peaks.

Long-line telephone circuits normally correct for speech asymmetry, and music seldom contains unbalanced waveforms. Thus, asymmetrical modulation peaks are mainly caused by live or tape-recorded voice programs originating locally or over relatively short telephone lines. Since the main purpose of this device is to distribute nonsymmetrical energy equally without disturbing symmetrical sources, the remaining modulation problem is thereby removed. Local voice program levels can be raised to equal those of other programs without danger of overmodulation.

This unit also offers another advantage for users of compressor, limiter, or uniform-level type amplifiers. With reduced peak energy, limiting will take place only during higher average modulation levels. When speech clippers are employed, the unit removes the low-frequency bounce normally produced by dc shifts in clipped nonsymmetrical waves.

14-3. LIMITER AMPLIFIERS

The limiting amplifier, or compression amplifier, is a very important link in a broadcast installation. However, its effect may not be advantageous if the wrong operational interpretation is given to the main purpose for which it is designed. This type of amplifier, as designed for use in a broadcast installation, is intended as a peak limiting device, the amount of gain reduction being a function of the program peak amplitude. In order to prevent material reduction in the dynamic range of the signal, the peak gain reduction should not be more than 3 to 5 dB. A broadcast limiting amplifier, therefore, should not be considered as a volume limiter, but as a peak limiter intended to prevent adjacent-channel interference and overloading of transmitter components.

In some installations, a special agc amplifier is used for automatic announcer override of musical background. In this application (which can be considered a small step toward automation), the normal program reaches the agc input at −35 dB, for example. The announce circuit has added amplification ahead of the agc amplifier so that the signal is at a level of −20 dB, for example. Since the agc tends to hold the output constant, about 15 dB of program reduction occurs so that the announce portion is at normal level and gain riding of background is not necessary. The reader should bear in mind that the main limiter amplifier has a normal limiting of only 3 to 5 dB and is not to be confused with the operation previously noted, where the actual compression of the announce portion may be as much as 15 to 25 dB.

It is true that doubling the output power of a transmitter raises the signal intensity 3 dB. It is also true that the limiter amplifier raises the signal level about 3 dB on program peaks. To those familiar with volume indications of program circuits, however, this 3-dB increase on speech or music is of small consequence. As far as the transmitter is concerned, the operator should think of this amplifier as a protective device to limit peaks caused by line transmission and those program peaks that escape the action of the control-room operator.

That the primary purpose of a limiting amplifier may be defeated by erroneous operation is a very important fact for the broadcast operator to know. Seriously detrimental effects will result if this amplifier is operated as a volume-compression device to attempt to provide a coverage area greater than a given power and transmitter location warrant. The attack time of peak limiting (about 0.001 second) is determined by a resistor-capacitor charging circuit with the inherent characteristics of a low-pass filter. At high frequencies, and where the duration of the peak is short compared to this operating time, a portion of the peak energy will escape limiting action. If the average signal level is so high that a great amount of compression takes place at all times, a larger amount of adjacent-channel interference will result, defeating one of the main purposes of the amplifier.

This has been quite noticeable in practice when the program content consists of music from dance orchestras of brass instruments where high peak powers at high frequencies are prevalent. A limiting amplifier operated properly for broadcast service will show about 3 to 5 dB of intermittent gain reduction as indicated by the peak-reading meter used to show the amount of program peak compression. The operator must realize that for certain types of programs, such as symphonies, liturgical music, and operas, the average audio signal may be very low over a period of time—even with limiting amplifiers in use. Dynamic range is just as important to high-fidelity transmission of these types of programs as is the frequency range.

Another consideration is the recovery time, or time required to restore the gain to normal after a peak has momentarily reduced the gain. Optimum recovery time varies for different types of program material. Piano music, for example, sounds unnatural when the recovery time is too short, because the effect is similar to inadequate damping of the strings after they are struck or to holding the sustaining pedal too long on the loud notes. As the recovery time is lengthened, however, the gain will be reduced (in effect) a greater proportion of the total time, and unnatural transmission of certain passages in music will result. If limiting amplifiers are operated properly and not subjected to more than the specified amount of peak load, however, they will serve their primary purpose quite satisfactorily without introducing undesirable effects.

When thinking of a compression amplifier as a means of increasing the service area of a transmitter, keep in mind the known facts concerning the psychological differences that exist in listening habits for various types of programs. A lower relative signal level is tolerable for dance music, news broadcasts, etc., where the average audio level is high over a period of time. In this case where listeners well outside the primary service area of the station may be numerous, the maximum amount of peak limiting may be used to help raise the signal-to-noise ratio at the receiving point. However, symphony broadcasts, choral music, certain liturgical music, opera, etc., where the average audio signal may be very low over a period of time, will appeal only to those listeners who are adequately served with strong carrier signals. In the interests of preserving the original dramatic effects of this type of program, it simply is not technically feasible for a broadcaster to attempt to set a fixed value of coverage area for all types of program material. Similarly, the engineer responsible for the transmission of programs should not attempt to operate all equipment in the same manner regardless of the type of programs being transmitted.

14-4. ANTENNA IMPEDANCE AND TUNING (A-M)

When an antenna is properly tuned, the reactive component is cancelled, leaving only the resistive component. The power input is the magnitude of the square of the rf current times the antenna resistance in which the current is measured. The antenna resistance is determined during the original installation and is part of the data filed with the FCC. This should be checked at periodic intervals.

There are two basic methods of measuring the antenna impedance, the rf-bridge method and the substitution method. In general, a greater degree of skill is necessary in using the rf bridge than in using the substitution method. However, the bridge method is more accurate and reliable. It requires becoming thoroughly acquainted with the particular bridge used. It is highly desirable to obtain initial guidance from an experienced user of the bridge.

Bridge Method

Fig. 14-5 shows the equivalent circuit of a shielded rf bridge such as is used for antenna work. The rf impedance is measured either by a substitution method on the bridge (not to be confused with the substitution method without a bridge as described later) or directly by the unity-ratio bridge method.

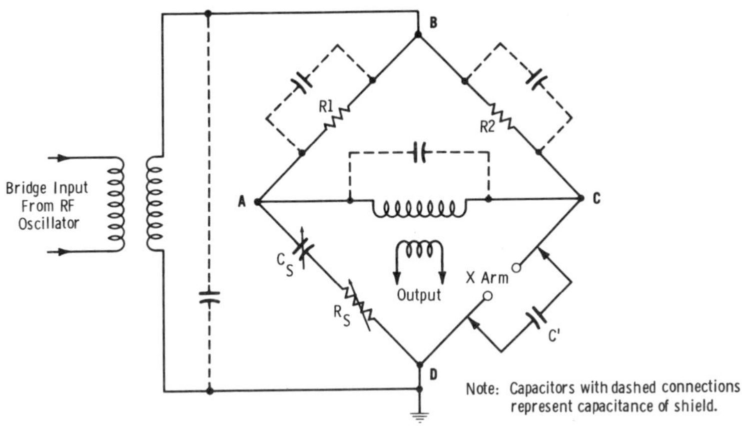

Fig. 14-5. Equivalent circuit of shielded rf bridge.

Method 1 (Bridge-Substitution Method)—The unknown impedance is placed in parallel with C_SR_S. Then R_S is adjusted to zero, and C_S is adjusted to some convenient value. The bridge is then balanced (zero output at the output terminals) with a convenient resistance and capacitance in the X arm. The unknown impedance is then removed. The resistive and reactive components are now determined from the changes required in the calibrated C_S and R_S controls to restore the balance. Obviously, in the use of any bridge, the manufacturer's instructions and precautions must be followed carefully.

Method 2 (Unity-Ratio Bridge Method)—The bridge is first balanced with R_S shorted to ground. The balance is made by adjusting C_S or by the use of an external capacitance from terminal A or C to ground as required. The unknown impedance is then connected to the X arm. The bridge is again balanced by adjusting C_SR_S to match the unknown impedance, indicating the resistive and reactive component values. With this particular type of bridge, when the unknown impedance is inductive, it must be made capacitive by shunting it with a known capacitance (shown in Fig. 14-5 as C′) so that balance may be obtained by the capacitance in the standard (S) arm.

Refer to Fig. 14-6 (T section of a single-antenna tuning unit). The reactance values for the circuit can be calculated as follows:

$$X1 = +\sqrt{R1R2}$$

$$X2 = +\sqrt{R1R2}$$

$$X3 = -\sqrt{R1R2}$$

where,

X1 is the reactance of L1,
X2 is the reactance of L2,
X3 is the reactance of C,
R1 is the line impedance,
R2 is the antenna resistance.

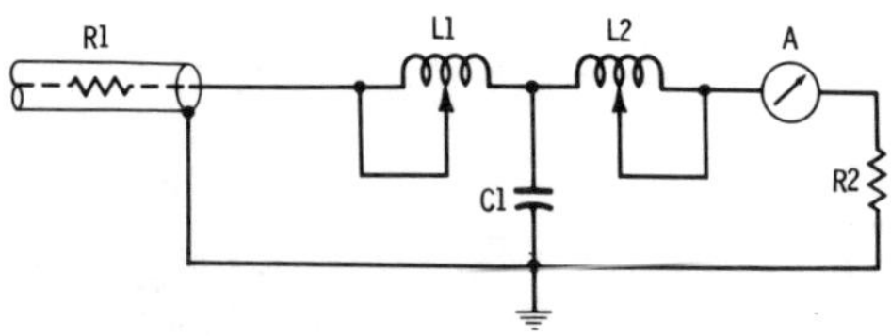

Fig. 14-6. Representation of single T section, antenna tuning unit.

First, it is necessary to determine the impedance of the antenna at the operating frequency so that the proper values of the inductance and capacitance arms may be computed. Impedance measurement of the antenna is made directly at the antenna-tower input terminal with the line disconnected. Usually, measurements are made at frequencies above and below the operating frequency, and the resistance at the operating frequency is determined from a graph of antenna resistance versus frequency.

As an example, suppose a tower has a height of 190 electrical degrees and 28 ohms of resistance and +j25 ohms of reactance. The transmission-line impedance is 70 ohms. Substitution of these values in the preceding equations gives:

$$\sqrt{(70)(28)} = \sqrt{1960} = 44 \text{ ohms (approx)}$$

Therefore, X1, X2, and X3 should have a value of 44 ohms at the operating frequency.

The antenna reactance in this case is +j25 ohms. It is assumed that this reactance is a part of the value of X2. Therefore, subtract the 25 ohms of positive antenna reactance from 44 ohms, giving 19 ohms to be obtained in X2. When X1 and X3 have been adjusted to 44 ohms, an impedance match is produced between the transmission line and the antenna resistance. Adjustment of X2 to 19 ohms cancels the antenna reactance, which would cause a loss of efficiency and a high vswr.

The rf impedance bridge is used in making these adjustments by connecting it across the input of the matching network with the transmission line disconnected. When measurements indicate that an input impedance of 70 ohms has been obtained (with antenna connected), the transmission line is reconnected for the final check before applying power from the transmitter.

Care should be taken that the capacitive branch uses capacitors having a high enough current rating to withstand the current at that point. Usually two or more capacitors of the proper value (to total the required value) are used to handle a current larger than one could safely carry.

Substitution Method

When an impedance bridge cannot be obtained for the necessary length of time, a method commonly known as the substitution method is used to tune the antenna system. Fig. 14-7 shows the arrangement of coupling circuit and tuning unit necessary for this procedure.

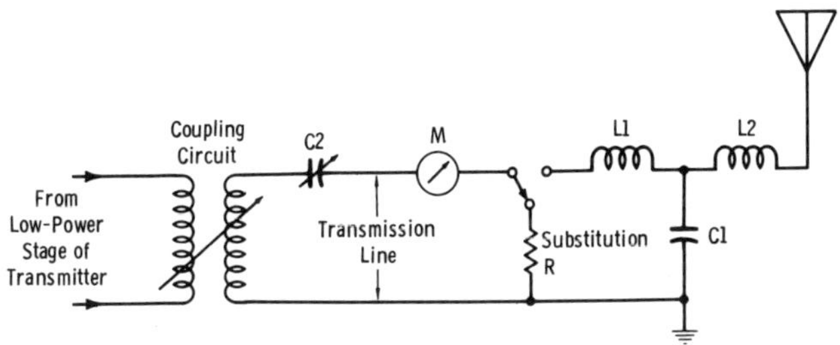

Fig. 14-7. Substitution method for adjusting antenna system.

It may be seen that it is necessary to switch the line from the tuner to a substitution resistance, which should equal the line impedance, so that a change in line current indicated by the meter (M) is noted when the switching action occurs. The tuner is properly adjusted when no change occurs.

Obviously, full power must not be applied to the line, since the voltage rating of a nonreactive resistor is comparatively low. Therefore, connections must be made from a low-power stage in the transmitter, or an accurate rf oscillator may be employed.

Capacitor C is first tuned for a maximum current reading in the line meter (M) with the switch in the resistor position. The tuner input is then switched in, and C is adjusted again for maximum line current. The change in the capacitance of C indicates whether the antenna circuit reactance is capacitive or inductive. If it is necessary to increase the capacitance, the load is capacitively reactive. Conversely, if the capacitor must be decreased to increase the line current, the load is inductively reactive. If no change in C is necessary, the load is resistive.

Final Check of Antenna Match

Whichever method of tuning has been employed, a final check of match conditions should be made before full power is applied to the circuits. The measuring equipment is removed, and low-range thermal milliammeters are inserted at each end of the ungrounded conductor of the line. Sufficient power is then applied from a low-power stage of the transmitter so that the meters give an adequate reading. When the tuning adjustments are correct, these readings agree within 15 percent, showing a proper feeding match between the line and tower networks.

14-5. FIELD-STRENGTH MEASUREMENT TECHNIQUES (A-M)

The inverse field is the unattenuated ground wave (the ground-wave field intensity that would exist considering only the effect of distance) and is used for comparison purposes. The rms value of this field is the radius of a circle (Fig. 14-8) which has the same area as the pattern formed by all the inverse-field-strength values at 1 mile in all horizontal directions from the antenna. The inverse field is found by taking measurements of the actual attenuated field with a field-strength meter; this information is then used to determine the unattenuated field strength at 1 mile in the direction of each radial. The rms value is then:

$$E_o = \sqrt{\frac{E_{10}^2 + E_{20}^2 + \ldots + E_{360}^2}{36}}$$

where,

E_o is the field strength,
E_{10} is the field strength at an azimuth angle of 10° (unattenuated),
E_{20} is the field strength at an azimuth angle of 20°, etc.

Less computation is involved if the actual field is first plotted on polar paper and a polar planimeter used to determine the equivalent-area circle.

The unattenuated ground wave is inversely proportional to the distance and proportional to the square root of the power. Thus if 100 mV/m is produced at 1 mile with 1 kW, 5 kW will produce $\sqrt{5}$ times 100 mV/m, or 223.5 mV/m from the same antenna.

Fig. 14-9 is a fundamental block diagram of a field-strength meter. A loop antenna into which a known oscillator amplitude may be injected for calibration of the meter is normally employed for a-m measurements. The measurement of the received signal may be taken from the calibrated attenuator and/or gain controls. Most of the later meters for standard a-m frequencies are designed for direct reading in millivolts per meter or microvolts per meter. It is obviously essential to become thoroughly familiar with the equipment before field-strength measurements are attempted.

The following requirements govern the taking and submission of data on the field intensity produced. First, have available a sufficient quantity of log-

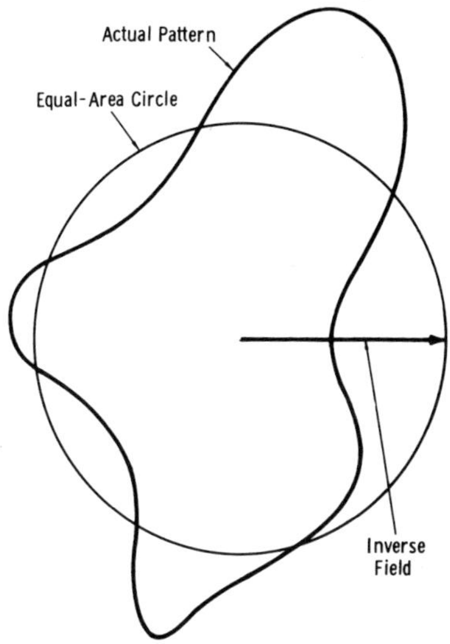

Fig. 14-8. Inverse field.

log graph paper. For a direct match to the FCC Documentary Graphs use K&E Ground-Wave Field Intensity paper No. 61729. Other graph papers that may be used are Dietzgen 3 × 3 cycle log No. 340-L33, or K&E No. 359-120. The steps to be followed are prescribed by the FCC as follows (be sure to check the current rules):

1. Beginning as near to the antenna as possible without including the induction field and to provide for the fact that a broadcast antenna is not a point source of radiation (not less than one wavelength or 5 times the vertical height in the case of a single-element antenna or 10 times the spacing between the elements of a directional antenna),

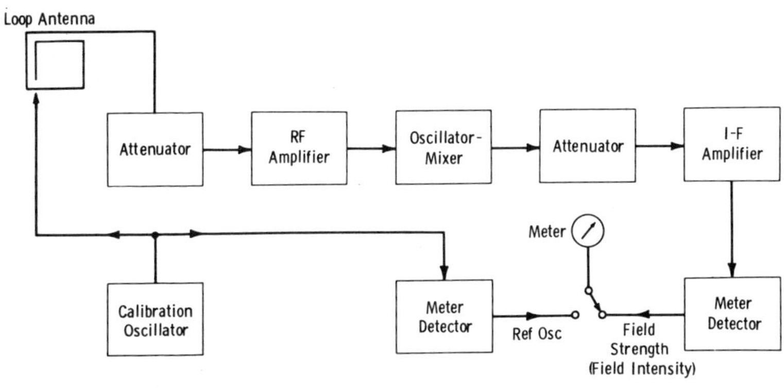

Fig. 14-9. Block diagram of a field-strength meter.

measurements shall be made on eight or more radials, at intervals of approximately one-tenth mile up to 2 miles from the antenna, at intervals of approximately one-half mile from 2 miles to 6 miles from the antenna, at intervals of approximately 2 miles from 6 miles to 15 or 20 miles from the antenna, and few additional measurements if needed at greater distances from the antenna. When the antenna is rurally located and unobstructed measurements can be made, there shall be as many as 18 or 20 measurements on each radial. However, when the antenna is located in a city where unobstructed measurements are difficult, measurements shall be made on each radial at as many unobstructed locations as possible, even though the number of intervals is considerably less than stated above, particularly within 2 miles of the antenna. In cases when it is not possible to obtain accurate measurements at the closer distances (even out to 5 or 6 miles due to the character of the intervening terrain), the measurements at greater distances should be made at closer intervals. (It is suggested that "wave tilt" measurements may be made to determine and compare locations for taking field-intensity measurements, particularly to determine that there are no abrupt changes in ground conductivity or that reflected waves are not causing abnormal intensities.)

2. The required data should be plotted for each radial in accordance with either of the two following methods:
   (A) Using log-log coordinate paper, plot the field intensity as the ordinate and the distance as the abscissa.
   (B) Using semilog coordinate paper, plot the field intensity times distance as the ordinate on the log scale and the distance as the abscissa on the linear scale.
3. Regardless of which of the methods is employed, the proper curve to be drawn through the points plotted shall be determined by comparison with the curves in FCC Paragraph 73.184 as follows: Place the sheet on which the actual points have been plotted over the appropriate graph in FCC paragraph 73-184. Hold it to the light if necessary and adjust until the curve most closely matching the points is found. This curve should then be drawn on the sheet on which the points were plotted, together with the inverse-distance curve corresponding to that curve. The field at 1 mile for the radial concerned shall be the ordinate on the inverse-distance curve at 1 mile.

   Fig. 14-10 shows a typical graph; in this instance the inverse field at 1 mile is found to be 210 mV/m. Dot this in on the 1-mile abscissa and the 210-mV/m ordinate. Since the inverse field varies inversely with the distance, place another dot on the 10-mile abscissa at the 21-mV/m ordinate (one-tenth of 210). The entire inverse curve may then be drawn with a straightedge through these two points.
4. When all radials have been analyzed, a curve shall be plotted on polar-coordinate paper from the fields obtained which gives the inverse-

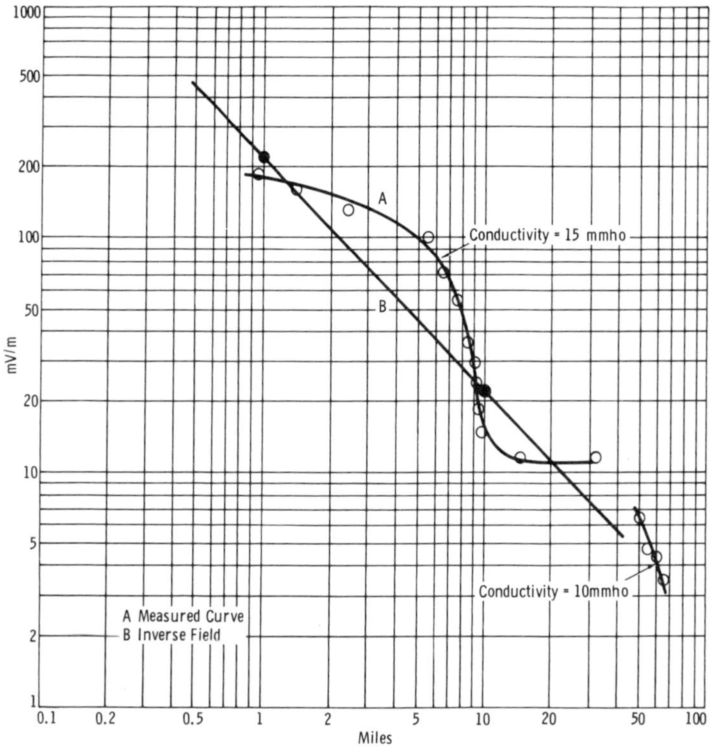

Fig. 14-10. Example of field-strength plot.

distance field pattern at 1 mile. The radius of a circle whose area is equal to the area bounded by this pattern is the effective field.

5. While making the field-intensity survey, the output power of the station shall be maintained at the licensed power as determined by the direct method. To do this it is necessary to determine accurately the total antenna resistance (the resistance variation method, the substitution method, or the bridge method is acceptable) and to measure the antenna current by an ammeter of acceptable accuracy.

The complete data taken in conjunction with the field-intensity measurements should be recorded, including the following:

1. Tabulation by number of each point of measurement to agree with the map required in item 2 below, the date and time of each measurement, the field intensity (E), the distance from the antenna (D), and the product of the field intensity and distance (ED) (if data for each radial are plotted on semilogarithmic paper, see above) for each point of measurement.

2. Map showing each point of measurement numbered to agree with tabulation required above.
3. Description of method used to take field-intensity measurements.
4. The family of theoretical curves used in determining the curve for each radial properly identified by conductivity and dielectric constants.
5. The curves drawn for each radial and the field-intensity pattern.
6. Antenna resistance measurement:
   (A) Antenna resistance at operating frequency.
   (B) Description of method employed.
   (C) Tabulation of complete data.
   (D) Curve showing antenna resistance versus frequency.
7. Antenna current or currents maintained during field-intensity measurements.
8. Description, accuracy, date, and by whom each instrument was last calibrated.
9. Name, address, and qualifications of the engineer making the measurements.
10. Any other pertinent information.

FIELD INTENSITY MEASUREMENT LOG

POINT 1 (Location)

AZIMUTH ANGLE	DISTANCE MILES	MV/M @ 1 mile Specified	MV/M @ 1 mile Obtained	Actual mv/m measured
12°	1.8	200	150	30
		POINT 2 (Location)		
14°	2.0	180	140	18
- - -	- - -	- - -	- - -	- - -
		POINT 10 (Location)		
40°	2.7	90	76	6.2

Fig. 14-11. Example of field-strength log for directional antenna.

A station license may require periodic field-strength measurements (particularly in the case of directional arrays). Even if it is not specifically required, it is a good practice to make these measurements at least quarterly. The following data are required:

1. Monitor-point identification. On the log for field-strength measurements, this may be only an identifying number. However, a complete description for each point must be on file at the station, usually a

duplicate of the original proof-of-performance data filed with the FCC for license application.
2. Specified unattenuated field strength at 1 mile (inverse field).
3. The inverse field at 1 mile *obtained.*
4. Actual received field strength at monitoring point logged.

An example of a field-strength log is shown in Fig. 14-11.

It is a good practice to make a complete list of components employed in the antenna system, such as meter types and calibration, types and manufacture of coils and capacitors, etc., along with the actual bridge measurements of reactance for the portions of the coils and (variable) capacitors used. This allows quick replacement and adjustment in case of damage by electrical storms or other causes. Keep a record of all dial settings on a matching or phasing unit. Keep all components and relay contacts clean.

14-6. PROOF OF PERFORMANCE (A-M)

The important characteristics of any modern broadcast installation are an adequate frequency range to convey as much of the original sound as possible, low noise and distortion levels necessary for the required dynamic range, and dependability of performance. One of the most important pieces of auxiliary equipment in the transmitting plant is the instrument used to determine noise and/or distortion over the usable frequency range. Several manufacturers supply such equipment, and most stations are equipped with a means of checking noise and distortion.

Adjustments and maintenance must be performed to assure that the following FCC requirements are met. Always check the latest revisions of FCC rules.

1. The total audio-frequency distortion from microphone terminals to the antenna output, including the microphone amplifier, must not exceed 5 percent harmonics (voltage measurements of arithmetical sum or rss) when modulating from 0 to 84 percent, and not over 7.5 percent harmonics (voltage measurements of arithmetical sum or rss) when modulating 85 to 95 percent. (Distortion shall be measured with modulating frequencies of 50, 100, 400, 1000, 5000, and 7500 Hz up to the tenth harmonic or 16,000 Hz, or any intermediate frequency that readings on these frequencies indicate is desirable.)
2. The audio-frequency transmitting characteristics of the equipment from the microphone terminals (including microphone amplifier unless microphone frequency correction is included, in which event proper allowance shall be made accordingly) to the antenna output must not depart more than 2 dB from the response at 1000 Hz between 100 and 5000 Hz.
3. The carrier shift (current) at any percentage of modulation must not exceed 5 percent.

4. The carrier hum and extraneous noise (exclusive of microphone and studio noise) level (unweighted rss) must be at least 45 dB below 100-percent modulation for the frequency band of 30 to 20,000 Hz.

Refer to Fig. 2-14 (and associated text in Chapter 2) for a typical audio-oscillator feed used for a broadcast proof of performance. Remember that this proof requires the signal-generator output to feed a studio microphone-preamplifier input.

Fig. 14-12 shows a block diagram of a typical noise and distortion meter, which is also used to check frequency response. In measuring harmonic distortion, the following takes place:

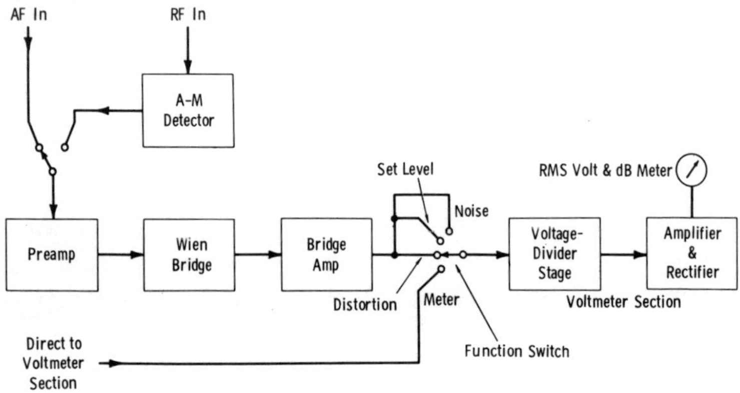

Fig. 14-12. Block diagram of noise-distortion meter.

A. The amplitude of a single-frequency sine wave is measured.
B. A tuning circuit is adjusted to suppress the fundamental frequency. This is done by a sharply tuned circuit and bridge-balance control to achieve at least 80 dB of suppression at the fundamental frequency.
C. The remaining measured amplitude is the total harmonic distortion.

Normally the transmitter output measurement is taken from a special output of the modulation monitor designed for this purpose. However, an a-m detector may also be provided for direct off-the-air measurement as shown in Fig. 14-12.

Always be sure of the "back-to-back" characteristics of the audio oscillator and noise-distortion meter before taking complete proofs or after tube changes or other servicing of the measuring equipment. The direct combination of these instruments should indicate a frequency response flat within 2 dB from 30 to 15,000 Hz. The distortion should measure less than 0.2 percent from 60 Hz to 20 kHz and not more than 0.35 percent from 20 to 50 Hz. The noise should measure at least 70 dB below the output level of the audio oscillator. If it does not, substitute new tubes one at a time (allowing good warmup time) or follow the manufacturers' instructions for necessary adjustments to bring the equipment within the above specifications or the specifications for the particular equipment used.

For frequency-response runs, always consider the back-to-back response of the measuring equipment in calibration of the readings for the input level at the studio. For example, if the back-to-back response is down 2 dB at 5000 Hz relative to 1000 Hz (reference frequency), then feed an input two decibels higher to the studio at this frequency.

Frequency-Response Runs

Fig. 14-13 shows a typical setup for a complete proof-of-performance run. The FCC rules require an overall response (from studio microphone input to transmitter output) flat within 2 dB from 100 to 5000 Hz. They further require that measurements be made at 25, 50, 85, and 100 percent (or the highest attainable) modulation from 50 to 7500 Hz. A step-by-step procedure follows.

1. With the audio oscillator set at 1000 Hz, feed −50 dBm to a studio microphone input (Fig. 14-13). If the oscillator meter indicates zero dBm at the adjusted gain, the calibrated attenuators will total 50 dB. Usually the oscillator meter has a full-scale reading of +15 to +17 dBm. It is advisable to adjust the oscillator gain to obtain, say, +10 dBm and set the attenuators for a total of 60 dB to result in a −50

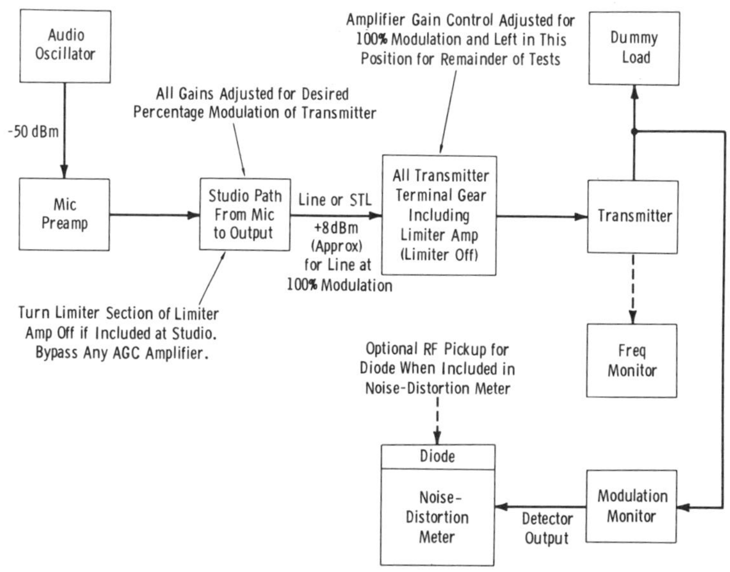

Fig. 14-13. Block diagram of setup for a-m proof of performance.

dBm output. Usually the back-to-back measurement previously described gives the best results with this type of operation, particularly when distortion and noise measurements are taken. Usually there are three sets of calibrated attenuators—tens, units, and tenths—so that precise readings may be made.

OVERALL AUDIO FREQUENCY RESPONSE DATA

25% MODULATION

Hz	30	50	100	400	1000	5000	7500
(1)							
(2)							
(3)							

50% MODULATION

Hz	30	50	100	400	1000	5000	7500
(1)							
(2)							
(3)							

85% MODULATION

Hz	30	50	100	400	1000	5000	7500
(1)							
(2)							
(3)							

100% (or %) MODULATION

Hz	30	50	100	400	1000	5000	7500
(1)	60	60	60	60	60	60	60
(2)	58.2	59.1	59.7	59.9	60	61	58
(3)	+1.8	+0.9	+0.3	+0.1	0	−1	+2

(1) RECORD THE ATTENUATOR READING FOR THE 1000 Hz REFERENCE SIGNAL IN EACH SPACE IN THIS ROW.

(2) RECORD THE ATTENUATOR READINGS FOR THE SPECIFIED FREQUENCIES IN THIS ROW.

(3) RECORD THE AUDIO FREQUENCY RESPONSE VARIATION IN THIS ROW WHICH IS OBTAINED BY SUBTRACTING ROW (2) FROM ROW (1). THESE FINAL FIGURES ARE TO BE USED IN PLOTTING THE GRAPHS.

Form No. AFP-1

Engineer ______________________

Date ______________________

Courtesy National Association of Broadcasters

Fig. 14-14. Form for recording frequency-response data.

2. Adjust the associated faders and master gain control for reference line output (an indication of zero on the studio-line VU meter). As discussed in Chapter 2, this is actually +4 to +12 VU. All faders and the master gain control should be in approximately the normal operating positions. The term "approximately" must be used since the microphone-fader setting will in most cases actually be somewhat

higher due to the high peak factor of speech waves compared to the sine-wave rms value.

3. Be sure that any agc amplifier is bypassed (patched around) and that if a limiting amplifier is employed at the studio, the limiter section is turned off.
4. At the transmitter, adjust the line or limiter amplifier to obtain 100-percent modulation. The limiter section of the limiter amplifier must be off.
5. Record the oscillator attenuator reading. Refer to Fig. 14-14 for an example of tabulated response data in the form suggested by the NAB (National Association of Broadcasters). Note that the attenuator reading for 100-percent modulation at 1000 Hz is simply recorded along the entire top row. Assume this to be 60 for illustrative purposes. Copy the figure 60 also in row 2 for 1000 Hz.
6. Tune the oscillator to 30 Hz, and readjust the attenuators (if necessary) to again obtain 100-percent modulation. Record the new attenuation figure in row 2 under 30 Hz. Repeat this procedure for the other frequencies listed.
7. Fill in row 3 by subtracting the readings in row 2 from those in row 1. This is a record of the response variation. Note that, for example, if it was necessary to reduce the attenuation 1.8 dB at 30 Hz, this indicates that a 1.8-dB *higher level* was required at 30 Hz relative to 1000 Hz to obtain 100-percent modulation. Therefore the response of the system is *down* 1.8 dB at this frequency relative to the reference frequency, and is so plotted on the graph (Fig. 14-15).
8. The entire process is repeated at the lower percentages of modulation required. Two methods may be used:
   (A) The audio-oscillator attenuators may be raised in value to obtain the new (lower) percentage modulation, or
   (B) The faders on the console may be adjusted to obtain the new percentage, retaining the −50 dBm input to the microphone preamplifier at the reference frequency of 1000 Hz.

Remember that the basic idea for a proof run is to prove that the system can be brought within specifications by the proper adjustments or servicing. A proof is a record of the performance as it exists at the time of measurement. This is no guarantee that a single tube in the system may not deteriorate the next hour, day, or week to result in different performance. However, the more often such tests are made and proper remedial steps taken to correct deficiencies, the better the overall results will be.

When a system fails to meet response specifications, first ascertain whether the trouble is at the studio, at the transmitter, or in the line. When the transmitter is at a separate location from the studio, feed the audio oscillator directly into the line. If the trouble persists, run a characteristic curve on the line (or stl) itself to isolate the fault or determine if the result is cumulative. If the trouble is at the studio, it is a simple matter to run studio-only frequency-response checks.

Noise Level

Hum and noise must be at least 45 dB below the level representing 100-percent modulation between 30 Hz and 20 kHz. The reference frequency is 1000 Hz, and this measurement may actually be performed in Step 5 of the preceding procedure by removing the oscillator (after the reference level has been established) and terminating the microphone-amplifier input with a resistor equal to its input impedance. Leave all gains as originally set, but be sure that no other source faders are open. The noise-distortion meter is then placed in the noise mode and its sensitivity increased to

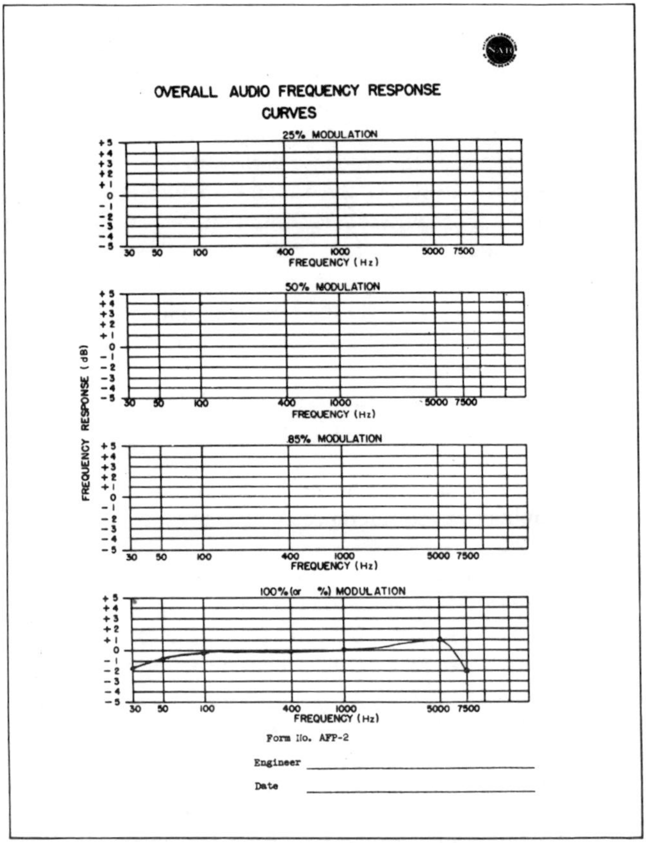

Courtesy National Association of Broadcasters

Fig. 14-15. Form for plotting frequency-response curves.

obtain the noise level in the unmodulated carrier. The microphone input level should be no lower than −50 dBm.

The input tubes of preamplifiers are the most common cause of noise. If an oscilloscope is available, connect it to the scope terminals of the noise-distortion meter, and determine if the noise is caused by a hum component. Many amplifier or preamplifier power supplies employ "hum" potentiometers which should be adjusted while the noise measurement is watched. Use the same trace-down procedure for hum and noise as that previously described for running down the cause of an improper frequency response.

Audio-Frequency Harmonic Distortion

Fig. 14-16 illustrates the NAB-suggested forms for recording the audio-frequency harmonic content, which is normally measured in terms of percentage. The measurements must be taken at the modulation percentages indicated and at the specified single sine-wave frequencies. The noise-distortion meter must be capable of measuring throughout the harmonic spectrum required by the FCC. The harmonic distortion must not exceed 5 percent up to 84-percent modulation, or 7.5 percent for modulation percentages greater than 84 percent. A typical procedure for the measurement of audio-frequency harmonic distortion follows.

1. The reference modulation at the reference frequency (1000 Hz) is obtained as in the frequency-response procedure. The noise-distortion meter is then placed in the set-level position, and the meter is adjusted to 100-percent calibration.
2. The noise-distortion meter is then placed in the distortion position, and the tuning and bridge-balance controls are adjusted for minimum reading. The sensitivity is next increased in steps of 10 dB, and these controls are readjusted for minimum each time.
3. The percentage is read directly on the most sensitive scale possible to obtain a minimum reading by use of the tuning and balance controls.

Excessive harmonic distortion can be caused by tubes, low power-supply voltages, or overdrive of any amplifier or chain. Again, this is a case of tracing down the source of distortion first to the transmitter only, line or stl, or studio. Also, if excessive noise is present, the distortion measurement is apt to be high because of the noise level. It is important to keep the microphone input level from the audio oscillator no lower than −50 dBm.

Carrier Shift

Carrier shift is a change in the average value of the modulated rf carrier compared to the average value of the unmodulated carrier. A shift in the upward direction is called a positive carrier shift; a shift downward is a negative carrier shift. Excessive carrier shift results in unwanted harmonics and additional sideband frequencies with consequent interference on adjacent channels.

Fig. 14-17 illustrates a convenient form for use in recording the carrier-shift data. A tone of 400 Hz is required. Assume, for example, that the rectified unmodulated carrier measures 1 volt. This is recorded in row 1. Under modulation, the voltage drops to 0.95 volt. This is recorded under the first reading in row 2. The difference voltage is 0.05 volt, recorded in row 3. The ratio of the number in row 3 to the corresponding number in row 1 is 0.05/1, or 0.05. This is 5-percent carrier shift (negative).

In practice, the rf input meter on the modulation monitor (which is normally adjusted to 100 percent in operation) may be used to measure this carrier shift. In the preceding example, if the modulated carrier exhibits a 5-percent negative carrier shift under modulation, the input meter will indicate 95 percent rather than 100 percent.

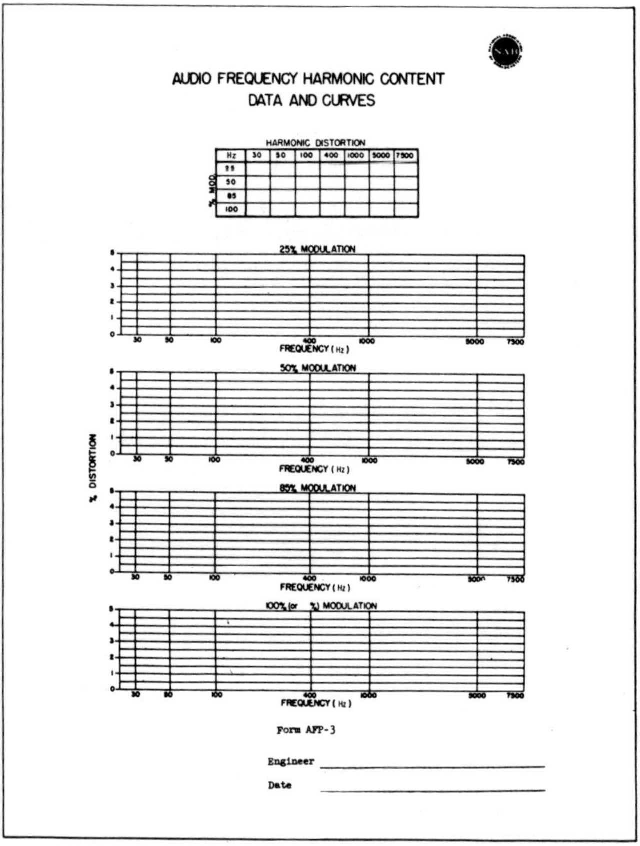

Courtesy National Association of Broadcasters

Fig. 14-16. Form for recording harmonic-distortion data.

CARRIER SHIFT AND COMBINED
NOISE AND HUM DATA

CARRIER SHIFT DATA (at 400 Hz)

% MOD.	25	50	85	100
(1)	1.00			
(2)	0.95			
(3)	0.05			
(4)	5%			

(1) RECORD DC VOLTMETER READING WITHOUT MODULATION IN EACH SPACE IN THIS ROW.

(2) RECORD DC VOLTMETER READINGS WITH MODULATION IN THIS ROW.

(3) SUBTRACT ROW (2) FROM ROW (1) AND RECORD DIFFERENCE IN THIS ROW.

(4) COMPUTE CARRIER SHIFT BY EQUATION: $\frac{\text{ROW (3)}}{\text{ROW (1)}} \times 100$, AND RECORD RESULTS IN THIS ROW.

COMBINED NOISE AND HUM READING

dB	%

Form No. AFP-4

Engineer ______________________

Date ______________________

Courtesy National Association of Broadcasters

Fig. 14-17. Form for recording carrier-shift and noise and hum data.

The maximum carrier shift at any of the specified degrees of modulation must be less than 5 percent in either the positive or negative direction. Excessive carrier shift may result from any of the following: overmodulation (check accuracy of modulation monitor with oscilloscope), improper grid bias, poor grid-bias supply regulation, poor plate-supply regulation, defective power-supply filters, faulty neutralization, or improper rf excitation.

Spurious Radiations and RF Harmonics

Spurious radiations and rf harmonics must be kept at a minimum and must never be of sufficient amplitude to cause undue interference to other services. Fig. 14-18 is a suggested form for these measurements and is self-explanatory.

Applicable FCC Rules (always check latest rulings) are as follows:

1. Any emission appearing on a frequency removed from the carrier by between 15 kHz and 30 kHz inclusive shall be attenuated at least 25 dB below the level of the unmodulated carrier.
2. Any emission appearing on a frequency removed from the carrier by more than 30 kHz and up to and including 75 kHz shall be attenuated at least 35 dB below the level of the unmodulated carrier.
3. Any emission appearing on a frequency removed from the carrier by more than 75 kHz shall be attenuated at least [43 + 10Log_{10} (power, in watts)] dB below the level of the unmodulated carrier, or 80 dB, whichever is the lesser attenuation.

1. ON A GENERAL COVERAGE COMMUNICATIONS TYPE RECEIVER SLOWLY SCAN THE RADIO SPECTRUM FROM 540 kHz TO 30 MHz FOR ANY INDICATION OF SPURIOUS EMISSIONS (OTHER THAN HARMONIC RADIATION) AND RECORD RESULTS BELOW.

FREQUENCY	DESCRIPTION AND INTENSITY OF EMISSION

2. USING THE SAME RECEIVER, MAKE OBSERVATIONS ON HARMONICALLY RELATED FREQUENCIES UP TO AND INCLUDING THE 15 TH HARMONIC, NOTING IN THE BOX BELOW THE S METER OR AUDIBLE RESULTS FOR EACH HARMONIC.

HARMONIC	S-METER READING OR AUDIBILITY RATING
2nd	
3rd	
4th	
5th	
6th	
7th	
8th	
9th	
10th	
11th	
12th	
13th	
14th	
15th	

FORM AFP-4A

ENGINEER ______________

DATE ______________

Courtesy National Association of Broadcasters

Fig. 14-18. Form for recording spurious-radiation and harmonic data.

14-7. PROOF OF PERFORMANCE (FM MONO)

The same technique as that previously described for a-m is used for fm proof of performance; the essential difference is the broadened and more stringent requirements of performance. The frequency-response measurement may be taken either with or without de-emphasis. It is a good engineering practice to run this check both ways. This proves that the station de-emphasis network in the monitor circuit is essentially complementary to the 75-microsecond pre-emphasis.

Fig. 14-19 is an example of tabulated data obtained by measuring the transmitter output without de-emphasis. Just as in a-m measurements, the entry in row 1 is the attenuator setting for the 1000-Hz reference frequency to obtain the specified modulation. Recorded in row 2 are the actual settings required at the specified frequencies to maintain the reference modulation. The numbers in row 3 are obtained by taking the difference between each entry in row 1 and the corresponding entry in row 2. Note

.... 25% MODULATION

Hz	50	100	400	1000	5000	10,000	15,000
(1)	60	60	60	60	60	60	60
(2)	57	58	59	60	69	74	76
(3)	+3	+2	+1	0	-9	-14	-16

(1) Record attenuator reading for the 1000-Hz reference signal in this row.
(2) Record the attenuator readings for the specified frequencies in this row.
(3) Record the AF response variations, which are obtained by subtracting row (2) from row (1), in this row. These final figures are to be used in plotting the graphs.
NOTE: The figures inserted in this example indicate pre-emphasis measurement.

Fig. 14-19. Example of data from frequency-response check of fm transmitter.

that as in the a-m measurements the plus attenuator setting indicates that the response is lower. These figures are plotted on the standard 75-μs pre-emphases curve (Fig. 14-20). Note in this example that the original run, where the 1000-Hz reference is 0 dB, results in the values for frequencies of 5000 Hz and 10,000 Hz falling outside the tolerable limits. Note, however, that at 5000 Hz (farthest out of the range) a shift of −0.8 dB brings this measurement within the limits. Therefore a new curve may be drawn (as shown by solid dots) with an axis shift of −0.8 dB, and the proof is satisfactorily completed. If this shift had resulted in other points falling outside the limits, remedial measures would have been needed to correct the fault. Measurements must be taken at least at the frequencies specified in Fig. 14-19 and for modulation percentages of approximately 25, 50, and 100 percent. The response is then run with standard de-emphasis, and the overall curve is drawn as shown in Fig. 14-15 for a-m, except for the extended frequency range.

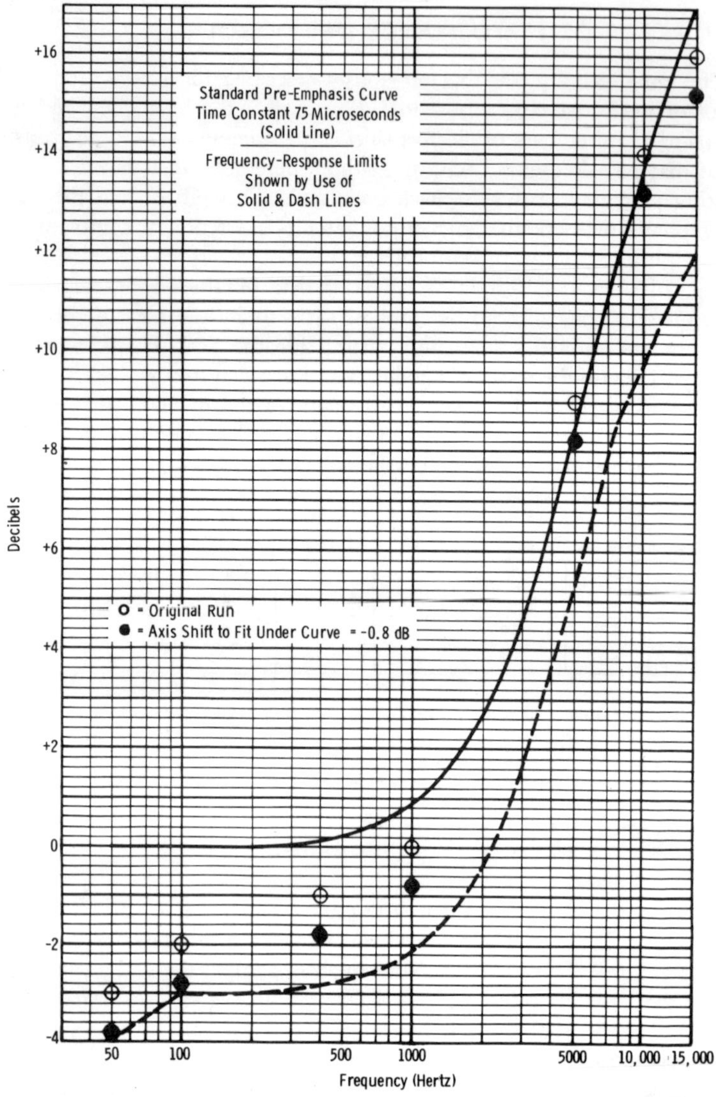

Fig. 14-20. Sample plot of frequency-response data for fm transmitter.

Audio-frequency harmonic distortion is measured in the same manner as described for a-m, except for the following: Distortion must be measured for fundamental frequencies of 50, 100, 400, 1000, and 5000 Hz and 25-, 50-, and 100-percent modulation. In addition, at 100-percent modulation, distortion frequencies of 10,000 and 15,000 Hz must also be measured. The measurements must include harmonics to 30,000 Hz. The maximum allowable distortion (measured through a standard 75-μs de-emphasis circuit) is as follows:

50-100 Hz: 3.5 percent
100-7500 Hz: 2.5 percent
7500-15,000 Hz: 3.0 percent

The output noise level for fm is measured in two categories, fm noise and a-m noise. The method of measuring fm noise is the same as that described for a-m in the preceding section. This includes any noise in the entire system that would result in frequency modulation of the carrier. Just as in a-m, fm noise is measured in dB below the level corresponding to 100-percent modulation, which for fm broadcast is a frequency swing of ±75 kHz. This measurement must be made with standard 75-μs de-emphasis. The indicating instrument must have ballistic characteristics similar to those of a standard VU meter. The fm noise must be at least 60 dB below 100-percent modulation, with a 75-μs de-emphasis circuit employed.

It is necessary to obtain the a-m noise level in terms of what would correspond to 100-percent amplitude modulation of the transmitter, but it is obviously not possible to amplitude modulate the fm transmitter to 100 percent. Some other means must be used to calibrate the noise meter.

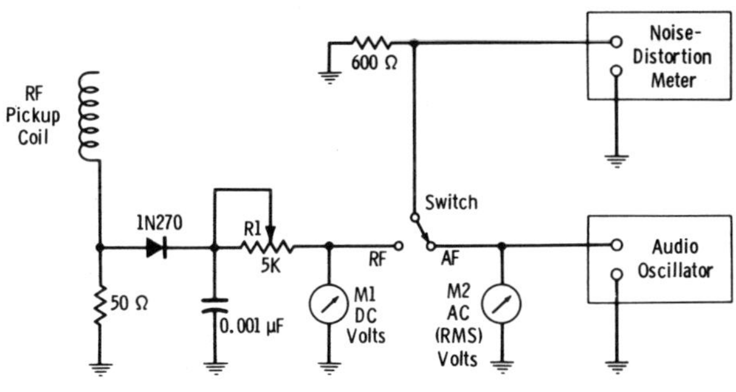

Fig. 14-21. One method of measuring a-m noise level of fm transmitter.

The arrangement shown in Fig. 14-21 may be used for this purpose. A diode rectifier rectifies a small amount of rf energy from the output of the transmitter. A dc voltage proportional to the carrier output of the transmitter appears across the 600-ohm resistor with the switch in the rf position. Should the carrier output be amplitude modulated, there would also appear an ac voltage that would be proportional to the percentage of modulation. If the carrier were amplitude modulated 100 percent, this ac voltage (rms) would be equal to 0.707 times the dc voltage. (The rectifier is a peak-voltage rectifier.) Thus an external calibration of the noise meter can be set up. This is done with an audio oscillator having a 600-ohm output adjusted so that there appears across the output an ac voltage equal to 0.707 times the rectified dc voltage. This voltage is fed into the noise meter, and the latter is adjusted for full-scale deflection. Thus calibrated, it is ready for use in measuring the a-m modulation level that appears across the output of the diode rectifier. The actual steps follow.

1. A diode rectifier is coupled to the output of the transmitter (Fig. 14-21). In some transmitters, one-half of the audio-monitor coupling links may be used. Adjust R1 to obtain a convenient reading, such as 1 volt, on M1.
2. With the switch in the rf position, measure the dc output voltage of of the diode rectifier by means of voltmeter M1.
3. Throw the switch to the af position, and adjust the output of the audio oscillator so that voltmeter M2 indicates an ac voltage equal to 0.707 times the dc voltage just measured. The noise meter is then adjusted to zero dB. (The reference is now set.)
4. Return the switch to the rf position, and read the noise level indicated by the noise meter. This is the measurement of the a-m noise level. This level must be at least 50 dB below 100-percent modulation, with a 75-μs de-emphasis circuit employed.

Fig. 14-22 illustrates an alternate method of measuring a-m noise without the conversion factor of 0.707 described previously. Only one voltmeter is required. The procedure is as follows:

1. Open S2 and close S1. Adjust the trimmer capacitor (if used) and R1 to obtain a convenient voltmeter reading, for example 2 volts.
2. Open S1 and close S2. Adjust R2 to obtain exactly the same voltage.
3. Calibrate the noise meter for 0 dB reference.
4. Open S2 and close S1. Take the noise reading according to the instructions for using the noise-distortion meter.

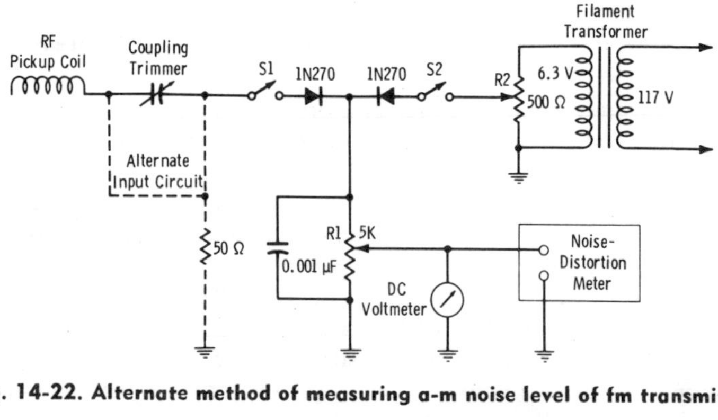

Fig. 14-22. Alternate method of measuring a-m noise level of fm transmitter.

Spurious emissions are measured essentially as previously outlined for a-m transmitters. The results must be as shown in Table 14-1.

Table 14-1. Spurious-Emission Limitations (FM)

Frequency Difference From Carrier	Required Level
120 to 240 kHz	At least 25 dB below unmodulated carrier
240 to 600 kHz	At least 35 dB below unmodulated carrier
More than 600 kHz	At least [43 + 10 $\log_{10}$ (power in watts)] dB below level of unmodulated carrier, or 80 dB, whichever is the lesser attenuation

14-8. PROOF OF PERFORMANCE, MULTIPLEXED TRANSMITTERS (SCA & STEREO)

The basic principles of SCA and compatible stereo have been covered in previous chapters. This section is concerned with engineering standards, transmitter adjustments, and proof-of-performance measurements.

Engineering Standards for Subsidiary Communications Multiplex Operations

The following standards for SCA operation by fm stations are from paragraph 73.319 of the FCC Rules and Regulations.

A. Frequency modulation of SCA subcarriers shall be used.

B. The instantaneous frequency of SCA subcarriers shall at all times be within the range of 20 to 75 kHz, provided, however, that when the station is engaged in stereophonic broadcasting the instantaneous frequency of SCA subcarriers shall at all times be within the range 53 to 75 kHz.

C. The arithmetic sum of the modulation of the main carrier by SCA subcarriers shall not exceed 30 percent, provided, however, that when the station is engaged in stereophonic broadcasting the arithmetic sum of the modulation of the main carrier by the SCA subcarriers shall not exceed 10 percent.

D. The total modulation of the main carrier, including SCA subcarriers, shall meet the requirements of 73.268 (85 to 100 percent modulation total).

E. Frequency modulation of the main carrier caused by the SCA subcarrier operation shall, in the frequency range 50 to 15,000 Hz, be at least 60 dB below 100 percent modulation, provided, however, that when the station is engaged in stereophonic broadcasting, frequency modulation of the main carrier by the SCA subcarrier operation shall, in the frequency range 50 to 53,000 Hz, be at least 60 dB below 100 percent modulation.

Measuring Cross Talk

The following is the RCA in-plant procedure for testing transmitters employing subchannel SCA before shipment. It appears here through the courtesy of RCA.

The measurement of special multiplex parameters should begin with main-to-subchannel cross talk. To set the reference, the subchannel should be modulated 100 percent (±7.5 kHz deviation) with a 400-Hz tone. Set the distortion analyzer for noise measurement, and adjust the meter deflection to 0 dB. The filter is set to pass a band of frequencies from 50 to 15,000 Hz (Fig. 14-23).

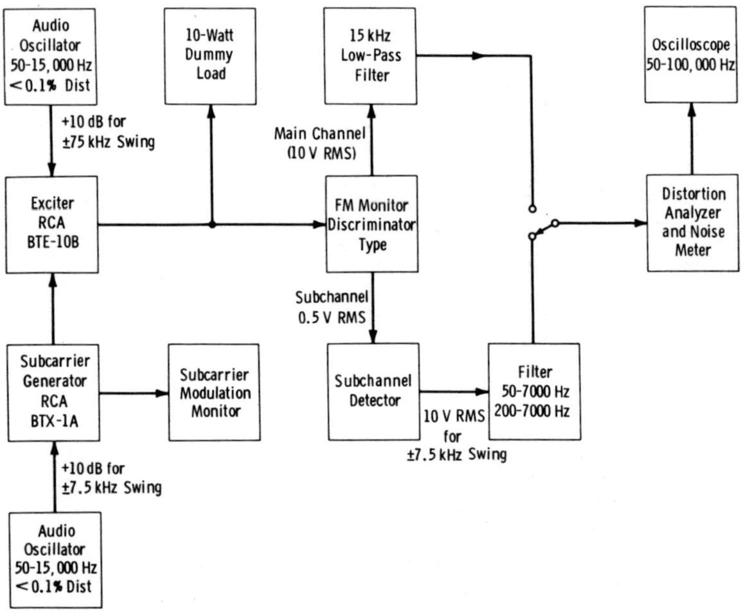

Fig. 14-23. Test setup for checking exciter and subcarrier-generator performance.

Now remove the subcarrier modulation and apply modulation to the main channel using several frequencies ranging from 50 to 15,000 Hz. Adjust each frequency to 85-percent modulation if one subchannel is used. (If two subchannels are used, the main channel should be modulated 70 percent instead.) Read the cross talk for every frequency on the distortion analyzer.

Next, modulate the main channel again with 400 Hz at 85 percent, and carefully adjust all multiplier stages of the exciter to give a minimum cross-talk reading. Be sure not to detune the tuned circuits too much. Also touch up the monitor tuning (rf and i-f coils and discriminator), and be sure that the monitor and separate subcarrier adapter, if they are used, are fed with the right amplitudes and are not overdriven.

If it is necessary, repeat the steps previously indicated; however, a slight adjustment of the first tripler coil should correct any excessive cross talk on 67 kHz. By observing the waveform oscilloscope, make sure that the meter indication of the distortion analyzer represents the components required. Certain beat frequencies generated in the monitor may give a false impression. It should be noted that the above adjustment is greatly simplified because only three tuned circuits may contribute to cross talk. In some instances, it may be advantageous to shift the modulator grid-tuning capacitor slightly to give 2 to 3 dB better cross-talk reduction. The subcarrier modulator itself requires no tuning (in the RCA system).

Sub-to-main channel cross talk is measured in a similar way. This time the 0-dB reference is 100 percent modulation (±75 kHz deviation) by a 400-Hz tone on the main channel. This modulation then is removed, and tones from 50 to 6000 Hz are applied to the subcarrier generator, giving 100 percent (±7.5 kHz) modulation on the subchannel. Slight cross talk may result from improper balancing of the main-channel modulator tubes. Therefore, the modulator-grid tuning, after an initial setting for maximum swing, should ultimately be tuned for minimum sub-to-main channel cross talk. This setting requires only a very slight change. The exciter multiplier tuning has practically no effect on sub-to-main channel cross talk.

To measure intersubcarrier cross talk, one proceeds in the same way, setting a reference level for the channel in which cross talk is being measured by first modulating it with a 400-Hz tone at 100 percent modulation (±7.5 kHz).

The cross talk measurement for the stereo subchannel is made by the same procedure as before, except that the main channel is modulated 90 percent instead of 100 percent (81 percent if SCA is included with stereo operation). The subchannel is amplitude modulated, and the suppressed-carrier sidebands frequency modulate the main channel.

To check the requirement in FCC paragraph 73.322(c), a "standard stereo receiver" or a reliable station receiving-type stereo monitor may be used. However, it is also possible to check the subcarrier generator itself. An oscilloscope connected to a suitable point in the adder stage (addition of 38-kHz subchannel and 19-kHz pilot) may be used. The scope must have dual-beam provisions. The procedure is as follows:

1. Connect one scope input to the 19-kHz grid of the adder stage and the other input to the 38-kHz grid.
2. Be sure that no modulation occurs (no audio signal).
3. Since the adder follows the balanced modulator, it is necessary (in order to obtain the 38-kHz subcarrier) to unbalance this circuit. This has no effect on the measurement.

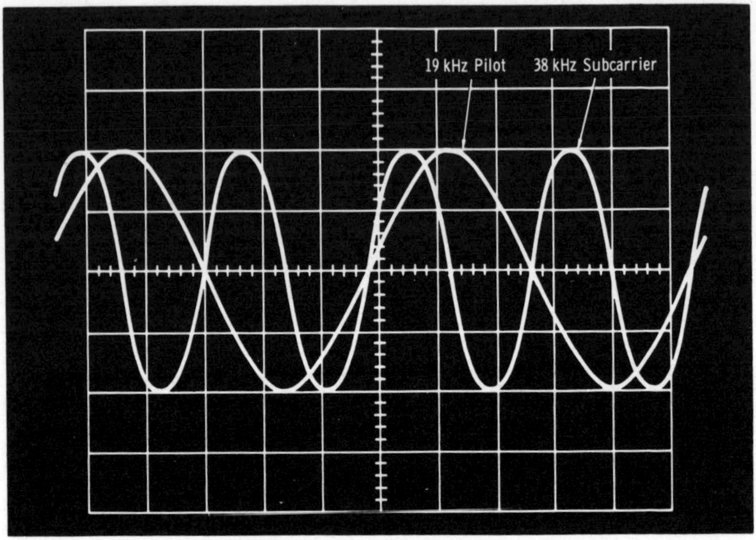

(A) Subcarrier and pilot subcarrier.

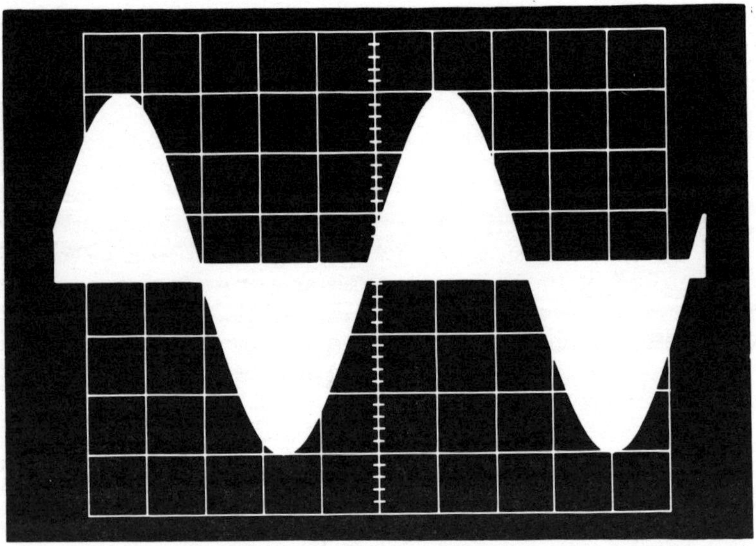

(B) Proper detected main-channel output.

Fig. 14-24.

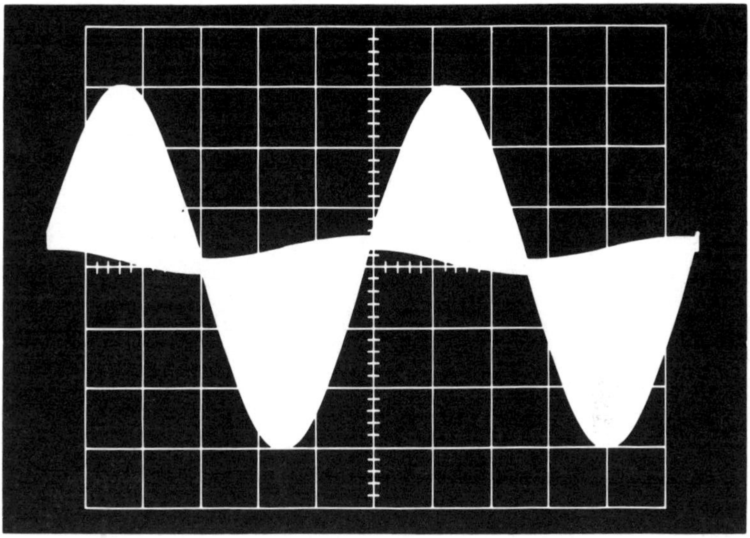

(C) Detected main-channel output in which L − R sidebands lag L + R.

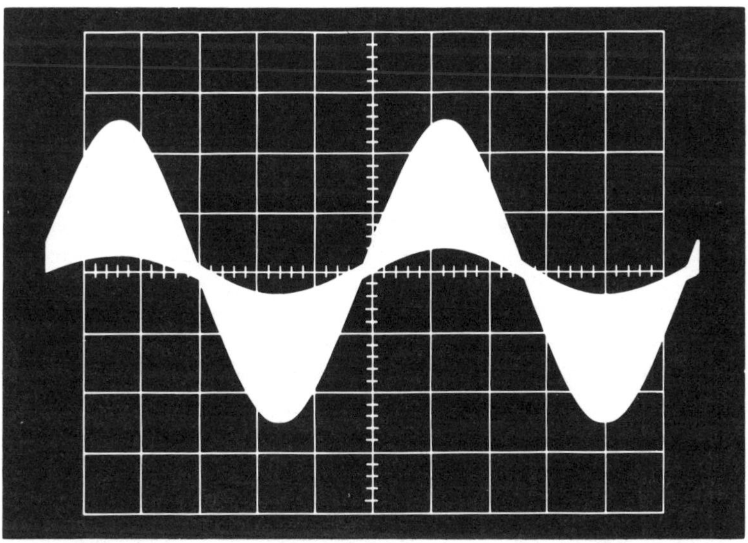

(D) Detected main-channel output where L − R channel lacks sufficient gain.

Stereo waveforms.

4. See Fig. 14-24A. Note that each time the pilot crosses the zero axis, the subcarrier, which is the second harmonic of the pilot, crosses the zero axis in the positive direction.

The detector output waveform from the receiver should appear as in Fig. 14-24B. The waveform is a composite of the L + R signal and the L − R sideband signal for a 400-Hz left-only input signal. The zero axis should be straight for all modulating frequencies.

Fig. 14-24C indicates that although the amplitudes of L + R and L − R are correct, the L − R sidebands *lag* L + R. In this particular case the lag is 6°, which is twice that allowed by the FCC. If the tilt is in the opposite direction, the L − R sidebands *lead* the L + R signal. In general, the phase is within the 3° tolerance if the amplitude of either half of the waveform is 9 times the measurable height of tilt. Thus, if the scope gain is adjusted to allow 4.5 cm of deflection for half of the waveform, the tilt amplitude should be no more than 0.5 cm.

An improper amplitude (insufficient gain) of the L − R sideband signal is indicated by the waveform of Fig. 14-24D. The phase is correct. If the bottom trace bowed upward instead of dipping, excessive gain of the L − R sideband signal would be indicated.

All stereo transmitters are accompanied by detailed instructions, and the maintenance engineer is obligated to become thoroughly familiar with the necessary adjustments of his particular installation to bring it within FCC specifications on proof runs. These procedures naturally vary according to the circuitry.

For example, the BTE 10B exciter in the above example is a tube-type unit (of which many are still in use) requiring more adjustments than the more recent RCA type BTE 15A all solid-state unit (Fig. 14-25).

Modulation of the temperature-compensated basic on-frequency oscillator is achieved by applying the composite stereo or SCA signals from the BTS 1B and BTX 1B generators, respectively, to a pair of push-pull varicap diodes which are coupled to the frequency-determining resonant circuit of the basic oscillator. The output of the basic oscillator is isolated from the following buffer amplifier by a 10-dB resistive attenuator. Thus, the stability and modulation characteristics of the basic direct fm oscillator are not disturbed by following rf power amplifiers.

The output of the buffer amplifier, approximately 500 mW, is used to drive the 15-Watt, three-stage rf amplifier as well as the binary divider chain in the afc circuit. The basic oscillator, buffer amplifier, and afc circuit are mounted inside a shielded enclosure. The rf power amplifier is also completely shielded.

Automatic frequency control (afc) for the on-frequency basic oscillator is achieved by taking a sample of the buffer output and applying it to a chain of 14 divide-by-two frequency dividers. A low-frequency reference crystal operating at 1/1024 of the desired output frequency has its fre-

Courtesy RCA

Fig. 14-25. RCA BTE 15A fm exciter.

quency divided by 16 in a binary chain. Integrated circuits operating in the saturated mode are used in both binary dividing chains. The outputs from the reference and basic-oscillator binary dividers are phase compared in a time-sharing IC comparator. The output of the circuit, which represents the afc error voltage, is filtered and applied to another pair of varicap diodes coupled to the basic-oscillator tuned circuit. Thus, the basic oscillator is phase locked to the 1024th harmonic of the oven-controlled reference crystal.

An off-frequency detector is incorporated in the design of the BTE 15A fm exciter. When the basic oscillator frequency is not phase locked to the reference crystal, an ac component appears in the afc output. This voltage is rectified to operate a relay the contacts of which can be used to turn off the fm transmitter.

Two multimeters are located on the hinged door in front of the regulated power supply section. One of these meters is used to indicate power supply and operating voltages within the exciter and 15-watt rf amplifier. The second meter is a peak-reading voltmeter that is used to indicate key modulating signals.

The rf power output of the BTE 15A can be continuously adjusted from 7 to 15 watts by means of a front-panel control. The primary power is turned on with a circuit breaker. The rf output is turned on with a front-panel switch or by jumping contacts available on the rear of the unit. The exciter will tolerate load mismatches from short circuit to open circuit without damage to the output transistor. Another safety feature prevents turning on the 41-kHz SCA subcarrier when the stereo generator is in the stereo mode.

In the Model BTS 1B stereo generator, the left and right input channels are identical, each having resistive input terminations, isolating transformers, 15-kHz low-pass filters, and an operational amplifier for obtaining pre-emphasis. The pre-emphasis is convertible from 75 to 50 microseconds in the field, or can be removed entirely. The left and right channels can be matched to within ½ percent gain difference and ½° phase difference from 30 to 15,000 Hz, including the 15-kHz low-pass filters. These filters are less than 0.5 dB down at 15 kHz, and more than 50 dB down at 19 kHz and above. This insures an absolute minimum of disturbance to the pilot carrier and subcarrier regions by the program material.

The pre-emphasized and filtered left and right audio signals are applied to a switching modulator which alternately switches between the two audio channels. The balanced and symmetrical 38-kHz switching signal is derived from a buffered 38-kHz output of a bistable multivibrator. The negligible amount of second harmonic (76 kHz) in the 38-kHz switching signal assures a minimum of interference to a 67-kHz SCA channel. The 76-kHz crystal-controlled signal driving the binary divider assures a frequency-stable 38-kHz stereo subcarrier.

The output of the switching modulator, along with the sinusoidal pilot (less than 1 percent distortion), is applied to a phase-linear filter to remove the third and all higher-order harmonic components of the switching signal. The complete composite stereo signal, or a left or right monaural signal, is selected by relays and applied to the input of an operational amplifier. The output of this amplifier is then applied to the wideband input of the BTE 15A fm exciter.

Switching between monaural right, monaural left, or stereo may be accomplished by front-panel push buttons on the BTS 1B or by momentary remote-control contact closures. The selected mode is indicated by front-panel lamps. Left, right, and composite program outputs are also applied to a peak-reading meter on the main frame of the fm exciter.

The Model BTX 1B SCA generator, using all hermetically sealed metal-cased integrated circuits and transistors, is designed to operate on either the 41-kHz or 67-kHz SCA channels. The audio input is applied to a resistive terminating pad and then to an isolating transformer before being amplified. An optional 5-kHz low-pass filter may be inserted in the input to prevent higher-order lower sidebands of the 67-kHz subcarrier from penetrating the upper regions of the stereophonic spectrum.

The audio amplifier includes an active pre-emphasis network which may be easily changed from 75 microseconds to 50 or 150 microseconds or adjusted for a flat response. The audio sensitivity of the BTX 1B is sufficiently high that line amplifiers are not required.

The processed audio input signal is then applied as modulation to a direct-fm SCA generator that includes a temperature-compensating circuit for extreme frequency stability. A vernier center-frequency control is available on the front panel.

Following this generator are a series diode muting gate, a buffer amplifier, and a wideband low-pass filter to remove subcarrier harmonics. The total harmonic content of the subcarrier output is less than 1 percent, and the incidental a-m is less than 5 percent peak with 10 percent subcarrier modulation. The output of the low-pass filter is applied to another buffer amplifier and output level control for application to the multiplex input of the BTE 15A fm exciter.

A sample of the pre-emphasized audio is used to drive a peak-reading multimeter on the main frame of the exciter. Automatic muting of the subcarrier is accomplished in the following manner. A portion of the pre-emphasized audio is applied to a variable-gain amplifier and, with an adjustable time constant, peak detected to operate a Schmitt trigger circuit. The output of the Schmitt trigger is shaped with a low-pass filter and used to turn on or off the series diode muting gate. When audio is applied to the input of the BTX 1B, the muting diode gate is turned on to allow the subcarrier output to appear. In the absence of audio, the Schmitt trigger pauses for a selected time interval before turning off the diode muting gate. The subcarrier envelope rise and fall times are constant and so chosen as to minimize clicks and pops in an SCA multiplex receiver. The amount of Schmitt-trigger delay is adjustable with a front-panel control. With this control, subcarrier muting can be adjusted to occur from 0.5 to 5 seconds after the audio input is removed. Two transistors are used to operate front-panel lamps to indicate the on-off status of the subcarrier. Also, a front-panel switch provides manual control of the subcarrier output or the use of the automatic muting feature. The subcarrier also can be turned on or off remotely.

Stereo Adjustments and Proof of Performance

Fig. 14-26 presents a suggested setup at the studio for stereo proof-of-performance measurements. The audio generator, step attenuator, and VU meter are usually incorporated in one unit. The balanced Y pad splits the common audio signal into two paths for simultaneous left and right channel (microphone input) feeds. Switches S1 and S2 allow either left only, right only, or left plus right feeds. Switch S3 allows either normal or reverse polarity feed to the right-channel input. Note that in the normal position, the A output of the signal generator feeds terminals 1 and 3 of the input transformers, while the B output feeds terminals 2 and 4. Thus the two inputs are being fed with a common in-phase signal, where $L = R$. With S3 in the reverse position, the polarity of feed to the right channel input is 180° out of phase with the left channel input, or $L = -R$. This is necessary for certain adjustments and measurements as described later.

When only the transmitter is to be tested to prepare for overall proof-of-performance measurements, the left and right inputs are to the inputs of the stereo generator. A typical setup at the transmitter for overall proof-of-performance runs is illustrated in Fig. 14-27.

Audio frequency-response and noise-distortion measurements are taken exactly as outlined earlier for mono fm, except that the procedure is repeated for the opposite channel. This simply means that the left and right channels must individually meet requirements of the FCC for the various modulation levels just as in mono fm specifications. We are then ready to take the remaining measurements which have to do with the composite signal: separation and cross talk between channels.

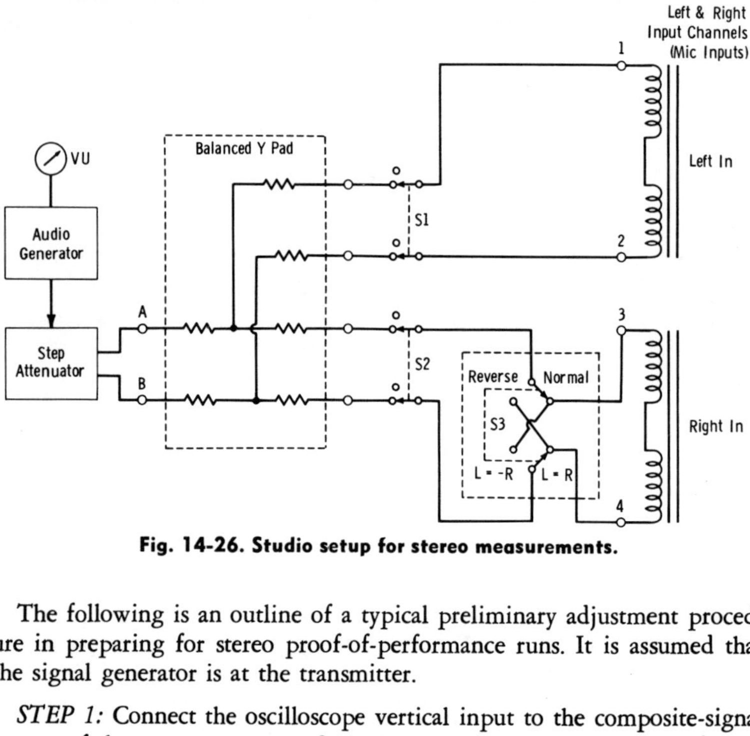

Fig. 14-26. Studio setup for stereo measurements.

The following is an outline of a typical preliminary adjustment procedure in preparing for stereo proof-of-performance runs. It is assumed that the signal generator is at the transmitter.

STEP 1: Connect the oscilloscope vertical input to the composite-signal output of the stereo generator. Sync the scope internally. Apply a left-only signal of 400 Hz at the specified input level of the stereo generator. Turn the pilot amplitude control to minimum.

STEP 2: The pattern on the scope should be as in Fig. 14-28A. If the base line is not flat (Fig. 14-28B), adjust the left-channel (L + R) gain control (sometimes termed "separation" control) to obtain a flat base line.

STEP 3: Apply a right-only 400-Hz signal and repeat the above adjustment for the right-channel gain control. Sometimes both controls are used; sometimes only an L + R gain control is incorporated. The point is that when the two gains are equal, the base line will be flat.

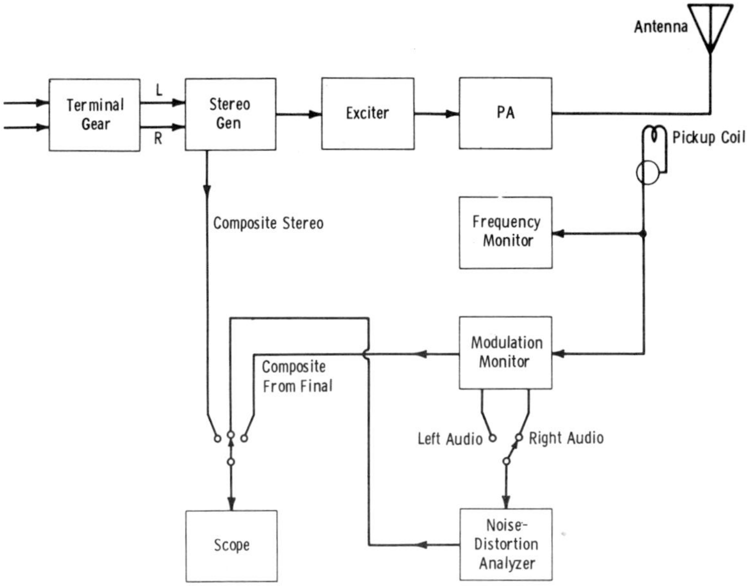

Fig. 14-27. Transmitter setup for stereo measurements.

> CAUTION: When a sampling type of modulator is used, a large number of odd-order harmonics (3rd, 5th, etc.) of the 38-kHz switching rate exist. For this reason, an output filter is employed to confine component frequencies below 75 kHz. Sometimes an output-filter phase-linearity control is incorporated to set the driving impedance of the filter for maximum phase-shift linearity over the range of 50 Hz to 53 kHz. Misadjustment of this control will cause a left-only or right-only signal to appear very similar to the waveform of Fig. 14-28B. If it is not possible to obtain a perfectly flat base line by adjusting the L + R and/or L − R gains, investigate this possibility. More recent solid-state stereo generators employing the sampling tye of modulation have fixed filtering networks that require no adjustment. The same problem can occur, of course, with a component change or failure within the filtering network.

STEP 4: Reverse the polarity of the right-channel input (Fig. 14-26), and feed both channels so that L = −R. Turn the pilot amplitude control to normal. With the scope time base set at about 200 μs/cm, the pattern of Fig. 14-29A should be obtained. This is termed the L − R "butterfly" and can be used to adjust the pilot phase control accurately. Remember that proper phase is indicated when a zero crossing of the pilot signal occurs at precisely the same time as the positive-going zero crossing of the (suppressed) 38-kHz subcarrier (Fig. 14-24A).

STEP 5: Expand the time base of the scope to around 50 μs/cm. Trigger the scope externally with the signal-generator frequency or the 19-kHz

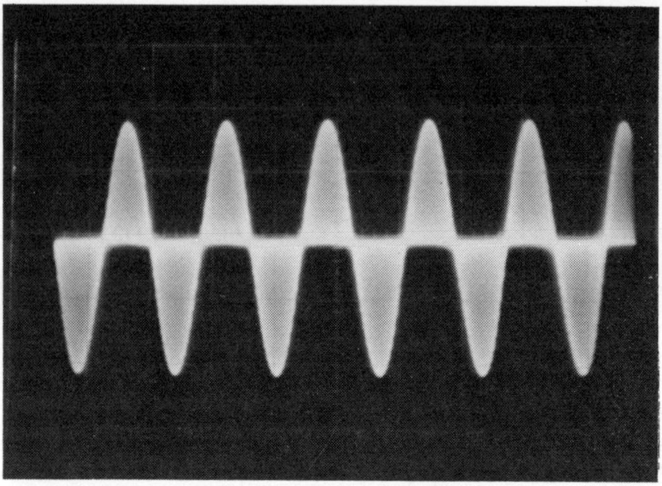

(*A*) *L* + *R and L* – *R gains equal.*

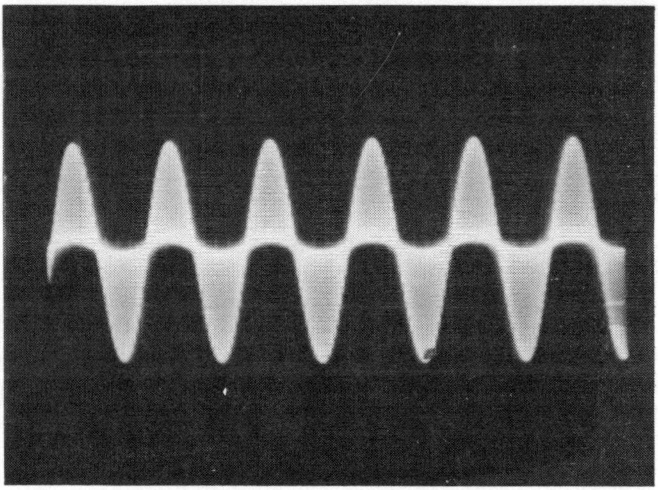

(*B*) *L* + *R and L* – *R gains unequal.*

Fig. 14-28. Setting separation controls.

pilot. If the pilot phase is correct, a perfect diamond pattern such as that of Fig. 14-29B will be displayed. This indicates proper zero crossings of the pilot and subcarrier sine waves. If the phase is not correct, the four crossings do not form a diamond pattern, and a distorted presentation such as that of Fig. 14-29C is obtained. Adjust the pilot phase control so that the pattern of Fig. 14-29B is obtained. Always recheck the pilot gain after pilot phase adjustments are made.

STEP 6: Move the scope to the modulation monitor position. The scope display from the composite output of the modulation monitor should be identical to that obtained from the composite output of the stereo generator. If a different pilot phase is indicated, adjust the pilot phase control in the monitor to agree with the phase shown at the composite output of the stereo generator, or follow specific instructions of the monitor manufacturer.

Measuring Separation

Feed a signal of 1 kHz to either the left or right (only) input. With the modulation monitor set in the Total position, modulate the transmitter 100 percent. Switch the modulation meter to the channel opposite that being fed; the modulation indication should be at least 29.7 dB below the 100-percent modulated channel (actually 90 percent with 10-percent pilot). This condition should prevail over a frequency range of 50 Hz to 15 kHz.

Measuring Cross Talk

The left and right input channels of the stereo generator are connected in parallel and *in phase.* The frequencies at which cross talk is to be measured modulate the main (L + R) channel without modulation of the subchannel (L − R). With the main channel at 90-percent modulation (10-percent pilot modulation), cross talk of the main channel into the subchannel is then indicated on the modulation monitor in the L − R position. The percent modulation in this channel should be a maximum of 1 percent, indicating −40 dB of cross talk.

Next, reverse the phase of the signal to the right channel, and set L − R to 90-percent modulation. Cross talk of the subchannel into the main channel is now indicated by the L + R position of the modulation monitor.

Problems in Stereo Proof of Performance

The problems of frequency response, fm and a-m noise of the fm carrier, and distortion are no different from those previously described for the mono fm transmitter, except that the number of measurements is essentially doubled. As in mono fm, poor performance is isolated to transmitter only, lines or stl from studio to transmitter, or the studio.

The cross-talk specification is usually the most difficult to meet. This measurement should first be made on the transmitter alone to check adjustments of the stereo generator, exciter and power amplifiers, and antenna system. If the cross-talk measurement deteriorates between the stereo-generator output and the final-stage output, check for proper circuit tuning and stage neutralization.

After the stereo generator and following rf stages have been optimized, move the signal generator to the terminal gear where the stl or wire lines normally feed. Unequal phase shifts and/or gains through agc or limiting amplifiers can cause major cross-talk problems. Gain must be kept to within one percent and phase to within one degree if cross-talk specifications are to be met.

Next, the signal generator is moved to the input of the stl or lines at the studio. Thus any additional factors in this area will influence the measurements. The final step is to feed the left- and right-channel microphone inputs for the regular overall proof-of-performance measurements.

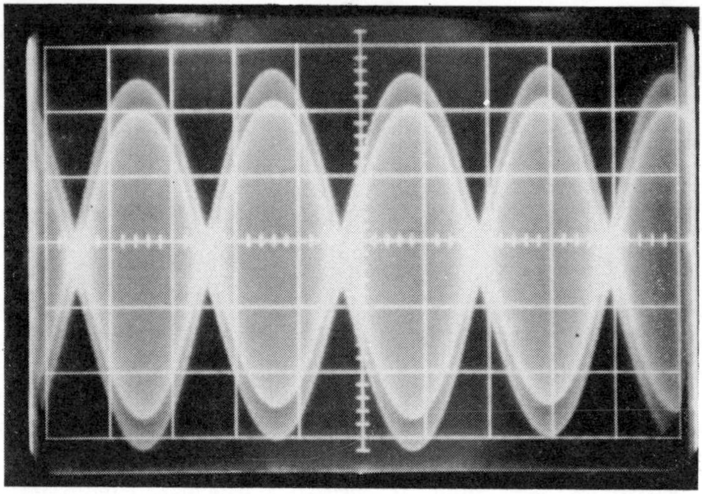

(A) Time base of 200 μs/cm.

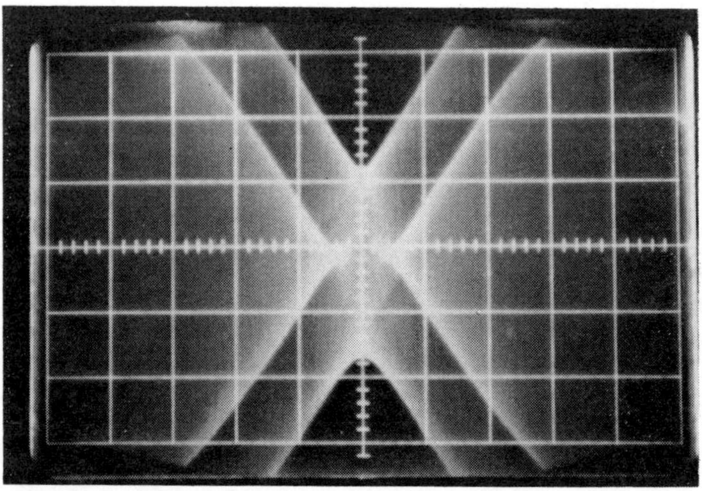

(B) Time base of 50 μs/cm, correct phase.

Fig. 14-29. L − R display

14-9. PREVENTIVE-MAINTENANCE SCHEDULES

Of course, the primary purpose of any preventive-maintenance schedule is to reduce as much as possible the likelihood of a failure of any component part of the broadcasting installation during the broadcast day. Regular maintenance schedules are in effect at most broadcast stations and do much to increase the useful life of equipment and anticipate many tube and parts failures that would otherwise occur.

Preventive maintenance on any sort of equipment may be defined as a systematic series of operations performed periodically on the equipment in order to prevent breakdowns. This type of maintenance may be divided into two phases: work performed while the equipment is functioning, and work performed during the normal shutdown periods. This discussion is concerned only with preventive maintenance procedures carried out during the shutdown period.

The importance of preventive maintenance cannot be overestimated. The owners of a broadcast station depend on its being on the air every second of its scheduled periods of transmission. It is very important that the personnel maintain the equipment so that lapses in the transmission will be kept to a minimum.

Cleanliness of the equipment is of the utmost importance since collections of dust and dirt can cause a number of troubles. This is particularly true in the higher-power stages of transmitters, since accumulation of foreign matter over a period of time reduces the effectiveness of the insula-

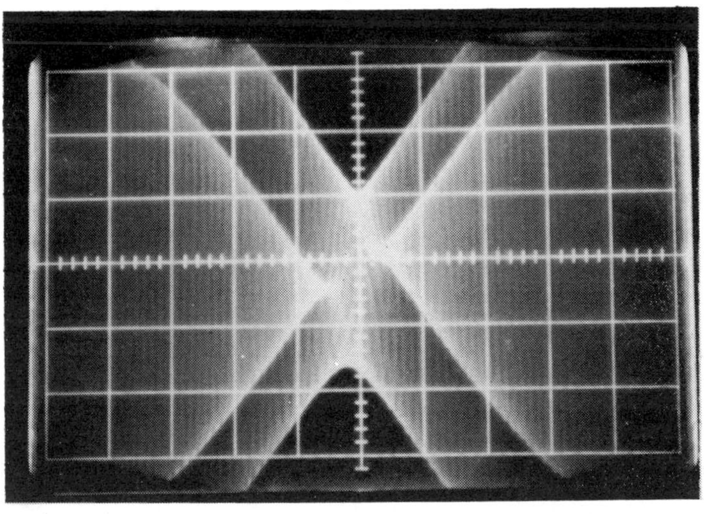

(C) Time base of 50 μs/cm, incorrect phase.

for setting pilot phase.

tion to a point where leakage currents and arcovers are common. High-voltage contacts have an extreme tendency to collect dirt (this is the principle used in electronic smoke eliminators), and the higher relative humidity existing in summer or in southern locations tends to aggravate this characteristic. A dusting and clean-up procedure, then, is desirable at a transmitter plant. A source of dry air under pressure is a common means of blowing out dirt, dead insects, and the like from inaccessible corners and from variable tuning capacitors. Insulators, safety gaps, etc., should be polished with a dry cloth. Carbon tetrachloride may be used to loosen excessive dirt and grime.

Transmitter Maintenance Schedule

So that preventive maintenance is effective, it must be performed at regular intervals; that is, certain portions of the equipment must be inspected for certain things every day in older-type equipment, while other parts of the equipment need be inspected only weekly or monthly. The following is a comprehensive maintenance schedule that may be considered a guide to anyone desiring to set up a means of preventing breakdowns. Naturally, this listing may include items that are unnecessary at a particular station, but it has been compiled with the thought that very precaution should be taken.

A. DAILY

1. Hourly read all meters and check power-tube filament voltages.
2. Check air-cooled anode temperatures. Check water temperature of water-cooled tubes if used.
3. Check for correct cabinet temperature of air around high-voltage rectifiers.
4. After shutdown, make a general inspection for overheated components, such as capacitors, inductors, transformers, relays, and blowers.
5. Investigate any peculiarities of meter readings.
6. In case of overloads, examine safety gaps and transmitter components for arc pits, etc. Clean and polish surfaces where arcs have occurred. Reset gaps if necessary. Investigate causes of outages.
7. In the event of lightning or heavy static discharges, inspect the transmission line, terminating equipment, and the antenna including the gaps. Polish pitted surfaces.
8. If gas-filled coaxial line is used, check pressure.

B. WEEKLY (In addition to items in daily inspection)

1. Immediately after shutdown, check antenna-terminating components for signs of overheating.
2. Clean antenna-tuning apparatus. Check for arc pits, etc. Clean and polish gaps and adjust if necessary.

3. Test antenna-monitor rectifier tubes.
4. Calibrate remote antenna meters against meters at the antenna.
5. Clean transmitter using a vacuum cleaner.
6. Clean component parts of transmitter.
   A. Brush terminal boards.
   B. Clean insulators with carbon tetrachloride.
   C. Clean power tubes and high-voltage rectifiers with tissue and alcohol or distilled water.
7. Check filament voltages and dc voltages at the sockets of all tubes that are not completely metered by panel meters.
8. Check air-flow interlocks for proper operation. Check all door interlocks for proper operation.
9. Check operation of grounding switches. Examine mechanical operation and electrical contacts.
10. Inspect blowers for loose impellers, free rotation, and sufficient oil.
11. Inspect relays for proper mechanical and electrical operation. If necessary, clean and adjust components.
12. Inspect air filters; clean if excessive dirt has accumulated.
13. Check all sphere and needle gaps. Clean any pits or dirt. Check gap spacings.
14. Check filter-bank surge resistors with an ohmmeter.
15. Check any power-tube series resistors with an ohmmeter.
16. Check power-change switches if used; check for serious arcing during day-night antenna changeover if used.
17. Make general performance checkup (distortion, noise, and frequency response). Observe modulated waveform on cro.
18. Check neutralization by disabling crystal oscillator and observing grid currents.
19. Check proper operating voltage for pure-tungsten-filament tubes. Determine lowest permissible voltage as follows:
    A. A-m transmitters—distortion and carrier-shift checks.
    B. Fm transmitters—decrease filament voltage until output begins to drop.
    C. Operate filaments approximately 1 percent above the filament voltage determined in A or B.
20. If water cooling is employed, check entire system for any signs of leakage and for electrical leakage.
21. Check pressure of any gas-filled capacitors.

C. MONTHLY (In addition to daily and weekly items)

1. Make detailed inspection of all transmitter components using whatever tests seem advisable.
2. Clean and inspect tube-socket contacts and tube pins.
3. Clean or replace air filter. Brush dirt from blower impellers, canvas boots, etc.

4. Clean and adjust all relay contacts. Clean pole faces on contactors. Replace badly worn contacts.
5. Oil the blower motors (carefully).
6. Operate all spare vacuum tubes for a minimum of two hours under normal operating conditions. Clean up any gassy tubes as described in a later section on large power tubes.
7. Operate all spare mercury-vapor rectifiers normally, after first applying filament voltage only for a minimum of 30 minutes. Store the tubes upright.
8. Inspect all variable-inductor contacts for tension, signs of overheating, and dirt. Clean and adjust as required. Carbon tetrachloride or crocus cloth may be used for cleaning. Do not use emery cloth.
9. Check for proper operation of time delays, notching relays, and any automatic-control systems.
10. Clean audio-equipment (console, etc.) attenuator and low-level switching contacts with cleaner; wipe off excess.
11. Check tubes in station monitoring equipment, such as frequency monitor, modulation monitor, etc.
12. Clean switches in monitoring equipment with cleaner.

D. QUARTERLY (In addition to the preceding)

1. Lubricate tuning motors and inspect for ease of rotation.
2. Check all indicating meters. Check ac filament voltmeters with an accurate dynamometer-type meter.
3. Check all connections and terminals for tightness.
4. Inspect any flexible cables to door connections.
5. Inspect and lubricate if necessary any flexible drive cables.
6. Inspect, clean, and service (if necessary) all switches (voltmeter selector switches, push-button switches, control switches, etc.).
7. Clean transmission-line insulators, and take up slack if open-wire lines are used.
8. Check oil circuit breakers, if used, for sufficient oil and loose or defective parts.

E. SEMIANNUALLY (In addition to the preceding)

1. Test transformer oil for breakdown, and filter it if necessary. (This is done by the power company.)
2. Check protective overload relays or circuit breakers for correct operation.
   A. Ac overload relays may be checked by shorting the high-voltage transformer secondary.
   B. Dc overload relays may be checked by shorting the dc through the relay in the circuit protected by the relay.
3. In fm installations, check the accuracy of the modulation monitor as described below.

Measuring Frequency Deviation by the Bessel-Zero Method

An fm wave may be resolved into its carrier and sideband-frequency components, the amplitudes of which vary as Bessel functions of the modulation index. The modulation index is defined as the ratio of the peak deviation from the center frequency to the frequency of the modulating signal:

$$m = \frac{\Delta F}{f}$$

where,

m is the modulation index,
ΔF is the peak deviation in Hz,
f is the modulating frequency in Hz.

At certain values of m, the carrier amplitude becomes zero; that is, all the energy is transmitted in the sidebands. Fig. 14-30 shows relative carrier amplitude as a function of modulation index. Table 14-2 gives the modulation indexes necessary to produce successive carrier nulls. If the modulating frequency (f) is properly chosen, one of the carrier-null points can be made to correspond to the desired maximum deviation. For example, the peak deviation for an SCA subcarrier is ±7.5 kHz. For the first null, the modulating frequency (f) must be:

$$f = \frac{\Delta F}{m} = \frac{7.5 \text{ kHz}}{2.4} = 3.125 \text{ kHz}$$

Obviously, for the main carrier with a ΔF of ±75 kHz, the modulating frequency must be 31.25 kHz to use the first null. Since this modulating frequency is outside the passband of the audio system, the second null must be used, and the modulating frequency becomes f = 75 kHz/5.52 = 13.586 kHz, which is practical. However, the alternate method described later may be used.

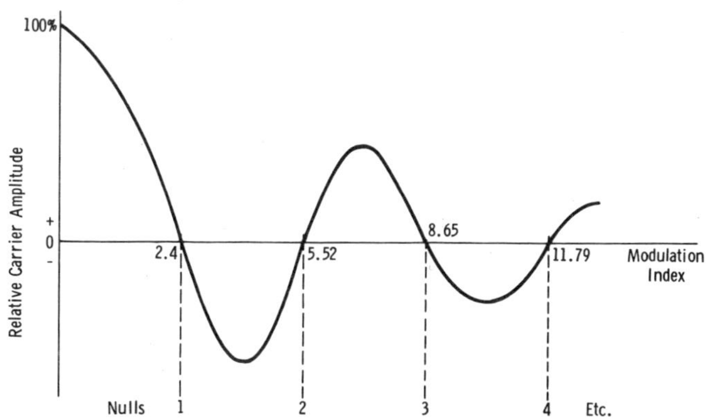

Fig. 14-30. Variation of carrier amplitude with modulation index.

Table 14-2. Modulation Indexes Required for Carrier Null

Carrier Null	Modulation Index
1	2.405
2	5.52
3	8.654
4	11.792
5	14.931
6	18.071
7	21.212
8	24.353
9	27.494
10	30.635

Procedure

1. Tune a communications receiver (using the narrowest i-f selectivity possible to avoid beats with other sideband frequencies) to the unmodulated carrier (Fig. 14-31).
2. Peak the bfo in the receiver to the unmodulated carrier.
3. Assume the main carrier deviation (75 kHz) is to be measured. Apply 13.586 kHz to the transmitter from the audio signal generator; set the output of the generator to zero.
4. Increase the signal-generator amplitude slowly from zero. The sideband frequencies and their beat notes are produced in the headphones. As the amplitude is increased, note in particular the gradual attenuation of the beat note produced by the carrier. The first null exists when this beat first disappears.
5. Further increase of the input signal amplitude will cause the beat to reappear. When the second null occurs, the modulation monitor should indicate 100-percent modulation (accuracy is normally within 5 percent by this method). Otherwise, a suitable VU meter placed across the modulator-input terminals may be calibrated in percentage modulation.

Alternate Procedure—This procedure uses the fact that frequency-multiplier stages increase both the center frequency and the frequency deviation by the same factor. For example, a transmitter incorporating a reactance tube may have its oscillator frequency increased 18 times before its application to the power-output stage.

1. Couple a small portion of the modulated-oscillator output to the receiver (dash line in Fig. 14-31).
2. Tune the receiver to the oscillator frequency. If the transmitter has a typical 18-times multiplication factor, this frequency is the assigned carrier frequency divided by 18, which is somewhere between 5 and 6 MHz.

3. The peak deviation desired at the modulator stage is 75 kHz divided by 18, or 4.166 kHz. Therefore adjust the signal generator for 4166/2.4 = 1736 Hz. (The accuracy of measurement depends on the oscillator calibration accuracy and the operator technique in setting the dial.)

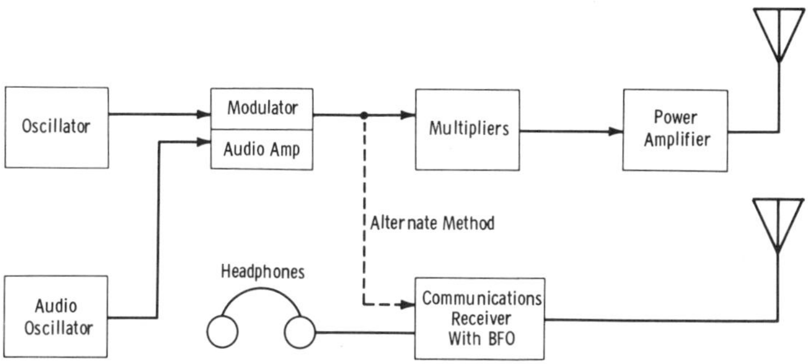

Fig. 14-31. Test setup for measurement of frequency deviation by the Bessel-zero method.

4. Gradually increase the input signal from zero until the first null occurs. This amplitude represents 100-percent modulation.

14-10. PREVENTIVE MAINTENANCE, BASIC INFORMATION

In the preceding material, some general facts about preventive maintenance were presented together with schedules showing when the different operations should be performed. Inasmuch as some of these operations deal with apparatus that can easily be damaged unless proper care is exercised, certain procedures should be followed so that no damage results from the periodic inspections and so that if repairs to the apparatus are necessary, they can be properly made. Remember that the information in the following pages is general, and it may happen that some manufacturers recommend specific procedures for their products. Of course, these procedures should be followed.

The reasons why preventive-maintenance operations are followed are obvious. It might be desirable, however, for the technicians who are responsible for this maintenance work to remember that the procedures discussed in the following pages are designed with the following objectives in mind:

1. Combat the detrimental effects of dirt, dust, moisture, water, and the ravages of weather on the equipment.
2. Keep the equipment in condition to insure uninterrupted operation for the longest possible period of time.

3. Maintain the equipment so that it always operates at maximum efficiency.
4. Prolong the useful life of the equipment.

The actual work performed during the application of the preventive-maintenance schedule items can be divided into six types of operations. Throughout this section, the lettering system for the six operations is as follows.

Feel (F)—The feel operation is extensively used to check rotating machinery (such as blower motors, drive motors, and generators) for overheated bearings. Feeling may indicate the need for lubrication or the existence of some other type of defect. The normal operating temperature is that which will permit placing the bare hand in contact with the motor-bearing cover for 5 seconds without discomfort. The feel operation also is applied to items other than rotating machinery; the feel operation for these is explained in the discussion of each specific item. It is important that the feel operation be performed as soon as possible after shutdown and always before any other maintenance.

Inspect (I)—Inspection is probably the most important of all the preventive-maintenance operations. If more than one technician is available to do this work, choose the most observant, since careful observation is necessary to find defects in the functioning of moving parts and other abnormal conditions. To carry out the inspection operation most effectively, make every effort to become thoroughly familiar with normal operating conditions and to learn to recognize and identify abnormal conditions immediately.

Inspection consists of carefully observing all parts in the equipment. Notice characteristics such as color, placement, and state of cleanliness. Inspect for the following conditions:

1. Overheating, as indicated by discoloration, blistering, or bulging of the part or surface of the container; leakage of insulating compounds; and oxidation of metal-contact surfaces.
2. Placement, by observing that all leads and cabling are in their original positions.
3. Cleanliness, by carefully examining all recesses in the units for accumulation of dust, especially between connecting terminals. Parts, connections, and joints should be free of dust, corrosion, and other foreign matter. In tropical and high-humidity locations, look for fungus and mildew.
4. Tightness, by testing any connection or mounting which appears to be loose by slightly pulling on the wire or feeling the lug or terminal screw.

Tighten (T)—Any movement of the equipment caused by transportation or by vibrations from moving machinery may result in loose connections which are likely to impair the operation. The importance of firm mountings and connections cannot be overemphasized; however, *never* tighten screws, bolts, and nuts unless it is definitely known that they are loose. Fittings that are tightened beyond the pressure for which they were designed will be damaged or broken. When tightening, always be certain to use the correct type and size of tool.

Clean (*C*)—When the schedule calls for a cleaning operation, it does not mean that every item which bears that identifying letter must be cleaned each time it is inspected. Clean parts only when inspection shows it necessary. The cleaning operation performed on each part is described later.

Adjust (*A*)—Adjustments are made only when necessary to restore normal operating conditions. Specific types of adjustments are described later.

Lubricate (*L*)—Lubrication means the addition of oil or grease to form a film between two surfaces that slide against each other, in order to prevent mechanical wear from friction. Generally, lubrication is performed only on motors and bearings.

> NOTE: When a part is suspected of impending failure, even after protective maintenance operations have been performed, immediately notify the person in charge, who will see that the condition is corrected by repair or replacement before a breakdown occurs.

Suggested List of Tools Necessary for Relay and Commutator Maintenance

Several items listed on the preventive-maintenance schedule require work of a special and somewhat delicate nature. This work includes cleaning and repairing relay contacts, cleaning plugs and receptacles, polishing commutators, and adjusting motor and generator brushes. To do the work properly, special supplies and specially constructed tools are necessary. A suggested list is given in Table 14-3.

Construction of Relay and Commutator Tools

Crocus-cloth, canvas-cloth, and sandpaper sticks are constructed in the following manner:

1. First prepare a length of wood 3¾ inches long, ⅜ inch wide, and 1/16 inch or less thick (Fig. 14-32A). Cut one piece of crocus cloth 2½ inches long and 1 inch wide.
2. Fold the crocus cloth as in Fig. 14-32A and cement it to the stick. Note that both sides of the stick are covered. Place the stick in a vise, press it, and wait until the cement hardens. Cut off the crocus cloth that extends past the edge of the stick.
3. Obtain three pieces of wood that measure 8 inches long, 1 inch wide, and approximately ¼ inch thick. Cut one piece of crocus cloth, one

piece of No. 0000 sandpaper, and one piece of canvas cloth, each 5¼ inches long and 1 inch wide.

4. Fold the long, narrow pieces of crocus cloth, sandpaper, and canvas cloth prepared in step 3 as shown in Fig. 14-32B and cement one of them to each of the three sticks. Note that in this case the fold is over one end of the stick rather than over the side. Place the stick in a vise, press, and wait until the cement hardens.

Table 14-3. Suggested List of Special Maintenance Tools and Supplies

Quantity	Item
1	Nonmagnifying dental mirror
1	Cleaning brush, 2-inch
1	Canvas-cloth strip
1	Small relay crocus-cloth stick
1	Relay-contact burnishing tool
1	Fine-cut file
1	Brush seating stone
1	Commutator polishing stone
1	Canvas-cloth stick
1	Sandpaper-covered stick
1	Brush, cleaning, 1-inch
1	Carbon tetrachloride, quart can
2	Cement, household, tube
1	Cloth, canvas, 2x4 feet
1	Cloth or canvas strip, 2x6-inch, cut from sheet
1	Cloth, lint-free, package
6	Crocus-cloth sheets
1	Crocus-cloth strip, ¾x6-inch, cut from sheet
1	Lubricant, petroleum jelly, container
6	Sandpaper sheets, No. 0000
6	Sandpaper sheets, No.00
1	Sandpaper No. 0000, ¾x6-inch, cut from sheet
1	Sandpaper strip, No. 00, ¾x6-inch, cut from sheet
1	Stick, crocus-cloth, large
50	Tags, small marker

Use and Care of Tools

The proper care of tools is as necessary as the proper care of radio equipment. Any effort or time spent in caring for tools is worthwhile. Clean them when necessary, and always store them so that they are easily accessible. The following information will be helpful in using and caring for the tools listed.

Crocus-Cloth Stick—The crocus-cloth sticks are used to clean the contacts of relays in the radio equipment.

Large Commutator Sticks—Commutator sticks with coverings of sandpaper or canvas are used for cleaning the commutators of electric motors and generators.

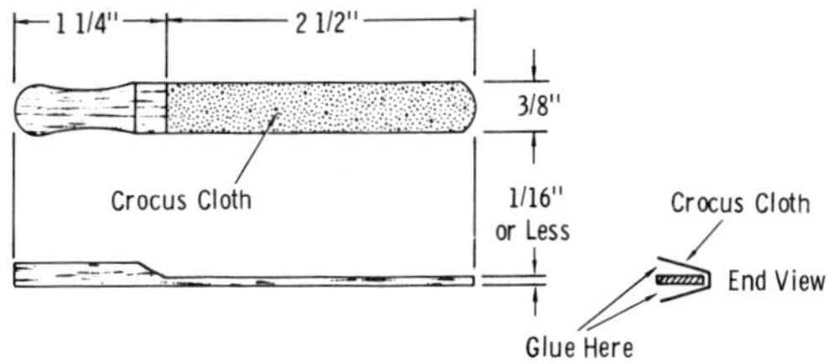

(*A*) *Stick for cleaning relay contacts.*

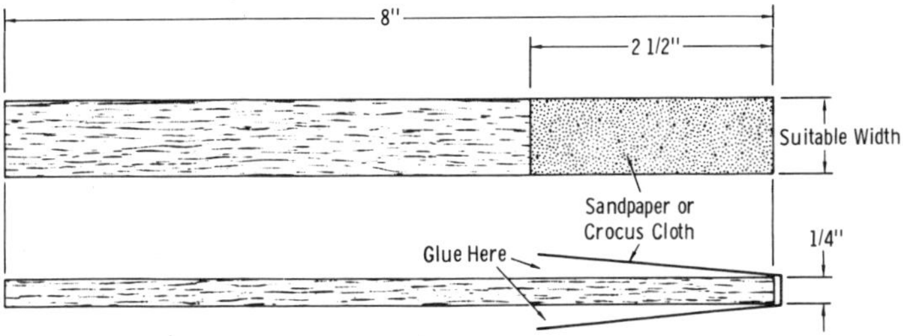

(*B*) *Stick for cleaning commutators.*

Fig. 14-32. Method of preparing crocus-cloth, canvas-cloth, and sandpaper sticks for relays and commutators.

Commutator Dressing Stone—The dressing stone is used to dress a commutator in case of emergency.

Brush Seating Stone—The seating stone is used when a set of new brushes is installed in an alternator or exciter. Only a very limited application of the seating stone is required to seat the average set of brushes.

Electric Soldering Iron—The use of the soldering iron is generally known. Remember to keep the tip properly tinned and shaped.

Allen Wrenches—Allen wrenches are used to tighten or remove setscrews on fan pulleys, motor pulleys, etc. Keep these small wrenches in the container provided. After use, wipe them off with an oily rag and return them to their proper place.

Diagonal-Cutting Pliers—Diagonal pliers are used to cut copper wire (no larger than No. 14) when working in small places. Do not cut iron wire with the diagonals.

Gas Pliers—Gas pliers are used to hold round tubing, round studs, or any other round metal objects that do not have screwdriver slots or flat sides for wrenches.

Long-Nose Pliers—Long-nose pliers are used to hold and dent small wires and to grip very small parts. They are generally used on delicate apparatus.

Adjustable-End Wrenches—Adjustable-end wrenches are designed to remove or hold bolts, studs, and nuts of various sizes. Keep the adjusting nut free from dirt and sand, and oil these wrenches frequently.

Nut-Driver Wrenches—Nut-driver wrenches are used to remove or install nuts of various sizes. Choose a wrench that fits the nut snugly.

Screwdrivers—Screwdrivers of different sizes are important tools and must be kept in good condition. Select the proper size for the job. Never force a screw; if undue resistance is felt, examine the threads for damage and replace the screw if necessary.

Shorting Bar—The shorting bar must be constructed at the station. Obtain a piece of wood about 15 inches long and 1 inch thick. Fasten a piece of copper, brass rod, or tubing securely to one end of the stick in such a manner that the rod extends 12 inches beyond the end of the stick. Solder a piece of heavy, flexible wire about 18 inches long to the metal rod at the point where it is fastened to the stick. When using the shorting bar, always attach the free end of the wire to a *good* ground connection *before* making contact with the terminal to be grounded.

14-11. PREVENTIVE-MAINTENANCE TECHNIQUES

Suggested preventive-maintenance procedures for various types of equipment and components are described in the following paragraphs.

Vacuum Tubes

The purpose of tube maintenance is to prevent tube failures caused by loose or dirty connections and to maintain the tubes in a clean operating condition. Certain types of vacuum tubes, especially those used in high-voltage circuits, operate at high temperatures. Careless contact with the bare hands or arms causes severe burns. Keep a pair of asbestos gloves handy. Otherwise, sufficient time must be allowed for the tubes to cool before handling.

Maintenance of vacuum tubes involves making minor adjustments and cleaning. Because of their high operating potentials, tubes used in high-voltage circuits require more frequent inspection and cleaning than tubes used in low-voltage circuits. Loose coupling at the terminals of high-voltage tubes will result in pitting and corrosion of the terminals. Loose connections cause poor electrical contact and lower the operational efficiency of the unit.

Apply maintenance to vacuum tubes only when necessary; too-frequent handling may result in damage to the tube terminals and connections. As a rule, vacuum tubes need little maintenance; therefore, when the program calls for maintenance, but inspection shows that the tubes do not require it, omit the operation. It is advisable, however, to clean the glass envelopes of the tubes and remove dust or dirt accumulations in the immediate area. The object of the maintenance program is to keep the tubes free from dirt, oil deposits, and corrosion.

For maintenance purposes, vacuum tubes are divided into two groups, transmitting-type tubes and receiving-type tubes. Maintenance procedures for vacuum tubes differ according to their types. Certain maintenance operations that must be performed on transmitting-type tubes may be omitted for receiving-type tubes. Transmitting-type tubes are those used in transmitters, modulators, and high-voltage rectifier units. Because of their physical construction, they require careful inspection and cleaning during maintenance.

Five procedures are required in the performance of maintenance of vacuum tubes: feel, inspect, tighten, clean, and adjust. The procedures involved depend on the type of tube being maintained. Transmitting tubes may require the application of all the procedures, although the procedures for receiving tubes are limited by the tube types.

The following procedures are employed for the maintenance of vacuum tubes. (CAUTION: Discharge all high-voltage capacitors before performing any maintenance operations. Avoid burns by allowing a sufficient time for tubes to cool before handling.) These operations should be performed 5 to 10 minutes after the power has been removed from the tubes.

Feel (*F*)—This operation should be applied only to high-voltage tubes, such as those used in transmitters, modulators, and high-voltage rectifier units. Feel the grid, plate, and filament terminals of the tubes for excessive heat. Practice will determine the temperature to be accepted as normal. For example, when two grid terminals are felt, one should not be warmer than the other. The development of excessive heat at the terminals indicates poor connections.

Inspect (*I*)—This maintenance operation is applicable to all types of vacuum tubes and should be performed after the tubes have had sufficient time to cool.

1. Inspect the glass or metal envelopes of tubes for accumulations of dust, dirt, and grease. Inspect the tube caps and connector clips for dirt and corrosion. Inspect the complete tube assembly and socket for dirt and corrosion. Check the tube caps to determine if any are loose. For glass tubes, check the glass envelope to determine whether or not it has become loosened from the tube base. Replace the tubes that have loose grid caps or envelopes. If replacement is impossible, do not attempt to clean or handle the tube, but operate the tube as it is, providing that its operation is normal. Enter the tube condition in the log so that replacement can be made at the earliest possible time.
2. Examine the spring clips that connect the grid, plate, and filament caps for looseness. Also examine all leads connected to these clips for poorly soldered or loose connections. These leads should be free of frayed insulation and broken strands. When removing clips from loosened grid caps, extreme care must be exercised, particularly if corrosion exists. Never try to force or pry a grid clip from the grid cap of a tube since damage to the tube or grid cap may result. If the grid cap is loose and it is necessary to remove the grid clip, first loosen

the tension of the clip by spreading it open; then gently remove (do not force) the clip from the tube cap.

3. Inspect the tubes to be sure they are secure in their sockets. Certain types of receiving tubes are mechanically fastened with tube spring locks; others have sockets that lock the tube in place. Inspect by turning the tube in a clockwise direction in its socket until it is locked in place. This type of socket is generally used for transmitting-type tubes. However, the firmness with which the tube is held in place depends on the tension of the terminals in the socket. These terminals are the spring type (contact springs) and must have sufficient tension to make good contact against the tube prongs. The tension can be tested by grasping the tube and turning it counterclockwise and then clockwise to its original position. If the tube seems to snap into place as it is turned, the spring tension of the socket terminals is firm enough; however, if the tension seems weak, the terminals may be tightened or adjusted as explained later in the tube maintenance procedure under Adjust.
4. Inspect all metal tubes for signs of corrosion and looseness of mounting. Many receiving-type tubes have keyways in the center of the tube bases. These keyways sometimes become broken, and this has a tendency to loosen the tube in its socket. Do not attempt to replace tubes having broken keyways unless it is absolutely necessary and it is possible to replace the tube correctly in its proper position. Inspect the sockets of metal tubes for cracks or breaks. Do not force metal tubes into their sockets. If they are hard to insert, examine the tube pins for signs of corrosion or solder deposits.

Tighten (*T*)—In this operation, take care not to overtighten tube sockets, tube clamps, and tube-socket insulators. Porcelain sockets and standoff insulators crack due to heat expansion if they are excessively tightened. Also be careful when tightening the tube caps of high-voltage tubes. Use the proper screwdriver or tool; if the tool should slip it might fall against the glass envelope and ruin a good tube.

Tighten all tube connections, terminals, sockets, and standoff insulators that were found to be loose during the inspection procedure. When tightening tube sockets having standoff insulators, determine before tightening if the fiber washers between the socket and the standoff insulators are intact. If these fiber spacers are cracked or missing, replace them before tightening the tube socket. Tightening the socket without these spacer washers breaks or cracks the porcelain tube socket.

Clean (*C*)—In the performance of this procedure, clean only where it is necessary. Do not remove tubes for cleaning purposes unless it is impossible to clean them in their original positions. If the tube must be removed, exercise caution. Do not attempt to clean the envelopes if they are located in an out-of-the-way place; in this case remove them for cleaning. When tubes are removed for cleaning, replace them immediately afterward. Do not leave them where they can be broken.

1. Clean the entire tube with a clean, dry cloth if the glass envelope is excessively dirty. Then wipe the glass envelope with a cloth moistened with water and polish after cleaning with a clean, dry cloth. Do not wipe metal tubes with a cloth moistened with water, since this causes the metal body of the tube to rust. Use a cleaning agent if the tube is excessively dirty because of oil deposits. Generally, metal tubes having oil deposits on their envelopes can be cleaned successfully by polishing them dry with a clean, dry cloth. The oil film remaining on the metal body of the tube prevents rusting. To remove oiliness, corrosion, or rust from tube envelopes, moisten a clean cloth with cleaning agent and apply to the area affected until it is clean. Wipe the envelopes dry with a clean, dry cloth.
2. Clean the grid and plate caps, if necessary, with a piece of No. 0000 sandpaper or crocus cloth. Wrap the paper around the cap and gently run it along the surface. Excessive pressure is unnecessary; neither is it necessary to grip the cap tightly.
3. When the tube sockets are cleaned and the contacts are accessible, fine sandpaper may be used if there is corrosion on the contacts. Clean the contacts thoroughly after sandpapering. Clean the area surrounding the tube sockets with a brush and a clean, dry cloth; this prevents dust and dirt from being blown back on the tube envelopes when the unit is put back into operation.

Adjust (A)—When performing this operation, care must be taken to place all leads and terminals as close as possible to their original positions.

1. Adjust all leads and tube connections. Check to determine if the leads are resting on the glass envelope of the high-voltage tubes; if they are, redress the leads so that the proper spacing is obtained. Examine all leads connecting to the tube caps. These should not be so tight that they barely reach the caps of the tubes. If this condition is found, redress these leads so that enough "play" is obtained. Adjust all the grid clamps so that the proper tension is obtained. To increase the tension of tube clamps, close the spring clamps slightly with a pair of long-nose pliers until the proper tension is obtained. Do not flatten the clamps.
2. The tube sockets for transmitting-type tubes should be adjusted if the tube is found loose in its mounting. The terminals of these sockets are spring-tensioned so that they may be adjusted to increase the pressure against the tube pins. To adjust these contacts, simply bend them toward the center of the socket until the correct tension is obtained. Do not apply too much pressure to the spring contacts; they may be broken from their mountings in the porcelain socket.

3. Any difficulty in removing or inserting metal tubes can usually be remedied easily. Remove the metal tube and examine the tube pins to determine if solder or corrosion has accumulated on them. Remove the solder deposits with a penknife; then polish the pins with fine sandpaper. Do not use a soldering iron to remove solder deposits; this makes them worse—the solder is built up on the pins rather than removed. To remove corrosion, use fine sandpaper, but never use it unless it is absolutely necessary. Saturate a small piece of cleaning cloth with a light lubricating oil or petroleum jelly and wipe the tube pins. Remove the excess oil from the pins by wiping them almost dry with a clean, dry cloth. If these procedures are followed, no difficulty should be experienced in removing or reinserting metal tubes. Do not force metal tubes into their sockets. Do not pry or wiggle them loose, since this damages the prongs of the socket and results in the intermittent operation of the unit employing them.

Special Instructions for Transmitting-Type Tubes

In addition to the preceding, high-power transmitter tubes and mercury-vapor rectifiers require special consideration as follows:

Mercury-Vapor Rectifier—As soon as possible after receipt of a new batch of mercury-vapor rectifiers, they should be placed in the tube sockets of the transmitter *without* the anode lead connected. Filament voltage should be applied and maintained for at least 30 minutes to distribute the mercury properly. These tubes should then be placed upright in a rugged container and protected from any jarring or tipping that would splatter the mercury. Should this occur, they must again be seasoned before the application of plate voltage. Applying anode voltage to an unseasoned mercury-vapor rectifier will cause severe arc-backs in the tube.

Unless proper precautions are taken, a major portion of lost airtime will be due to faulty rectifiers. These tubes should be observed whenever possible during each operating day. A good mercury-vapor rectifier is characterized by a healthy, clear-blue glow. A greenish-yellow color usually indicates a faulty tube or one which will soon cause trouble.

Due to the importance of foreseeing such trouble and due to the lack of familiarity of the average operator with testing methods for this type of tube, the reader should become familiar with the maintenance procedure illustrated by Fig. 14-33. The cathode-ray oscilloscope provides a convenient check. An isolation transformer of at least 300 volt-amperes rating should be used together with a series current-limiting resistor of 50 ohms as shown. The mercury-vapor rectifier tube is left in its regular socket with its regular plate-cap connection removed. The secondary of the isolation transformer is then connected in series with the resistor to the rectifier plate, and the other lead is connected to the filament center tap. The vertical-deflection plates of the oscilloscope are connected directly across the tube in the same manner. With the scope self-synchronized with the 60-Hz power line and power applied to the filament of the tube being checked, the scope pattern will show both the nonconducting half of the ac cycle and the conduction half which gives the dc potential. The sharp peak at the start of conduction reveals the condition of the tube under operating

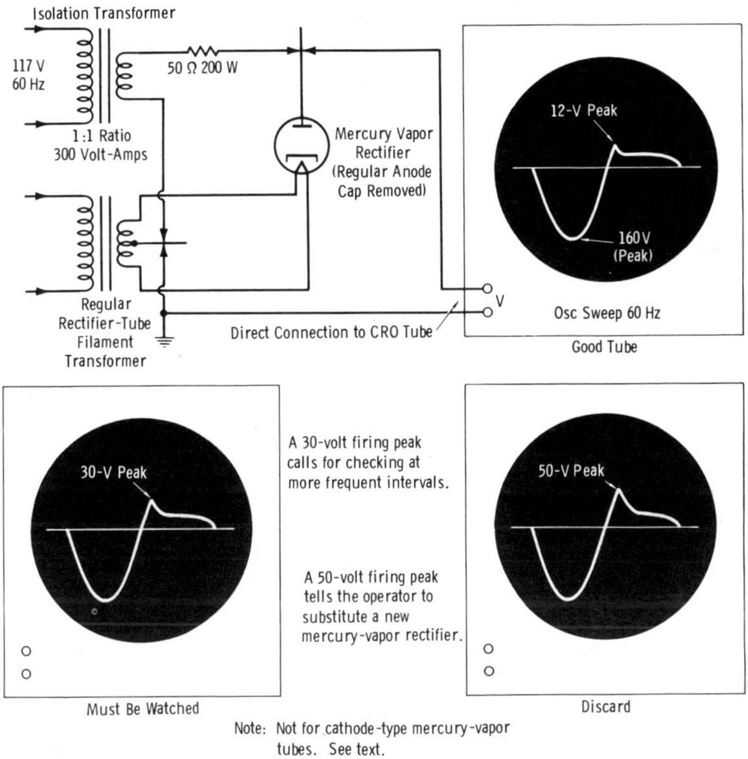

Fig. 14-33. Method of checking condition of mercury-vapor rectifier tube.

conditions. A good tube will fire at between 10 and 20 volts, as indicated by the amplitude of this peak on a calibrated screen. A tube approaching the end of its useful life will require a higher firing voltage and will break into conduction later in the conducting interval. When this breakdown peak reaches from 30 to 40 volts, the tube must be tested at more frequent intervals, preferably once a week. When the firing peak reaches close to 50 volts, the tube must be replaced with a new rectifier. Operators following this procedure will greatly minimize off-the-air time caused by rectifier arc-backs and otherwise defective tubes.

Always remember that mercury-vapor rectifiers must have their filaments operated at normal voltage for a minimum of 30 minutes and then be stored *upright* to prevent the mercury from splashing back on the envelope and elements. Tubes which have been accidentally jarred must again be preheated before application of the anode potential.

CAUTION: This type of simple check is possible only for gas tubes (such as the 866 and 8008) in which the filament directly faces the plate. Certain types of tubes (such as the 673) employ an indirect (cathode) heater which provides an extra element between filament and plate. This projection is similar in action to a grid, since it is connected to one side of the filament and is either positive or negative with respect to the plate, depending on which way the filament transformer is connected. The best way to field-test this type of tube is described in an RCA bulletin, *Pulse Method of Testing Hot-Cathode Gas Tubes,* Application Note AN-157. If this Bulletin is not included in your instruction book, write to RCA Electronic Components, Harrison, N.J. 07029 and request a copy. The circuitry is more involved, but it is worth your time.

Large Power Tubes—Larger-type power tubes, such as those used in modulator and final stages in a transmitter of 5 kW or more, also require special treatment on receipt from the factory and at 4- to 6-month intervals thereafter. They should be placed in the transmitter and only filament voltage applied for 30 minutes. Low plate voltage should then be applied for about 15 minutes. This processing materially aids in preventing gas formation within the tube—a common occurrence if such measures are not followed.

Occasionally a tube will develop a small amount of gas while in storage. The RCA recommendations for their 892R tube are as follows. With the tubes in the power amplifier, apply a low plate voltage without modulation. After a few minutes, apply 1000-Hz tone modulation, gradually increasing the percentage of modulation. If no gas flashes occur after 15 minutes of full tone operation, remove the modulating signal, apply high power, and then repeat the process.

If gas flashes occur during the process, go back to the low-power position with no modulation and repeat. Allow the tube to run for a considerable length of time with a low percentage of modulation, and then repeat the foregoing procedure. A tube that develops a small amount of gas in modulator service may usually be cleaned up by operating it in the power amplifier.

Forced-Air Systems

In recent years, the elimination of the water-cooling system has been accomplished for transmitters having as much as a 50-kW rating by the development of forced-air cooling systems. The control circuits for these systems are greatly simplified, consisting primarily of an air-interlock damper, which prevents the application of filament and plate voltages until the normal air-flow pressure is present, and a blower-motor "keep-alive" relay, which is a time-delay relay that keeps the blower motors functioning 4 to 7 minutes after the filament voltage is removed.

The maintenance of forced-air systems is simpler than that of water systems, but it is just as important for trouble-free operation. The canvas air ducts should be cleaned about once a month by removing them, turning them inside out, and using a vacuum cleaner to remove the accumulated dirt. While these ducts are removed, a cloth may be used to clean between the fins of the tube, especially against the tube anode. Care must be taken not to damage the mercury air-flow switches mounted on the blower housing. These switches prevent the application of filament and plate voltages until the proper air flow is present. Both sides of the air-flow vanes (half-circle discs used to operate the mercury switch) should be wiped clean with a cloth or chamois, and a small wire brush may be used to clean the corners of the fan blades. A vacuum cleaner then may be used to pick up any dust from inside the bottom of the blower frames.

After this cleaning procedure, the blowers should be started to check the air-flow vanes for proper operation of the mercury switches. Then the canvas ducts should be replaced and the over-all operation checked.

Capacitors

Capacitors are vital components in transmitters. The following paragraphs outline procedures for their maintenance.

High-Voltage Capacitors—Because of their high operating potentials, high-voltage capacitors must be kept clean at all times to prevent losses and arcing. Dirt, oil deposits, or any other foreign matter must not be allowed to accumulate on the high-voltage terminals of these capacitors. All leads and terminal connections must be inspected periodically for signs of looseness and corrosion, and the porcelain insulators must be inspected for cracks and breaks. CAUTION: To avoid severe electrical shock in case of bleeder failures, always discharge high-voltage capacitors before maintenance.

Low-Voltage Capacitors, Oil-Filled—Low-voltage, oil-filled capacitors require the same care as the high-voltage type, although the frequency of the maintenance operation is not so critical. The terminals and connections of these capacitors should be given the same careful inspection as those of the high-voltage types. The leads of these capacitors are usually not as rugged as those on the high-voltage capacitors, and they should be inspected more closely for poorly soldered connections.

Tubular Capacitors—These capacitors are of the low-voltage paper type and are generally used in low-voltage circuits for coupling and bypassing. They should be inspected and cleaned whenever the chassis in which they are located is removed for maintenance. The only maintenance requirement for these capacitors is inspection of the tubular body of the capacitor for bulging, excessive swelling of the capacitors, and signs of wax leakage. Inspect the terminal leads (pigtail type) of the capacitors for firmness of contact at their respective points of connection. Never use a cloth to clean this type of capacitor; this may result in damage to the surrounding circuits. Dirt and dust are brushed from the capacitor and surrounding area with a small, soft brush.

Mica Capacitors—Mica capacitors require very little maintenance other than being kept free from dust and oil. Two types of mica capacitors are generally used, the high-voltage and low-voltage types. The low-voltage types are inspected whenever maintenance is performed on the chassis of the unit in which they are located. The capacitors are inspected for body cracks caused by excessive heat, and their leads (pigtail type) are inspected for firmness of contact at their points of connection. The terminals of the high-voltage types must be inspected for tightness and corrosion, firmness of mounting, and body conditions. The body of this type of capacitor is made of a ceramic material, and care must be exercised when tightening the mountings. The bodies of the capacitors are easily kept clean with a dry, clean cloth. For satisfactory operation, the terminals must be free from dirt and corrosion at all times. Take care when tightening the terminals of these capacitors, as excessive pressure damages or cracks the ceramic case where the terminals are coupled to the body of the capacitor.

Trimmer Capacitors—In very damp climates, trimmer capacitors must be inspected often. Moisture, if allowed to accumulate on the plates of the capacitors, causes erratic operation of the unit using the capacitors. In certain cases where high voltage is used, serious damage to the capacitors results. A minute amount of moisture is all that is necessary to short-circuit the plates of the capacitor and cause abnormal operation. When such conditions are encountered, the capacitor must be thoroughly dried with a small portable heater. A cleaning cloth used to dry the plates of the capacitors may throw the plates out of alignment when the cloth is inserted between them. In cases where the plates of the capacitors are very closely spaced, use a magnifying glass to locate the exact position of the moisture beads existing between the plates. Due to the sheen of the capacitor plates, very minute particles of moisture cannot always be detected by the naked eye.

Capacitor Maintenance

The following suggested procedures may be followed in the maintenance of the various types of capacitors already described.

Feel (F)—Feel the terminals of the high-voltage filter capacitors. These should be fairly cool. Excessive heat probably indicates losses due to loose, dirty, or corroded terminal connections. Feel the sides of oil-filled and electrolytic capacitors. These should be cool or slightly warm. If they are very warm or hot, excessive internal leakage is indicated. Capacitors in this condition can fail at any time and should be reported for immediate replacement.

Inspect (I)—Inspect the general condition of all capacitors, regardless of type. Inspect for broken, frayed, or loose terminals, leads, and connections. Inspect the condition of the terminals of the high-voltage capacitors.

Check these for dirt, corrosion, and looseness. Inspect the bodies of the capacitors for signs of excessive bulging and oil leakage. Inspect the plates of the tuning capacitors for dirt and corrosion. Check all capacitor shafts, bushings, bearings, and couplings for looseness or binding.

Tighten (*T*)—Tighten all loose terminals, connections, and terminal leads on all types of capacitors. Tighten all capacitor mountings and standoff insulators. Tighten all loose shaft couplings and bushings.

Clean (*C*)—Special attention should be given to all high-voltage capacitors to insure that they not only are kept clean but are free from moisture. Thoroughly clean the insulators, terminals, and leads of high-voltage capacitors. When they are extremely damp due to high humidity, these capacitors frequently must be wiped dry with a clean, absorbent cloth to prevent arcovers and breakdown of insulation. Remove the terminals that appear to be either corroded or dirty; also remove those causing power losses due to high-resistance connections. Clean them with crocus cloth which is either dry or moistened with cleaning fluid. After cleaning the terminals, polish them dry with a clean, dry cloth. Replace all the connections after cleaning, making certain that good electrical contact is obtained. Low-voltage capacitors require little attention. However, all insulated bushings and supports should be kept clean and free from foreign matter.

Adjust (*A*)—Adjust all leads if necessary. This requires the redressing of leads which may have been dislocated during the maintenance procedure. If capacitor leads are stretched too tightly, redress or replace them to obtain the correct placement.

Resistors

Resistors may be divided for maintenance purposes into two groups: those resistors easily detachable and known as *ferrule-type resistors* and those with soldered terminals and known as *pigtail-type resistors.* CAUTION: Do not touch power resistors immediately after the power has been removed. They are usually hot, and severe burns can result from contact with them.

Feel (*F*)—The springiness of ferrule clips may be ascertained when the ferrule-type resistor is removed. Insufficient pull at the clip may be an indication of a loose connection and poor electrical contact.

Inspect (*I*)—It is important to inspect all types of resistors for blistering or discoloration, since these are indications of overheating. Inspect the leads, clips, and metallic ends of the resistors and their adjacent connections for corrosion, dirt, dust, looseness, and broken strands in the connecting wires; also inspect the firmness of mounting.

Tighten (*T*)—Tighten all resistor mountings and connections found to be loose. If the tension at the end clips has decreased, it is common practice to press the clip ends together by hand or with a pair of pliers. The hand method is preferred because the pliers may bend the clip or damage the contact surface.

Clean (*C*)—Dirty or corroded connections of ferrule-type resistors can be cleaned by using a brush or cloth dipped in cleaning fluid. If the condition persists, use crocus cloth moistened with cleaning fluid. It may be necessary to sandpaper the resistors lightly with fine-grade sandpaper, such as No. 0000. Always wipe them clean with a dry cloth before replacing them. Vitreous resistors connected across high voltage should be kept clean at all times to prevent leakage or flashovers between terminals. They can be wiped clean with a dry cloth or a cloth moistened with cleaning fluid. If cleaning fluid is used, polish the resistors with a dry, clean cloth.

Pigtail-Type Resistors—Maintenance of pigtail-type resistors is limited to an inspection of the soldered connections. These connections may break if the soldering is faulty or if the resistors are located in a place subject to vibration. The recommended practice is to slide a small insulated stick lightly over the connections and to inspect them visually for solidness. If connections are noticeably weak or loose, resolder them immediately. Discolored or chipped resistors indicate possible overloads. Although replacement is recommended, resistors in this condition may last indefinitely. The pigtail-type connections should be dusted with a brush or with an air blower.

Fuses

A fuse consists of a strip of fusible metal inserted in an electrical circuit. If the current increases beyond a safe value, the metal melts, thus interrupting the current. Fuses vary in size and rating depending on the circuits in which they are used. Two types of fuses are used: renewable and nonrenewable. The first type is designed so that the fuse link, or element contained within the fuse cartridge, may be removed and replaced when blown. The second type, however, is constructed so that the fuse element is permanently sealed within the fuse housing. When a fuse blows, an attempt must be made to determine the reason for its failure and to make corrections before a new fuse is installed.

Renewable Type—The renewable-type fuse assembly consists of a housing or cartridge of insulating material with a threaded metal cap (ferrule) at each end. As a precaution against damage, the fuse element, or link, is placed inside the cartridge, or housing, and is held in position by the two end caps, or ferrules. When a fuse is placed in service, the two ends of the fuse cartridge are slid into spring contacts mounted on the fuse block. This places the fuse in the circuit to be protected.

Nonrenewable Type—When nonrenewable fuses are blown, they must be discarded. Certain types of nonrenewable fuses are removed by unscrewing and withdrawing the cap screws holding them in place. When they are removed, the fuse and cap screw are separated by pulling them apart. The glass fuses are easily removed for inspection. Care must be taken to see that the fuse ends and holding clips are kept clean and tight. If they are not, overheating will result and make replacement necessary.

Inspect (*I*)—Inspect the fuse caps for evidence of overheating and corrosion. Inspect the fuse clips for dirt, loose connections, and proper tension.

Tighten (*T*)—Tighten the end caps, the fuse clips, and connections to the clips on replaceable fuses if they are loose. The tension of the fuse clips may be increased by pressing the sides closer together. Fuse caps should be hand tightened only. Excessive tightening results in difficulty in removing them.

Clean (*C*)—Clean all fuse ends and fuse clips with fine sandpaper when needed; wipe with a clean cloth after cleaning. If it becomes necessary to use a file to remove deep pits in the clips, fuse ends, or contacts, always finish with fine sandpaper in order to leave a smooth contact surface. As a final step, wipe the surface clean with a clean, dry cloth.

Bushings and Insulators

Bushings and insulators are extremely important elements in electrical circuits, especially when located in high-voltage circuits where insulation breakdown is most common. Most of the high-voltage insulators are constructed of ceramic material with highly glazed surfaces. Exercise extreme care when working near these insulators, because they are easily chipped or broken.

Inspect (*I*)—Thoroughly inspect all high-voltage insulators and bushings for moisture, dust, and other accumulated foreign matter. Unless they are both clean and dry, leakage or arc-overs will occur and damage them permanently. Check for chipped surfaces, hairline cracks, carbonized arc-over paths, and other surface defects that may make the insulator unserviceable. Insulators in this condition should be reported to the person in charge for replacement.

Tighten (*T*)—Feedthrough bushings and standoff and other insulators should be tightened if they have loose mountings or supports. Tighten these insulators with care because the gaskets absorb only a small amount of pressure before permitting the insulator to break.

Clean (*C*)—Cleaning operations are similar to those outlined for tubes. Use a clean cloth (dampened with cleaning fluid if necessary) to remove dust, dirt, or other foreign matter. Always polish with a dry, absorbent cloth after cleaning.

Relays

The various types of relays may be classified as follows: overload relays, time-delay relays, and magnetic contactors. Relays require a certain amount of preventive maintenance, which *must never be performed except when absolutely necessary.* Certain types are completely encased in dustproof and moistureproof cases. These require little maintenance other than a periodic inspection.

To service the relay contacts, several types of tools are necessary. Each of these has a special function, as described in the following paragraphs.

Burnishing Tool—This tool is used on relays that have extremely hard contacts; it is not a file. A contact should not be burnished unless it is found to be pitted or oxidized, and then not burnished more than necessary to restore a clean, smooth surface. The original shape of the contact must be retained.

Small Fine-Cut File—This file is to be used only on the larger contacts when they have become badly burned or pitted and a replacement is not available. This tool is not to be used on silver-plated contacts or on the contacts of telephone-type relays. Do not use the file more than necessary to remove the pit. The original shape of the contact must be preserved. After filing, No. 0000 sandpaper may be applied to the contact and followed by crocus cloth to obtain a smooth finish on the contact surface. A clean, dry cloth is used for the final polishing.

No. 0000 Sandpaper Stick—This tool is made in the same way as the crocus-cloth stick, except that sandpaper is used instead of crocus cloth. The use of sandpaper is limited, as is the use on the fine-cut file, to the treatment of badly burned or pitted contacts on the larger relays. Sandpaper is not used on silver-plated contacts, except under extreme circumstances; when it is used it should be followed by crocus cloth. All contacts should be polished after sanding with a clean, dry cloth.

Crocus Cloth—This maintenance aid is available in two forms—as a tool and as a strip of material. It serves a twofold purpose: It may be used to remove corrosion from all relay contacts, or it may be applied to the contacts following the use of the fine-cut file and No. 0000 sandpaper. Neither the file nor sandpaper leaves a finish smooth enough for proper relay operation. Use crocus cloth to polish the surface of the contact. The choice between the stick and the piece of cloth depends on the accessibility of the contacts. If the location of the relay and the position of the contacts permit the use of the crocus-cloth stick, it should be used; otherwise, the strip of crocus cloth must be used. The crocus cloth and tool are used as illustrated in Figs. 14-34 and 14-35. In both cases, the maintenance aid is inserted between the contacts and is drawn through them while the contacts are pressed together with the fingers.

Maintenance of relays requires that they be inspected periodically and preventive-maintenance measures performed if necessary. The inspection procedure requires that the terminals be inspected for looseness, dirt, and corrosion. Contacts may have become loosened because of the jarring of the equipment during shipment. The contacts may become dirty or corroded due to climatic conditions. Relay contacts must never be sandpapered or filed unless this procedure is absolutely necessary for the normal operation of the relay unit. A relay is considered normal if:

1. The relay assembly is free from dirt, dust, and other foreign matter.
2. The contacts are not burned, pitted, or corroded.
3. The contacts are properly aligned and correctly spaced.

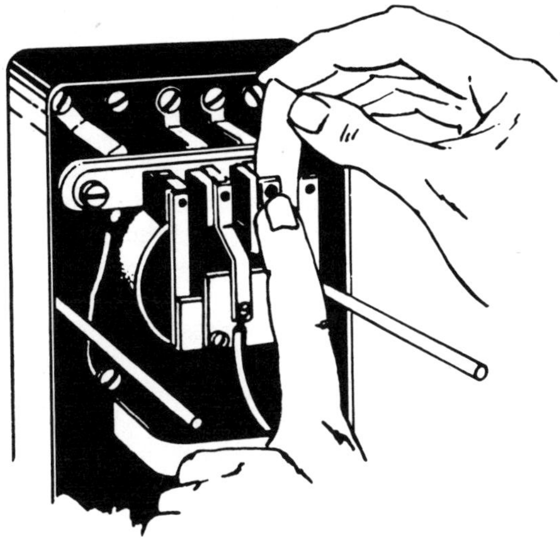

Fig. 14-34. Use of crocus-cloth strip for polishing relay contacts.

4. The contact springs are in good condition.
5. The moving parts travel freely and function in a satisfactory manner. The solenoids of plunger-type relays must be free from obstructions.
6. The connections to the relay are tight.
7. The wire insulation is not frayed or torn.
8. The relay assembly is securely mounted.
9. The coil shows no sign of overheating.

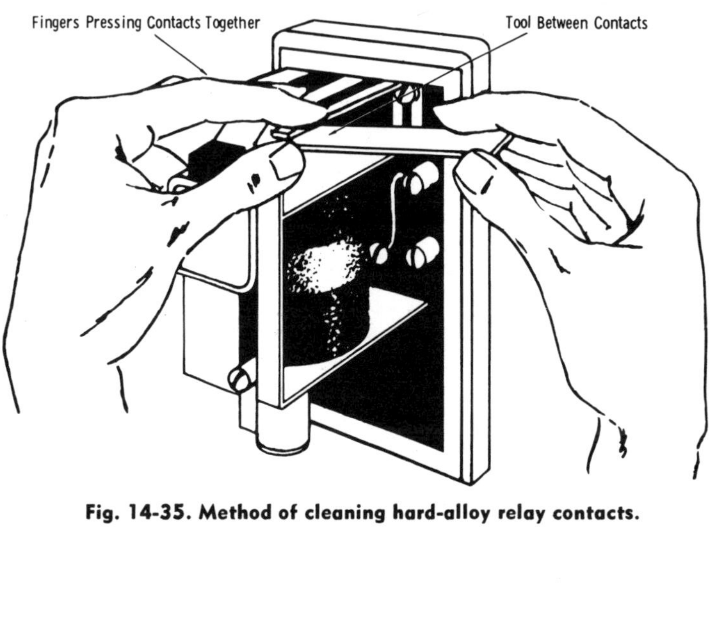

Fig. 14-35. Method of cleaning hard-alloy relay contacts.

A relay is considered abnormal if it fails to meet any of the above-mentioned requirements. The following are the procedures used in the maintenance of relay units.

Inspect (I)—Using the check list given above, inspect the relays to determine abnormal conditions. If the contacts are not readily accessible, they may be examined with the aid of a flashlight and mirror. Many of the relays can be inspected and cleaned without being removed from their mountings or without being taken apart. Mechanical action of the relays should be checked to make certain that the moving and stationary contacts come together in a definite manner and that they are directly in line with each other. The armature or plunger mechanism should move freely without binding or dragging. Be careful during inspection not to damage or misalign the relay mechanism. Relays that require the removal of the cover for complete inspection may be found enclosed in glass, Bakelite, or metal cases. Relays must never be taken apart unless it is absolutely necessary. Exercise care if they must be taken apart for maintenance purposes. When disassembling relays, tag all leads as they are removed. This insures that the leads are returned to their proper terminals after the maintenance procedure is completed.

Tighten (T)—Tighten all mounting screws found to be loose, but do not apply enough force to damage the screw or to break the part that it holds. Do not start screws with their threads crossed. If a screw does not turn easily, remove it and start again. Relay coils can be tightened by inserting, if possible, a small wooden or paper wedge between the coil and the core of the relay. This prevents chatter of the relay. Tighten any and all loose connections. Also tighten the mounting of the relay assembly if it is loose. When replacing glass or Bakelite covers over relay cases, take care not to overtighten the screw cap holding the cover over the relay.

Clean (C)—Clean the exterior of the relay with a dry cloth. If it is very dirty, clean with a cloth or brush dipped in cleaning fluid; then wipe the surface with a dry cloth. If loose connections are found, they should be inspected. Remove and clean connections that inspection reveals are either dirty or corroded.

The relay service aid is a narrow piece of folded cloth or canvas. It serves a twofold purpose: it is suitable for polishing a clean surface, and it is used as a follow-up to a crocus cloth. It is also intended to remove the grains that come off the crocus cloth and adhere to the contact surface. The cloth is used as shown in Fig. 14-34.

Hard Contacts—Hard-alloy contacts are cleaned by drawing a strip of clean wrapping paper between them while they are held together. It may be necessary in some cases to moisten the paper with cleaning fluid. Corroded, burned, or pitted contacts must be cleaned with the crocus-cloth strip or the burnishing tool as shown in Fig. 14-35.

Solid-Silver Contacts—Dirty solid-silver contacts are easily cleaned with a brush dipped in cleaning fluid. After they are cleaned, the contacts are polished with a clean, dry cloth. Note that the brown discoloration that is found on silver and silver-plated relay contacts is silver oxide and is a good conductor. It should be left alone unless the contacts must be cleaned for some other reason. It may be removed at any time by a cloth moistened in cleaning fluid.

Dress corroded contacts first with crocus cloth, using either the stick or the strip of crocus material. When all of the corrosion has been removed, wipe with a clean cloth moistened in cleaning fluid and polish with a piece of folded cloth. Make certain that the shape of the contacts has not been altered from the original.

Burned or pitted contacts may be resurfaced, if necessary, with No. 0000 sandpaper, making certain that the original shape of the contacts is not changed. Next, smooth the surface of the contacts with crocus cloth until a high polish is obtained. Wipe thoroughly with a clean cloth to remove the abrasive remaining on the contacts. When contacts are very badly burned or pitted and a replacement is not available, use a small fine-cut file and No. 0000 sandpaper.

Silver-Plated Contacts—Dirty silver-plated contacts are cleaned with a cloth or brush dipped in cleaning fluid. After they are cleaned, the contacts are polished with a dry cloth.

Dress corroded contacts first with crocus cloth, using either the stick or strip of crocus material. The work must be done very carefully so as not to remove an excessive amount of silver plating. When all of the corrosion has been removed, polish with a clean, dry cloth. Make certain that the shape of the contacts has not been altered.

Dress burned or pitted contacts with crocus cloth until the burned or pitted spots are removed. This may require an appreciable amount of time and energy, but it is preferable to using a file or sandpaper. If the crocus cloth does not remove the burns or the pits, use the sandpaper tool very carefully. When sandpaper is used, it must be followed with crocus cloth to polish the contacts, and then with a cloth moistened in cleaning fluid. The contacts are then polished with a clean, dry cloth.

Never use highly abrasive materials, such as emery cloth, coarse sandpaper, or Carborundum paper for servicing relay contacts, since damage to the contacts will result.

Adjust (A)—Adjust relay contacts after cleaning if necessary. The contacts should close properly when the plunger is hand operated. Adjust the relay springs if necessary, but do not tamper with them unless it is absolutely necessary. These springs are factory adjusted and maintain a certain given tension; they rarely get out of adjustment. If the spring tension must be changed, exercise care when doing so. The adjustment of current-control relays is usually accomplished by turning calibrated knobs to the desired setting or by turning a knurled adjustment sleeve which has a calibrated scale mounted adjacent to it. The adjustments should not be changed from their original factory setting except in cases of emergency. Overload relays must never be adjusted unless the person in charge has been notified and has sanctioned the adjustment.

Shapes of Relay Contacts—Relay contacts have varied shapes (Fig. 14-36) depending on their size and application. In some instances, both contacts are flat; in others, one contact is convex although its mate is flat. The original shape of a contact must be retained during cleaning. If burning or pitting has distorted the contact so that it must be reshaped, the original shape must be restored. It is essential that the maintenance personnel familiarize themselves with all details of the relays by examining them while the relays are in good condition.

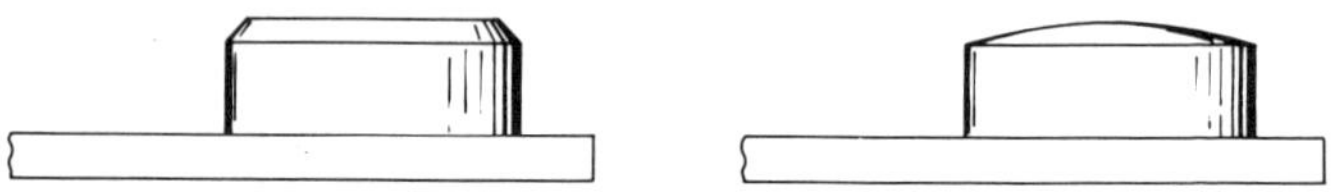

(*A*) *Flat.* (*B*) *Convex.*

Fig. 14-36. Relay-contact shapes.

Sparks and "Key Thump" Suppression

In some composite equipment (equipment not commercially designed), the engineer encounters "key clicks" or excessive sparking of relay or switch contacts. If this occurs, capacitor-resistor suppressor circuits placed in shunt with the contacts will aid materially in reducing the problem and increasing the life of the contacts.

Exact values of capacitance and resistance to use are best determined by trial. The resistance should be as high and the capacitance as low as is effective for the circuit in question. When the controlled current is pulsating or alternating, start with a capacitance of about 0.04 μF and a series resistance of 1000 ohms. If these values are not effective, gradually increase the capacitance and decrease the resistance for each capacitance until the desired suppression is achieved. When sparking occurs across contacts in a noninductive dc circuit, start the test runs with 0.5 μF capacitance and 200 ohms resistance. The optimum values will not normally be beyond these limits.

Switches

For the purpose of maintenance, switches may be classified into two general groups: those with contacts that are readily accessible, and those with contacts that are completely encased. The basic maintenance operations of inspection, cleaning, adjusting, and lubrication are applicable only to the first group. Because of the enclosed construction of the second group, no maintenance can be applied except to make a mechanical test of their operation.

Accessible-Contact Switches—This group consists of knife-blade switches, start-stop push-button switches, and high-voltage shorting bars. With the exception of the shorting bars, all of these switches consist of blades that are mechanically inserted into spring contacts.

Inspect (I) all the terminal connections of each individual switch for tightness and cleanliness. Check the mounting of the switch for firmness. Operate the mechanism of the switch and see if the parts move freely. Observe the stationary spring contacts to determine if they have lost tension and if they are making good electrical contact.

Tighten (T) all loose mountings and connections properly. If inspection shows that the fixed contacts have lost tension, tighten them with the fingers or pliers. Tighten every loose connection or terminal.

If inspection shows that any terminal, connection, or section of the switch is dry, dusty, corroded, or pitted, *clean* (c) the part with a dry, clean cloth. If the condition is more serious, moisten the cloth with cleaning fluid and rub vigorously. Surfaces which have been touched with the bare hands must be thoroughly cleaned with a cloth moistened in cleaning fluid and then polished with a clean cloth. The points of contact with the moving blade are naturally those which most often show signs of wear. Examine these points very carefully to insure that both sides of each blade, as well as the contact surfaces of the clips, are spotlessly clean at all times. Crocus cloth moistened with cleaning fluid usually produces this condition; however, if this is not sufficient, No. 0000 or No. 000 sandpaper may be used. Always polish clean after the sandpapering operation.

Adjust (A) if necessary. Some of the switches have a tendency to fall out of alignment because of loosening of the pivot. In most cases, tightening the screw on the axis of motion corrects this condition.

Lubricate (L) when necessary. If binding is noted during inspection of the operation of the switch, apply a drop of instrument oil with a toothpick to the point of motion or rotation. Do not allow oil to run into the electrical contacts, since a film of oil may cause serious damage or a poor contact. Lubrication of switches is not recommended unless serious binding is noticed.

Nonaccessible-Contact Switches—Under this heading are included all the remaining switches not discussed above. Interlock switches, toggle switches, meter-protective push buttons, and selector switches have been designed so that it is impossible to get at the contacts without breaking the switch assemblies. The only maintenance possible is to check the operation of the switch and, if something abnormal is detected, to notify the person in charge immediately so that a spare may be obtained and a replacement made as soon as possible. Do not lubricate these switches under any circumstances.

Generators and Motors

Certain preventive-maintenance procedures must be applied to motors and generators if proper functioning and dependable performance are to be obtained. There are three principal causes that contribute to faulty operation of this type of equipment: accumulation of dirt, dust, or other foreign matter on the windings and moving parts of the equipment; lack of sufficient lubrication on bearings and other moving parts; and improper adjustments or damaged parts. Given proper care, motors and generators provide long and efficient service. In addition to the techniques given in the following paragraphs, additional maintenance instructions covering specific motors or generators will be found in the manufacturer's instruction books. The maintenance techniques that follow apply to the motors and generators used at the transmitter for standby power or in field use.

Feel (*F*)—The bearing and the housings can be tested by feeling them to determine overheated conditions. An accepted test, except in very hot climates, is to hold the bare hand in contact with the bearing or housing for a period of at least 5 seconds. If the temperature can be tolerated for this length of time, the bearing temperature may be considered normal. Overheating may indicate lack of sufficient lubrication, a damaged bearing surface, or, in rare situations, an accumulation of dirt in the field windings.

Inspect (*I*)—Each motor and generator exterior, and any other visible parts, must be inspected for dirt and signs of mechanical looseness or defects. Wherever wires are exposed, see that all connections are tight and in good condition and that the insulation is not frayed. Inspect the motor ends for excess oil and the mounting for loose bolts. Wherever possible and practicable, feel the pulleys, belts, and mechanical couplings to insure that the proper tension or tightness is present. (Naturally, this must *not* be done while the machine is in motion.)

Tighten (*T*)—Any mounting, connection, or part found to be loose must be properly tightened. If any internal part such as a commutator segment or an armature coil is loose, notify the person in charge, and repair the part immediately or replace it at the first opportunity. Operation under these conditions will cause considerable damage in a very short period of time.

Clean (*C*)—Carefully wipe the exterior, base, and mountings of each motor or generator with an oiled cloth in order to leave a thin, protective film of oil on the surfaces. If available, use an air blower or hand bellows to blow the dust and dirt out if inspection shows that the windings are dusty or dirty.

If inspection of the commutator and brushes shows that cleaning is necessary, the accepted cleaning practice is as follows. Lift or remove the most accessible brush assembly, and press a piece of canvas cloth folded to the exact width of the commutator against the commutator; then run the motor for about 1 minute, exerting the necessary pressure. If the condition persists because the commutator has been burned or pitted, use a piece of fine sandpaper (No. 0000), preferably mounted on the commutator cleaning stick. While exerting the necessary pressure, rotate the motor for approximately 1 minute. Stop the motor and wipe around the commutator bars with a clean cloth. It may be necessary to polish the commutator with a piece of canvas, as explained previously. Identical maintenance procedures apply to slip rings.

Transformers and Choke Coils

Some transformers are enclosed in metal housings; others are not. However, similar maintenance techniques are applicable to all of them.

Inspect (*I*)—Carefully inspect each transformer and choke for general cleanliness, for tightness of mounting brackets and rivets, for solid terminal connections, and for secure connecting lugs. The presence of dust, dirt, and moisture between terminals of high-voltage transformers and chokes may cause flashovers. In general, overheating in wax- or tar-impregnated transformers or coils is indicated by the presence of insulating compound on the outside or around the base of a transformer or coil. If this condition is encountered, immediately notify the person in charge.

Tighten (*T*)—Properly tighten mounting lugs, terminals, and rivets found to be loose.

Clean (*C*)—All metal-encased transformers can be easily cleaned by wiping the outer casings with a cloth moistened with cleaning fluid. Clean the casing and the immediate area surrounding the transformer base. Clean any connections that are dirty or corroded. This operation is especially important on high-voltage transformers and coils. It is very important that all transformer terminals and bushings be kept clean at all times and examined regularly.

Variable Transformers

Variable transformers, as a rule, are sturdily built and are protected so that very little maintenance other than regular inspection is required.

Inspect (*I*)—Carefully inspect the exterior for signs of dirt and rust. Inspect the mounting of each variable transformer to determine if it is securely mounted. Inspect all connections for looseness, corrosion, and dirt. Check the slip rings for signs of corrosion or dirt.

Clean (*C*)—The perforated casing of each variable transformer as well as the area surrounding the base must be cleaned regularly. If the slip rings need cleaning, disassemble the unit and clean them with a cloth moistened in cleaning fluid, and then polish with a clean, dry cloth. If the dirty condition persists, use crocus cloth and rub vigorously. Again polish with a clean cloth. Reassemble the unit; then reinstall it, reconnecting all terminals.

Lubricate (*L*)—If the shaft shows signs of binding or if it squeaks, apply a few drops of household oil to the front and rear bearings. Rotate the control shaft back and forth several times to insure an equal distribution of the lubricant in the front and rear bearings.

Rheostats and Potentiometers

Rheostats and potentiometers fall into two main groups for maintenance purposes: those which have the resistance winding and the sliding contact open and accessible, and those which, by construction, have their inner parts totally enclosed. In the latter group very little maintenance can be performed, since opening and removing the metal case may damage the unit.

Inspect (*I*)—The mechanical condition of each rheostat must be inspected regularly. The control knob should be tight on the shaft. Inspect the contact arm and resistor winding for cleanliness and good electrical contact. Check the rheostat assembly and mounting screws for firmness, the sliding arm for proper tension, and the insulating body of the rheostat for cracks, chipped places, and dirt.

Tighten (*T*)—Tighten carefully any part of the rheostat or potentiometer assembly found to be loose.

Clean (*C*)—The rheostat or potentiometer assembly is easily cleaned by using a soft brush and then polishing with a soft, clean cloth. If additional cleaning is needed, or if the windings show signs of corrosion or grease, the brush may be dipped in cleaning fluid and brushed over the winding and contacts. Use a clean cloth to remove the film that remains after the cleaning fluid has evaporated. If the contact point of the sliding arm is burned or pitted, it is a good practice to place a piece of folded crocus cloth between the contact and the winding and then to slide the arm over the crocus cloth a number of times. When cleaning the winding, do not exert excessive pressure, or damage will result.

Adjust (*A*)—If the tension of the sliding contact is insufficient, an adjustment can be made with long-nose pliers. A slight bending of the rotating piece in the proper direction restores the original tension.

Lubricate (*L*)—Apply lubrication only when necessary; that is, when binding or squeaking is noticed. One or two drops of instrument oil applied to the bearings with a toothpick is sufficient. Since the slightest flow of oil into the winding or the sliding-arm contact may cause serious damage, lubrication must be applied very carefully and only to the bearings. Wipe off all excess oil.

Terminal Boards and Connecting Panels

Little preventive maintenance is required on terminal boards and connecting panels. The following paragraphs contain some suggested procedures.

Inspect (*I*)—Carefully inspect terminal boards for cracks, breaks, dirt, loose connections, and loose mountings. Examine each connection for mechanical defects, dirt, corrosion, or breakage.

Tighten (*T*)—All loose terminals, screws, lugs, and mounting bolts should be tightened properly. Use the proper tools for the tightening procedure, and do not overtighten, or the assembly may crack or break.

Clean (*C*)—If a connection is corroded or rusty, it is necessary to disconnect it completely. Clean each part individually and thoroughly with cloth or crocus cloth moistened with cleaning fluid. All the contact surfaces should be immaculate for good electrical contact. Replace and tighten the connection after it has been thoroughly cleaned.

Air Filters

Air filters are placed in blowers and ventilating ducts to remove dust from the air before it is drawn into and circulated through the ventilating system. Some filters are impregnated with oil and some are filled with cut strands of glass to facilitate the filtering action. The following procedures cover their maintenance.

Inspect (*I*)—The filter should be inspected for any large accumulation of dirt and for lack of oil. Note whether the filter is mounted correctly and whether the retaining clips are in place. Improperly assembled filter elements allow unfiltered air to leak around the edges and thus permit dust to enter the ventilating system.

Tighten (*T*)—Tighten the retaining clips if they are loose, and readjust the filter in its mounting.

Clean (*C*)—Usually the filters are easily accessible and may be taken out after the removal of the cover plate. The general procedure is as follows. Mark the outside of the filter before removing it from the air duct. Before washing it, tap its edges against the wall or on the ground to remove as much dirt as possible. Wash the filter in gasoline, using a brush to remove dirt from the steel wool. After the filter has been washed, place it face down on two supports. Allow the filter to drain and dry thoroughly before lubricating it.

Lubricate (*L*)—Lubricate or recharge the filter element by dipping it in a bath of oil. For temperatures above 20°F, use SAE-10 oil. Allow the filter to drain thoroughly, intake side down, before it is put into use. While the filter is draining, keep it away from places where sand or dirt is being blown through the air. Always replace a filter with its intake side facing the incoming air flow.

Cabinets

The cabinets that house the various components of the installation are generally constructed of sheet metal. Suggestions for their care follow.

Inspect (*I*)—The outside and inside of each cabinet must be inspected. Check the door hinges (if any), the ventilator mountings, the panel screws, and the zero-setting of the meters. Examine the pilot-light covers for cracks and breaks. Occasionally remove the covers and see if the pilot-light bulbs are secure in their sockets. Inspect the control panels for loose knobs and switches.

Adjust (*A*)—Adjust the zero setting of meters if they are incorrect. Follow the specific instructions given in the subsection on meters.

Clean (*C*)—Clean each cabinet, including the control panel, outside and inside with a clean, dry cloth. Also clean the meter glasses and control knobs with a clean, dry cloth.

Lubricate (*L*)—Door hinges and latches need little lubrication, but if inspection reveals that they are becoming dry, apply a small amount of instrument oil. Remove excess oil with a clean, dry cloth.

Meters

Meters require very little maintenance. They are extremely delicate instruments and must be handled very carefully, and, because they are precision instruments, they cannot be repaired in the field. A damaged meter should be replaced with a spare and the defective meter returned to the maker for repair and calibration.

Inspect (*I*)—Inspect the leads and connections to the meter. Check for loose, dirty, and corroded connections and for cracked or broken cases and meter glasses. Since the movement of a meter is extremely delicate, its accuracy is seriously affected if the case or glass is broken and dirt and water filter through. If the climate is damp, it is only a matter of time until enough moisture seeps through a crack to ruin the meter movement.

Tighten (*T*)—Tighten all loose connections and screws. Any loose meter wires should be inspected for dirt or corrosion before they are tightened. The tightening of meter connections requires a special technique because careless handling can easily crack the meter case. To prevent breakage, firmly hold the hexagonal nut under the connecting lug while the outside nut is tightened. This permits the tightening of the connection without increasing the pressure of the head of the stud against the inside of the meter case.

Clean (*C*)—Meter cases are usually made of hard, highly polished Bakelite and can be cleaned with a dry cloth. If cleaning is difficult, the cloth may be dampened with cleaning fluid. Dirty connections may be cleaned with a small, stiff brush dipped in cleaning fluid or with a small piece of cloth dipped in solvent. Remember that solvents do not remove all dirt from hard surfaces. Some of the dirt remains in a softened state and must be removed with a damp cloth. Corroded connections are cleaned by sanding them lightly with a very fine grade of sandpaper, such as No. 0000. After they are cleaned, the connections are wiped carefully with a clean cloth.

Adjust (*A*)—Normally, all meters should indicate zero when the equipment is turned off. The procedure for setting a meter to zero is not difficult. The tool required is a thin-bladed screwdriver. Before deciding that a meter needs adjusting, tap the meter case lightly with the tip of one finger. This helps the needle overcome the slight friction that sometimes exists at the pointer bearings and prevents an otherwise normal unit from coming to rest at zero. If an adjustment is needed, insert the tip of the screwdriver in the slotted screw head located below the meter glass, and slowly turn the adjusting screw until the pointer rests at zero. Observe the following precautions: View the meter face and pointer full on and not from either side. Avoid turning the zero-adjust screw too far, since the meter pointer may be bent against the stop peg, or the spring may be damaged. Zero adjustments should not be made for several minutes after shutdown. Always remember that meters are delicate instruments.

Pilot Lights

Pilot lights are used to indicate that power has been applied to a circuit or that a circuit is ready for the application of power. They are easily removed and replaced. The colored pilot-light covers must be removed carefully, lest they be dropped and broken. The maintenance of pilot lights presents no special difficulty, but the following instructions are given for general guidance.

Inspect (*I*)—Inspect the pilot-light assembly for broken or cracked pilot-light shields, loose bulbs, bulbs with loose bases, loose mounting screws, and loose, dirty, or corroded connections.

Tighten (*T*)—Tighten all mounting screws, and resolder any loose connections. If the connections are dirty or corroded, they should be cleaned before they are soldered. Loose bulbs should be tightly screwed into their bases. Broken or cracked pilot-light shields may sometimes be temporarily repaired by joining the broken or cracked pieces with a narrow piece of friction tape. Replace them as soon as possible; also replace broken or burned-out pilot-light bulbs as soon as possible. While the removal of a bulb may sometimes be difficult, the process is simplified by folding a small piece of friction tape over the top of the bulb and pressing it firmly from the two sides. After the tape is attached, the bulb can be unscrewed and removed from the socket. The socket connections are, of course, inspected when the bulb is out. A new bulb can usually be replaced with the fingers, but if difficulty is experienced, use friction tape to grip the glass envelope of the bulb.

Clean (*C*)—The pilot-light shield, the base assembly, and the glass envelope of the light bulb can be cleaned with a clean, dry cloth. Clean the accumulated dust or dirt from the interior of the socket base with a small brush. Corroded socket contacts or connections can be cleaned with a piece of cloth or a brush dipped in cleaning fluid. The surfaces are then polished with a dry cloth. Clean contacts and connections are important.

Plugs and Receptacles

There are two main types of plugs and receptacles used to interconnect the various components. The first type of plug is used with a coaxial line and consists of a metal shell with a single pin in the center and insulated from the shell. When the plug is inserted into the receptacle, this pin is gripped firmly by a spring connector. There is a knurled metal ring around the plug that is screwed onto the corresponding threads on the receptacle.

The second type of plug is used for connecting multiconductor cables. The plug usually consists of a number of pins insulated from the shell and inserted into a corresponding number of female connectors in the receptacle, although in some cases the female connectors are in the plug, and the male connectors are in the receptacle. This type of plug usually has two small pins or buttons that are mounted on a spring inside the shell and protrude through the shell. When the shell is properly oriented and placed in the receptacle, one of these pins springs up through a hole in the receptacle, firmly locking the plug and receptacle together. When it becomes necessary to remove the plug, the other pin is simply depressed and the plug removed.

The connections between all plugs and their cables are made inside the plug shell. The cable conductor may be soldered to the pin, or there may be a screw holding the wire to the pin. Remove the shell if it is necessary to get at these connections for repair or inspection. Loosen the screws if there is a clamp holding the cable to the shell. In some cases, it is found that the shell and plug body are both threaded; in this case the shell may simply be unscrewed. Usually there are several screws holding the shell; these are removed and the shell is pulled off.

Inspect (*I*)—Each of the following parts should be inspected for the items indicated.

1. The part of the cable that was inside the shell for dirt and cracked or burned insulation.
2. The conductor or conductors and their connection to the pins for broken wires, bad insulation, and dirty, corroded, broken, or loose connections.
3. The male or female connectors in the plug for looseness in the insulation, damage, dirt, or corrosion.
4. The plug body for damage to the insulation and for dirt or corrosion.
5. The shell for damage, such as dents or cracks, and dirt or corrosion.
6. The receptacle for damaged or corroded connectors, cracked insulation, and proper electrical connection between the connectors and the leads.

Tighten (*T*)—Perform the necessary operations to correct the following conditions.

1. Any looseness of the connectors in the insulation, if possible; if not, replace the plug.
2. Any loose electrical connections. Resolder if necessary.

Clean (*C*)—Perform the necessary cleaning operations on the following items.

1. The cable, using a cloth and cleaning fluid.
2. The connectors and connections, using a cloth and cleaning fluid. Use crocus cloth to remove corrosion.
3. The plug body and shell, using a cloth and cleaning fluid; use crocus cloth to remove corrosion.

4. The receptacle, using a cloth and cleaning fluid if necessary. Corrosion can be removed with crocus cloth.

Adjust (*A*)—Adjust the connectors for proper contact if they are the spring type.

Lubricate (*L*)—Lubricate the plug and receptacle with a thin coat of petroleum jelly if they are difficult to connect or remove. The type of plug with the threaded ring may especially require this treatment.

14-12. MEETING EMERGENCIES AT THE TRANSMITTER

"We're off the air!" This is a declaration that invariably causes a state of panic for the newcomer to a transmitter operating job. In many instances, he is alone with the responsibility of correcting the trouble as quickly as possible to avoid loss of revenue for his employer. The highest efficiency in correcting trouble will come with more experience at the particular installation. However, the operator who can visualize general circuit theory in relation to the particular circuits with which he is concerned will find a logical and natural sequence for locating the fault. The main requirement, quite naturally, is to become thoroughly acquainted with the circuits used. He should be able to draw a good general functional diagram of all circuits from memory, and be able to draw a block diagram of the operating sequence for starting relays and protective relays in the power-control circuits. It is obvious that confidence and peace of mind can be achieved only with a complete familiarity with all circuits and their relation to the overall performance of the transmitter.

There is one piece of equipment at the transmitter installation that should be the central focusing point for the operator's first attention when trouble occurs. This is the modulation monitor; it has an rf-input meter that reads a definite place on the scale for normal operation, and, of course, it has the percentage-modulation indicator. The purpose of observing this instrument will be evident in the following discussions.

At the first interruption of the program, or the occurrence of noise or distortion in the monitoring loudspeaker, the modulation monitor should be observed. Assume that the rf-input meter is at the normal point on the scale; this means that the trouble is not in the rf section, because any trouble there would cause some deviation in the rf input to the monitor.

The following is a procedure to use when the program suddenly stops:

1. If the rf-input meter shows normal and the modulation meter shows modulation taking place, the trouble is obviously in the monitoring line or amplifier, and the station is *not* off the air.
2. If the rf input is normal and no modulation is shown on the meter, the trouble is either in the audio section of the transmitter, in the line amplifier, in the program line from the studio to the transmitter, or at the studio.

3. Call studio control to ascertain the condition at that point. If everything there is normal, check the line by patching the line into the monitor amplifier or spare amplifier to see if the program is coming into the transmitter. If not, notify control to feed the program on the spare line, and call the local test board of the telephone company. If the program is coming in satisfactorily from the line, use a spare line amplifier to feed the transmitter. If the regular line amplifier is working normally, then the trouble obviously lies in the audio section of the transmitter. Usually any trouble there will be indicated by abnormal plate-current meter readings, and, of course, tube trouble is the most common source of program interruption.

The same procedure should be used when noise or distortion occurs; first check with the studio, then check the line, the line amplifier, and the audio section of the transmitter. If all speech-input tube currents are zero, then the trouble is in the associated power supply. Most likely, the trouble again is due to a tube, and the tube should be changed on indication of abnormal plate current. Next in line come bleeder resistors, taps from a bleeder supply, and line-to-plate circuits of tubes. Power-supply component parts usually show a visual indication of damage, such as smoking, unless they have opened.

If, at the first indication of trouble, a glance at the modulation monitor shows zero or low rf input, then the trouble lies in the rf stages of the transmitter. The operator must accordingly proceed to check for the trouble by observing all rf-circuit meter indications. Observation of plate-current and grid-current meters aids in quickly determining which of the stages is faulty.

When the transmitter is shut down by relay operation in the control circuits, the cause of the failure can be traced quickly if the operator is familiar with the relay sequence and functions. Control circuits are divided into two functional purposes, those which control circuits to the primaries of power supplies and those with protective functions. Pilot lights are often associated with the various relays to show when they are open or closed. As stated before, the sequence of operation should be committed to memory. The filament power supply, for example, will not operate until the cooling-motor contactors have functioned to supply the cooling medium (water or air) to the tubes. After the filament contactor has applied filament voltage, the plate-voltage contactor will not operate until the time-delay relay has functioned, etc.

Rectifier tubes of the mercury-vapor type usually arc backward several times before expiring. When arcback indicators are used, the faulty tube may be observed quickly and changed immediately. Other troubles in high-voltage power supplies nearly always show signs of physical deterioration.

Short circuits which cause a quick tripping of overload relays are the most difficult troubles to locate. In some difficult troubles of this kind, overload relays have been strapped out of the circuits, and limiting resistors put in the current fuse box to limit the amount of current. The circuits were then visually observed for arcing with doors open and interlock switches short-circuited. This is a dangerous procedure, however, and should be left to the more experienced operators. Any unusual procedure of this kind should be carried out by *two or more men.*

This all may be summarized into the most important factor: Be familiar with the transmitter, and know what indications would be for the most common sources of trouble, such as tubes and power supplies, in the various circuits.

Be Mentally Prepared

We have discussed the art of being mentally prepared for emergencies in the section on studio emergency techniques. Obviously, the same sort of preparedness is an important factor in mastering technical emergencies at the transmitter.

The first step, then, is to become familiar with the circuits. The second step is to become familiar with the locations of circuit components. Where is the main rectifier time-delay relay? Where are the interlock contactors? Where is the modulator bias relay? Take the complete (not simplified) schematic "behind the doors" of your transmitter, and visualize each component in the physical layout in relation to the schematic. In most modern commercial transmitters of vertical-chassis construction, the schematic numbers are stamped either directly on the parts or on the chassis adjacent to the particular component. For example, if a certain coupling capacitor is shown on the schematic as C102, this number will appear on the capacitor itself or on the chassis adjacent to it. In some cases, only the tube numbers (V1, V2, etc.) and transformer numbers, (T1, T2, etc.) are shown. The parts associated with a particular tube may be oriented easily by tracing their connections to the proper tube socket or transformer terminals. When the operator has become thoroughly familiar with the equipment and feels at ease with the first steps, he is ready for the all-important third step, getting mentally prepared for trouble.

One important example of mental prepardeness: Will *reduced power* help to stay on the air, either until the end of the day's schedule or until a sustaining program comes along to avoid loss of revenue? This will, of course, be determined by the nature of the trouble. Any capacitor, resistor, tube, transformer, insulator, or power lead that would undergo less strain because of a power reduction obviously might hold a while longer by this means. Also, in the case of tripping of overload relays in rapid succession when the source of the trouble is still not apparent, reducing the power should be the very first step in attempting to stay on the air. This is advantageous, too, when help may be needed in cases of the more serious type. The station may be kept on the air until help arrives to carry out emergency operations.

Transmitter Emergency Procedure

The following example routines are applicable to any station, whether locally or remotely controlled.

1. The first duty of the transmitter operator is to minimize down time. If you have a failure and are not able to get back on the air in a reasonable length of time (within 1 or 2 minutes), advise the studio operator of the condition. The studio operator will then stand by to make telephone calls to get in touch with the chief engineer or provide any other assistance the transmitter operator may request.
2. Operators not on duty and listening at home may call in to offer help, but they should not call the transmitter. They should call the studio operator, who will keep them advised.
3. In no event should more than 2 minutes elapse before advising the studio of a failure. Advise them sooner if you feel that there might be a chance of not getting back on in 2 minutes.
4. At the first sign of a possible failure, the filaments of the auxiliary transmitter should be turned on and preliminary adjustments made so that it can go on the air if the main transmitter fails.

NOTE: If trouble develops on the antenna side of the directional relays, where the antenna system is common to both transmitters, it will not help to change transmitters. It is quite possible, however, that components that will not hold on full power will hold on less power.

5. Make it an important habit to note the time of any interruption to the program, whether transmitter failure or lack of program from the studio. This will serve also as a double check on studio or network failures.
6. Try to be mentally prepared to meet an emergency. Inspect relays and lightning gaps often during an electrical storm. In this way, you may see part of the trouble developing and can replace them or have temporary clip leads ready to put in the circuit.
7. In case of time lost due to failure at the transmitter, report to the control room the time the program stopped and the time it started again.
8. Fill out a transmitter operating-room report form.

Lightning

Mental preparedness at the transmitter is rigorously tested during electrical storms. In spite of the many provisions in modern transmitter installations to help protect the equipment from heavy lightning surges (ball lightning gaps, automatic carrier-off relays, etc.), the great majority of time lost in well-maintained stations is due to lightning.

There seems to be no rhyme or reason to some troubles that develop during storms. Regardless of critically spaced lightning gaps at the bases of the towers and both ends of transmission lines, lightning has been known to open or short interlock circuits, power supplies, speech-input equipment, etc. The shortest path to ground relative to the antenna circuit is definitely *not* a 100-percent rule for lightning.

It is true, however, that a majority of failures due to electrical storms occur in the antenna or directional phasing equipment. The trouble usually may be recognized immediately from blackened or smoking parts. When a line or antenna current meter is damaged by lightning, the face is nearly always so black as to be unreadable. Even when switch-blade shunts are kept on meters, lightning charges are apt to damage a transmission-line current meter by a heavy current arc from coil to magnet to ground. It is necessary to remove the leads from the meter to remove the short that remains after the initial strike. Relays in the antenna circuits are another common source of failure when severe lightning surges occur. It is very important that every transmitter operator have various sizes of clip jumpers handy to strap around any such failure. If only the relay holding coil is opened, the relay may be blocked shut by some kind of prop or weight, depending on the type of relay and the method of mounting.

Cracked insulator bowls on top of the tuning house will cause continued arcing when power is restored, resulting in tripping of overload relays. Look for arcing inside the bowl.

> CAUTION: Electrical storms are hazardous to a transmitter operator when work on the antenna system is necessary during the height of the storm. Never touch anything under these conditions until the tower has been well grounded by a "hot stick" or other arrangement; hang the metal connection onto the tower lead as close to the actual tower base as possible. Fig. 14-37 illustrates such an arrangement. Grounding sticks of this type are an important and necessary item for any transmitter operating room or tuning house.

Control Circuits

Over a period of several years, an operator will be faced with the situation of quick tripping of overload relays, finally resulting in a complete shutdown of the transmitter. Modern transmitters employ an automatic return circuit so that three to five overloads must occur in rapid succession before the power is removed, requiring a notching relay to be reset by hand. Also, some form of visual indication is provided to show which general section is being overloaded, such as the power amplifier, modulator, etc. The exact cause of the overload, however, is often hard to locate unless visible arcing serves to indicate the source. Past history of the particular installation and observation of parts with voltage applied are helpful in most such instances.

Where arcing occurs, the sense of hearing usually is able to locate the approximate vicinity. On opening the doors, signs of arcing, such as burned spots on the frame immediately adjacent to a coil or capacitor corner, should be visible.

The first emergency measure is to reduce power. If the carrier will hold on reduced power, the chief engineer or supervisory personnel may then be consulted to determine the best possible course of action from then on. When overloads must be traced down by strapping interlock circuits and opening the doors for visual observation with voltage applied, extreme care must be taken to avoid contact with high voltages.

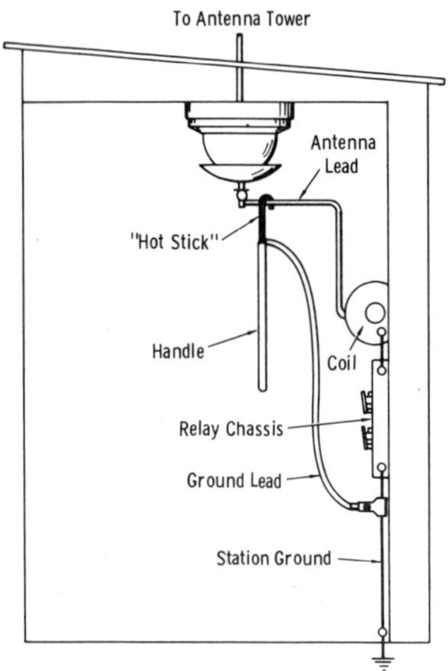

Fig. 14-37. Safety device for grounding high-voltage points.

Power-amplifier overloads may, of course, be caused by trouble in the antenna system. Any fault that would cause grounding of the transmission-line center conductor or antenna (unless shunt fed) would cause the final stage to overload and operate the overload relay. Such faults may usually be found by visual observation. Look for blackened meter faces, smoking capacitors, or a collapsed static-drain choke causing a tower to be grounded for rf and dc.

Field mice have been known to get into antenna tuning houses and meet their end by becoming a direct rf path to ground on the tuning-component chassis. When examining any part of the transmitter or antenna system for possible sources of overloads, look around for such possibilities as mice, bugs between plates of tuning capacitors, etc. The occurrence is more common than the newcomer might suspect.

Rectifier line overloads that often trip circuit breakers or blow line fuses are almost always caused by a defective mercury-vapor rectifier tube which arcs back. Modern transmitters employ arcback indicators which will give a visual check on any tube in which current has passed in the reverse direction. If tubes are visible from the front of the transmitter through the panel doors, a tube so afflicted will have its blue haze extinguished momentarily at the time of the arcback, or a distinct flash will be apparent within the envelope. If the rectifiers are not visible, the entire set of mercury-vapor tubes should be replaced with new ones. Remember that the new tubes must have been preheated for at least 30 minutes to assure that all mercury has been removed from the tube elements.

In some cases, the relay is at fault. If, for example, a holding coil should open, as sometimes happens, the simplest procedure is to prop or tie the relay shut manually. When this cannot be done conveniently, the proper terminal board numbers should be jumpered to complete the circuit, or the jumper may be used at the relay contacts in some instances.

An example of jumpering terminal-board connections to meet emergencies in control circuitry can be seen by reviewing Fig. 9-36, Chapter 9. Assume that the filaments have been turned on by closing the main line switch. Power is now also applied to timing relay K1, which should cause green indicating light IS1 to light after the customary 30 seconds of time delay. This will occur if:

1. Relay K1 is in working order.
2. Door interlocks S1 and S2 are closed.
3. Green indicating light IS1 is in working order.

Now assume that after one minute the green light (indicating that high voltage can be applied) has not come on, although the doors are closed (closing S1 and S2) and it should do so. Pressing momentary-contact switch IS1 fails to apply plate voltage, so you can assume that either the time-delay relay has not operated or a door interlock switch is open. Note from Fig. 9-36 that you can jumper terminals 11-10 and 11-11 to complete the interlock circuit even with the doors open.

If this still does not allow application of plate voltage, you can jumper terminals 11-11 and 11-12 to bypass the time-delay contacts. In case of extreme emergency where it is necessary to get on the air in a hurry, and doubt exists as to which circuit is faulty, a jumper from terminal 11-10 to terminal 11-12 completes the entire circuit of timing relay and door interlocks.

14-13. DUTIES OF THE CHIEF ENGINEER

The chief engineer normally assumes the entire responsibility for all technical matters concerning his installation. Exceptions are the following duties sometimes assumed by a consulting engineering firm.

1. Filing of the original engineering sections of forms submitted to the FCC on an initial application for a broadcast station.
2. Design and supervision of the initial adjustment of a directional antenna system as may be required by the FCC.

An engineering consultant is often kept on a retainer fee by the station management for the purpose of alerting the station to legal changes in engineering requirements and to assist the chief engineer in meeting the legal requirements for changes in the technical installation.

The type of work performed by the chief engineer can vary from the small station where he performs all actual maintenance and repair to the larger stations where his duties are mainly administrative. He normally handles scheduling for technical operations and important preventive maintenance and test measurements. He should have a current copy of the FCC Rules and Regulations available and at least be sufficiently familiar with their format to know where to look for reference in cases such as the following.

The engineer in charge of the FCC field office for the district in which the station is located must be notified when indicating instruments such as frequency monitors, modulation monitors, antenna meters (such as the common-input meter for directional-antenna systems), etc., are temporarily out of service during an emergency. A record of the telegram sent should be kept for reference. The same notification must be given on the restoration of such devices to service.

Notification must also be given when the power must be reduced below specified tolerances for some reason (such as failure of the antenna transmission line or other component under authorized power). Check current rules for the maximum allowable time of operation on low power before notification is made.

Compliance with FCC requirements on tower painting and lighting, technical logs, and all other matters pertaining to the technical operation of the station is normally the responsibility of the chief engineer. He will see that the local FAA (Federal Aviation Administration) office is notified immediately when a flasher beacon at the top of the tower installation is inoperative. In some cases, notification is required if any other tower light is off, particularly when the location is on or near a main artery of air traffic. A record of the notification must be kept in the technical log, which should have a designated space for tower-light condition and time of observation, as well as the time of FAA notification when required.

The chief engineer should have on file an up-to-date equipment and component list from a sufficient number of sources so that he can immediately take under study any special requirement proposed by management. In some stations, special problems occur almost daily, and the chief engineer must find the quickest and most practical solution at the lowest possible cost.

In the case of a new installation or expansion of an existing facility requiring new construction, the chief engineer must be able to direct the electrical and construction contractors in drawing up the initial technical plans. He must have a sufficient knowledge of heat dissipation to assist the air-conditioning engineer in properly installing an adequate system if it is required.

He should have available the latest revised FCC Tariff No. 198 giving regulations, minimum costs, and leasing times for various grades of service, etc., concerning channels for program transmission in connection with radio broadcasting. This enables him to be prepared to come up with a quick estimate on the technical costs of a special event and provides a basis for judging whether certain classes of lines should be ordered instead of using mobile equipment that is either owned by the station or available for lease. This information is available from the American Telephone & Telegraph Co. Also, it is a good idea for the chief engineer to become acquainted with the supervisor of the local AT&T toll test and telephone company test board and be familiar with the extent of prior notification necessary for the various classes of line service. This requirement varies somewhat with local conditions.

He should be sure that all emergency numbers, such as police, fire, power and light company emergency service, AT&T toll test, and telephone company test are plainly posted at the studio and transmitter operating positions. He should also be sure that fire extinguishers of the proper type for electronic equipment are installed and that they are recharged at the required intervals.

The chief engineer must be acquainted with the FCC requirements for proof-of-performance measurements, including the technical standards that must be met and the dates when the appropriate forms must be filed. For example, a license-renewal application must be filed at least 90 days prior to the license expiration date. A complete proof of performance must be run in the four-month period preceding the date of filing application. Thus, if a station license expires October 1, the renewal application must be filed not later than July 2. (NOTE: Always check current FCC Rules and Regulations.) The proof-of-performance runs must then be made in the period of March 3 to July 2. It is advisable to run such tests about three times every year at intervals such that the tests occur at the proper time in the renewal year. Remember that proof-of-performance data must be filed with the application for license to cover the construction permit (CP) for new or modified facilities.

The chief engineer must also be familiar with FCC requirements for checks on existing directional-antenna system patterns, the frequency with which such checks must be made, and the proper recording of all related information.

When preparing the technical portion of an application for renewal of the license, be sure that data entered in the form are in agreement with the log (where involved) and that such data indicate compliance with the FCC rules and the license specifications.

It is good practice for the chief engineer to check the readings of meters periodically under normal operating conditions. Compare these with the log, the present license, and the previous application for the license or license renewal. This helps to eliminate errors, to ascertain that the observed data are typical, and to determine if the variations from licensed values are within the limits permitted.

Carefully file all original copies of station-frequency checks made by the frequency-measuring service, and be sure that this information has been properly logged. Preserve all records of such items as meter calibration for any meter replacement in the final stage or antenna system with the full details required by the FCC.

All of the foregoing condenses to this: Know exactly what the FCC and the specific station license require, and see that the proper steps are taken to meet these requirements. Make equipment-performance measurements more often than required by the FCC. Measurements made on a regular basis and a complete record of steps taken to remedy any deficiency go a long way toward assuring the radio inspector that the station is in competent engineering hands.

EXERCISES

Q14-1. Would you expect the studio VU meter and the transmitter modulation meter to "peak" the same on program waves at normal modulation?

Q14-2. Name the two basic methods of measuring antenna impedance.

Q14-3. How must transmitter output power be determined when field-intensity surveys are made?

Q14-4. How is transmitter output power determined by the (A) direct method (B) indirect method?

Q14-5. Must the limiter or agc amplifier installed in the normal transmission path be used in proof-of-performance measurements?

Q14-6. The stereophonic transmission standards state: "The stereophonic subcarrier shall be the second harmonic of the pilot subcarrier and shall cross the time axis with a positive slope simultaneously with each crossing of the time axis by the pilot subcarrier." Why is it necessary to standardize this time relationship?

Q14-7. The stereophonic standards also state: "At the instant when only a positive left signal is applied, the main channel modulation shall cause an upward deviation of the main carrier frequency, and the stereophonic subcarrier and its sidebands signal shall cross the time axis simultaneously and in the same direction." What stereo characteristic does this standard affect?

Q14-8. Review paragraphs L and M of the stereophonic standards (see Section 10-6, Chapter 10). What stereo characteristic do these paragraphs of the standard affect?

Q14-9. In stereo transmission, do the terms "separation" and "cross talk" define the same characteristic?

Q14-10. When audio is weak or lost entirely at the transmitter, where is the first place to look?

APPENDIX A

Reference Tables

Table A-1. Common Logarithms

N	0	1	2	3	4	5	6	7	8	9	Proportional Parts 1	2	3	4	5	6	7	8	9
10	0000	0043	0086	0128	0170	0212	0253	0294	0334	0374	4	8	12	17	21	25	29	33	37
11	0414	0453	0492	0531	0569	0607	0645	0682	0719	0755	4	8	11	15	19	23	26	30	34
12	0792	0828	0864	0899	0934	0969	1004	1038	1072	1106	3	7	10	14	17	21	24	28	31
13	1139	1173	1206	1239	1271	1303	1335	1367	1399	1430	3	6	10	13	16	19	23	26	29
14	1461	1492	1523	1553	1584	1614	1644	1673	1703	1732	3	6	9	12	15	18	21	24	27
15	1761	1790	1818	1847	1875	1903	1931	1959	1987	2014	3	6	8	11	14	17	20	22	25
16	2041	2068	2095	2122	2148	2175	2201	2227	2253	2279	3	5	8	11	13	16	18	21	24
17	2304	2330	2355	2380	2405	2430	2455	2480	2504	2529	2	5	7	10	12	15	17	20	22
18	2553	2577	2601	2625	2648	2672	2695	2718	2742	2765	2	5	7	9	12	14	16	19	21
19	2788	2810	2833	2856	2878	2900	2923	2945	2967	2989	2	4	7	9	11	13	16	18	20
20	3010	3032	3054	3075	3096	3118	3139	3160	3181	3201	2	4	6	8	11	13	15	17	19
21	3222	3243	3263	3284	3304	3324	3345	3365	3385	3404	2	4	6	8	10	12	14	16	18
22	3424	3444	3464	3483	3502	3522	3541	3560	3579	3598	2	4	6	8	10	12	14	15	17
23	3617	3636	3655	3674	3692	3711	3729	3747	3766	3784	2	4	6	7	9	11	13	15	17
24	3802	3820	3838	3856	3874	3892	3909	3927	3945	3962	2	4	5	7	9	11	12	14	16
25	3979	3997	4014	4031	4048	4065	4082	4099	4116	4133	2	3	5	7	9	10	12	14	15
26	4150	4166	4183	4200	4216	4232	4249	4265	4281	4298	2	3	5	7	8	10	11	13	15
27	4314	4330	4346	4362	4378	4393	4409	4425	4440	4456	2	3	5	6	8	9	11	13	14
28	4472	4487	4502	4518	4533	4548	4564	4579	4594	4609	2	3	5	6	8	9	11	12	14
29	4624	4639	4654	4669	4683	4698	4713	4728	4742	4757	1	3	4	6	7	9	10	12	13
30	4771	4786	4800	4814	4829	4843	4857	4871	4886	4900	1	3	4	6	7	9	10	11	13
31	4914	4928	4942	4955	4969	4983	4997	5011	5024	5038	1	3	4	6	7	8	10	11	12
32	5051	5065	5079	5092	5105	5119	5132	5145	5159	5172	1	3	4	5	7	8	9	11	12
33	5185	5198	5211	5224	5237	5250	5263	5276	5289	5302	1	3	4	5	6	8	9	10	12
34	5315	5328	5340	5353	5366	5378	5391	5403	5416	5428	1	3	4	5	6	8	9	10	11
N	0	1	2	3	4	5	6	7	8	9	Proportional Parts 1	2	3	4	5	6	7	8	9

Table A-1. Common Logarithms—cont

N	0	1	2	3	4	5	6	7	8	9	PP 1	PP 2	PP 3	PP 4	PP 5	PP 6	PP 7	PP 8	PP 9
35	5441	5453	5465	5478	5490	5502	5514	5527	5539	5551	1	2	4	5	6	7	9	10	11
36	5563	5575	5587	5599	5611	5623	5635	5647	5658	5670	1	2	4	5	6	7	8	10	11
37	5682	5694	5705	5717	5729	5740	5752	5763	5775	5786	1	2	3	5	6	7	8	9	10
38	5798	5809	5821	5832	5843	5855	5866	5877	5888	5899	1	2	3	5	6	7	8	9	10
39	5911	5922	5933	5944	5955	5966	5977	5988	5999	6010	1	2	3	4	5	7	8	9	10
40	6021	6031	6042	6053	6064	6075	6085	6096	6107	6117	1	2	3	4	5	6	8	9	10
41	6128	6138	6149	6160	6170	6180	6191	6201	6212	6222	1	2	3	4	5	6	7	8	9
42	6232	6243	6253	6263	6274	6284	6294	6304	6314	6325	1	2	3	4	5	6	7	8	9
43	6335	6345	6355	6365	6375	6385	6395	6405	6415	6425	1	2	3	4	5	6	7	8	9
44	6435	6444	6454	6464	6474	6484	6493	6503	6513	6522	1	2	3	4	5	6	7	8	9
45	6532	6542	6551	6561	6571	6580	6590	6599	6609	6618	1	2	3	4	5	6	7	8	9
46	6628	6637	6646	6656	6665	6675	6684	6693	6702	6712	1	2	3	4	5	6	7	7	8
47	6721	6730	6739	6749	6758	6767	6776	6785	6794	6803	1	2	3	4	5	5	6	7	8
48	6812	6821	6830	6839	6848	6857	6866	6875	6884	6893	1	2	3	4	4	5	6	7	8
49	6902	6911	6920	6928	6937	6946	6955	6964	6972	6981	1	2	3	4	4	5	6	7	8
50	6990	6998	7007	7016	7024	7033	7042	7050	7059	7067	1	2	3	3	4	5	6	7	8
51	7076	7084	7093	7101	7110	7118	7126	7135	7143	7152	1	2	3	3	4	5	6	7	8
52	7160	7168	7177	7185	7193	7202	7210	7218	7226	7235	1	2	2	3	4	5	6	7	7
53	7243	7251	7259	7267	7275	7284	7292	7300	7308	7316	1	2	2	3	4	5	6	6	7
54	7324	7332	7340	7348	7356	7364	7372	7380	7388	7396	1	2	2	3	4	5	6	6	7
55	7404	7412	7419	7427	7435	7443	7451	7459	7466	7474	1	2	2	3	4	5	5	6	7
56	7482	7490	7497	7505	7513	7520	7528	7536	7543	7551	1	2	2	3	4	5	5	6	7
57	7559	7566	7574	7582	7589	7597	7604	7612	7619	7627	1	2	2	3	4	5	5	6	7
58	7634	7642	7649	7657	7664	7672	7679	7686	7694	7701	1	1	2	3	4	4	5	6	7
59	7709	7716	7723	7731	7738	7745	7752	7760	7767	7774	1	1	2	3	4	4	5	6	7
N	0	1	2	3	4	5	6	7	8	9	1	2	3	4	5	6	7	8	9

(PP = Proportional Parts)

Table A-1. Common Logarithms—cont

N	0	1	2	3	4	5	6	7	8	9	Proportional Parts 1	2	3	4	5	6	7	8	9
60	7782	7789	7796	7803	7810	7818	7825	7832	7839	7846	1	1	2	3	4	4	5	6	6
61	7853	7860	7868	7875	7882	7889	7896	7903	7910	7917	1	1	2	3	4	4	5	6	6
62	7924	7931	7938	7945	7952	7959	7966	7973	7980	7987	1	1	2	3	3	4	5	6	6
63	7993	8000	8007	8014	8021	8028	8035	8041	8048	8055	1	1	2	3	3	4	5	5	6
64	8062	8069	8075	8082	8089	8096	8102	8109	8116	8122	1	1	2	3	3	4	5	5	6
65	8129	8136	8142	8149	8156	8162	8169	8176	8182	8189	1	1	2	3	3	4	5	5	6
66	8195	8202	8209	8215	8222	8228	8235	8241	8248	8254	1	1	2	3	3	4	5	5	6
67	8261	8267	8274	8280	8287	8293	8299	8306	8312	8319	1	1	2	3	3	4	5	5	6
68	8325	8331	8338	8344	8351	8357	8363	8370	8376	8382	1	1	2	3	3	4	4	5	6
69	8388	8395	8401	8407	8414	8420	8426	8432	8439	8445	1	1	2	2	3	4	4	5	6
70	8451	8457	8463	8470	8476	8482	8488	8494	8500	8506	1	1	2	2	3	4	4	5	6
71	8513	8519	8525	8531	8537	8543	8549	8555	8561	8567	1	1	2	2	3	4	4	5	5
72	8573	8579	8585	8591	8597	8603	8609	8615	8621	8627	1	1	2	2	3	4	4	5	5
73	8633	8639	8645	8651	8657	8663	8669	8675	8681	8686	1	1	2	2	3	4	4	5	5
74	8692	8698	8704	8710	8716	8722	8727	8733	8739	8745	1	1	2	2	3	4	4	5	5
75	8751	8756	8762	8768	8774	8779	8785	8791	8797	8802	1	1	2	2	3	3	4	5	5
76	8808	8814	8820	8825	8831	8837	8842	8848	8854	8859	1	1	2	2	3	3	4	5	5
77	8865	8871	8876	8882	8887	8893	8899	8904	8910	8915	1	1	2	2	3	3	4	4	5
78	8921	8927	8932	8938	8943	8949	8954	8960	8965	8971	1	1	2	2	3	3	4	4	5
79	8976	8982	8987	8993	8998	9004	9009	9015	9020	9025	1	1	2	2	3	3	4	4	5
80	9031	9036	9042	9047	9053	9058	9063	9069	9074	9079	1	1	2	2	3	3	4	4	5
81	9085	9090	9096	9101	9106	9112	9117	9122	9128	9133	1	1	2	2	3	3	4	4	5
82	9138	9143	9149	9154	9159	9165	9170	9175	9180	9186	1	1	2	2	3	3	4	4	5
83	9191	9196	9201	9206	9212	9217	9222	9227	9232	9238	1	1	2	2	3	3	4	4	5
84	9243	9248	9253	9258	9263	9269	9274	9279	9284	9289	1	1	2	2	3	3	4	4	5
N	**0**	**1**	**2**	**3**	**4**	**5**	**6**	**7**	**8**	**9**	**1**	**2**	**3**	**4**	**5**	**6**	**7**	**8**	**9**
											Proportional Parts								

Table A-1. Common Logarithms—cont

N	0	1	2	3	4	5	6	7	8	9	Proportional Parts								
											1	2	3	4	5	6	7	8	9
85	9294	9299	9304	9309	9315	9320	9325	9330	9335	9340	1	1	2	2	3	3	4	4	5
86	9345	9350	9355	9360	9365	9370	9375	9380	9385	9390	1	1	2	2	3	3	4	4	5
87	9395	9400	9405	9410	9415	9420	9425	9430	9435	9440	0	1	1	2	2	3	3	4	4
88	9445	9450	9455	9460	9465	9469	9474	9479	9484	9489	0	1	1	2	2	3	3	4	4
89	9494	9499	9504	9509	9513	9518	9523	9528	9533	9538	0	1	1	2	2	3	3	4	4
90	9542	9547	9552	9557	9562	9566	9571	9576	9581	9586	0	1	1	2	2	3	3	4	4
91	9590	9595	9600	9605	9609	9614	9619	9624	9628	9633	0	1	1	2	2	3	3	4	4
92	9638	9643	9647	9652	9657	9661	9666	9671	9675	9680	0	1	1	2	2	3	3	4	4
93	9685	9689	9694	9699	9703	9708	9713	9717	9722	9727	0	1	1	2	2	3	3	4	4
94	9731	9736	9741	9745	9750	9754	9759	9763	9768	9773	0	1	1	2	2	3	3	4	4
95	9777	9782	9786	9791	9795	9800	9805	9809	9814	9818	0	1	1	2	2	3	3	4	4
96	9823	9827	9832	9836	9841	9845	9850	9854	9859	9863	0	1	1	2	2	3	3	4	4
97	9868	9872	9877	9881	9886	9890	9894	9899	9903	9908	0	1	1	2	2	3	3	4	4
98	9912	9917	9921	9926	9930	9934	9939	9943	9948	9952	0	1	1	2	2	3	3	4	4
99	9956	9961	9965	9969	9974	9978	9983	9987	9991	9996	0	1	1	2	2	3	3	3	4
N	0	1	2	3	4	5	6	7	8	9	1	2	3	4	5	6	7	8	9
											Proportional Parts								

Table A-2. Natural Trigonometric Functions

Degrees	Sin	Cos	Tan	Cot	
0° 00′	0.0000	1.0000	0.0000	∞	90° 00′
10	.0029	1.0000	.0029	343.77	50
20	.0058	1.0000	.0058	171.89	40
30	.0087	1.0000	.0087	114.59	30
40	.0116	.9999	.0116	85.940	20
50	.0145	.9999	.0145	68.750	10
1° 00′	0.0175	0.9998	0.0175	57.290	89° 00′
10	.0204	.9998	.0204	49.104	50
20	.0233	.9997	.0233	42.964	40
30	.0262	.9997	.0262	38.188	30
40	.0291	.9996	.0291	34.368	20
50	.0320	.9995	.0320	31.242	10
2° 00′	0.0349	0.9994	0.0349	28.636	88° 00′
10	.0378	.9993	.0378	26.432	50
20	.0407	.9992	.0407	24.542	40
30	.0436	.9990	.0437	22.904	30
40	.0465	.9989	.0466	21.470	20
50	.0494	.9988	.0495	20.206	10
3° 00′	0.0523	0.9986	0.0524	19.081	87° 00′
10	.0552	.9985	.0553	18.075	50
20	.0581	.9983	.0582	17.169	40
30	.0610	.9981	.0612	16.350	30
40	.0640	.9980	.0641	15.605	20
50	.0669	.9978	.0670	14.924	10
4° 00′	0.0698	0.9976	0.0699	14.301	86° 00′
10	.0727	.9974	.0729	13.727	50
20	.0756	.9971	.0758	13.197	40
30	.0785	.9969	.0787	12.706	30
40	.0814	.9967	.0816	12.251	20
50	.0843	.9964	.0846	11.826	10
5° 00′	0.0872	0.9962	0.0875	11.430	85° 00′
10	.0901	.9959	.0904	11.059	50
20	.0929	.9957	.0934	10.712	40
30	.0958	.9954	.0963	10.385	30
40	.0987	.9951	.0992	10.078	20
50	.1016	.9948	.1022	9.7882	10
6° 00′	0.1045	0.9945	0.1051	9.5144	84° 00′
10	.1074	.9942	.1080	9.2553	50
20	.1103	.9939	.1110	9.0098	40
30	.1132	.9936	.1139	8.7769	30
40	.1161	.9932	.1169	8.5555	20
50	.1190	.9929	.1198	8.3450	10
7° 00′	0.1219	0.9925	0.1228	8.1443	83° 00′
10	.1248	.9922	.1257	7.9530	50
20	.1276	.9918	.1287	7.7704	40
30	.1305	.9914	.1317	7.5958	30
40	.1334	.9911	.1346	7.4287	20
50	.1363	.9907	.1376	7.2687	10
	Cos	Sin	Cot	Tan	Degrees

Table A-2. Natural Trigonometric Functions—cont

Degrees	Sin	Cos	Tan	Cot	
8° 00′	0.1392	0.9903	0.1405	7.1154	82° 00′
10	.1421	.9899	.1435	6.9682	50
20	.1449	.9894	.1465	6.8269	40
30	.1478	.9890	.1495	6.6912	30
40	.1507	.9886	.1524	6.5606	20
50	.1536	.9881	.1554	6.4348	10
9° 00′	0.1564	0.9877	0.1584	6.3138	81° 00′
10	.1593	.9872	.1614	6.1970	50
20	.1622	.9868	.1644	6.0844	40
30	.1650	.9863	.1673	5.9758	30
40	.1679	.9858	.1703	5.8708	20
50	.1708	.9853	.1733	5.7694	10
10° 00′	0.1736	0.9848	0.1763	5.6713	80° 00′
10	.1765	.9843	.1793	5.5764	50
20	.1794	.9838	.1823	5.4845	40
30	.1822	.9833	.1853	5.3955	30
40	.1851	.9827	.1883	5.3093	20
50	.1880	.9822	.1914	5.2257	10
11° 00′	0.1908	0.9816	0.1944	5.1446	79° 00′
10	.1937	.9811	.1974	5.0658	50
20	.1965	.9805	.2004	4.9894	40
30	.1994	.9799	.2035	4.9152	30
40	.2022	.9793	.2065	4.8430	20
50	.2051	.9787	.2095	4.7729	10
12° 00′	0.2079	0.9781	0.2126	4.7046	78° 00′
10	.2108	.9775	.2156	4.6382	50
20	.2136	.9769	.2186	4.5736	40
30	.2164	.9763	.2217	4.5107	30
40	.2193	.9757	.2247	4.4494	20
50	.2221	.9750	.2278	4.3897	10
13° 00′	0.2250	0.9744	0.2309	4.3315	77° 00′
10	.2278	.9737	.2339	4.2747	50
20	.2306	.9730	.2370	4.2193	40
30	.2334	.9724	.2401	4.1653	30
40	.2363	.9717	.2432	4.1126	20
50	.2391	.9710	.2462	4.0611	10
14° 00′	0.2419	0.9703	0.2493	4.0108	76° 00′
10	.2447	.9696	.2524	3.9617	50
20	.2476	.9689	.2555	3.9136	40
30	.2504	.9681	.2586	3.8667	30
40	.2532	.9674	.2617	3.8208	20
50	.2560	.9667	.2648	3.7760	10
15° 00′	0.2588	0.9659	0.2679	3.7321	75° 00′
10	.2616	.9652	.2711	3.6891	50
20	.2644	.9644	.2742	3.6470	40
30	.2672	.9636	.2773	3.6059	30
40	.2700	.9628	.2805	3.5656	20
50	.2728	.9621	.2836	3.5261	10
	Cos	Sin	Cot	Tan	Degrees

Table A-2. Natural Trigonometric Functions—cont

Degrees	Sin	Cos	Tan	Cot	
16° 00′	0.2756	0.9613	0.2867	3.4874	**74° 00′**
10	.2784	.9605	.2899	3.4495	**50**
20	.2812	.9596	.2931	3.4124	**40**
30	.2840	.9588	.2962	3.3759	**30**
40	.2868	.9580	.2994	3.3402	**20**
50	.2896	.9572	.3026	3.3052	**10**
17° 00′	0.2924	0.9563	0.3057	3.2709	**73° 00′**
10	.2952	.9555	.3089	3.2371	**50**
20	.2979	.9546	.3121	3.2041	**40**
30	.3007	.9537	.3153	3.1716	**30**
40	.3035	.9528	.3185	3.1397	**20**
50	.3062	.9520	.3217	3.1084	**10**
18° 00′	0.3090	0.9511	0.3249	3.0777	**72° 00′**
10	.3118	.9502	.3281	3.0475	**50**
20	.3145	.9492	.3314	3.0178	**40**
30	.3173	.9483	.3346	2.9887	**30**
40	.3201	.9474	.3378	2.9600	**20**
50	.3228	.9465	.3411	2.9319	**10**
19° 00′	0.3256	0.9455	0.3443	2.9042	**71° 00′**
10	.3283	.9446	.3476	2.8770	**50**
20	.3311	.9436	.3508	2.8502	**40**
30	.3338	.9426	.3541	2.8239	**30**
40	.3365	.9417	.3574	2.7980	**20**
50	.3393	.9407	.3607	2.7725	**10**
20° 00′	0.3420	0.9397	0.3640	2.7475	**70° 00′**
10	.3448	.9387	.3673	2.7228	**50**
20	.3475	.9377	.3706	2.6985	**40**
30	.3502	.9367	.3739	2.6746	**30**
40	.3529	.9356	.3772	2.6511	**20**
50	.3557	.9346	.3805	2.6279	**10**
21° 00′	0.3584	0.9336	0.3839	2.6051	**69° 00′**
10	.3611	.9325	.3872	2.5826	**50**
20	.3638	.9315	.3906	2.5605	**40**
30	.3665	.9304	.3939	2.5386	**30**
40	.3692	.9293	.3973	2.5172	**20**
50	.3719	.9283	.4006	2.4960	**10**
22° 00′	0.3746	0.9272	0.4040	2.4751	**68° 00′**
10	.3773	.9261	.4074	2.4545	**50**
20	.3800	.9250	.4108	2.4342	**40**
30	.3827	.9239	.4142	2.4142	**30**
40	.3854	.9228	.4176	2.3945	**20**
50	.3881	.9216	.4210	2.3750	**10**
23° 00′	0.3907	0.9205	0.4245	2.3559	**67° 00′**
10	.3934	.9194	.4279	2.3369	**50**
20	.3961	.9182	.4314	2.3183	**40**
30	.3987	.9171	.4348	2.2998	**30**
40	.4014	.9159	.4383	2.2817	**20**
50	.4041	.9147	.4417	2.2637	**10**
	Cos	**Sin**	**Cot**	**Tan**	**Degrees**

Table A-2. Natural Trigonometric Functions—cont

Degrees	Sin	Cos	Tan	Cot	
24° 00′	0.4067	0.9135	0.4452	2.2460	**66° 00′**
10	.4094	.9124	.4487	2.2286	**50**
20	.4120	.9112	.4522	2.2113	**40**
30	.4147	.9100	.4557	2.1943	**30**
40	.4173	.9088	.4592	2.1775	**20**
50	.4200	.9075	.4628	2.1609	**10**
25° 00′	0.4226	0.9063	0.4663	2.1445	**65° 00′**
10	.4253	.9051	.4699	2.1283	**50**
20	.4279	.9038	.4734	2.1123	**40**
30	.4305	.9026	.4770	2.0965	**30**
40	.4331	.9013	.4806	2.0809	**20**
50	.4358	.9001	.4841	2.0655	**10**
26° 00′	0.4384	0.8988	0.4877	2.0503	**64° 00′**
10	.4410	.8975	.4913	2.0353	**50**
20	.4436	.8962	.4950	2.0204	**40**
30	.4462	.8949	.4986	2.0057	**30**
40	.4488	.8936	.5022	1.9912	**20**
50	.4514	.8923	.5059	1.9768	**10**
27° 00′	0.4540	0.8910	0.5095	1.9626	**63° 00′**
10	.4566	.8897	.5132	1.9486	**50**
20	.4592	.8884	.5169	1.9347	**40**
30	.4617	.8870	.5206	1.9210	**30**
40	.4643	.8857	.5243	1.9074	**20**
50	.4669	.8843	.5280	1.8940	**10**
28° 00′	0.4695	0.8829	0.5317	1.8807	**62° 00′**
10	.4720	.8816	.5354	1.8676	**50**
20	.4746	.8802	.5392	1.8546	**40**
30	.4772	.8788	.5430	1.8418	**30**
40	.4797	.8774	.5467	1.8291	**20**
50	.4823	.8760	.5505	1.8165	**10**
29° 00′	0.4848	0.8746	0.5543	1.8040	**61° 00′**
10	.4874	.8732	.5581	1.7917	**50**
20	.4899	.8718	.5619	1.7796	**40**
30	.4924	.8704	.5658	1.7675	**30**
40	.4950	.8689	.5696	1.7556	**20**
50	.4975	.8675	.5735	1.7437	**10**
30° 00′	0.5000	0.8660	0.5774	1.7321	**60° 00′**
10	.5025	.8646	.5812	1.7205	**50**
20	.5050	.8631	.5851	1.7090	**40**
30	.5075	.8616	.5890	1.6977	**30**
40	.5100	.8601	.5930	1.6864	**20**
50	.5125	.8587	.5969	1.6753	**10**
31° 00′	0.5150	0.8572	0.6009	1.6643	**59° 00′**
10	.5175	.8557	.6048	1.6534	**50**
20	.5200	.8542	.6088	1.6426	**40**
30	.5225	.8526	.6128	1.6319	**30**
40	.5250	.8511	.6168	1.6212	**20**
50	.5275	.8496	.6208	1.6107	**10**
	Cos	**Sin**	**Cot**	**Tan**	**Degrees**

Table A-2. Natural Trigonometric Functions—cont

Degrees	Sin	Cos	Tan	Cot	
32° 00′	0.5299	0.8480	0.6249	1.6003	**58° 00′**
10	.5324	.8465	.6289	1.5900	**50**
20	.5348	.8450	.6330	1.5798	**40**
30	.5373	.8434	.6371	1.5697	**30**
40	.5398	.8418	.6412	1.5597	**20**
50	.5422	.8403	.6453	1.5497	**10**
33° 00′	0.5446	0.8387	0.6494	1.5399	**57° 00′**
10	.5471	.8371	.6536	1.5301	**50**
20	.5495	.8355	.6577	1.5204	**40**
30	.5519	.8339	.6619	1.5108	**30**
40	.5544	.8323	.6661	1.5013	**20**
50	.5568	.8307	.6703	1.4919	**10**
34° 00′	0.5592	0.8290	0.6745	1.4826	**56° 00′**
10	.5616	.8274	.6787	1.4733	**50**
20	.5640	.8258	.6830	1.4641	**40**
30	.5664	.8241	.6873	1.4550	**30**
40	.5688	.8225	.6916	1.4460	**20**
50	.5712	.8208	.6959	1.4370	**10**
35° 00′	0.5736	0.8192	0.7002	1.4281	**55° 00′**
10	.5760	.8175	.7046	1.4193	**50**
20	.5783	.8158	.7089	1.4106	**40**
30	.5807	.8141	.7133	1.4019	**30**
40	.5831	.8124	.7177	1.3934	**20**
50	.5854	.8107	.7221	1.3848	**10**
36° 00′	0.5878	0.8090	0.7265	1.3764	**54° 00′**
10	.5901	.8073	.7310	1.3680	**50**
20	.5925	.8056	.7355	1.3597	**40**
30	.5948	.8039	.7400	1.3514	**30**
40	.5972	.8021	.7445	1.3432	**20**
50	.5995	.8004	.7490	1.3351	**10**
37° 00′	.6018	.7986	.7536	1.3270	**53° 00′**
10	.6041	.7969	.7581	1.3190	**50**
20	.6065	.7951	.7627	1.3111	**40**
30	.6088	.7934	.7673	1.3032	**30**
40	.6111	.7916	.7720	1.2954	**20**
50	.6134	.7898	.7766	1.2876	**10**
38° 00′	0.6157	0.7880	0.7813	1.2799	**52° 00′**
10	.6180	.7862	.7860	1.2723	**50**
20	.6202	.7844	.7907	1.2647	**40**
30	.6225	.7826	.7954	1.2572	**30**
40	.6248	.7808	.8002	1.2497	**20**
50	.6271	.7790	.8050	1.2423	**10**
39° 00′	0.6293	0.7771	0.8098	1.2349	**51° 00′**
10	.6316	.7753	.8146	1.2276	**50**
20	.6338	.7735	.8195	1.2203	**40**
30	.6361	.7716	.8243	1.2131	**30**
40	.6383	.7698	.8292	1.2059	**20**
50	.6406	.7679	.8342	1.1988	**10**
	Cos	**Sin**	**Cot**	**Tan**	**Degrees**

Table A-2. Natural Trigonometric Functions—cont

Degrees	Sin	Cos	Tan	Cot	
40° 00′	0.6428	0.7660	0.8391	1.1918	**50° 00′**
10	.6450	.7642	.8441	1.1847	**50**
20	.6472	.7623	.8491	1.1778	**40**
30	.6494	.7604	.8541	1.1708	**30**
40	.6517	.7585	.8591	1.1640	**20**
50	.6539	.7566	.8642	1.1571	**10**
41° 00′	0.6561	0.7547	0.8693	1.1504	**49° 00′**
10	.6583	.7528	.8744	1.1436	**50**
20	.6604	.7509	.8796	1.1369	**40**
30	.6626	.7490	.8847	1.1303	**30**
40	.6648	.7470	.8899	1.1237	**20**
50	.6670	.7451	.8952	1.1171	**10**
42° 00′	0.6691	0.7431	0.9004	1.1106	**48° 00′**
10	.6713	.7412	.9057	1.1041	**50**
20	.6734	.7392	.9110	1.0977	**40**
30	.6756	.7373	.9163	1.0913	**30**
40	.6777	.7353	.9217	1.0850	**20**
50	.6799	.7333	.9271	1.0786	**10**
43° 00′	0.6820	0.7314	0.9325	1.0724	**47° 00′**
10	.6841	.7294	.9380	1.0661	**50**
20	.6862	.7274	.9435	1.0599	**40**
30	.6884	.7254	.9490	1.0538	**30**
40	.6905	.7234	.9545	1.0477	**20**
50	.6926	.7214	.9601	1.0416	**10**
44° 00′	0.6947	0.7193	0.9657	1.0355	**46° 00′**
10	.6967	.7173	.9713	1.0295	**50**
20	.6988	.7163	.9770	1.0235	**40**
30	.7009	.7133	.9827	1.0176	**30**
40	.7030	.7112	.9884	1.0117	**20**
50	.7050	.7092	.9942	1.0058	**10**
45° 00′	0.7071	0.7071	1.0000	1.0000	**45° 00′**
	Cos	**Sin**	**Cot**	**Tan**	**Degrees**

Table A-3. Decibel Table

dB	Current or Voltage Ratio		Power Ratio		dB	Current or Voltage Ratio		Power Ratio	
	Gain	Loss	Gain	Loss		Gain	Loss	Gain	Loss
0	1.000	1.0000	1.000	1.0000	5.5	1.884	.5309	3.548	.2818
.1	1.012	.9886	1.023	.9772	5.6	1.905	.5248	3.631	.2754
.2	1.023	.9772	1.047	.9550	5.7	1.928	.5188	3.715	.2692
.3	1.035	.9661	1.072	.9333	5.8	1.950	.5129	3.802	.2630
.4	1.047	.9550	1.096	.9120	5.9	1.972	.5070	3.890	.2570
.5	1.059	.9441	1.122	.8913	6.0	1.995	.5012	3.981	.2512
.6	1.072	.9333	1.148	.8710	6.1	2.018	.4955	4.074	.2455
.7	1.084	.9226	1.175	.8511	6.2	2.042	.4898	4.169	.2399
.8	1.096	.9120	1.202	.8318	6.3	2.065	.4842	4.266	.2344
.9	1.109	.9016	1.230	.8128	6.4	2.089	.4786	4.365	.2291
1.0	1.122	.8913	1.259	.7943	6.5	2.113	.4732	4.467	.2239
1.1	1.135	.8810	1.288	.7762	6.6	2.138	.4677	4.571	.2188
1.2	1.148	.8710	1.318	.7586	6.7	2.163	.4624	4.677	.2138
1.3	1.161	.8610	1.349	.7413	6.8	2.188	.4571	4.786	.2089
1.4	1.175	.8511	1.380	.7244	6.9	2.213	.4519	4.898	.2042
1.5	1.189	.8414	1.413	.7079	7.0	2.239	.4467	5.012	.1995
1.6	1.202	.8318	1.445	.6918	7.1	2.265	.4416	5.129	.1950
1.7	1.216	.8222	1.479	.6761	7.2	2.291	.4365	5.248	.1905
1.8	1.230	.8128	1.514	.6607	7.3	2.317	.4315	5.370	.1862
1.9	1.245	.8035	1.549	.6457	7.4	2.344	.4266	5.495	.1820
2.0	1.259	.7943	1.585	.6310	7.5	2.371	.4217	5.623	.1778
2.1	1.274	.7852	1.622	.6166	7.6	2.399	.4169	5.754	.1738
2.2	1.288	.7762	1.660	.6026	7.7	2.427	.4121	5.888	.1698
2.3	1.303	.7674	1.698	.5888	7.8	2.455	.4074	6.026	.1660
2.4	1.318	.7586	1.738	.5754	7.9	2.483	.4027	6.166	.1622
2.5	1.334	.7499	1.778	.5623	8.0	2.512	.3981	6.310	.1585
2.6	1.349	.7413	1.820	.5495	8.1	2.541	.3936	6.457	.1549
2.7	1.365	.7328	1.862	.5370	8.2	2.570	.3890	6.607	.1514
2.8	1.380	.7244	1.905	.5248	8.3	2.600	.3846	6.761	.1479
2.9	1.396	.7161	1.950	.5129	8.4	2.630	.3802	6.918	.1445
3.0	1.413	.7079	1.995	.5012	8.5	2.661	.3758	7.079	.1413
3.1	1.429	.6998	2.042	.4898	8.6	2.692	.3715	7.244	.1380
3.2	1.445	.6918	2.089	.4786	8.7	2.723	.3673	7.413	.1349
3.3	1.462	.6839	2.138	.4677	8.8	2.754	.3631	7.586	.1318
3.4	1.479	.6761	2.188	.4571	8.9	2.786	.3589	7.762	.1288
3.5	1.496	.6683	2.239	.4467	9.0	2.818	.3548	7.943	.1259
3.6	1.514	.6607	2.291	.4365	9.1	2.851	.3508	8.128	.1230
3.7	1.531	.6531	2.344	.4266	9.2	2.884	.3467	8.318	.1202
3.8	1.549	.6457	2.399	.4169	9.3	2.917	.3428	8.511	.1175
3.9	1.567	.6383	2.455	.4074	9.4	2.951	.3388	8.710	.1148
4.0	1.585	.6310	2.512	.3981	9.5	2.985	.3350	8.913	.1122
4.1	1.603	.6237	2.570	.3890	9.6	3.020	.3311	9.120	.1096
4.2	1.622	.6166	2.630	.3802	9.7	3.055	.3273	9.333	.1072
4.3	1.641	.6095	2.692	.3715	9.8	3.090	.3236	9.550	.1047
4.4	1.660	.6026	2.754	.3631	9.9	3.126	.3199	9.772	.1023
4.5	1.679	.5957	2.818	.3548	10.0	3.162	.3162	10.000	.1000
4.6	1.698	.5888	2.884	.3467	10.1	3.199	.3126	10.23	.09772
4.7	1.718	.5821	2.951	.3388	10.2	3.236	.3090	10.47	.09550
4.8	1.738	.5754	3.020	.3311	10.3	3.273	.3055	10.72	.09333
4.9	1.758	.5689	3.090	.3236	10.4	3.311	.3020	10.96	.09120
5.0	1.778	.5623	3.162	.3162	10.5	3.350	.2985	11.22	.08913
5.1	1.799	.5559	3.236	.3090	10.6	3.388	.2951	11.48	.08710
5.2	1.820	.5495	3.311	.3020	10.7	3.428	.2917	11.75	.08511
5.3	1.841	.5433	3.388	.2951	10.8	3.467	.2884	12.02	.08318
5.4	1.862	.5370	3.467	.2884	10.9	3.508	.2851	12.30	.08128

Table A-3. Decibel Table—cont

dB	Current or Voltage Ratio Gain	Current or Voltage Ratio Loss	Power Ratio Gain	Power Ratio Loss	dB	Current or Voltage Ratio Gain	Current or Voltage Ratio Loss	Power Ratio Gain	Power Ratio Loss
11.0	3.548	.2818	12.59	.07943	15.5	5.957	.1679	35.48	.02818
11.1	3.589	.2786	12.88	.07762	15.6	6.026	.1660	36.31	.02754
11.2	3.631	.2754	13.18	.07586	15.7	6.095	.1641	37.15	.02692
11.3	3.673	.2723	13.49	.07413	15.8	6.166	.1622	38.02	.02630
11.4	3.715	.2692	13.80	.07244	15.9	6.237	.1603	38.90	.02570
11.5	3.758	.2661	14.13	.07079	16.0	6.310	.1585	39.81	.02512
11.6	3.802	.2630	14.45	.06918	16.1	6.383	.1567	40.74	.02455
11.7	3.846	.2600	14.79	.06761	16.2	6.457	.1549	41.69	.02399
11.8	3.890	.2570	15.14	.06607	16.3	6.531	.1531	42.66	.02344
11.9	3.936	.2541	15.49	.06457	16.4	6.607	.1514	43.65	.02291
12.0	3.981	.2512	15.85	.06310	16.5	6.683	.1496	44.67	.02239
12.1	4.027	.2483	16.22	.06166	16.6	6.761	.1479	45.71	.02188
12.2	4.074	.2455	16.60	.06026	16.7	6.839	.1462	46.77	.02138
12.3	4.121	.2427	16.98	.05888	16.8	6.918	.1445	47.86	.02089
12.4	4.169	.2399	17.38	.05754	16.9	6.998	.1429	48.98	.02042
12.5	4.217	.2371	17.78	.05623	17.0	7.079	.1413	50.12	.01995
12.6	4.266	.2344	18.20	.05495	17.1	7.161	.1396	51.29	.01950
12.7	4.315	.2317	18.62	.05370	17.2	7.244	.1380	52.48	.01905
12.8	4.365	.2291	19.05	.05248	17.3	7.328	.1365	53.70	.01862
12.9	4.416	.2265	19.50	.05129	17.4	7.413	.1349	54.95	.01820
13.0	4.467	.2239	19.95	.05012	17.5	7.499	.1334	56.23	.01778
13.1	4.519	.2213	20.42	.04898	17.6	7.586	.1318	57.54	.01738
13.2	4.571	.2188	20.89	.04786	17.7	7.674	.1303	58.88	.01698
13.3	4.624	.2163	21.38	.04677	17.8	7.762	.1288	60.26	.01660
13.4	4.677	.2138	21.88	.04571	17.9	7.852	.1274	61.66	.01622
13.5	4.732	.2113	22.39	.04467	18.0	7.943	.1259	63.10	.01585
13.6	4.786	.2089	22.91	.04365	18.1	8.035	.1245	64.57	.01549
13.7	4.842	.2065	23.44	.04266	18.2	8.128	.1230	66.07	.01514
13.8	4.898	.2042	23.99	.04169	18.3	8.222	.1216	67.61	.01479
13.9	4.955	.2018	24.55	.04074	18.4	8.318	.1202	69.18	.01445
14.0	5.012	.1995	25.12	.03981	18.5	8.414	.1189	70.79	.01413
14.1	5.070	.1972	25.70	.03890	18.6	8.511	.1175	72.44	.01380
14.2	5.129	.1950	26.30	.03802	18.7	8.610	.1161	74.13	.01349
14.3	5.188	.1928	26.92	.03715	18.8	8.710	.1148	75.86	.01318
14.4	5.248	.1905	27.54	.03631	18.9	8.810	.1135	77.62	.01288
14.5	5.309	.1884	28.18	.03548	19.0	8.913	.1122	79.43	.01259
14.6	5.370	.1862	28.84	.03467	19.1	9.016	.1109	81.28	.01230
14.7	5.433	.1841	29.51	.03388	19.2	9.120	.1096	83.18	.01202
14.8	5.495	.1820	30.20	.03311	19.3	9.226	.1084	85.11	.01175
14.9	5.559	.1799	30.90	.03236	19.4	9.333	.1072	87.10	.01148
15.0	5.623	.1778	31.62	.03162	19.5	9.441	.1059	89.13	.01122
15.1	5.689	.1758	32.36	.03090	19.6	9.550	.1047	91.20	.01096
15.2	5.754	.1738	33.11	.03020	19.7	9.661	.1035	93.33	.01072
15.3	5.821	.1718	33.88	.02951	19.8	9.772	.1023	95.50	.01047
15.4	5.888	.1698	34.67	.02884	19.9	9.886	.1012	97.72	.01023

Note: For values from 20 to 100 dB, see next page.

Table A-3. Decibel Table—cont

dB	Current or Voltage Ratio		Power Ratio
	Gain	Loss	
20.0	10.00 Use the same numbers as 0-20 dB, but shift point one step to the right. Thus since 10 dB = 3.162, 30 dB = 31.62.	.1000 Use the same numbers as 0-20 dB, but shift point one step to the left. Thus since 10 dB = .3162, 30 dB = .03162.	10.00 This column repeats every 10 dB instead of every 20 dB.
40.0	100 Use the same numbers of 0-20 dB, but shift point two steps to the right. Thus since 10 dB = 3.162, 50 dB = 316.2.	.01 Use the same numbers as 0-20 dB, but shift point two steps to the left. Thus since 10 dB = .3162, 50 dB = .003162.	20.00 This column repeats every 10 dB instead of every 20 dB.
60.0	1000 Use the same numbers as 0-20 dB, but shift point three steps to the right. Thus since 10 dB = 3.162, 70 dB = 3162.	.001 Use the same numbers as 0-20 dB, but shift point three to the left. Thus since 10 dB = .3162, 70 dB = .0003162.	30.00 This column repeats every 10 dB instead of every 20 dB.
80.0	10,000 Use the same numbers as 0-20 dB, but shift point four steps to the right. Thus since 10 dB = 3.162, 90 dB = 31620.	.0001 Use the same numbers as 0-20 dB, but shift point four steps to the left. Thus since 10 dB = .3162, 90 dB = .00003162.	40.00 This column repeats every 10 dB instead of every 20 dB.
100.0	100,000	.00001	50.00

APPENDIX **B**

Answers to Exercises

CHAPTER 1

A1-1. (A) The rf carrier is increased and decreased in amplitude by the applied signal. (B) The rf carrier remains constant in amplitude, but the frequency is varied by the applied signal.

A1-2. 535 kHz to 1605 kHz.

A1-3. Lowest: 540 kHz. Highest: 1600 kHz.

A1-4. (A) Lower sideband = 880 kHz. (B) Upper sideband = 900 kHz.

A1-5. (A) Minimum, 10 kW; maximum, 50 kW.
(B) Minimum, 250 watts; maximum, 250 watts nighttime, 1 kW daytime.

A1-6. 88 to 108 MHz.

A1-7. 200 kHz.

A1-8. Antenna height (Review Section 1-4).

A1-9. The pilot subcarrier is used in stereophonic fm broadcasting as a control signal for fm stereo receivers.

A1-10. 23 to 53 kHz.

CHAPTER 2

A2-1.

$$\begin{array}{r} 4 \\ 7 \\ 8 \\ 9 \\ \hline 28 \end{array} \qquad \begin{array}{r} -10 \\ -\ 6 \\ \hline -16 \end{array}$$

28 − 16 = 12 (Answer)

A2-2.

$$\begin{array}{r} 21 \\ 5 \\ 2 \\ \hline 28 \end{array} \qquad \begin{array}{r} -30 \\ -16 \\ \hline -46 \end{array}$$

−46 + 28 = −18 (Answer)

A2-3. $5 - (-16) = 21$

A2-4. $(4/5)(3) = 12/5$

A2-5. $\frac{9/3}{32} = \frac{3}{32}$

or:

$$\frac{9}{32(3)} = \frac{9}{96} = \frac{3}{32}$$

A2-6. $1/7 + 1/3 + 1/2 = 6/42 + 14/42 + 21/42 = 41/42$

A2-7. $1/4 - 1/6 = 3/12 - 2/12 = 1/12$

A2-8. $(2/3)(1/5) = 2/15$

A2-9. $(3/4)(6/10) = 18/40 = 9/20$

A2-10. $2/3 \div 3/5 = (2/3)(5/3) = 10/9 = 1\frac{1}{9}$

A2-11. $30/4 - 15/3 + 180 - 5$

$= 7.5 - 5 + 180 - 5$

$= 187.5 - 5 - 5$

$= 187.5 - 10 = 177.5$

A2-12. $\dfrac{48 - 47.5}{47.5}(100)$

$= \dfrac{0.5}{47.5}(100)$

$= 0.0105(100) = 1.05\%$

A2-13. $10^0 = 1$

$4^0 = 1$

$2^0 = 1$

A2-14. 5^{-8}

A2-15. (A) 4^{15}

(B) 4^{15}

(C) 4^{-15}

(D) 4^{-12}

A2-16. 9/16

A2-17. $\sqrt{6'95'73}$ Step 1

```
      2
    √6'95'73          Step 2
      4
      ----
      2 95
```

```
      2  6
    √6'95'73          Step 3
      4
 46 ) 2 95
      2 76
    --------
        19 73
```

```
      2  6  3
    √6'95'73          Step 4
      4
 46 ) 2 95
523 )   19 73
        15 69
    ---------
         4 04   Remainder
```

So: $\sqrt{69{,}573} = 263$

A2-18. $\dfrac{(4)(5)(2)10^{3-12+6}}{(5)(4)(10^{16-7})}$

$= \dfrac{40(10^{-3})}{20(10^{9})}$

$= \dfrac{2(10^{-3})}{10^{9}} = 2(10^{-3-9}) = 2(10^{-12})$

A2-19. (A) 3
(B) 2
(C) 1
(D) 0
(E) −4

A2-20. 1.7376

A2-21. Log 1.24 = 0.0934
Log 246 = 2.3909
Total = 2.4843

Antilog 2.4843 = 305 (Answer)

A2-22. Log 961 = 2.9827
Log 224 = 2.3502
Difference = 0.6325

Antilog 0.6325 = 4.29 (Answer)

A2-23. Log 638 = 2.8048
Exponent = 5
Product = 14.0240

Antilog 14.024 = $1.057(10^{14})$ (Answer)

A2-24. B = X/A

A = X/B

A2-25. R = 14.54

r = 39°54′

The resultant is $14.54\ \angle 39°54'$

A2-26. $20\ \angle 30° = 20(\cos 30° + j \sin 30°)$

$= 20(0.866 + j0.5)$

$= 17.32 + j10$

A2-27. $\sqrt{30^2 + 50^2}$

$= \sqrt{900 + 2500}$

$= \sqrt{3400} = 58.3$ (Magnitude)

$\tan r = 50/30 = 1.6667$

Then $r = 59°02'$ (Approx)

Complete answer: $58.3 \angle 59°02'$

A2-28. 1 milliwatt (mW) in 600 ohms.

A2-29. $100/150 = 0.66 = 1.8$ dB loss.

A2-30. 10^4

CHAPTER 3

A3-1. The R1-R2 voltage divider gives a voltage of 3 volts at the Q1 base. See Fig. B-1; this is the dc equivalent circuit. Note that the feedback shown in this diagram is dc feedback only; C2 and R4 provide a bypass around R6 to frequencies in the passband. Coupling capacitors C1 and C3 prevent the dc operating point from being influenced by source and load characteristics.

Since the voltage at the base of Q1 is 3 volts, V_E of Q1 is $+3 - 0.6 = 2.4$ volts. The current through R6 to produce this 2.4 volts is $2.4/1000 = 2.4$ mA.

At Q2 there will be a 0.6-volt difference between emitter and base, so V_B of $Q2 = 6 - 0.6 = 5.4$ volts. This is also the voltage at the collector of Q1. The current through R3 to drop 0.6 volt is $0.6/1000 = 0.6$ mA, and almost all of this current goes to the Q1 collector. The emitter current of Q1 will also be about 0.6 mA.

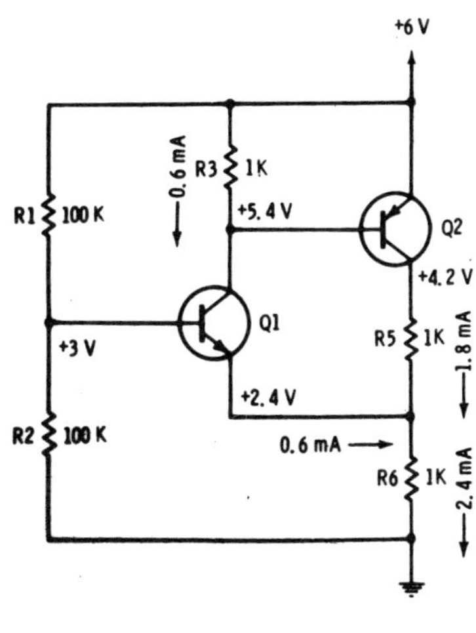

Fig. B-1. Dc equivalent of Fig. 3-69.

The current through R6 is 2.4 mA. Since 0.6 mA of this current comes from the emitter of Q1, $2.4 - 0.6 = 1.8$ mA must come through R5. The voltage drop across R5 is then $(1.8)(1) = 1.8$ volts. The voltage at the collector of Q2 is the sum of the voltage drops across R5 and R6, or $2.4 + 1.8 = 4.2$ volts.

A3-2. (A) Positional Weight → 16 8 4 2 1

Binary → 1 0 0 0 0

= 10000.00 binary

(B) $\text{Bcd} = \underbrace{0001}_{1}\quad\underbrace{0110}_{6}$

$= 1 \quad 6 = \text{decimal } 16$

(C) Binary $\underbrace{010}_{2}\underbrace{000}_{0}$

= 20 octal

A3-3. (A) 1011. (B) 13.

A3-4. (A) 1000111000. (B) 1070.

A3-5. 41.25

A3-6. (A) A *and* B. (B) A *or* B.

A3-7. (A) Diode-transistor logic. (B) Resistor-transistor logic. (C) Resistor-capacitor-transistor logic. (D) Transistor-transistor logic.

A3-8. See Fig. B-2. Note that the truth table starts with binary number zero (0000) and increases by adding 1 each time until the number 15 (1111) is reached. This assures that no combination is omitted.

A3-9. See Fig. B-3.

A3-10. See Fig. B-4.

A3-11. No. A reset pulse can be made to coincide with the binary equivalent of an odd number, restarting the count to obtain division by an odd integer.

CHAPTER 4

A4-1. Only at the expense of frequency discrimination (Fig. 4-13).

A4-2. A "figure eight" pattern.

AND

A	B	C	D	H
0	0	0	0	0
0	0	0	1	0
0	0	1	0	0
0	0	1	1	0
0	1	0	0	0
0	1	0	1	0
0	1	1	0	0
0	1	1	1	0
1	0	0	0	0
1	0	0	1	0
1	0	1	0	0
1	0	1	1	0
1	1	0	0	0
1	1	0	1	0
1	1	1	0	0
1	1	1	1	1

Fig. B-2. Truth table for AND portion of Fig. 3-52A.

A4-3. Three. One terminal is common. The left and common feed the left channel. The right and common feed the right channel.

A4-4. Yes. The left and right terminals are paralleled to feed a single channel.

A4-5. Single.

OR

E	F	G	OUT TO Q1 BASE
0	0	0	0
0	0	1	1
0	1	0	1
0	1	1	1
1	0	0	1
1	0	1	1
1	1	0	1
1	1	1	1

Fig. B-3. Truth table for OR portion of Fig. 3-52A.

NOT

INPUT	OUTPUT
1	0
0	1

Fig. B-4. Truth table for NOT portion of Fig. 3-52A.

A4-6. 0.25 inch. (Actually 0.246 plus or minus 0.002 in.)

A4-7. 0.24 inch.

A4-8. Track width is 80 mils plus slight "spill over" at edges. Center-to-center spacing is 160 mils.

A4-9. Track width = 37 mils. Center-to-center track spacing = 71 mils.

A4-10. To compensate for the nonlinear B-H curve of the magnetic tape.

CHAPTER 5

A5-1. No. Practically all tape cartridge systems require an external bulk eraser to erase the tape. Also, some very-slow-speed reel-to-reel recorders, such as those used for logging, do not have an erase head.

A5-2. No. It supplies only sufficient torque to maintain correct tape tension between the capstan and take-up reel.

A5-3. The capstan motor and pinch-roller assembly.

A5-4. 3¾ and 7½ in/s.

A5-5. As the frequency is decreased, the magnetic field energy decreases at a 6 dB/octave rate and falls into the noise level.

A5-6. 7½ in/s.

A5-7. (A) 1000 Hz. (B) 150 Hz. (C) 8000 Hz.

A5-8. The lower track of a monophonic system, and the lowest track of a three-track stereo system.

A5-9. Left channel: upper track. Right channel: middle track.

A5-10. So that the proper speed is established prior to the instant at which the capstan pinch roller engages the tape.

CHAPTER 6

A6-1. The balanced circuit requires an H pad.

Attenuation required = 32 dB.

Refer to Fig. 6-6B for configuration. From Table 6-2:

$$K_1 = 0.950$$

$$K_2 = 19.8$$

$$R1 = \frac{(600 + 150)(0.950) + (600 - 150)}{2}$$

$$= \frac{1162}{2} = 581 \text{ ohms}$$

$$R2 = \frac{(600 + 150)(0.950) - (600 - 150)}{2}$$

$$= \frac{262}{2} = 131 \text{ ohms}$$

$$R3 = \frac{600 + 150}{(2)(19.8)} = \frac{750}{39.6} = 18.9 \text{ ohms}$$

Final solution:

$$\frac{R1}{2} = 291 \text{ ohms (nearest EIA value} = 300 \text{ ohms)}$$

$$\frac{R2}{2} = 65.5 \text{ ohms (nearest EIA value} = 68 \text{ ohms)}$$

$$R3 = 18.9 \text{ ohms (nearest EIA value} = 18 \text{ ohms)}$$

A6-2. The loss when the attenuator is placed fully clockwise (maximum volume).

A6-3. 6 dB.

A6-4. Zero insertion loss in both cases when operated between like impedances.

A6-5. The output level required for a s/n ratio of 60 dB is $-80 + 60 = -20$ dBm. Since the gain of the amplifier is 40 dB, then the input level required for a −20-dBm output is $-20 + (-40) = -60$ dBm.

A6-6. (A) Max input level $= +18 - 40 = -22$ dBm
(B) s/n ratio $= +18 - (-80) = 98$ dB

A6-7. The required attenuation is 25 dB. From Table 6-3, R1 = 5000 ohms and $\frac{R1}{2}$ = 2500 ohms. Two 300-ohm resistors are added across the amplifier input terminals with the center junction of the resistors grounded (see the balanced-configuration diagram in Table 6-3). Note that this bridging impedance precludes adding any further bridging circuits to this line.

A6-8. (A) $R = \frac{3R_s}{5} = \frac{450}{5} = 90$ ohms

(B) $\text{Min loss} = 20 \log 4$
$= (20)(0.602)$
$= 12$ dB

Therefore the amplifier input level is −72 dBm if the microphone delivers −60 dBm.

A6-9. With a −72-dBm input and a gain of 40 dB, the output level is −32 dBm. Then the s/n ratio is 80 − 32 = 48 dB.

A6-10. 8/4 = 2 watts for each speaker.

CHAPTER 7

A7-1. The color coding of the lead that is "positive" relative to the other leads. The positive wire of each microphone must have the same connection to corresponding terminals of the preamplifiers. This is true for either stereo or mono broadcasting.

A7-2. Ambient sound versus direct sound.

A7-3. If only quad discs are used, no further encoding is necessary. If four microphones or four-track (four-head) tape is employed, encoding is necessary.

A7-4. It converts the original four-channel sound to two channels in such a manner that the original four channels can be separated by a decoding device.

A7-5. Left and right speakers reproducing the same magnitude of in-phase signal. For quad sound, this applies only to the left-front and right-front speakers.

A7-6. No.

A7-7. Yes.

A7-8. Yes.

A7-9. No.

CHAPTER 8

A8-1. 947-952 MHz.

A8-2. Yes. For remote-pickup and mobile service, see Chart 8-1. For stl's, see answer A8-1.

A8-3. 0.6 of the first-Fresnel-zone radius.

A8-4. Path of most concentrated energy bounded by a path where an additional half-wavelength in the path length causes phase cancellation.

A8-5. Because of strong reflections from a smooth surface.

A8-6. Fm.

A8-7. Coaxial cable.

A8-8. (A) 50 dB
(B) 60 dB

CHAPTER 9

A9-1. The angle subtended by a section of the circumference a circle equal in length to the radius of that circle. Mathematically:

$$1 \text{ Radian} = \frac{180°}{\pi} = 57.32°$$

A9-2. $\text{Radians} = (\text{Degrees})\left(\frac{\pi}{180°}\right)$

A9-3. π radians, or 3.14.

A9-4. (A) 500 meters, or 1640 feet. (B) 187.5 meters, or 615 feet.

A9-5. Ground-wave signals.

A9-6. 10 mV/m.

A9-7. The class-C stage used to amplify carrier power (unmodulated) employs high-Q (narrow-band) tuned amplifiers for the carrier frequency. The linear rf amplifier must have a bandwidth to include all sideband frequencies, which calls for lower-Q tuned circuits.

A9-8. 22.5 percent over the unmodulated value of antenna current.

A9-9. 85 percent (approximately).

A9-10. (A) 15 kW.
(B) 11.25 kW.
(C) 10.3125 kW.

CHAPTER 10

A10-1. Almost directly proportional to the number of stacked bays. See Table 10-2.

A10-2. (A) $10 \log 6 = (10)(0.7782) = 7.78$ dB
(B) $\sqrt{6} = 2.44$

A10-3. Approximately proportional to the number of bays divided by 2. See Table 10-4.

A10-4. (A) Amplitude. (B) Amplitude and frequency.

A10-5. The frequency response (after pre-emphasis) is made proportional to the reciprocal of the modulating frequency ($1/f_m$).

A10-6. Same as modulation index. (A) 5. (B) 3750.

A10-7. (A) 1 radian $= 57.3°$, so $(5)(57.3) = 286.5°$
(B) $(3750)(57.3) = 215{,}000°$ (approx)

A10-8. ± 75 kHz$/4 = \pm 18.75$ kHz

A10-9. Frequency modulated.

A10-10. Amplitude modulated.

A10-11. L − R, suppressed carrier.

A10-12. 23 kHz to 53 kHz.

CHAPTER 11

A11-1. The hearing sense is not "flat." (Review Figs. 2-16 and 11-5, with associated text.) This simply emphasizes that if the operator groups performers around a microphone so that lows and highs are in balance with a "flat" monitoring system, a flat frequency response in the remainder of the system is desirable. Bass and treble controls in the receiver compensate for room acoustics and individual listener preference.

A11-2. The compression meter, which shows the amount of compression or limiting taking place.

A11-3. Around 32 percent, or about −10 dB from program peaks.

A11-4. In practice, three basic types: (A) acoustical, Fig. 11-11; (B) magnetic tape, Figs. 11-12 and 11-13; and (C) electrosonic delay lines. The latter type is encountered often in professional recording studios but very seldom in broadcast installations.

A11-5. Position the piano so that the keyboard hammer line is more on axis to the microphone, keeping the same distance.

A11-6. Use the minimum number possible for adequate coverage. Reposition performers around one microphone if possible to obtain balance.

A11-7. The FCC says the station licensee. In the final analysis, the operator on duty has control over "loudness."

CHAPTER 12

A12-1. Remote programs require more microphones to accentuate the ratio of direct to reflected sound and to prevent the ambient noise level from destroying the program.

A12-2. Use an isolation coil as shown in Fig. 12-3. Use the balanced-line side of the transformer for the line, with 10K bridging resistors in each side. Connect the public-address input to the unbalanced side of the transformer. The pa amplifier normally has sufficient gain to allow this bridging. It may be necessary to use the microphone input on the pa amplifier.

A12-3. Many remote amplifiers have an adjustable multiplier for the VU meter, allowing zero reference to be +4, +8, or +12 VU. Use the highest level possible (such as +12) for the amplifier feeding the program line. This normally overrides such interference.

A12-4. Use of an equalizer at the studio will normally provide satisfactory quality for voice (only) transmission.

CHAPTER 13

A13-1. Place a 47K resistor in series with the vom, and take this into account in the reading.

A13-2. A broken shield or ground wire.

A13-3. No. The playback response should be the inverse of the recording characteristic.

A13-4. Be sure the IC leads are not folded against the board on the soldered side. Otherwise, damage to the board will result.

A13-5. Clean and demagnetize the heads.

A13-6. Head-to-tape contact, tilt, height, tangency, azimuth.

A13-7. Yes. See Section 13-9.

A13-8. All transmitters and receivers in the same system are tuned to the same frequency.

A13-9. Total harmonic distortion (thd).

A13-10. Be sure to use an input signal which is below the maximum input rating of the amplifier into which the signal is inserted.

CHAPTER 14

A14-1. No. The VU meter has a full-wave rectifier and a specially damped movement. The modulation meter is a device to measure either negative or positive modulation at a given time, and it is a semipeak device rather than rms.

A14-2. The bridge method and the substitution method. Details are covered in Section 14-4.

A14-3. By the direct method.

A14-4. (A) Antenna current squared times antenna impedance.

(B) $E_p \times I_p \times F$

where,

E_p is the plate voltage of the final stage,
I_p is the plate current of the final stage,
F is an efficiency factor furnished by the manufacturer.

A14-5. If used, the limiting or agc action must be removed.

A14-6. The relationship provides indentification of the left and right channels. If the second harmonic should cross the time axis with a negative slope at the transmitter and a positive slope at the receiver, the "left" channel as transmitted would be reproduced in the "right" channel of the receiver, and conversely.

A14-7. It maintains the relationship of the left-plus-right signal to the left-minus-right signal. If this standard varied from station to station (deviation upward in some transmitters and downward in others), the left and right channels would be reversed in the receiver, depending on the station being received.

A14-8. Paragraphs L and M give the FCC requirements for channel separation. Paragraph L specifies the required amplitude response of the two channels, and paragraph M specifies the phase response. With a steady-state signal in the left channel only, signals of the same frequency and equal amplitudes (within 3½ percent) exist in the left-plus-right and left-minus-right channels. The phase difference between the left-plus-right and left-minus-right signals must not exceed $\pm 3°$ between 50 Hz and 15 kHz.

The above may be summarized by stating that for proper channel separation, the left-plus-right and left-minus-right channels must have the same frequency response, and all frequencies in both channels must arrive at the receiving matrix at the same time (in the proper phase relationship).

The FCC note following paragraph M specifies when further measurements in checking channel separation are required. For a given signal in the left channel only, adjustment is satisfactory if the signal recovered at the receiver (or receiver-type monitor) in the right channel is attenuated at least 29.7 dB. This must be true throughout the frequency range of 50 Hz to 15 kHz. If this check is negative, further measurements of the frequency and phase response must be taken.

A14-9. No. Paragraphs N and O of the standards specify transmitter requirements in checking channel separation with respect to cross talk from the left-minus-right channel into the left-plus-right channel (stereo subchannel into main channel) and cross talk from the left-plus-right channel into the left-minus-right channel (main channel into stereo subchannel).

A14-10. The modulation meter. See Section 14-12.

Index